TABLE OF CONTENTS

NEW PRODUCTS

In this Optics Guide, we are proud to introduce many new products. Our selection of stock optical components has been greatly expanded to include meniscus lenses, sapphire lenses, many new beamsplitters, more laser diode optics, high energy laser optics and many new coatings. We offer, by far, the largest selection of fully specified, stock optical components available today.

Additionally, we would like to draw your attention to the following major Melles Griot innovations:

OPTICAL TABLES

We have taken the first fresh look at Optical table requirements in many years. You will find our products a significant improvement over prior technology. Our new tables are accompanied by a vastly expanded range of optical hardware components. See pages 403-500.

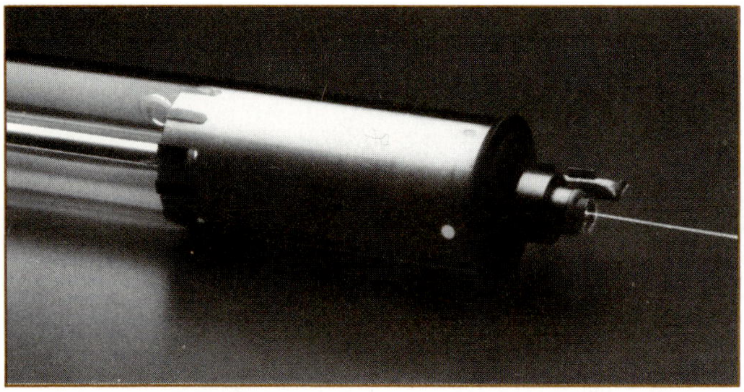

GREEN AND IR HeNe LASERS

Yes, HeNe's! We stock HeNe lasers operating at green (543 nm) and infrared (1.5 micron) wavelengths. These, we believe, open to many new applications the convenience and reliability of proven HeNe technology. See pages 345-360.

INFRARED POLARIZERS

Wire grid polarization scaled down to near-infrared wavelengths offers significant advantages over conventional dichroic sheet polarizers. We believe this product, never before available, will be of great significance in communication technology.

See pages 283-308 for these products and our vastly expanded range of polarization components, including calcite polarizing prisms.

CARBON DIOXIDE LASERS

We have adapted our proven and reliable approach to HeNe lasers in producing this new range of sealed tube, waveguide CO_2 lasers. Providing an output of 7 watts at 10.6 microns, these lasers will find many uses in R&D and medical applications. See pages 396-402.

FUNDAMENTAL OPTICS

IMAGING PROPERTIES OF SIMPLE LENS AT FINITE CONJUGATE RATIO

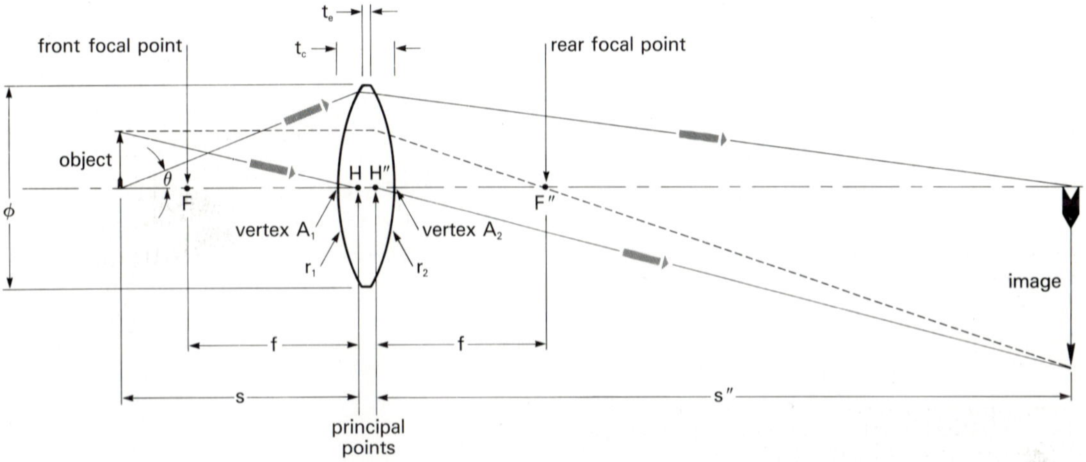

Note location of object and image relative to front and rear focal points.

ϕ = Lens diameter

r_1 = Radius of curvature of 1st surface (positive if center of curvature is to right)

r_2 = Radius of curvature of 2nd surface (negative if center of curvature is to left)

r_2 = $-r_1$ for symmetric lens

m = s″/s = magnification = conjugate ratio, said to be infinite if either s″ or s is infinite

θ = Arctan (ϕ/2s)

s = Object distance, positive for object (whether real or virtual) to the left of principal point H

s″ = Image distance (s and s″ are collectively called conjugate distances, with object and image in conjugate planes), positive for image (whether real or virtual) to the right of the principal point H″

t_c = Center thickness

t_e = Edge thickness

f = Effective focal length (EFL), may be positive (as shown or negative. f represents both $\overline{FH}$ and $\overline{H''F''}$, assuming lens to be surrounded by medium of index 1.0

In this section we take you through the fundamentals of optics and introduce many basic formulas which you will find useful in the selection and use of optical components. We specify our stock components very precisely so that, understanding the definitions presented here, you may select optics with confidence.

Contents **Page Number**

FOCAL LENGTHS AND PRINCIPAL SURFACES

FOCAL LENGTH

There are two distinct types of focal lengths associated with every lens or lens system. Most important of these is the Effective Focal Length (EFL) or Equivalent Focal Length (denoted f on the drawings which follow), which determines magnification and hence the image size. You will notice that f appears frequently in the lens formulas, in addition to being the quantity listed as focal length in the tables of standard lenses. Unfortunately, f is measured with reference to points (the principal points) which are usually inside the lens, and so the meaning of f is not immediately apparent when a lens is visually inspected.

The second type of focal length relates the focal plane positions directly to landmarks on the lens surfaces (namely the vertices) which are immediately recognizable. It is not simply related to image size, but is especially convenient for use when one is concerned about correct lens positioning or mechanical clearances. Examples of this second type of focal length are the Front Focal Length (FFL, denoted f_f on the drawing that follows) and the Back Focal Length (BFL, denoted f_b on the drawing that follows).

The convention in all of the drawings (with the exception of a single deliberately reversed ray in the following drawing) is that light travels from left to right.

FOCAL POINT (F OR F″)

Rays which pass through or originate at either focal point must be, on the opposite side of the lens, parallel to the optical axis. This fact is the basis for locating both focal points.

PRIMARY PRINCIPAL PLANE (OR SURFACE)

Let us imagine that rays originating at the front focal point F (and therefore parallel to the optical axis after emergence from the opposite side of the lens) are singly refracted at some imaginary surface, instead of twice refracted (once at each lens surface) as actually happens. There is a unique imaginary surface at which this can happen. To locate this unique surface, consider a single ray traced from the air on one side of the lens, through the lens and into the air on the other side. The ray is broken into three segments by the lens. Two of these are external (in the air), and the third is internal (in the glass). The external segments can be extended to a common point of intersection (certainly near, and usually within, the lens). The principal surface is the locus of all such points of intersection of extended external ray segments. The principal surface is perpendicular to the optical axis at the principal points and its curvature is negligible in the paraxial limit (defined on pg. 7) — hence the more common language "Principl Plane."

SECONDARY PRINCIPAL PLANE (OR SURFACE)

Defined analogously to the primary principal plane, but for a collimated beam incident from the left and focused to the rear focal point F_0 on the right. Rays in that part of the beam nearest the axis can be thought of as once refracted at the secondary principal surface, instead of being refracted by both lens surfaces.

PRIMARY PRINCIPAL POINT (H)

The intersection of the primary principal plane with optical axis.

SECONDARY PRINCIPAL POINT (H″)

The intersection of the secondary principal plane with the optical axis.

PRIMARY VERTEX (A_1)

The intersection of the primary lens surface with the optical axis.

SECONDARY VERTEX (A_2)

The intersection of the secondary lens surface with the optical axis.

EFFECTIVE FOCAL LENGTH (f)

Assuming that the lens is surrounded by air or vacuum (refractive index 1.0), this is both the distance from the front focal point (F) to the primary principal point (H) and the distance from the secondary principal point (H_0) to the rear focal point (F_0). Later we use (f_o) to designate the paraxial effective focal length for the design wavelength (λ_o).

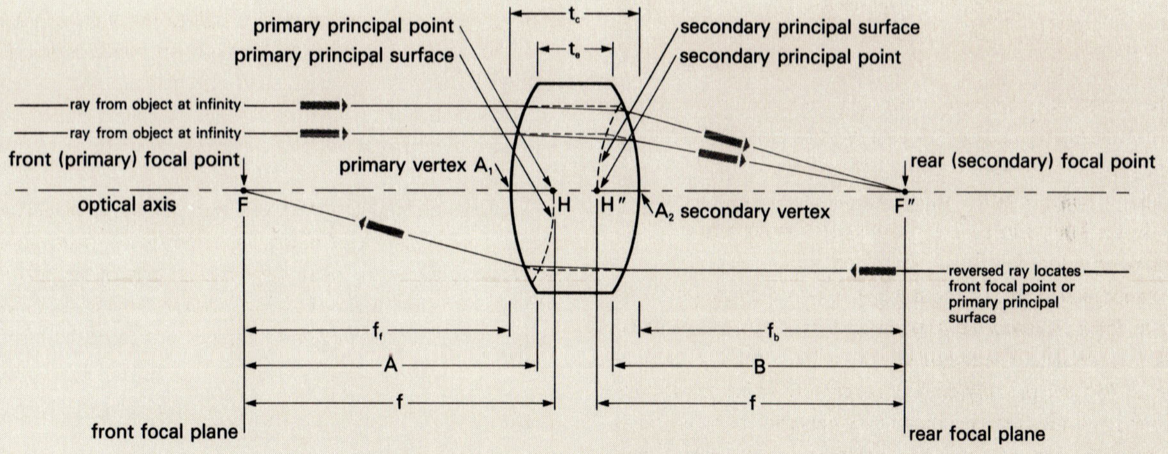

f = Effective focal length (EFL), may be positive (as shown) or negative

f_f = Front focal length (FFL)	A = Front focus to front edge distance	t_e = Edge thickness
f_b = Back focal length (BFL)	B = Rear edge to rear focus distance	t_c = Center thickness

FRONT AND BACK FOCAL LENGTHS of a lens having spherical surfaces and surrounded by air. Under these conditions, distances labeled f are equal whether or not the lens is symmetric, but distances f_f and f_b are equal *only* if the lens is symmetric. In the paraxial limit (see text), the curvature of the principal surfaces may be neglected.

FRONT FOCAL LENGTH (f_f)

The distance from the front focal point (F) to the primary vertex (A_1).

BACK FOCAL LENGTH (f_b)

The distance from the secondary vertex (A_2) to the rear foca point (F").

EDGE-TO-FOCUS DISTANCES (A AND B)

A is the distance from the front focus to the front edge of the lens. B is the distance from the rear edge of the lens to the rear focus. Both distances are presumed to be always positive.

If you need help with the use of formulas presented in this Optics Guide, our Applications Engineers will be pleased to assist you.

PRINCIPAL POINT LOCATIONS

The following sketches of various standard lens shapes indicate approximately where the principal points fall in relation to the lens surfaces. The exact positions depend on the index of refraction of the lens material, and on the lens radii, and can be found by formula. In extreme meniscus lens shapes (short radii or steep curves) it is possible that both principal points fall outside the lens boundaries. For symmetric lenses the principal points divide that part of the optical axis between the vertices into three approximately equal segments. For plano lenses one principal point is at the curved vertex, and the other is approximately one-third of the way to the plane vertex.

If a symmetric convex lens was continuously transformed (both radii becoming infinite, changing sign, and returning to finite values while maintaining symmetry) into a symmetric concave lens, the principal points would remain approximately stationary during this process. In contrast, the focal points would change sides of the lens.

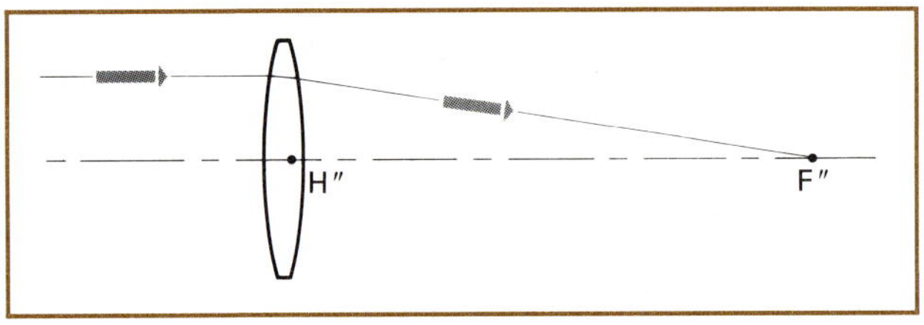

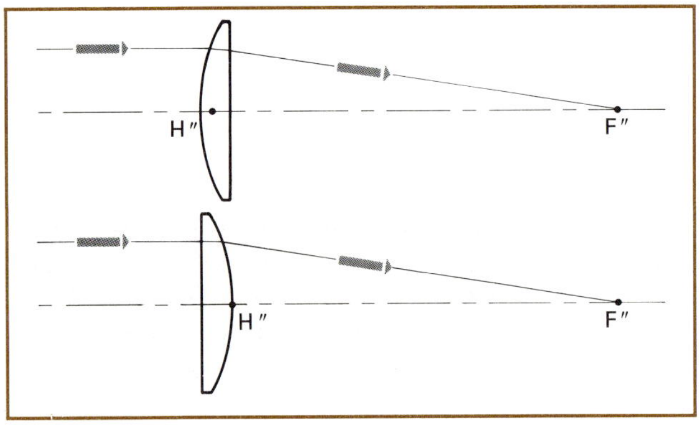

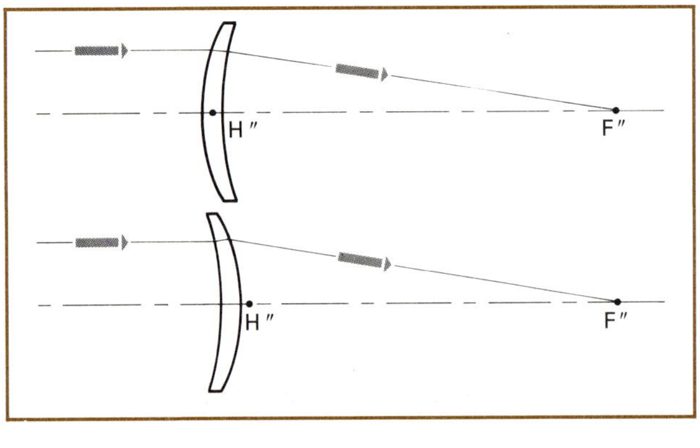

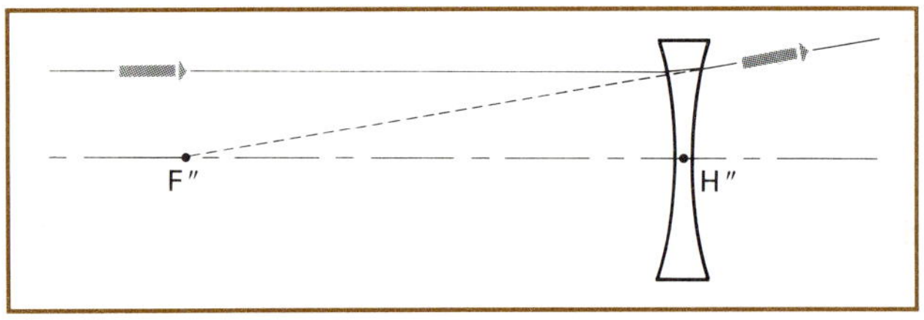

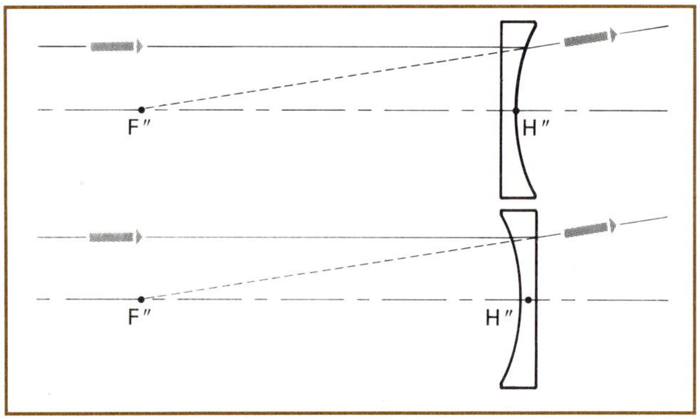

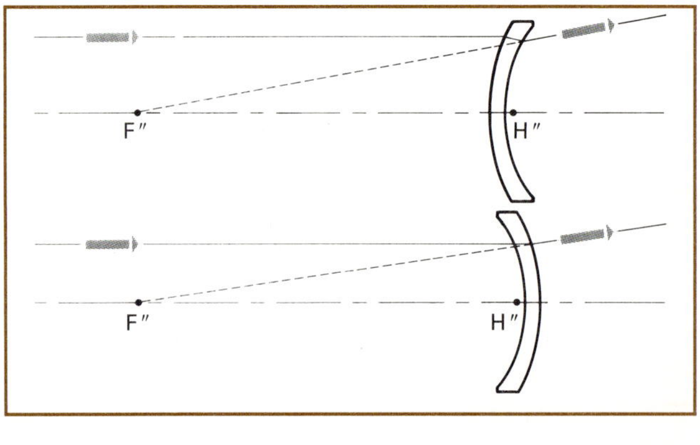

PARAXIAL FORMULAS

PARAXIAL FORMULAS FOR LENSES IN AIR

The formulas which follow are based on the behavior of paraxial rays, which are rays always very close to, and very nearly parallel to, the optical axis. In this region lens surfaces are always very nearly normal to the optical axis, and hence all angles of incidence and refraction are small. As a result, the sines of the angles of incidence and refraction (as used in Snell's Law) can be approximated by the angles themselves (in radian measure).

The paraxial formulas do not include effects of spherical aberration such as might be experienced by a marginal ray — a ray passing through the lens near its edge or margin. All effective focal length values (f) tabulated in this catalog are paraxial values, corresponding to the paraxial formulas.

The formulas which follow are paraxial and are valid for both thick and thin lenses unless otherwise noted. The refractive index of the lens glass, denoted n, is the ratio of the speed of light in vacuum to the speed of light in the lens glass.

SIGN CONVENTION

The validity of the lens formulas that follow is dependent on adherence to the sign convention shown here:

- s is + for object to left of H
- s is – for object to right of H
- s″ is + for image to right of H″
- s″ is – for image to left of H″
- r_1 is + for center of curvature to right of surface
- r_1 is – for center of curvature to left of surface
- r_2 is + for center of curvature to right of surface
- r_2 is – for center of curvature to left of surface
- A_1H is + for first principal point H to right of vertex A_1 (interior to lens)
- A_1H is – for first principal point H to left of vertex A_1 (exterior to lens)
- A_2H'' is + for secondary principal point H″ to right of secondary vertex A_2 (H″ exterior to lens)
- A_2H'' is – for secondary principal point H″ to left of secondary verdex A_2 (H″ interior to lens)

Other authors may use a different sign convention for conjugate distances, for focal length, and in their aberration formulas.

Focal Length

$$f = m\frac{(s + s'')}{(m + 1)^2}$$

where (s + s″) is the approximate object-to-image distance.

$$f = \frac{sm}{m + 1}$$

$$f = \frac{s + s''}{m + 2 + \frac{1}{m}}$$

The Gaussian form of "the lens formula" is

$$\frac{1}{f} = \frac{1}{s} + \frac{1}{s''},$$

where s and s″ are the conjugate distances.
The Newtonian form of "the lens formula" is

$$xx'' = f^2,$$

where x=(s–f), and x″=(s″–f).

$$\frac{1}{f} = (n - 1)\left(\frac{1}{r_1} - \frac{1}{r_2}\right) + \frac{(n - 1)^2}{n}\frac{t_c}{r_1 r_2},$$

where n is the refractive index, t_c is the center thickness, and the sign convention previously given for the radii r_1 and r_2 applies. For thin lenses, $t_c \cong 0$, and for plano lenses either r_1 or r_2 is infinite. In either case the second term of the above equation vanishes, and we are left with the familiar Lens Maker's Formula

$$\frac{1}{f} = (n - 1)\left(\frac{1}{r_1} - \frac{1}{r_2}\right)$$

Surface Sagitta and Radius of Curvature

$$r^2 = (r - s)^2 + \left(\frac{d}{2}\right)^2$$

$$s = r - \sqrt{r^2 - \left(\frac{d}{2}\right)^2} > 0$$

$$r = \frac{s}{2} + \frac{d^2}{8s}$$

An often useful approximation is to neglect s/2.

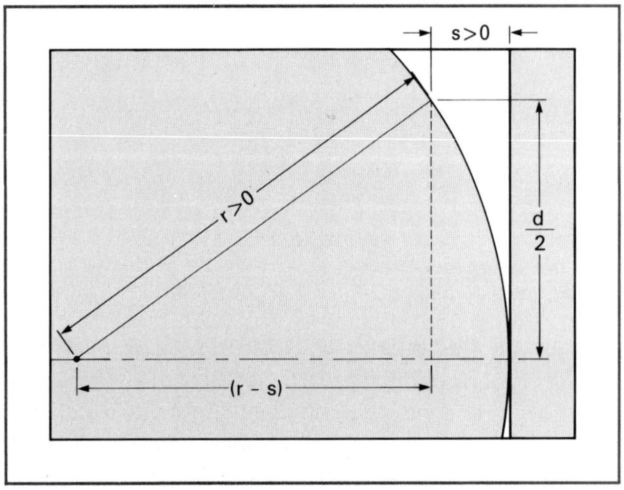

SURFACE SAGITTA and radius of curvature.

Symmetric Lens Radii ($r_2 = -r_1$)

With center thickness constrained,

$$r_1 = (n-1)\left[f \pm \sqrt{f^2 - \left(\frac{ft_c}{n}\right)}\right]$$

$$= (n-1)f\left[1 \pm \sqrt{1 - \left(\frac{t_c}{nf}\right)}\right]$$

where in the first form the + sign is chosen for the square root if f is positive, but the − sign must be used if f is negative. In the second form the + sign must be used regardless of the sign of f. With edge thickness constrained the equation for r_1 becomes transcendental:

$$\frac{1}{f} = \frac{2(n-1)}{r_1} - \frac{(n-1)^2}{nr_1^2}\left[t_c + 2r_1\left\{1 - \cos\left(\arcsin\frac{\phi}{2r_1}\right)\right\}\right],$$

where ϕ is the lens diameter. This equation can be solved by numerical methods.

Plano Lens Radius

r_2 is infinite, so

$$r_1 = (n-1)f.$$

Principal Point Locations (signed distances from vertices)

$$A_2H'' = \frac{-r_2t_c}{n(r_2 - r_1) + t_c(n-1)},$$

$$A_1H = \frac{-r_1t_c}{n(r_2 - r_1) + t_c(n-1)}, \text{ and}$$

where the above sign convention applies.

For symmetric lenses ($r_2 = -r_1$),

$$A_1H = -A_2H''$$

$$= \frac{r_1t_c}{2nr_1 - t_c(n-1)}.$$

If either r_1 or r_2 is infinite, "l' Hôpital's Rule" from calculus must be used. Thus, for plano-convex lenses in the correct orientation (explained later),

$$A_1H = 0 \text{ and}$$

$$A_2H'' = -\frac{t_c}{n}.$$

For flat plates let $r_1 \rightarrow \infty$ in a symmetric lens, thus obtaining $A_1H = A_2H'' = t_c/2n$. These results are useful in connection with the paraxial lens combination formulas which follow.

Hiatus or Interstitium (principal point separation)

$$HH'' = t_c\left\{1 - \frac{f}{n}\left[\frac{1}{f} - \frac{(n-1)^2t_c}{n\,r_1r_2}\right]\right\},$$

which in the thin lens approximation (exact for plano lenses) becomes

$$HH'' = t_c\left(1 - \frac{1}{n}\right).$$

This result also applies to flat plates.

Object-to-First Principal Point Distance

(approximate lens-to-object distance, often simply called object distance) s is one of the two conjugate distances.

$$s = \frac{s''}{m}$$

$$= f\left(\frac{1}{m} + 1\right)$$

$$= \frac{s''f}{s'' - f}$$

Second Principal Point-To-Image Distance

(approximate lens-to-image distance, often simply called image distance) s″ is one of the two conjugate distances.

$$s'' = sm$$
$$= f\,(m + 1)$$
$$= \frac{sf}{s - f}$$

Sum of Conjugate Distances

(approximate object-to-image distance)

$$(s + s'') = f\left(m + \frac{1}{m} + 2\right)$$

Focal Ratio, f-Number, or Relative Aperture

is strictly defined as f/ϕ, where ϕ is the clear aperture diameter. It is sometimes incorrectly used in the broader sense as meaning either s/ϕ or s''/ϕ, whichever is smaller. Either s or s″ may be referred to as a working distance. **Ray f-number** is f/2h, where h is the height (measured from the optical axis) at which a ray enters a lens.

Numerical Aperture (N.A.)

N.A. = $n\sin\theta$, where $\theta = \phi/2s$, and in this formula only, n is the object space refractive index. Since f-number was previously defined as s/ϕ, it is clear the N.A. and f-number are related by the formula:

$$\text{f-number} = \frac{1}{2\ \text{N.A.}}\ .$$

> **FOR QUICK APPROXIMATIONS**
>
> Much time and effort can be saved by ignoring the differences between f, f_b, and f_f in these formulas (assume $f=f_b=f_f$), by thinking of s as the lens-to-object distance, by thinking of s″ as the lens-to-image distance, and by thinking of the sum of conjugate distances s + s″ as being the object-to-image distance. These approximations are especially good at large f-numbers. Only at f-numbers approaching 1.0 is it necessary to be very careful about associating these quantities with their exact meanings, a process which necessarily involves considerable labor.

Solid Angle

The solid angle subtended by a lens, for an observer situated at an on-axis image point, is

$$\Omega = 2\pi\,(1 - \cos\theta)$$
$$= 4\pi \sin^2\left(\frac{\theta}{2}\right)\ ,$$

where this result is in steradians, and where

$$\theta = \arctan\left(\frac{\phi}{2\,s''}\right)$$

is the apparent angular radius of the lens clear aperture. For an observer at an on-axis object point, use s instead of s″. To convert from steradians to the more intuitive "sphere" units, simply divide Ω by 4π. If the Abbe sine condition (discussed later) is known to apply θ may be calculated using the arc sine function instead of the arc tangent.

Back Focal Length

$$f_b = f + A_2H''$$
$$= f - \frac{r_2 t_c}{n(r_2 - r_1) + t_c\,(n - 1)}\ ,$$

where the sign convention presented above applies to A_2H'' and to the radii. If r_2 is infinite, "l' Hôpital's Rule" from calculus must be used, whereby

$$f_b = f - \frac{t_c}{n}.$$

Front Focal Length

$$f_f = f - A_1H$$
$$= f + \frac{r_1 t_c}{n(r_2 - r_1) + t_c\,(n - 1)}\ ,$$

where the sign convention presented above applies to A_1H and to the radii. If r_1 is infinite, "l' Hôpital's Rule" from calculus must be used, whereby

$$f_f = f - \frac{t_c}{n}.$$

Edge-To-Focus Distances

For positive lenses,

$$A = f_f + s_1, \text{ and}$$

$$B = f_b + s_2,$$

where s_1 and s_2 are the sagittas of the first and second surfaces. Bevel is neglected.

Magnification or Conjugate Ratio

$$m = \frac{s''}{s} = \frac{f}{s - f} = \frac{s'' - f}{f}.$$

PARAXIAL FORMULAS FOR LENSES IN ARBITRARY MEDIA

These formulas allow for the possiblity of distinct and completely arbitrary refractive indices for the object space medium (refractive index n), lens (refractive index n′), and image space medium (refractive index n″). Refractive index is the ratio of the speed of light in vacuum to the speed of light in a medium. In this situation the EFL (effective focal length, formerly denoted f) assumes two distinct values, namely f in object space and f″ in image space. It is additionally necessary to distinguish the principal points from the nodal points. The lens serves both as a lens and as a window separating the object space and image space media.

The situation of a lens immersed in a homogenous fluid is included as a special case (n = n″). This case is of considerable practical importance. The two values f and f″ are again equal, so that the lens combination formulas are applicable to systems immersed in a common fluid. The general case (two different fluids) is more difficult, and must be approached by ray tracing on a surface-by-surface basis.

LENS CONSTANT (k)

This is a number which appears frequently in the formulas which follow. Notice that it is an explicit function of the complete lens prescription (both radii, t_c, n′) and both media indices (n and n″). This dependence is implicit anywhere that k appears.

$$k = \frac{n' - n}{r_1} + \frac{n'' - n'}{r_2} - \frac{t_c(n' - n)(n'' - n')}{n' r_1 r_2}$$

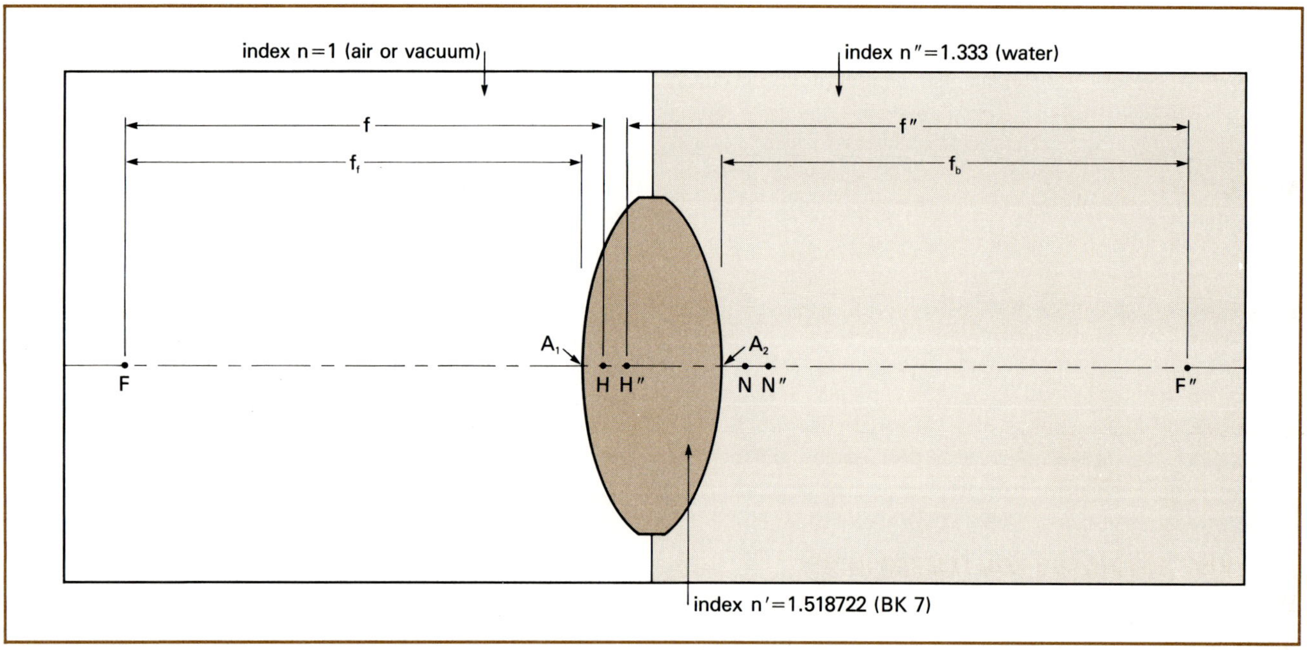

SYMMETRIC LENS WITH DISPARATE OBJECT AND IMAGE SPACE INDICES.

Effective Focal Lengths

$$f = \frac{n}{k} \qquad f'' = \frac{n''}{k}$$

Lens Formula (Gaussian form)

$$\frac{n}{s} + \frac{n''}{s''} = k$$

Lens Formula (Newtonian form)

$$xx'' = ff'' = \frac{nn''}{k^2} ,$$

where $x = s - f$ and $x'' = s'' - f''$.

Principal Point Locations

$$A_1H - \frac{nt_c}{k}\left(\frac{n''-n'}{n'r_2}\right)$$

$$A_2H'' = \frac{-n''t_c}{k}\left(\frac{n'-n}{n'r_1}\right)$$

Object To First Principal Point Distance

$$s = \frac{ns''}{ks''-n''},$$

Second Principal Point To Image Distance

$$s'' = \frac{n''s}{ks-n}$$

Magnification

$$m = \frac{s''}{s}$$

Lens Maker's Formula

$$\frac{n}{f} = \frac{n''}{f''} = k$$

Nodal Point Locations

$$A_1N = A_1H + HN$$

$$A_2N'' = A_2H'' + HN$$

Separation of Nodal Point From Corresponding Principal Point

$HN = H''N'' = (n'' - n) / k$, positive for N to the right of H and N'' to the right of H''.

Back Focal Length

$$f_b = f'' + A_2H''$$

Front Focal Length

$$f_f = f - A_1H$$

Focal Ratios

f/ϕ and f''/ϕ, where ϕ is the diameter of the clear aperture of the lens.

Numerical Apertures

$$n \sin\theta \qquad n'' \sin\theta''$$

where $\theta = \arctan\left(\frac{\phi}{2s}\right)$ where $\theta'' = \arctan\left(\frac{\phi}{2s''}\right)$

Solid Angles (in steradians)

$$\Omega = 2\pi(1 - \cos\theta) \qquad \Omega'' = 2\pi(1 - \cos\theta'')$$

$$= 4\pi\sin^2\left(\frac{\theta''}{2}\right) \qquad = 4\pi\sin^2\left(\frac{\theta}{2}\right)$$

where $\theta = \arctan\left(\frac{\phi}{2s}\right)$ where $\theta'' = \arctan\left(\frac{\phi}{2s''}\right)$

To convert from steradians to spheres, simply divide by 4π.

PHYSICAL SIGNIFICANCE OF THE NODAL POINTS

A ray directed at the primary nodal point N of a lens appears to emerge from the secondary nodal point N'' without change of direction. Conversely, a ray directed at N'' appears to emerge from N without change of direction. At infinite conjugate ratio, if a lens is rotated about a rotational axis orthogonal to the optical axis at the secondary nodal point (i.e., if N'' is the center of rotation) the image remains stationary during the rotation. This fact is the basis for the nodal slide method for measurement of nodal point location. The nodal points coincide with their corresponding principal points when the image space and object space refractive indices are equal ($n = n''$), making the nodal slide method the most precise method of principal point location.

LENS COMBINATION FORMULAS

PARAXIAL LENS COMBINATION FORMULAS

Here are some formulas that will help in selecting combinations of standard catalog lenses to perform more complex imaging tasks. Many of these tasks cannot be performed with a single lens. Any number of lenses can be dealt with by considering the combination of two lenses and then combining this "combination" with the next lens, and so on, until all the lenses in the system are accounted for. The utility of this procedure, which includes location of both combination principal points, is threefold:

1. Combination conjugate distances can be related by the simple lens formula

   $$\frac{1}{f} = \frac{1}{s} + \frac{1}{s''} ,$$

   greatly speeding the preliminary design of systems employing such combinations.
2. Proper element spacing to achieve specified combination characteristics can be found. For example, these formulas allow standard lenses to be combined to obtain focal lengths or f-numbers, either between, or totally different from, catalog values. Similarly, multi-element optical systems can be mathematically modeled, using standard lenses, saving time and money.
3. Necessary component and position tolerances can be estimated.

COMBINATION FOCAL LENGTH (EFL)

For *all* cases, whether separations are large or small, and where f_1 and f_2 values of either sign are permitted,

$$f = \frac{f_1 f_2}{f_1 + f_2 - d} .$$

This may be more familiar in the form

$$\frac{1}{f} = \frac{1}{f_1} + \frac{1}{f_2} - \frac{d}{f_1 f_2} .$$

Notice that the formula is symmetric with respect to interchange of the lenses (end-for-end rotation of the combination) at constant d. The next two formulas are not.

COMBINATION FOCAL POINT LOCATION

For *all* cases,

$$s_2'' = \frac{f_2 (f_1 - d)}{f_1 + f_2 - d}$$

COMBINATION SECONDARY PRINCIPAL POINT LOCATION

For *all* cases,

$$z = s_2'' - f.$$

SYMBOLS

- f = combination focal length (EFL), positive if combination final focal point falls to right of combination secondary principal point, negative otherwise.
- f_1 = focal length (EFL) of first element.
- f_2 = focal length (EFL) of second element.
- d = distance from secondary principal point of first element to primary principal point of second element (positive if primary principal point is to right of the secondary principal point, negative otherwise).
- s_2'' = distance from secondary principal point of second element to final combination focal point (location of final image for object at infinity to left), positive if the focal point is to the right of second element secondary principal point.
- z = distance to combination secondary principal point measured from secondary principal point of second element, positive if combination secondary principal point is to right of secondary principal point of second element.

Note:

These formulas are paraxial and apply to coaxial combinations of both thick and thin lenses immersed in any fluid with refractive index independent of position. They assume that light propagates from left to right through an optical system.

GENERALIZATION FROM COMBINATIONS TO SYSTEMS

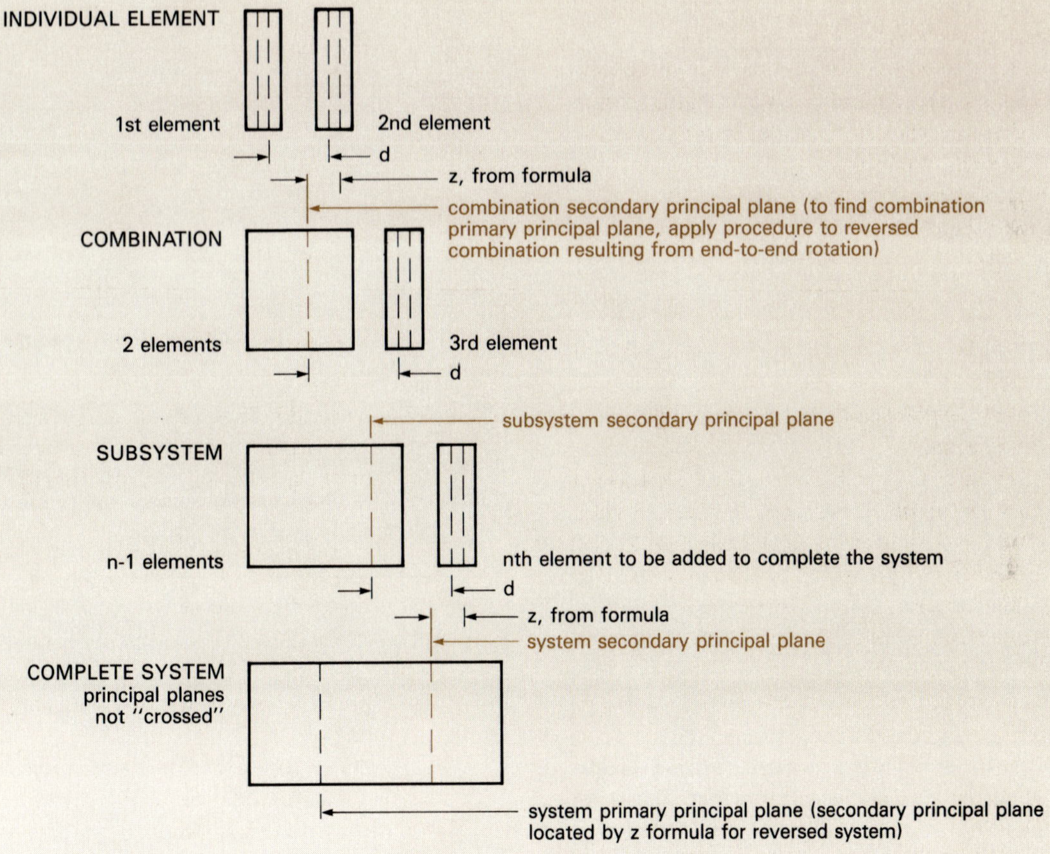

Lens combinations or systems may exhibit "crossed" principal planes as single lenses cannot.

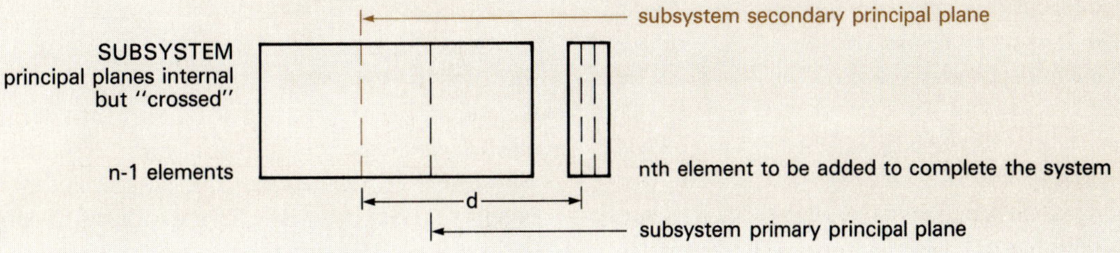

It is also possible for a lens combination or system to exhibit principal planes which are far removed from the system. When such systems are themselves combined, negative d-values may arise. Probably the simplest example of a negative d-value situation is a pair of meniscus lenses with steep surface vertices in contact, as shown in the following drawing.

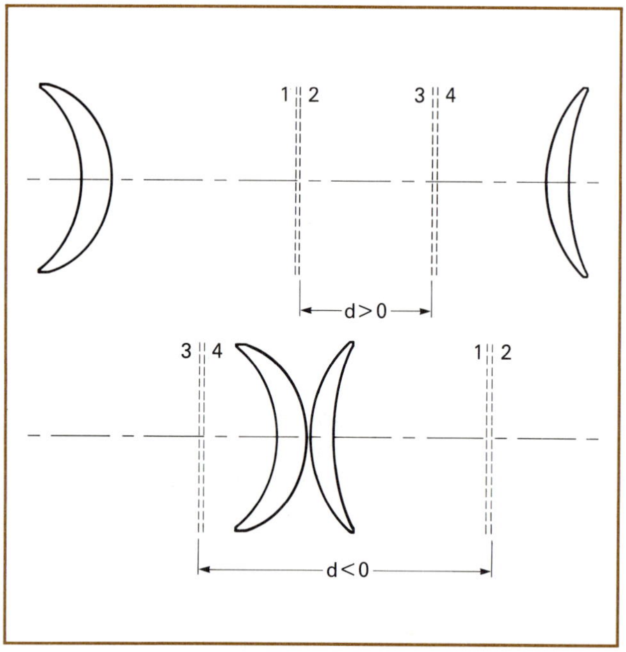

NEGATIVE d-VALUE example is shown using extreme meniscus form lenses having external principal planes. When the lenses are brought into contact, a negative d-value situation can occur. (Drawing not to scale.)

The paraxial combination formulas are useful for simple lenses in predicting where images can be expected to fall in the paraxial limit. However, they tell nothing about image quality or location of the plane of best focus. The paraxial combination formulas are much more useful in connection with well-corrected achromats, whose paraxial and full-aperture properties should not differ significantly.

There is no simple way of combining the aberrations of individual system elements, or of predicting image quality. More complete understanding of the behavior of a lens combination requires exact surface-by-surface trigonometric ray tracing at full aperture, usually at several field angles.

COMBINATION EXAMPLES

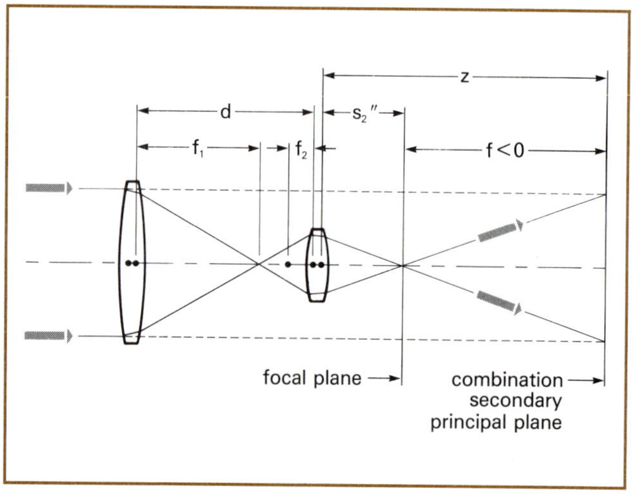

PAIR OF POSITIVE LENSES separated by distance d greater than $f_1 + f_2$. Symmetry of the lenses is not required. f is *negative*, while both s_2'' and z are positive.

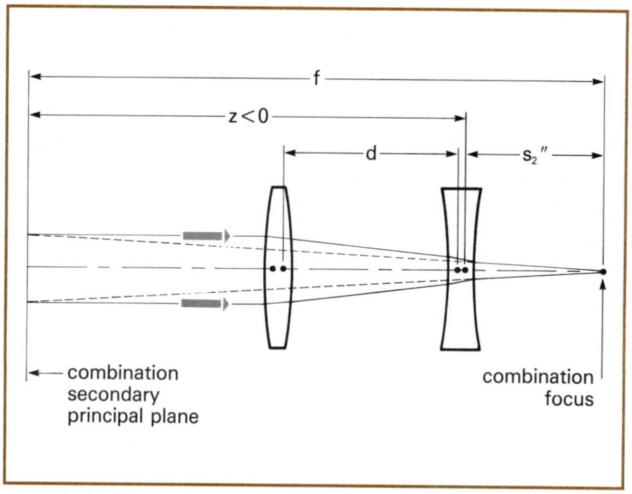

TELEPHOTO COMBINATION. Symmetry of the lenses is not required. The EFL, and hence the image size, can be made much larger than the distance from combination to image would suggest. Suppose that f_1 is positive and $f_2 = -f_1/2$, for example. Then f is negative for d less than $f_1/2$, infinite for $d = f_1/2$ (Galilean telescope or beam expander), and positive for d larger than $f_1/2$. To make the example even more specific, cat-

alog lenses 01 LDX 189 and 01 LDK 021, with d = 78.2mm, will yield $s_2'' = 2.0$ meters, f = 5.2 meters, and z = - 3.2 meters. Hand-held in sunlight, a solar image of 45.4mm diameter is easily produced, while the comination secondary principal plane is 3.2 meters overhead.

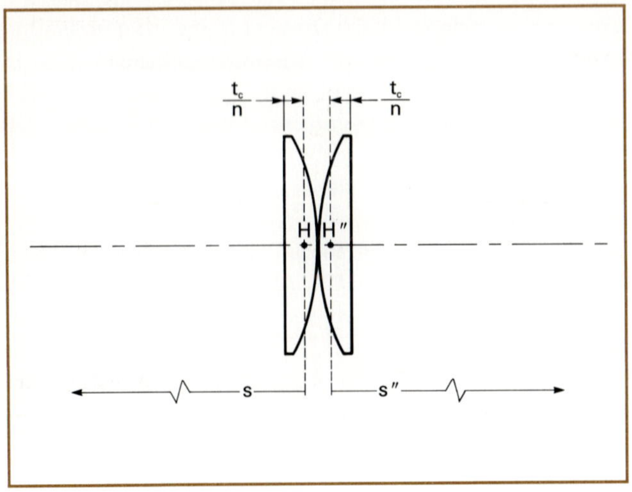

PAIR OF IDENTICAL PLANO LENSES in air, used as a condenser, with convex vertices in contact. The lenses could also be plano aspherics. Because of contact, d = 0, f = $f_1/2 = f_2/_2$, $f_1/_2 = s_2''$, and z = 0. The second element secondary principal point and combination secondary principal point coincide at H″, at depth t_c/n beneath the vertex of the plano surface of the second element, where t_c is the element center thickness and n is the refractive index of the element. By symmetry, the combination primary principal point is similarly located in the first element. Combination conjugate distances must be measured from these points.

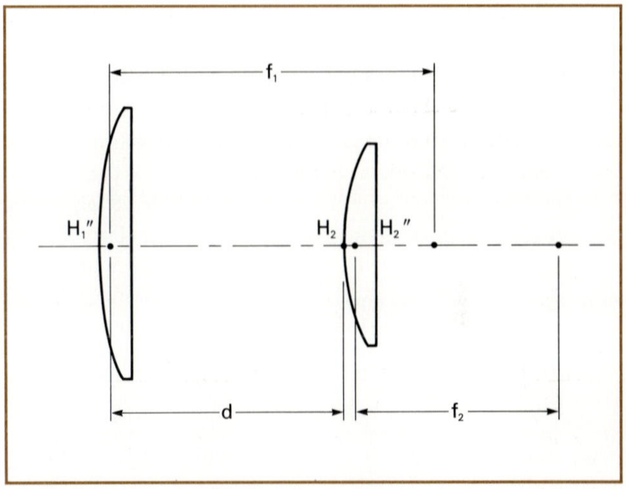

ACHROMATIC COMBINATIONS. Air-spaced lens combinations can be made achromatic, in the sense that the derivative of focal length with respect to wavelength can be made to vanish at the design wavelength and kept small at adjacent wavelengths, even though both elements are made of the same material. This fact is the basis for Huygens and Ramsden eyepieces. Achievement of achromatism in the present sense requires, in the thin-lens approximation, only that

$$d = \frac{(f_1 + f_2)}{2}.$$

This approximation is also adequate for most thick-lens situations. The signs of f_1, f_2, and d are unrestricted, while d must have a value which guarantees the existence of an air space. Element shapes are unrestricted, and hence can be chosen to compensate other aberrations after the above criterion has been met.

COST EFFECTIVE DESIGN USING CATALOG LENSES

There are many benefits to the potential lens user in selecting a combination of simple, off-the-shelf, catalog lenses to perform quite complex imaging tasks. Ready availability of prototypes, lower design costs, lack of tooling costs and lower prices when you go into production are some of these benefits.

We have all our lens prescriptions filed on magnetic disc and readily accessible to lens design programs. Our Applications Engineers will be pleased to assist you in the selection and evaluation of lens combination systems.

ABERRATIONS OF LENSES

The direction of a light ray after refraction at the interface between two homogeneous, isotropic media of differing index of refraction is given by Snell's Law: $n_1 \sin \theta_1 = n_2 \sin \theta_2$.

Here θ_1 is the angle of incidence, θ_2 is the angle of refraction, and both angles are measured from the surface normal as is shown in the accompanying figure. In theory, to determine the imaging characteristics of a lens of lens system, we could trace many rays through the system, applying Snell's Law at each interface. In practice, we can estimate the image quality of most optical systems by using certain approximations. The sine functions in Snell's Law can be expanded in infinite Taylor series, i.e.,

$$\sin \theta_1 = \theta_1 - \theta_1^3/3! + \theta_1^5/5! - \theta_1^7/7! + \theta_1^9/9! - \ldots$$

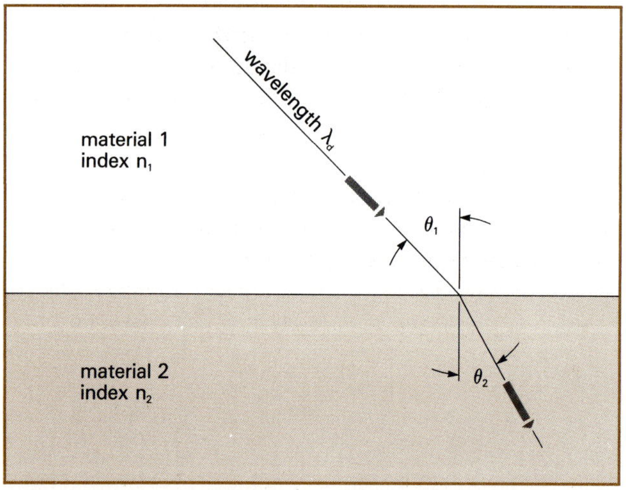

REFRACTION OF LIGHT at a dielectric boundary.

The first approximation is to simply replace all sine functions with their arguments, i.e., replace $\sin \theta_1$ with θ_1 itself and so on. This is called "first order theory" or paraxial theory because we are only using the first terms of the sine expansions. All the formulas given in the preceding pages are paraxial results. While it would seem that this is an oversimplification, in fact paraxial results are of great utility in determining system focal length, magnification, conjugate distances, etc., and are in fact nearly always used in the first stages of designing an optical system. A perfect optical system (one without any aberrations) would form its image at the point and to the size indicated by paraxial theory, so aberrations are really a measure of how the image differs from the paraxial prediction.

The next step we could take in improving the approximation is to include the third order terms in the sine expansions. The results of doing this were investigated extensively by Seidel, and third order lens aberrations are therefore called the Seidel aberrations. Seidel put the aberrations of an optical system into several different classifications. In monochromatic light they are spherical aberration, astigmatism, field curvature, coma and distortion. In polychromatic light there are also chromatic aberration and lateral color. Seidel developed a formalism for approximation of each of these aberrations without the necessity of actually tracing large numbers of rays using all the terms in the sine expansions. In addition, it is possible to assess the aberrative contributions of each individual surface in an optical system using the Seidel aberrations. While in reality, aberrations occur in combination rather than alone, this system of classifying them makes analysis much simpler, and gives a good description of the image quality of many optical systems.

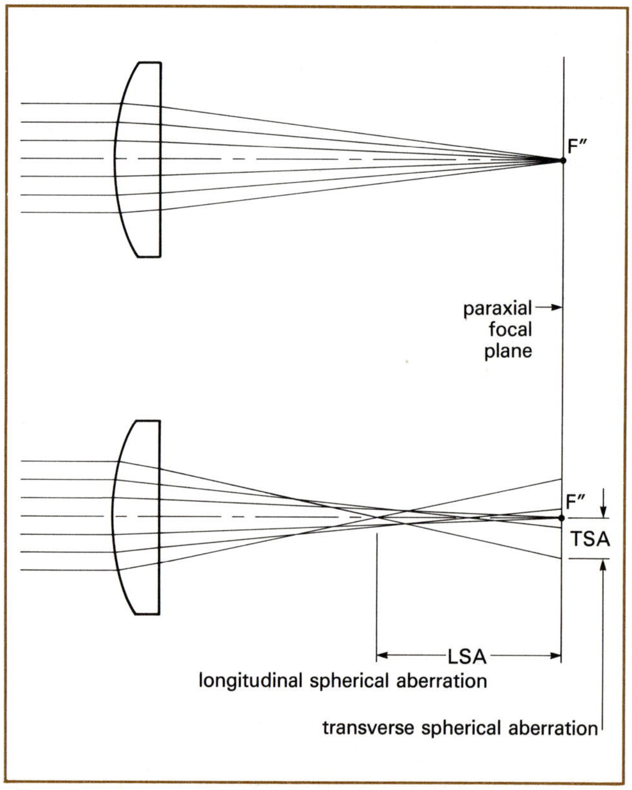

SPHERICAL ABERRATION of a plano-convex lens.

SPHERICAL ABERRATION

In the figure we see how an aberration free lens focuses incoming collimated light. All the rays pass through the focal point F″. In the second part of the figure is shown the situation more normally encountered in single lenses. The farther from the optical axis the ray enters the lens, the nearer to the lens it focuses. The distance along the optical axis between the intercept of the rays which are nearly on the optical axis (paraxial rays) and rays which went through the edge of the lens (marginal rays) is called longitudinal spherical aberration (LSA). The difference in the height at which these rays intercept the paraxial focal plane is called transverse spherical aberration (TSA). As can be seen, these quantities are related by (transverse aberration) = (longitudinal aberration) x tan u″.

Spherical aberration is dependent on the shape of the lens, its orientation and the conjugate ratio at which it is being used. Decision parameters, for choosing the best lens shape and orientation for a given task, are presented later in this chapter. In general, simple positive lenses have undercorrected spherical aberration; negative lenses usually have overcorrected spherical aberration. By combining positive and negative elements with nearly equal but opposite spherical aberration, we can greatly correct a system. This is done in doublets, where a positive and a negative lens are cemented together; the result being a lens with very little spherical aberration.

ASTIGMATISM

A schematic view of an optical system imaging an off axis point is shown in the figure. Rays which lie in the plane containing both the optical axis and the object point are called meridional or tangential rays. Rays not in this plane are referred to as skew rays. The chief, or principal ray goes from the object point through the center of the aperture of the lens system. The plane perpendicular to the tangential plane that contains the chief ray is called the sagittal or radial plane. In the figure we see that tangential rays from the object come to a focus closer to the lens than do rays in the saggital plane. When we evaluate the image at the tangential conjugate we see a line in the sagittal direction. A line in the tangential direction is formed at the sagittal conjugate. In between these conjugates the image is either an elliptical or circular blur. The separation of these conjugates is astigmatism.

The amount of astigmatism in a lens depends on lens shape only when there is an aperture in the system that is not in contact with the lens itself. (In all optical systems there is an aperture or stop, although in many cases it is simply the clear aperture of the lens element itself.) Astigmatism depends strongly on conjugate ratio.

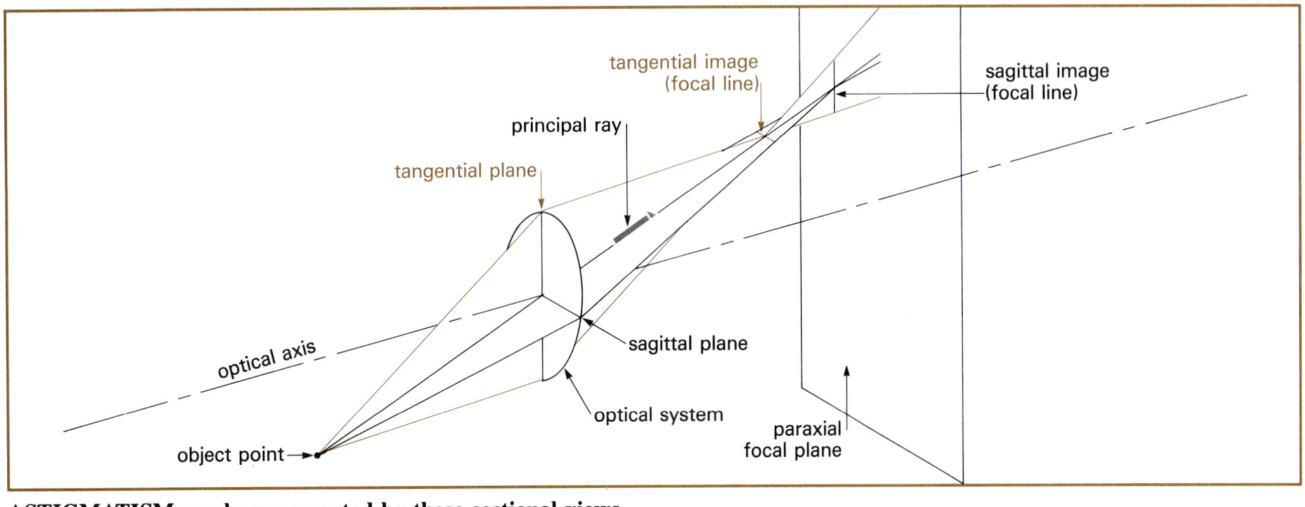

ASTIGMATISM can be represented by these sectional views.

COMA

Coma is defined to be the variation of magnification (conjugate ratio) with aperture. This effect is somewhat analogous to spherical aberration, but here rays from the edge of the lens come to a focus closer to the optical axis than do those passing through the center of the lens. In the second part of the figure we see the relationship between the ray's position at the lens aperture and its final position in the image plane. As can be seen, the rays which enter the lens opposite each other are imaged at the same point. Furthermore, rays from both the tangential and sagittal plane are both imaged in the tangential plane. If we construct the image of several concentric circles of rays passing through the lens we see that they are imaged as circles, but that they now lie tangent to a 60° angle.

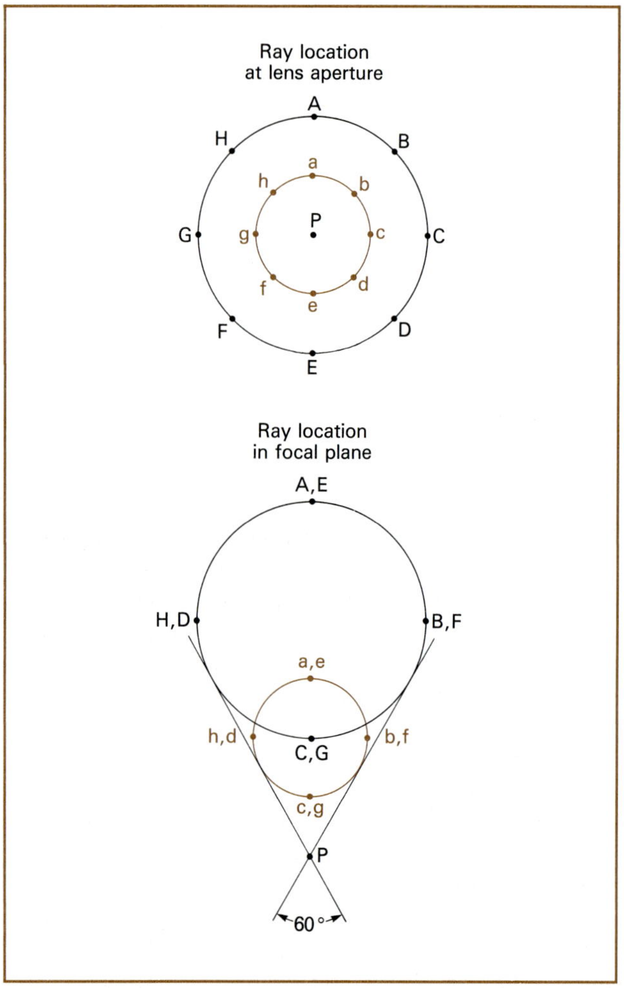

COMA can be illustrated by the focal position of rays compared with their pupil coordinates.

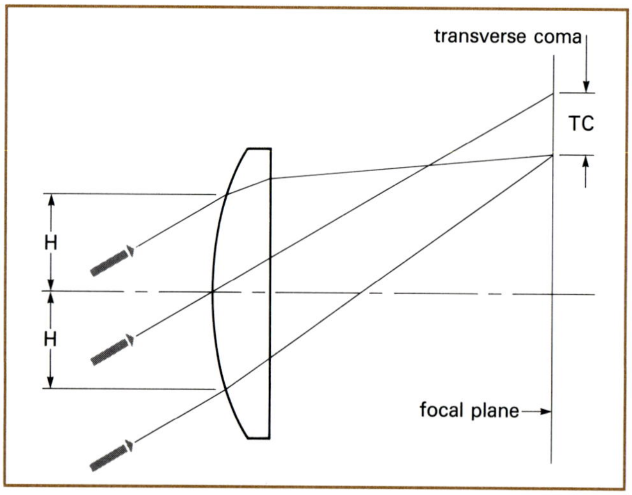

TRANSVERSE COMA.

Coma, again like spherical aberration, depends on lens shape, lens orientation and magnification.

FIELD CURVATURE

Even in the absence of astigmatism there is a tendency of optical systems to image better on curved surfaces rather than flat planes. This effect is called field curvature. In the presence of astigmatism

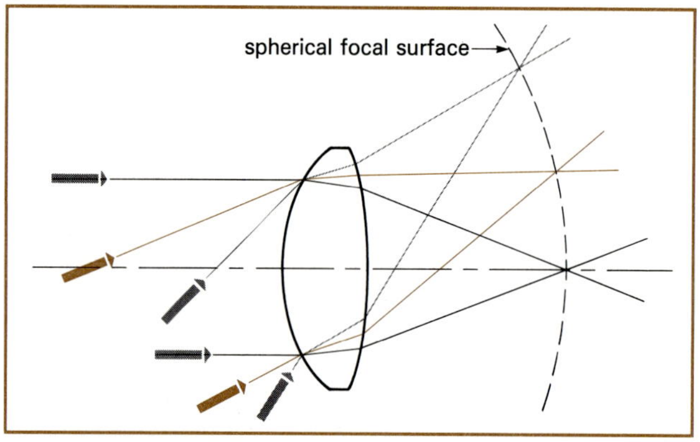

FIELD CURVATURE.

this problem is compounded, since there are two separate astigmatic focal surfaces corresponding to the tangential and sagittal conjugates.

Positive lens elements usually have inward curving fields, while negative lenses have outward curving fields. Curvature of field can be corrected to some extent by combining positive and negative lens elements, as is done when using LAO series achromats and their corresponding LFF series field flatteners.

DISTORTION

When the image of an off-axis point is not formed at the location predicted by the paraxial equations, the image is said to be distorted. This aberration is distinct from coma (where rays from an off-axis point failed to meet perfectly in the image plane). Distortion means that even if a perfect off-axis point image is formed, its location on the image plane is not correct. Furthermore, the amount of distortion usually increases with increasing image height. The effect of this can be seen in the figure which shows the object and two different kinds of distortion. Distortion does not lower system resolution; it simply means that the image shape does not exactly correspond to the shape of the object. Distortion is defined to be separation of the actual image point from the paraxially predicted location on the image plane, and can be expressed either as an absolute value or a percentage of the paraxial image height.

It should be apparent that a lens or lens system has opposite types of distortion depending on whether it is used forwards or backwards. This means that if a lens were used both to make a photograph, and then used in reverse to project it, there would be no distortion in the final screen image. Also, perfectly symmetrical optical systems at 1:1 magnification have no distortion (or coma for that matter).

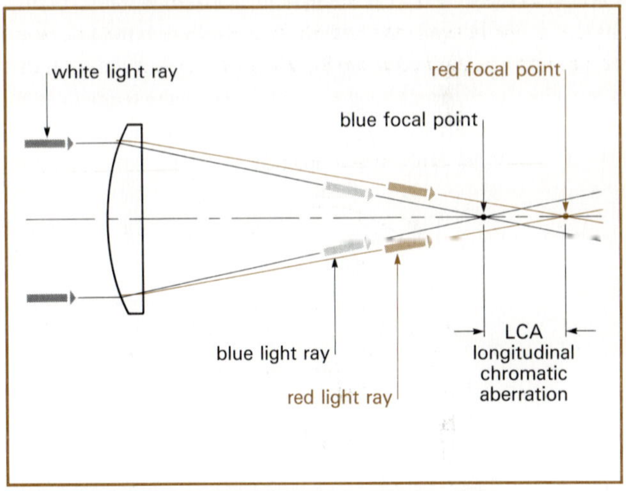

LONGITUDINAL CHROMATIC ABERRATION.

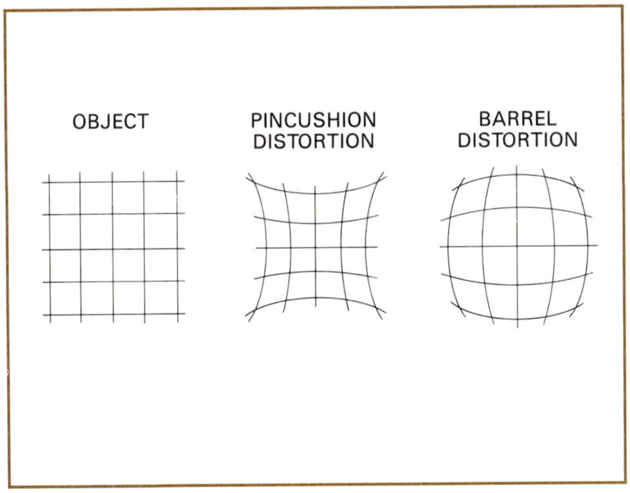

DISTORTION is illustrated by these two image types.

CHROMATIC ABERRATION

The index of refraction of a material is a function of wavelength. The exact relationship between index and wavelength is given in the chapter on materials along with tables for some materials. If we go back to the equation for Snell's Law (see page 16), it can be seen that light rays of different wavelengths or colors will be refracted at different angles, since the index is not a constant. The figure shows the result when polychromatic collimated light is incident on a positive lens element. Because the index of refraction is higher for shorter wavelengths, these are focused closer to the lens than are longer wavelengths. Longitudinal chromatic aberration is the axial distance from the nearest to the farthest focal point.

As in the case of spherical aberration, positive and negative elements have opposite signs of chromatic aberration. Once again, by combining elements of nearly opposite aberration to form a doublet, we can partially correct chromatic aberration. It is necessary to use two glasses with different dispersion characteristics, so that the weaker negative element can balance the aberration of the stronger positive element. Dispersion is defined as the amount that index of refraction varies with wavelength over some internal.

LATERAL COLOR

The figure shows the chief ray of an optical system consisting of a simple positive lens and a separate aperture. Because of the change in index with wavelength, blue light is refracted more strongly than red light; the result being that the rays intercept the image plane at different heights. Stated simply, the magnification depends on color. Lateral color is strongly dependent on the location of the system stop; for a simple lens it is zero when the stop is in contact with the lens.

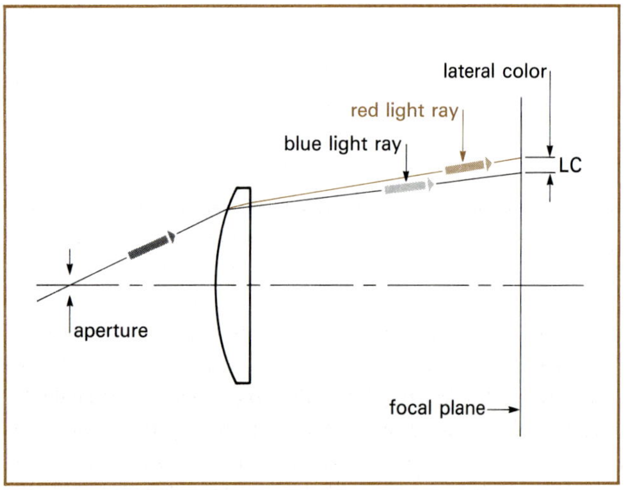

LATERAL COLOR is the difference in image height between red and blue rays.

HIGH ORDER ABERRATIONS

Having considered using only the third order term in Snell's Law, we might now ask what the results of including the fifth order term would be. Unfortunately, inclusion of this term considerably complicates matters and it is no longer possible to separate and quantify the aberrations as neatly as in the third order case. The advantage of being able to determine the magnitude of aberrations by simple formulas is not available when fifth order terms are taken into account. For many optical systems, the third order term is all that may be needed to quantify the aberrations. In highly corrected systems or those having large apertures or a large angular field-of-view, third order theory is inadequate. In these cases exact trigonometric ray tracing is essential.

Variation of Aberrations with Aperture, Field Angle, and Image Height

Aberration	Aperture (ϕ)	Field Angle (θ)	Image Height (y)
Lateral Spherical	ϕ^3	—	—
Longitudinal Spherical	ϕ^2	—	—
Coma	ϕ^2	θ	y
Astigmatism	—	θ^2	y^2
Field Curvature	—	θ^2	y^2
Distortion	—	θ^3	y^3
Chromatic	—	—	—

The purpose of lenses is to form images. It is natural to wonder how well they perform this function. Simple lenses typically aberrate badly at full aperture, so diffraction theory is useless. In such a case, how can you specify performance? The answer is simple, *turn to page 40 for an introduction to MTF.*

PREFERRED CONJUGATE RATIOS

Most lenses and lens systems are designed for optimum performance at particular conjugate ratios (ratios of image distance to object distance), and there is some loss of resolution in using such systems and lenses at other than the intended conjugate ratio. Microscope objectives are an example of this, and the intended ratios are often engraved upon them. In this sense, asymmetric lenses are not reversible devices.

At any conjugate ratio the singlet shape which minimizes spherical aberration is called "best-form" because spherical aberration is often most important, especially near the optical axis. The criterion for best-form at any conjugate ratio is that marginal rays be equally deviated at both lens surfaces. This is also the criterion for minimum surface reflectance loss. Happily, it turns out, for both infinite and unit conjugate ratios, that absolute coma is nearly minimized for the best-form shape. For wide field applications best-form is not the optimum singlet shape, especially at infinite conjugate ratio, where it yields maximum field curvature. Normally a doublet or triplet is chosen for wide field applications. As the conjugate ratio is varied, the ideal singlet lens shape also varies. Ideal lens shape is also affected by the position of any stop (not in contact or near-contact) with which the lens may be used.

UNIT CONJUGATE RATIO

Periscope relay lenses, copying lenses, and the condenser lenses of many simple illumination systems normally operate at unit conjugate ratio ($s = s'' = 2f$). At this conjugate ratio best-form is the symmetric convex lens. Not only is spherical aberration minimized, but coma, distortion, and lateral chromatic aberration exactly cancel out. These last results presume that the lens is its own aperture stop, and must be modified if a remote stop is present. These results are true regardless of material index or wavelength, and explain the importance of symmetric convex lenses.

For demanding requirements at specific conjugate ratios and focal ratios, a computer-optimized optical design will be preferred in many applications. Melles Griot will be pleased to design and manufacture optical systems and subsystems to your specific requirements.

INFINITE CONJUGATE RATIO

Camera lenses usually operate at, or close to, infinite conjugate ratio ($s = \infty$, $s'' = f''$). In the figure which follows it is assumed that collimated light is incident from the left and focused to the right. The curves have all been calculated for refractive index 1.518722 (BK 7 at the mercury green line, 546.1mm). The Coddington shape factor q is defined as

$$q = \frac{r_2 + r_1}{r_2 - r_1},$$

and is a function only of the radii of curvature of the lens surfaces. It can be seen from this figure that the plano-convex shape ($q = 1$), with convex side toward the infinite conjugate, is a very good shape for use at infinite conjugate ratio. It exhibits near-minimum spherical aberration, near-minimum total transverse aberration, and near-zero coma. Reversing the rays, the same lens in the same orientation is suitable for producing a collimated beam from a point source. These facts, along with ease of manufacture, explain the importance of plano-convex and near-plano-convex lenses.

Focal length is held constant at 100mm as shape is varied via the Coddington shape factor q. The data shown applies specifically to a ray f-number (f/2h, where h is the entrance ray height from the axis) of 5.0. Off-axis data is shown only for a field angle of 11 degrees. Lens thickness is zero (thin lens approximation). All data is paraxial in the sense that all aberration values are measured either within or with reference to the paraxial focal plane, which is not the plane of best geometric focus. All aberration values are measured in millimeters. Aberration values at the plane of the best geometric focus will differ significantly from those shown here.

This figure also demonstrates the existence of an asymmetric shape, corresponding to a q-value of about 0.7426 for this material and wavelength, which truly is the best singlet shape for on-axis imaging. This shape, for which spherical aberration is minimized, is called "best form" and simultaneously minimizes transverse spherical aberration at both the paraxial focal plane and the plane of best geometric focus. Best form shapes are used in Laser Line Focusing Singlet lenses (see page 120).

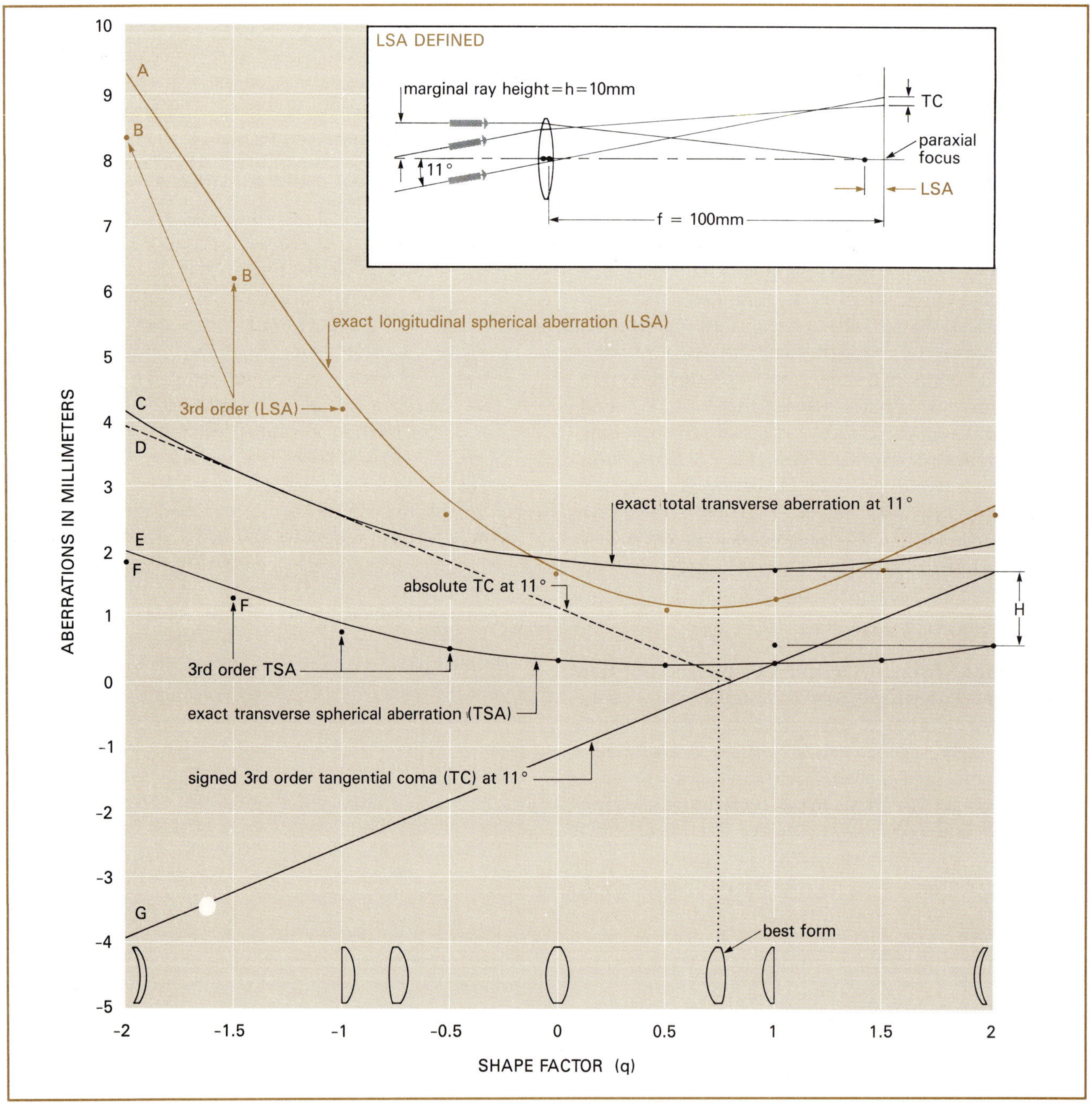

Aberrations of Positive Singlets at Infinite Conjugate Ratio as a Function of Shape.

A: Longitudinal spherical aberration (LSA) is defined in the inset to the figure. It is the distance in front of the paraxial focal plane at which the marginal ray crosses the optical axis, and is measured from the paraxial focal plane. LSA is the only variable shown which is measured out of the paraxial focal plane instead of within it, and LSA is therefore shown in brown instead of black. The solid brown curve is exact LSA, calculated from exact trigonometric ray traces.

B: The isolated brown points show the behavior of 3rd order LSA, for comparison with the exact curve. Agreement is excellent for q-values between 0 and 1.

C: Total transverse aberration is the major dimension (edge-to-edge) of the off-axis blur patch on the paraxial focal plane for collimated light incident from the left at a field angle of 11 degrees. It is exact in the sense that it is computed from exact trigonometric ray traces. It includes contributions from spherical aberration, coma, astigmatism, and curvature of field. Notice that this curve has its minimum at the same q-value as the minima of curves A and E.

D: Absolute TC at 11 degrees (the dashed straight line) is the absolute value of signed 3rd order tangential coma (TC) at 11 degrees. See discussion of curve G below.

E: Exact transverse spherical aberration (TSA) is the diameter of the on-axis blur circle at the paraxial plane, computed from exact trigonometric ray traces. At the present f-number (5.0) the spherical aberration is almost pure 3rd order, and it can therefore be said that the same ray traces would yield geometric blur circle diameters at the plane of best geometric focus about 4 times smaller than on the paraxial plane. These smaller values correspond to the geometric term in the hybird spot size formula which we present later in this text. At lower f-numbers, where higher order spherical aberration sets in, the ratio of geometric spot diameter at the paraxial focal plane to geometric spot diameter at the plane of best geometric focus is significantly different from 4.

F: The isolated points are calculated from 3rd order approximations for comparison with the exact curve E.

G: Signed 3rd order tangential coma (TC) is defined except for sign in the inset to the figure, and is again measured within the paraxial focal plane. The reference point for TC is the intercept of the chief (central) ray with the paraxial focal plane. TC is measured from this reference point to the intercept of the off-axis marginal ray with the paraxial focal plane. The applicable sign convention is that TC is positive if measured upward from the reference point, and negative if measured downward. TC is negative in the inset, in agreement with curve G for a q-value of 0.0, since the example lens chosen for the inset is symmetric. For clarity, the magnitudes of TC and LSA are exaggerated in the inset.

H: The distance H applies to a q-value of 1.0 only, i.e., to plano convex lenses in the preferred orientation (at infinite conjugate ratio). At this q-value TC and TSA are approximately equal, so that their sum is twice TSA. H measures the difference between exact total transverse aberration and the sum of TC and TSA. Thus, for the chosen field angle of 11 degrees, H is identified with both astigmatism and curvature of field. H is considerably larger than the sum of TC and TSA, demonstrating that the 3rd order formula for TC is of little practical use for field angles this large. For this reason we do not publish the 3rd order TC formula. The exact total transverse aberration (see discussion of curve C above) is a direct measure of the size of the focal blur patch, and is therefore of immediate practical importance. TSA, TC, and (TSA+TC) are unrelated to the size of the blur patch at this field angle.

Our Applications Engineers can supply you with paraxial properties and performance characteristics of any of our lenses as applied to your particular application.

DETAIL OF FOCAL REGION (Inset: previous page)

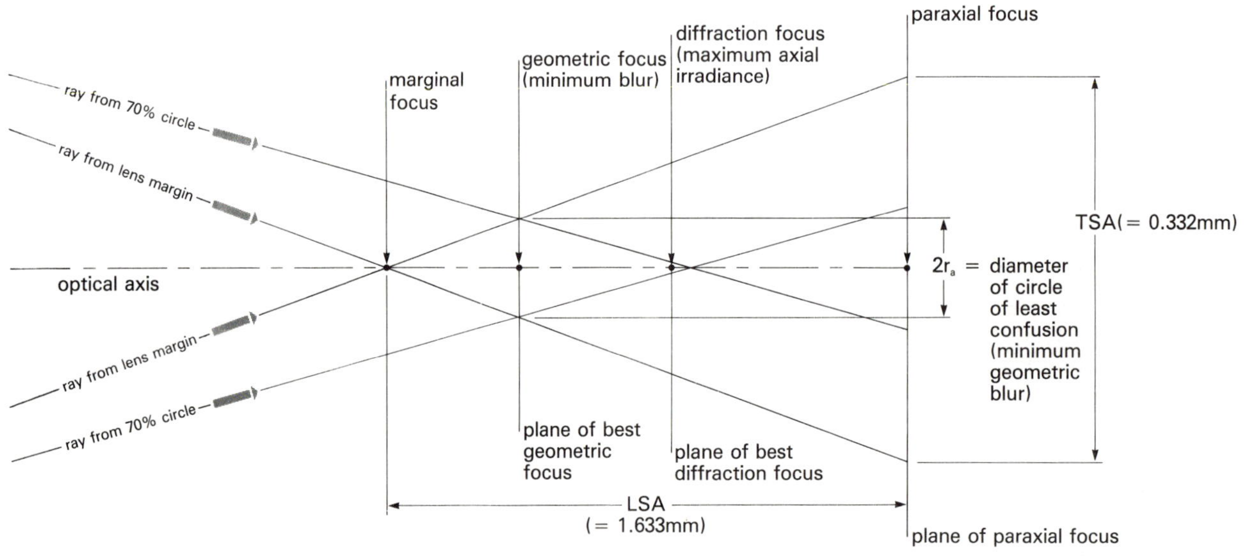

Exaggerated for clarity. Horizontal and vertical scales are different.

DETAIL OF FOCAL REGION IN THE INSET TO THE PREVIOUS FIGURE.

The focal region is shown in meridional section. The figure is approximately to longitudinal scale in the 3rd order thin lens limit. The f-number has been exaggerated for clarity (1.0 here versus 5.0 in the previous figure), with the result that transverse scale is also exaggerated.

The plane of best geometric focus is defined by the intersection of rays from the lens margin with rays arriving from a circle having a diameter of $1/\sqrt{2}$ times (or approximately 70% of) the lens diameter. For pure 3rd order spherical aberration this occurs at a plane located at 3/4 of the distance from the paraxial to the marginal focus. At this plane the circle of least confusion is found. Alternatively, it is said that the geometric blur circle has its minimum diameter at this plane. For pure 3rd order spherical aberration, this minimum diameter is exactly 1/4 of the diameter (TSA) of the geometric blur at the paraxial focal plane.

The plane of best diffraction focus passes through that point at which the axial irradiance is maximum. In the limit of pure 3rd order spherical aberration, this point is midway between the marginal and paraxial foci.

For the example shown, LSA = 1.633mm and TSA = 0.332mm.

These ratios 3/4 and 1/4, and the relative positions of the foci, are correct only for pure 3rd order spherical aberration. They change significantly for singlets at low f-number (with the onset of significant higher order spherical aberration).

LENS COMBINATIONS

The previous section examined the aberrations of single lenses. We can achieve much better correction in an optical system by using more than one element; this section presents some guidelines for that selection. Once again, we will examine the two cases of an infinite conjugate ratio system and a unit conjugate ratio system.

INFINITE CONJUGATE RATIO

As shown in the previous discussion, the best form singlet lens for use at infinite conjugate ratios is generally nearly plano-convex. The figure shows a plano-convex lens (01 LPX 023) with incoming collimated light at 546.1nm wavelength. This drawing, including the rays traced through it, is shown to exact scale. As can be seen, the marginal ray (ray f-number 1.5) strikes the paraxial focal plane significantly off the optical axis.

To improve this situation we use a two element system. The second part of the figure shows a precision achromat (01 LAO 014), which consists of a positive crown element cemented to a negative meniscus flint element. This is drawn to the same scale as the plano convex lens. No spherical aberration can be discerned in the lens at this scale. Of course, not all of the rays pass exactly through the paraxial focal point, but in this case the departure is measured in microns, rather than in millimeters as in the case of the plano-convex lens. Additionally, chromatic aberration (not shown) is corrected much better corrected in the doublet. It is important to recognize that even with monochromatic light the doublet's performance is superior.

We can get a feeling for the f-number at which singlet performance becomes unacceptable from this drawing. The ray with f-number 7.5 practically intercepts the paraxial focal point, and the f/3.75 ray is not far off. This drawing can be scaled to fit a plano-convex lens of any focal length, and thereby to estimate the magnitude of its spherical aberration, although lens thickness does affect the results slightly.

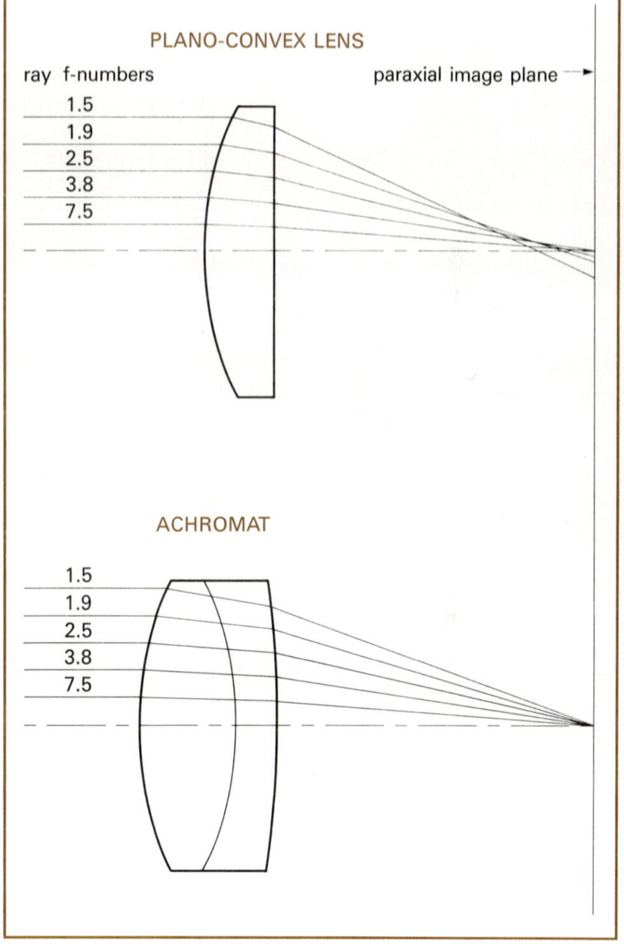

UNIT CONJUGATE RATIO

The figure shows three possible systems for use at unit conjugate ratio. All are shown to the same scale and using the same ray f-numbers with light of 546.1nm wavelength. The first system is a symmetric biconvex lens (01 LDX 027) already shown to be the best form singlet in this application. Clearly, there is significant spherical aberration present in this lens at f/2.7. Not until f/13.3 does the ray closely approach the paraxial focus.

A dramatic improvement in performance is gained by going to two identical plano-convex lenses with convex surfaces facing and nearly in contact. Those shown are both 01 LPX 081. They yield almost exactly the same focal length as the biconvex lens. If we were to split the biconvex lens down the middle, we would have two identical plano-convex lenses, each working at infinite conjugate ratio, but with the convex surface toward the focus, rather than vice-versa. This orientation is opposite to that we have already shown to be optimum for this shape lens. If we reverse these lenses, we have the system just described and get much better correction of spherical aberration.

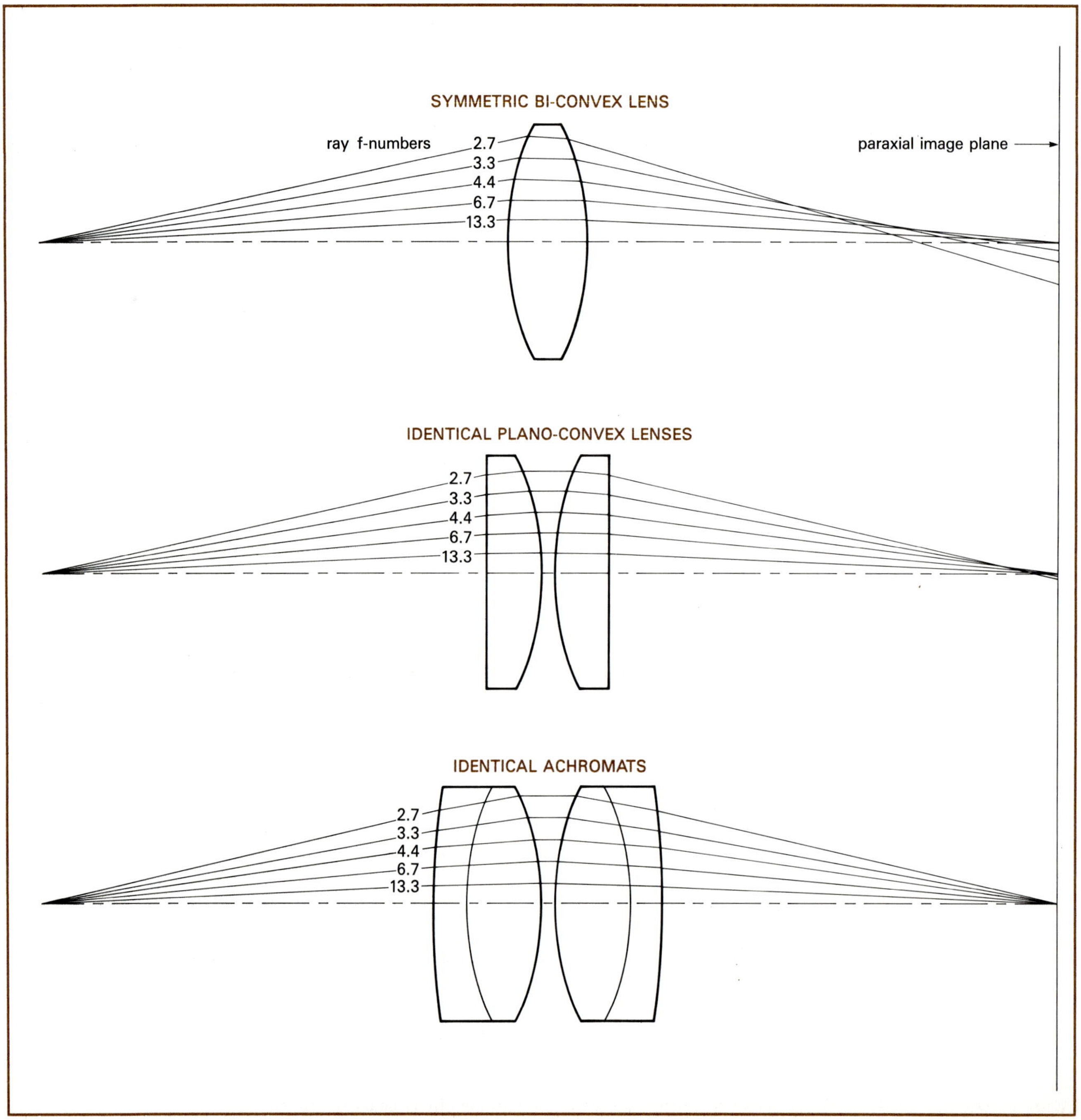

SYMMETRIC BI-CONVEX LENS
ray f-numbers
2.7
3.3
4.4
6.7
13.3
paraxial image plane
IDENTICAL PLANO-CONVEX LENSES
2.7
3.3
4.4
6.7
13.3
IDENTICAL ACHROMATS
2.7
3.3
4.4
6.7
13.3

We saw in the previous examples that an achromat is superior in performance to a singlet when used at infinite conjugate ratio and at low f-numbers. Since the unit conjugate case can be thought of as two lenses, each working at infinite conjugate ratio, the next natural step is to replace the plano-convex singlets with achromats, yielding a four element system. The third part of the figure shows such a system (composed of two 01 LAO 037's). Once again, spherical aberration is not evident at this scale, even in the f/2.7 ray.

SELECTION OF LENS SHAPE

The following discussion will help you to select either a plano-convex or a symmetric-convex lens according to its intended use. These results are independent of f-number provided it is high enough to validate 3rd order theory, yet not so high as to be diffraction limited. Small field angles are also assumed; at large field angles other aberrations, especially field curvature, become prominent.

The object distance at which the symmetric convex lens and plano-convex lens (with convex surface facing the real object) exhibit equal spherical aberration according to 3rd order theory is about

$$s_2 = f\,\frac{8(n^2 - 1)}{4n^2 - n - 6}\ .$$

The subscript 2 serves to identify this special object distance in the accompanying illustration. The magnification corresponding to s_2 is

$$m_2 = \frac{f}{s_2 - f}\ .$$

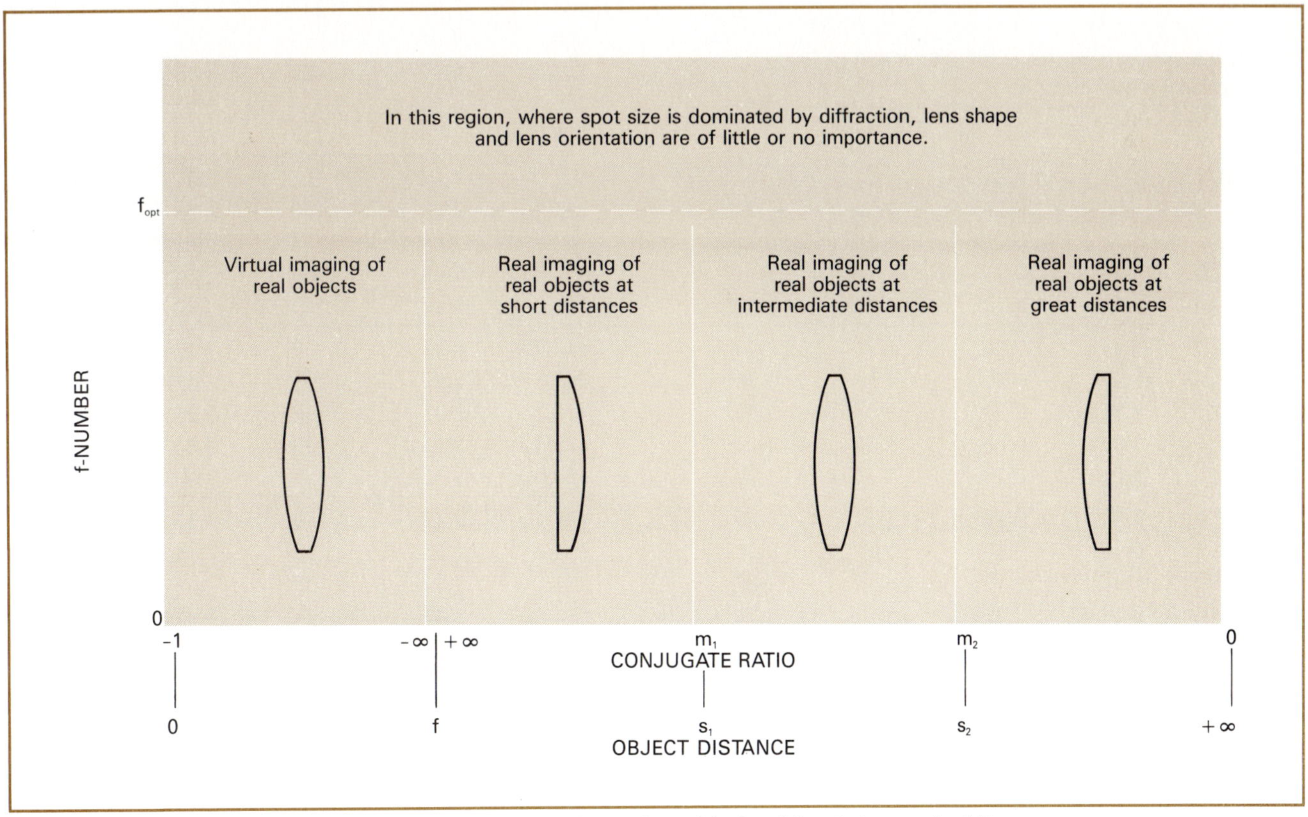

LENS SHAPE PREFERENCES to minimize spherical aberration with the object being to the left.

At the reciprocal of this magnification, namely

$$m_1 = \frac{1}{m_2} = \frac{s_2 - f}{f},$$

the symmetric convex lens and plano-convex lens again exhibit equal spherical aberration, but now the plane surface must face the object. The corresponding object distance, namely

$$s_1 = f\left(\frac{m_1 + 1}{m_1}\right) = f\frac{8(n_2 - 1)}{4n_2 + n - 2},$$

is only slightly larger than f. These results are implicitly wavelength dependent, through the refractive index n. Values for both BK 7 and synthetic fused silica are tabulated for several spectrum lines below.

Decision Parameters for BK 7 Simple Lenses (As Functions of Wavelength)

λ (nm)	s_2/f	m_2	m_1	s_1/f
351.1	5.66	0.215	4.66	1.21
363.8	5.71	0.212	4.71	1.21
365.0	5.71	0.212	4.71	1.21
442.0	5.94	0.202	4.94	1.20
457.9	5.98	0.201	4.98	1.20
465.8	5.99	0.200	4.99	1.20
472.7	6.01	0.200	5.01	1.20
476.5	6.01	0.200	5.01	1.20
488.0	6.03	0.199	5.03	1.20
496.5	6.05	0.198	5.05	1.20
501.7	6.06	0.198	5.06	1.20
514.5	6.08	0.197	5.08	1.20
532.0	6.10	0.196	5.10	1.20
546.1	6.12	0.195	5.12	1.20
632.8	6.22	0.192	5.22	1.19
694.3	6.27	0.190	5.27	1.19
820.0	6.35	0.187	5.35	1.19
904.0	6.39	0.186	5.39	1.19
1064.0	6.46	0.183	5.46	1.18

Note the useful extended UV range of Synthetic Fused Silica Lenses.

Decision Parameters for Synthetic Fused Silica Simple Lenses (As Functions of Wavelength)

λ (nm)	s_2/f	m_2	m_1	s_1/f
185.0	5.05	0.247	4.05	1.25
200.0	5.44	0.225	4.44	1.23
253.7	6.49	0.182	5.49	1.18
266.0	6.68	0.176	5.68	1.18
325.0	7.36	0.157	6.36	1.16
337.1	7.47	0.155	6.47	1.15
351.1	7.58	0.152	6.58	1.15
363.8	7.68	0.150	6.68	1.15
365.0	7.68	0.150	6.68	1.15
442.0	8.12	0.140	7.12	1.14
457.9	8.19	0.139	7.19	1.14
465.8	8.22	0.138	7.22	1.14
472.7	8.25	0.138	7.25	1.14
476.5	8.26	0.138	7.26	1.14
488.0	8.30	0.137	7.30	1.14
496.5	8.33	0.136	7.33	1.14
501.7	8.35	0.136	7.35	1.14
514.5	8.39	0.135	7.39	1.14
532.0	8.44	0.134	7.44	1.13
546.1	8.48	0.134	7.48	1.13
632.8	8.68	0.130	7.68	1.13
694.3	8.79	0.128	7.79	1.13
820.0	8.96	0.126	7.96	1.13
904.0	9.06	0.124	8.06	1.12
1064.0	9.22	0.122	8.22	1.12

DIFFRACTION EFFECTS

Quite apart from the aberrations we have previously discussed, all optical systems are subject to a fundamental physical effect called diffraction. In all light beams some energy is spread outside the region which would be expected from considerations of rectilinear propagation.

This spreading can be explained by the wave nature of light. "Huygens Principle" states that each point on a propagating wavefront is an emitter of secondary wavelets. The locus of expanding wavelets forms the propagating wave. Interference between the secondary wavelets gives rise to a fringe pattern which rapidly decreases in intensity with increasing angle from the initial direction of propagation. Huygens Principle nicely describes diffraction, but rigorous explanation demands a detailed study of wave theory.

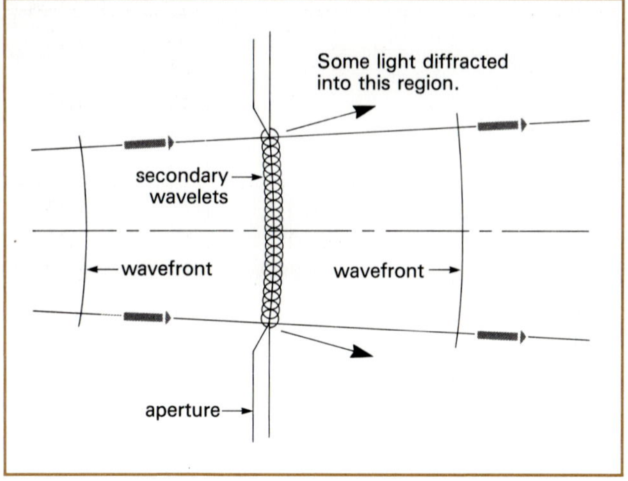

HUYGENS PRINCIPLE states that each point on a propagating wavefront is an emitter of secondary wavelets.

Diffraction effects are traditionally classified into either Fresnel or Fraunhofer types. Fresnel diffraction is primarily concerned with what happens to light in the immediate neighborhood of a diffracting object — with the object illuminated by a point source usually at a finite distance. Fraunhofer diffraction is primarily concerned with diffracted light sensed at infinite distances from the diffracting object with the illumination usually coming from an infinitely distant point source (plane wave illumination). A lens or lens system of finite positive focal length with plane wave input maps the far-field diffraction pattern of the aperture onto the focal plane, and so it is Fraunhofer diffraction which determines the limiting performance of optical systems. More generally, at any conjugate ratio, the far-field pattern is mapped onto the secondary conjugate plane, so that Fraunhofer diffraction is again applicable.

This mapping is most easily understood from a purely geometric viewpoint. Let θ be the angle in object space between any ray and the optical axis. The distance within the secondary conjugate plane by which this ray will miss the secondary conjugate point is simply $s'' \tan(\theta)$, where s'' is the secondary conjugate distance. Thus angles in the far field are transformed into distances in the image plane. This relationship is quite apparent for the chief ray, but in the absence of aberration is valid for all rays.

CIRCULAR APERTURE

Fraunhofer Diffraction at a circular aperture is of considerable importance in that it dictates the fundamental limits of performance for circular lenses. The diffraction pattern resulting from a uniformly illuminated circular aperture consists of a central bright region, known as the Airy Disc, surrounded by a number of fainter rings. Each ring is separated by a circle of zero intensity. The irradiance distribution in this pattern can be described by

$$I_x = I_o \left[\frac{2\,J_1(x)}{x} \right]^2 ,$$

where I_o = peak irradiance in image,

$$J_1(x) = x \sum_{n=1}^{\infty} (-1)^{n+1} \frac{x^{2n-2}}{(n-1)!\,n!\,2^{2n-1}} ,$$

$J_1(x)$ = Bessel function of the first kind of order unity,

$$x = \frac{\pi D}{\lambda} \sin\theta \text{ , where}$$

λ = wavelength,
D = aperture diameter, and
θ = angular radius from pattern maximum.

This useful formula shows the far field irradiance distribution from a uniformly illuminated circular aperture of diameter, D.

SLIT APERTURE

A slit aperture is mathematically simpler and is useful in relation to cylindrical optical elements. The irradiance distribution in the diffraction pattern of a uniformly illuminated slit aperture is described by

$$I_x = I_o \left[\frac{\sin x}{x}\right]^2$$

where,

I_o = peak irradiance in image ,

$$x = \frac{\pi \text{ w } \sin\theta}{\lambda},$$

λ = wavelength ,

w = slit width , and

θ = angular deviation from pattern maximum.

ENERGY DISTRIBUTION TABLE

Following is a table showing the major features of the pure (unaberrated) Fraunhofer diffraction patterns of circular and slit apertures. The table shows the location, relative intensity, and percentage of total pattern energy corresponding to each ring or band in the patterns. It is especially convenient ot characterize positions in either pattern with the same variable x. This variable is related to field angle in the circular aperture case by

$$\sin\theta = \frac{\lambda x}{\pi D},$$

where D is the aperture diameter. For a slit aperture this relationship is given by

$$\sin\theta = \frac{\lambda x}{\pi w},$$

where w is the slit width. π has its usual meaning, and we presume that D, w, and λ are all in the same units (preferably mm).

Linear instead of angular field positions are simply found from

$$r = s'' \tan(\theta),$$

where s″ is the secondary conjugate distance. This last result is often seen in a different form which we now derive for the circular aperture case.

Let U″ denote the angular radius of the lens clear aperture as viewed from the secondary conjugatee point, and suppose that the secondary principal surface is a sphere centered on the secondary conjugate point, a situation of which the Abbe sine condition (discussed elsewhere in this text) is a special case. Then D, s″, and U″ are related by

$$\frac{D/2}{s''} = \sin U''$$

Energy Distribution in the Diffraction Pattern of a Circular or Slit Aperture

Ring or Band	Circular Aperture			Slit Aperture		
	Position (x)	Relative Intensity (I_x/I_o)	Energy in Ring (%)	Position (x)	Relative Intensity (I_x/I_o)	Energy in Band (%)
Central Maximum	0.0	1.0	83.8	0.0	1.0	90.3
First Dark	1.22	0.0		1.00	0.0	
First Bright	1.64	0.0175	7.2	1.43	0.0472	4.7
Second Dark	2.23	0.0		2.00	0.0	
Second Bright	2.68	0.0042	2.8	2.46	0.0165	1.7
Third Dark	3.24	0.0		3.00	0.0	
Third Bright	3.70	0.0016	1.5	3.47	0.0083	0.8
Fourth Dark	4.24	0.0		4.00	0.0	
Fourth Bright	4.71	0.0008	1.0	4.48	0.0050	0.5
Fifth Dark	5.24	0.0		5.00	0.0	

Note: position variable (x) is defined in the text.

Solving for s″ and substituting above we find

$$r=\frac{D}{2\sin U''}\tan\theta$$

In the present situation,

$$\sin\theta=\frac{\lambda x}{\pi n''D},$$

where n″ is the image space refractive index. From consideration of a right triangle of unit hypotenuse it follows that

$$\tan\theta=\frac{\left(\frac{\lambda x}{\pi n''D}\right)}{\sqrt{1-\left(\frac{\lambda x}{\pi n''D}\right)^2}}$$

In the small angle approximation the denominator is unity (the tangent equals the sine). Substituting in the last formula for r above, we find

$$r=\frac{D}{2\sin U''}\cdot\frac{\lambda x}{\pi n''D}$$

$$=\frac{\lambda x}{2\pi n''\sin U''}$$

$$=\frac{\lambda x}{(2\pi)(N.A.)},$$

where (N.A.) denotes the numerical aperture. This result depends both on the generalized Abbe sine condition and the small angle approximation. These formulas reappear later in this text, with x assigned the particular value 1.22π which corresponds to the first minimum of the circular aperture diffraction pattern.

The accompanying graph shows the form of both circular and slit aperture diffraction patterns when plotted on the same normalized scale. Aperture diameter has been set equal to slit width so that for both patterns the correspondence between x-values and angular deviations in the far field is the same.

APPLICATION TO REAL LENSES

In most practical situations aberration effects predominate over diffraction effects. However, for some imaging situations, aberrations can be minimized to the point at which diffraction effects are more significant. Such a lens is said to be "diffraction limited." No further improvement in performance can be made beyond the diffraction limit; all that can be done is to increase the aperture.

An example is shown in the computer generated diffraction point spread function. This was calculated for a real lens (01 LAO 123) in monochromatic light at 546.1nm. Any section through the symmetry axis of the point spread function closely resembles the irradiance profile for the diffraction pattern of a circular aperture. This is true for all Melles Griot precision optimized achromats (LAO series) when used with uniform, axial, collimated, monochromatic input.

For all real lenses there exists an optimum aperture. If used at a larger aperture, aberrations will limit image quality while if the lens is stopped down, diffraction effects predominate. This is discussed in more detail on page 36.

For demanding imaging requirements at specific conjugate ratios and focal ratios, a computer-optimized optical design will be preferred in many applications. Melles Griot will be pleased to design and manufacture optical systems and sub-systems to your specific requirements.

CIRCULAR APERTURE

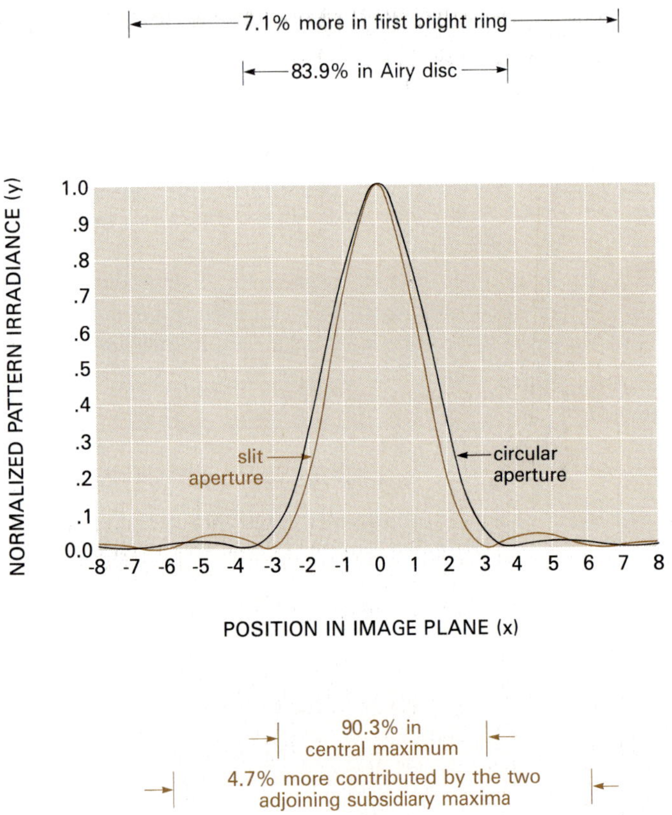

$y_c = \left(\frac{2J_1(x)}{x}\right)^2$, where $J_1(x) = x \sum_{n=1}^{\infty} (-1)^{n+1} \frac{x^{2n-2}}{(n-1)!n!2^{2n-1}}$

Note: $J_1(x)$ is the Bessel function of the first kind of order unity.

$x = \frac{\pi}{\lambda} D \sin\theta$

λ = Wavelength

D = Aperture diameter

θ = Angular radius from pattern maximum

$y_s = \left(\frac{\sin x}{x}\right)^2$, where $x = \frac{\pi}{\lambda} w \sin\theta$

λ = Wavelength

w = Slit width

θ = Angular deviation direction of pattern maximum

SLIT APERTURE

FRAUNHOFER DIFFRACTION PATTERN of a singlet slit superimposed on the Fraunhofer Diffraction Pattern of a circular aperture.

GAUSSIAN BEAMS

Most laser based optical systems **do not** have uniform pupil irradiance. The output beam of a laser operating in the TEM_{00} mode has a Gaussian irradiance profile.

Apodization, or nonuniformity of pupil irradiance, alters diffraction patterns. If your pupil irradiance is nonuniform, the formulas and results above do not apply.

For a discussion of Gaussian Beam Optics see page 339.

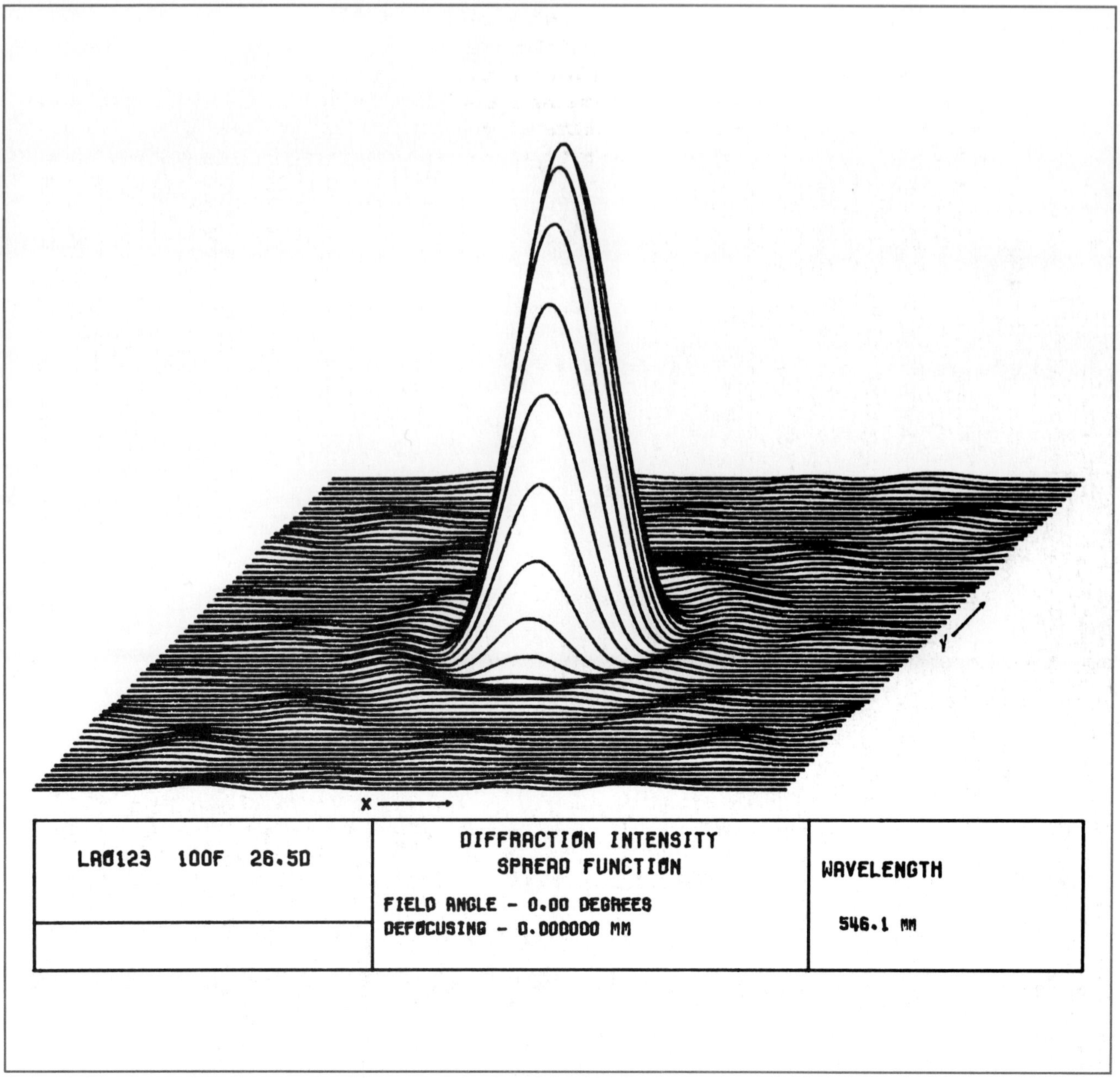

POINT SPREAD FUNCTION OF A REAL ACHROMAT BY MERCURY GREEN LIGHT. The lens is the precision optimized achromat 01 LAO 123. The point spread function for this achromat has been calculated by rigorous diffraction theory at full aperture, on axis.

QUALITY CONTROL

Melles Griot has always believed that quality comes first. That's why we invest heavily in equipment and skilled personnel to ensure that we control every specified parameter in the Optics Guide.

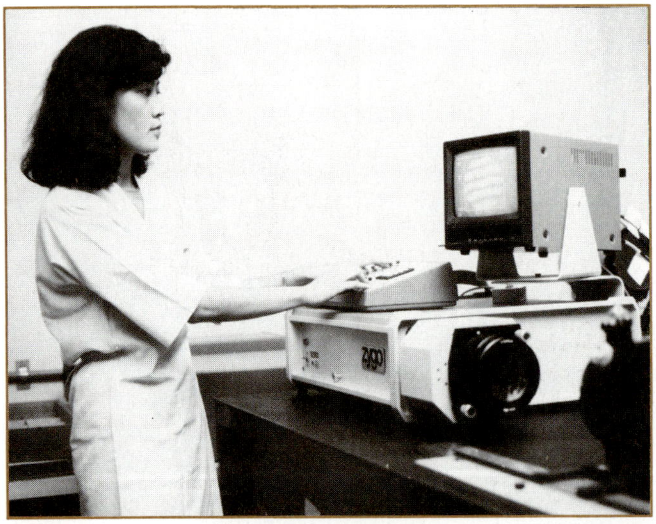

Melles Griot Q.C. standards and procedures are controlled world-wide to a common level. Our inspectors in the USA, Taiwan, Europe, and Japan, all use the same methods and equipment. Our statistical sampling procedures assure you of reliability and consistency.

OPTICAL SPECIFICATIONS

In this section discuss the main optical parameters you will wish to specify. Throughout the Optics Guide we use these specifications in the way they are defined here.

Contents **Page Number**

SPOT SIZE

SPOT SIZE

The size of the image of a distant axial source, whether this is real (like a star) or virtual (like a laser beam waist), is indicative of the finest detail that a lens can resolve. It is also of great practical importance in that many lenses are required to collect energy or collimate it. In this situation the design requirement is to focus the maximum possible energy into the smallest possible spot.

The system designer would like to have an idea how well any particular lens will perform this task. Unfortunately it is not so simple, as there are three separate mechanisms contributing to image spread. They are diffraction, aberration and tolerances.

Diffraction is a physical limitation which must be accepted, but can be perfectly quantified.

Aberration is a design limitation which can be minimized for the specific application. Even a well chosen singlet can have minimal aberration in certain situations.

Tolerances are a manufacturing limitation which cannot be overlooked. They can contribute to image spread in many ways. Poor surface quality can give rise to scattered light which will cause veiling glare in the image plane. Decentration in compound lenses can reduce the actual performance introducing a pseudo-aberration. All Melles Griot lenses are carefully toleranced and specified to minimize possible problems due to manufacturing limitations.

In practice most lenses are either aberration or diffraction limited so it is possible to calculate the spot size. There is, of course, a region in which both effects are important. Consideration of this region leads us into the concept of "optimum aperture." For every imaging situation there exists an aperture which gives the best performance. Larger apertures will be aberration limited and smaller apertures will be diffraction limited.

In the limit of small angles and simple lenses a reasonable approximation to spot size can be derived by adding the spherical aberration and diffraction terms. While this is not at all rigorous, it does provide a simple, practical way to select lenses.

MINIMUM ON-AXIS SPOT SIZE AND OPTIMUM APERTURE

The assumption that the aberration and diffraction angular radii are simply additive takes the form

$$\theta = \theta_d + \theta_a ,$$

where θ is the total on-axis spot angular radius, θ_d is the diffraction contribution, and θ_a is the aberration contribution. The corresponding linear spot radii are additive only in the small angle limit, since $r_d = s'' \tan(\theta_d)$, $r_a = s'' \tan(\theta_a)$, and $r = s'' \tan(\theta)$

The following discussion is restricted to the special case of a simple lens immersed in a medium of constant refractive index

$n = n''$. Define

$$N = \frac{n'}{n} ,$$

where n' is the refractive index of the lens. Using the third order approximation of perfectly centered thin lens aberration theory, it can be shown that for several special lens shapes and conjugate ratios of practical interest

$$\theta_a = \left(\frac{1}{2}\right) K(N) \left(\frac{D}{f}\right)^3 ,$$

where K(N) is an explicit function of index ratio N and lens shape. This result fails for low f-numbers, but is very accurate where we shall apply it. Without yet knowing any formulas for K(N), it is possible to derive several useful results. Numerically it can be shown that these results do not differ significantly, except at low f-numbers, from corresponding results based upon exact trigonometric ray traces instead of third order approximations.

Using the third order formula for θ_a, the total on-axis spot angular radius becomes

$$\theta = \left(\frac{1}{2}\right) K(N) \left(\frac{D}{f}\right)^3 + \frac{1.22\lambda}{nD} .$$

The diffraction term (the second term) is an inverse function of the diameter D, while the spherical aberration term (the first term) increases as the cube of D. D may be regarded either as the lens diameter or the diameter of a perfectly centered stop in close proximity. An optimum diameter (D_{opt}) or a corresponding optimum focal ratio $(f/D)_{opt}$ must exist, for which the angular radius of the image spot or blur circle is minimized.

From calculus

$$D_{opt} = \left[\frac{(2)(1.22)}{3} \cdot \frac{\lambda f^3}{nK(N)} \right]^{1/4} ,$$

and

$$\left(\frac{f}{D} \right)_{opt} = \left[\frac{3}{(2)(1.22)} \cdot \frac{fnK(N)}{\lambda} \right]^{1/4} .$$

If we now let $D = \alpha D_{opt}$, we discover that

$$\left(\frac{\theta}{\theta_d} \right) = \frac{\alpha^4}{3} + 1 .$$

In the special case of $\alpha = 1$ (or $D = D_{opt}$) this reduces to $(\theta/\theta_d) = 4/3$ exactly, a result which simplifies the calculation of θ under optimum conditions.

The preceding results are quite general, applying to any simple lens shape and any conjugate ratio for which an appropriate index function K(N) can be derived. Furthermore these optima occur at values for which the role of higher order spherical aberration is insignificant. Hence third order formulas are highly accurate in the present context.

The following explicit formulas for K(N), in a few specific cases, have been derived for the image surface of best geometric focus at full aperture. This image surface is located at 3/4 of the distance from the paraxial to the marginal conjugate. **These results are not paraxial.** To obtain third order paraxial results, multiply K(N) by 4.0.

API: Asymmetric positive spherical best form lens, infinite conjugate ratio, shortest radius surface facing the infinite conjugate. Best form means that the radius ratio has been chosen to minimize spherical aberration at this wavelength and conjugate ratio.

$$K_{api}(N) = \frac{1}{32(N-1)^2} \left[N^2 - \frac{N(2N+1)^2}{4(N+2)} \right]$$

PPI: Plano positive lens, infinite conjugate ratio, spherical curved surface facing the infinite conjugate.

$$K_{ppi}(N) = \frac{1}{32(N-1)^2} \left(N^2 - 2N + \frac{2}{N} \right)$$

SPI: Symmetric positive spherical lens, infinite conjugate ratio.

$$K_{spi}(N) = \frac{1}{32(N-1)^2} \left(N^2 - N - \frac{1}{4} + \frac{1}{2N} \right)$$

SPU: Symmetric positive spherical lens, unit conjugate ratio. The symmetric shape is best form (minimizes spherical aberration) at unit conjugate ratio and all wavelengths.

$$K_{spu}(N) = \frac{1}{128} \left(\frac{N}{N-1} \right)^2 \text{ and } s'' = 2f.$$

SMI: Spherical concave mirror infinite conjugate ratio only.

$$K_{smi} = \frac{1}{128} , \text{ independent of N.}$$

CYLINDRICAL LENSES AND MIRRORS

The formulas for the index function K(N) can be applied without modification to the corresponding cylindrical cases. Notice, however, that the factor 1.22 appearing in the formulas for θ, θ_d, D_{opt}, and $(f/D)_{opt}$ must be replaced by 1.00 before application to cylindrical cases. D_{opt} might then be better denoted w_{opt}, the optimum full width of a perfectly centered slit (rectangular aperture) in contact or near-contact. In the spherical surface cases θ is the angular radius of an irradiance contour circle containing 83.8% of the image spot energy. In the cylindrical surface cases, θ is one-half of the angular separation of a pair of symmetrically displaced irradiance contour lines which bound approximately 90% of the image stripe energy. The contour lines are symmetrical about the stripe centerline.

ON-AXIS MINIMUM SPOT SIZE TABLE

The table includes data for both lenses and mirrors, and for both spherical and circular cylindrical surface shapes. A wavelength of 546.1nm is assumed. The lenses are assumed to be made of BK 7, which has a refractive index of $n' = 1.518722$ at this wavelength. It is also assumed that $n = n'' = 1$. For elements with cylindrical surfaces, D_{opt} should be interpreted as w_{opt}, the optimum full width of a slit in contact or near-contact. Likewise for these elements, the spot radii should be interpreted as one-half of the full width of the image stripe.

Optimum f-numbers, on- and off-axis, are sometimes a convenient measure of lens or lens system quality, at least for systems having equal EFL values. Other systems can be compared by scaling. The smaller the scaled optimum f-number, the higher the quality.

Simple lens performance can be substantially improved by using materials of high refractive index. Such materials have the further advantage of being better matched to single layer MgF_2 antireflection coatings. In most applications high index materials have other disadvantages, including reduced transmittance, vulnerability to chemical staining, and vulnerability to thermal shock. Cemented doublet achromats offer performance overwhelmingly superior to the best high-index singlets even under optimum conditions for the singlets.

On-Axis Minimum Spot Size at Best Geometric Focus

Conjugate Ratio	Element Type	Element Shape	Surface Shape	f (mm)	D_{opt} (mm)	$(f/D)_{opt}$	Angular Spot Radius (radians)	Linear Spot Radius (microns)
∞	lens, BK 7	asymmetric best form	spherical	10	1.63	6.15	5.46×10^{-4}	5.46
				100	9.14	10.9	9.72×10^{-5}	9.72
				1000	51.4	19.5	1.73×10^{-5}	17.3
∞	lens, BK 7	plano	spherical	10	1.60	6.26	5.56×10^{-4}	5.56
				100	8.99	11.1	9.88×10^{-5}	9.88
				1000	50.5	19.8	1.76×10^{-5}	17.6
∞	lens, BK 7	symmetric	spherical	10	1.45	6.90	6.13×10^{-4}	6.13
				100	8.15	12.3	1.09×10^{-4}	10.9
				1000	45.8	21.8	1.94×10^{-5}	19.4
1.0	lens, BK 7	symmetric	spherical	10	1.60	6.23	5.54×10^{-4}	11.1
				100	9.02	11.1	9.84×10^{-5}	19.7
				1000	50.7	19.7	1.75×10^{-5}	35.0
∞	mirror	concave	spherical	10	2.75	3.64	3.24×10^{-4}	3.24
				100	15.4	6.48	5.75×10^{-5}	5.75
				1000	86.8	11.5	1.02×10^{-5}	10.2
					w_{opt} (mm)			1/2 stripewidth (microns)
∞	lens, BK 7	plano	circular cylindrical	10	1.52	6.58	4.97×10^{-4}	4.97
				100	8.55	11.7	8.51×10^{-5}	8.51
				1000	48.1	20.8	1.51×10^{-5}	15.1
∞	mirror	concave	circular cylindrical	10	2.61	3.83	2.79×10^{-4}	2.79
				100	14.7	6.81	4.96×10^{-5}	4.96
				1000	82.6	12.1	8.81×10^{-6}	8.81

OFF-AXIS SPOT SIZE UNDER OPTIMUM CONDITIONS

The accompanying 12 curves show the growth of spot size with field angle for both plano-convex and symmetric convex lenses, at the conjugate ratios for which these shapes are best form. Results are shown for both types in focal lengths of 10, 100, and 1000mm. Precisely centered stops of the appropriate diameter D_{opt} are presumed in contact with each lens. D_{opt} is calculated at zero field angle, and subsequently applied to all higher field angles. Growth of spot size with field angle is caused by coma, astigmatism, and curvature of field. The effect of pure field curvature is seen as the difference between corresponding curves (flat field and curved field) for the same focal length. Spot sizes off-axis on the natural curved image surface (curved field) are much smaller than those on the flat image surface of best on-axis focus (flat field).

Because of coma, the off-axis spot is asymmetric, and the definition of spot radius must be specific. The first term is a diffraction term calculated from D_{opt}. The second term is the radius of a circle (not an irradiance contour) which according to purely geometrical optics would contain 83.8% of the total image energy. This geometric radius is determined by tracing a uniform lattice of about 200 rays through the lens to the image surface, and finding the size of the circle most likely to contain 83.8% of the ray intersections with this surface.

The Airy disc radius ($1.22\lambda/D_{opt}$) has been chosen as the unit of measurement for off-axis angular spot radius, so that the approach to the diffraction limit on-axis would be clear. The Airy disc radius is different for each lens, and it therefore appears that the long focal length lens does not perform as well off-axis as the shorter focal length lenses. This appearance is deceptive, for the Airy disc of the large focal length lens is extremely small. If consistent units of angular radius (such as 10^{-4} radian) had been used for all lenses, the largest lens would be seen to offer the best off-axis performance. The ratio θ/θ_{do} is the ratio of the total off-axis spot angular radius to the on-axis diffraction term alone (Airy disc radius) for $D = D_{opt}$.

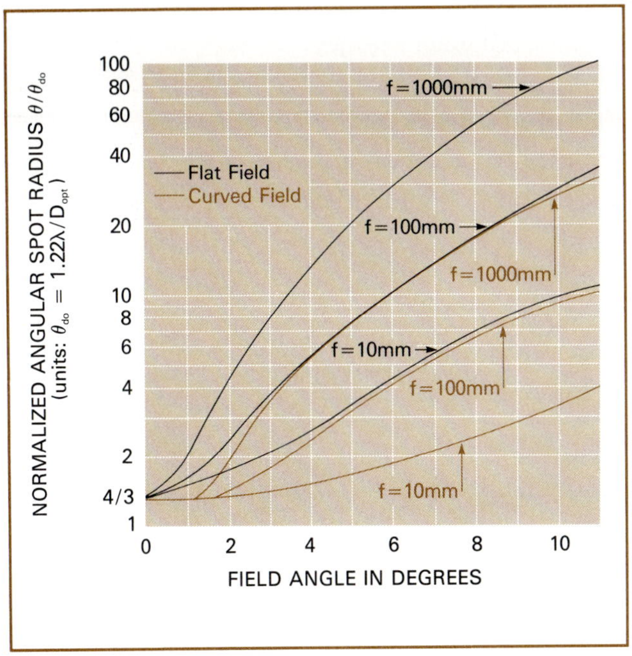

GROWTH OF SPOT SIZE WITH FIELD ANGLE: plano-convex spherical lenses at infinite conjugate ratio.

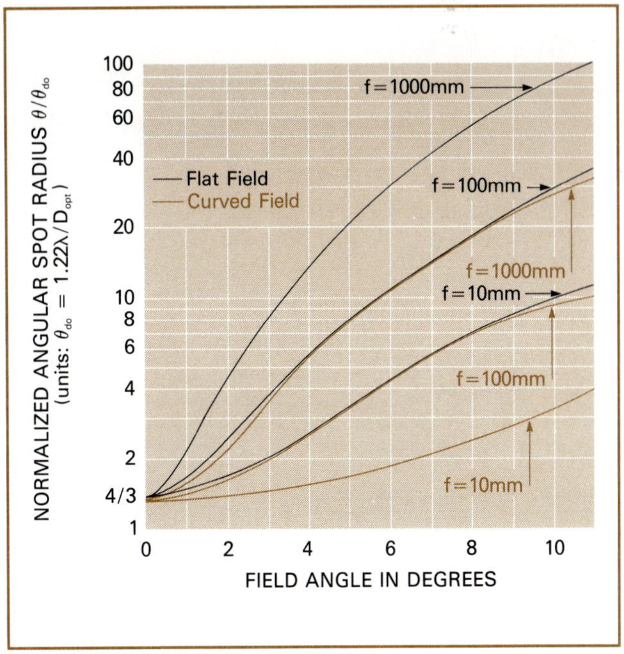

GROWTH OF SPOT SIZE WITH FIELD ANGLE: symmetric convex spherical lenses at unit conjugate ratio.

MODULATION TRANSFER FUNCTION

Modulation Transfer Function (MTF) is a quantitative measure of image quality. It is the best such measure yet devised, and is far superior to any of the classical resolution criteria. From MTF the information content of an image can be deduced. MTF describes the ability of a lens or system, as a function of spatial frequency, to transfer object contrast to the image. MTF curves can be associated with the subsystems which make up a complete electro-optical or photographic system. With MTF data available, it can be seen whether overall system expectations are feasible.

Bar chart resolution testing of lens systems is deceptive, because almost 20% of the energy arriving at a lens system from a bar chart is modulated at 3rd harmonic and higher harmonic frequencies of the bar chart fundamental frequency. Consider instead a sine wave chart, in the form of a positve transparency on which transmittance varies one dimensionally in a precisely sinusoidal fashion. Such charts are difficult to make, but there exist electronic methods by which experimental results duplicating those possible with a truly perfect sine wave chart can be achieved. Assume that the transparency is viewed against a uniformly illuminated background. The maximum and minimum transmittances are T_{max} and T_{min}, respectively. A lens system under test forms a real image of the sine wave chart, and the spatial frequency of the image is u cycles per millimeter. Corresponding to the transmittances T_{max} and T_{min} are the image irradiances I_{max} and I_{min}. By analogy with Michelson's definition of "visibility" of interference fringes, the "contrast" or "modulation" of the chart and image are respectively defined as

$$M_c = \frac{T_{max} - T_{min}}{T_{max} + T_{min}}$$

and

$$M_i = \frac{I_{max} - I_{min}}{I_{max} + I_{min}}$$

The modulation transfer function of the optical system at spatial frequency u is then defined to be

$$\text{MTF} = \text{MTF(u)} = M_i/M_c.$$

The graph of MTF versus u is a modulation transfer function curve. MTF is defined only for lenses or systems having positive focal length and forming real images.

MTF can be measured by a variety of methods, and instruments are commercially available which automatically measure MTF curves. MTF curves change as field angle positions are varied. They also change with conjugate ratio. In a system with astigmatism or coma, different MTF curves are obtained corresponding to various azimuths in the image plane through a single image point. For cylindrical lenses only one azimuth is meaningful. MTF curves can be either polychromatic or monochromatic.

Polychromatic curves show the effect of any chromatic aberration that may be present. For a well-corrected achromatic system, polychromatic MTF can be computed by weighted averaging of monochromatic MTF's at a single image surface. Most instruments measure polychromatic MTF directly.

PERFECT CIRCULAR LENS

In diffraction limited performance, the monochromatic MTF of a circular aperture (perfect aberration free spherical lens) at arbitrary conjugate ratio is given by the formula

$$\text{MDMTF(x)} = \frac{2}{\pi}\left[\arccos(x) - x\sqrt{1 - x^2}\right],$$

where radian mode is intended for the arc cosine function, and x is the normalized spatial frequency

$$x = \frac{u}{u_{ic}},$$

where u is the absolute spatial frequency, and u_{ic} is the incoherent diffraction cutoff spatial frequency. There are several formulas for u_{ic} including

$$u_{ic} = \frac{1.22}{r_d}$$

$$= \frac{n'' D \sqrt{1 - \left(\frac{1.22\lambda}{n''D}\right)^2}}{\lambda s''}$$

$$= \frac{2n'' \sin u'' \sqrt{1 - \left(\frac{1.22\lambda}{n''D}\right)^2}}{\lambda}$$

$$= \frac{2n'' \sin(u'')}{\lambda}$$

$$= \frac{n'' D}{\lambda s''},$$

where r_d is the linear spot radius in the case of pure diffraction (Airy disc radius), D is the diameter of the lens clear aperture (or of a stop in near-contact), λ is the wavelength, s″ is the secondary conjugate distance, u″ is the largest angle between any ray and the optical axis at the secondary conjugate point, and the product n″ sin (u″) is by definition the image space numerical aperture, n″ is the image space refractive index. It is essential that D, λ, and s″ have consistent units (usually mm, in which case u and u_{ic} will be in cycles per millimeter). The relationship

$$\sin(u'') = \frac{D}{2s''}$$

implies that the secondary principal surface is a sphere centered upon the secondary conjugate point. This means that the lens is completely free of spherical aberration and coma, and in the special case of inifinite conjugate ratio (s″ = f″) the Abbe Sine Condition is satisfied. Also at infinite conjugate ratio this leads to the familiar textbook result that

$$u_{ic} = n''\frac{D}{\lambda f}.$$

PERFECT RECTANGULAR LENS

In diffraction limited performance, the monochromatic MTF of a rectangular aperture (perfect aberration less cylindrical lens) at arbitrary conjugate ratio is given by the formula

$$\text{MDMTF}(x) = (1-x),$$

where x is again the normalized spatial frequency u/u_{ic}, where in the present cylindrical case

$$u_{ic} = \frac{1}{r_d},$$

and r_d must be interpreted as one-half the full width of the central stripe of the diffraction pattern measured from first minimum to first minimum. Notice how this formula differs (by a factor 1.22) from the corresponding formula in the circular aperture case. For both the circular and rectangular apertures, the result

$$u_{ic} = \frac{2n''\sin(u'')}{\lambda}$$

applies. The remaining three expressions for u_{ic} in the circular aperture case can be applied to the present rectangular aperture case provided that two substitutions are made. Everywhere the constant 1.22 formerly appeared, it must be replaced by 1.00. Also the aperture diameter D must now be replaced by the aperture width w. The relationship sin (u″) = w/2s″ means that the secondary principal surface is a circular cylinder centered upon the secondary conjugate line. In the special case of infinite conjugate ratio this gives the result that for cylindrical lenses the incoherent cutoff frequency is

$$u_{ic} = n''\frac{w}{\lambda f}.$$

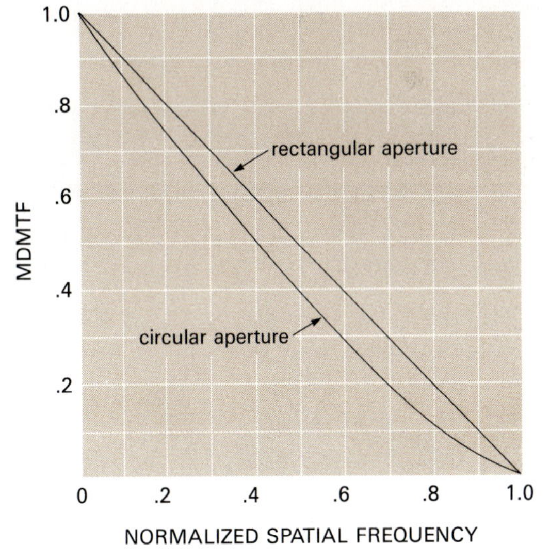

MDMTF(x) vs x. These curves (and the formulas they represent) precisely quantify an important and fundamental property common to all electromagnetic, de Broglie, and acoustic waves. All objects become more difficult to image, and ultimately become impossible to image, as their dimensions diminish toward the radiation wavelength.

IDEAL PERFORMANCE AND REAL LENSES

For both formulas above, the x-intercept and MDMTF-intercept are at unity (1.0). MDMTF(x) for the rectangular case is a straight line between these intercepts. For the circular case MDMTF(x) is a curve which dips slightly below the straight line.

All real cylindrical monochromatic MTF curves fall on or below the straight MDMTF(x) line. Similarly all real spherical and monochromatic MTF curves fall on or below the circular MDMTF(x) curve. Thus the two MDMTF(x) curves represent the ultimate in optical performance. Optical element or system quality is measured by how closely the real MTF curve approaches the corresponding MDMTF(x) curve from below, and it is possible for real MTF curves to approach the MDMTF(x) curves very closely.

MTF is an extremely sensitive measure of image degradation. To illustrate this consider a lens having a quarter wavelength of spherical aberration. This aberration, barely discernible by eye, would reduce the MTF by as much as 0.2 at the midpoint of the spatial frequency range.

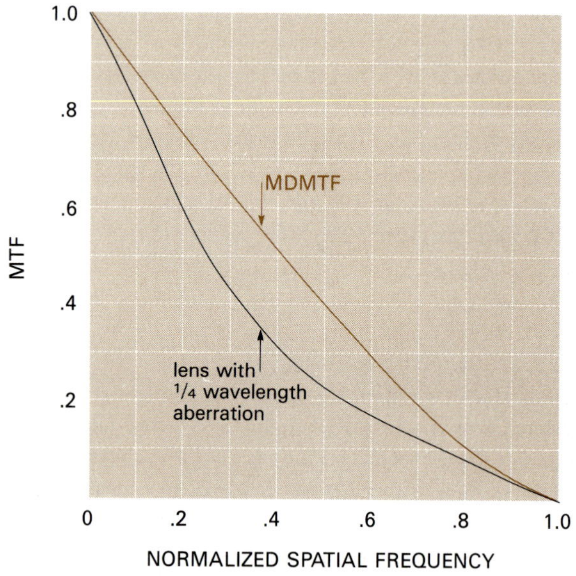

MTF as a function of normalized spatial frequency.

A WORD OF CAUTION

It is tempting to take component MTF's and multiply them together to determine the performance of a complete system. There are circumstances in which this is legitimate but in most cases it is not.

For the multiplication of two imaging subsystem MTF's to be legitimate or to yield a correct overall system MTF, it is first necessary that the first imaging subsystem be "isoplanatic." This means that the point spread function must be rotationally and translationally invariant throughout the entire field shared by the subsystems. In practice this usually demands that the first subsystem must be diffraction limited throughout the shared field. Secondly, the point spread function of the first system must be smaller than that of the second (in the object space of the second). Thus it is obviously wrong to multiply the incoherent MTF's of telescope objective and eyepieces, for the objective point spread function is the larger, the intermediate image formed by the objective therefore appears as a partially coherent virtual object to the eyepiece, and it is therefore unreasonable to hope that the product MTF can describe telescope operation. When criteria for legitimate multiplication are satisfied by imaging subsystems, it is necessary prior to multiplication to mathematically transform the MTF's to a common image surface.

In general, we advise against combining MTF's in this way.

SURFACE ACCURACY

When attempting to specify how closely an optical surface conforms to its intended shape a measure of surface accuracy is needed. Common optical industry practice in the USA, which is adhered to by Melles Griot, is to combine coarse and fine form errors stating one value for surface accuracy. The combination of form errors cannot exceed this value. Errors in radius, surface asymmetry and local deviations are all taken into account. Surface accuracy is a "peak-to-peak" value, as is wavefront distortion, which is discussed later.

Surface accuracy is defined as twice the maximum deviation of the surface under test from its ideal mathematical reference surface. Deviation has its customary statistical meaning here. The reference surface is defined by the ideal equation with coefficients which best approximate the surface under test. It is permitted to take up a position, embedded in the surface, so as to minimize the deviation. Specifications are normally given in fractions of a wavelength for either the mercury green line (546.1nm.), or the helium neon laser line (632.8nm).

It should be pointed out that a practice which is used by some suppliers (not Melles Griot) is to define surface accuracy as equal to the maximum deviation. This is not a peak-to-peak value. **Do not be mislead.**

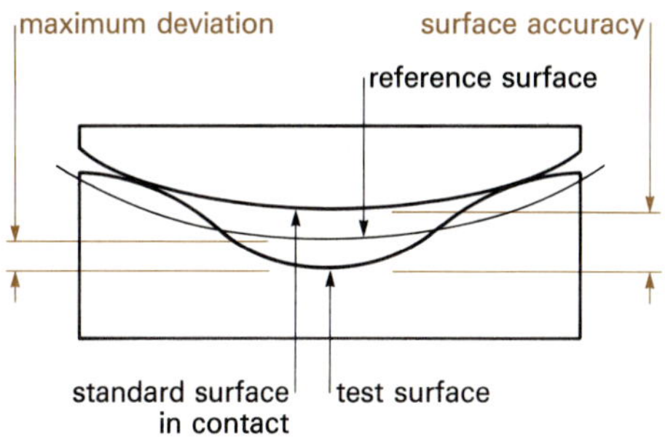

DEFINITION OF SURFACE ACCURACY.

SURFACE FLATNESS

Surface flatness is simply surface accuracy with respect to a plane reference surface. It is used extensively in mirror specifications.

POWER AND IRREGULARITY

During manufacture, a precision component is frequently compared with a test plate having an accurate polished surface which is the inverse of the surface under test. The two surfaces are brought together and viewed in near monochromatic light. Newton's rings (interference fringes caused by the near contact of the surfaces) are seen. The number of rings indicates the difference in radius between the surfaces. This is known as "power," or sometimes "figure." It is measured in "rings" which are equivalent to half wavelengths.

Beyond their number the rings may exhibit crookedness indicating non-uniform shape differences. The crookedness may be local to one small area of the surface or may be in the form of non-circular fringes over the whole aperture. All such non-uniformities are known collectively as "irregularity."

DIN FORM ERRORS

The DIN (Deutsche Industrie Norm), specification DIN 3140 Sheet 5, refers to allowable surface accuracy errors measured by interferometric testing of optical surfaces. The DIN identification for allowable form or surface accuracy errors is:

3M (ΔM)F where,

- 3 = the general identification number referring to form errors.
- M = coarse form error, stated in number of fringes allowed. Coarse form error is power or the difference in radius from an appropriate test plate.
- ΔM = regular form errors, stated in the number of fringes allowed. Regular form errors are the maximum difference between the numbers of fringes in orthogonal directions. These are asymmetrical form errors such as astigmatism or cylindrical errors.
- F = Fine form errors, stated in the number of fringes allowed. Fine form errors correspond directly with fringe irregularity.

For example:

3/2 (0.5) 0.5

specifies a surface accuracy allowing 1 wave (2 fringes) of power, 1/4 wave (0.5 fringe) of asymmetry and 1/4 wave (0.5 fringe) irregularity.

WAVEFRONT DISTORTION

Sometimes the best specification for an optical component is its effect on the emergent wavefront. This is particularly true for windows and retroreflectors.

Wavefront distortion is the total, peak-to-peak, deformation (in the direction of propagation) of the emergent wavefront from its intended shape. Specifications are normally given in fractions of a wavelength for the helium neon laser line (632.8nm.).

Consider a perfectly plane monochromatic wavefront, incident in a direction normal to the face of a window. Deviations from perfect flatness in both surfaces of the window and, more importantly, inhomogeneities of the index within the window, will all cause a deformation of that incident wavefront during transmission through the window. Likewise in a retroreflector each of the faces plus the material will affect the emergent wavefront. Also collimating lenses whose prime purpose is to produce a plane wavefront are best specified in terms of the maximum error in the emergent wavefront.

INTERFEROMETER MEASUREMENTS

Measurements of wavefront distortion are made using a laser interferometer. In the interferometers we use, the wavefront from a helium neon laser ($\lambda = 632.8$nm) is expanded and then divided into a reference wavefront and a test wavefront by using a partially transmitting reference surface. The reference wavefront is reflected back to the interferometer and the test wavefront is transmitted through the surface to the test element. The reference surface is a known flat or spherical surface whose surface error is on the order of $\lambda/20$.

The test wavefront is reflected back to the interferometer either from the surface being tested or from another $\lambda/20$ reference surface, in this case passing twice through the element under test. The reference and test wavefronts are recombined at the interferometer where constructive and destructive interference takes place between the two wavefronts. A difference in optical paths of the two wavefronts is caused by any error present in the test element and any tilt of one wavefront relative to the other. The fringe pattern so produced is projected to a viewing screen or a camera system if a permanent record of the pattern is needed.

A slight tilt of the test wavefront to the reference wavefront produces a set of fringes whose parallelism and straightness depend on the amount that the test wavefront is altered by the element under test. The distance between successive fringes (usually measured dark band to dark band) represents one wavelength difference in optical path traveled by the two wavefronts. In both surface testing and transmitted wavefront testing, the test wavefront travels through the error in the test piece twice. Therefore one fringe spacing represents one half wavelength of surface or transmission error of the test element.

Determination of convexity or concavity of the test element error can be made if the zero order direction of the interference cavity, the space between the referrence and test surfaces, is known. The zero order direction is the direction of the center of tilt between the reference and test wavefronts.

Fringes which curve around the center of tilt (zero order) are convex, caused by a high area on the test surface. Conversely, fringes which curve away from the center of tilt (zero order) are concave, caused by a low area on the test surface.

A fringe pattern in which a tilt of the reference wavefront with respect to the test wavefront is introduced, and the zero order direction is known, contains information as to the amount of error in the test element and the direction of the error, concave or convex. Normally about six fringes are introduced; increasing the angle between wavefronts decreases the spacing between interference fringes.

Wavefront distortion testing is used for the testing of achromats, windows, filters, beamsplitters, prism, retroreflectors, and many other transmitting elements. This testing method is most nearly consistent with the way these optical elements or assemblies are used.

A flat, partially transmitting reference surface is used for most wavefront testing together with a flat or concave spherical reference mirror to reflect the test wavefront after it is transmitted by the test element.

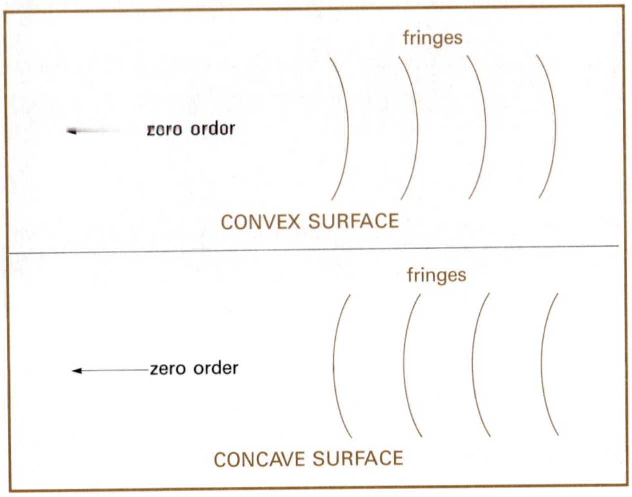

CONVEXITY OR CONCAVITY is determined by the direction of the zero order fringe in relation to the curvatures of the fringes.

INTERFEROGRAM INTERPRETATION

A straight, equally spaced line grid is the fringe pattern produced by a perfect reference surface and a perfect wavefront with a tilt introduced between the two. The deviation of a fringe from this straight line grid is a measure of the wavefront error of the element under test.

Test elements whose surface accuracy or wavefront distortion specification is $\lambda/10$ or worse can be evaluated by observing the fringe pattern at the interferometer. Since one fringe separation is equal to $\lambda/2$, an error of 0.2 fringe corresponds to an error of $\lambda/10$. This can easily be estimated by eye. If there is any doubt or if the specification is better than $\lambda/10$, the fringe pattern is photographed and an accurate evaluation is made from the interferogram.

To evaluate the interferogram a straight, equally spaced line grid with adjustable spacing to fit to the fringe pattern on the photograph so that all the fringe error is to one side of the grid lines. The maximum deviation of the fringe pattern from the grid is measured and divided by two times the average spacing to give the number of wavelengths of error. The average fringe spacing is the same as the line grid spacing.

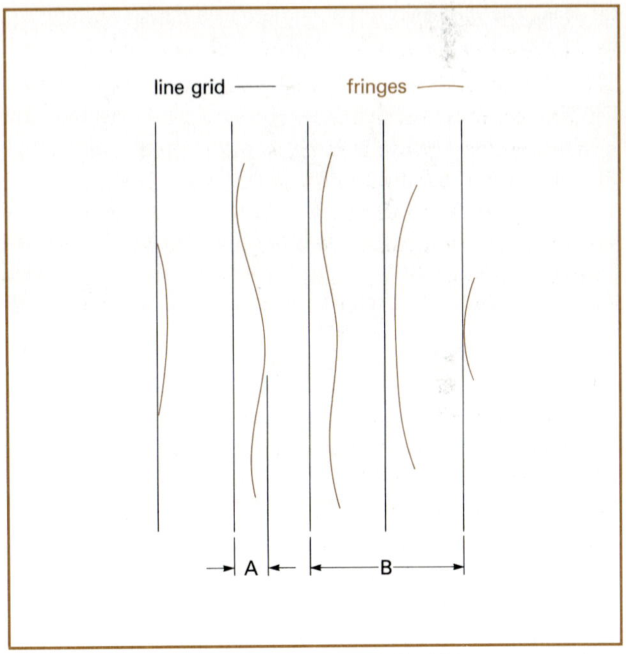

EVALUATION OF AN INTERFEROGRAM by comparison with a line grid. A represents the maximum deviation of the fringe pattern. B represents one wavelength. Thus the wavefront error = A/B wavelengths.

We routinely check wavefront distortion on the components for which this parameter is specified.

CENTRATION

Centration is an often misunderstood specification of optical components. It arises from the fact that an optical element usually has two axes of symmetry.

OPTICAL AND MECHANICAL AXES

The optical axis is of great importance in relation to the imaging formulas presented earlier. It is defined, for a simple lens, as the straight line joining the centers of curvature of the two surfaces of the lens. The mechanical axis is determined during manufacture by the centerline of the edging machine. The rotational axis during edge grinding becomes the mechanical axis of the lens when mounted by its edge. Ideally the optical and mechanical axes would coincide. Centration is the tolerance within which they meet this ideal.

Centration is defined as the upper limit on the angular deviation of a ray which is incident precisely along the mechanical axis of a lens. As a lens is rotated about its mechanical axis the focal point will describe an orbit. Centration is the maximum angular radius of this orbit.

Decentration may be in the form of an angular and/or a linear separation of the axes. Both of these will have a similar effect in enhancing the asymmetrical aberrations, coma and astigmatism. Centration, as defined, takes both types of error into account.

DOUBLETS AND TRIPLETS

To achieve a given centration specification for a doublet is more difficult than for a singlet since each element must be individually centered to a tighter specification and the two optical axes must be carefully aligned during the cementing process. The situation is even more complex for triplets in which three optical axes must be aligned to result in a maximum angular deviation within the centration specification.

One method which is sometimes used to obtain precise centration in compound lenses is to align the elements optically and then edge the combination.

CYLINDRICAL OPTICS

Centration, as defined, is meaningless for cylindrical optics and must be modified to take account of the one dimensional imaging. Centration, for cylindrical optics, is defined as the upper limit on the angle between the focal line and the projection of the mechanical centerplane on the focal plane.

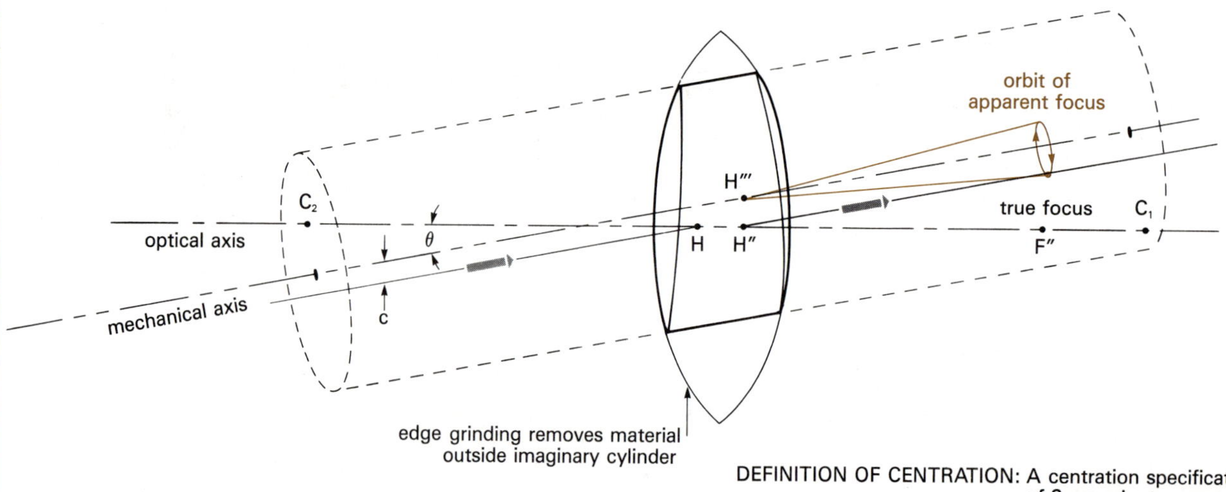

DEFINITION OF CENTRATION: A centration specification of 3 arc minutes means that the angular radius of the orbit of apparent focus cannot exceed 3 minutes.

DEFINITION OF CENTRATION

COSMETIC SURFACE QUALITY

Cosmetic surface quality describes the degree of perfection found in the surface of an optical component. It defines the state of polish, freedom from scratches and digs, and the edge treatment of components. These factors are important, not only to enhance the appearance of the component, but because they can have a serious adverse effect on performance due to the scattering of light. Scattering can be particularly important in laser applications due to the intensity of the incident illumination. Unwanted diffraction patterns caused by scratches can lead to problems and scattering of high energy laser radiation can cause damage. Overspecifying cosmetic surface quality can be costly though. Melles Griot components offer a high degree of perfection but are specified at reasonable levels according to their intended application.

U.S. MIL SPEC – SURFACE QUALITY

The U.S. Military Surface Quality Specification, MIL-0-13830A, defines the surface quality standards which are used in this catalog and to which Melles Griot adheres. The surface quality standards include those for scratches, digs, grayness, edge chips and cemented interfaces. We compare our products with scratch and dig standards which were manufactured according to the U.S. military drawing C7641866 Rev L and our inspection areas are equipped with lighting which meets the requirements of MIL-0-13830A.

Inspection of polished optical surfaces for scratches and digs is accomplished by visual comparison to scratch and dig standards. A part is rejected if there are any scratches present which are more visible than the specified maximum or if there are any digs present which are larger in diameter than the specified maximum.

The scratch and dig designation for a component or assembly is specified by two numbers; the first defining the allowable maximum scratch visibility and the second referring to the allowable maximum dig diameter, separated by a hyphen.

For example:

80-50 represents a commonly acceptable cosmetic standard.

10-10 represents a tight standard suitable for demanding laser applications.

SCRATCHES

A scratch is defined as any marking or tearing of a polished optical surface and the maximum allowed scratch widths (reference dimensions only) are as follows:

80 scratch = 8 microns width
60 scratch = 6 microns width
40 scratch = 4 microns width
20 scratch = 2 microns width
10 scratch = 1 micron width

If maximum visibility scratches are present, their combined lengths cannot exceed a quarter of the part diameter. Even with maximum visibility scratches present, many combinations of smaller scratch sizes and lengths are allowed on a polished surface.

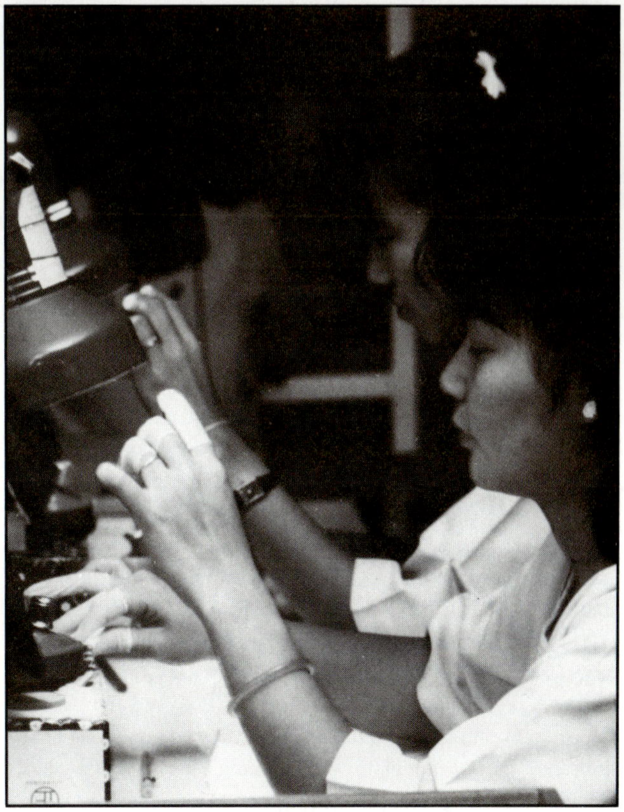

DIGS

A dig is defined as a pit or small crater on the polished optical surface.

Digs are defined by their diameters. The diameter of an irregularly shaped dig is 1/2 x (length plus width).

50 dig = 0.5mm diameter
40 dig = 0.4mm diameter
30 dig = 0.3mm diameter
20 dig = 0.2mm diameter
10 dig = 0.1mm diameter

Smaller diameter digs are allowed with maximum diameter digs as shown below:

1. Clear aperture diameter 0-2mm, 1 maximum dig allowed, and the maximum sum of dig diameters is 2 times the maximum dig diameter.
2. Clear aperture diameter 20-40mm, 2 maximum digs allowed, and the maximum sum of dig diameters is 4 times the maximum dig diameter.
3. Clear aperture diameter 40-60mm, 3 maximum digs allowed, and the maximum sum of dig diameter is 6 times the maximum dig diameter.

Digs of a specified maximum size of 10, or 0.1mm diameter, must be separated edge to edge by at least 1mm.

EDGE CHIPS

Lens edge chips are only allowed outside the clear aperture of the lens. The clear aperture is 90% of the lens diameter unless otherwise specified. Chips smaller than 0.5mm are ignored and those larger than 0.5mm are stoned or ground so that there is no shine to the chip. The sum of the widths of chips larger than 0.5mm cannot exceed 10% of the lens perimeter.

Prism edge chips outside the clear aperture of the prism are allowed. If the prism leg dimension is 25.4mm or less, chips may extend inward 1.0mm from the edge. If the prism leg dimension is larger than 25.4mm, chips may extend inward 2.0mm from the edge. Chips smaller than 0.5mm are ignored, and those larger than 0.5mm must be stoned, or ground, leaving no shine to the chip. The sum of the widths of chips larger than 0.5mm cannot exceed 10% of the length of the edge on which they occur.

CEMENTED INTERFACES

A cemented interface is considered a lens surface and specified surface quality standards apply. Edge separation at a cemented interface cannot extend into the element more than half the distance to the element clear aperture up to a maximum of 1.0mm. The sum of edge separations larger than 0.5mm cannot exceed 10% of the element perimeter.

BEVELS

Although bevels are not specified in MIL-0-13830A, our standard shop practice specifies that element edges be beveled to a width of 0.25 to 0.5mm at an angle of 45° ± 15°. Edges meeting at angles of 135° or larger are not beveled.

COATING DEFECTS

Defects caused by the coating of an optical element, such as coating scratches, voids or pinholes and dust or stain under the coating, are considered with the scratch and dig specification for that element. Coating defects are allowed if their size is within the stated scratch and dig tolerance.

DIN SURFACE INSPECTION

The DIN (Deutsche Industrie Norm) specification, DIN 3140, SHEET 7, corresponds generally to the U.S. military specification for surface quality or scratch and dig. The surface imperfections specified are chips, scratches, digs, pits, crazes, and spots.

The identification number for the specification of surface imperfections is 5. A complete surface imperfection specification is:

5/A x S where,

A = the number of permitted surface imperfections and,

S = a number specifying the edge length of a surface imperfection of square area. For imperfections other than square, a square area is approximated.

Normally, subdivision of surface imperfections is permitted allowing larger numbers of smaller imperfections than the specified number and size. If no subdivision is allowed, the A x S term of the identification number is in brackets.

5/(A x S)

The table gives the number and size of permitted imperfections. To use this table look up the S value in the left hand column and read accross the row to determine the size and number of permitted smaller imperfections.

DIN Imperfection Acceptance Chart

	Size of Imperfections, S (mm)			
	Number of Permitted Imperfections			
A	2.5 x A	6.3 x A	16 x A	40 x A
0.0063	0.0040	0.0025	—	—
0.010	0.0063	0.0040	0.0025	—
0.016	0.010	0.0063	0.0040	0.0025
0.025	0.016	0.010	0.0063	0.0040
0.040	0.025	0.016	0.010	0.0063
0.063	0.040	0.025	0.016	0.010
0.10	0.063	0.040	0.025	0.016
0.16	0.10	0.063	0.040	0.025
0.25	0.16	0.10	0.063	0.040
0.40	0.25	0.16	0.10	0.063
0.63	0.40	0.25	0.16	0.10
1.0	0.63	0.40	0.25	0.16
1.6	1.0	0.63	0.40	0.25
2.5	1.6	1.0	0.63	0.40
4.0	2.5	1.6	1.0	0.63
6.3	4.0	2.5	1.6	1.0
10.	6.3	4.0	2.5	1.6

For example:

5/3 x 0.63 Identification number

- 5 identifies surface imperfection specification.
- 3 = Number of allowed surface imperfections without subdivision.
- 0.63 = Maximum size of imperfections allowed in mm.

Three 0.63mm imperfections are permitted or from the table, 3 x 2.5 = 8 imperfections of size 0.40mm are permitted. Combinations are allowed such as 2 imperfections of size 0.63mm and 1 x 6.3 = 6 imperfections of size 0.25mm. All possible combinations are allowed as long as the total area of all surface imperfections is not greater than the area corresponding to the specified identification number. Accumulation of surface imperfections in a small area, however is not allowed. There cannot be more than two maximum size imperfections within a circle 0.2 times the clear aperture diameter.

The size of permitted edge chips is specified as follows:

5/A x S; RB where,

- R = the symbol for edge chip specification and,
- B = the maximum distance any chip may extend in from the edge of the part.

For example:

5/8 x 0.40; R0.5

- 5 identifies surface imperfection specification.
- 8 = permitted number of imperfections without subdivision
- 0.40 = maximum size of allowed imperfections in mm.
- R identifies edge chip specification.
- 0.5 = distance chips may extend in from edge in mm.

Subdivision is allowed since the 8 x 0.40 term is not in brackets. The size of permitted scratches in addition to imperfections is specified as follows:

5/A x S; KC x D where,

- K = the symbol designating a scratch specification and,
- C = the maximum number of permitted scratches and,
- D = the maximum width of allowed scratches.

Scratch length is not specified.

For example:

5/3 x 0.25; K2 x 0.01

- 5 identifies the surface imperfection specification.
- 3 = number of allowed imperfections without subdivision.
- 0.25 = maximum size of allowed imperfections in mm.
- K identifies the scratch specification.
- 2 = number of permitted scratches
- 0.01 = maximum width of permitted scratches in mm.

Despite the attempts of the MIL and DIN specifications to attempt to quantify cosmetic surface quality, measurements are still highly subjective. However, Melles Griot uses skilled inspectors to ensure that our optics meet the standard specified.

SAPPHIRE

Sapphire is a highly superior optical material. It is extremely hard and resistant to chemical attack, it exhibits high transmittance from 150 nm to 6 microns wavelength, and it has a high thermal conductivity. This equipment is used to produce optical quality synthetic sapphire.

Melles Griot introduces in this catalog a unique selection of stock sapphire optics including windows, plano convex and plano concave lenses.

OPTICAL MATERIALS

In this section we present the important physical properties of common optical materials and the formulas from which they may be derived. We then list these properties in detail for the majority of optical materials used in Melles Griot products.

Contents **Page Number**

PROPERTIES OF OPTICAL MATERIALS

The most important properties of lens materials are their internal and external transmittances, surface reflectances, and refractive indices. The formulas which connect these variables in the on-axis case are now presented.

When considering an image forming lens system, the external transmittance is of paramount concern. This neglects multiple reflections between lens surfaces and therefore transmittance measured with an integrating sphere, will be slightly higher. Let T_e denote the desired external irradiance (not amplitude) transmittance, T_i denote the corresponding internal transmittance, t_1 denote the single pass transmittance of the first surface, and t_2 denote the single pass transmittance of the second surface. Then

$$T_e = t_1 t_2 T_i = t_1 t_2 e^{-\mu x},$$

where e is the base of the natural system of logarithms, μ is the absorption coefficient of the lens material, and x is the lens center thickness (elsewhere denoted t_c). This allows for the possibility that the lens surfaces might have unequal transmittances (i.e., that perhaps one is coated and the other is not). Suppose now that both surfaces are uncoated. Then

$$t_1 t_2 = 1 - 2r + r^2,$$

where

$$r = \left(\frac{n - 1}{n + 1} \right)^2$$

is the single surface, single pass irradiance reflectance at normal incidence as given by the Fresnel formula. The refractive index n must be calculated from the material dispersion formula or be otherwise known. The present results are monochromatic. Both μ and n are functions of wavelength.

To calculate either T_i or T_e for a lens at any wavelength of interest, first find the value of the absorption coefficient μ. Typically the internal transmittance T_i is tabulated as a function of wavelength for two distinct thicknesses x_1 and x_2, and μ must be found from these. Thus

$$\bar{\mu} = -\frac{1}{2} \left[\frac{\ln T_i (x_1)}{x_1} + \frac{\ln T_i (x_2)}{x_2} \right]$$

where the bar denotes averaging. In portions of the spectrum where absorption is strong, a value for T_i is typically given only for the lesser thickness. Then

$$\mu = -\frac{1}{x} \ln T_i \, .$$

When it is necessary to find transmittance at wavelengths other than those for which T_i is tabulated, proceed by linear interpolation.

The on-axis T_e value is normally the most useful, but some applications require that transmittance be known along other ray paths, or that it be averaged over the entire lens surface. The method outlined above is easily extended to encompass such cases. The values of t_1 and t_2 must in general be found from the complete Fresnel formulas for arbitrary angles of incidence. t_1 and t_2 will generally be unequal because the angles of incidence will be different at the two surfaces. The distance x becomes the surface-to-surface distance along a particular ray, and must be found by ray tracing. It is necessary to account separately for the s- and p-planes of polarization, and it is usually sufficient to average the results for both planes at the end of the calculation.

Most of our singlet lenses are made from BK 7 grade A, fine annealed optical glass rather than loosely specified "optical crown."

MATERIALS USED IN MELLES GRIOT LENSES

Most Melles Griot simple lenses are made of either synthetic fused silica or BK 7 grade A fine annealed. Only a few other materials are used. The following table identifies the materials used in the manufacture of Melles Griot lens products. Some of these materials are also used in prisms and other products. Melles Griot reserves the right to make material changes without prior notice.

Glass type designations and physical constants are the same as published by Schott Glass Technologies, Inc. Melles Griot occasionally uses corresponding glasses made by other glass manufacturers, but only when this does not result in a significant change in optical properties. Melles Griot Optical Crown is not a Schott material. Literature from any of the synthetic fused silica manufacturers will be helpful in further appreciating the capabilities of this material.

Physical constant values which follow are reasonable means based on historical experience. Individual material specimens may deviate from these means. Materials having tolerances more restrictive than those published in the following pages, or materials traceable to specific manufacturers, are available only on special request and at increased cost.

The quality of performance of optical lenses and prisms is partly predicated on the quality of material used. No amount of skill during manufacture can eradicate striae, bubbles, inclusions or variations in index. Melles Griot takes considerable care in its material selection. Only first class optical materials from reputable glass manufacturers are used. The result is reliable, repeatable, consistent performance.

Materials used in the manufacture of Melles Griot Lenses

Materials	Lens Product Numbers
Synthetic Fused Silica, UV grade	01 LQP, 01 LQD 01 LQS 01 LQN, 01 LQC.
Synthetic Fused Silica, optical quality	01 LQF, 01 LQB, 01 LQT, 01 LQR, and half of 01 CMP series.
BK 7, grade A fine annealed	01 LPX, 01 LDX, 01 LPK, 01 LDK, 01 LCP, 01 LCN, 01 LMN, 01 LCF, 01 LMP, 01 LFS, 01 LFF.
LaSF 9, grade A fine annealed	only 01 LPX 401, 01 LPX 405, 01 LPX 407, 01 LPX 411, 01 LPX 413, 06 LMS
BaK 1, grade A fine annealed	01 LPX 415, 01 LPX 421, and 01 LPX 423.
Optical Crown	01 LAG.
Pyrex	half of 01 CMP series.
SF 11, grade A fine annealed	06 LXP.
SK 11/SF 5, grade A fine annealed	06 LAI.
Sapphire	01 LSK, 01 LSX.
Optical Silicon	06 LSM.
Proprietary Glass Combinations	01 LAU, 01 LAO 01 LCD, 01 LAT, 01 LBX, 04 OAS, 04 OAP, 04 EHY, 04 EPP, 04 ERA 001, 04 EWA, 04 EWR 001, 04 EWP 001, 04 ECW 001, 06 GLC, 06 GBX, 06 GLR, 09 LSL, 09 LCM, and 09 LBM.

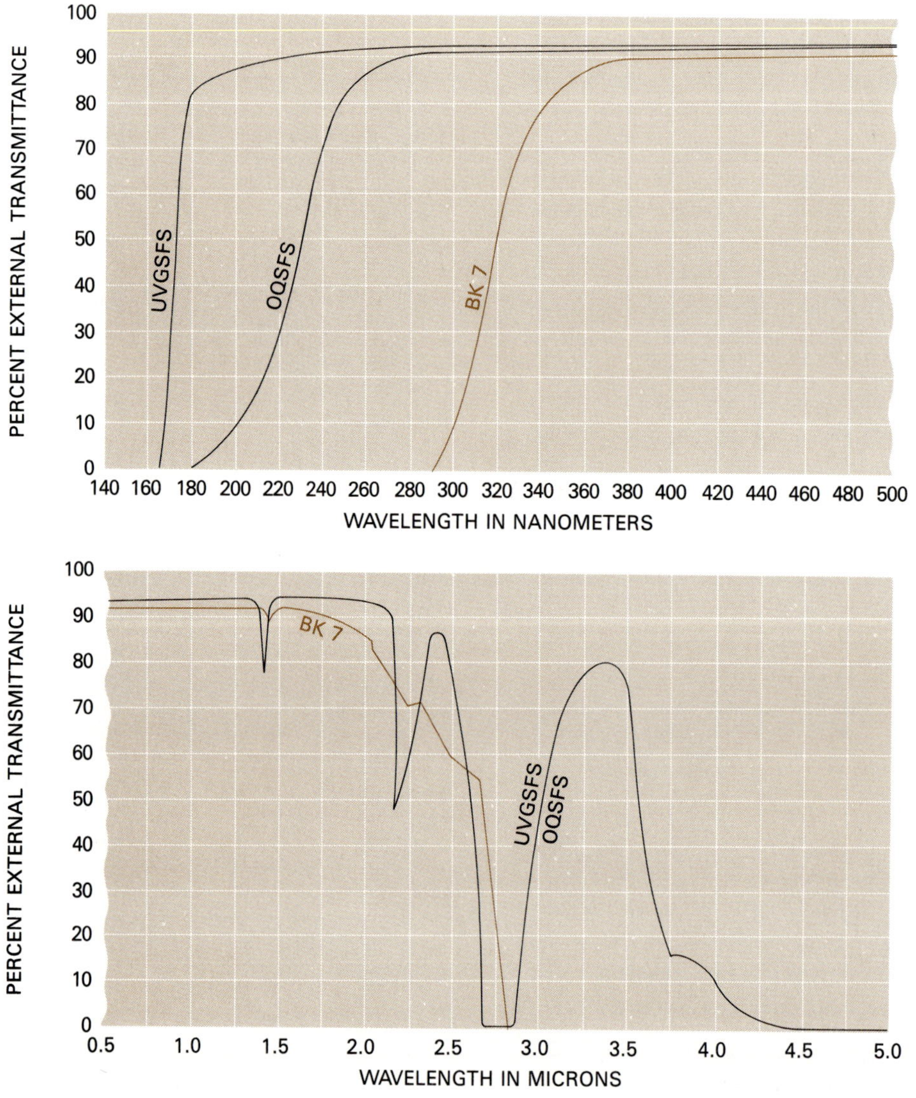

COMPARISON OF UNCOATED EXTERNAL TRANSMITTANCES for ultraviolet grade synthetic fused silica (UVGSFS), optical quality synthetic fused silica (OQSFS) and a common optical glass (BK 7), all of 10mm thickness.

SYNTHETIC FUSED SILICA

Fused silica is an ideal optical material for many applications. It is extremely transparent over a wide spectral range, has a low coefficient of thermal expansion, and is resistant to scratching and thermal shock.

Synthetic fused silica (amorphous silicon dioxide) is formed by chemical combination of silicon and oxygen. It is not to be confused with fused quartz which is made by crushing and melting natural crystals. Synthetic fused silica is far purer than fused quartz. This increased purity assures higher ultraviolet transmission and freedom from striae or inclusions.

When compared with glass or fused quartz, fused silica lenses offer a number of advantages:

1. Greater UV and IR transmission.
2. Low coefficient of thermal expansion, providing stability and resistance to thermal shock over large temperature excursions.
3. Wider thermal operating range.
4. Increased hardness and resistance to scratching.
5. Much higher resistance to radiation darkening from UV, X-rays, gamma rays, and neutrons.

Manufacture of fused quartz by crushing and melting natural crystalline quartz, or by fusing silica sands, results in a granular microstructure and bubble entrapment. Microstructure and impurities lead to local index variations, and contribute, along with bubbles and opaque particles, to reduced transmission throughout the spectrum. The synthetic fused silica materials used by Melles Griot are manufactured by flame hydrolysis to extremely high standards. The resultant material is colorless, non-crystalline, and has an impurity content of only about one part per million. Controlling the purity of the reactants and conditions of reaction assures the high quality of the synthetic fused silica from which our lenses are made.

Optical quality synthetic fused silica lenses are ideally suited for applications in energy-gathering and imaging systems in the mid-UV, visible ad near-IR spectral regions. The low dispersion of fused silica reduces chromatic aberration.

Melles Griot **UV grade** synthetic fused silica is selected from among the best of the commercially available UV grades of synthetic fused silica. It offers both the highest transmission (especially in the deep ultraviolet), and very low fluorescence levels (approximately 0.1% that of fused natural quartz excited at 254nm). UV grade synthetic fused silica does not fluoresce in response to wavelengths longer than 290nm. In deep UV applications, UV grade synthetic fused silica is an ideal choice. Its tight index tolerance assures highly predictable lens specifications.

Synthetic Fused Silica Constants

Abbe constant: 67.8 $\pm$ 0.5

Change of refractive index with temperature (0 to 700°C): 1.28 x 10^{-5}/°C

Homogeneity (Maximum index variation over 10cm aperture): 2 x 10^{-5}

Continuous operating temperature: Maximum 900°C

Coefficient of thermal expansion: 5.5 x 10^{-7}/°C

Thermal conductivity (100°C): 0.177 cal g^{-1} °C^{-1}

Dispersion formula at 20°C(*), (λ in microns):

$$n^2 - 1 = \frac{0.6961663\lambda^2}{\lambda^2 - (0.0684043)^2} + \frac{-.4079426\lambda^2}{\lambda^2 - (0.1162414)^2} + \frac{0.8974794\lambda^2}{\lambda^2 - (9.896161)^2}$$

The following table shows the refractive index of a typical UV grade synthetic fused silica versus wavelength at 20°C. To obtain index for optical quality synthetic fused silica, round the values off to the fourth decimal place. Accuracy of index:

$\pm 3 \times 10^{-5}$.

Glass transmittances are affected by thermal history after manufacture, as well as by the manufacturing process. Depending upon the manufacturer and subsequent thermal processing (coating, annealing, or tempering) it is possible for any optical glass, including BK 7, to show internal transmittance reductions of several percent across the entire spectrum with external transmittance correspondingly affected. Transmittance of all glasses is especially uncertain at wavelengths approaching the water absorption band at 2.7 microns.

Synthetic fused silica also shows batch-to-batch transmittance variations, especially in the deep ultraviolet and infrared. These variations are related to manufacture and impurity content rather

(*)Malitson, I.H. "Interspecimen Comparison of the Refractive Index of Fused Silica," *Journal of the Optical Society of America*, Vol. 55, No. 10, October 1965, pp. 1205-1209.

Refractive Index of Synthetic Fused Silica

Wavelength (nm)	Index of Refraction	Wavelength (nm)	Index of Refraction
180.0	1.58529	532.0	1.46071
190.0	1.56572	546.1	1.46008
200.0	1.55051	587.6	1.45846
213.9	1.53431	589.3	1.45840
226.7	1.52275	632.8	1.45702
230.2	1.52008	643.8	1.45670
239.9	1.51337	656.3	1.45637
248.3	1.50840	694.3	1.45542
265.2	1.50003	706.5	1.45515
275.3	1.49591	786.0	1.45356
280.3	1.49404	820.0	1.45298
289.4	1.49099	830.0	1.45282
296.7	1.48873	852.1	1.45247
302.2	1.48719	904.0	1.45170
330.3	1.48054	1014.0	1.45024
340.4	1.47858	1064.0	1.44963
351.1	1.47671	1100.0	1.44920
361.1	1.47513	1200.0	1.44805
365.0	1.47454	1300.0	1.44692
404.7	1.46962	1400.0	1.44578
435.8	1.46669	1500.0	1.44462
441.6	1.46622	1550.0	1.44402
457.9	1.46498	1660.0	1.44267
476.5	1.46372	1700.0	1.44217
486.1	1.46313	1800.0	1.44087
488.0	1.46301	1900	1.43951
496.5	1.46252	2000	1.43809
514.5	1.46156	2100	1.43659

than to subsequent history. In the ultraviolet these variations have been attributed to uncontrollable fluctuations in metallic impurity content at the parts per billion level. Ultraviolet transmittance is the basis for the classifications "UV grade" and "optical quality." Specification of UV grade assures that a specimen is represented on the highest curve. Transmittance curves for optical quality may fall anywhere between the UV grade curve and the curve identified as optical quality. This latter curve is actually the lower limiting curve.

Infrared batch-to-batch transmittance variations in synthetic fused silica are attributable to fluctuations in OH chemical bond content, are most pronounced at wavelengths near and beyond the water absorption band at 2.7 microns, and are normally uncontrolled because ultraviolet transmittance is generally regarded as more important. High infrared transmittance can be assured by appropriate manufacturing controls, but only at the sacrifice of ultraviolet transmittance.

Visible spectrum batch-to-batch transmittance variations in synthetic fused silica are insignificant.

The high ultraviolet internal transmittance of UV grade synthetic fused silica is correlated with a visible internal transmittance that is so high as to be beyond the traditional methods of measurement. It is instead necessary to measure optical signal attenuation in fibers drawn of the material.

Following is a semilogarithmic comparison of the internal transmittances of UV grade synthetic fused silica and BK 7 glass. It is evident from this graph that UV grade synthetic fused silica averages about two orders of magnitude less absorption loss than BK 7 across the visible spectrum. In a sample thickness of 10 millimeters the internal transmittance of UV grade synthetic fused silica differs from unity only in the 5th decimal place. The high internal transmittance of such a material can be exploited by both transmission at Brewster's angle for the appropriate linear polarization, and with the assistance of high-efficiency antireflection coatings such as HEBBAR™ or one of the laser line "V"-coats. With these coatings it is possible to achieve external transmittances of 98.5% and 99.5%, respectively. Synthetic fused silica and HEBBAR™ are especially well suited for each other in visible spectrum applications.

The internal transmittance of UV grade synthetic fused silica shows a pronounced dip at 950nm, while the data for BK 7 give no hint of a corresponding feature. It should be understood that BK 7 and UVGSFS are manufactured by very different processes. One of the many differences in these materials is that UVGSFS has a much higher content of OH chemical bond ("hydroxyl content") than does BK 7. The dip in UVGSFS transmittance may be attributed to OH bond resonance.

We offer the widest selection of stock lenses made from synthetic fused silica. This truly superior optical material should be considered for all demanding laser applications.

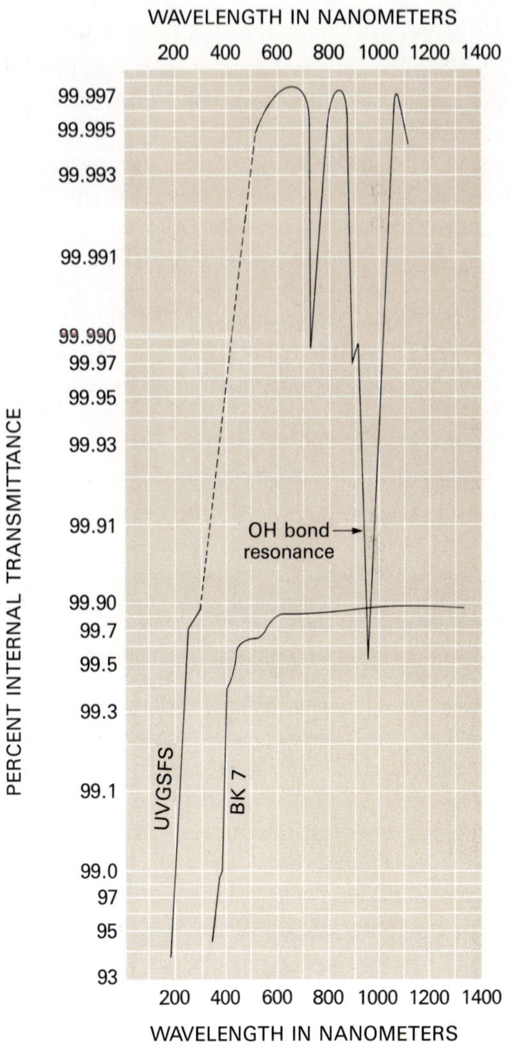

SEMILOGARITHMIC COMPARISON OF INTERNAL TRANSMITTANCES of UV grade synthetic fused silica and BK 7.

INTRODUCTION TO OPTICAL GLASSES

There are nearly 250 different optical glasses listed in the Schott Optical Glass catalog. For the lens designer, the most important difference between each of these glasses is the value of the index of refraction, and the dispersion (rate of change of index with wavelength). Typically, an optical glass is specified by giving its index of refraction at a wavelength in the middle of the visible spectrum, usually 587.56nm (the helium d-line), and by the Abbe v-value, defined to be $v_d = n_d - 1 / n_F - n_C$. The designations F and C stand for 486.1nm and 656.3nm, respectively. As can be seen, v_d is a measure of how index of refraction varies with wavelength; the smaller v_d is, the faster the rate of change. Glasses are roughly divided into two categories, crowns and flints. Crown glasses are those with $n_d < 1.60$ and $v_d > 55$ or $n_d > 1.60$ and $v_d > 50$. The others are flint glasses.

The refractive index of a glass in the wavelength range of 365 to 1014nm can be calculated from the following formula;

$$n^2 = A_0 + A_1\lambda^2 + A_2\lambda^{-2} + A_3\lambda^{-4} + A_4\lambda^{-6} + A_5\lambda^{-8}$$

Here λ, the wavelength, must be in microns, and the constants A_0 through A_5 are given by the manufacturer for each different glass. We have tabulated these constants in the tables given on page 59. This equation yields a value for index that is accurate to $\pm 3 \times 10^{-6}$ between 400 and 750nm, and accurate to $\pm 5 \times 10^{-6}$ from 750 to 1014nm.

For the lens manufacturer the mechanical and chemical properties of glasses are important. Different polishing techniques and special handling may be called for depending on whether the glass is hard or soft, or extremely sensitive to acid or alkali. To quantify the chemical properties of glasses, ratings are given to each glass in four different categories; climatic resistance, stain resistance, acid resistance and alkali resistance.

Water vapor in the air can cause a cloudy film to appear on the surface of some optical glasses. Climatic resistance expresses the susceptibility of a glass to such a process. In this test, the glass is placed in a water vapor saturated environment and subjected to a temperature cycle which alternately causes condensation and evaporation. The glass is given a rating from 1 to 4 depending on the amount of surface scattering induced by the test. A rating of 1 indicates little or no change after 7 days of exposure; a rating of 4 means a significant change occurred in less than 30 hours.

Stain resistance expresses the glass' resistance to mildly acidic water solutions, such as might be encountered in fingerprints or perspiration. In this test, a few drops of a mild acid is placed on the glass; a colored stain, due to interference effects, will appear if the glass starts to decompose. A rating from 1 to 5 is given to each glass, depending on the time before onset of the stain. A rating of 1 indicates no observed stain in 100 hours of exposure; a rating of 5 means staining occurred in less than 0.2 hours.

Acid resistance quantifies the resistance of a glass to stronger acidic solutions. Acid resistance can have particular importance to the lens manufacturer since acidic solutions are typically used to strip coatings from glass or to separate cemented elements. In this case a rating from 1 to 4 or 51 to 53 is given. Ratings 1 to 4 indicate progressively less resistance to a pH 0.3 acid solution, and values from 51 to 53 are used for glasses with too little resistance to be tested with such a strong solution.

Alkali resistance is also important to the lens manufacturer, since the polishing process usually takes place in an alkaline solution. A two digit number is used to designate the alkali resistance of optical glasses; the first number, from 1 to 4, indicates the length of time before the onset of any surface change in the glass, and the second digit reveals the extent of that change.

The most important mechanical property of a glass for the lens manufacturer is the "microhardness." To test this, a precisely specified diamond scribe is placed on the glass' surface under a known force. The size of the resultant indentation is then measured. Two different tests, the Knoop and the Vickers michrohardness test, measure the hardness of a polished surface and a freshly fractured surface, respectively.

OPTICAL GLASS

A borosilicate crown glass, BK 7 is the material used in a large fraction of Melles Griot products. There are several reasons for this. BK 7 performs well in the chemical tests discussed earlier; this means that special treatment of the glass during polishing is not necessary. BK 7 is a relatively hard glass, it does not scratch easily, and can be handled without special precautions. The bubble and inclusion content of BK 7 is very low; bubble and inclusion content cross-section totals less than 0.029mm^2 per 100cm^3. Another important characteristic of BK 7 is its excellent transmittance, down to 350nm. A variant of BK 7, designated UBK 7, has transmission down to nearly 300nm. This special glass is useful in applications requiring the high index of refraction, and desirable chemical properties of BK 7, but demanding transmission deeper into the ultraviolet. Because of these characteristics, BK 7 is used widely throughout the optics industry.

THREE SCHOTT GLASSES

The following tables list the most important optical and physical constants for the three Schott optical glasses BK 7, SF 11 and LaSF 9. These are the materials used in the majority of Melles Griot simple lens products. Most of the values are calculated from the dispersion formula given earlier, using the constants shown in the table. Values marked with an "x" are extrapolations, since they fall outside the wavelength interval for which the dispersion formula coefficients were determined. Especially in the wavelength regions where index has been extrapolated, either index or internal transmittance values for corresponding glasses by other manufacturers may differ from the values shown. In recent years, Schott Glass Technologies has begun the introduction of experimental melt-to-melt mean indices at four specific infrared wavelengths into their optical glass catalog. It must be emphasized that these wavelengths (1060.0, 1529.6, 1970.1 and 2325.4nm) occur in what we identify as the extrapolation region for the dispersion formula. For these wavelengths, deviations of refractive index from formula predictions are both real and significant. In the tables, we have identified the corresponding indices "m," for measured, rather than "x." Internal transmittance values shown are melt-to-melt experimental means, and may be affected by thermal history (coating, annealing or tempering operations) after manufacture.

Physical Constants of Three Schott Glasses

	Glass Type		
	BK 7	SF 11	LaSF 9
Melt-to-Melt Mean Index Tolerance	±0.001	±0.001	±0.002
Homogeneity Within Melt	$\pm 1 \times 10^{-4}$	$\pm 1 \times 10^{-4}$	$\pm 1 \times 10^{-4}$
Striae Grade (MIL-G-174-A)	A	A	A
Stress Birefringence, nm/cm, yellow light	10	10	10
Abbe Factor (v_d)	64.17	25.76	32.23

Physical Constants of Three Schott Glasses (continued)

	Glass Type		
	BK 7	SF 11	LaSF 9
Constants of Dispersion Formula			
A_0	2.2718929	3.0539614	3.3051830
A_1	$-1.0108077 \times 10^{-2}$	$-1.1580432 \times 10^{-2}$	$-1.3857059 \times 10^{-2}$
A_2	1.0592509×10^{-2}	3.9199816×10^{-2}	3.5921736×10^{-2}
A_3	2.0816965×10^{-4}	2.9462812×10^{-3}	2.6740381×10^{-3}
A_4	$-7.6472538 \times 10^{-6}$	$-2.0371019 \times 10^{-4}$	$-1.9764177 \times 10^{-6}$
A_5	4.9240991×10^{-7}	2.7633569×10^{-5}	1.9381052×10^{-6}
Density (g/cm^{-3})	2.51	4.74	4.53
Coefficient of linear Thermal Expansion (α)			
-30° to +70° (per °C)	7.1×10^{-6}	6.1×10^{-6}	7.6×10^{-6}
+20° to +300° (per °C)	8.3×10^{-6}	6.9×10^{-6}	8.6×10^{-6}
Transformation Temperature	559 °C	503 °C	704 °C
Young's Modulus (dynes/mm^2)	8.10×10^9	6.60×10^9	1.07×10^{10}
Modulus of Rigidity (dynes/mm^2)	3.4×10^9	2.7×10^9	4.2×10^9
Poisson's Ratio	0.208	0.237	0.285
Climate Resistance	2	1	1
Stain Resistance	0	0	0
Acid Resistance	1	1	1
Alkali Resistance	2.0	1.2	1.0
Knoop Hardness	520	380	520

Refractive Index of Three Schott Glasses

Wavelength λ (nm)	Refractive Index, n BK 7	SF 11	LaSF 9	Fraunhofer Designation	Source	Spectral Region
351.1	1.53894x	—	—	—	Ar laser	UV
363.8	1.53648x	—	—	—	Ar laser	UV
404.7	1.53024	1.84211	1.89841	h	mercury arc	violet
435.8	1.52669	1.82518	1.88463	g	mercury arc	blue
441.6	1.52611	1.82259	1.88249		HeCd laser	blue
457.9	1.52462	1.81596	1.87697		Ar laser	blue
465.8	1.52395	1.81307	1.87455		Ar laser	blue
472.7	1.52339	1.81070	1.87256		Ar laser	blue
476.5	1.52310	1.80945	1.87150		Ar laser	blue
480.0	1.52283	1.80834	1.87056	F′	cadmium arc	blue
486.1	1.52238	1.80645	1.86896	F	hydrogen arc	blue
488.0	1.52224	1.80590	1.86850		Ar laser	blue
496.5	1.52165	1.80347	1.86643		Ar laser	green
501.7	1.52130	1.80205	1.86522		Ar laser	green
514.5	1.52049	1.79880	1.86244		Ar laser	green
532.0	1.51947	1.79480	1.85900		Nd laser	green
546.1	1.51872	1.79190	1.85650	e	mercury arc	green
587.6	1.51680	1.78472	1.85026	d	helium arc	yellow
589.3	1.51673	1.78446	1.85003	D	sodium arc	yellow
632.8	1.51509	1.77862	1.84491		HeNe laser	red
643.8	1.51472	1.77734	1.84378	C′	cadmium arc	red
656.3	1.51432	1.77599	1.84259	C	hydrogen arc	red
694.3	1.51322	1.77231	1.83933		Ruby laser	red
786.0	1.51106	1.76559	1.83332			IR
821.0	1.51037	1.76360	1.83152			IR
830.0	1.51021	1.76312	1.83109		GaAlAs laser	IR
852.1	1.50981	1.76202	1.83009	s	cesium arc	IR
904.0	1.50894	1.75971	1.82798		GaAs laser	IR
1014.0	1.50731	1.75579	1.82435	t	mercury arc	IR
1060.0	1.50669m	1.75444	1.82307		Nd laser	IR
1300.0	1.50371x	1.74888x	1.81767		InGaAsP laser	IR
1500.0	1.50130x	1.74525x				
1550.0	1.50068x	1.74440x	1.81309x			
1970.1	1.49500m	1.73843m	1.80727m		mercury arc	
2325.4	1.48929m	1.73294m	1.80158m		mercury arc	

Internal Transmittance of Three Schott Glasses

Wavelength λ (nm)	Internal Transmittance (%) BK 7 thickness (mm) 5	25	SF 11 thickness (mm) 5	25	LaSF 9 thickness (mm) 5	25
300	0.26	—	—	—	—	—
310	0.59	0.07	—	—	—	—
320	0.81	0.35	—	—	—	—
330	0.93	0.70	—	—	—	—
340	0.97	0.86	—	—	—	—
350	0.986	0.93	—	—	—	—
360	0.992	0.96	—	—	0.11	—
370	0.995	0.974	—	—	0.39	—
380	0.996	0.980	0.13	—	0.60	0.08
390	0.998	0.989	0.46	0.02	0.76	0.26
400	0.948	0.991	0.73	0.21	0.84	0.41
420	0.998	0.993	0.93	0.69	0.88	0.63
440	0.999	0.994	0.97	0.87	0.91	0.75
460	0.999	0.994	0.986	0.93	0.959	0.81
480	0.999	0.995	0.990	0.96	0.968	0.86
500	0.999	0.996	0.995	0.976	0.977	0.89
540	0.999	0.996	0.998	0.990	0.986	0.93
580	0.999	0.996	0.998	0.992	0.991	0.958
620	0.999	0.997	0.998	0.992	0.995	0.975
660	0.990	0.997	0.999	0.993	0.996	0.980
700	0.999	0.998	0.999	0.994	0.998	0.988

The precision of the published mean refractive indices, and of the dispersion formula coefficients, is appropriate for "restricted tolerance" materials. These materials are prepared by special procedures which result in highly repeatable melt-to-melt indices. The ± 0.001 melt-to-melt mean index tolerance applies to ordinary optical glasses for which these procedures are not used.

OPTICAL CROWN GLASS

Optical crown glass is a low index, commercial quality glass in which index of refraction, transmittance and homogeneity are not controlled as carefully as in optical quality glasses such as BK 7. Optical crown is a suitable material in applications where component tolerances are fairly loose or as a substrate material for mirrors. Transmittance characteristics for optical crown are shown in the graph; relevant properties of optical crown are tabulated below.

Refractive Index of Optical Crown Glass

Wavelength (nm)	Refractive Index,n	Fraunhofer Designation	Source	Spectral Region
435.8	1.53394	g	mercury arc	blue
480.0	1.52960	F′	cadmium arc	blue
486.1	1.52908	F	hydrogen arc	blue
546.1	1.52501	e	mercury arc	green
587.6	1.52288	d	helium arc	yellow
589.0	1.52280	D_2	sodium arc	yellow
643.8	1.52059	C′	cadmium arc	red
656.3	1.52015	C	hydrogen arc	red

OPTICAL CROWN GLASS CONSTANTS

Abbe Factor:

$$v_e = \frac{n_e - 1}{n_{F'} - n_{C'}} = 58.3 \quad v_d = \frac{n_d - 1}{n_F - n_C} = 58.8$$

Dispersion: $(n_F - n_c) = 0.0089$

Glass type designation: B270

Density: 2.55 g cm^{-3} at 23 °C

Transformation temperature: 521 °C

Softening temperature: 708 °C

Coefficient of linear expansion:
(20 to 300 °C) = 93.3×10^{-7} °C^{-1}

Specific heat: C_p (20 to 100 °C) = 0.184 cal. g^{-1} °C^{-1}

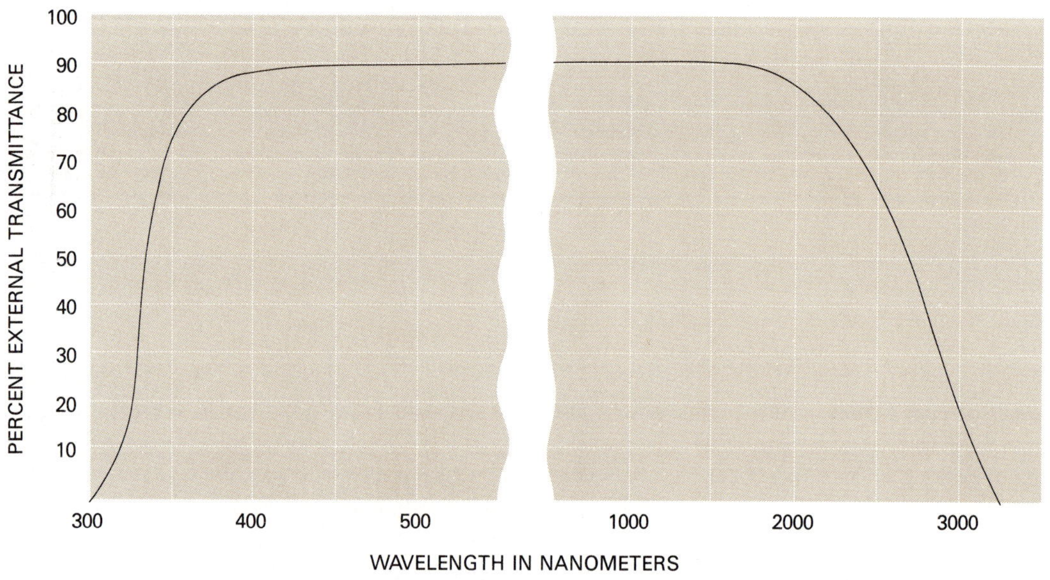

EXTERNAL TRANSMITTANCE FOR OPTICAL CROWN GLASS of 10mm thickness.

PYREX

Pyrex is a special type of borosilcate glass which has a low thermal expansion coefficient. Pyrex is well suited for applications where high temperature, thermal shock or resistance to chemical attack are primary considerations. The main drawback to pyrex is that it is typically less homogeneous, and contains more striae and bubbles than optical glasses such as BK 7. Pyrex is ideally suited to such tasks as low expansion mirror substrates, condenser lenses for high power illumination systems or windows in high temperature environments. Because of its low cost and excellent thermal stability, pyrex is the standard material used in test plates and optical flats. As can be seen in the accompanying graph, pyrex's transmission extends both into the ultraviolet and well into the infrared. Index of refraction in pyrex varies considerably from batch to batch, typical values are tabulated below.

PYREX CONSTANTS

Glass type designation: Pyrex 7740
Density: 2.23 g cm^{-3} at 25 °C
Young's Modulus: 5.98 x 10^9 dynes/mm^2
Poisson's Ratio: 0.20
Specific Heat at 25 °C: 0.17 cal/g °C
Coefficient of linear expansion:
(0 to 300 °C) = 3.25 x 10^{-6} °C
Abbe Factor: v_d = 66

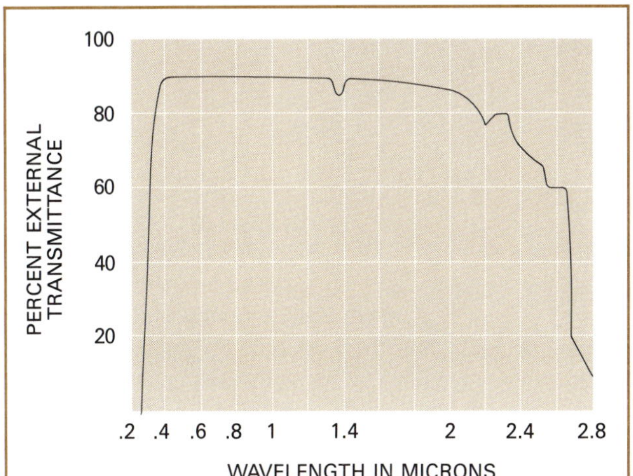

EXTERNAL TRANSMITTANCE FOR PYREX of 8mm thickness.

Refractive Index of Pyrex

Wavelength (nm)	Refractive Index, n	Fraunhofer Designation	Source	Spectral Region
486.1	1.479	F	hydrogen arc	blue
514.5	1.477		argon laser	green
546.1	1.476	e	mercury arc	green
587.6	1.474	d	sodium arc	yellow
643.8	1.472	C′	cadmium arc	red

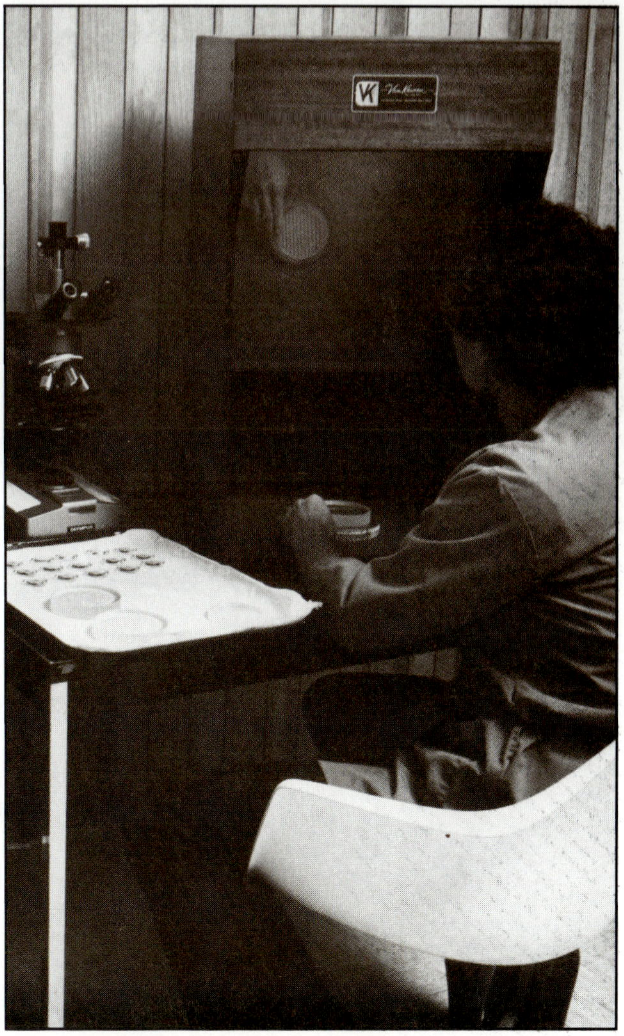

SAPPHIRE

Sapphire is in many ways a superior window material. Because of its extreme surface hardness, sapphire can only be scratched by a few substances (such as diamond or boron nitride) other than itself. Chemically inert and insoluble in almost everything except at highly elevated temperatures, sapphire can be cleaned with impunity. For example, even hydrogen fluoride fails to attack sapphire at temperatures below 300 °C. Sapphire exhibits high internal transmittance all the way from 150 nanometers in the vacuum ultraviolet out to 6000 nanometers in the middle infrared. External transmittance of sapphire is shown in the accompanying graph. Because of its great strength (Young's modulus is roughly 5 times higher than fused silica and most glasses), sapphire windows can safely be made much thinner than windows of other common dielectrics. Sapphire windows are therefore useful even at wavelengths which are very close to its transmission limits. Because of the exceptionally high thermal conductivity of sapphire, thin windows can be very effectively cooled by forced air or other methods. Conversely, sapphire windows can easily be heated to prevent condensation.

Sapphire is single crystal aluminum oxide or Al_2O_3. Because of its hexagonal crystalline structure, sapphire exhibits anisotropy in many optical and physical properties. The exact characteristics of an optical component made from sapphire depend on the orientation of the optic axis or c-axis relative to the surface of the element. Sapphire exhibits birefringence; a difference in index of refraction in orthogonal directions. The difference in index is 0.008 between light traveling along the optic axis and light traveling perpendicular to it. Malitson determined a dispersion relationship for the ordinary ray in sapphire. This formula, along with the appropriate constants and a table of some calculated values, is shown below, (λ in microns):

$$n^2-1=\frac{A_1\lambda^2}{\lambda^2-\lambda_1^2}+\frac{A_2\lambda^2}{\lambda^2-\lambda_2^2}+\frac{A_3\lambda^2}{\lambda^2-\lambda_3^2}\quad\text{, where}$$

A_1 = 1.023798

A_2 = 1.058264

A_3 = 5.280792

λ_1^2 = 0.00377588

λ_2^2 = 0.0122544

λ_3^2 = 321.3616.

*Malitson, I.H., "Refraction and Dispersion of Synthetic Sapphire," *Journal of the Optical Society of America,* Vol. 52, No. 12, Dec. 1967, pg. 1377.

The transmission of sapphire is limited primarily by losses due to surface reflections. The high index of sapphire makes magnesium fluoride almost an ideal single-layer antireflection coating. When a single-layer of magnesium fluoride is deposited on sapphire and optimized for 550nm, total transmission of a sapphire component can be kept above 98% throughout the entire visible spectrum.

SAPPHIRE* CONSTANTS

Density: 3.98 g cm^{-3} at 25 °C

Young's Modulus: *3.7 x 10^{10} dynes/mm^2

Poisson's Ratio: *–0.02

Moh hardness: 9 (by definition)

Softening Point: 1800 °C

Specific Heat at 25 °C: 0.18 cal/g °C

Coefficient of linear expansion:

(0 °C to 500 °C) ° 7.7 x 10^{-6} °C

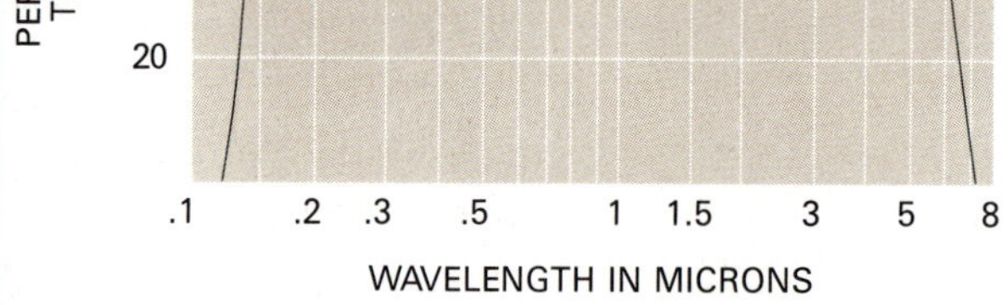

EXTERNAL TRANSMITTANCE FOR SAPPHIRE of 1mm thickness.

*Sapphire is anisotropic in many of its properties which require tensor description. These values are averages over many directions.

Refractive Index of Sapphire

Wavelength (nm)	Refractive Index, n
265.2	1.8337
275.3	1.8272
280.3	1.8243
289.4	1.8185
296.7	1.8159
302.2	1.8135
330.3	1.8030
340.4	1.7999
351.1	1.7970
361.1	1.7945
365.0	1.7836
404.7	1.7858
435.8	1.7812
441.6	1.7805
457.9	1.7785
476.5	1.7765
486.1	1.7756
488.0	1.7754
496.5	1.7746
514.5	1.7731
532.0	1.7718
546.1	1.7708
587.6	1.7682
589.3	1.7691
632.8	1.7660
643.8	1.7655
656.3	1.7649
694.3	1.7635
706.5	1.7630
786.0	1.7606
820.0	1.7597
830.0	1.7594
852.1	1.7589
904.0	1.7577
1014.0	1.7555

Refractive Index of Sapphire (continued)

Wavelength (nm)	Refractive Index, n
1064.0	1.7545
1100.0	1.7539
1200.0	1.7522
1300.0	1.7505
1400.0	1.7488
1500.0	1.7471
1550.0	1.7462
1660.6	1.7443
1700.0	1.7436
1800.0	1.7417
1900.0	1.7398
2000.0	1.7377

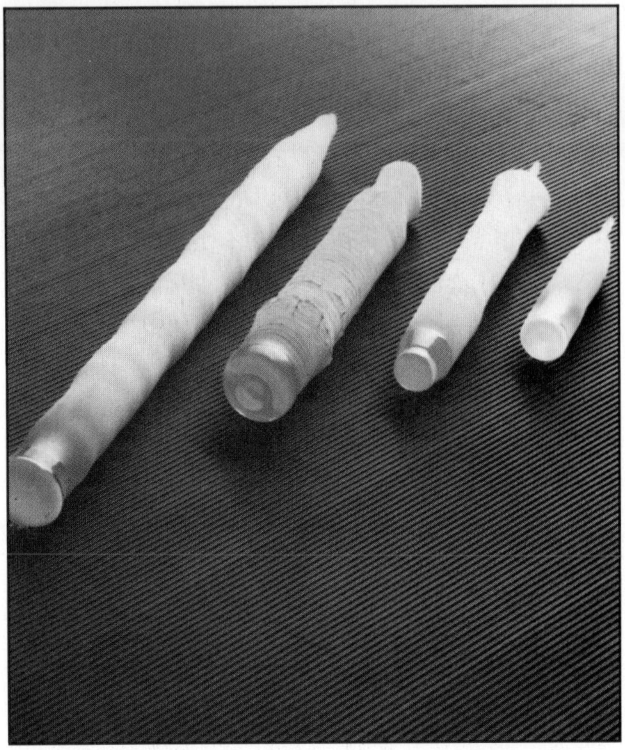

SILICON

Silicon, the material at the heart of semiconductor electronic technology, is also a useful material for transmissive optics at infrared wavelengths. With its high index of refraction and durability, silicon can be preferable over optical glass in the region where their transmission bands overlap.

Silicon crystals for optical use are prepared in much the same manner as silicon substrates for semiconductor applications. In a technique known as the Czochralski pulling process, high purity silicon is melted in a graphite or fused silica crucible. Melting is typically accomplished using radio frequencies. To begin crystallization, a small seed crystal is introduced into the melt. The seed crystal acts as a center for further crystallization to occur. This crystal is slowly withdrawn while being rotated from the molten silicon. Much attention is paid to any possible sources of impurities during the process; a positive presure of helium or argon is kept over the melt to minimize diffusion of impurities from the melt into the crystal. Rotation of the crystal distributes any impurities evenly. While electronics applications usually call for a single crystal substrate, polycrystalline silicon is adequate for most optical uses. There is slightly increased scattering at the grain boundaries in polycrystalline silicon; this effect can become important in high power laser applications.

The refractive index of silicon as a function of wavelength is given in the formula below along with the appropriate constants. This formula is for single crystal silicon; the refractive index of adequately pure polycrystalline silicon should be essentially the same. The temperature coefficient of refractive index is $1.5 \times 10^{-4}/°C$.

Dispersion formula for silicon* (λ in microns):

$$n = A + BL + CL^2 + D\lambda^2 + E\lambda^4$$

$A = 3.41696$

$B = 0.138497$

$C = 0.013924$

$D = -2.09 \times 10^{-5}$

$E = 1.48 \times 10^{-7}$

$L = (\lambda^2 - 0.028)^{-1}$

*Salzberg, C.D., "Infrared Refractive Indexes of Silicon, Germanium and Modified Selenium Glass," *Journal of the Optical Society of America*, Vol., 47, No. 3, Mar. 1957, p. 244.

SILICON CONSTANTS

Density: 2.33 g cm^{-3} at 18 °C
Young's Modulus: 1.067×10^{10} dynes/mm^{-2}
Poisson's Ratio: 0.44
Moh Hardness: 7
Specific Heat at 27 °C: .171 cal/g °C
Coefficient of linear expansion at 300 °C: $2.33 \times 10^{-6}/°C$

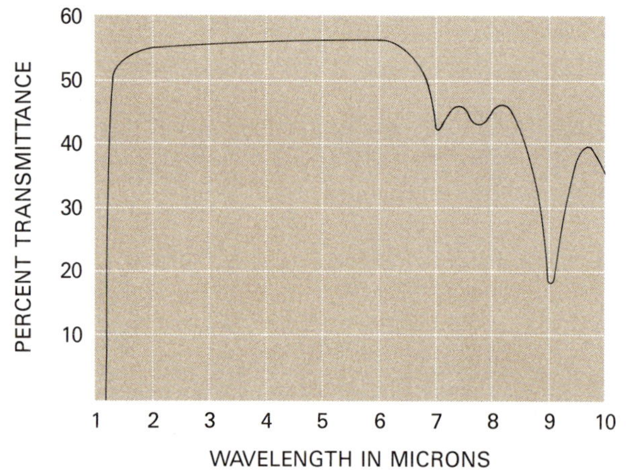

EXTERNAL TRANSMITTANCE FOR SILICON of 2mm thickness.

Silicon singlets for use with laser diodes at 1300 and 1550nm are offered on page 123.

Refractive Index of Silicon

Wavelength (nm)	Refractive Index, n
1250	3.5131
1260	3.5115
1270	3.5099
1280	3.5083
1290	3.5060
1300	3.5053
1310	3.5039
1320	3.5024
1330	3.5011
1340	3.4997
1350	3.4984
1360	3.4971
1370	3.4959
1380	3.4947
1390	3.4935
1400	3.4923
1410	3.4912
1420	3.4901
1430	3.4890
1440	3.4979
1450	3.4869
1460	3.4859
1470	3.4849
1480	3.4839
1490	3.4830
1500	3.4821
1510	3.4812
1520	3.4803
1530	3.4794
1540	3.4785
1550	3.4777
1560	3.4769
1570	3.4761
1580	3.4753
1590	3.4745
1600	3.4738
1610	3.4730
1620	3.4723

Refractive Index of Silicon

Wavelength (nm)	Refractive Index, n
1630	3.4716
1640	3.4709
1650	3.4702
1660	3.4696
1670	3.4689
1680	3.4682
1690	3.4676
1700	3.4670
1710	3.4664
1720	3.4658
1730	3.4652
1740	3.4646
1750	3.4641
1760	3.4635
1770	3.4629
1780	3.4624
1790	3.4619
1800	3.4614
1810	3.4609
1820	3.4604
1830	3.4599
1840	3.4594
1850	3.4589
1860	3.4584
1870	3.4580
1880	3.4575
1890	3.4571
1900	3.4566
1910	3.4562
1920	3.4558
1930	3.4554
1940	3.4550
1950	3.4546
1960	3.4542
1970	3.4538
1980	3.4534
1990	3.4530
2000	3.4526

ZERODUR®

For applications demanding the ultimate in thermal stability and thermal shock resistance, a "zero expansion" material such as Schott ZERODUR® is the best choice. Low expansion glass-ceramics of this type are made by combining a negative thermal expansion coefficient crystalline phase with a positive coefficient amorphous phase in such proportions as to result in an expansion coefficient of almost zero over a large temperature interval. The reason that negative thermal expansion coefficient materials exist is that as temperature varies, so also does the ratio of the energies in vibrational modes which are parallel to or orthogonal to chemical bond directions. This in turn is because of the quantization of vibrational energy levels, and the fact that allowed levels for orthogonal modes need not fall at the same energies. Thus, zero and negative expansion coefficients are large scale manifestations of quantum theory. ZERODUR® contains 70% to 78% crystalline phase by weight. Because of the amount of scattering due to the crystalline content, glass-ceramics such as ZERODUR® are not used for transmissive optics. They are, however, clear enough for visual inspection of bubble and inclusion content.

A reasonable upper bound on the absolute value of the expansion coefficient of low expansion ZERODUR® glass-ceramic at room temperature is $0.15 \times 10^{-6}/°C$. The coefficient can be much smaller, and can have either sign. By design, these materials typically exhibit a change in the sign of the coefficient of linear thermal expansion near room temperature. A comparison of the thermal expansion coefficients of ZERODUR® and fused silica is shown in the graph. As can be seen, ZERODUR® is markedly superior over a large temperature range. ZERODUR® makes ideal mirror substrates for such stringent applications as multiple-exposure holography, holographic interferometry, interferometry generally, manipulation of moderately powerful laser beams, and spaceborne imaging systems.

ZERODUR® CONSTANTS

Density: 2.53 g cm^{-3} at 25 °C
Young's Modulus: 9.1×10^9 dynes/mm^2
Poisson's Ratio: 0.24
Specific Heat at 25 °C: .196 cal/g °C
Coefficient of linear expansion:
(20 °C to 300 °C) = $0.05 \pm .10 \times 10^{-6}$ °C
Abbe Factor: $v_d = 66$

ZERODUR® is a registered trademark of Schott Glass Technologies.

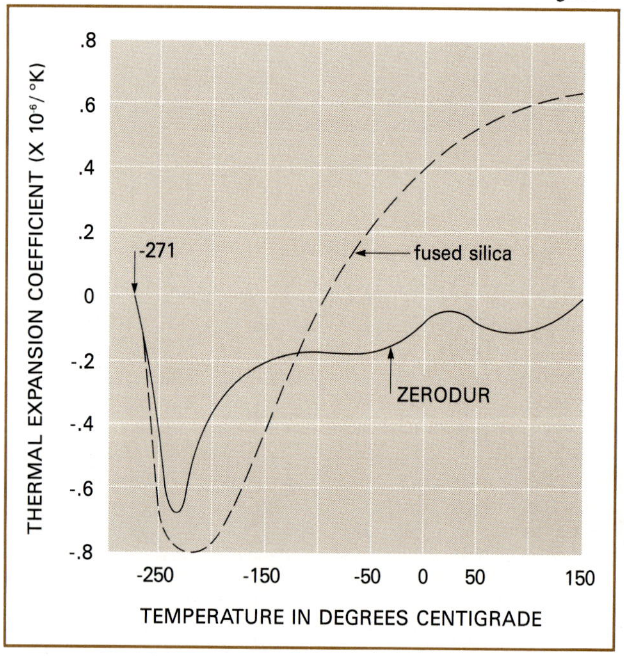

COMPARISON OF THERMAL EXPANSION COEFFICIENTS OF ZERODUR® AND FUSED SILICA.

ZERODUR® mirrors of 1/20 wave flatness are offerred on pages 186 and 187.

OPTICAL COATINGS

In this section we discuss the various types of coating which can be applied to an optical component. Using our simple Product Number Suffix system makes it easy for you to specify coated optics.

Contents **Page Number**

COATINGS

Coatings are applied to optical surfaces in order to modify their reflectance/transmittance properties. Various types of coating are available from Melles Griot including antireflection coatings, high reflection coatings, partial reflection coatings and wavelength selective coatings (filters).

We are able to apply coatings designed to operate within a wavelength region from 200 to 1600nm. Coating types include metals, single layer dielectrics, multilayer dielectrics and hybrids. Some coatings are for use at a specific wavelength while others cover a broad spectral range.

Some coatings are specific to a particluar product but most can be applied to any component. Our coating suffix system facilitates your selection and specification of coating options. Simply add the appropriate coating suffix to the product number and we'll take care of the rest. We will supply a coating to meet the stated specifications regardless of the substrate material. Many coated components are carried in stock, but if your particular choice is not, we will take an uncoated part and coat it. You benefit from the large volume of coating we do by our ability to combine your order with others and thereby offer a reasonable piece price. For some less popular coatings we regretfully have to charge more.

For those working with high energy lasers we offer a separate range of very high durability coatings. The special demands of high energy lasers (HELs) make it necessary to control carefully, not only the coating, but the substrate and the entire manufacturing cycle. In this way we can reduce the probability of laser induced damage to a level at which we are prepared to guarantee these optics. HEL coatings are restricted to certain substrates and are listed separately (see pages 203-222).

COATING DESIGN

We strive to offer standard coatings for most common applications and each of our coatings is carefully specified on the following pages. However there are still many requirements for which we do not have a "standard solution." If yours is one of these situations, we are interested in working with you. Our in-house engineers will match coating performance to your special requirements. We use microcomputer based coating design programs. Working closely with production, our engineers are able to optimize your performance with a cost-effective design.

COATING PRODUCTION

Melles Griot has over 500m^2 of coating facilities in California and 760m^2 of coating facilities in the Republic of China. We have three 1.2m box coaters, two 900mm box coaters, one 600mm bell jar coater and three 450mm bell jar coaters. Positive pressurized filtration systems are used together with ultrasonic cleaning and vapor drying to assure you of low scatter, less staining and superior adhesion. Our coating chambers are equipped with electron beam guns to evaporate high melting point materials and we use a mixture of optical and quartz crystal monitoring techniques depending on the nature of the job.

We handle a high volume of coating work and are very confident of our ability to meet your needs. If you don't see the coating you need on the following pages, please call us for technical assistance.

QUALITY CONTROL

Our inspection system meets the stringent demands of MIL-1-45208A and our spectrophotometers are calibrated to MIL-C-45662 requirements. We can provide complete environmental and photometric testing to MIL-C-675 and MIL-M-13508. All are firm assurances of dependability and accuracy. Also, our people are well trained, careful and attentive to detail; they know that our coatings must be as good as they possibly can be.

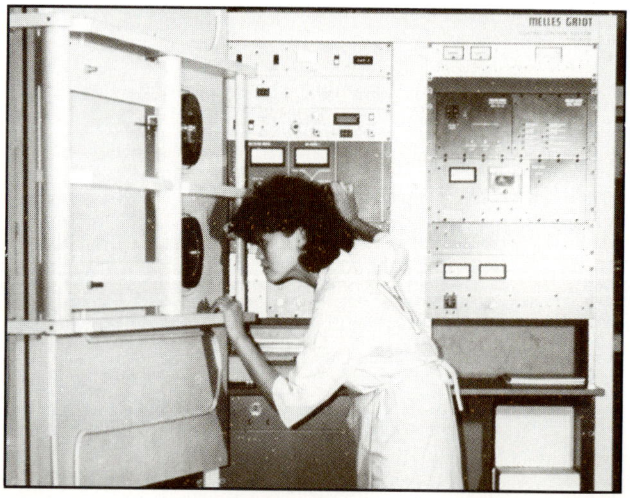

ANTIREFLECTION COATINGS

About 4% of visible light is reflected at any uncoated air-to-glass surface of a lens, prism, or other optical component. For designs using more than a few components, losses in light level can accumulate rapidly. More important are the corresponding losses in image contrast or modulation due to weak reflected "ghost" images superimposed on the desired image. Such unwanted images are often defocused beyond recognition so that contrast reduction (rather than confusion) is their primary effect.

SINGLE LAYER COATINGS

Surface reflectance depends on the difference in refractive indices for air and glass. For visible light this is approximately the difference between 1.0 and 1.5. By selecting a material with an index of refraction that falls between these two values (near the geometric mean), and by applying a coating of approximately 1/4-wavelength optical thickness to the surface, it is possible to reduce the reflectance significantly. Magnesium fluoride (MgF_2), a durable substance with a thin film refractive index of 1.38 at 550 nanometers, is the most commonly used coating material for single-layer antireflection (AR) coatings.

Instead of one reflection at the air-glass interface, there is a first reflection at the air-MgF_2 interface followed by a second reflection at the MgF_2-glass interface. These reflections interfere destructively, resulting in a minimum reflectance within the visible spectrum at normal incidence of 1.5% or less, as compared with the original 4% for the uncoated glass surface. From the principle of energy conservation, confirmed experimentally, there must be a corresponding increase in overall transmission of a coated surface, including both interfaces.

Such coatings work extremely well over a wide range of wavelengths and angles of incidence, despite the fact that the theoretical idea of zero reflectance is achieved by a film of 1/4-wavelength optical thickness only for normal incidence, and only if the refractive index of the coating material is exactly the geometric mean. In fact, the single layer ¼-wave thickness MgF_2 coating designed for normal incidence makes its most significant contribution to the transmission of steep surfaces, where most rays are incident at very large angles.

MULTILAYER COATINGS

By combining multi-layers of high, medium and low index materials it is possible to cause the reflected and transmitted beams to combine selectively in order to achieve an improved performance over a single layer coating. The improvement can be in broadening the spectral bandwidth of the coating or in almost nullifying the reflectance at a particular wavelength.

Our HEBBAR™ (High Efficiency BroadBand AntiReflection) coatings give a very low reflectance over a broad spectral bandwidth. The normal (0°) coating typically gives less than 0.5% reflectance over a range of 440 to 660nm. This coating can be modified for use at 45° or can be shifted to cover a wavelength range from 750 to 1100nm.

To obtain the ultimate in transmittance at a single wavelength, it is possible to deposit a "V-coat" which will reduce the reflectance to less than 0.25%, and typically closer to 0.1% for that one wavelength at normal incidence only.

COATING FORMULAS (SINGLE LAYER)

Because of the practical importance and wide usage of single layer coatings, especially at oblique incidence, it is valuable to have the formulas from which coating reflectance curves (as functions of either wavelength or angle of incidence) can be calculated.

Coating Dispersion Formula

For optical purposes a thin film may be considered to be perfectly homogeneous. The refractive index of MgF_2, whether amorphous or crystalline, is therefore connected to density via the Lorentz-Lorenz formula, where for the amorphous phase the crystalline ordinary and extraordinary indices of refraction may be averaged.

The formulas for crystalline MgF_2 are respectively

$$n_{co} = 1.36957 + \frac{3.5821 \times 10^{-3}}{(\lambda - 0.14925)} \text{ and}$$

$$n_{ce} = 1.381 + \frac{3.7415 \times 10^{-3}}{(\lambda - 0.14947)}$$

for the ordinary and extraordinary rays, where λ is the wavelength in microns.

For the average of the ordinary and extraordinary indices of refraction,

$$n = n(\lambda) = \frac{1}{2}(n_{co} + n_{ce})$$

The value 1.38 is the universally accepted amorphous film index for MgF_2 at a wavelength of 550 nanometers.

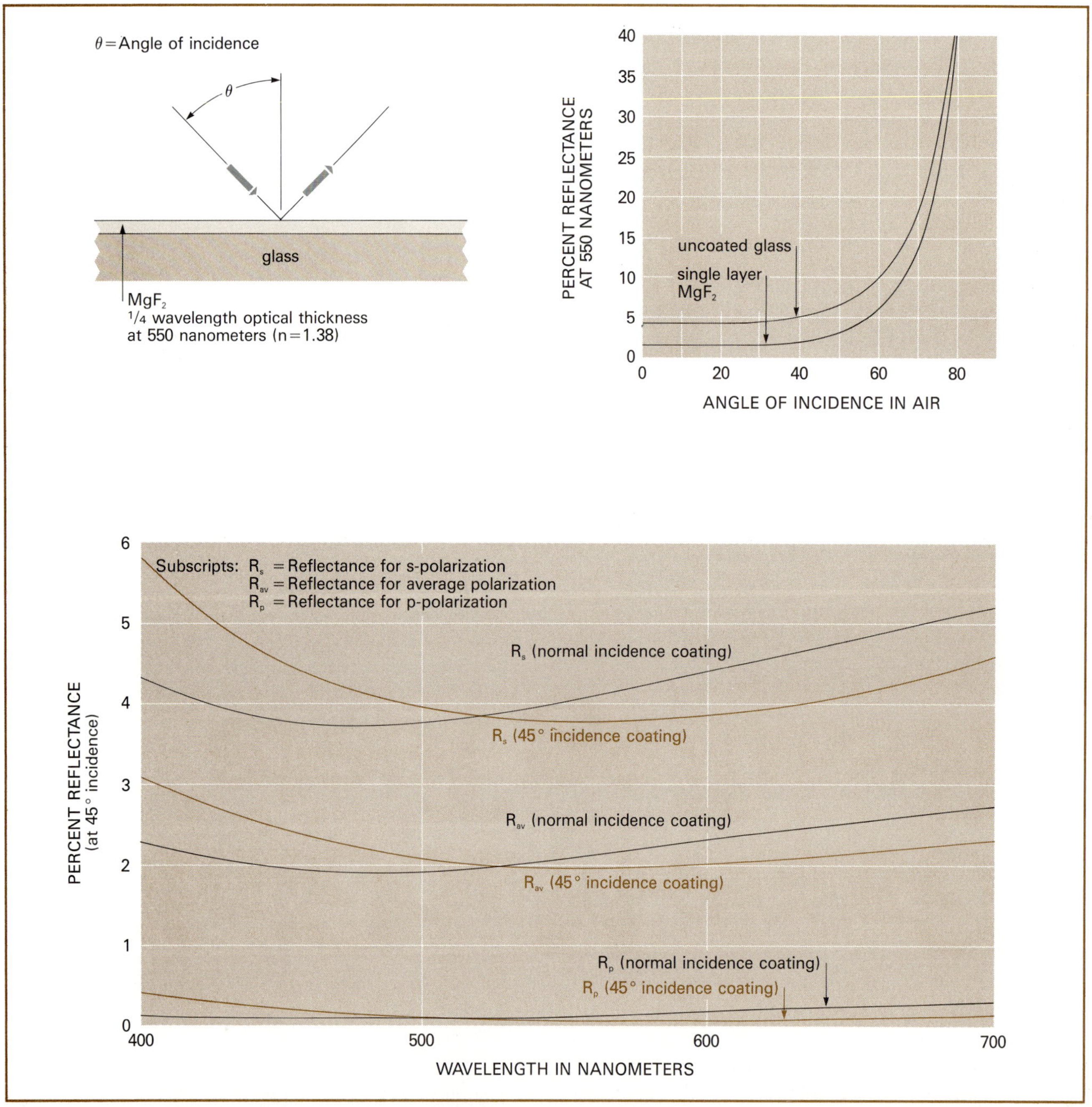

REFLECTION LOSS from a coated surface as a function of angle of incidence, state of polarization and wavelength.

Coated Surface Reflectance at Normal Incidence

Suppose that the coating is accurately of quarter wave optical thickness for some wavelength λ. Let n_o denote the refractive index of the external medium at this wavelength (1.0 for air or vacuum), and let n_f and n_s, respectively, denote the film and substrate indices. Then, for normal incidence at this wavelength, the single-pass irradiance reflectance of the coated surface can be shown to be

$$R = \left(\frac{n_o n_s - n_f^2}{n_o n_s + n_f^2} \right)^2 ,$$

regardless of the polarization state of the incident radiation.

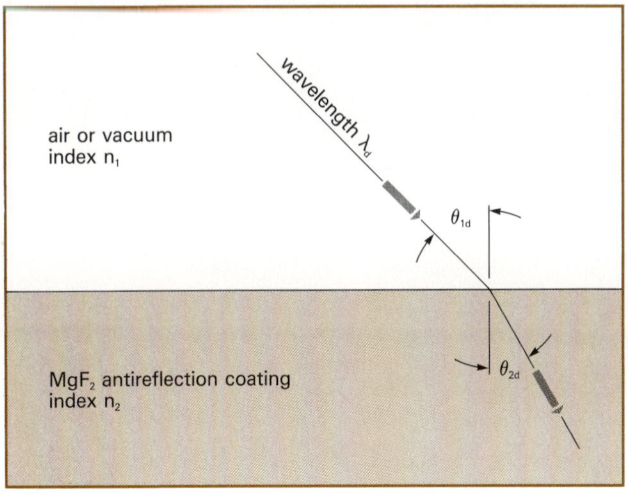

Coated Surface Reflectance at Oblique Incidence

At oblique incidence the situation is more complex. Let n_1, n_2, and n_3, respectively, represent the wavelength dependent refractive indices of the external medium (air or vacuum), coating film, and substrate. Assume the coating exhibits a reflectance extremum of the first order for some wavelength λ_d and angle of incidence θ_{1d} in the external medium. The coating is completely specified when θ_{1d} and λ_d are known. One may then identify n_2 with the film index n_f (= 1.38 for MgF_2 at 550nm). The extremum is a minimum if n_2 is less than n_3 and a maximum if n_2 exceeds n_3. The same formulas apply in either case. The formulas which follow lead to reflectance of the coated surface for a completely arbitrary angle of incidence θ_1 and arbitrary wavelength λ_1.

Corresponding to the angle of incidence θ_{1d} is an angle of refraction in the film:

$$\theta_{2d} = \arcsin \left(\frac{\sin \theta_{1d}}{n_2 (\lambda_d)} \right) .$$

As θ_1 is reduced from θ_{1d} to zero, the reflectance extremum shifts in wavelength from λ_d to λ_n, where the subscript n denotes normal incidence.

This wavelength is given by the equation

$$\lambda_n = \frac{n_2 (\lambda_n)}{n_2 (\lambda_d)} \frac{\lambda_d}{\cos \theta_{2d}} ,$$

Corresponding to the arbitrary angle of incidence θ_1 and arbitrary wavelength λ_1 are angles of refraction in the coating and substrate, given by

$$\theta_2 = \arcsin \left(\frac{n_1 \lambda_1 \sin \theta_1}{n_2(\lambda_1)} \right)$$

and

$$\theta_3 = \arcsin \left(\frac{n_1 \lambda_1 \sin \theta_1}{n_3(\lambda_1)} \right) .$$

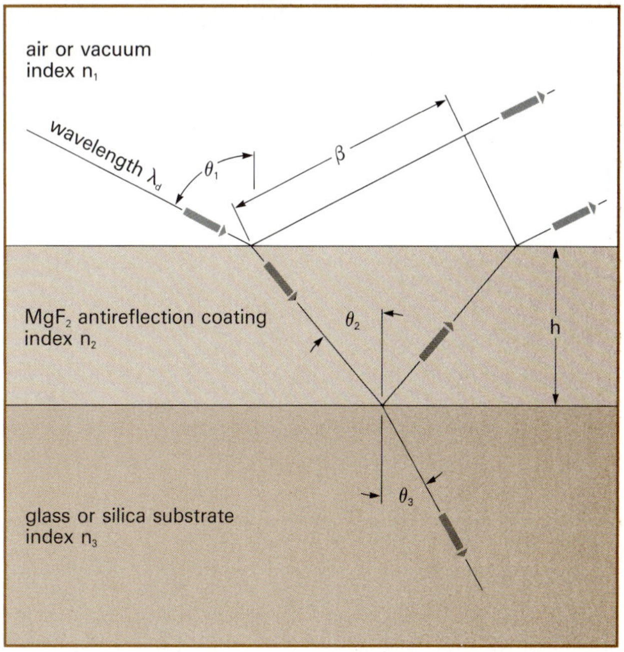

Following are formulas for the single interface amplitude reflectances for both the p- and s-polarizations:

$$r_{12p} = \frac{n_2 \cos \theta_1 - n_1 \cos \theta_2}{n_2 \cos \theta_1 + n_1 \cos \theta_2} ,$$

$$r_{23p} = \frac{n_3 \cos \theta_2 - n_2 \cos \theta_3}{n_3 \cos \theta_2 + n_2 \cos \theta_3} ,$$

$$r_{12s} = \frac{n_1 \cos \theta_1 - n_2 \cos \theta_2}{n_1 \cos \theta_1 + n_2 \cos \theta_2} , \text{ and}$$

$$r_{23s} = \frac{n_2 \cos \theta_2 - n_3 \cos \theta_3}{n_2 \cos \theta_2 + n_3 \cos \theta_3} .$$

The subscript "12p," for example, means that the formula gives the amplitude reflectance for the p-polarization at the interface between the first and second media.

The corresponding irradiance reflectances for the coated surface, accounting for both interferences and the phase differences between the reflected waves, are given by

$$R_p = \frac{r_{12p}^2 + r_{23p}^2 + 2r_{12p}r_{23p} \cos (2\beta)}{1 + r_{12p}^2 r_{23p}^2 + 2r_{12p}r_{23p} \cos (2\beta)} , \text{ and}$$

$$R_s = \frac{r_{12s}^2 + r_{23s}^2 + 2r_{12s}r_{23s} \cos (2\beta)}{1 + r_{12s}^2 r_{23s}^2 + 2r_{12s}r_{23s} \cos (2\beta)} .$$

where β is the phase difference (in the external medium) between waves reflected from the first and second surfaces of the coating.

$$\beta = \frac{2\pi}{\lambda_1} n_2 (\lambda_1) h \cos \theta_2 .$$

The cosines must be in radians. The average reflectance is given by

$$\overline{R} = \frac{1}{2} (R_p + R_s) .$$

With these formulas reflectance curves can be calculated as functions of either wavelength λ_1 or angle of incidence θ_1.

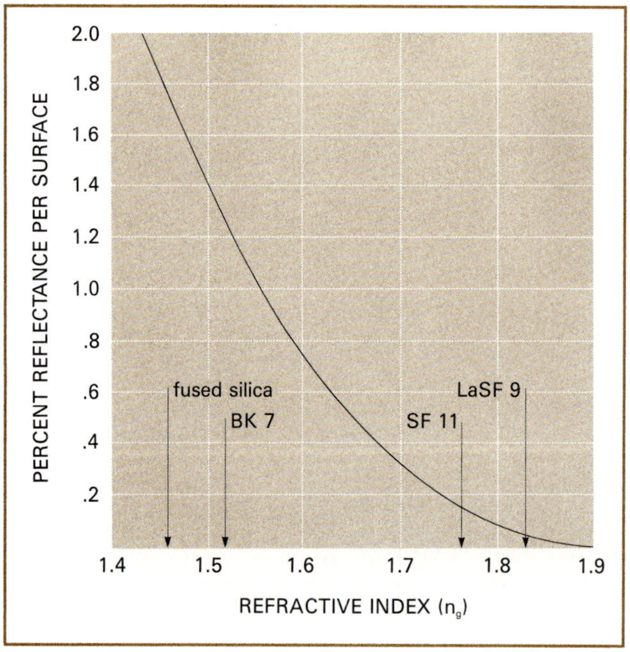

REFLECTANCE OF COATED SURFACE of index n_g when coated with a quarter wavelength of magnesium flouride (index = 1.38).

Special coatings are no problem to us — call our Application Engineers to discuss your coating requirements.

SINGLE LAYER ANTIREFLECTION COATINGS

MAGNESIUM FLUORIDE COATINGS

Magnesium Fluoride (MgF_2) is commonly used for single layer antireflection coatings because of its almost ideal refractive index (1.38 at 550nm) and its high durability. Although we can routinely apply antireflection coatings of other materials, we do offer a standard, economical MgF_2 coating which can be applied to any of our optical components. It is optimized for 550nm and normal incidence but, as can be seen from the reflectance curve, it is extremely insensitive to wavelength and incidence angle. Many of our standard lenses are held in stock already coated with MgF_2. Our precision optimized achromats (01 LAO series) are normally supplied with this coating.

Should you wish to specify a different wavelength and incidence angle, it is no problem to shift the reflectance curve. Please bear in mind, however, that additional delivery time is needed for special coatings, and that care should be taken in selecting the quantity of items coated to maximize the efficiency of the coating run. Partially filled chambers result in higher unit prices.

Single layer antireflection coatings are routinely available for almost any angle of incidence and any wavelength between 200nm in the ultraviolet and 1.6 microns in the infrared. To obtain such coatings, simply specify (for each surface of each part) the precise wavelength and angle of incidence for which reflectance is to be minimized. As the 1.6 micron wavelength is approached, the angle of incidence range becomes restricted to near-normal incidence. This is because of practical limitations on physical coating thickness. It is usually inadvisable to request a MgF_2 coating for any wavelength greater than 1.6 microns. While thicker MgF_2 coatings are possible, they tend to exhibit crazing, poor adhesion, and significantly increased scattering.

Single layer antireflection coatings for use on very steeply curved or short radius surfaces should be specified for an angle of incidence approximately half as large as the largest angle of incidence encountered by the surface. Depending on the specific application, determination of the best wavelength for use in a coating specification may require both ray tracing and energy tracing of the optical system in its anticipated environment.

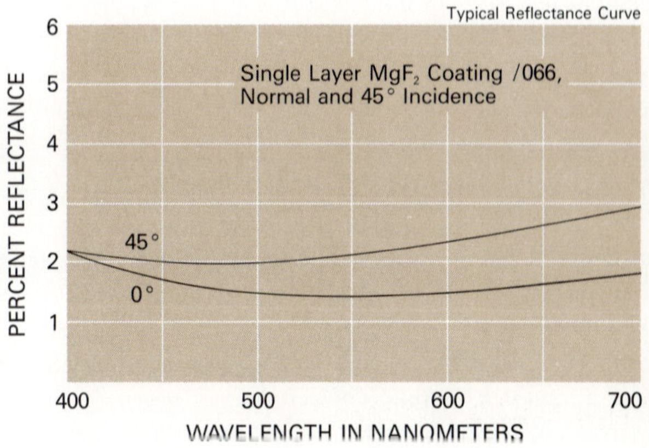

The effectiveness of MgF_2 as an antireflection coating is increased dramatically with increasing refractive index of the component material (see the graph on the previous page). This means that, for use on high index materials, there is often little point in using more complex coatings. The reflectance curve above is for MgF_2 on BK 7 optical glass.

Single Layer MgF_2 Antireflection Coating; Normal Incidence

Wavelength Range (nm)	Maximum Reflectance on BK 7 (%)	Maximum Reflectance on Fused Silica (%)	COATING SUFFIX
500-600	1.5	1.75	/066

To order this coating append coating suffix to product number and specify which surfaces are to be coated.

If you need a coating for another wavelength or angle of incidence, please do not use the /066 suffix, but instead specify the conditions separately.

HEBBAR™ COATINGS

Our HEBBAR™ (High Efficiency BroadBand AntiReflection) coatings give a very low reflectance over a broad spectral bandwidth. They are multilayer films, comprising alternate layers of various index materials, combined in such a way to reduce the overall reflectance to an extremely low level for the broad spectral range covered.

These coatings exhibit a characteristic, double minimum reflectance curve covering a range of some 300nm in wavelength. Over this entire range, the reflectance does not exceed 1.0% and is more typically below 0.6%. Within a more limited spectral range on either side of the central peak, the reflectance can be held well below 0.4%. HEBBAR™ coatings are somewhat insensitive to angle of incidence. The effect of increasing the angle of incidence however, is to shift the curve to slightly shorter wavelengths and to increase the long wavelength reflectance slightly. Both normal and 45° incidence are shown in the curves. One effect of this insensitivity is to make the coatings extremely useful for high numerical aperture (low f-number) lenses or steeply curved surfaces. In these cases, the incidence angle varies significantly over the aperture.

We offer four standard versions of HEBBAR™ and many of our components are carried in stock with a HEBBAR™ coating. The most commonly used version covers most of the visible spectrum (415 to 700nm) and is optimized for normal incidence. A 45° incidence version of this coating is also available. Two other HEBBAR™ coatings cover different spectral ranges. An infrared shifted version covers the range from 700 to 1100nm. We also offer an overlapping version covering the range from 780 to 850nm, and specifically optimized for extremely low reflectance over the range of laser diode wavelengths (780 to 850nm). At 632.8nm our /078 coating will typically reflect less than 0.3% at 0°±15° and less than 0.4% where incident angle is increased to 30°.

These coatings are not intended for use with high energy lasers although they are quite suitable for diagnostic applications. (See also page 213 for details of our guaranteed, Broadband High Energy Laser Window Coatings.)

For coating types /077, /078, and /079, reflectance values apply to indices 1.47 to 1.55 only. Other indices, while having their own designs, will have reflectance values approximately 20% higher for incidence angles 0°±15° and 25% higher for incidence angles 0°±30°. For example, 0.4% absolute becomes 0.5% absolute.

HEBBAR™ Coatings

Wavelength Range (nm)	Maximum Reflectance (%)	Angle of Incidence (deg)	COATING SUFFIX
440-660	0.6 Abs	0 - 15	/078
415-685	0.4 Ave	0 - 15	
415-700	1.0 Abs	0 - 15	
415-680	1.0 Abs	0 - 30	
632.8	0.3 Abs	0 - 15	
632.8	0.4 Abs	0 - 30	
750-1100	0.6 Abs	0 - 15	/077
700-1100	0.4 Ave	0 - 15	
700-1150	1.0 Abs	0 - 15	
700-1100	1.0 Abs	0 - 30	
780-850	0.25 Abs	0 - 15	/076
780-850	0.40 Abs	0 - 30	
725-875	1.0 Abs	0 - 30	
425-670	0.6 Ave	45	/079
425-670	1.0 Abs	45	
440-670	0.4 Abs	30 - 45	
440-670	1.0 Abs	45 - 50	

To order append coating suffix to product number and specify which surfaces to be coated.

For your convenience we show typical reflectance curves for the coatings both for their intended angle of incidence and for 45 degrees. (Normal in the case of the 45 degree coating.)

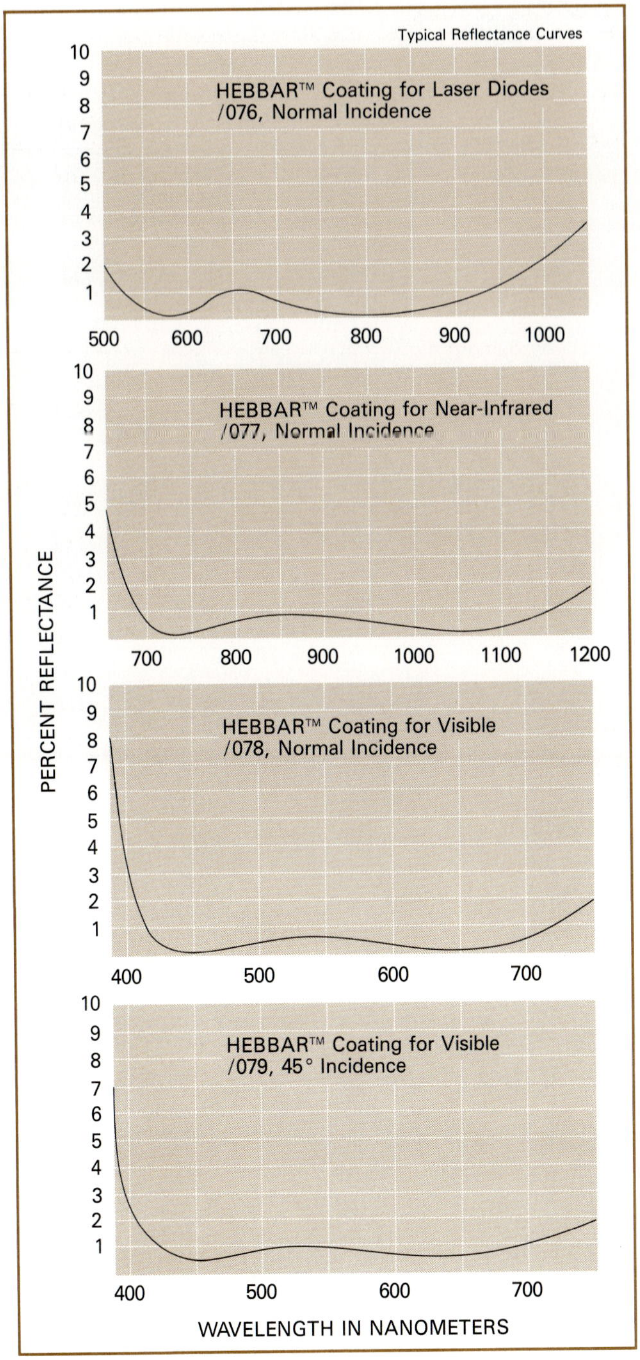
Typical Reflectance Curves
HEBBAR™ Coating for Laser Diodes
/076, Normal Incidence
HEBBAR™ Coating for Near-Infrared
/077, Normal Incidence
HEBBAR™ Coating for Visible
/078, Normal Incidence
HEBBAR™ Coating for Visible
/079, 45° Incidence
PERCENT REFLECTANCE
WAVELENGTH IN NANOMETERS

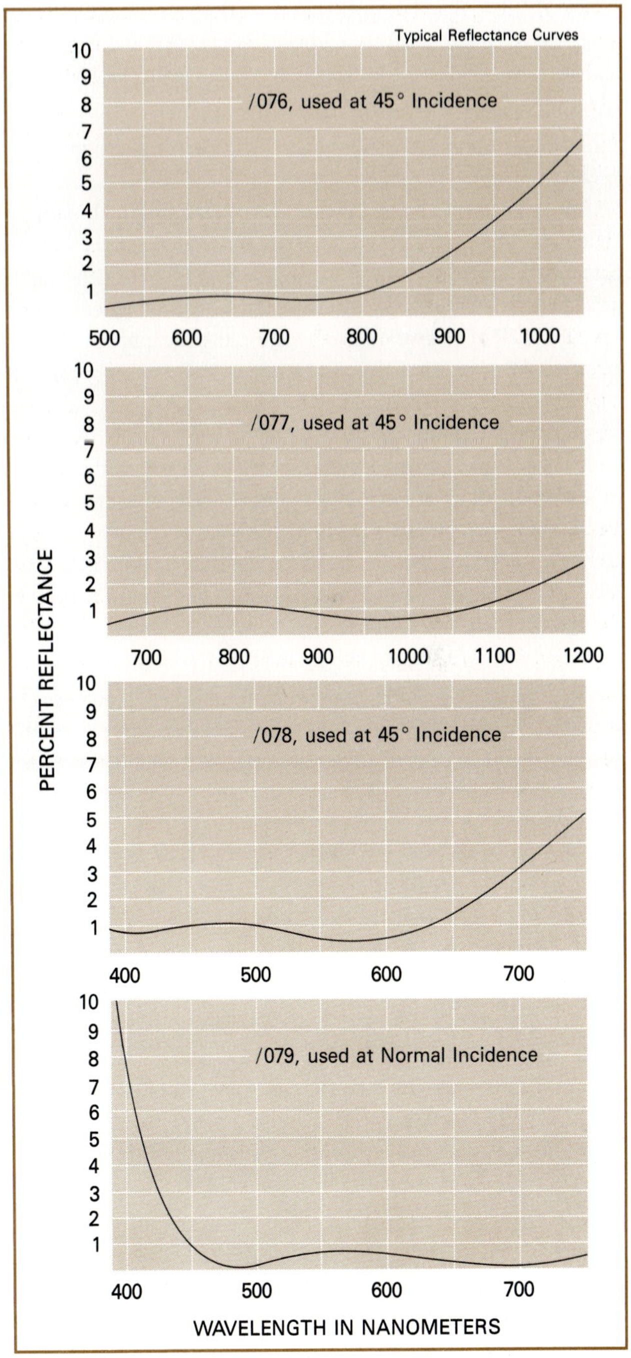
Typical Reflectance Curves
/076, used at 45° Incidence
/077, used at 45° Incidence
/078, used at 45° Incidence
/079, used at Normal Incidence
PERCENT REFLECTANCE
WAVELENGTH IN NANOMETERS

V-COATINGS

"V"-coatings are multilayer antireflection coatings which reduce the reflectance of a component to near-zero for one very specific wavelength. Our "V"-coatings are intended for use at normal incidence, and when used in this way will achieve maximum reflectances of not more than 0.25% at their design wavelength.

They are extremely sensitive to both wavelength and angle of incidence. For example, a "V"-coating intended for the helium neon wavelength (632.8nm) when used at 30° incidence will reflect about 0.8%. At 45° incidence the same coating will reflect over 2.5%. If your application involves other than normal angles of incidence or high numerical aperture (low f-number) optics, it may be better to use a HEBBAR™ coating.

Experience shows that the maximum reflectance typically achieved by these coatings is often closer to 0.1% than the 0.25% we specify. Using "V"-coatings on fused silica optics can therefore provide exceptionally high external transmittances.

The typical reflectance curve illustrated is for a "V"-coating on BK 7 optical glass. Clearly the performance of such a coating is highly dependent on the refractive index of the component material. However, we specify all our coatings by their performance figures and not by their design. This means that we will change the design to suit the material being coated. In this way we make ordering coatings simple: you just select the specification you want to achieve, tell us what to put it on, and we do the rest. That does mean that your reflectance curve may vary from the one shown on this page, but the coating will meet your requirements.

These coatings are not intended for use with high energy lasers although the YAG coatings are suitable for many medium power applications including diagnostics.

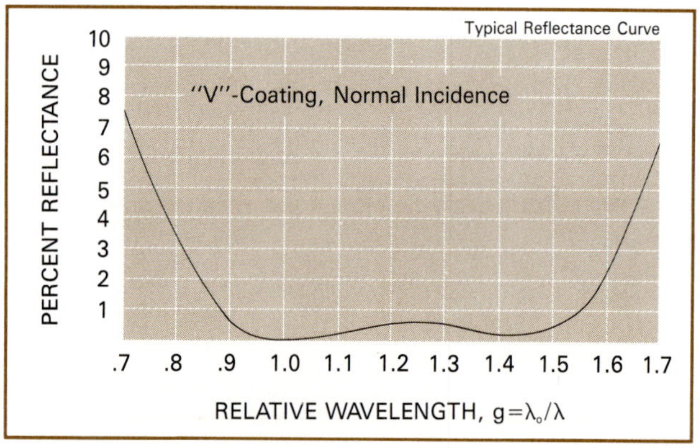

"V"-Coatings; Normal Incidence

Wavelength (nm)	Laser Type	Maximum Reflectance (%)	COATING SUFFIX
351	Ar ion	0.25	/105
364	Ar ion	0.25	/107
442	HeCd	0.25	/111
458	Ar ion	0.25	/112
466	Ar ion	0.25	/113
473	Ar ion	0.25	/114
476	Ar ion	0.25	/115
488	Ar ion	0.25	/116
496	Ar ion	0.25	/117
502	Ar ion	0.25	/118
514	Ar ion	0.25	/119
532	YAG 2nd harm.	0.25	/122
543	HeNe	0.25	/121
633	HeNe	0.25	/123
694	Ruby	0.25	/124
780	GaAlAs	0.25	/163
830	GaAlAs	0.25	/166
850	GaAlAs	0.25	/167
904	GaAs	0.25	/125
1064	YAG	0.25	/126
1300	InGaAsP	0.25	/168
1523	HeNe	0.25	/169
1550	InGaAsP	0.25	/169

To order append coating suffix to product number and specify which surfaces to be coated.

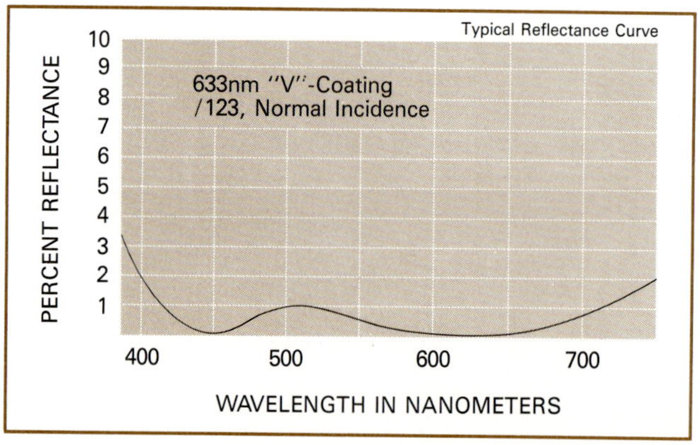

HIGHREFLECTION COATINGS

It was previously shown that about 4% of incident light is reflected from any uncoated glass surface at normal incidence. As the angle of incidence is increased, so also is the reflectance, although polarization effects start to occur (see the discussion on page 174 for a full explanation). In order to increase the reflectivity of a surface, high reflection coatings may be applied. Multilayer stacks of 15 to 35 layers are common but even greater numbers of layers may be needed for same applications. These coatings may be applied to the outside of a component, such as a flat piece of glass, to produce a "first-surface" mirror. They may also be applied to an internal surface, of a prism for example, to produce a "second-surface" mirror.

The coatings we offer may be applied to many of our components to increase the reflectivity. Prism faces, cylindrical lenses and plano concave lenses are examples of common applications for high reflection coatings in addition to the more usual flat mirrors.

We offer a variety of high reflection coatings for use at any specific angle of incidence or range of angles you may require. Our coatings fall into two categories, metallic and dielectric. Coatings in either category will meet or exceed all applicable military specifications.

METALLIC COATINGS

These are used primarily for mirrors and for prism surfaces. Metallic coatings can in turn be overcoated with dielectric films to increase the reflectance over a desired range of wavelengths or angles of incidence. In this case the metallic coating is said to be enhanced.

By overcoating metallic coatings with hard dielectric layers of 1/2 wavelength optical thickness, abrasion and tarnish resistance can be increased. The optical properties are only slightly affected, assuming that the dielectric has high internal transmittance at the wavelengths of interest. Depending on the dielectric used, such overcoated metals are sometimes referred to as durable, protected or hard coatings.

Metals which we routinely supply as high reflection coatings include:

Aluminum	Ultraviolet-enhanced aluminum
Protected aluminum	Gold
Enhanced aluminum	Silver (second surface only)

DIELECTRIC COATINGS

If the index of a dielectric quarter wave optical thickness film is lower than the index of the substrate, reflectance of the coated surface will be reduced at normal incidence. The reflectance of this film will, in fact, be quite low over a wide range of angles of incidence.

If the index of the quarter wave film is instead higher than the index of the substrate, reflectance of the coated surface will be substantially increased.

If both types of quarter wave films are applied, alternating between high and low index, very high levels of reflectance may be obtained. These stacks achieve their very high reflectance in a narrow wavelength interval surrounding the wavelength for which the thickness of the high index plus the low index layers is a halfwave. The reflectance achieved is very sensitive to the film index ratio, and to the number of periods or pairs of high and low index layers. Increase either and you increase the reflectance.

The width of the high reflectance wavelength interval is also determined by the film index ratio. The higher the ratio the wider the interval. There are two ways in which the wavelength interval can be made larger than the limit set by index ratio. First, it is possible to deposit stacks for slightly different wavelengths on top of each other. Second, the interval can be made wider by systematically perturbing the layer thicknesses in a single stack to be different from precisely quarterwave optical thickness values at one wavelength. Computer optimization programs exist for arriving at efficient multilayer designs. The reflectances of such films can easily be made to exceed the highest metallic reflectances over very large wavelength intervals, spanning almost the entire visible spectrum for example. Such films are effective for both the s- and p-polarizations, and over a wide range of angles of incidence.

Depending on the materials chosen for the multilayer, it is also possible for the durability and abrasion-resistance of such films to be superior to that of metallic films. It is a natural characteristic of multilayer dielectric broadband reflecting coatings that their reflectance values diminish at oblique incidence.

See page 173 for Mirrors.

METALLIC HIGH-REFLECTION COATINGS

We offer six forms of standard metallic high reflection coating. All of these coatings are formed by vacuum deposition. The following is a brief summary of the properties of these metals, giving some indication of where and why they are used. They are convenient in that they can be used at any angle of incidence. These coatings may be applied to most of the optical components listed. To do so simply append the coating suffix number to the component product number.

Metallic reflective coatings are delicate and require care during cleaning. Dielectric overcoats substantially improve abrasion resistance but they are not impervious to abrasive cleaning techniques. If it is necessary to clean these coatings, care is required. Clean, dry pressurized gas can be used for blowing off loose particles, then clean deionized water, mild detergent and alcohol may be used. Gentle cleaning with a swab is recommended.

ALUMINUM

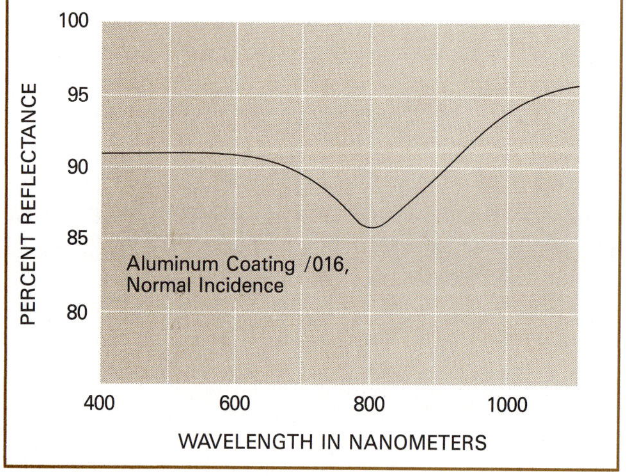

Aluminum is the most widely used metal for reflecting films, offering consistently high reflectance throughout the visible, near-infrared, and near-ultraviolet regions of the spectrum. While silver exhibits slightly higher reflectance than aluminum through most of the visible spectrum, the advantage is temporary because of tarnishing due to oxidation. Aluminum also oxidizes, though more slowly, and its oxide is tough and corrosion-resistant. Oxidation significantly reduces aluminum reflectance in the ultraviolet, and the oxide causes slight scattering throughout the spectrum.

PROTECTED ALUMINUM

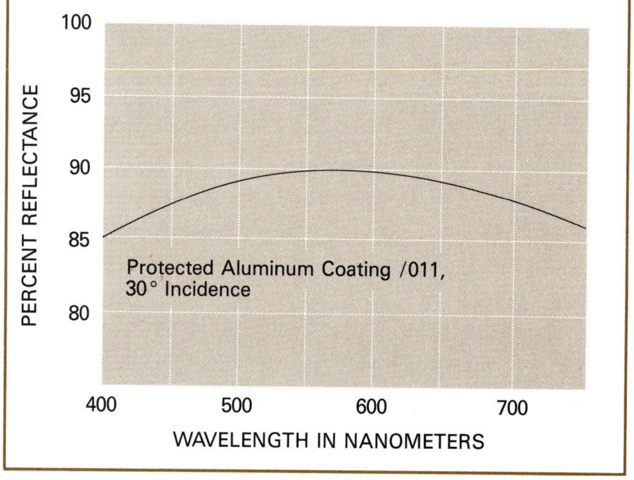

Protected aluminum is the very best general purpose metallic coating for use as an external reflector in the visible and near-infrared spectrum. All of our mirrors are coated with protected aluminum unless we specify otherwise or you specifically request a different coating. The Coating Suffix need not be appended. Protected aluminum is aluminum overcoated with a dielectric film of silicon monoxide (Si0) of 1/2-wavelength optical thickness at 550nm. The protective film arrests oxidation and helps to maintain the initial high reflectance. It is hard enough to protect the aluminum from minor abrasions.

Please call to discuss your OEM mirror coating requirements.

ENHANCED ALUMINUM

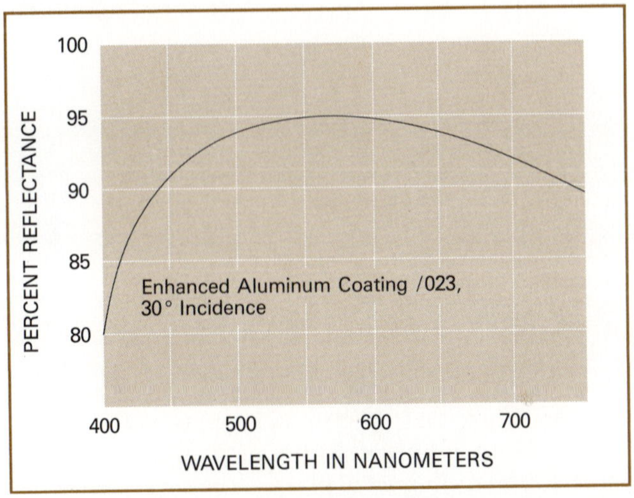

By overcoating the aluminum with a multilayer dielectric film, reflectance is increased over a wide range of wavelengths. The durable enhancing multilayer produces a peak reflectance of 95% with an average across the visible spectrum of 93%. This coating is well suited to applications requiring the durability and reliability of protected aluminum, but with higher reflectance in the mid-visisble regions. The reflectance of enhanced aluminum peaks between 530 and 580nm, and is high from 400 to 800nm.

ULTRAVIOLET-ENHANCED ALUMINUM

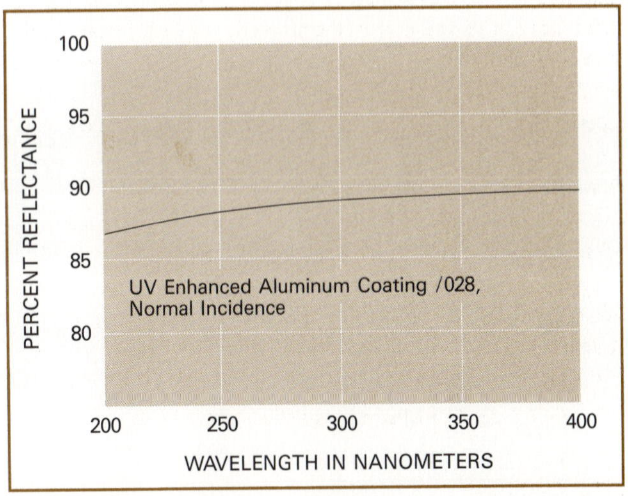

By applying a film of an ultraviolet-transmitting dielectric (usually MgF_2), the reflectance of pure bare aluminum can be preserved in the ultraviolet. The dielectric layer prevents oxidation of the aluminum surface and provides abrasion resistance. While the resulting surface is not as abrasion-resistant as our protected aluminum, this coating may be cleaned with care. Reflectance averages over 88% from 200nm to 400nm and over 85% throughout the visible spectrum. This coating can be applied to all of our mirrors with the exception of the paraboloidal reflectors and ellipsoidal reflectors.

SILVER

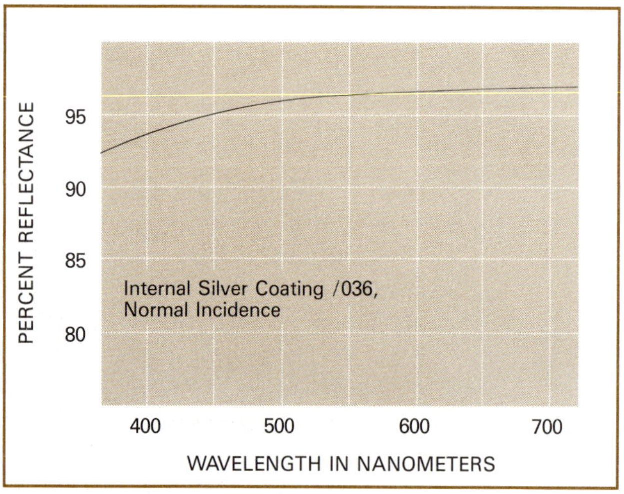

Through most of the visible and in the near-infrared spectrum, silver has higher reflectance than aluminum, at least for a short time following deposition. Because the advantage is temporary, silver is rarely used in external reflection. Instead, silver is used almost exclusively in internal reflection. Oxidation and tarnish are prevented by over-coating the external surface with an additional layer of either Inconel or copper. The Inconel or copper layers are subsequently painted to increase abrasion resistance. In this way, the high initial reflectance of silver is indefinitely preserved.

Silver is frequently used in the near-infrared (the interval containing the neodymium and gallium arsenide laser lines) because it avoids the small dip in reflectance which aluminum exhibits in this interval. In the near-ultraviolet, silver has very low reflectance, and aluminum is a preferable choice. From the visible into the middle-infrared spectrum, silver offers the highest internal reflectance available from a metallic coating. Silver has less effect than aluminum on the polarization state in these regions of the spectrum.

All our mirrors may be coated with gold for use in the infrared.

GOLD

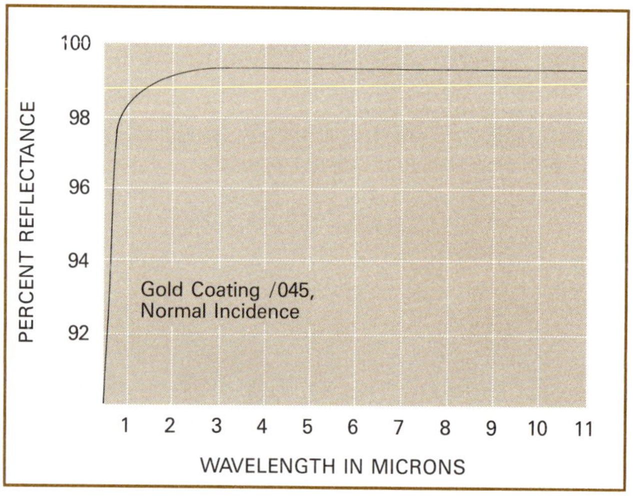

Combining good tarnish resistance with consistently high reflectance throughout the near-, middle-, and far-infrared, gold is by far the most widely used reflecting material in these regions. While it is possible to construct multilayer films which may surpass the reflectance of gold at specific wavelengths, the useful range of gold is unequaled. Gold is especially effective in the control of thermal radiation. Bare gold is soft and scratches easily, so bare gold mirrors should be cleaned only by flow-washing with solvents and clean water or by blowing the surface clean with a low-pressure stream of dry air.

Metallic High Reflection Coatings

Coating Type	Wavelength Range (nm)	Average Reflectance (%)	COATING SUFFIX
Aluminum	200-IR	90	/016
Protected Aluminum	400-800	87	/011
Enhanced Aluminum	450-750	93	/023
UV-Enhanced Aluminum	200-400	88	/028
Silver (internal)	400-IR	98	/036
Gold	700-IR	99	/045

To order append coating suffix to product number.

MAXBRIte™ COATING

The MAXBRIte™ (Multilayer All-dielectic Xerophilous Broadband Reflecting Interference) coating is, without a doubt, the very best broadband mirror coating available anywhere. It reflects well over 99% of any incident visible laser radiation. Furthermore, it is durable and can withstand high levels of illumination without damage.

This coating exhibits exceptionally high reflectance for both the s- and p-polarizations throughout most of the visible spectrum. For most applications it is superior to any metallic or enhanced metallic coatings.

At the most important visible laser wavelengths and for angles of incidence as high as 45 degrees the average of s- and p-reflectances exceeds 99%.

This coating is suitable for instrumental and external laser beam manipulation tasks. It is the ideal choice for use with tunable dye and parametric oscillator systems. The structural design of this coating is such that flatness specifications as tight as 1/20 wavelength can be maintained using low expansion substrate materials.

Because of the wide general applications of this superior coating, we carry a number of our better mirrors in stock already coated with MAXBRIte™.

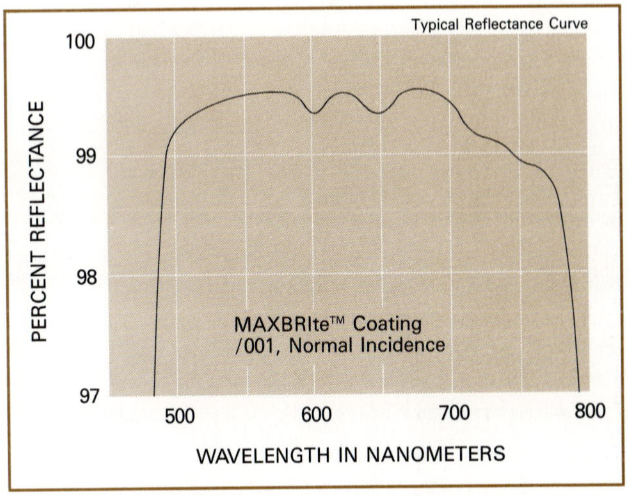

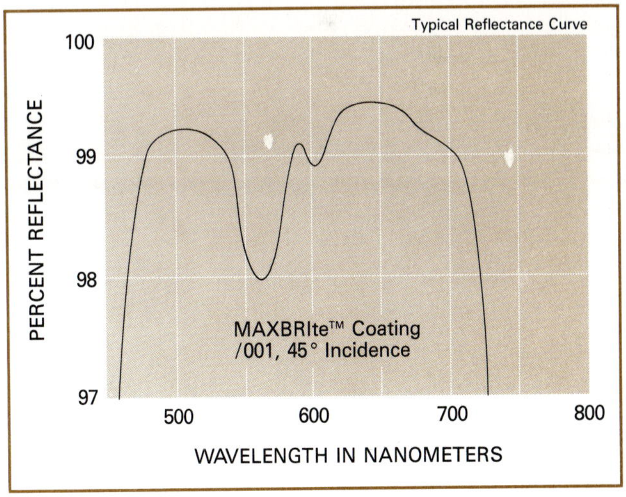

MAXBRIte™ Coating

Wavelength Range (nm)	Minimum Reflectance (%)	Angle of Incidence (deg)	COATING SUFFIX
480-700	99	0 ± 45	/001

To order append coating suffix to product number.

LASERLINE MAX-R COATINGS

These multilayer coatings are stacks intended to achieve the highest possible reflectances at specific laser line wavelengths and at particular angles of incidence. At these wavelengths and angles, laser line MAX-R coatings outperform MAXBRIte™. For limited ranges of angle of incidence (centered about the intended angles) the reflectances of these coatings will meet or exceed our published specifications at the design wavelength. We offer coatings for angles of incidence of 0 and 45 degrees. The layer thicknesses differ for these two angles so the coatings cannot be used interchangeably. Laser line MAX-R coatings are intended for external beam manipulation applications. Specified coating reflectances apply to the p-polarization (except at normal incidence, where the p- and s-polarization states are indistinguishable). Reflectances for the s-polarization generally exceed those for the p-polarization.

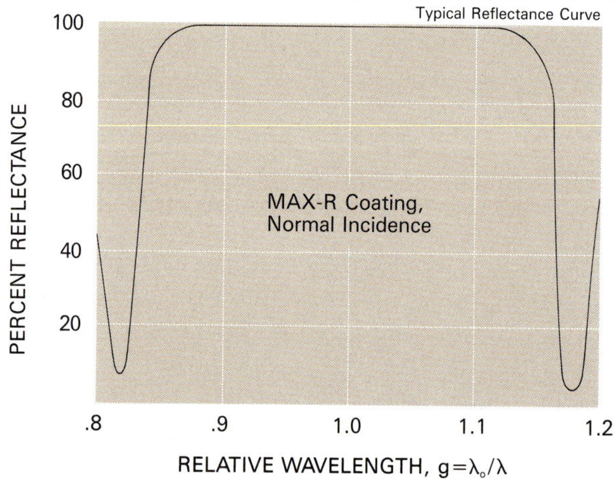

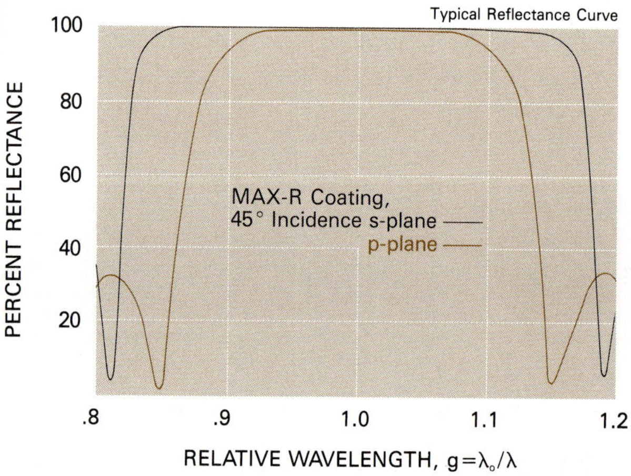

We offer a range of optical coating services. Ask for your free copy of our brochure.

These coatings are intended for low to medium power applications. We also offer a range of guaranteed HEL Mirror Coatings for use with high energy lasers (see pages 206-209).

Laser Line MAX-R Coatings; Normal Incidence

Wavelength (nm)	Laser Type	Minimum Reflectance R_p (%) 0° Incidence	0° ± 15 Incidence	COATING SUFFIX
351	Ar ion	98.0	95.0	/205
364	Ar ion	98.0	95.0	/207
442	HeCd	99.3	99.0	/209
458	Ar ion	99.5	99.3	/211
466	Ar ion	99.5	99.3	/213
473	Ar ion	99.5	99.3	/215
476	Ar ion	99.5	99.3	/217
488	Ar ion	99.5	99.3	/219
496	Ar ion	99.5	99.3	/221
502	Ar ion	99.5	99.3	/222
514	Ar ion	99.5	99.3	/223
532	YAG (2nd harm.)	99.5	99.3	/225
543	HeNe	99.5	99.3	/226
633	HeNe	99.5	99.3	/229
694	Ruby	99.3	99.0	/231
700	GaAlAs	99.3	99.0	/233
830	GaAlAs	99.3	99.0	/237
850	GaAlAs	99.3	99.0	/238
904	GaAs	99.3	99.0	/239
1064	YAG	99.2	99.0	/241
1300	InGaAsP	99.2	99.0	/245
1523	HeNe	99.2	99.0	/247
1550	InGaAsP	99.2	99.0	/247

To order append coating suffix to product number.

Laser Line MAX-R Coatings; 45° Incidence

Wavelength (nm)	Laser Type	Minimum Reflectance R_p (%) 45° Incidence	45° ± 10 Incidence	COATING SUFFIX
351	Ar ion	97.0	95.0	/255
364	Ar ion	97.0	95.0	/257
442	HeCd	99.0	98.0	/259
458	Ar ion	99.3	98.0	/261
466	Ar ion	99.3	98.5	/263
473	Ar ion	99.3	98.5	/265
476	Ar ion	99.3	98.5	/267
488	Ar ion	99.3	98.5	/269
496	Ar ion	99.5	98.5	/271
502	Ar ion	99.5	98.5	/272
514	Ar ion	99.5	98.5	/273
532	YAG (2nd harm.)	99.5	98.5	/275
543	HeNe	99.5	98.5	/276
633	HeNe	99.5	98.5	/279
694	Ruby	99.0	98.5	/281
780	GaAlAs	99.0	98.5	/283
830	GaAlAs	99.0	98.5	/287
850	GaAlAs	99.0	98.5	/288
904	GaAs	99.0	98.5	/289
1064	YAG	99.0	98.0	/291
1300	InGaAsP	99.0	98.5	/295
1523	HeNe	99.2	9.0	/297
1550	InGaAsP	99.0	98.5	/297

To order append coating suffix to product number.

We can produce MAX-R coatings for other wavelengths on special request.

PARTIAL REFLECTION COATINGS

When it is necessary to split an optical beam into two parts, the best way to do this is with a partial reflection coating. These coatings, again, can be made of metal, dielectric, or a combination of the two.

The essential trade-off in selecting a partial reflection coating is between efficiency and spectral flatness. At one end of the scale, an all-dielectric coating can be controlled to provide a precisely known reflectance to transmittance (R/T) ratio with very little absorption loss. The R/T ratio will, however, be extremely sensitive to wavelength, angle of incidence and polarization. Metal coatings, by comparison, can be spectrally flat but do have absorption losses, often higher than 30%.

Melles Griot offers a wide selection of partial reflection coatings to suit every need. These coatings are provided only as finished products. Since it is inappropriate to apply them to other components, we do not list them here with generalized coating suffixes. Instead, we refer you to the relevant section of the Optics Guide for more details.

One of the most commonly used partial reflection coatings is our hybrid coating which is supplied only as a cube beamsplitter. This is a combination of metal all-deielectric which is moderately efficient and exceptionally flat. All-dielectric plate beamsplitters are offered with R/T ratios of 0.5, 1.0, and 2.0 (see pages 236-237).

Cube beamsplitters are offered only for equal splitting but they are available with both hybrid and all-dielectric coatings (see pages 238-239).

Neutral density filters manufactured by Melles Griot are glass plates on which is deposited a coating of a metallic alloy. This alloy is chosen because of its excellent spectral neutrality. Absorption in the coating can be as high as 30% but the reflected and transmitted beams will be in the same ratio over a broad wavelength range. These neutral density filters, up to a density of 1.0, make good beamsplitters for some applications (see pages 248-252).

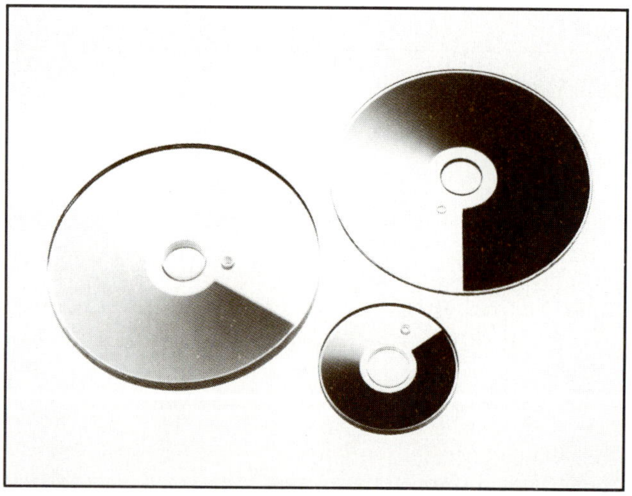

Our linear wedge neutral density filters offer a simple means of varying the R/T ratio in an optical system (see pages 253-254).

We also offer a selection of pellicle beamsplitters with a variety of coating options (see page 240-241).

High energy lasers demand much more of coatings. We offer a selection of HEL beamsplitters and output couplers having highly durable coatings for use with these lasers (see pages 214-219).

For situations in which it is necessary to split or combine beams with regard to the plane of polarization, we offer a range of coatings which are polarization selective. These coatings, which are optimized for particular laser wavelengths, are supplied in the form of polarizing cube beamsplitters (see pages 294-295). A selection of HEL polarizer coatings is also offered on flat glass substrates for use with YAG and Ruby lasers (see page 222).

WAVELENGTH SELECTIVE COATINGS (FILTERS)

Melles Griot supplies many types of wavelength selective coating and all of these are listed in detail in the section on filters.

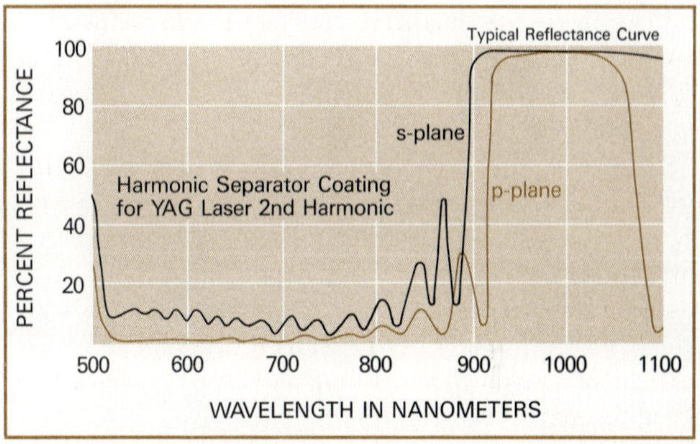

We, of course, list many narrow and broad band interference filters (see pages 270-281).

Our hot and cold mirror coatings are extremely useful in selecting wheer to allow heat into and out of a system. These are supplied on flat glass substrates but concave cold mirrors are also available (see pages 246-247).

We also offer a choice of HEL harmonic separators. These are coatings designed to separate, by selective filtering, the harmonics of YAG and Ruby from their fundamental frequencies (see pages 220-221).

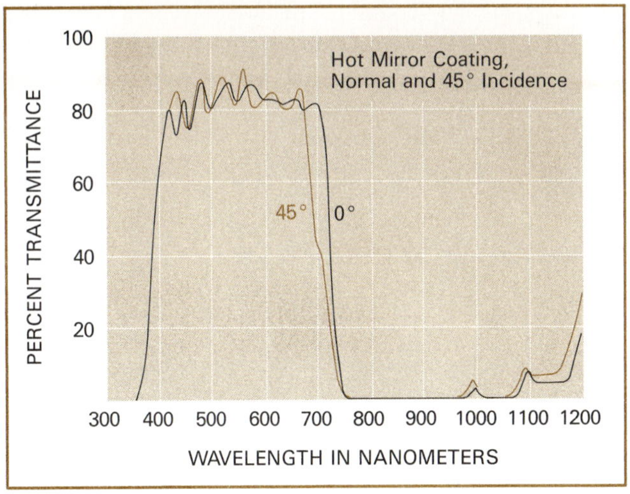

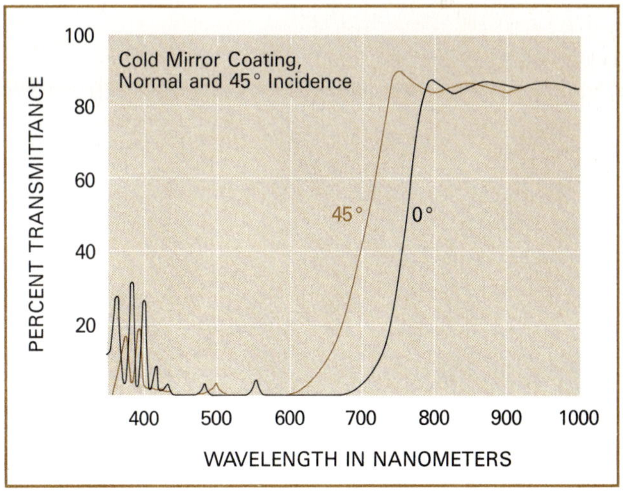

See the filter section starting on page 243.

OPTICAL COATING

Melles Griot is one of the most prolific suppliers of precision optical coatings in the world. Our Irvine, California coating facility is equipped with modern coating design software, six fully equipped coating chambers and a selection of spectrophotometers, refractometers, and other Q.C. equipment.

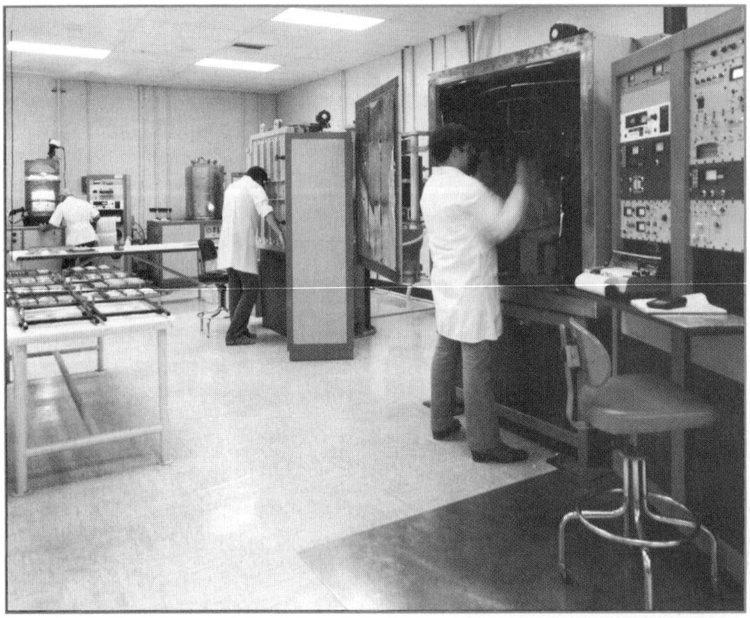

In addition to providing a wide selection of catalog coating options on any of our components, Melles Griot is able to design and provide special coatings to your specifications.

If you would like us to coat your optics or supply coated optics to your special requirements, please contact your nearest office and provide your specification.

SINGLET LENSES

In this section we describe our broad selection of simple single element, spherical section lenses. These are manufactured from the best optical grade materials including BK 7 optical glass, synthetic fused silica, sapphire and silicon.

The lenses are normally supplied from stock uncoated. If you require an antireflection coating, simply append a coating suffix as described in the coatings section of this Optics Guide.

These lenses find a wide variety of uses. They may be combined into more complex, multi-element systems and our Applications Engineers will be pleased to assist you in designing such systems.

Contents **Page Number**

PLANO-CONVEX GLASS LENSES

These lenses have positive focal lengths, converge incident light, and form both real images (as might be focused on a piece of paper) and virtual images (as are seen *through* the lenses if they are used as magnifiers). They closely approach best-form for infinite and near-infinite conjugate ratios, with the preferred focal plane being that adjacent to the plano surface. We have earlier tabulated "decision parameters" which enable you to make the best choice between plano-convex and symmetric-convex lenses for your particular wavelength and conjugate ratio. (See page 27-28.)

Plano-convex spherical glass lenses are widely used in telescopes, collimators, optical transceivers, magnifiers, radiometers, and condensers. In this last application, which concerns transferring light with maximum efficiency from source to subject, pairs of plano-convex lenses are used in the convex-side-facing-convex-side orientation (thus minimizing spherical aberration). Condenser conjugate ratios can be varied by pairing plano-convex lenses of different focal lengths.

A large variety of single and multilayer antireflection coatings is available for these lenses. See the preceding section on coatings (page 69).

Fused silica plano-convex lenses are also available. (See page 110.)

All tabulated lens properties (here and later) are nominal with tolerances corresponding to tolerances on f_o and n_o.

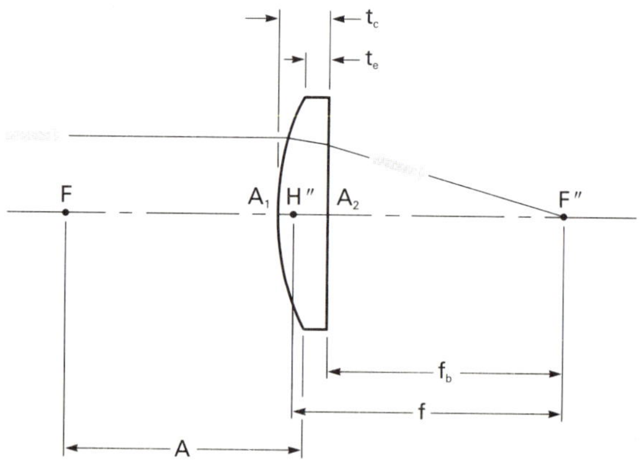

01 LPX

SPECIFICATIONS: PLANO-CONVEX GLASS LENSES

Paraxial Focal Length (f_o): ±5%
Design Index (n_o): ±0.001 (±0.002 for LaSF 9)
Centration: 3 arc minutes
Diameter (ϕ): ±0.25mm (±0.5mm if $\phi \geq$ 100mm)
Center Thickness (t_c): ±0.2mm
Materials: Grade A Fine Annealed optical glass.
Cosmetic Surface Quality: Pitch polished, 80-50 scratch and dig
Circumference is ground and both edges have a slight protective bevel.

Diameters of all lenses may be reduced ("edged") by us to meet your requirements. Larger diameters and focal lengths are available on special request.

Plano-Convex Glass Lenses

f_o (mm)	ϕ (mm)	Glass Type	Design Wavelength λ_o (mm)	Design Index n_o	t_c (mm)	t_e (mm)	A_1H (mm)	A_2H'' (mm)	f_b (mm)	A (mm)	f_o/ϕ	PRODUCT NUMBER
1.0	1.5	LaSF 9	587.6	1.8502	0.8	0.3	0	-0.43	0.57	1.45	0.67	01 LPX 401
1.2	1.5	LaSF 9	587.6	1.8502	0.8	0.5	0	-0.43	0.77	1.53	0.80	01 LPX 405
1.2	2.0	LaSF 9	587.6	1.8502	0.9	0.1	0	-0.43	0.77	2.02	0.60	01 LPX 407
1.6	2.0	LaSF 9	587.6	1.8502	0.8	0.4	0	-0.43	1.17	2.04	0.80	01 LPX 411
1.6	2.5	LaSF 9	587.6	1.8502	0.9	0.1	0	-0.43	1.17	2.42	0.64	01 LPX 413
2.5	2.8	BaK 1	587.6	1.5725	1.5	0.4	0	-0.95	1.55	3.63	0.89	01 LPX 415
3.5	2.2	BK 7	587.6	1.5168	2.0	1.6	0	-1.32	2.18	3.87	1.59	01 LPX 417
3.5	2.8	BK 7	587.6	1.5168	2.0	1.3	0	-1.32	2.18	4.16	1.25	01 LPX 419
3.5	3.6	BaK 1	587.6	1.5725	2.0	0.9	0	-1.27	2.23	4.62	0.97	01 LPX 421
4.0	4.2	BaK 1	587.6	1.5725	1.9	0.5	0	-1.21	2.79	5.38	0.95	01 LPX 423
6.0	3.8	BK 7	587.6	1.5168	1.7	1.0	0	-1.12	4.88	6.65	1.58	01 LPX 425
6.0	4.0	BK 7	587.6	1.5168	1.5	0.8	0	-0.99	5.01	6.73	1.50	01 LPX 427
8.0	3.0	BK 7	587.6	1.5168	2.0	1.7	0	-1.32	6.7	8.3	2.67	01 LPX 429
8.0	4.0	BK 7	546.1	1.5187	2.5	2.0	0	-1.6	6.4	8.5	2.00	01 LPX 001
10.0	5.0	BK 7	546.1	1.5187	2.6	2.0	0	-1.7	8.3	10.6	2.00	01 LPX 003
10.0	8.0	BK 7	546.1	1.5187	3.9	2.0	0	-2.6	7.4	11.9	1.25	01 LPX 005
12.5	7.5	BK 7	587.6	1.5168	2.2	1.0	0	-1.5	11.0	13.7	1.67	01 LPX 435
12.7	6.3	BK 7	546.1	1.5187	2.8	2.0	0	-1.8	10.9	13.5	2.02	01 LPX 007
13.0	7.5	BK 7	546.1	1.5187	2.0	0.9	0	-1.32	11.7	14.1	1.73	01 LPX 008
15.0	8.0	BK 7	546.1	1.5187	3.1	2.0	0	-2.0	13.0	16.1	1.88	01 LPX 009
15.0	10.0	BK 7	546.1	1.5187	3.8	2.0	0	-2.5	12.5	16.8	1.50	01 LPX 011
17.0	6.0	BK 7	546.1	1.5187	2.5	2.0	0	-1.6	15.4	17.5	2.83	01 LPX 013
18.0	8.0	BK 7	546.1	1.5187	2.9	2.0	0	-1.9	16.1	18.9	2.25	01 LPX 015
18.0	15.0	BK 7	546.1	1.5187	5.7	1.9	0	-3.8	14.2	21.8	1.20	01 LPX 017
20.0	10.0	BK 7	546.1	1.5187	3.3	2.0	0	-2.2	17.8	21.3	2.00	01 LPX 019
20.0	12.0	BK 7	546.1	1.5187	3.9	2.0	0	-2.6	17.4	21.9	1.67	01 LPX 021
21.0	15.0	BK 7	546.1	1.5187	5.0	2.0	0	-3.3	17.7	24.0	1.40	01 LPX 023
22.0	9.0	BK 7	546.1	1.5187	2.9	2.0	0	-1.9	20.1	22.9	2.44	01 LPX 025
22.0	10.0	BK 7	546.1	1.5187	3.1	1.9	0	-2.0	20.0	23.2	2.20	01 LPX 027
25.0	12.0	BK 7	546.1	1.5187	3.5	2.0	0	-2.3	22.7	26.5	2.08	01 LPX 029
25.0	14.0	BK 7	546.1	1.5187	4.0	1.9	0	-2.6	22.4	27.1	1.79	01 LPX 035
25.4	6.3	BK 7	546.1	1.5187	2.4	2.0	0	-1.6	23.8	25.8	4.03	01 LPX 037
25.4	25.4	BK 7	546.1	1.5187	11.4	1.7	0	-7.5	17.9	35.1	1.00	01 LPX 041
26.0	13.5	BK 7	546.1	1.5187	3.8	2.0	0	-2.5	23.5	27.8	1.93	01 LPX 043
28.6	12.7	BK 7	546.1	1.5187	3.4	2.0	0	-2.2	26.4	30.0	2.25	01 LPX 045

continued

Plano-Convex Glass Lenses (continued)

f_o (mm)	ϕ (mm)	Glass Type	Design Wavelength λ_o (mm)	Design Index n_o	t_c (mm)	t_e (mm)	A_1H (mm)	A_2H'' (mm)	f_b (mm)	A (mm)	f_o/ϕ	PRODUCT NUMBER
30.0	11.7	BK 7	546.1	1.5187	3.2	2.1	0	-2.1	27.9	31.1	2.56	01 LPX 047
30.0	18.5	BK 7	546.1	1.5187	5.0	2.0	0	-3.3	26.7	33.0	1.62	01 LPX 049
31.7	25.4	BK 7	546.1	1.5187	7.9	1.9	0	-5.2	26.5	37.7	1.25	01 LPX 055
32.7	15.8	BK 7	546.1	1.5187	3.9	1.9	0	-2.6	30.1	34.7	2.07	01 LPX 057
34.0	17.0	BK 7	546.1	1.5187	4.2	2.0	0	-2.8	31.2	36.2	2.00	01 LPX 059
34.0	22.0	BK 7	546.1	1.5187	5.8	1.9	0	-3.8	30.2	37.9	1.55	01 LPX 061
35.0	20.0	BK 7	546.1	1.5187	4.2	1.2	0	-2.8	32.2	38.0	1.75	01 LPX 063
35.0	30.0	BK 7	546.1	1.5187	9.9	2.0	0	-6.5	28.5	42.9	1.17	01 LPX 064
36.0	15.0	BK 7	546.1	1.5187	3.6	2.0	0	-2.4	33.6	37.6	2.40	01 LPX 065
38.1	12.7	BK 7	546.1	1.5187	3.0	2.0	0	-2.0	36.1	39.1	3.00	01 LPX 067
38.1	30.0	BK 7	546.1	1.5187	8.8	1.9	0	-5.8	32.3	45.0	1.27	01 LPX 077
40.0	10.0	BK 7	546.1	1.5187	2.6	2.0	0	-1.7	38.3	40.6	4.00	01 LPX 079
40.0	20.0	BK 7	546.1	1.5187	4.5	1.9	0	-3.0	37.0	42.6	2.00	01 LPX 081
40.0	22.4	BK 7	546.1	1.5187	5.3	2.0	0	-3.5	36.5	43.3	1.79	01 LPX 083
40.0	30.0	BK 7	546.1	1.5187	8.4	2.0	0	-5.5	34.5	46.4	1.33	01 LPX 084
43.0	13.0	BK 7	546.1	1.5187	3.0	2.0	0	-2.0	41.0	44.0	3.31	01 LPX 085
44.5	12.7	BK 7	546.1	1.5187	2.4	1.5	0	-1.6	42.9	45.4	3.50	01 LPX 087
47.0	11.0	BK 7	546.1	1.5187	2.6	2.0	0	-1.7	45.3	47.6	4.27	01 LPX 101
50.0	6.6	BK 7	587.6	1.5168	2.1	1.9	0	-1.4	48.6	50.2	7.58	01 LPX 441
50.0	15.0	BK 7	546.1	1.5187	3.1	2.0	0	-2.0	48.0	51.1	3.33	01 LPX 105
50.0	20.0	BK 7	546.1	1.5187	4.0	2.0	0	-2.6	47.4	52.0	2.50	01 LPX 107
50.0	27.0	BK 7	546.1	1.5187	5.8	2.0	0	-3.8	46.2	53.8	1.85	01 LPX 108
50.0	29.0	BK 7	546.1	1.5187	6.4	2.0	0	-4.2	45.8	54.4	1.72	01 LPX 109
50.0	30.0	BK 7	546.1	1.5187	6.8	2.0	0	-4.5	45.5	54.8	1.67	01 LPX 110
50.0	40.0	BK 7	546.1	1.5187	11.3	1.9	0	-7.4	42.6	59.4	1.25	01 LPX 111
50.8	25.4	BK 7	546.1	1.5187	5.2	1.9	0	-3.4	47.4	54.1	2.00	01 LPX 113
50.8	38.1	BK 7	546.1	1.5187	10.0	1.9	0	-6.6	44.2	58.9	1.33	01 LPX 115
55.0	10.0	BK 7	546.1	1.5187	2.4	2.0	0	-1.6	53.4	55.4	5.50	01 LPX 119
60.0	22.4	BK 7	546.1	1.5187	4.1	2.0	0	-2.7	57.3	62.1	2.68	01 LPX 123
60.0	30.0	BK 7	546.1	1.5187	5.8	1.9	0	-3.8	56.2	63.9	2.00	01 LPX 125
60.0	42.0	BK 7	546.1	1.5187	10.1	1.9	0	-6.7	53.3	68.2	1.43	01 LPX 127
60.0	47.0	BK 7	546.1	1.5187	12.6	1.9	0	-8.3	51.7	70.7	1.28	01 LPX 129
63.5	25.4	BK 7	546.1	1.5187	4.5	2.0	0	-3.0	60.5	66.0	2.50	01 LPX 133
63.5	38.1	BK 7	546.1	1.5187	8.0	1.9	0	-5.3	58.2	69.6	1.67	01 LPX 135
68.0	5.1	BK 7	587.6	1.5168	2.0	1.9	0	-1.3	66.7	68.1	13.33	01 LPX 443
70.0	30.0	BK 7	546.1	1.5187	5.2	2.0	0	-3.4	66.6	73.2	2.33	01 LPX 137
70.0	40.0	BK 7	546.1	1.5187	7.9	1.9	0	-5.2	64.8	76.0	1.75	01 LPX 139
70.0	60.0	BK 7	546.1	1.5187	18.6	2.7	0	-12.2	57.8	85.9	1.17	01 LPX 141

continued

Plano-Convex Glass Lenses (continued)

f_o (mm)	ϕ (mm)	Glass Type	Design Wavelength λ_o (mm)	Design Index n_o	t_c (mm)	t_e (mm)	A_1H (mm)	A_2H'' (mm)	f_b (mm)	A (mm)	f_o/ϕ	PRODUCT NUMBER
75.0	60.0	BK 7	546.1	1.5187	17.0	2.9	0	-11.2	63.8	89.1	1.25	01 LPX 142
76.2	25.4	BK 7	546.1	1.5187	4.1	2.0	0	-2.7	73.5	78.3	3.00	01 LPX 143
76.2	50.8	BK 7	546.1	1.5187	12.1	2.9	0	-8.0	68.2	85.4	1.50	01 LPX 145
80.0	15.0	BK 7	546.1	1.5187	2.7	2.0	0	-1.8	78.2	80.7	5.33	01 LPX 147
80.0	22.4	BK 7	546.1	1.5187	3.5	2.0	0	-2.3	77.7	81.5	3.57	01 LPX 149
80.0	30.0	BK 7	546.1	1.5187	4.8	2.0	0	-3.2	76.8	82.8	2.67	01 LPX 153
80.0	42.0	BK 7	546.1	1.5187	7.7	2.0	0	-5.1	74.9	85.7	1.90	01 LPX 155
80.0	50.0	BK 7	546.1	1.5187	10.4	2.0	0	-6.8	73.2	88.4	1.60	01 LPX 157
85.0	35.0	BK 7	546.1	1.5187	5.6	2.0	0	-3.7	81.3	88.6	2.43	01 LPX 159
88.9	50.8	BK 7	546.1	1.5187	10.6	3.0	0	-7.0	81.9	96.5	1.75	01 LPX 161
90.0	40.0	BK 7	546.1	1.5187	6.5	2.0	0	-4.3	85.7	94.5	2.25	01 LPX 163
90.0	60.0	BK 7	546.1	1.5187	13.8	2.9	0	-9.1	80.9	100.9	1.50	01 LPX 165
95.0	50.0	BK 7	546.1	1.5187	9.7	2.9	0	-6.4	88.6	101.8	1.90	01 LPX 167
100.0	6.6	BK 7	587.1	1.5168	3.0	2.9	0	-2.0	98.0	100.1	15.15	01 LPX 445
100.0	15.0	BK 7	546.1	1.5187	2.5	2.0	0	-1.6	98.4	100.5	6.67	01 LPX 169
100.0	22.4	BK 7	546.1	1.5187	3.2	2.0	0	-2.1	97.9	101.2	4.46	01 LPX 177
100.0	30.0	BK 7	546.1	1.5187	4.2	2.0	0	-2.8	97.2	102.2	3.33	01 LPX 178
100.0	40.0	BK 7	546.1	1.5187	6.0	2.0	0	-4.0	96.0	104.0	2.50	01 LPX 179
100.0	52.0	BK 7	546.1	1.5187	9.9	2.9	0	-6.5	93.5	107.0	1.92	01 LPX 181
100.0	60.0	BK 7	546.1	1.5187	12.5	2.9	0	-8.2	91.8	109.6	1.67	01 LPX 183
100.0	80.0	BK 7	546.1	1.5187	21.6	2.8	0	-14.2	85.8	118.8	1.25	01 LPX 187
105.0	25.0	BK 7	546.1	1.5187	3.4	1.9	0	-2.2	102.8	106.5	4.20	01 LPX 198
110.0	30.0	BK 7	546.1	1.5187	4.0	2.0	0	-2.6	107.4	112.0	3.67	01 LPX 201
120.0	30.0	BK 7	546.1	1.5187	3.8	2.0	0	-2.5	117.5	121.8	4.00	01 LPX 207
120.0	60.0	BK 7	546.1	1.5187	10.6	2.9	0	-7.0	113.0	127.7	2.00	01 LPX 209
120.0	80.0	BK 7	546.1	1.5187	17.4	2.8	0	-11.5	108.5	134.6	1.50	01 LPX 211
120.8	100.0	BK 7	546.1	1.5187	28.8	3.6	0	-19.0	101.0	145.2	1.20	01 LPX 215
125.0	30.0	BK 7	546.1	1.5187	3.7	1.9	0	-2.4	122.6	126.8	4.17	01 LPX 217
127.0	25.4	BK 7	546.1	1.5187	3.2	2.0	0	-2.1	124.9	128.2	5.00	01 LPX 219
127.0	50.8	BK 7	546.1	1.5187	8.0	2.9	0	-5.3	121.7	132.1	2.50	01 LPX 221
127.0	76.2	BK 7	546.1	1.5187	15.0	2.9	0	-9.9	117.1	139.1	1.67	01 LPX 223
130.0	80.0	BK 7	546.1	1.5187	16.0	2.9	0	-10.5	119.5	143.1	1.63	01 LPX 225
130.0	95.0	BK 7	546.1	1.5187	22.3	2.7	0	-14.7	115.3	149.6	1.37	01 LPX 227
135.0	25.0	BK 7	546.1	1.5187	3.1	2.0	0	-2.0	133.0	136.1	5.40	01 LPX 229
140.0	77.0	BK 7	546.1	1.5187	13.9	2.9	0	-9.2	130.8	151.0	1.82	01 LPX 235
150.0	22.4	BK 7	546.1	1.5187	2.8	2.0	0	-1.8	148.2	150.8	6.70	01 LPX 237
150.0	30.0	BK 7	546.1	1.5187	3.5	2.0	0	-2.3	147.7	151.5	5.00	01 LPX 238
150.0	42.0	BK 7	546.1	1.5187	4.9	2.0	0	-3.2	146.8	152.9	3.57	01 LPX 239

continued

Plano-Convex Glass Lenses (continued)

f_o (mm)	ϕ (mm)	Glass Type	Design Wavelength λ_o (mm)	Design Index n_o	t_c (mm)	t_e (mm)	A_1H (mm)	A_2H'' (mm)	f_b (mm)	A (mm)	f_o/ϕ	PRODUCT NUMBER
150.0	52.0	BK 7	546.1	1.5187	7.4	2.9	0	-4.9	145.1	154.5	2.88	01 LPX 241
150.0	60.0	BK 7	546.1	1.5187	9.0	3.0	0	-5.9	144.1	156.0	2.50	01 LPX 242
150.0	80.0	BK 7	546.1	1.5187	14.0	2.9	0	-9.2	140.8	161.1	1.87	01 LPX 243
150.0	105.0	BK 7	546.1	1.5187	24.2	3.8	0	-15.9	134.1	170.4	1.43	01 LPX 245
152.4	25.4	BK 7	546.1	1.5187	3.0	2.0	0	-2.0	150.4	153.4	6.00	01 LPX 247
152.4	50.8	BK 7	546.1	1.5187	7.2	3.0	0	-4.7	147.7	156.6	3.00	01 LPX 249
160.0	10.0	BK 7	546.1	1.5187	2.1	1.9	0	-1.4	158.6	160.2	16.00	01 LPX 255
160.0	29.0	BK 7	546.1	1.5187	3.3	2.0	0	-2.2	157.8	161.3	5.52	01 LPX 257
177.0	50.8	BK 7	546.1	1.5187	6.6	3.0	0	-4.3	172.7	180.6	3.48	01 LPX 259
185.0	75.0	BK 7	546.1	1.5187	10.6	3.0	0	-7.0	178.0	192.6	2.47	01 LPX 263
196.0	15.0	BK 7	546.1	1.5187	2.3	2.0	0	-1.5	194.5	196.3	13.07	01 LPX 265
200.0	22.4	BK 7	546.1	1.5187	2.6	2.0	0	-1.7	198.3	200.6	8.93	01 LPX 267
200.0	30.0	BK 7	546.1	1.5187	3.1	2.0	0	-2.0	198.0	201.1	6.67	01 LPX 268
200.0	40.0	BK 7	546.1	1.5187	3.9	2.0	0	-2.6	197.4	201.9	5.00	01 LPX 269
200.0	50.0	BK 7	546.1	1.5187	6.0	2.9	0	-4.0	196.0	203.1	4.00	01 LPX 277
200.0	77.0	BK 7	546.1	1.5187	10.3	2.9	0	-6.8	193.2	207.4	2.60	01 LPX 279
200.0	90.0	BK 7	546.1	1.5187	13.2	2.9	0	-8.7	191.3	210.3	2.22	01 LPX 280
200.0	130.0	BK 7	546.1	1.5187	26.6	3.7	0	-17.5	182.5	222.9	1.54	01 LPX 281
250.0	30.0	BK 7	546.1	1.5187	2.9	2.0	0	-1.9	248.1	250.9	8.33	01 LPX 282
250.0	42.0	BK 7	546.1	1.5187	3.7	2.0	0	-2.4	247.6	251.7	5.95	01 LPX 285
250.0	50.0	BK 7	546.1	1.5187	4.4	2.0	0	-2.9	247.1	252.4	5.00	01 LPX 286
250.0	140.0	BK 7	546.1	1.5187	24.3	3.8	0	-16.0	234.0	270.5	1.79	01 LPX 283
250.0	145.0	BK 7	546.1	1.5187	25.9	3.7	0	-17.1	232.9	272.2	1.72	01 LPX 289
254.0	50.8	BK 7	546.1	1.5187	5.5	3.0	0	-3.6	250.4	256.5	5.00	01 LPX 301
254.0	101.6	BK 7	546.1	1.5187	14.1	3.9	0	-9.3	244.7	264.2	2.50	01 LPX 303
300.0	47.0	BK 7	546.1	1.5187	3.8	2.0	0	-2.5	297.5	301.8	6.38	01 LPX 305
300.0	66.0	BK 7	546.1	1.5187	6.5	3.0	0	-4.3	295.7	303.5	4.55	01 LPX 307
300.0	95.0	BK 7	546.1	1.5187	10.4	3.0	0	-6.8	293.2	307.4	3.16	01 LPX 309
330.0	145.0	BK 7	546.1	1.5187	20.0	3.9	0	-13.2	316.8	346.1	2.28	01 LPX 311
400.0	25.0	BK 7	546.1	1.5187	2.4	2.0	0	-1.6	398.4	400.4	16.00	01 LPX 313
450.0	145.0	BK 7	546.1	1.5187	15.5	4.0	0	-10.2	439.8	461.5	3.10	01 LPX 336
500.0	25.0	BK 7	546.1	1.5187	2.3	2.0	0	-1.5	498.5	500.3	20.00	01 LPX 319
500.0	30.0	BK 7	546.1	1.5187	2.4	2.0	0	-1.6	498.4	500.4	16.67	01 LPX 318
500.0	45.0	BK 7	546.1	1.5187	3.0	2.0	0	-2.0	498.0	501.0	11.11	01 LPX 321
500.0	50.0	BK 7	546.1	1.5187	3.2	2.0	0	-2.1	497.9	501.2	10.00	01 LPX 322
600.0	42.0	BK 7	546.1	1.5187	2.7	2.0	0	-1.8	598.2	600.7	14.29	01 LPX 323

continued

Plano-Convex Glass Lenses (continued)

f_o (mm)	ϕ (mm)	Glass Type	Design Wavelength λ_o (mm)	Design Index n_o	t_c (mm)	t_e (mm)	A_1H (mm)	A_2H'' (mm)	f_b (mm)	A (mm)	f_o/ϕ	PRODUCT NUMBER
600.0	95.0	BK 7	546.1	1.5187	6.6	3.0	0	-4.3	595.7	603.6	6.32	01 LPX 327
700.0	25.0	BK 7	546.1	1.5187	2.2	2.0	0	-1.4	698.6	700.2	28.00	01 LPX 329
700.0	30.0	BK 7	546.1	1.5187	2.3	2.0	0	-1.5	698.5	700.3	23.33	01 LPX 331
700.0	50.0	BK 7	546.1	1.5187	2.9	2.0	0	-1.9	698.1	700.9	14.00	01 LPX 332
800.0	30.0	BK 7	546.1	1.5187	2.3	2.0	0	-1.5	798.5	800.3	26.67	01 LPX 341
800.0	50.0	BK 7	546.1	1.5187	2.8	2.0	0	-1.8	796.2	800.8	16.00	01 LPX 343
1000.0	25.0	BK 7	546.1	1.5187	2.1	1.9	0	-1.4	998.6	1000.2	40.00	01 LPX 333
1000.0	30.0	BK 7	546.1	1.5187	2.2	2.0	0	-1.4	998.6	1000.2	33.33	01 LPX 345
1000.0	47.0	BK 7	546.1	1.5187	2.5	2.0	0	-1.6	998.4	1000.5	21.28	01 LPX 334
1000.0	50.0	BK 7	546.1	1.5187	2.6	2.0	0	-1.7	998.3	1000.6	20.00	01 LPX 347
1000.0	80.0	BK 7	546.1	1.5187	4.5	3.0	0	-3.0	997.0	1001.5	12.50	01 LPX 335
1000.0	145.0	BK 7	546.1	1.5187	9.0	3.9	0	-5.9	994.1	1005.1	6.90	01 LPX 337

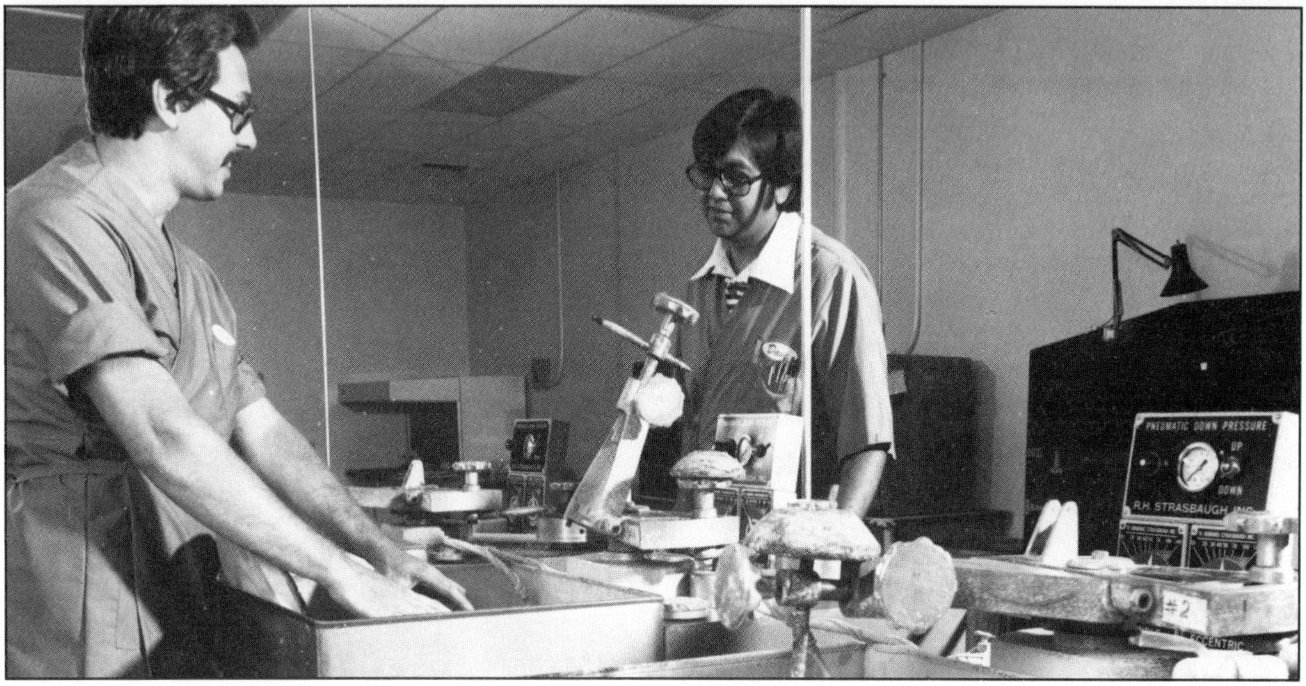

SYMMETRIC-CONVEX GLASS LENSES

Like plano-convex lenses, these have positive focal lengths and form both real and virtual images. Symmetry is assured by finishing both surfaces with the same tool. Because of symmetry, several aberrations almost exactly cancel at unit conjugate ratio ($s=s''=2f$). These include coma, distortion, and lateral chromatic aberration. At this conjugate ratio, spherical aberration is the minimum possible with a single element, and symmetric-convex lenses are by definition best-form. These aberrations grow as conjugate ratios depart from unity.

Symmetric-convex lenses are recommended for virtual imaging of real objects, and for positive conjugate ratios from about 0.2 up to about 5. Beyond 5 (and between 0 and 0.2) plano-convex lenses are preferable. These values are only approximate, and are wavelength sensitive. To make the best choice between symmetric-convex and plano-convex lenses for your particular choice of wavelength and conjugate ratio, consult the earlier tabulated values of "decision parameters" (see alphabetical index). Requirements for short focal length (and hence short overall package length) sometimes outweigh requirements for image sharpness. Since both surfaces contribute to the power of symmetric lenses, they have shorter focal lengths than plano-convex lenses of equal diameter and surface radius.

A large variety of single and multilayer antireflection coatings is available for these lenses. See the preceding section on coatings (page 69).

Diameters of all lenses may be reduced ("edged") by us to meet your requirements. Larger diameters and focal lengths are available on special request.

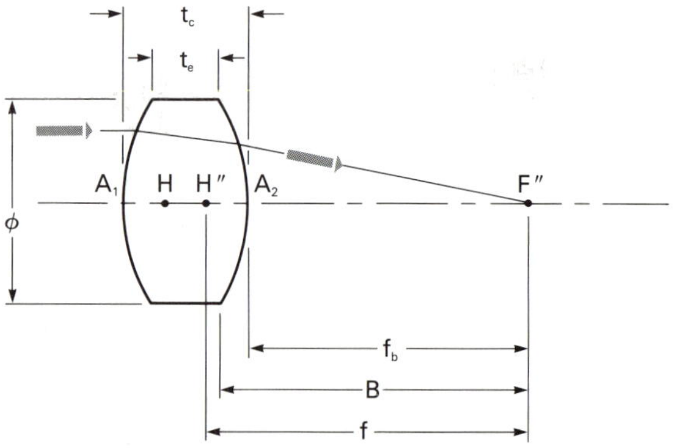

01 LDX

SPECIFICATIONS: SYMMETRIC-CONVEX GLASS LENSES

Paraxial Focal Length (f_o): ±5%
Design Index (n_o): ±0.001
Centration: 3 arc minutes
Diameter (ϕ): ±0.25mm (±0.5mm if $\phi \geq$ 100mm)
Center Thickness (t_c): ±0.2mm
Material: BK 7 Grade A Fine Annealed
Cosmetic Surface Quality: Pitch polished, 80-50 scratch and dig
Circumference is ground and both edges have a slight protective bevel.

Fused silica symmetric-convex spherical lenses are also available. Fused silica provides extreme transparency over a wide spectral range, has a low coefficient of thermal expansion, and is resistant to scratching and thermal shock. By combining fused silica optics with our laser line "V"-coatings, external transmittances of 99.5% can be obtained. (See page 112.)

Symmetric-Convex Glass Lenses

f_o (mm)	ϕ (mm)	Design Wavelength λ_o (mm)	Design Index n_o	t_c (mm)	t_e (mm)	A_1H (mm)	f_b (mm)	A=B (mm)	f_o/ϕ	PRODUCT NUMBER
4.5	5.0	587.6	1.5168	2.0	0.4	0.7	3.8	4.6	0.90	01 LDX 401
7.0	5.0	546.1	1.5187	3.0	2.0	1.1	5.93	6.4	1.40	01 LDX 001
8.0	6.0	546.1	1.5187	3.2	2.0	1.1	6.9	7.5	1.33	01 LDX 003
10.0	5.0	546.1	1.5187	2.6	2.0	0.9	9.1	9.4	2.00	01 LDX 005
10.0	7.5	546.1	1.5187	3.5	2.0	1.2	8.8	9.5	1.33	01 LDX 007
10.0	10.0	546.1	1.5187	4.8	1.9	1.7	8.3	9.7	1.00	01 LDX 009
12.7	6.3	546.1	1.5187	2.8	2.0	1.0	11.7	12.1	2.02	01 LDX 013
14.2	12.5	546.1	1.5187	5.0	2.0	1.8	12.4	13.9	1.14	01 LDX 015
15.0	6.0	587.6	1.5168	1.7	1.1	0.6	14.4	14.7	2.50	01 LDX 403
15.0	7.5	546.1	1.5187	2.9	2.0	1.0	14.0	14.5	2.00	01 LDX 017
15.0	14.5	546.1	1.5187	5.9	2.0	2.1	12.9	14.9	1.03	01 LDX 019
18.0	13.0	546.1	1.5187	4.4	2.0	1.5	16.5	17.7	1.38	01 LDX 023
20.0	15.0	546.1	1.5187	4.9	2.0	1.7	18.3	19.8	1.33	01 LDX 025
20.0	21.0	546.1	1.5187	8.2	2.0	2.9	17.1	20.2	0.95	01 LDX 027
22.0	20.0	546.1	1.5187	6.9	2.0	2.4	19.6	22.1	1.10	01 LDX 029
25.0	22.4	546.1	1.5187	7.3	1.9	2.5	22.5	25.2	1.12	01 LDX 035
25.4	12.7	546.1	1.5187	3.6	2.0	1.2	24.2	25.0	2.00	01 LDX 039
25.4	20.0	546.1	1.5187	6.1	2.0	2.1	23.3	25.4	1.27	01 LDX 041
25.4	25.4	546.1	1.5187	9.0	2.0	3.2	22.2	25.8	1.00	01 LDX 043
28.6	12.7	546.1	1.5187	3.4	2.0	1.1	27.5	28.2	2.25	01 LDX 049
30.0	22.4	546.1	1.5187	6.3	2.0	2.2	27.8	30.0	1.34	01 LDX 055
30.0	30.0	546.1	1.5187	10.3	2.0	3.6	26.4	30.5	1.00	01 LDX 057
31.0	20.0	546.1	1.5187	5.3	2.0	1.8	29.2	30.8	1.55	01 LDX 059
31.7	25.4	546.1	1.5187	7.3	2.0	2.5	29.2	31.9	1.25	01 LDX 061
31.7	35.0	546.1	1.5187	12.9	1.9	4.6	27.1	32.6	0.91	01 LDX 063
33.0	10.0	546.1	1.5187	2.7	2.0	0.9	32.1	32.5	3.30	01 LDX 065
35.0	19.5	546.1	1.5187	4.7	2.0	1.6	33.4	34.8	1.79	01 LDX 067
38.1	25.4	546.1	1.5187	6.3	2.0	2.1	36.0	38.1	1.50	01 LDX 069
38.1	31.7	546.1	1.5187	8.9	2.0	3.1	35.0	38.5	1.20	01 LDX 077
38.1	38.1	546.1	1.5187	12.4	1.9	4.3	33.8	39.0	1.00	01 LDX 079
40.0	22.4	546.1	1.5187	5.1	1.9	1.7	38.3	39.9	1.79	01 LDX 083
40.0	25.0	546.1	1.5187	5.9	1.9	2.0	38.0	40.0	1.60	01 LDX 085
45.0	38.0	546.1	1.5187	10.4	2.0	3.6	41.4	45.7	1.18	01 LDX 087
50.0	22.4	546.1	1.5187	4.5	2.0	1.5	48.5	49.7	2.23	01 LDX 101
50.0	25.0	546.1	1.5187	5.1	2.0	1.7	48.3	49.8	2.00	01 LDX 103

continued

Symmetric-Convex Glass Lenses (continued)

f_o (mm)	ϕ (mm)	Design Wavelength λ_o (mm)	Design Index n_o	t_c (mm)	t_e (mm)	A_1H (mm)	f_b (mm)	A=B (mm)	f_o/ϕ	PRODUCT NUMBER
50.0	30.0	546.1	1.5187	6.5	2.0	2.2	47.8	50.1	1.67	01 LDX 104
50.0	31.0	546.1	1.5187	6.8	1.9	2.3	47.7	50.1	1.61	01 LDX 105
50.0	42.0	546.1	1.5187	11.2	1.9	3.8	46.2	50.8	1.19	01 LDX 107
50.8	38.1	546.1	1.5187	9.3	1.9	3.2	47.6	51.3	1.33	01 LDX 113
50.8	50.8	546.1	1.5187	16.8	2.8	5.9	44.9	51.9	1.00	01 LDX 115
50.8	60.0	546.1	1.5187	23.7	2.8	8.5	42.3	52.7	0.85	01 LDX 117
54.0	32.0	546.1	1.5187	6.7	1.9	2.3	51.7	54.1	1.69	01 LDX 119
60.0	22.4	546.1	1.5187	4.0	1.9	1.3	58.7	59.7	2.68	01 LDX 121
60.0	50.0	546.1	1.5187	12.9	2.0	4.4	55.6	61.1	1.2	01 LDX 123
63.5	40.0	546.1	1.5187	8.3	1.9	2.8	60.7	63.9	1.59	01 LDX 125
63.5	50.0	546.1	1.5187	13.2	3.0	4.5	59.0	64.1	1.27	01 LDX 127
65.0	35.0	546.1	1.5187	6.7	2.0	2.2	62.8	65.1	1.86	01 LDX 129
65.0	42.0	546.1	1.5187	8.8	1.9	3.0	62.0	65.5	1.55	01 LDX 133
70.0	30.0	546.1	1.5187	5.2	2.0	1.7	68.3	69.9	2.33	01 LDX 135
70.0	42.0	546.1	1.5187	8.3	2.0	2.8	67.2	70.4	1.67	01 LDX 137
70.0	60.0	546.1	1.5187	16.4	2.8	5.6	64.4	71.2	1.17	01 LDX 139
75.0	25.0	546.1	1.5187	4.0	2.0	1.3	73.7	74.7	3.00	01 LDX 141
76.2	38.1	546.1	1.5187	6.7	2.0	2.2	74.0	76.3	2.00	01 LDX 143
76.2	50.8	546.1	1.5187	11.5	2.9	3.9	72.3	76.6	1.50	01 LDX 145
80.0	22.4	546.1	1.5187	3.5	2.0	1.2	78.8	79.6	3.57	01 LDX 149
80.0	47.0	546.1	1.5187	8.9	2.0	3.0	77.0	80.5	1.70	01 LDX 157
82.0	29.0	546.1	1.5187	4.5	2.0	1.5	80.5	81.8	2.83	01 LDX 159
88.9	25.4	546.1	1.5187	3.8	2.0	1.3	87.6	88.5	3.50	01 LDX 161
100.0	22.4	546.1	1.5187	3.2	2.0	1.1	98.9	99.6	4.46	01 LDX 167
100.0	30.0	546.1	1.5187	4.2	2.0	1.4	98.6	99.7	3.33	01 LDX 168
100.0	42.0	546.1	1.5187	6.3	2.0	2.1	97.9	100.1	2.38	01 LDX 169
100.0	50.0	546.1	1.5187	8.2	2.0	2.7	97.3	100.4	2.00	01 LDX 171
101.6	38.1	546.1	1.5187	5.5	2.0	1.8	99.8	101.5	2.67	01 LDX 177
101.6	50.8	546.1	1.5187	9.3	3.0	3.1	98.5	101.6	2.00	01 LDX 179
101.6	63.5	546.1	1.5187	12.9	2.9	4.3	97.3	102.3	1.60	01 LDX 181
110.0	35.0	546.1	1.5187	4.7	2.0	1.6	108.4	109.8	3.14	01 LDX 183
125.0	25.0	546.1	1.5187	3.2	2.0	1.1	123.9	124.5	5.00	01 LDX 185
125.0	40.0	546.1	1.5187	5.1	2.0	1.7	123.3	124.9	3.13	01 LDX 187
127.0	50.8	546.1	1.5187	8.0	3.0	2.7	124.3	126.8	2.50	01 LDX 189
150.0	22.4	546.1	1.5187	2.8	2.0	0.9	149.1	149.5	6.70	01 LDX 201

continued

Symmetric-Convex Glass Lenses (continued)

f_o (mm)	ϕ (mm)	Design Wavelength λ_o (mm)	Design Index n_o	t_c (mm)	t_e (mm)	A_1H (mm)	f_b (mm)	A=B (mm)	f_o/ϕ	PRODUCT NUMBER
150.0	30.0	546.1	1.5187	3.5	2.0	1.2	148.8	149.5	5.00	01 LDX 202
150.0	42.0	546.1	1.5187	4.8	1.9	1.6	148.4	149.8	3.57	01 LDX 203
150.0	50.0	546.1	1.5187	6.1	2.0	2.0	148.0	150.0	3.00	01 LDX 205
150.0	66.0	546.1	1.5187	10.1	2.9	3.4	146.6	150.2	2.27	01 LDX 207
150.0	75.0	546.1	1.5187	12.2	2.9	4.1	145.9	150.6	2.00	01 LDX 209
150.0	95.0	546.1	1.5187	18.0	2.8	6.0	144.0	151.5	1.58	01 LDX 211
150.0	100.0	546.1	1.5187	20.8	3.9	7.0	143.0	151.4	1.50	01 LDX 213
175.0	100.0	546.1	1.5187	18.2	3.9	6.1	168.9	176.1	1.75	01 LDX 215
200.0	22.4	546.1	1.5187	2.6	2.0	0.9	199.1	199.4	8.93	01 LDX 217
200.0	30.0	546.1	1.5187	3.1	2.0	1.0	199.0	199.5	6.67	01 LDX 218
200.0	42.0	546.1	1.5187	4.1	2.0	1.4	198.6	199.7	4.76	01 LDX 219
200.0	50.0	546.1	1.5187	5.0	2.0	1.7	198.3	199.8	4.00	01 LDX 222
200.0	75.0	546.1	1.5187	9.8	2.9	3.3	196.7	200.2	2.67	01 LDX 221
200.0	100.0	546.1	1.5187	16.3	3.9	5.4	194.6	200.8	2.00	01 LDX 223
250.0	42.0	546.1	1.5187	3.7	2.0	1.2	248.8	249.6	5.95	01 LDX 225
250.0	75.0	546.1	1.5187	8.4	2.9	2.8	241.2	250.0	3.33	01 LDX 229
250.0	100.0	546.1	1.5187	13.7	3.9	4.6	245.4	250.4	2.50	01 LDX 235
300.0	42.0	546.1	1.5187	3.4	2.0	1.1	298.9	299.6	7.14	01 LDX 237
400.0	42.0	546.1	1.5187	3.1	2.0	1.0	399.0	399.5	9.52	01 LDX 243
500.0	30.0	546.1	1.5187	2.4	2.0	0.8	499.2	499.4	16.67	01 LDX 245
500.0	42.0	546.1	1.5187	2.8	1.9	0.9	499.1	499.5	11.90	01 LDX 247
500.0	50.0	546.1	1.5187	3.2	2.0	1.1	498.9	499.5	10.00	01 LDX 248
500.0	80.0	546.1	1.5187	6.1	3.0	2.0	498.0	499.5	6.25	01 LDX 249
600.0	42.0	546.1	1.5187	2.7	2.0	0.9	599.1	599.5	14.29	01 LDX 255
600.0	47.0	546.1	1.5187	2.9	2.0	1.0	599.0	599.5	12.77	01 LDX 257
1000.0	30.0	546.1	1.5187	2.2	2.0	0.7	999.3	999.4	33.33	01 LDX 258
1000.0	42.0	546.1	1.5187	2.4	2.0	0.8	999.2	999.4	23.81	01 LDX 259
1000.0	47.0	546.1	1.5187	2.5	2.0	0.8	999.2	999.4	21.28	01 LDX 261
1000.0	50.0	546.1	1.5187	2.6	2.0	0.9	999.1	999.4	20.00	01 LDX 262
2000.0	42.0	546.1	1.5187	2.2	2.0	0.7	1999.3	1999.4	47.62	01 LDX 263

Please call us to discuss your high volume OEM requirements for lenses.

PLANO-CONCAVE GLASS LENSES

This lens type has a negative focal length, diverges collimated incident light, and forms only virtual images which are seen through the lens. It is often used to expand light beams, or to increase focal lengths in existing systems.

The plano-concave lens offers very nearly minimum spherical aberration at negative-infinite or near-infinite conjugate ratios, with the concave surface facing the longest conjugate distance. A collimated beam incident on the curved surface will be transformed into a divergent beam emergent from the plane surface. The divergent beam will appear to emerge from a smaller (more nearly diffraction-limited) virtual point source than if the plane surface had faced toward the collimated beam.

Reversing the rays in the above description, a convergent beam is transformed with least additional spherical aberration into a collimated beam if the convergent beam is incident on the plane surface. The word "additional" is used as the convergent beam may already contain spherical aberration, and depending on its sign, the plano-concave lens may either increase or decrease the total for a lens combination. Thus, in some applications, the opposite orientation of an element may reduce the overall aberration of a lens combination.

A variety of single and multilayer antireflection coatings is available for these lenses. See the preceding section on coatings (page 69). When a reflective coating is deposited on the concave lens surface an excellent concave mirror is formed. A variety of reflective and partially reflective coatings will also be found in the coating suffix system.

Diameters of all lenses may be reduced ("edged") by us to meet your requirements.

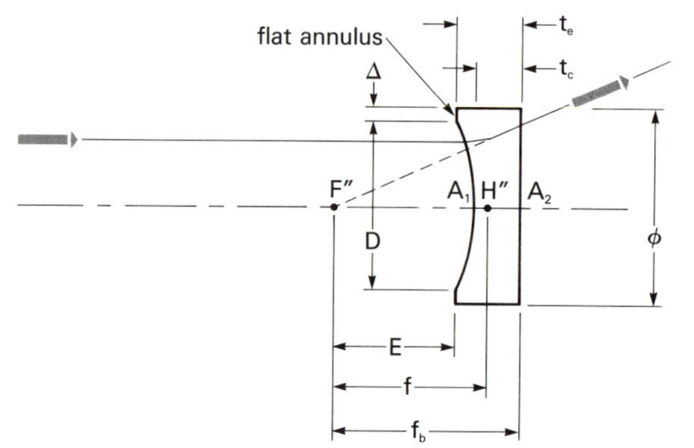

01 LPK

SPECIFICATIONS: PLANO-CONCAVE GLASS LENSES

Paraxial Focal Length (f_o): ±5%
Design Wavelength (λ_o): 546.1nm
Design Index (n_o): 1.5187 ±0.001
Centration: 3 arc minutes
Diameter (ϕ): ±0.25mm
Clear Aperture Diameter (D): $D=\phi-2\Delta$
Nominal Flat Annulus Width (Δ): +0%, -70%
$\Delta=0.2$mm for $\phi\leq40$mm
$\Delta=0.4$mm for $\phi>40$mm
Center Thickness (t_c): ±0.2mm
Material: BK 7 Grade A Fine Annealed
Cosmetic Surface Quality: Pitch polished, 80-50 scratch and dig
Circumference and flat annulus are ground, and both edges have a slight protective bevel.

Fused silica plano-concave spherical lenses are also available. Fused silica provides greater UV and IR transmission, an increased thermal operating range, increased resistance to scratching, and much higher resistance to radiation darkening. *(See page 114.)*

Plano-Concave Glass Lenses

f_o (mm)	ϕ (mm)	t_c (mm)	t_e (mm)	A_1H (mm)	A_2H'' (mm)	f_b (mm)	E (mm)	f_o/D	D (mm)	PRODUCT NUMBER
-10.0	5.0	2.0	2.3	0	-1.3	-11.3	9.0	-2.17	4.6	01 LPK 002
-15.0	8.0	2.0	3.0	0	-1.3	-16.3	13.3	-1.97	7.6	01 LPK 004
-20.0	10.0	2.0	3.2	0	-1.3	-21.3	18.1	-2.08	9.6	01 LPK 001
-25.0	25.0	2.0	10.9	0	-1.3	-26.3	15.5	-1.02	24.6	01 LPK 003
-29.0	18.0	2.0	4.8	0	-1.3	-30.3	25.5	-1.65	17.6	01 LPK 005
-30.0	20.0	2.0	5.5	0	-1.3	-31.3	25.8	-1.53	19.6	01 LPK 007
-38.1	25.4	2.0	6.5	0	-1.3	-39.4	33.0	-1.52	25.0	01 LPK 011
-40.0	21.0	2.0	4.7	0	-1.3	-41.3	36.6	-1.94	20.6	01 LPK 013
-50.0	20.0	2.0	3.9	0	-1.3	-51.3	47.4	-2.55	19.6	01 LPK 015
-50.0	22.4	2.0	4.4	0	-1.3	-51.3	46.9	-2.27	22.0	01 LPK 017
-50.0	30.0	2.0	6.6	0	-1.3	-51.3	44.7	-1.69	29.6	01 LPK 019
-50.0	40.0	2.0	11.2	0	-1.3	-51.3	40.1	-1.26	39.6	01 LPK 021
-50.0	47.0	2.0	16.1	0	-1.3	-51.3	35.2	-1.08	46.2	01 LPK 023
-63.5	40.0	2.0	8.6	0	-1.3	-64.8	56.2	-1.60	39.6	01 LPK 027
-75.0	15.0	2.0	2.7	0	-1.3	-76.3	73.6	-5.14	14.6	01 LPK 029
-75.0	50.0	3.0	11.8	0	-2.0	-77.0	65.2	-1.52	49.2	01 LPK 033
-85.0	22.0	2.0	3.3	0	-1.3	-86.3	83.0	-3.94	21.6	01 LPK 035
-100.0	22.4	2.0	3.2	0	-1.3	-101.3	98.1	-4.55	22.0	01 LPK 041
-100.0	30.0	2.0	4.2	0	-1.3	-101.3	97.2	-3.38	29.6	01 LPK 043
-100.0	42.0	2.0	6.3	0	-1.3	-101.3	95.1	-2.43	41.2	01 LPK 045
-100.0	50.0	2.0	8.2	0	-1.3	-101.3	93.1	-2.03	49.2	01 LPK 047
-125.0	30.0	2.0	3.7	0	-1.3	-126.3	122.6	-4.22	29.6	01 LPK 049
-125.0	50.0	3.0	7.8	0	-2.0	-127.0	119.1	-2.54	49.2	01 LPK 055
-150.0	22.4	2.0	2.8	0	-1.3	-151.3	148.5	-6.82	22.0	01 LPK 057
-150.0	30.0	2.0	3.4	0	-1.3	-151.3	147.9	-5.07	29.6	01 LPK 058
-150.0	40.0	2.0	4.6	0	-1.3	-151.3	146.5	-3.79	39.6	01 LPK 059
-150.0	50.0	3.0	7.0	0	-2.0	-152.0	145.0	-3.05	49.2	01 LPK 061
-200.0	30.0	2.0	3.1	0	-1.3	-201.3	198.2	-6.76	29.6	01 LPK 062
-200.0	50.0	2.0	5.0	0	-1.3	-201.3	196.3	-4.07	49.2	01 LPK 064
-200.0	42.0	2.0	4.1	0	-1.3	-201.3	197.3	-4.85	41.2	01 LPK 063
-600.0	50.0	2.0	3.0	0	-1.3	-601.3	598.3	-12.20	49.2	01 LPK 066
-600.0	95.0	3.0	6.6	0	-2.0	-602.0	595.4	-6.37	94.2	01 LPK 065

SYMMETRIC-CONCAVE GLASS LENSES

The symmetric-concave lens has a negative focal length, diverges collimated incident light, and forms only virtual images which are seen through the lens. It is often used to expand light beams or to increase focal lengths in existing systems, and is normally used in combination with other lenses. Among the many devices utilizing symmetric-concave lenses are laser beam expanders, optical character readers, viewers, and projection systems.

A variety of single and multilayer antireflection coatings is available for these lenses. See the preceding section on coatings (page 69).

Diameters of all lenses may be reduced ("edged") by us to meet your requirements.

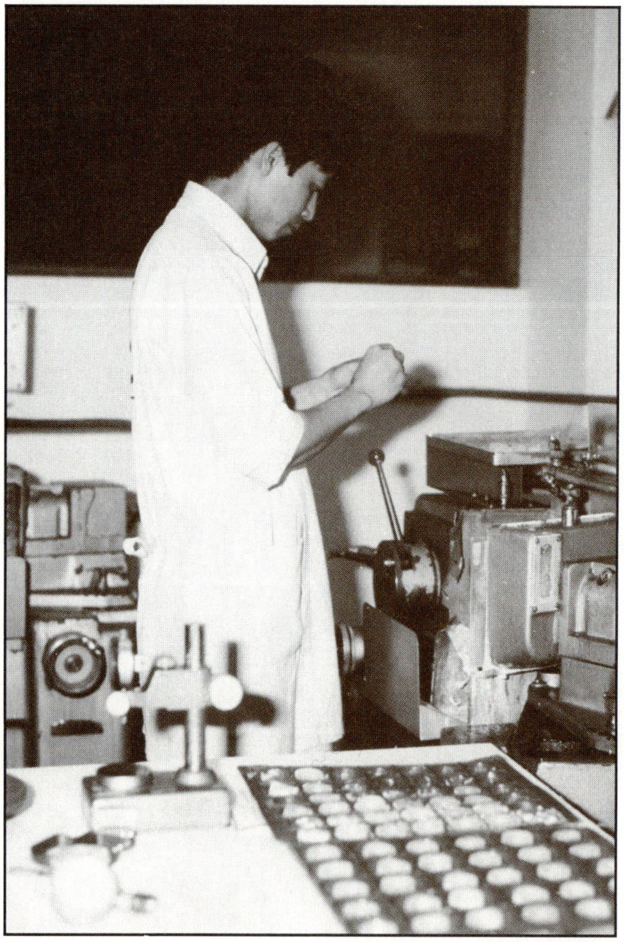

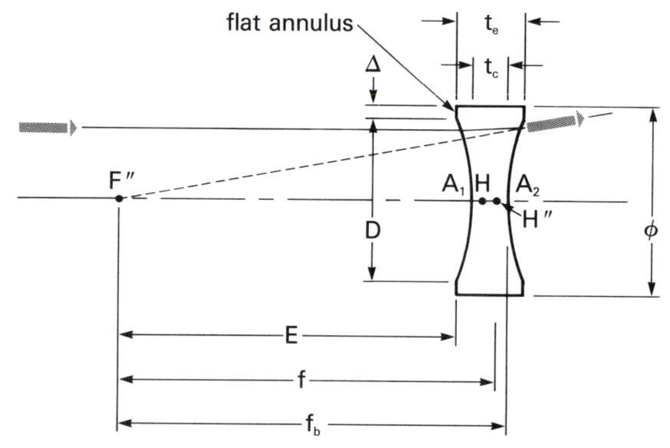

01 LDK

SPECIFICATIONS: SYMMETRIC-CONCAVE GLASS LENSES

Paraxial Focal Length (f_o): ±5%
Design Wavelength (λ_o): 546.1nm
Design Index (n_o): 1.5187±0.001
Centration: 3 arc minutes
Diameter (ϕ): ±0.25mm
Clear Aperture Diameter (D): $D=\phi-2\Delta$
Nominal Flat Annulus Width (Δ): +0%, -70%
  $\Delta=0.2$mm for $\phi\leq40$mm
  $\Delta=0.4$mm for $\phi>40$mm
Center Thickness (t_c): ±0.2mm
Material: BK 7 Grade A Fine Annealed
Cosmetic Surface Quality: Pitch polished, 80-50 scratch and dig
Circumference and flat annulus are ground, and both edges have a slight protective bevel.

Symmetric concave fused silica lenses are also available *(see page 116).*

Symmetric-Concave Glass Lenses

f_o (mm)	ϕ (mm)	t_c (mm)	t_e (mm)	A_1H mm)	A_2H_o (mm)	f_b (mm)	E (mm)	f_o/D	D (mm)	PRODUCT NUMBER
-5.0	5.0	2.0	3.0	0.6	-0.6	-5.62	3.1	-1.09	4.6	01 LDK 001
-8.0	6.0	2.0	2.9	0.6	-0.6	-8.6	6.2	-1.43	5.6	01 LDK 003
-12.0	12.0	2.0	4.8	0.6	-0.6	-12.6	9.2	-1.03	11.6	01 LDK 005
-20.0	20.0	2.0	6.8	0.6	-0.6	-20.6	16.2	-1.02	19.6	01 LDK 007
-25.0	25.0	2.0	8.1	0.6	-0.6	-25.6	20.6	-1.02	24.6	01 LDK 009
-30.0	13.0	2.0	3.3	0.7	-0.7	-30.7	28.0	-2.38	12.6	01 LDK 012
-30.0	20.0	2.0	5.1	0.7	-0.7	-30.7	27.1	-1.53	19.6	01 LDK 013
-30.0	30.0	2.0	9.4	0.7	-0.7	-30.7	25.0	-1.01	29.6	01 LDK 015
-35.0	20.0	2.0	4.7	0.7	-0.7	-35.7	32.3	-1.79	19.6	01 LDK 016
-40.0	22.4	2.0	4.9	0.7	-0.7	-40.7	37.2	-1.82	22.0	01 LDK 017
-50.0	22.4	2.0	4.3	0.7	-0.7	-50.7	47.5	-2.27	22.0	01 LDK 019
-50.0	30.0	2.0	6.3	0.7	-0.7	-50.7	46.6	-1.69	29.6	01 LDK 018
-50.0	42.0	2.0	10.5	0.7	-0.7	-50.7	44.4	-1.21	41.2	01 LDK 021
-70.0	20.0	2.0	3.3	0.7	-0.7	-70.7	68.0	-3.57	19.6	01 LDK 023
-100.0	30.0	2.0	4.1	0.7	-0.7	-100.7	97.6	-3.38	29.6	01 LDK 024
-100.0	42.0	2.0	6.1	0.7	-0.7	-100.7	96.6	-2.43	41.2	01 LDK 025
-100.0	50.0	2.0	7.9	0.7	-0.7	-100.7	95.8	-2.03	49.2	01 LDK 026
-150.0	22.4	2.0	2.8	0.7	-0.7	-150.7	148.3	-6.82	22.0	01 LDK 027
-150.0	30.0	2.0	3.4	0.7	-0.7	-150.7	148.0	-5.07	29.6	01 LDK 028
-150.0	42.0	2.0	4.7	0.7	-0.7	-150.7	147.3	-3.64	41.2	01 LDK 029
-150.0	50.0	2.0	5.9	0.7	-0.7	-150.7	146.7	-3.05	49.2	01 LDK 031
-150.0	95.0	3.0	17.5	1.0	-1.0	-151.0	140.7	-1.59	94.2	01 LDK 033
-200.0	30.0	2.0	3.1	0.7	-0.7	-200.7	198.2	-6.76	29.6	01 LDK 034
-200.0	50.0	2.0	4.9	0.7	-0.7	-200.7	197.2	-4.07	49.2	01 LDK 036
-250.0	42.0	2.0	3.6	0.7	-0.7	-250.7	247.8	-6.07	41.2	01 LDK 035
-300.0	42.0	2.0	3.4	0.7	-0.7	-300.7	298.0	-7.28	41.2	01 LDK 037
-400.0	42.0	2.0	3.0	0.7	-0.7	-400.7	398.1	-9.71	41.2	01 LDK 039
-500.0	42.0	2.0	2.8	0.7	-0.7	-500.7	498.2	-12.14	41.2	01 LDK 041
-600.0	42.0	2.0	2.7	0.7	-0.7	-600.7	598.3	-14.56	41.2	01 LDK 043
-600.0	50.0	2.0	3.0	0.7	-0.7	-600.7	598.3	-12.20	49.2	01 LDK 045

Lens holders are available to suit all Melles Griot lenses. *(See page 452.)*

MENISCUS GLASS LENSES

THEORY OF MENISCUS LENSES

A meniscus lens is defined as having one surface convex and one surface concave. A meniscus lens is most often used in conjunction with another lens to produce a system of a longer or shorter focal length than the original lens. For example, a positive meniscus lens can be placed after a plano-convex lens to shorten its focal length and decrease its f-number.

Meniscus lenses can be designed with a shape that results in no additional spherical aberration or coma being introduced into the system. This shape is called the aplanatic form. It is possible to achieve the aplanatic form only when the lens is operated at a magnification that is numerically equal to the reciprocal of the index of refraction of the material from which it is made. Referring to the drawing, third order optics show that a positive aplanatic meniscus lens has a front radius given by

$$r_1 = s/(n+1),$$

where

$$\text{magnification} = -1/n = -s''/s \text{ or } s'' = -s/n.$$

The second radius of a positive aplanatic meniscus is given by

$$r_2 = s'' - t_c.$$

Thus, the second radius is concentric with the image point. In the case of the negative aplanatic meniscus, the first radius is concentric with the (virtual) object, so

$$r_1 = s.$$

The second radius is given by

$$r_2 = n(s-t)/(n+1).$$

The aplanatic lens form, although very useful because of its aberration properties, is rather restrictive in its application since it must be used under specific conditions (a particular magnification and object distance) to realize its full advantages. In general, it would be more useful to have a meniscus lens that would perform satisfactorily over a wide range of object distances or magnifications. For a general form meniscus lens, third order aberration theory gives the total transverse spherical aberration (TSA) of the meniscus and preceding element as

$$TSA = TSA_o(s''/s) + TSA_m,$$

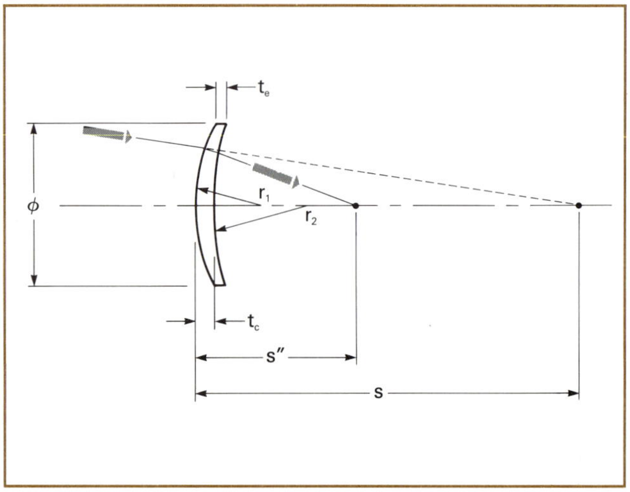

APLANATIC MENISCUS LENS with index of refraction, n.

where TSA_o is the transverse spherical aberration of the original system, and TSA_m is the transverse spherical aberration of the meniscus lens itself. In the case of the positive meniscus lens, the transverse spherical aberration of the original system is actually reduced by a factor equal to the magnification of the meniscus element. Since the focal length of the system is shortened by this same amount, the angular aberration of the system is unchanged. The total spherical aberration of the system is only increased by the amount TSA_m. Thus, the positive meniscus can be used to shorten the focal length and lower the f-number of a system without adding much aberration if the term TSA_m is small. Of course, in the case of the aplanatic meniscus, the term TSA_m is equal to zero. In the case of the negative meniscus lens, the aberration of the original system is increased an amount equal to the magnification. Since the focal length of the system also increases by the same factor, there is again no change in angular resolution. By examining third order, thin lens aberration results, it is possible to select a more or less optimal shape for a general purpose meniscus lens. In making this selection, it is important that the aberration contributed by the lens have a low absolute value so that the term TSA_m in the preceding equation is small. This means designing the lens to work at a fairly low magnification, somewhat near a value equal to the reciprocal of the index of refraction of the lens material. The meniscus lenses presented here were all designed somewhere in the range between 1/1.6 to 1/2.5 magnification with a nominal value of 1/1.75 (we express the magnification as a fraction with a decimal denominator so that the conjugate ratio, and its relationship to the refractive index, is readily apparent). Lenses designed for a lower magnification (between 1/3 to 1/10 times) would have had an unacceptably large aberration term, TSA_m. At magnifications lower than about 1/10 power the plano-convex or plano-concave form is nearly optimal, so in this case lenses from the 01 LPX or 01 LPK series can be used. Thus, the nominal magnification of 1/1.75 represents a compromise between a value low enough to make the lenses useful as auxiliary elements, but not too low so as to make their aberrations unacceptably large.

USING MENISCUS LENSES

Meniscus lenses are employed to greatest advantage when they are used to shorten the focal length of a doublet or a plano-convex lens already working at an infinite conjugate ratio. To determine which particular meniscus lens to use in this situation, it is necessary to know either the desired focal length of the final system (including the meniscus lens), or the focal length of the original lens. For positive meniscus lenses, the relationship between the meniscus lens focal length, f_m, original lens focal length, f_o, and the combination focal length, f, is given by

$f=4f_o/7$, and

$f_m=4f_o/3$.

For systems using negative meniscus lenses, the focal lengths are related by

$f=7f_o/4$, and

$f_m=-7f_o/3$.

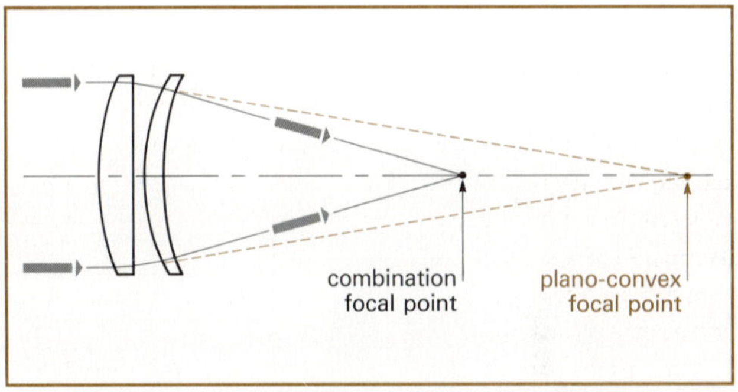

A MENISCUS LENS BEING USED TO SHORTEN THE FOCAL LENGTH OF A PLANO-CONVEX LENS. The focal length of the plano-convex lens alone is indicated, as is the combination focal length of the system including the meniscus lens. The meniscus lens should always be oriented with the concave surface towards the focal point, a shown above.

In most cases, there will not be lenses available with focal lengths exactly matching those predicted by the formulas. In these cases, lenses with the closest focal length to those derived from the formulas should be chosen. One can then adjust the element spacing to achieve the correct magnification. To determine the proper spacing, one solves the combination focal length formula for the value d;

$f=f_of_m/(f_o+f_m-d)$.

In this formula, f_o and f_m are the values for the actually available lenses, and f is calculated from one of the previous formulas.

The judicious use of a positive meniscus lens can result in a system with significantly better performance than would otherwise be achieved with a single component, especially in systems working with monochromatic light. The table compares the angular resolu-

Resolution of plano-convex meniscus combinations

	Combination		Equivalent Plano-Convex	
Focal Ratio	PRODUCT NUMBERS	Angular Resolution (mrad)	PRODUCT NUMBER	Angular Resolution (mrad)
1.5	01 LPX 209 & 01 LMP 011	4.02	01 LPX 141	16.08
2.0	01 LPX 249 & 01 LMP 019	1.80	01 LPX 161	7.47
2.5	01 LPX 277 & 01 LMP 023	1.04	01 LPX 209	4.07
3.0	01 LPX 301 & 01 LMP 031	0.56	01 LPX 242	2.22

tion of a single plano-convex lens (operating at infinite conjugate and with monochromatic light) with a system composed of a plano-convex lens and a meniscus lens having the same f-number and effective focal length.

As can be seen, the plano-convex/meniscus lens combinations have about four times better resolution than the plano-convex lens acting alone. This is because the resolution in a plano-convex lens is roughly proportional to the reciprocal of the f-number cubed. To achieve a given focal length and f-number in the two lens systems, one starts with a plano-convex lens of higher f-number, and hence, better resolution. Since the meniscus lens doesn't add much aberration (it works near the aplanatic condition), the total system resolution is nearly that of the original lens. It is possible to use more than one meniscus lens at a time to compound these effects; one might construct a system consisting of a plano-convex lens followed by two different meniscus lenses. For example, a system consisting of an 01 LPX 277 an 01 LMP 023 and an 01 LMP 011 has a focal length of 64.5mm, operates at f/1.3, and has an angular resolution of 2.3mrad. In contrast, a single plano-convex lens (01 LPX 129) having nearly the same effective focal length and f-number has an angular resolution of 25.1mrad, over ten times worse. To determine the proper elements needed for such a system, one simply applies the previous formulas. The focal length of the first meniscus lens is found as before; to calculate the focal length of the second meniscus lens, the formulas are used again, but the focal length of the first meniscus/plano-convex lens combination is used as the value for f_o.

A variety of single and multilayer coatings is available for these lenses. See the preceding section on lenses (page 69).

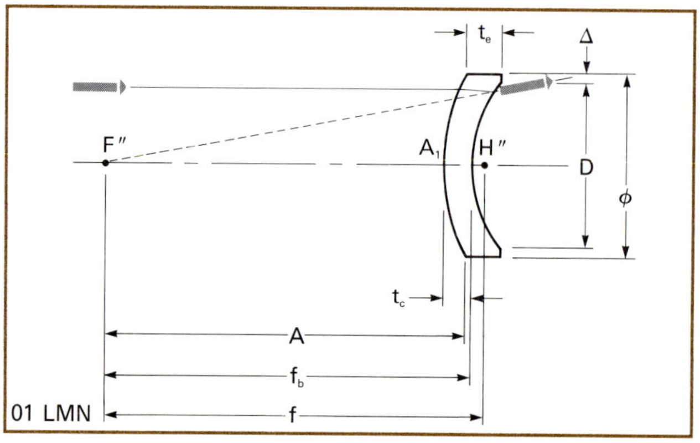

SPECIFICATIONS: MENISCUS GLASS LENSES

Paraxial Focal Length (f_o): ±5%
Design Wavelength: 546.07nm
Design Index (n_o): 1.5187 ±0.001
Centration: 3 arc minutes
Diameter (ϕ): ±0.25mm
Clear Aperture Diameter (D): $D=\phi-2\Delta$
Nominal Flat Annulus Width (Δ): +0%, -70%
  Δ=0.2mm for ϕ=25mm
  Δ=0.4mm for ϕ=50mm
Center Thickness (t_c): ±0.2mm
Material: BK 7, Grade A Fine Annealed
Cosmetic Surface Quality: Pitch polished, 80-50 scratch and dig
Circumference is ground and both edges have a slight protective bevel.

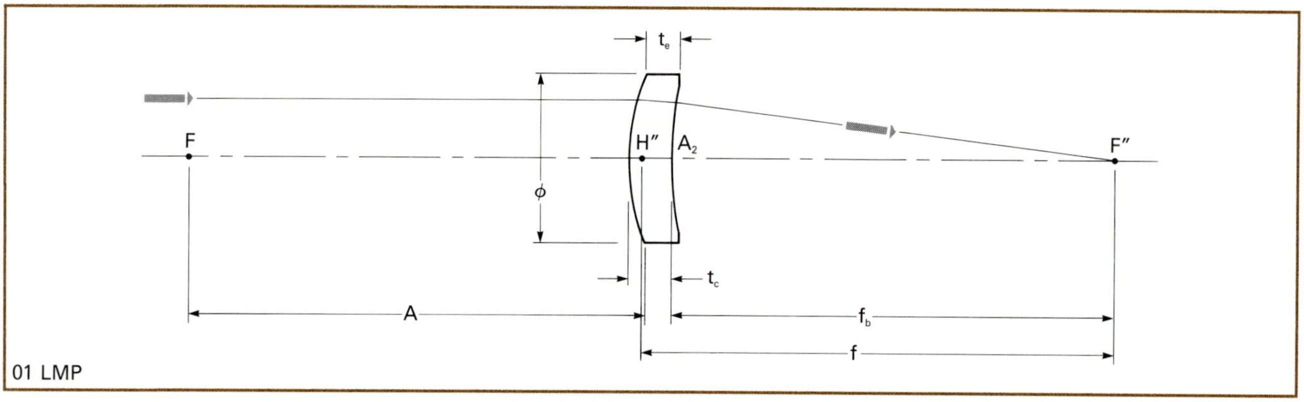

Meniscus Glass Lenses; Positive

f_o (mm)	ϕ (mm)	t_c (mm)	t_e (mm)	A_1H (mm)	A_2H'' (mm)	f_b (mm)	A (mm)	f_o/D	PRODUCT NUMBER
100.0	25.0	4.0	2.5	-1.5	-4.1	97.4	105.4	4.0	01 LMP 001
100.0	50.0	10.3	2.8	-3.8	-10.2	87.6	112.8	2.0	01 LMP 003
125.0	25.0	4.0	2.8	-1.9	-4.5	121.7	130.2	5.0	01 LMP 005
125.0	50.0	8.0	2.3	-3.7	-8.7	114.3	136.0	2.5	01 LMP 007
150.0	25.0	4.0	3.0	-2.3	-4.9	146.2	155.2	6.0	01 LMP 009
150.0	50.0	7.0	2.4	-3.9	-8.3	139.6	160.0	3.0	01 LMP 011
175.0	25.0	4.0	3.1	-2.7	-5.2	171.1	180.7	7.0	01 LMP 013
175.0	50.0	7.0	3.2	-4.6	-9.0	163.5	184.4	3.5	01 LMP 015
200.0	25.0	3.5	2.8	-2.7	-4.9	196.8	206.0	8.0	01 LMP 017
200.0	50.0	6.5	3.2	-4.9	-8.9	188.0	208.7	4.0	01 LMP 019
250.0	25.0	3.5	2.9	-1.3	-3.6	246.3	252.1	10.0	01 LMP 021
250.0	50.0	4.8	2.3	-1.7	-4.8	244.3	254.7	5.0	01 LMP 023
300.0	25.0	3.5	3.0	-1.5	-3.8	296.1	302.3	12.0	01 LMP 025
300.0	50.0	4.8	2.7	-2.1	-5.2	293.9	304.5	6.0	01 LMP 027
350.0	25.0	3.5	3.1	-1.8	-4.1	345.7	352.3	14.0	01 LMP 029
350.0	50.0	4.5	2.8	-2.3	-5.2	343.6	354.2	7.0	01 LMP 031
400.0	25.0	3.5	3.1	-2.0	-4.3	395.4	402.5	16.0	01 LMP 033
400.0	50.0	4.5	3.0	-2.6	-5.5	393.1	404.1	8.0	01 LMP 035
500.0	25.0	3.5	3.2	-2.5	-4.8	495.0	503.0	20.0	01 LMP 037
500.0	50.0	4.5	3.3	-3.3	-6.2	492.2	504.1	10.0	01 LMP 039
600.0	25.0	3.5	3.2	-1.2	-3.5	594.8	599.8	24.0	01 LMP 041
600.0	50.0	4.0	3.0	-1.4	-4.0	594.0	600.9	12.0	01 LMP 043
800.0	25.0	3.5	3.3	-1.6	-3.9	793.9	799.6	32.0	01 LMP 045
800.0	50.0	4.0	3.2	-1.8	-4.4	792.9	800.4	16.0	01 LMP 047
1000.0	25.0	3.5	3.3	-2.0	-4.3	992.9	999.5	40.0	01 LMP 049
1000.0	50.0	4.0	3.4	-2.3	-4.9	991.8	1000.1	20.0	01 LMP 051
1500.0	25.0	3.5	3.4	-3.0	-5.3	1490.7	1499.2	60.0	01 LMP 053
1500.0	50.0	4.0	3.6	-3.4	-6.0	1489.0	1499.4	30.0	01 LMP 055
2000.0	25.0	3.5	3.4	-4.0	-6.3	1988.5	1999.0	80.0	01 LMP 057
2000.0	50.0	4.0	3.7	-4.5	-7.2	1986.1	1998.6	40.0	01 LMP 059
4000.0	25.0	3.5	3.5	-8.0	-10.2	3981.3	3999.6	160.0	01 LMP 061
4000.0	50.0	4.0	3.9	-9.1	-11.7	3974.8	3996.2	80.0	01 LMP 063

All lenses may be edged to meet your diameter requirements.

Meniscus Glass Lenses; Negative

f_o (mm)	ϕ (mm)	t_c (mm)	t_e (mm)	A_1H (mm)	A_2H'' (mm)	f_b (mm)	A (mm)	D (mm)	f_o/D	PRODUCT NUMBER
-100.0	25.0	3.0	4.5	3.2	1.1	-99.1	97.0	24.6	-4.0	01 LMN 001
-125.0	25.0	3.0	4.2	3.4	1.4	-123.6	121.5	24.6	-5.0	01 LMN 005
-150.0	25.0	3.0	4.0	3.7	1.7	-149.4	147.3	24.6	-6.0	01 LMN 009
-150.0	50.0	3.0	7.6	3.7	1.7	-149.4	149.9	49.2	-3.0	01 LMN 011
-175.0	25.0	3.0	3.8	4.0	2.0	-174.4	172.3	24.6	-7.0	01 LMN 013
-175.0	50.0	3.0	6.8	4.0	2.0	-174.4	175.0	49.2	-3.5	01 LMN 015
-200.0	25.0	3.0	3.7	4.3	2.3	-199.5	197.4	24.6	-8.0	01 LMN 017
-200.0	50.0	3.0	6.3	4.3	2.3	-199.5	200.1	49.2	-4.0	01 LMN 019
-250.0	25.0	3.0	3.6	3.1	1.1	-248.5	245.9	24.6	-10.0	01 LMN 021
-250.0	50.0	3.0	5.4	3.1	1.1	-248.5	246.9	49.2	-5.0	01 LMN 023
-300.0	25.0	3.0	3.5	3.3	1.3	-298.3	295.6	24.6	-12.0	01 LMN 025
-300.0	50.0	3.0	5.0	3.3	1.3	-298.3	296.6	49.2	-6.0	01 LMN 027
-350.0	25.0	3.0	3.4	3.5	1.5	-348.0	345.3	24.6	-14.0	01 LMN 029
-350.0	50.0	3.0	4.7	3.5	1.5	-348.0	346.3	49.2	-7.0	01 LMN 031
-400.0	25.0	3.0	3.4	3.7	1.7	-397.2	394.5	24.6	-16.0	01 LMN 033
-400.0	50.0	3.0	4.5	3.7	1.7	-397.2	395.5	49.2	-8.0	01 LMN 035
-500.0	25.0	3.0	3.3	4.2	2.2	-496.5	493.9	24.6	-20.0	01 LMN 037
-500.0	50.0	3.0	4.2	4.2	2.2	-496.5	494.9	49.2	-10.0	01 LMN 039
-600.0	25.0	3.0	3.2	3.0	1.0	-596.8	593.9	24.6	-24.0	01 LMN 041
-600.0	50.0	3.0	4.0	3.0	1.0	-596.8	594.3	49.2	-12.0	01 LMN 043
-800.0	25.0	3.0	3.2	3.3	1.4	-795.6	792.7	24.6	-32.0	01 LMN 045
-800.0	50.0	3.0	3.7	3.3	1.4	-795.6	793.1	49.2	-16.0	01 LMN 047
-1000.0	25.0	3.0	3.1	3.7	1.7	-994.5	991.6	24.6	-40.0	01 LMN 049
-1000.0	50.0	3.0	3.6	3.7	1.7	-994.5	992.0	49.2	-20.0	01 LMN 051
-1500.0	25.0	3.0	3.1	4.5	2.6	-1491.5	1488.6	24.6	-60.0	01 LMN 053
-1500.0	50.0	3.0	3.4	4.5	2.6	-1491.5	1489.0	49.2	-30.0	01 LMN 055
-2000.0	25.0	3.0	3.1	5.4	3.4	-1986.9	1984.0	24.6	-80.0	01 LMN 057
-2000.0	50.0	3.0	3.3	5.4	3.4	-1986.9	1984.4	49.2	-40.0	01 LMN 059
-4000.0	25.0	3.0	3.0	8.8	6.8	-3969.9	3967.0	24.6	-160.0	01 LMN 061
-4000.0	50.0	3.0	3.2	8.8	6.8	-3969.9	3967.4	49.2	-80.0	01 LMN 063

Lens holders are available to suit all Melles Griot lenses. *(See page 452.)*

OPTICAL FABRICATION

The Melles Griot optical fabrication facility is located in the Hsinchu Science Based Industrial Park, Taiwan, R.O.C. It is one of the most modern optical facilities in the world, regularly producing over half a million optical components per year.

We employ precision, high-speed polishing methods to assure you of a superior and consistent standard of optical components.

Our flat work shop uses both continuous polishers and double-sided polishers to produce some of the best planar optics available.

The modern optical coating facility, staffed by graduate engineers, produces high quality multilayer coatings for laser applications.

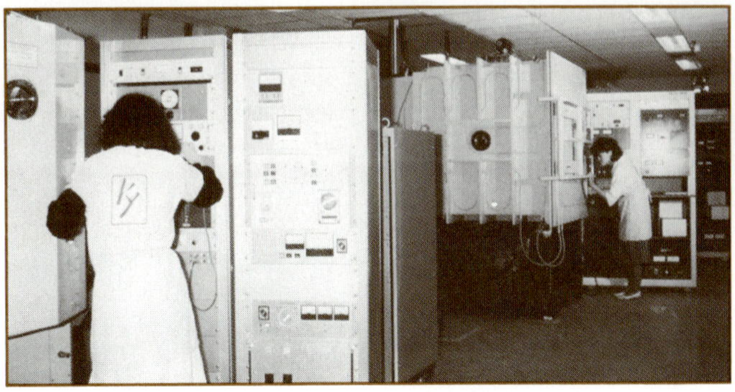

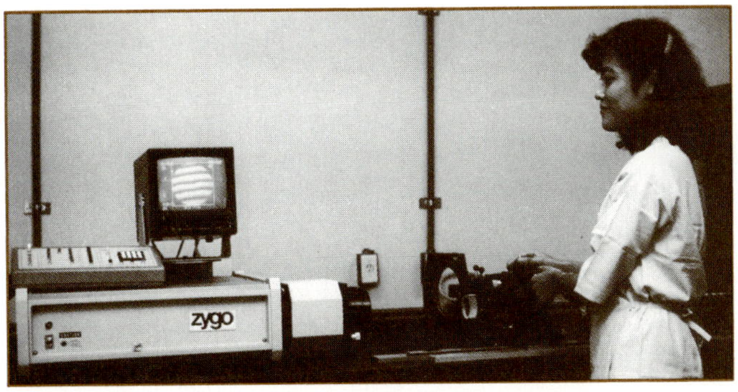

Our fully equipped quality control department assures you that your components will meet your demands first time and every time.

PLANO-CONVEX SYNTHETIC FUSED SILICA LENSES

Synthetic fused silica lenses are intended primarily (but not exclusively) for ultraviolet applications. Ultraviolet transmittance of synthetic fused silica is far superior to that of any glass. For example, in a sample thickness of 10mm, the wavelengths corresponding to 80% external transmittance are respectively 336nm, 243nm, and 175nm for BK 7, optical quality synthetic fused silica, and UV grade synthetic fused silica. Synthetic fused silica also has thermal properties superior to glasses. For a comprehensive presentation of the properties of this material, see the materials section on pages 51-68.

Synthetic fused silica lenses are offered in two distinct qualities of material. "UV grade" offers the highest possible ultraviolet and visible transmittance. Combining this material with suitable antireflection coatings (see earlier discussions of coatings; pages 69-87), external transmittances in excess of 99% can be achieved in 10mm sample thickness for any chosen visible wavelength. "Optical quality" synthetic fused silica exhibits considerable batch-to-batch variation in ultraviolet transmittance, and generally poorer ultraviolet transmittance, due to uncontrollable metallic impurity content at the parts per billion level. The transmittance data earlier presented for "optical quality" material is actually worst case data. Most specimens will show much better transmittance, but not nearly as good as UV grade. If you want a guarantee of consistently the very best ultraviolet transmittance, you must specify UV grade. For reasons explained elsewhere, cylindrical lenses are offered in UV grade exclusively.

A variety of antireflection coatings are available for synthetic fused silica lenses via the coating suffix system.

An edge grinding service is available if change of shape or size is desired.

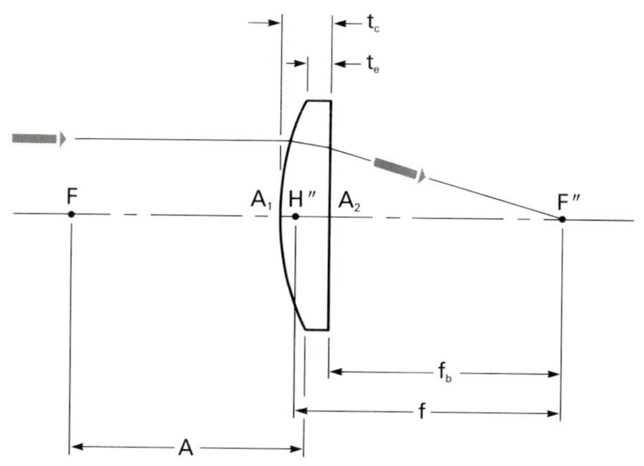

01 LQF/LQP

SPECIFICATIONS: PLANO-CONVEX SYNTHETIC FUSED SILICA LENSES

Paraxial Focal Length (f_o): ±5%
Design Wavelength (λ_o): 546.1nm
Design Index (n_o): 1.46008 ±0.0001 for Optical Quality
±0.00003 for UV Grade
Diameter (ϕ): ±0.15mm
Centration: 3 arc minutes
Center Thickness (t_c): ±0.2mm
Cosmetic Surface Quality: Pitch polished, 80-50 scratch and dig
Circumference is ground and both edges have a slight protective bevel.

Much of our business is in special optical components not listed in this catalog. Our optical designers are able to assist with your special needs and our production facilities in California and the Republic of China are ready to meet your quantity requirements.

Plano-Convex Synthetic Fused Silica Lenses

									PRODUCT NUMBER	
f_o (mm)	ϕ (mm)	t_c (mm)	t_e (mm)	A_1H (mm)	A_2H'' (mm)	f_b (mm)	A (mm)	f_o/ϕ	Optical Quality	UV Grade
10.0	5.0	2.7	2.0	0	-1.8	8.2	10.7	2.00	01 LQF 005	01 LQP 147
15.0	7.5	3.1	2.0	0	-2.1	12.9	16.1	2.00	01 LQF 011	01 LQP 155
25.0	10.0	3.1	2.0	0	-2.1	22.9	26.1	2.50	01 LQF 016	01 LQP 166
25.0	15.0	4.8	2.0	0	-3.3	21.7	27.8	1.67	01 LQF 023	01 LQP 177
50.0	25.0	5.7	2.0	0	-3.9	46.1	53.7	2.00	01 LQF 028	01 LQP 001
50.0	40.0	13.7	2.1	0	-9.4	40.6	61.6	1.25	01 LQF 036	—
75.0	25.0	4.4	2.1	0	-3.0	72.0	77.3	3.00	01 LQF 045	01 LQP 003
75.0	40.0	8.4	2.0	0	-5.8	69.2	81.4	1.88	01 LQF 055	—
75.0	50.0	13.8	3.1	0	-9.5	65.5	85.7	1.50	01 LQF 066	01 LQP 005
100.0	25.0	3.7	2.0	0	-2.5	97.5	101.7	4.00	01 LQF 078	01 LQP 007
100.0	40.0	6.6	2.0	0	-4.5	95.5	104.6	2.50	01 LQF 111	—
100.0	50.0	10.4	3.0	0	-7.1	92.9	107.4	2.00	01 LQF 116	01 LQP 009
100.0	75.0	22.5	3.1	0	-15.4	84.6	119.4	1.33	01 LQF 123	—
125.0	25.0	3.4	2.0	0	-2.3	122.7	126.4	5.00	01 LQF 128	01 LQP 011
150.0	25.0	3.1	2.0	0	-2.1	147.9	151.1	6.00	01 LQF 136	01 LQP 013
150.0	50.0	7.7	3.0	0	-5.3	144.7	154.7	3.00	01 LQF 145	01 LQP 015
150.0	75.0	14.1	3.0	0	-9.7	140.3	161.1	2.00	01 LQF 155	—
175.0	25.0	3.0	2.0	0	-2.1	172.9	176.0	7.00	—	01 LQP 017
175.0	50.0	7.0	3.0	0	-4.8	170.2	179.0	3.50	—	01 LQP 019
200.0	25.0	2.9	2.0	0	-2.0	198.0	200.9	8.00	01 LQF 222	01 LQP 021
200.0	50.0	6.5	3.0	0	-4.5	195.5	203.5	4.00	01 LQF 227	01 LQP 023
250.0	25.0	2.7	2.0	0	-1.8	248.2	250.7	10.00	01 LQF 235	01 LQP 025
250.0	50.0	5.8	3.1	0	-4.0	246.0	252.8	5.00	01 LQF 244	01 LQP 027
250.0	75.0	9.3	3.0	0	-6.4	243.6	256.3	3.33	01 LQF 249	—
300.0	25.0	2.6	2.0	0	-1.8	298.2	300.6	12.00	01 LQF 259	01 LQP 029
300.0	50.0	5.3	3.0	0	-3.6	296.4	302.3	6.00	01 LQF 277	01 LQP 033
500.0	25.0	2.3	2.0	0	-1.6	498.4	500.3	20.00	01 LQF 299	01 LQP 035
500.0	50.0	4.4	3.0	0	-3.0	497.0	501.4	10.00	01 LQF 337	01 LQP 037
500.0	75.0	6.1	3.0	0	-4.2	495.8	503.1	6.67	01 LQF 346	—
1000.0	25.0	2.2	2.0	0	-1.5	998.5	1000.2	40.00	—	01 LQP 039
1000.0	50.0	3.7	3.0	0	-2.5	997.5	1000.7	20.00	—	01 LQP 041

See page 55 for the exceptional properties of Synthetic Fused Silica.

SYMMETRIC-CONVEX FUSED SILICA LENSES

Symmetric-convex synthetic fused silica lenses provide extreme transparency over a wide spectral range. By combining these optics with our laser line "V"-coatings, external transmittances of 9.5% can be obtained.

Like plano-convex lenses, these have positive focal lengths and form both real and virtual images. Symmetry is assured by finishing both surfaces with the same tool. Because of symmetry, several aberrations almost exactly cancel at unit conjugate ratio ($s=s''=2f$). These include coma, distortion, and lateral chromatic aberration. At this conjugate ratio spherical aberration is the minimum possible with a single element, and symmetric-convex lenses are by definition best-form. These aberrations grow as conjugate ratios depart from unity.

Symmetric-convex lenses are recommended for virtual imaging of real objects, and for positive conjugate ratios from about 0.2 up to about 5. Beyond 5 (and between 0 and 0.2) plano-convex lenses are preferable. These values are only approximate, and are wavelength sensitive. To make the best choice between symmetric convex and plano-convex lenses for your particular choice of wavelength and conjugate ratio, consult the earlier tabulated values of "decision parameters."

A large variety of single and multilayer antireflection coatings is available for these lenses. See the preceding section on coatings (page 69).

Diameters of all lenses may be reduced ("edged") by us to meet your requirements. Larger diameters and focal lengths are available on special request.

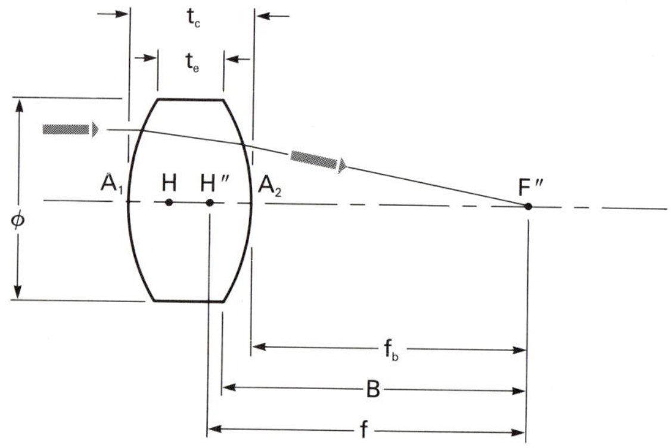

01 LQB/LQD

SPECIFICATIONS: SYMMETRIC-CONVEX SYNTHETIC FUSED SILICA LENSES

Paraxial Focal Length (f_o): ±5%
Design Wavelength (λ_o): 546.1nm
Design Index (n_o): 1.46008 ±0.0001 for Optical Quality
±0.00003 for UV Grade
Centration: 3 arc minutes
Diameter (ϕ): ±0.15mm
Center Thickness (t_c): ±0.2mm
Cosmetic Surface Quality: Pitch polished, 80-50 scratch and dig
Circumference is ground and both edges have a slight protective bevel.

Symmetric-Convex Synthetic Fused Silica Lenses

f_o (mm)	ϕ (mm)	t_c (mm)	t_e (mm)	A_1H (mm)	A_2H'' (mm)	f_b (mm)	A=B (mm)	f_o/ϕ	PRODUCT NUMBER Optical Quality	UV Grade
10.0	5.0	2.7	2.0	1.0	1.0	9.0	9.4	2.00	01 LQB 005	—
10.0	10.0	5.4	2.0	2.1	-2.1	7.9	9.6	1.00	01 LQB 011	—
25.0	10.0	3.1	2.0	1.0	-1.1	23.9	24.5	2.50	01 LQB 016	—
25.0	15.0	4.6	2.0	1.6	-1.6	23.4	24.7	1.67	01 LQB 023	—
25.0	25.0	10.2	2.1	3.8	-3.8	21.2	25.3	1.00	01 LQB 028	01 LQD 001
50.0	25.0	5.5	2.0	1.9	-1.9	48.1	49.8	2.00	01 LQB 036	01 LQD 003
50.0	40.0	11.6	2.0	4.1	-4.1	45.9	50.7	1.25	01 LQB 045	—
50.0	50.0	19.2	3.1	7.1	-7.1	42.9	51.0	1.00	01 LQB 055	01 LQD 005
75.0	25.0	4.3	2.0	1.5	-9.5	73.5	74.7	3.00	01 LQB 066	01 LQD 007
75.0	50.0	12.7	3.0	4.5	-4.5	70.5	75.4	1.50	01 LQB 111	01 LQD 009
100.0	25.0	3.7	2.0	1.3	-1.3	98.7	99.6	4.00	01 LQB 116	01 LQD 011
100.0	40.0	6.5	2.0	2.3	-2.3	97.7	100.0	2.50	01 LQB 123	—
100.0	50.0	10.1	3.0	3.5	-3.5	96.5	100.0	2.00	01 LQB 128	01 LQD 013
100.0	75.0	19.7	3.1	7.0	-7.0	93.0	101.3	1.33	01 LQB 136	—
150.0	25.0	3.1	2.0	1.1	-1.1	148.9	149.5	6.00	01 LQB 145	01 LQD 015
150.0	50.0	7.6	3.0	2.6	-2.6	147.4	149.7	3.00	01 LQB 155	01 LQD 017
150.0	75.0	13.6	3.0	4.7	-4.7	145.3	150.5	2.00	01 LQB 166	—
200.0	25.0	2.9	2.0	1.0	-1.0	199.0	199.4	8.00	01 LQB 178	01 LQD 019
200.0	50.0	6.4	3.0	2.2	-2.2	197.8	199.5	4.00	01 LQB 222	01 LQD 021
250.0	25.0	2.7	2.0	0.9	-0.9	249.1	249.4	10.00	01 LQB 227	—
250.0	50.0	5.7	3.0	2.0	-2.0	248.0	249.4	5.00	01 LQB 235	—
250.0	75.0	9.2	3.0	3.2	-3.2	246.8	249.9	3.33	01 LQB 244	—
300.0	25.0	2.6	2.0	0.9	-0.9	299.1	299.4	12.00	01 LQB 249	01 LQD 023
300.0	50.0	5.3	3.0	1.8	-1.8	298.2	299.3	6.00	01 LQB 259	01 LQD 025
500.0	25.0	2.3	2.0	0.8	-0.8	499.2	499.4	20.00	01 LQB 277	01 LQD 027
500.0	50.0	4.4	3.0	1.5	-1.5	498.5	499.2	10.00	01 LQB 299	01 LQD 029
500.0	75.0	6.1	3.0	2.1	-2.1	497.9	499.4	6.67	01 LQB 337	—
1000.0	25.0	2.2	2.0	0.8	-0.8	999.2	999.3	40.00	—	01 LQD 033
1000.0	50.0	3.7	3.0	1.3	-1.3	998.7	999.1	20.00	—	01 LQD 035

Our large inventory of stock lenses enables us to ship most items within 24 hours.

PLANO-CONCAVE SYNTHETIC FUSED SILICA LENSES

Plano-concave synthetic fused silica lenses provide greater UV and IR transmission and an increased thermal operating range. Like the glass lenses previously described, they have a negative focal length and arc often used to expand beams or to increase focal lengths.

The plano-concave lens offers very nearly minimum spherical aberration at negative-infinite or near-infinite conjugate ratios with the concave surface facing the longest conjugate distance. A collimated beam incident on the curved surface will be transformed into a divergent beam emergent from the plane surface. The divergent beam will appear to emerge from a smaller (more nearly diffraction-limited) virtual point source than if the plane surface had faced toward the collimated beam.

Reversing the rays in the above description, a convergent beam is transformed with least additional spherical aberration into a collimated beam if the convergent beam is incident on the plane surface. The word "additional" is used as the convergent beam may already contain spherical aberration, and depending on its sign, the plano-concave lens may either increase or decrease the total for a lens combination. Thus, in some applications, the opposite orientation of an element may reduce the overall aberration of a lens combination.

A variety of single and multilayer antireflection coatings is available for these lenses. See the preceding section on coatings (page 69). When a reflective coating is deposited on the concave lens surface an excellent concave mirror is formed. A variety of reflective and partially reflective coatings will also be found in the coating suffix system.

Diameters of all lenses may be reduced ("edged") by us to meet your requirements.

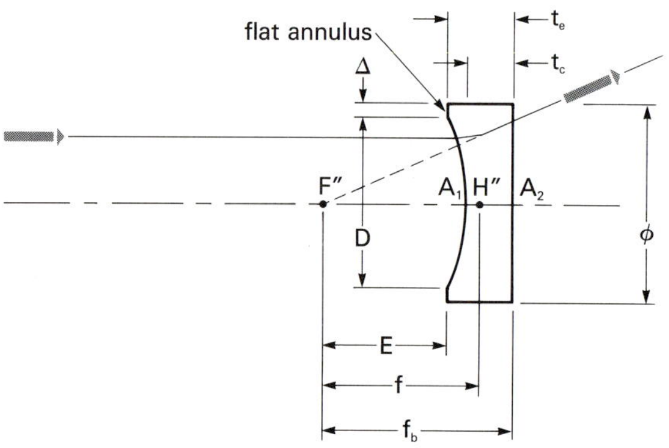

01 LQT/LQS

SPECIFICATIONS: PLANO-CONCAVE SYNTHETIC FUSED SILICA LENSES

Paraxial Focal Length (f_o): ±5%
Design Wavelength (λ_o): 546.1nm
Design Index (n_o): 1.46008 ±0.0001 for Optical Quality
±0.00003 for UV Grade
Centration: 3 arc minutes
Diameter (ϕ): ±0.15mm
Clear Aperture Diameter (D): $D = \phi - 2\Delta$
Nominal Flat Annulus Width (Δ): +0%, -70%
Δ=0.2mm for $\phi \leq$ 40mm
Δ=0.4mm for $\phi >$ 40mm
Center Thickness (t_c): ±0.2mm
Cosmetic Surface Quality: Pitch polished, 80-50 scratch and dig
Circumference and flat annulus are ground, and both edges have a slight protective bevel.

We offer a wide selection of optical hardware for use in your experimental and prototype systems (see pages 403-500).

Plano-Concave Synthetic Fused Silica Lenses

f_o (mm)	ϕ (mm)	t_c (mm)	t_e (mm)	A_1H (mm)	A_2H'' (mm)	f_b (mm)	E (mm)	f_o/ϕ	D (mm)	PRODUCT NUMBER Optical Quality	UV Grade
-10.0	5.0	2.0	2.6	0	-1.4	-11.4	8.8	-2.17	4.6	01 LQT 000	01 LQS 000
-15.0	7.5	2.0	3.0	0	-1.4	-16.4	13.4	-2.11	7.1	01 LQT 001	01 LQS 001
-25.0	10.0	2.0	3.0	0	-1.4	-26.4	23.3	-2.60	9.6	01 LQT 002	01 LQS 002
-50.0	25.0	2.0	5.6	0	-1.4	-51.4	45.8	-2.03	24.6	01 LQT 003	01 LQS 003
-75.0	25.0	2.0	4.3	0	-1.4	-76.4	72.1	-3.05	24.6	01 LQT 004	01 LQS 004
-100.0	25.0	2.0	3.7	0	-1.4	-101.4	97.7	-4.07	24.6	01 LQT 005	01 LQS 005
-100.0	50.0	3.0	10.1	0	-2.1	-102.1	91.9	-2.03	49.2	01 LQT 006	01 LQS 006
-150.0	25.0	2.0	3.1	0	-1.4	-151.4	148.3	-6.10	24.6	01 LQT 007	01 LQS 007
-150.0	50.0	3.0	7.5	0	-2.1	-152.1	144.5	-3.05	49.2	01 LQT 008	01 LQS 008
-200.0	25.0	2.0	2.8	0	-1.4	-201.4	198.5	-8.13	24.6	01 LQT 009	01 LQS 009
-200.0	50.0	3.0	6.3	0	-2.1	-202.1	195.7	-4.07	49.2	01 LQT 011	01 LQS 011
-250.0	25.0	2.0	2.7	0	-1.4	-251.4	248.7	-10.16	24.6	01 LQT 012	01 LQS 012
-250.0	50.0	3.0	5.7	0	-2.1	-252.1	246.4	-5.08	49.2	01 LQT 013	01 LQS 013
-300.0	25.0	2.0	2.5	0	-1.4	-301.4	298.8	-12.20	24.6	01 LQT 014	01 LQS 014
-300.0	50.0	3.0	5.2	0	-2.1	-302.1	296.8	-6.10	49.2	01 LQT 015	01 LQS 015
-500.0	25.0	2.0	2.3	0	-1.4	-501.4	499.0	-20.33	24.6	01 LQT 016	01 LQS 016
-500.0	50.0	3.0	4.3	0	-2.1	-502.1	497.7	-10.16	49.2	01 LQT 017	01 LQS 017
-1000.0	25.0	2.0	2.2	0	-1.4	-1001.4	999.2	-40.65	24.6	01 LQT 018	01 LQS 018
-1000.0	50.0	3.0	3.7	0	-2.1	-1002.1	998.4	-20.33	49.2	01 LQT 019	01 LQS 019

SYMMETRIC-CONCAVE SYNTHETIC FUSED SILICA LENSES

Symmetric-concave synthetic fused silica lenses may be used in any application demanding the superior properties of fused silica together with the characteristics of a negative lens.

The symmetric-concave lens has a negative focal length, diverges collimated incident light, and forms only virtual images which are seen through the lens. It is often used to expand light beams or to increase focal lengths in existing systems, and is normally used in combination with other lenses. Among the many devices utilizing symmetric-concave lenses are laser beam expanders, optical character readers, viewers, and projection systems.

A variety of single and multilayer antireflection coatings is available for these lenses. See the preceding section on coatings (page 69).

Diameters of all lenses may be reduced ("edged") by us to meet your requirements.

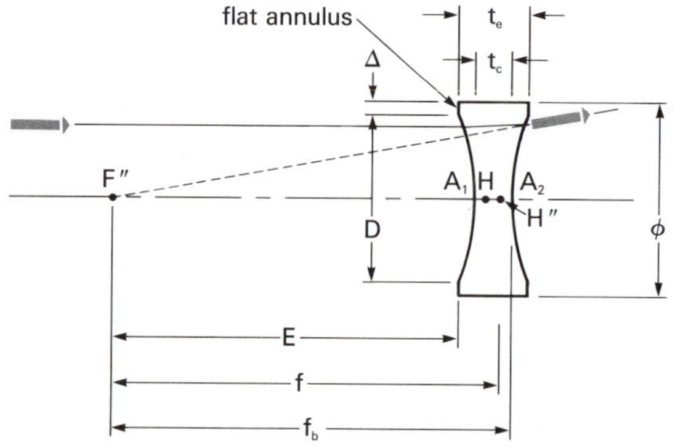

01 LQR/LQN

SPECIFICATIONS: SYMMETRIC-CONCAVE SYNTHETIC FUSED SILICA LENSES

Paraxial Focal Length (f_o): ±5%
Design Wavelength (λ_o): 546.1nm
Design Index (n_o): 1.46008 ±0.0001 for Optical Quality
±0.00003 for UV Grade
Centration: 3 arc minutes
Diameter (ϕ): ±0.15mm
Clear Aperture Diameter (D): $D = \phi - 2\Delta$
Nominal Flat Annulus Width (Δ): +0%, −70%
$\Delta = 0.2$mm for $\phi \leq 40$mm
$\Delta = 0.4$mm for $\phi > 40$mm
Center Thickness (t_c): ±0.2mm
Cosmetic Surface Quality: Pitch polished, 80-50 scratch and dig
Circumference and flat annulus are ground, and both edges have a slight protective bevel.

Call us with your special requirements. Our manufacturing facilities are ready to meet your needs.

Symmetric-Concave Synthetic Fused Silica Lenses

f_o (mm)	ϕ (mm)	t_c (mm)	t_e (mm)	A_1H (mm)	A_2H'' (mm)	f_b (mm)	E (mm)	f_o/ϕ	D (mm)	PRODUCT NUMBER Optical Quality	UV Grade
-10.0	5.0	2.0	2.6	0.7	-0.7	-10.7	8.4	-2.17	4.6	01 LQR 000	01 LQN 000
-15.0	7.5	2.0	2.9	0.7	-0.7	-15.7	13.2	-2.11	7.1	01 LQR 001	01 LQN 001
-25.0	10.0	2.0	3.0	0.7	-0.7	-25.7	23.2	-2.60	9.6	01 LQR 002	01 LQN 002
-50.0	25.0	2.0	5.3	0.7	-0.7	-50.7	47.0	-2.03	24.6	01 LQR 003	01 LQN 003
-75.0	25.0	2.0	4.2	0.7	-0.7	-75.7	72.6	-3.05	24.6	01 LQR 004	01 LQN 004
-100.0	25.0	2.0	3.6	0.7	-0.7	-100.7	97.9	-4.07	24.6	01 LQR 005	01 LQN 005
-100.0	50.0	3.0	9.7	1.0	-1.0	-101.0	94.7	-2.03	49.2	01 LQR 006	01 LQN 006
-150.0	25.0	2.0	3.1	0.7	-0.7	-150.7	148.1	-6.10	24.6	01 LQR 007	01 LQN 007
-150.0	50.0	3.0	7.4	1.0	-1.0	-151.0	145.8	-3.05	49.2	01 LQR 008	01 LQN 008
-200.0	25.0	2.0	2.8	0.7	-0.7	-200.7	198.3	-8.13	24.6	01 LQR 009	01 LQN 009
-200.0	50.0	3.0	6.3	1.0	-1.0	-201.0	196.4	-4.07	49.2	01 LQR 011	01 LQN 011
-250.0	25.0	2.0	2.7	0.7	-0.7	-250.7	248.7	-10.16	24.6	01 LQR 012	01 LQN 012
-250.0	50.0	3.0	5.6	1.0	-1.0	-251.0	246.7	-5.08	49.2	01 LQR 013	01 LQN 013
-300.0	25.0	2.0	2.5	0.7	-0.7	-300.7	298.4	-12.20	24.6	01 LQR 014	01 LQN 014
-300.0	50.0	3.0	5.2	1.0	-1.0	-301.0	296.9	-6.10	49.2	01 LQR 015	01 LQN 015
-500.0	25.0	2.0	2.3	0.7	-0.7	-500.7	498.5	-20.33	24.6	01 LQR 016	01 LQN 016
-500.0	50.0	3.0	4.3	1.0	-1.0	-501.0	497.4	-10.16	49.2	01 LQR 017	01 LQN 017
-1000.0	25.0	2.0	2.2	0.7	-0.7	-1000.7	998.6	-40.65	24.6	01 LQR 018	01 LQN 018
-1000.0	50.0	3.0	3.7	1.0	-1.0	-1001.0	997.7	-20.33	49.2	01 LQR 019	01 LQN 019

SAPPHIRE LENSES

Sapphire is a marvelous optical material. It has high transmission over an enormous spectral range, reaching from 150 nanometers to over 6 microns in the mid-infrared. It is extremely hard and can be cleaned repeatedly with no concern about damage or scratching. It is chemically inert and insoluble in almost everything except at extremely high temperatures. The resistance of sapphire to radiation browning and luminescence in a reactor environment makes this material useful for optical systems in in-core applications. The simple lenses described below will be of use to optical systems which must perform in difficult or hazardous conditions.

Because sapphire is a monotonic cubic crystal, it is slightly birefringent. The index difference between the ordinary and extraordinary rays (n_o–n_e) is typically less than 0.008. Stress birefringence also may be induced in sapphire by overly rigid mounting and constraining techniques. Further optical properties and a table of refractive index will be found in the materials section of the catalog. Please see page 64.

The raw aniostropic crystalline synthetic sapphire used in these lenses is manufactured by a newly developed technique which produces higher quality and better purity material than has previously been available. The single crystal growing technique is a modified Verneuil process which avoids contact between the molten material and the crucible during growth. Powdered (1 to 2 micron) aluminum oxide is fed through a downward burning gas flame. The molten material is collected at a seed crystal which has been placed at the tip of a corundum rod. By computer controlled motion of the corundum rod and the flow rate of the aluminum oxide, a single sapphire crystal may be grown to quite large dimensions.

Because of its great hardness, sapphire is a difficult material to use in optical fabrication. Diamond or boron nitride abrasive grains must be used for both the grinding and polishing processes. Historically polishing process has been the most difficult aspect of production; sleeks and haze have often been apparent in even the most carefully processed material. A new finishing technique has been developed for these lenses which results in a good, clean optical finish with a scratch and dig specification of at least 80-50.

Because of the relatively high index of sapphire (1.77) antireflection coatings will be found to be beneficial to reduction of Fresnel losses at the surface. HEBBAR™ and "V"-coatings are available for the lenses by affixing the Coating Suffix when ordering.

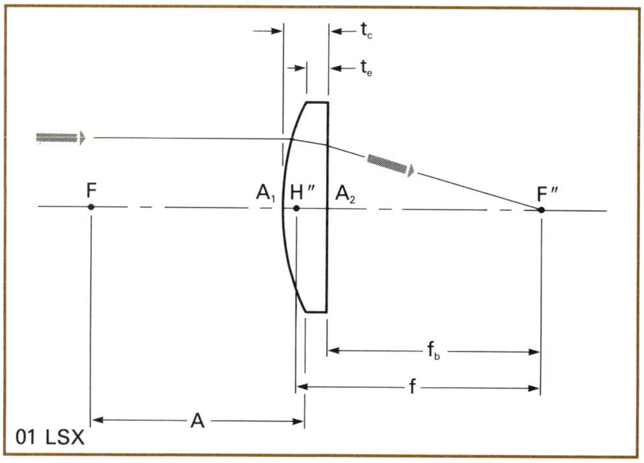

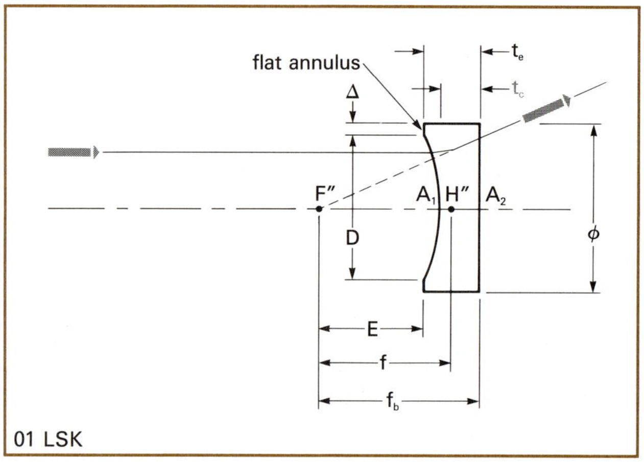

SPECIFICATIONS: SYNTHETIC SAPPHIRE LENSES

Paraxial Focal Length (f_o): ±5%
Design Wavelength (λ_o): 546.1nm
Design Index (n_o): ±0.001
Birefringence (n_e-n_o): 0.008 throughout the visible spectrum
Stress Birefringence: >1 fringe per cm thickness at 632.8nm
Centration: 3 arc minutes
Diameter (ϕ): ±0.25mm
Center Thickness (t_c): ±0.2mm
Material: Aniostropic crystalline synthetic sapphire (Al_2O_3)
Cosmetic Surface Quality: Diamond polished, 80-50 scratch and dig
Circumference is ground and both edges have a slight protective bevel.

Plano-Convex Synthetic Sapphire Lenses

f_o (mm)	ϕ (mm)	t_c (mm)	t_e (mm)	A_1H (mm)	A_2H'' (mm)	f_b (mm)	A (mm)	f_o/ϕ	PRODUCT NUMBER
5.0	5.0	2.9	2.0	0	-1.6	3.4	5.9	1.0	01 LSX 001
10.0	5.0	2.4	2.0	0	-1.4	8.6	10.4	2.0	01 LSX 003
10.0	10.0	3.8	2.0	0	-2.1	7.9	11.8	1.0	01 LSX 005
20.0	10.0	2.8	2.0	0	-1.6	18.4	20.8	2.0	01 LSX 007
20.0	20.0	5.7	2.0	0	-3.2	16.8	23.7	1.0	01 LSX 009
25.0	10.0	2.7	2.0	0	-1.5	23.5	25.7	2.5	01 LSX 011
25.0	25.0	6.6	2.0	0	-3.7	21.3	29.6	1.0	01 LSX 013
50.0	20.0	3.3	2.0	0	-1.9	48.1	51.3	2.5	01 LSX 015
50.0	25.0	4.1	2.0	0	-2.3	47.7	52.1	2.0	01 LSX 017
100.0	20.0	2.7	2.0	0	-1.5	98.5	100.7	5.0	01 LSX 019
100.0	25.0	3.0	2.0	0	-1.7	98.3	101.0	4.0	01 LSX 023

Plano-Concave Synthetic Sapphire Lenses

f_o (mm)	ϕ (mm)	t_c (mm)	t_e (mm)	A_1H (mm)	A_2H'' (mm)	f_b (mm)	E (mm)	f_o/ϕ	D	PRODUCT NUMBER
-5.0	5.0	2.0	2.8	0	-1.1	-6.1	3.3	-1.1	4.6	01 LSK 001
-10.0	5.0	2.0	2.4	0	-1.1	-11.1	8.7	-2.2	4.6	01 LSK 003
-10.0	10.0	2.0	3.7	0	-1.1	-11.1	7.4	-1.0	9.6	01 LSK 005
-20.0	10.0	2.0	2.8	0	-1.1	-21.1	18.3	-2.1	9.6	01 LSK 007
-20.0	20.0	2.0	5.5	0	-1.1	-21.1	15.6	-1.1	19.6	01 LSK 009
-25.0	10.0	2.0	2.6	0	-1.1	-26.1	23.5	-2.6	9.6	01 LSK 011
-25.0	25.0	2.0	6.4	0	-1.1	-26.1	19.7	-1.0	24.6	01 LSK 013
-50.0	20.0	2.0	3.3	0	-1.1	-51.1	47.8	-2.6	19.6	01 LSK 015
-50.0	25.0	2.0	4.0	0	-1.1	-51.1	47.1	-2.0	24.6	01 LSK 017
-100.0	20.0	2.0	2.6	0	-1.1	-101.1	98.5	-5.1	19.6	01 LSK 019
-100.0	25.0	2.0	3.0	0	-1.1	-101.1	98.1	-4.1	24.6	01 LSK 023

For the optical properties of sapphire see page 64.

LASER FOCUSING AND RECEIVER SINGLETS

These lenses are intended for focusing laser beams down to the smallest and most intense spots possible with spherical singlets. When used at optimum f-number these lenses will contribute peak-to-peak wavefront distortions less than $\lambda/10$ over the central 80% of the beam. They have spherical surfaces, and are best form (the shape which simultaneously minimizes spherical aberration and surface reflection loss) for each wavelength. Furthermore, these lenses are designed to accommodate the most common laser beam diameters at precisely optimum f-number. In accordance with the previous discussion of spot size, it follows that with uniform illumination of the intended aperture, the focal spot size at the 84% throughput contour approaches 4/3 times the diffraction limit. BK 7 is used for these singlets.

Applications of these lenses are wide ranging including: micromachining, pumping of other laser systems, production of nonlinear effects, production of stimulated Raman and Brillouin scattering, production of multiphoton processes, laser-induced damage resistance studies, aerosol particle counting and sizing, video disc recording and playback, and isotope separation.

The large lenses optimized for 47mm aperture are well suited for use as laser receiver and Raman backscatter receivers. They may be used with our 50mm round interference filters (see page 279) which have a 48mm coated apertuer specification.

The procedure for designing these lenses was related to the earlier discussion of optimum f-number and minimum spot size. For Gaussian laser beams it is convenient to choose the optimized aperture D_{opt} to be the diameter of the 99% throughput contour, which is 3/2 times the "$1/e^2$" irradiance contour diameter. If a central zone of diameter D_{opt} were subject to *uniform* plane wave illumination, the minimum singlet focal spot radius in microns would be approximately given by

$$r_{min} = \frac{4000}{3} \cdot \frac{1.22\lambda}{D_{opt}} \cdot f ,$$

where f, λ, and D_{opt} are all in millimeters. Such spot radii are shown in the accompanying table.

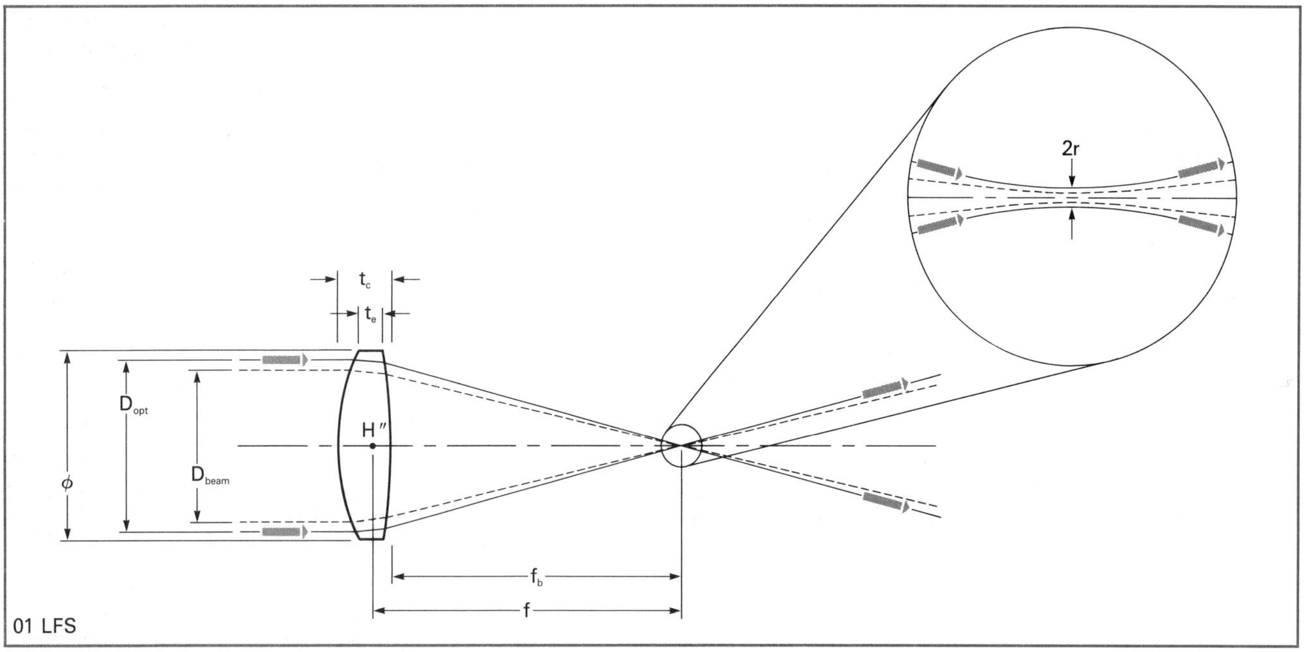

LASERLINE FOCUSING SINGLET. Showing the relationship between Gaussian beam size, diameter D_{opt} of the zone for which f-number is optimized, and diameter ϕ.

The lens diameters are oversized and have been chosen to allow for ease in handling and the use of retaining rings in mounting, thereby leaving the optimized area unobstructed. It is important that the input laser beam be accurately centered, and be of the approximate size for which the lens is intended. We have tabulated, in addition to D_{opt}, the "$1/e^2$" Gaussian laser beam diameters (D_{beam}) to which each singlet corresponds. These values assume 99% throughput for a circle of diameter D_{opt}. Our inventory contains singlets for the most common beam diameters of each laser type.

SPECIFICATIONS: LASER FOCUSING AND RECEIVER SINGLETS

Paraxial Focal Length (f): ±3%
Sphericity: λ/10 over 90% of diameter
Wavefront Distortion: <λ/10 over central 80% of diameter.
Centration: 3 arc minutes
Diameter (φ): ±0.15mm
Center Thickness (t_c): ±0.2mm
Materials: BK 7 Grade A, Fine Annealed
Surface Quality: Pitch polished, 40-20 scratch and dig
Circumference is ground, and lightly beveled.

Laser Focusing And Receiver Singlets

Wave-Length (mm)	Laser Type	D_{beam} ($1/e^2$) (mm)	D_{opt} (mm)	φ (mm)	f (mm)	f_b (mm)	t_c (mm)	t_e (mm)	Spot Radius (μm)	PRODUCT NUMBER
488.0	Ar	1.6	2.4	10	18	16.0	3.5	2.1	5.8	01 LFS 021
488.0	Ar	*	47.0	50	900	895.0	8.3	7.7	15.4	01 LFS 023
514.5	Ar	1.6	2.4	10	17	15.0	3.5	2.0	6.0	01 LFS 025
514.5	Ar	*	47.0	50	900	895.0	8.3	7.7	16.1	01 LFS 027
632.8	HeNe	0.7	1.0	5	5	3.2	3.1	1.5	5.2	01 LFS 035
632.8	HeNe	0.8	1.2	5	7	5.5	2.5	1.5	5.6	01 LFS 033
632.8	HeNe	1.0	1.5	5	9	7.7	2.2	1.5	6.0	01 LFS 037
632.8	HeNe	1.2	1.8	5	11	9.8	2.1	1.5	6.3	01 LFS 039
632.8	HeNe	2.0	3.0	10	22	20.0	3.2	2.0	7.5	01 LFS 041
632.8	HeNe	*	47.0	50	850	845.0	8.3	7.6	18.6	01 LFS 043

*Receiver Lenses; $1/e^2$ diameter not applicable.

We also offer a spatial filter with a selection of focusing optics suitable for Argon and Helium neon lasers. This may be attached directly to the mounting ring of a Melles Griot laser *(see page 362).*

LASER DIODE PLANO-CONVEX GLASS SINGLETS

These diffraction limited singlets have been optimized for use with laser diodes for focusing and collection applications.

Made from a high-index glass, these plano-convex singlets provide better levels of correction than are normally available over this wavelength range. Wavefront distortion is less than λ/5 at 830nm.

High-efficiency anti-reflection coatings are available for three specific wavelength ranges as noted below. These coatings will reduce reflection losses to less than 0.25% per surface at their design wavelengths.

For a complete range of laser diode optics see page 369.

COATING OPTIONS

The Coating Suffix system should be used to request antireflection coatings for these lenses. If no Coating Suffix is appended to the Product Number, the lenses will be supplied uncoated.

Antireflection Coatings; Normal Incidence

Wavelength (nm)	Maximum Reflectance (%)	COATING SUFFIX
780-830	0.25	/076
1300	0.25	/168
1550	0.25	/169

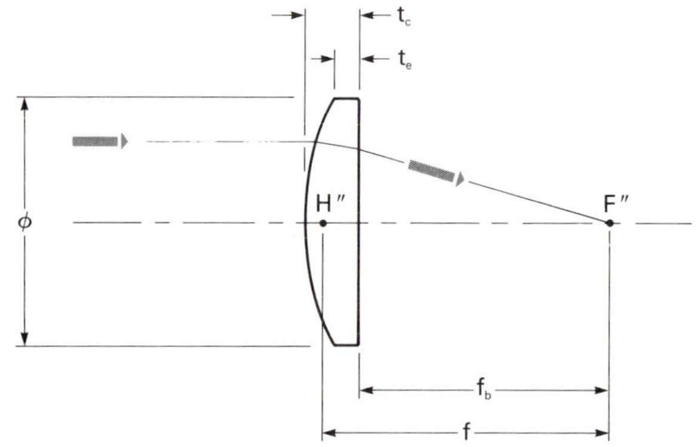

06 LXP

SPECIFICATIONS: LASER DIODE PLANO-CONVEX GLASS SINGLETS

Paraxial Focal Length: ±2%

Design Wavelength: 830nm

Wavelength Range: 780 to 1550nm

Refractive Index: $n_{830}=1.76312$, $n_{1300}=1.74888$, $n_{1550}=1.74525$

Wavefront Distortion: <λ/5 at 830nm

Centration: 3 arc minutes

Diameter (ϕ): ±0.25mm

Center Thickness (t_c): ±0.15mm

Materials: Schott SF 11, Grade A

Cosmetic Surface Quality: Pitch-polished, 80-50 scratch and dig

Coating: Must be specified by suffix

Circumference is ground and outside edges have a slight protective bevel.

Laser Diode Plano-Convex Glass Singlets

for λ=830nm						for λ=1300nm		for λ=1550nm		PRODUCT
f	f_b	f/ϕ	ϕ	t_c	t_e	f	f_b	f	f_b	NUMBER
15.0	13.9	5.0	3.0	2.0	1.9	15.3	14.1	15.4	14.2	06 LXP 001
25.0	23.6	5.0	5.0	2.5	2.3	25.5	24.0	25.6	24.2	06 LXP 003
50.0	48.3	6.3	8.0	3.0	2.8	51.0	49.2	51.2	49.5	06 LXP 005
65.0	63.0	6.5	10.0	3.5	3.3	66.2	64.2	66.6	64.6	06 LXP 007
110.0	107.8	4.0	15.0	4.0	3.7	112.1	109.8	112.6	110.3	06 LXP 009

Supplied uncoated. For antireflection coated lenses, append coating suffix from the coating options chart.

LASER DIODE SILICON SINGLETS

These silicon singlet lenses have been designed specifically for use with laser diodes at 1300 and 1550nm. At these wavelengths, the high index of refraction of silicon allows excellent aberration correction. All these lenses are diffraction limited and exhibit less than λ/10 of wavefront distortion at 1550nm.

Chromatic aberration cannot be corrected in singlets so these lenses have to be refocused when the wavelength domain is changed. These lenses are best-form (meniscus) for 1550nm wavelength and therefore minmize the effect of spherical aberration.

High-efficiency antireflection coatings are available for these lenses at two wavelengths (1300 and 1550nm). To request these coatings please append the coating suffix as specified below.

COATING OPTIONS

The Coating Suffix system should be used to request antireflection coatings for these lenses. If no Coating Suffix is appended to the Product Number the lenses will be supplied uncoated.

Antireflection coatings; Normal Incidence

Wavelength (nm)	Maximum Reflectance (%)	COATING SUFFIX
1300	0.25	/168
1550	0.25	/169

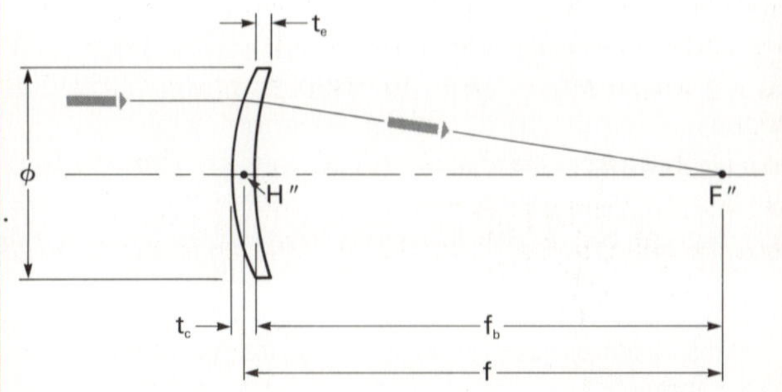

06 LSM

SPECIFICATIONS: LASER DIODE SILICON SINGLETS

Paraxial Focal Length: ±2%
Design Wavelength: 1550nm
Wavelength Range: 1300 to 1550nm
Wavefront Distortion: <λ/10 at 1550nm
Centration: 3 arc miniutes
Diameter (φ): ±0.25mm
Center Thickness (t_c): ±0.15mm
Material: Optical Grade Silicon
Cosmetic Surface Quality: Pitch-polished, 80-50 scratch and dig
Coating: Must be specified by suffix
Circumference is ground and outside edges have a slight protective bevel.

Laser Diode Silicon Singlets

for λ= 1550nm							for λ= 1300nm		PRODUCT
f	f_b	f/φ	φ	t_c	t_e	B	f	f_b	NUMBER
15.0	13.5	3.0	5	2.3	2.2	13.4	14.8	13.3	06 LSM 001
25.0	23.3	3.1	8	2.6	2.5	23.1	24.7	22.8	06 LSM 003
35.0	33.1	3.5	10	2.8	2.7	32.8	34.6	32.7	06 LSM 005
60.0	57.6	4.0	15	3.4	3.2	57.3	59.3	56.9	06 LSM 007
90.0	87.2	4.5	20	4.0	3.7	86.8	89.0	86.2	06 LSM 009
120.0	116.4	4.8	25	5.0	4.7	116.0	118.7	115.1	06 LSM 011

Supplied uncoated. For antireflection coated lenses, append coating suffix from the coating options chart.

COST-EFFECTIVE OPTICAL DESIGN

In optical systems engineering, there can be many advantages to using off-the-shelf components to build a compound lens system. Think about it!

- Lower design costs
- No tooling
- Rapid availability of prototypes
- Cost benefits of combining your production with existing high volume production.

That's why we've developed a new approach to optical system design.

Our engineers can combine any of our catalog lenses, vary the orientation and spacing, and determine system performance in terms of wavefront, spotsize or MTF. Furthermore, we can do this at any wavelength. We provide this information at a modest cost so that you can be assured that the system will work.

With this information, you can build prototypes, try alternatives and cut your development cycle drastically.

When your product's ready to go into production, you can be assured that we will be ready to meet your volume requirements at a competitive price. Why re-invent the lens for each new application — Try the Melles Griot approach — call us.

DOUBLETS AND TRIPLETS

Doublets and triplets are two- and three-element lenses, respectively. They are the simplest forms of multi-element lens which have been designed to improve on the performance available from singlets. In this section, we introduce our very important selection of computer-optimized cemented doublets and triplets.

These lenses have been used in many complex imaging systems with great success. They offer a stock solution to a wide variety of optical problems. Please don't hesitate to call our Applications Engineers for assistance in applying these lenses to your situation.

Contents **Page Number**

COMPUTER-OPTIMIZED ACHROMATIC DOUBLETS

Melles Griot offers two families of computer-optimized spherical achromats:

1. Precision Computer-Optimized Achromats (LAO series)
2. Spherical Achromats (LAU series)

The pages which follow explain optimized achromats, outline the optimization procedure, and present the characteristics of the two Melles Griot optimized achromat families. Modulation transfer function data is used to illustrate the advantages of the optional optimized field flatteners, and to document achromat superiority over singlets.

WHAT IS AN ACHROMATIC DOUBLET?

An achromatic doublet is a closely spaced, and often cemented, combination of positive and negative elements of differing refractive index. These elements are chosen so that the chromatic aberration is cancelled at two distinct and well separated wavelengths. These wavelengths are typically selected to be in the red and blue portions of the spectrum. At intermediate and nearby wavelengths, the focal length varies slightly from the value for the chosen red and blue wavelengths. This variation is negligible for most purposes, and a single focal length value is applied to the entire visible spectrum. Achromat focal lengths are specified for an intermediate wavelength, usually in the yellow or green portion of the spectrum. At this intermediate wavelength, the focal length is shorter by about 1 part in 2000 than it is at the selected red and blue wavelengths. By comparison, the focal length difference for a BK 7 singlet between these same red and blue wavelengths is typically 30 or 40 times larger than for an achromat. The 1 part in 2000 difference for the achromat is known as "secondary color" aberration, and is the result of fundamental glass property limitations. Positive achromatic doublets are usually composed of a positive low-index ("crown") and negative high-index ("flint") element.

It is generally true of achromats intended for use at infinite conjugate ratio that the most steeply curved surface (of the crown element) should face the infinite conjugate. The most nearly flat surface (of the flint element) should face the preferred focus. Often the surface facing the infinite conjugate is exposed to weather, abuse, and frequent cleaning. It is appropriate that this surface be on the harder and more stain resistant crown element.

Principles of achromat design are widely discussed in textbooks, and will not be presented here. With achromatization complete, considerable freedom remains with which to suppress monochromatic aberrations. Good achromats, and especially optimized achromats, closely satisfy the Abbe sine condition at infinite conjugate ratio. This means that they are very nearly free of both spherical aberration and coma. Optimized achromats are therefore vastly superior to simple singlet lenses for monochromatic applications at any visible wavelength. For example, they are well suited for low power laser beam manipulation.

WHAT IS A COMPUTER-OPTIMIZED ACHROMATIC DOUBLET?

A computer-optimized achromatic doublet is one which has been designed with the aid of a computer for the best possible performance under a certain set of conditions. In this procedure the computer is used to determine that particular combination of material dispersion curves, elemental thicknesses, and surface radii which will result in the best possible polychromatic field-averaged MTF curves (for the LAO series) or minimum possible field-averaged geometric blur sizes (for the LAU series). The best MTF and minimum blur criteria result in significantly different prescriptions.

The computer is also used to find that combination of manufacturing tolerances which must be satisfied if the lens is to perform in accordance with the full potential of the optimized prescription. Careful manufacture and testing assure that these tolerances are met.

WHY COMPUTER-OPTIMIZED ACHROMATS ARE FAR SUPERIOR TO SIMPLE LENSES FOR BOTH MONOCHROMATIC AND POLYCHROMATIC TASKS

1. In their correct orientation and at their intended (infinite) conjugate ratio, all Melles Griot computer-optimized achromats in the LAU and LAO series very closely satisfy the Abbe sine condition and are therefore almost completely free of both spherical aberration and coma.
2. Freedom from spherical aberration means that focal length is independent of aperture. Simple lenses require considerable refocusing as aperture is varied, but computer-optimized achromats do not.
3. Imagery with Melles Griot computer-optimized achromats of the LAU and LAO series is much sharper (fine and more contrasty) than the best possible simple lens imagery, *even under monochromatic conditions!*

   As a specific example, we may compare the performance of computer-optimized achromat 01 LAO 123 with the performance of a precisely corresponding (same focal length, same diameter) plano-convex BK 7 singlet in the preferred orientation. The spatial frequency at the intercept of the monochromatic (green) on-axis MTF curve with the MTF = 0.1 level is more than 15 times higher (and is possibly as much as 25 times higher) for the achromat than for the singlet. This frequency is indicative of the size of the finest detail that can be seen in the images. Thus the achromat can reveal details 15 to 25 times finer than the corresponding singlet can. It is also true polychromatically that the achromat can reveal details more than 15 times finer than the singlet can monochromatically.

   The area beneath the monochromatic on-axis MTF curve of this achromat is more than 20 times the area beneath the corresponding MTF curve of the corresponding singlet. Polychromatically, the ratio of areas is more than 13. Information content of the achromat image is correspondingly increased.
4. Optimized achromat imagery is typically sharper at full aperture than the best possible imagery with a simple lens of the same focal length, even if the simple lens is used at its D_{opt} value.
5. Optimized achromats have very high throughput, image brightness, and information rate advantages over simple lenses. D_{opt} for an achromat is much larger than D_{opt} for a simple lens of equal focal length. The advantage is proportional to the square of the ratio of these D_{opt} values. For many optimized achromats, D_{opt} approaches full aperture.
6. Freedom from coma in Melles Griot computer-optimized achromats means that their off-axis performance is far superior to that of simple lenses. Optimized spherical achromats can maintain almost uniform performance throughout a field of up to 2.5 degrees angular radius, while simple lens spot sizes under optimum conditions have been shown in graphs presented earlier to increase by as much as 4.5 times at 2.5 degrees.
7. Optimized achromats form good polychromatic images under circumstances in which simple lens imagery is totally unacceptable. Within the visible spectrum image position for an optimized achromat is typically 50 or more times less sensitive to wavelength change than is image position for a corresponding simple lens. Even for a really drastic wavelength change, as from the middle of the visible spectrum out to 1.06 microns in the infrared, optimized achromat focal length typically increases by amounts ranging from only 0.5% to 1.0%.
8. Melles Griot optimized achromats of the LAO series are so good as to be field flattener compatible, opening the door to even further dramatic increases in image quality. Computer-optimized companion field flatteners are available from stock.

COMPUTER-OPTIMIZED FIELD FLATTENERS

Field flatteners are lenses used near the focal or image plane of a companion element or system, where they may partially correct aberrations (principally but not exclusively classical field curvature), by reshaping the image surface without drastically affecting EFL, image size, and image location. Field flatteners are plano lenses, with their curved surfaces facing their companion achromatic doublets. Contrary to Petzval theorem predictions, optimized field flatteners may be either positive or negative. Field flatteners are unable to compensate either spherical aberration or coma, and therefore flatteners are most effective with lenses that are already exceptionally well corrected. The slightest trace of residual spherical aberration or coma drastically reduces field flattener effectiveness.

Because of their close proximity to the image plane, field flatteners must have a good surface finish and be kept scrupulously clean. Their surfaces and any dirt they may support are always in or near focus. Antireflection coatings, which make the surfaces more difficult to see, are highly desirable. Accordingly, all Melles Griot field flatteners are given HEBBAR™ coatings on both surfaces. Other coatings can be requested via the Coating Suffix system previously described.

Each optimized field flattener corresponds to a specific doublet, and will not function effectively with a different doublet. Melles Griot field flatteners are truly optimized. Their prescriptions are the result of optimization at infinite conjugate ratio with the doublet prescriptions serving as constraints. Such flattener prescriptions differ widely from Petzval theorem predictions. While slightly improved performance would have resulted from optimizing the 3-element system, instead of separately optimizing doublet and flattener, such a system optimization would have precluded use of the doublet alone. Melles Griot offers superb achromatic doublets for use alone, and offers the option of further significantly improving the field performance by use of a field flattener.

COATING OPTIONS

If you do not specify a particular coating you will receive the following coatings:

Coating Type	Lens Series
Single-layer MgF_2	LAO series
HEBBAR™	All field flatteners
Uncoated	LAU series

The Coating Suffix system should be used to request combinations of lenses and coatings different from those shown above. To obtain one of the listed coatings, simply append the Coating Suffix to the Product Number of the object to be coated, and indicate which surfaces are to be coated. (See the coatings section on page 69 for details.)

PRECISION COMPUTER-OPTIMIZED ACHROMATIC DOUBLETS

These are cemented spherical achromatic doublets which are optimized to have very good flat field MTF. These achromats are manufactured to test plate specifications, meaning that their surfaces are checked for power and irregularity by interference methods in direct comparison with precision master test plates. Tight tolerances are maintained in centration and in all dimensions. This assures that the achromats actually perform in accordance with the full potential of their optimized prescriptions. Positions of both focal points and both principal points are accurately known and are essentially independent of f-number, since spherical aberration is very well corrected. Instruments can be designed around these lenses with confidence that little mechanical adjustment will be required when the lenses are installed.

These Precision Optimized Achromats are specifically optimized for use at full aperture, alone, and on a flat image surface at infinite conjugate ratio (as demanded by the vast majority of applications). Companion optimized field flatteners are available off-the-shelf for most of these achromats. For the lowest f-numbers and shortest focal lengths, field flatteners are precluded by size requirements, or the presence at full aperture of some residual aberration.

Consistent with their design constraints and materials, these cemented achromats are as good as it is possible for cemented doublets to be. No better materials could have been chosen without both increasing costs and introducing other unfavorable properties. The principal residual aberrations are secondary color and curvature of field, both of which are due to material and doublet limitations. These aberrations are unimportant in tasks such as laser beam manipulation. The optional field flatteners allow field curvature to be largely eliminated. Spherical aberration is insignificant except at the very lowest f-numbers. Coma is insignificant. Within the specified 2.5 degree angular radius field, astigmatism plays only a minor role.

INCOHERENT MTF FOR A TYPICAL OPTIMIZED ACHROMAT

The graphs which follow serve to illustrate typical MTF data for the Precision Optimized Spherical Achromat series.

The first graph compares the performance of 01 LAO 123 with the performance of a BK 7 spherical plano-convex singlet of the same diameter (26.5mm) and focal length (100.0mm), demonstrating the overwhelming superiority of the achromat.

As can be seen, the advantage of the achromat remains overwhelming by comparison with the highest index singlets. The present comparison is with a BK 7 singlet. A doubling of the singlet geometric cutoff frequency would not be significant.

The second graph demonstrates the performance of the combination of achromat and field flattener by comparison with the achromat alone.

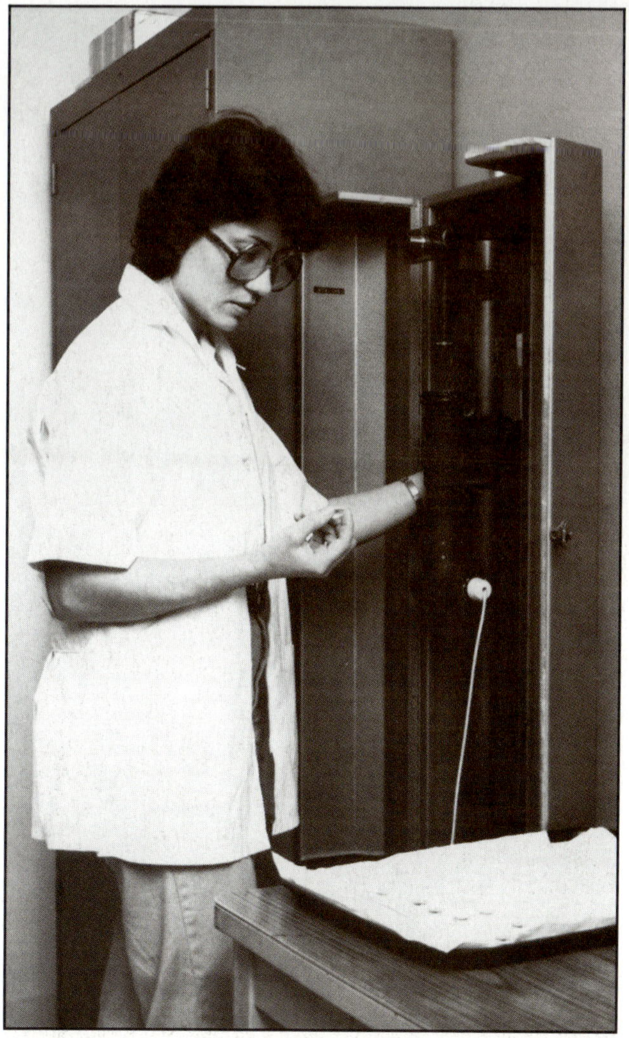

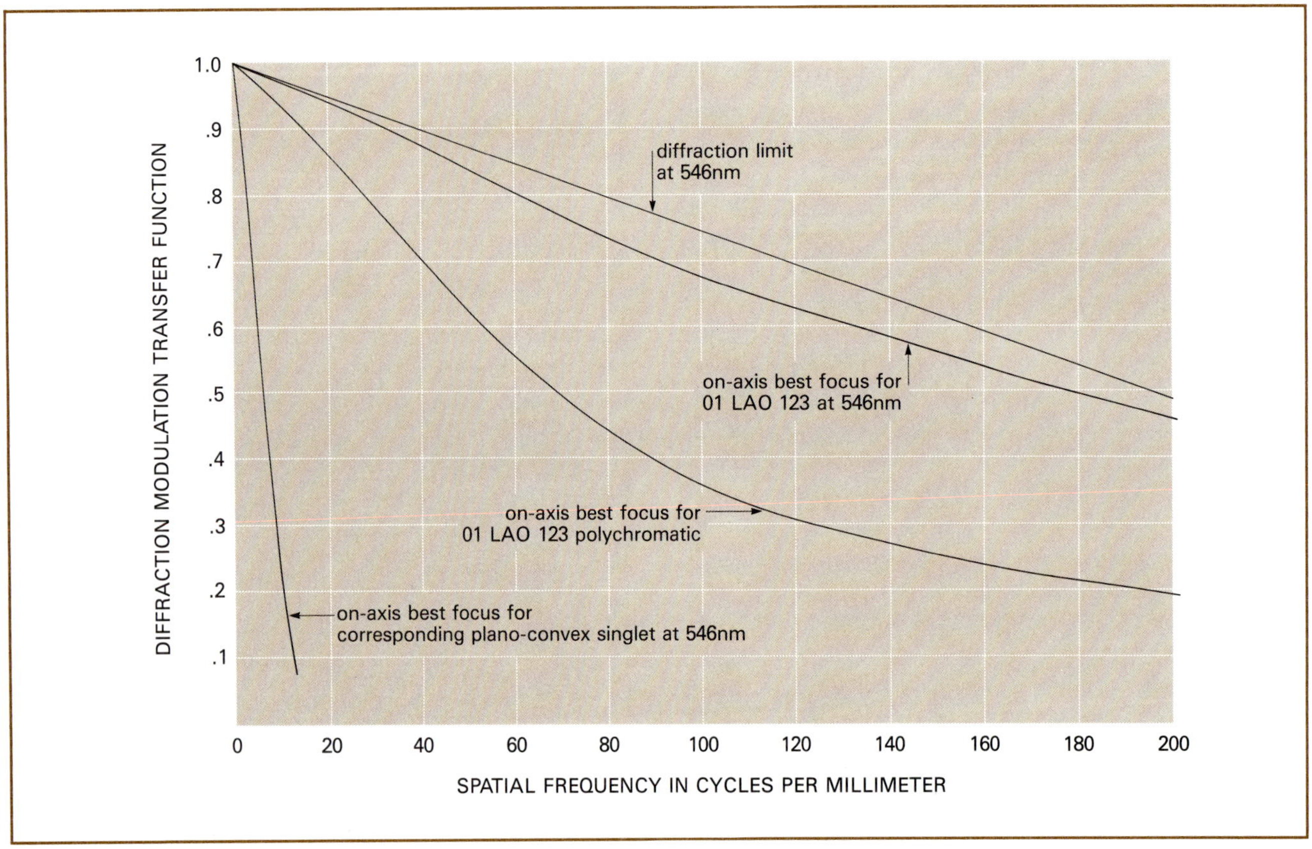

COMPARISON OF ON-AXIS VISIBLE MTF'S FOR ACHROMAT AND SINGLET

Both lenses have same diameter and same focal length. MTF's are evaluated on the respective planes of best on-axis focus, to show both lenses at their best. The achromat curves are diffraction MTF's, while the singlet curve is a geometric MTF. Because of the severity of spherical aberration in the singlet at this focal ratio (3.77 for both singlet and achromat), the singlet monochromatic and polychromatic curves are almost identical. In fact the same singlet curve suffices for all visible and infrared wavelengths out to 1.06 microns, with the geometric cut-off frequencies varying by less than 1/20th cycle per millimeter.

The area beneath the polychromatic achromat curve exceeds the area beneath the monochromatic singlet curve by more than a factor 13, while the area beneath the monochromatic achromat curve exceeds that beneath the monochromatic singlet curve by more than a factor 20. There are corresponding increases in image information content. Most of the difference between the two achromat curves can be attributed to the inevitable doublet secondary color aberration. Notice that the monochromatic curve for the achromat very closely approaches the diffraction limit curve in spite of the low focal ratio (3.77).

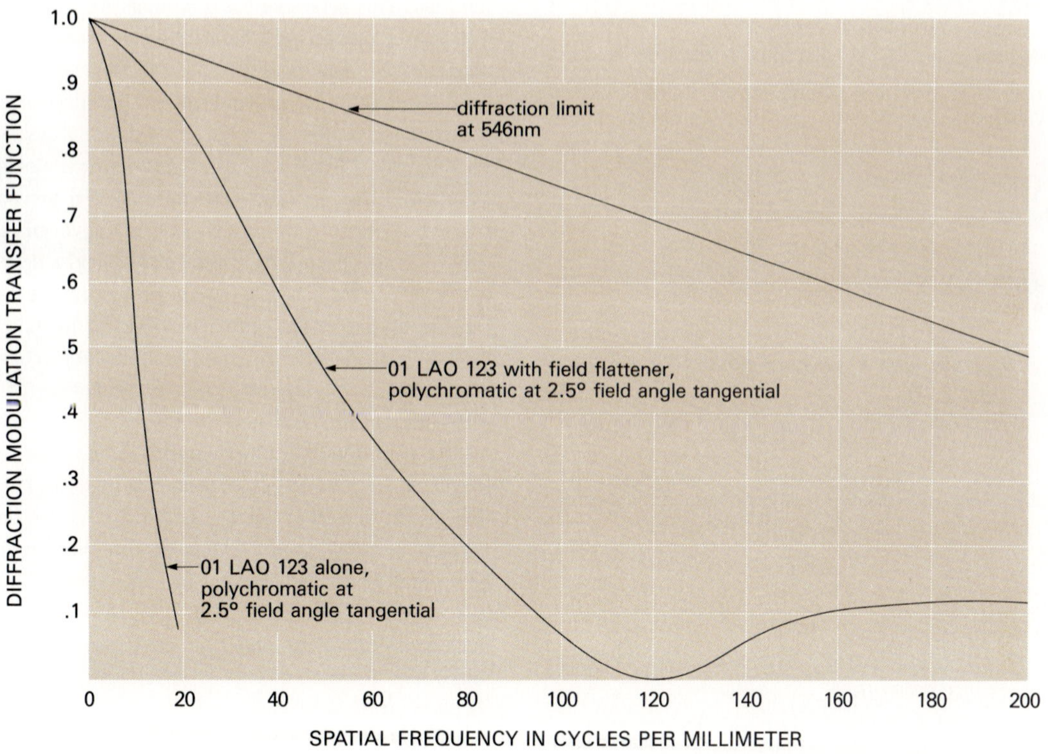

EFFECT OF FIELD FLATTENER ON POLYCHROMATIC MTF OF OPTIMIZED ACHROMATS

Field angle is 2.5°. The lower or flat field MTF curve shows the achromat at its worst, on the compromise plane of best field-averaged focus, at the field boundary, and for the orientation in the image plane for which astigmatism is most severe. Comparison with the singlet curve in the previous figure reveals that, even under these most unfavorable conditions, the achromat is better than the singlet at its very best. In fact the area beneath the worst case polychromatic curve of the achromat is more than 1.4 times the area under the best possible monochromatic curve for the singlet. This is what we mean by overwhelming achromat superiority.

The upper curve, for the combination of achromat and field flattener, is evaluated for a different compromise plane of best field-averaged focus for the combination. As before, all field angles up to and including 2.5° are included in the averaging process. Because the true focal surface of the combination is much more nearly flat than the focal surface of the achromat alone, there is considerable improvement in MTF at the field margin on the new compromise plane. In fact the flattener increases the area beneath the curve by a factor five, and there is a corresponding increase in image information content in the neighborhood of this field angle contour. It is presumed that the field flattener is correctly positioned. Ordinarily the position of the compromise plane is found experimentally. For purposes of MTF calculation the position of this plane is supplied by the optimization program.

OPTIMUM APERTURES FOR PRECISION OPTIMIZED ACHROMATS

An approximate value for D_{opt} can be found by graphing the sum of geometric and diffraction spot size terms, and looking for the value of D at which the curve shows a minimum. In this Guide we evaluate the geometric term at the plane of best on-axis focus with the position of this plane allowed to vary slightly as a function of D, as would be the case in an experimental determination of D_{opt}. The values found in this way are compared below with the corresponding plano-convex singlet values which were tabulated in the earlier discusion of spot size.

Comparison of D_{opt} values for Precision Optimized Achromats and Singlets of same EFL

f	D_{opt} (mm)	
(mm)	Achromat	Singlet
10	2.2	1.60
100	17.5	8.99
1000	>80.0	50.5

WAVELENGTH INSENSITIVITY OF PRECISION OPTIMIZED ACHROMAT SERIES

The following graphs show the variation of focal length and principal point position of a precision optimized achromat (01 LAO 123) as functions of wavelength between 0.3 and 1.2 microns. This achromat has a nominal focal length of 100mm at 546.1nm. Comparison focal length data for a plano-convex BK 7 singlet of the same nominal focal length is shown. Here it can be seen that the focal lengths for the shortest and longest of the three design wavelengths are only approximately equal, but that both focal lengths differ from the center design wavelength focal length by about 1 part in 2000. The optimization process has slightly relaxed the achromatization constraint in favor of further suppression of the monochromatic aberrations, and a superior achromat has resulted. The classical requirement of precise equality of focal lengths at the extreme design wavelengths is excessive, and impairs overall performance of un-optimized achromats.

Because of the superior imaging properties of the precision series achromats, there is often interest in using them outside the visible spectrum. We are frequently asked how much the focal length of an LAO series achromat changes in moving from the mid-visible spectrum out to the Nd:YAG laser line wavelength (1.064 microns) in the infrared. The answer is somewhere between 0.584% and 0.939% increase over the published green line nominal value. This range of variation results from the fact that several different glass type combinations are utilized in the series. For this same change of wavelength, variations in BFL and secondary principal point position also seem to be in the range of between 0.5% and 1.0% of their respective green line nominal values.

INFRARED MTF OF PRECISION OPTIMIZED SPHERICAL ACHROMAT SERIES

Studies of 01 LAO 123 (the example lens chosen for earlier graphic presentation of MTF data) show that the full aperture monochromatic MTF at 1.06 microns in the infrared is comparable to the visible polychromatic MTF. Infrared monochromatic MTF improves for wavelength approaching the visible, but does not reach MTF values as high as the monochromatic green line curve.

Our Applications Engineers will be happy to provide data on the properties of any of our achromats at any wavelength.

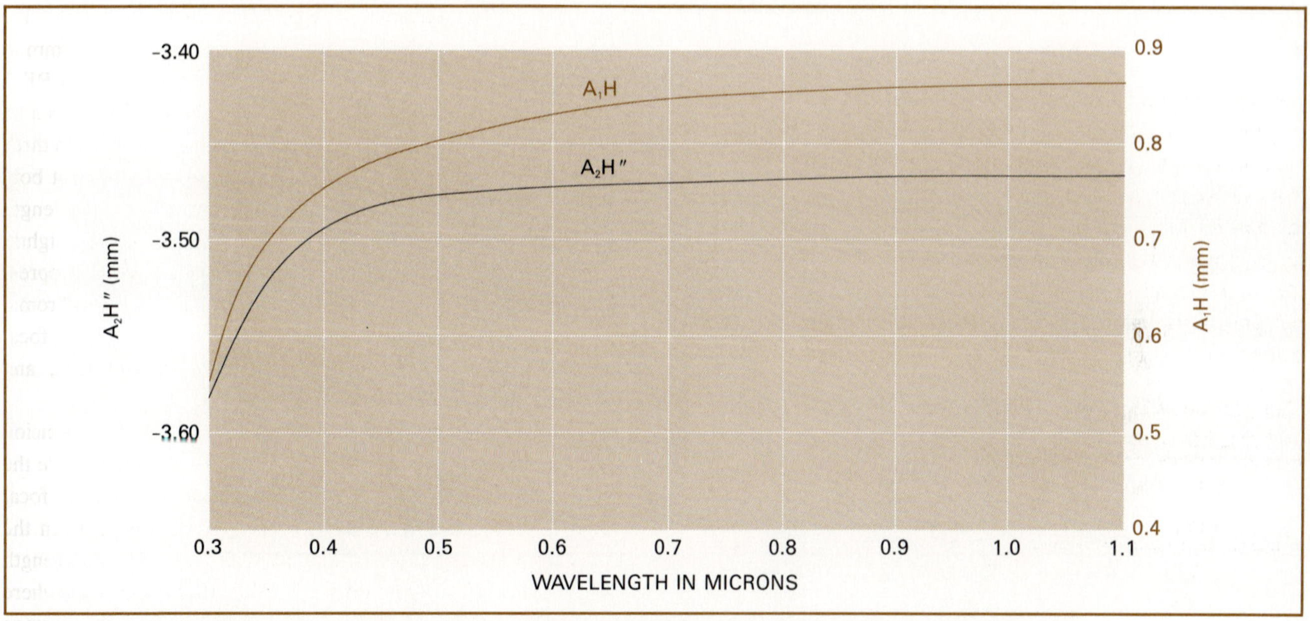

WAVELENGTH DEPENDENCE OF PRINCIPAL POINT POSITION for Nominal Achromat 01 LAO 123.

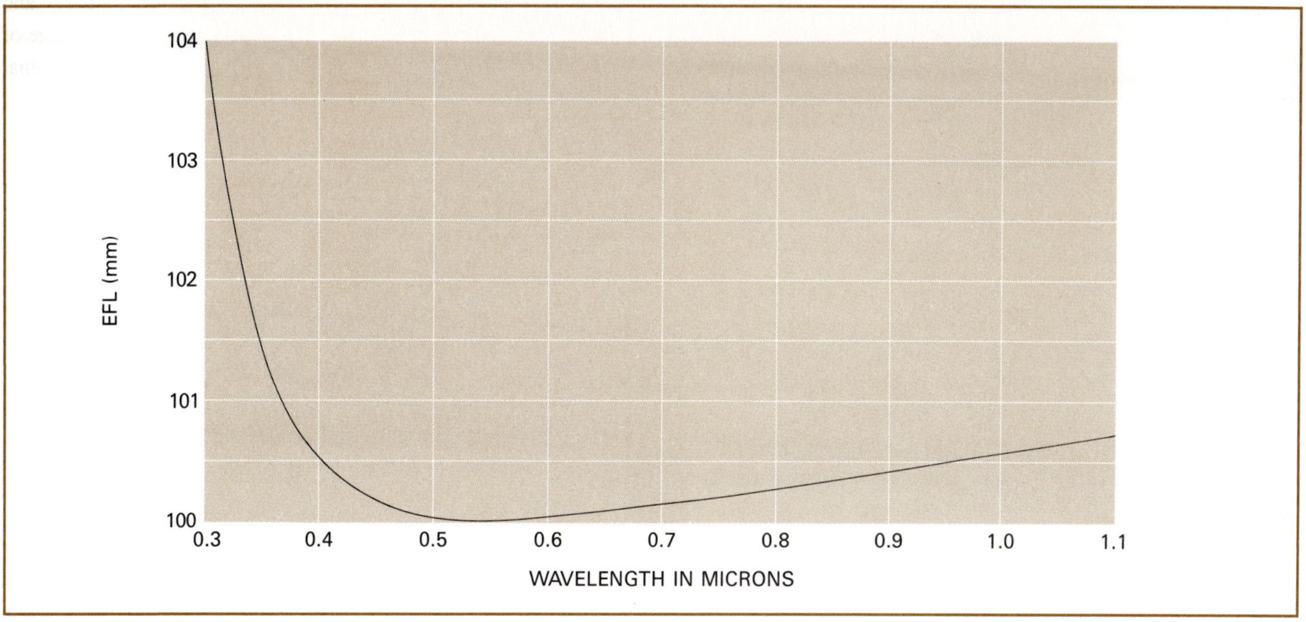

WAVELENGTH DEPENDENCE OF EFFECTIVE FOCAL LENGTH for Nominal Achromat 01 LAO 123.

PRECISION-OPTIMIZED ACHROMATS

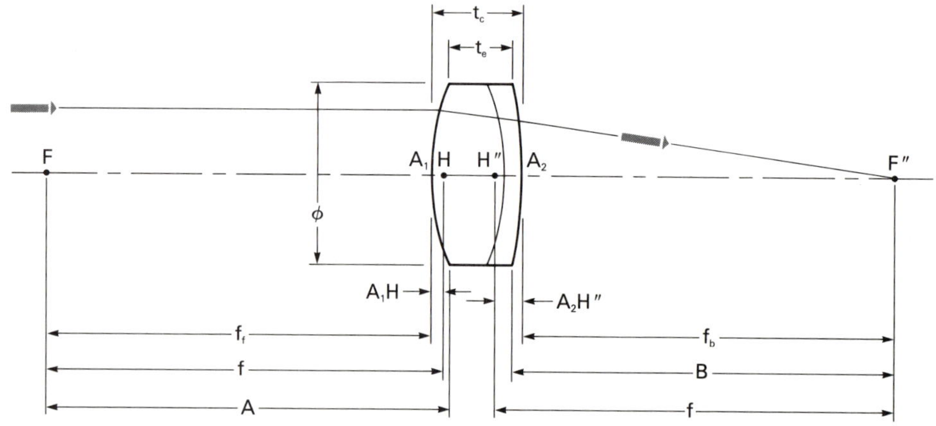

01 LAO

SPECIFICATIONS: PRECISION OPTIMIZED ACHROMATS

Paraxial Focal Length (f) at 546.1nm: $\pm 2\%$

Design Wavelengths: 480.0nm (blue), 546.1nm (green), 643.8nm (red)

Transmission Range: 400nm to 2500nm

Surface Accuracy (over 90% of aperture): For $(f/\phi) < 3.0$, spherical to within 0.5 wave at the mercury greenline (546.1nm). For $(f/\phi) \geq 3.0$, spherical to within 0.25 wave.

Centration: 3 arc minutes

Diameter (ϕ): ± 0.15mm

Center Thickness (t_c): ± 0.25mm

Materials: Grade A Fine Annealed crown and flint glasses

Cement: Ultraviolet cured synthetic polyester

Maximum Service Temperature: 90 °C (cement limitation)

Cosmetic Surface Quality: Pitch-polished, 60-40 scratch and dig

Coating: Single layer MgF_2

Correct Orientation: The most steeply curved (shortest radius) surface should face the infinite conjugate.

Circumference is ground and outside edges have a slight protective bevel.

For the benefit of those customers who may be able to exploit the advantages of curved field performance (with curved image sensors, fiber optics faceplates, refocusing, or by other means), we have tabulated the radii of the spheres which most closely approximate the true focal surface shape within the specified 2.5° angular radius field. If the focal surface curvature thus indicated is excessive for your application, you should instead use the appropriate companion field flattener.

Precision Optimized Achromats

f (mm)	ϕ (mm)	t_c (mm)	t_e (mm)	A_1H (mm)	A_2H'' (mm)	f_f (mm)	f_b (mm)	A (mm)	B (mm)	Focal Surface Radius (mm)	f/ϕ	PRODUCT NUMBER
10.0	6.0	4.4	3.5	0.4	-2.4	9.6	7.6	10	8	4.8	1.67	01 LAO 001
12.0	7.0	4.7	3.7	0.4	-2.6	11.6	9.4	12	10	5.2	1.71	01 LAO 004
16.0	8.0	4.9	3.9	0.4	-2.6	15.6	13.4	16	14	6.6	2.00	01 LAO 007
20.0	10.0	5.6	4.4	0.5	-3.0	19.5	17.0	20	17	8.1	2.00	01 LAO 011
21.0	14.0	8.1	5.8	0.7	-4.4	20.3	16.6	22	17	10.2	1.50	01 LAO 014
25.0	8.0	5.1	4.4	0.9	-2.4	24.1	22.6	25	23	10.5	3.12	01 LAO 018
25.0	12.0	5.8	4.0	1.1	-2.8	23.9	22.2	25	23	10.7	2.08	01 LAO 019
30.0	8.0	4.5	4.0	0.4	-2.4	29.6	27.6	30	28	12.0	3.75	01 LAO 022
30.0	12.5	5.3	4.1	0.5	-2.8	29.5	27.2	30	27	12.0	2.40	01 LAO 024
31.0	17.5	8.9	6.5	0.8	-4.7	30.1	26.3	32	27	12.3	1.77	01 LAO 028
36.0	16.0	6.9	5.2	0.6	-3.7	35.4	32.3	37	33	14.4	2.25	01 LAO 034
40.0	18.0	8.0	6.1	0.7	-4.3	39.3	35.7	41	36	16.0	2.22	01 LAO 037
40.5	13.0	4.6	3.5	0.6	-2.4	39.9	38.1	41	38	15.8	3.12	01 LAO 045
44.0	14.0	5.2	4.2	0.3	-2.9	43.7	41.1	45	41	17.2	3.14	01 LAO 047
48.0	16.0	5.2	4.0	0.3	-2.9	47.7	45.1	49	45	18.7	3.00	01 LAO 055
50.0	18.0	6.0	4.2	0.7	-3.1	49.3	46.9	51	47	19.5	2.78	01 LAO 059
56.0	19.0	5.9	4.2	0.7	-3.1	55.3	52.9	57	53	21.7	2.95	01 LAO 068
60.0	18.0	5.2	3.9	0.5	-2.7	59.5	57.3	61	58	23.4	3.33	01 LAO 077
60.0	30.0	12.5	8.4	1.6	-6.4	58.3	53.5	61	54	25.0	2.00	01 LAO 079
65.0	25.0	8.0	5.0	1.3	-4.0	63.7	61.0	66	62	25.7	2.60	01 LAO 083
73.0	17.0	5.2	4.3	0.3	-2.8	72.7	70.2	73	70	28.2	4.29	01 LAO 088
80.0	18.0	5.2	4.3	0.3	-2.9	79.7	77.1	81	77	30.8	4.44	01 LAO 111
80.0	31.5	9.5	6.1	1.3	-4.8	78.7	75.2	81	76	31.2	2.54	01 LAO 114
90.0	19.0	5.5	4.6	0.3	-3.1	89.7	86.9	91	87	34.6	4.74	01 LAO 117
90.0	25.0	6.5	4.4	1.1	-3.1	88.9	86.9	90	88	34.7	3.60	01 LAO 119
100.0	26.5	6.8	4.9	0.8	-3.5	99.2	96.5	101	97	38.4	3.77	01 LAO 123
100.0	30.0	8.7	6.3	1.1	-1.5	98.9	95.5	101	96	38.7	3.33	01 LAO 124
100.0	31.5	8.7	6.1	1.1	-4.5	98.9	95.5	101	96	38.7	3.17	01 LAO 126
120.0	24.0	5.5	4.4	0.4	-2.9	119.6	117.1	120	117	46.1	5.00	01 LAO 134
120.0	30.0	10.7	8.4	2.3	-4.7	117.7	115.3	119	116	47.2	4.00	01 LAO 135
120.0	40.0	11.2	7.6	1.4	-5.8	118.7	114.3	121	115	46.4	3.00	01 LAO 138
140.0	30.0	9.0	7.1	1.9	-4.0	138.1	136.0	139	137	54.0	4.66	01 LAO 139
140.0	40.0	10.1	7.1	1.2	-5.2	138.8	134.8	141	136	53.8	3.50	01 LAO 145
148.0	19.0	4.6	4.0	0.6	-2.3	147.4	145.7	148	146	56.0	7.79	01 LAO 148
160.0	30.0	6.5	5.2	0.5	-3.5	159.5	156.5	161	157	61.3	5.33	01 LAO 149

continued

Precision Optimized Achromats (continued)

f (mm)	ϕ (mm)	t_c (mm)	t_e (mm)	A_1H (mm)	A_2H'' (mm)	f_f (mm)	f_b (mm)	A (mm)	B (mm)	Focal Surface Radius (mm)	f/ϕ	PRODUCT NUMBER
160.0	31.5	6.5	5.1	0.5	-3.5	159.5	156.5	161	157	61.3	5.08	01 LAO 155
160.0	40.0	9.4	6.8	1.2	-4.8	158.8	155.2	161	156	61.2	4.00	01 LAO 159
175.0	25.0	5.5	4.7	0.5	-2.8	174.5	172.2	175	172	67.0	7.00	01 LAO 167
180.0	30.0	8.1	6.6	1.9	-3.4	178.1	176.6	179	180	68.8	6.00	01 LAO 169
180.0	40.0	9.6	7.2	1.4	-4.6	178.6	175.4	180	176	69.0	4.50	01 LAO 177
190.0	19.0	6.0	5.5	1.0	-2.7	188.9	187.3	189	187	72.6	10.00	01 LAO 188
200.0	30.0	8.1	7.1	0.7	-4.2	199.3	195.8	200	196	76.8	6.67	01 LAO 189
200.0	31.5	8.1	7.0	0.7	-4.2	199.3	195.8	200	196	76.8	6.35	01 LAO 222
200.0	40.0	10.6	8.8	0.6	-5.9	199.4	194.1	201	194	76.8	5.00	01 LAO 225
250.0	30.0	7.2	6.2	0.9	-3.6	249.1	246.5	250	247	94.5	8.33	01 LAO 226
250.0	31.5	7.2	6.2	0.9	-3.6	249.1	246.5	250	247	94.5	7.94	01 LAO 229
260.0	32.0	7.6	6.5	2.0	-3.7	259.0	256.3	260	257	98.3	8.12	01 LAO 235
260.0	50.0	12.3	9.8	1.7	-5.9	258.3	254.1	260	255	99.3	5.20	01 LAO 238
300.0	30.0	9.4	8.6	1.6	-4.3	298.4	295.7	299	296	114.6	10.0	01 LAO 239
300.0	31.5	9.4	8.6	1.6	-4.3	298.4	295.7	299	296	114.6	9.52	01 LAO 244
300.0	40.0	9.3	7.9	1.3	-4.5	298.7	295.5	300	296	113.6	7.50	01 LAO 248
300.0	50.0	14.7	12.5	2.3	-6.8	297.7	293.2	299	294	115.0	6.00	01 LAO 256
300.0	82.0	19.6	13.7	2.9	-9.4	297.1	290.5	301	292	115.7	3.66	01 LAO 267
350.0	40.0	7.0	5.9	0.7	-3.6	349.1	346.2	350	346	133.2	8.74	01 LAO 277
400.0	40.0	8.0	7.1	0.9	-4.0	399.1	396.0	400	396	152.5	10.00	01 LAO 288
400.0	63.0	12.0	9.4	1.8	-5.7	398.2	394.3	400	395	151.6	6.35	01 LAO 333
450.0	40.0	8.2	7.3	1.3	-3.8	448.7	446.2	449	446	170.5	11.25	01 LAO 339
500.0	63.0	12.0	10.2	1.3	-6.0	498.8	494.0	500	494	190.9	7.94	01 LAO 346
600.0	63.0	11.0	9.5	1.2	-5.5	598.8	594.5	600	595	228.5	9.52	01 LAO 349
700.0	50.0	10.0	8.9	2.6	-4.6	697.4	695.4	698	696	260.7	14.00	01 LAO 355
800.0	63.0	10.0	8.9	1.1	-5.0	798.9	795.0	800	795	303.8	12.70	01 LAO 357
1000.0	80.0	10.0	8.5	1.0	-5.0	999.0	995.0	1000	995	379.3	12.50	01 LAO 366
1000.0	150.0	24.6	18.7	3.2	−12.2	996.8	987.8	1001	989	377.3	6.67	01 LAO 367
1185.0	80.0	11.8	10.6	1.3	-5.9	1183.7	1179.1	1185	1179	449.5	14.81	01 LAO 369
2000.0	150.0	23.3	20.3	3.0	−11.5	1997.0	1988.5	1999	1989	750.0	13.33	01 LAO 379
3000.0	150.0	23.5	21.5	3.0	−11.6	2997.0	2988.4	2998	2982	1123.0	20.00	01 LAO 399

OPTICAL FABRICATION

The Melles Griot optical fabrication facility is located in the Hsinchu Science Based Industrial Park, Taiwan, R.O.C. It is one of the most modern optical facilities in the world, regularly producing over half a million optical components per year.

We employ precision, high-speed polishing methods to assure you of a superior and consistent standard of optical components.

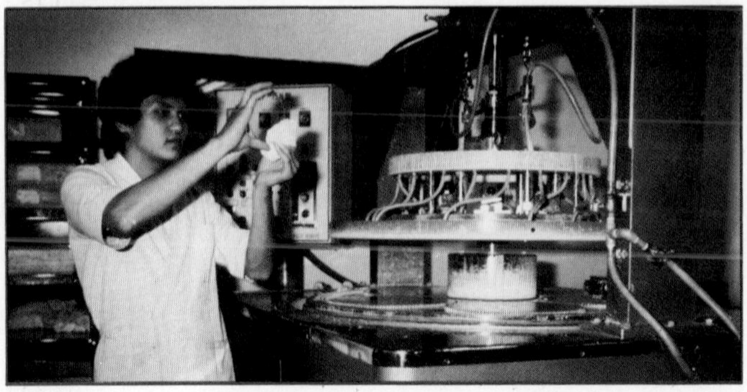

Our flat work shop uses both continuous polishers and double-sided polishers to produce some of the best planar optics available.

The modern optical coating facility, staffed by graduate engineers, produces high quality multilayer coatings for laser applications.

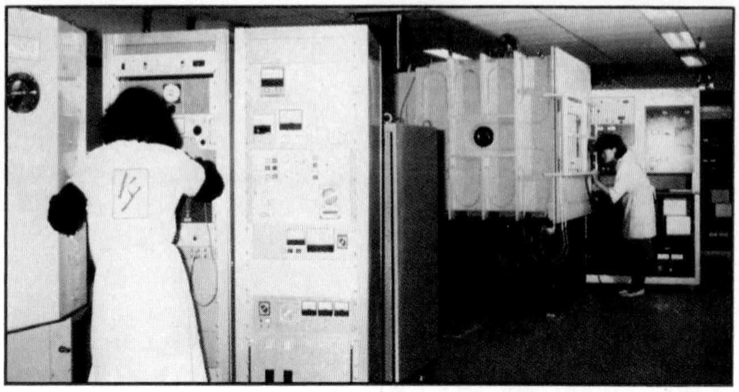

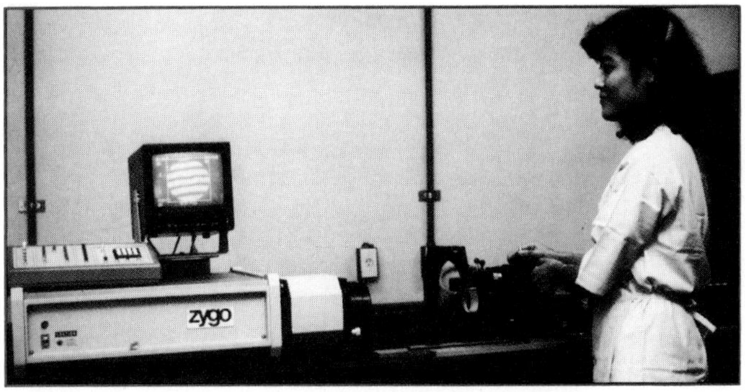

Our fully equipped quality control department assures you that your components will meet your demands first time and every time.

PRECISION OPTIMIZED APLANATIC MENISCUS LENSES

These are companion elements for the precision optimized spherical achromats. Aplanatic meniscus lenses may be used in conjunction with doublets to shorten focal length and reduce f-number. The theory of aplanatic meniscus lenses is described in the section on meniscus lenses (see page 104). Stated simply, one can use an aplanatic meniscus lens with a doublet to shorten its focal length without decreasing its performance. This capability significantly increases the utility of the doublet product line. As an example, an 01 LAO 028 used at f/1.8 (with a collimated input beam at 588nm) has a spot of size of about 74 microns. In contrast, a system consisting of an 01 LAO 059 and an 01 LAM 059 has the same focal length and f-number as the 01 LAO 028, but gives a spot size of approximately 5 microns under the same conditions. The reason for this dramatic difference in performance is that doublet performance is strongly dependent on f-number. In the case of the doublet/meniscus lens combination, the doublet itself is working at f/2.9; as a result, it will have significantly better performance than a doublet at f/1.8. The meniscus lens serves to lower the system f-number without lowering performance.

For use with a doublet, it is important to use an aplanatic meniscus lens rather than the type of meniscus lens listed on page 104. This is because the aberrations present in a doublet are very small; any additional aberrations introduced by the meniscus element can have a noticeable effect on overall resolution. In the case of simple lenses, the additional aberration from a meniscus lens is much less noticeable. For this same reason it is important to use the aplanatic meniscus lenses only with their companion doublets, and at the correct magnification.

USING THE APLANATIC MENISCUS LENSES

The product tables show each meniscus lens paired with a doublet from the 01 LAO series. The column headed f_c shows the final focal length of the doublet/meniscus lens combination. The f-number tabulated is also the value for the indicated doublet/meniscus lens combination. While the meniscus lenses can be used with any lens or lens system, as long as they are at the proper magnification, it is most convenient to use them with the doublets indicated. There are two major reasons for this. First, the meniscus lenses are the same diameter as their companion doublet, and second, they are designed to be positioned 1mm from the doublet.

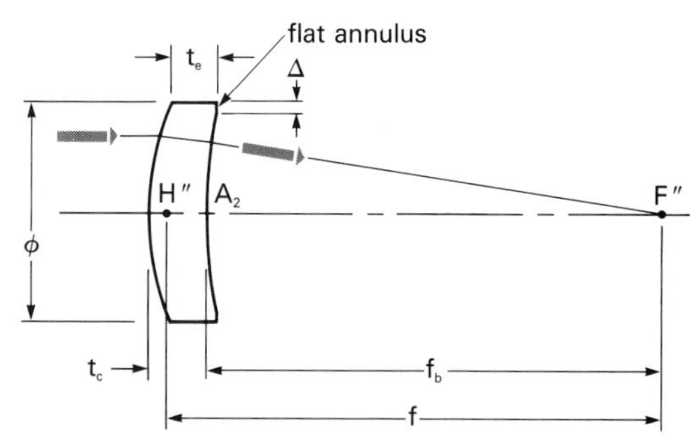

01 LAM

These two features make mounting and positioning of the aplanatic meniscus lenses fairly straightforward. To determine how to use these lenses under more general circumstances, one calculates the object and image distance using a magnification of $1/n = -0.59$ and the focal length, f_o, given in the product table. Because these are singlet lenses, there is no provision for color correction. This means that the meniscus lenses are used to greatest advantage in systems with monochromatic light.

SPECIFICATIONS: APLANATIC MENISCUS LENSES

Paraxial Focal Length (f_o): ±2%

Design Wavelength: 546.1nm

Surface Accuracy: 1λ at 546.1nm

Centration: 3 arc minutes

Clear Aperture: 90% of diameter

Diameter: ±0.15mm

Center Thickness: ±0.15mm

Material: SF 8 grade A fine annealed

Cosmetic Surface Quality: 60-40 scratch and dig

Coating: HEBBAR (for normal incidence at 550nm on both surfaces)

Correct Orientation: Concave surface facing towards focal point.

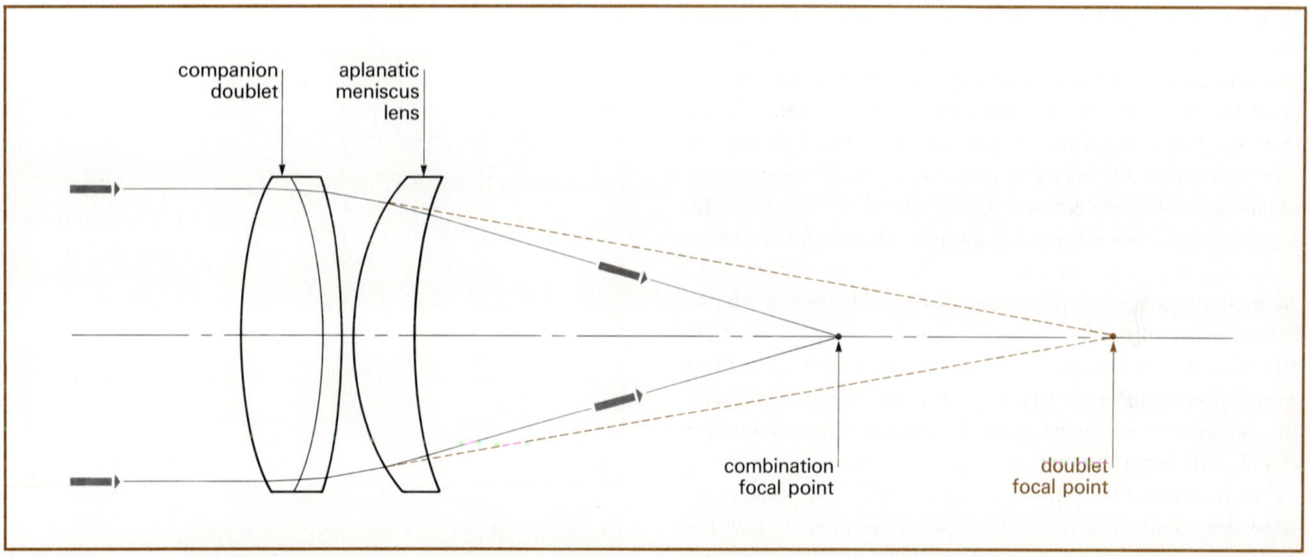

A MENISCUS LENS BEING USED TO SHORTEN THE FOCAL LENGTH OF A DOUBLET. The focal length of the doublet alone is shown, as is the combination focal length of the system including the meniscus lens. The meniscus lens should always be oriented with the concave surface towards the focal point, as shown here. When used with their companion doublets, the meniscus lens will be spaced 1mm from the doublet.

Precision Optimized Aplanatic Meniscus Lenses

						Combination		
f_o (mm)	ϕ (mm)	t_c (mm)	t_e (mm)	Δ (mm)	COMPANION DOUBLET	f_c (mm)	f_c/ϕ (mm)	PRODUCT NUMBER
73.5	18.0	4.0	3.0	0.5	01 LAO 059	30.1	1.67	01 LAM 059
116.2	18.0	4.0	3.4	0.5	01 LAO 111	47.6	2.64	01 LAM 111
130.2	19.0	4.0	3.4	0.5	01 LAO 117	53.3	2.81	01 LAM 117
146.4	31.5	6.0	4.7	0.5	01 LAO 126	60.0	1.90	01 LAM 126
175.2	40.0	7.0	5.2	0.5	01 LAO 138	71.2	1.78	01 LAM 138
214.6	19.0	4.0	3.7	0.5	01 LAO 148	87.9	4.63	01 LAM 148
233.6	31.5	6.0	5.2	0.5	01 LAO 155	95.7	3.03	01 LAM 155
290.7	40.0	7.0	6.0	0.5	01 LAO 225	119.1	2.98	01 LAM 225
362.8	31.5	6.0	5.5	0.5	01 LAO 229	148.6	4.72	01 LAM 229
434.4	82.0	11.0	8.0	1.0	01 LAO 267	178.0	2.17	01 LAM 267
507.8	40.0	7.0	6.4	0.5	01 LAO 277	208.1	5.20	01 LAM 277
580.2	63.0	9.0	7.7	1.0	01 LAO 333	237.7	3.77	01 LAM 333
723.6	63.0	9.0	7.9	1.0	01 LAO 346	296.5	4.71	01 LAM 346
1157.0	63.0	9.0	8.3	1.0	01 LAO 357	474.0	7.52	01 LAM 357

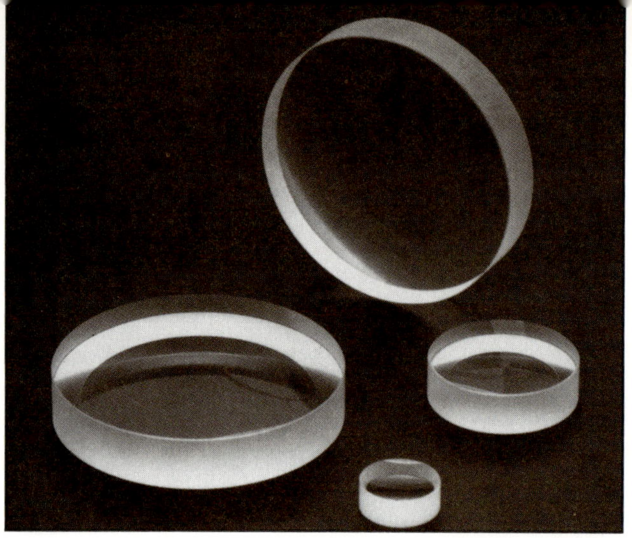

PRECISION OPTIMIZED FIELD FLATTENERS

These are companion elements for the precision optimized spherical achromats. Their operation was earlier explained in a more general context (see page 128). It is intended that doublets be operated at full aperture with their companion flatteners. Flatteners are offered only for those doublets with which flatteners are both reasonably sized and highly effective at doublet full aperture.

Field flattener MTF curves are clearly identified in the graphical presentation of MTF data for the example achromat 01 LAO 123. Interpretation of the MTF data is explained in that context.

The correct position for the optimized field flattener is with its flat surface in a plane 2mm ± 0.5mm toward the doublet from the experimentally determined plane of best focus for the combination. Within this plane the flattener should be centered on the doublet optical axis to within 2% of the flattener diameter for best results. The curved surface must face the doublet.

Field flatteners are supplied unmounted only. Their dimensions admit field angles up to 2.5 degrees (sometimes slightly more) at infinite conjugate ratio. Field angle range will be more restricted, and field flattener effectiveness will be diminished, at other conjugate ratios.

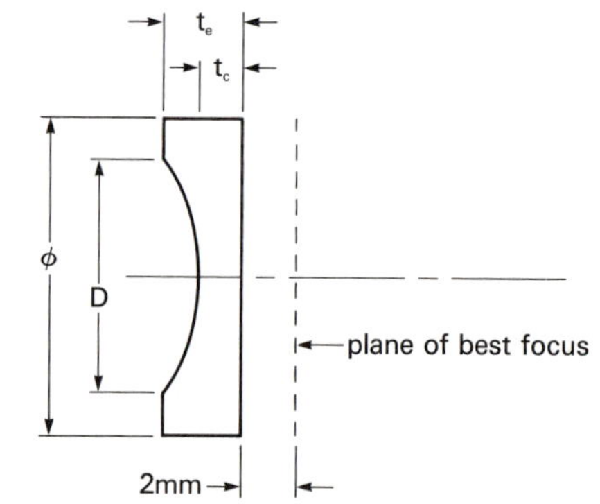

01 LFF

SPECIFICATIONS: PRECISION OPTIMIZED SPHERICAL FIELD FLATTENERS

Centration: 3 arc minutes

Diameter (ϕ): ±0.25mm

Diameter exceeds intended clear aperture by 4mm, to allow for possible retaining ring radial thickness of 2mm.

Center Thickness (t_c): ±0.2mm

Material: BK 7 Grade A Fine Annealed

Cosmetic Surface Quality: Pitch polished, 40-20 scratch and dig

Coating: HEBBAR™ for normal incidence at 550nm on both surfaces

Circumference and flat annulus are ground, and edges have a slight protective bevel.

Our Applications Engineers will be happy to analyze the performance of your proposed system.

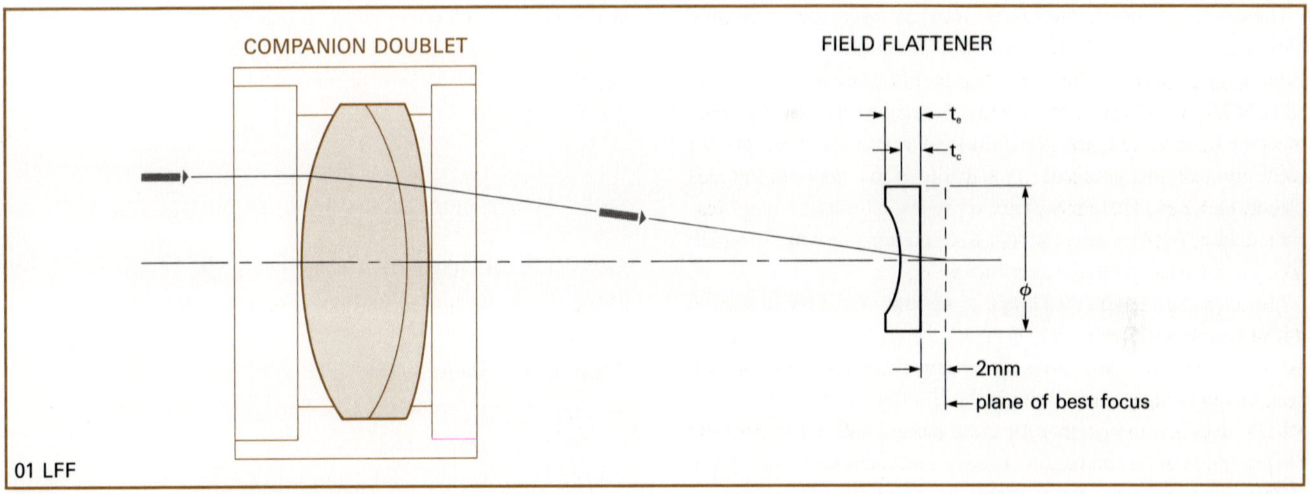

Precision Optimized Spherical Field Flatteners

ϕ (mm)	t_c (mm)	t_e (mm)	COMPANION DOUBLET PRODUCT NUMBER	PRODUCT NUMBER
10.4	3.0	3.9	01 LAO 088	01 LFF 088
11.0	3.0	4.0	01 LAO 111	01 LFF 111
11.0	3.0	3.8	01 LAO 114	01 LFF 114
11.9	3.0	4.1	01 LAO 117	01 LFF 117
12.7	3.0	4.2	01 LAO 123	01 LFF 123
12.7	3.0	4.1	01 LAO 126	01 LFF 126
14.5	3.0	4.2	01 LAO 134	01 LFF 134
14.5	3.0	4.2	01 LAO 138	01 LFF 138
16.2	3.0	4.4	01 LAO 145	01 LFF 145
16.9	3.0	4.6	01 LAO 148	01 LFF 148
18.0	3.3	4.5	01 LAO 155	01 LFF 155
18.0	3.0	4.6	01 LAO 159	01 LFF 159
19.3	3.0	4.6	01 LAO 167	01 LFF 167
19.7	3.0	4.6	01 LAO 177	01 LFF 177
20.6	4.0	5.7	01 LAO 188	01 LFF 188
21.5	3.0	4.809	01 LAO 222	01 LFF 222

continued

Precision Optimized Spherical Field Flatteners (continued)

ϕ (mm)	t_c (mm)	t_e (mm)	COMPANION DOUBLET PRODUCT NUMBER	PRODUCT NUMBER
21.5	3.0	4.9	01 LAO 225	01 LFF 225
25.8	4.0	6.3	01 LAO 229	01 LFF 229
26.7	3.0	5.4	01 LAO 235	01 LFF 235
26.7	3.0	5.4	01 LAO 238	01 LFF 238
30.2	3.0	5.5	01 LAO 244	01 LFF 244
30.2	4.0	6.6	01 LAO 248	01 LFF 248
30.2	4.0	6.6	01 LAO 256	01 LFF 256
30.2	3.0	5.4	01 LAO 267	01 LFF 267
34.6	4.0	6.9	01 LAO 277	01 LFF 277
38.9	4.0	7.3	01 LAO 288	01 LFF 288
38.9	4.0	7.2	01 LAO 333	01 LFF 333
43.3	5.0	8.5	01 LAO 339	01 LFF 339
47.7	5.0	8.9	01 LAO 346	01 LFF 346
56.4	6.0	10.6	01 LAO 349	01 LFF 349
73.9	8.0	14.0	01 LAO 357	01 LFF 357

SPHERICAL ACHROMATS

These are cemented achromatic doublets optimized at infinite conjugate ratio in accordance with a maximum energy through a small circle criterion. This criterion does not result in the best possible MTF curves. In contrast to the Precision Optimized Achromats previously presented, the present spherical achromats are not ideal for imaging purposes, though the Abbe sine condition is closely satisfied. These lenses are appropriate for use in radiometric and laser beam manipulation tasks, and imaging tasks in which MTF is not of paramount importance.

This series offers you the benefit of an optimized prescription, at a lower price than would be necessary for the best possible imagery. Glasses used in this series have been selected for low cost and ready availability.

MTF data is not available for these lenses, and would be inappropriate for the optimization criterion and tolerances used. Field flatteners are likewise not available for these achromats.

Achromats of this series are normally sold uncoated, but any of a variety of coatings can be requested via the Coating Suffix system previously described. There is considerable advantage to coatings for the external surfaces, since the reflectances of the two internal glass/cement interfaces are only about 15 parts per million and 3 parts per thousand respectively. These values presume normal incidence, and are found from the formula

$$R = \left(\frac{n_1 - n_2}{n_1 + n_2}\right)^2 ,$$

where n_1 is the refractive index on one side of an interface, and n_2 is the refractive index on the other side. The external surface reflectances without coatings are much larger, amounting to about 4% per surface.

SPECIFICATIONS: SPHERICAL ACHROMATS

Paraxial Focal Length (f) at 589.0nm: ±5%

Design Wavelengths: 486.1nm (blue), 589.0nm (yellow), and 656.3nm (red)

Transmission Range: 400 to 2500nm

Centration: 5 arc minutes

Diameter (ϕ): ±0.2mm

Center Thickness (t_c): ±0.25mm

Materials: Grade A Fine Annealed crown and flint glasses

Cement: Canada Balsam

Maximum Service Temperature: 65°C (cement limitation)

Cosmetic Surface Quality: Pitch-polished, 80-50 scratch and dig

Coating: None. See coating suffix system for coating options.

Correct Orientation: The most steeply curved (shortest radius) surface should face the infinite conjugate.

Circumference is ground and outside edges have a slight protective bevel.

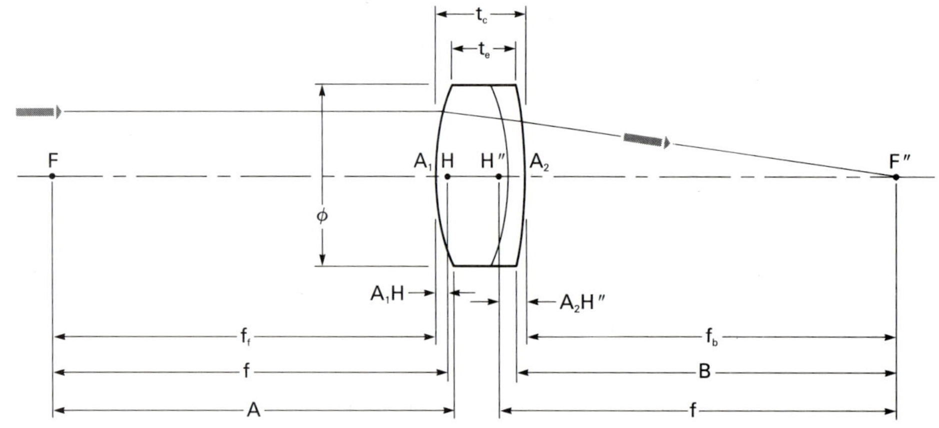

01 LAU

Spherical Achromats

f (mm)	ϕ (mm)	t_c (mm)	t_e (mm)	A_1H (mm)	A_2H'' (mm)	f_f (mm)	f_b (mm)	A (mm)	B (mm)	f/ϕ	PRODUCT NUMBER
10.0	5.0	3.7	2.9	0.8	−1.7	9.2	8.3	10	9	2.00	01 LAU 000
10.0	7.0	4.9	3.2	1.0	−2.4	9.0	7.6	10	8	1.43	01 LAU 126
12.0	6.0	4.4	3.5	1.0	−2.0	11.0	10.0	12	10	2.00	01 LAU 001
15.0	7.5	4.8	3.6	1.0	−2.2	14.0	12.8	15	13	2.00	01 LAU 128
20.0	10.0	5.9	4.3	1.2	−2.7	18.8	17.3	20	18	2.00	01 LAU 002
22.0	8.5	3.6	2.6	0.8	−1.5	21.2	20.4	22	21	2.59	01 LAU 134
25.0	5.1	3.5	3.2	0.9	−1.3	24.1	23.7	24	24	4.90	01 LAU 136
25.0	12.0	5.8	4.0	1.1	−2.8	23.9	22.2	25	23	2.08	01 LAU 004
30.0	12.0	5.8	4.3	1.3	−2.5	28.7	27.5	30	28	2.50	01 LAU 006
35.0	15.0	7.0	5.0	1.5	−3.1	33.5	31.9	35	33	2.33	01 LAU 007
40.0	15.0	7.0	5.3	1.5	−3.0	38.5	37.0	40	38	2.67	01 LAU 008
45.0	15.0	6.5	5.0	1.5	−2.7	43.5	42.3	45	43	3.00	01 LAU 016
50.0	12.0	5.4	4.5	1.3	−2.1	48.7	47.9	49	48	4.17	01 LAU 018
50.0	20.0	7.8	5.4	1.5	−3.6	48.5	46.4	50	47	2.50	01 LAU 022
60.0	20.0	7.2	5.2	1.5	−3.1	58.5	56.9	60	58	3.00	01 LAU 023
60.0	23.0	8.7	6.0	1.7	−3.9	58.5	56.1	60	57	2.61	01 LAU 144
65.0	6.3	5.0	4.8	1.2	−2.0	63.8	63.0	64	63	10.32	01 LAU 138
65.0	25.0	8.0	5.0	1.3	−4.0	63.7	61.0	66	62	2.60	01 LAU 024
70.0	25.0	8.0	5.3	1.3	−3.9	68.7	66.1	71	67	2.80	01 LAU 025
75.0	25.0	7.0	4.5	1.2	−3.4	73.8	71.6	76	72	3.00	01 LAU 026
80.0	25.0	7.0	4.6	1.2	−3.3	78.8	76.7	80	77	3.20	01 LAU 028
90.0	25.0	6.5	4.4	1.1	−3.1	88.9	86.9	90	88	3.60	01 LAU 029
100.0	25.0	7.0	5.1	1.4	−3.0	98.5	97.0	100	98	4.00	01 LAU 033
100.0	40.0	12.0	7.1	1.9	−6.0	98.1	94.0	102	96	2.50	01 LAU 034
120.0	43.5	13.0	8.2	2.0	−6.5	118.0	113.5	121	115	2.76	01 LAU 145
150.0	25.0	7.0	5.7	1.4	−3.0	148.6	147.0	149	147	6.00	01 LAU 036
180.0	57.5	17.0	11.4	3.4	−7.5	176.6	172.5	180	174	3.13	01 LAU 124
200.0	25.0	7.0	6.1	1.4	−3.0	198.6	197.0	199	197	8.00	01 LAU 044
200.0	50.0	11.0	7.2	2.1	−5.0	197.9	195.0	200	196	4.00	01 LAU 045
250.0	50.0	10.0	7.0	1.9	−4.4	248.1	245.6	250	247	5.00	01 LAU 046
300.0	50.0	10.0	7.5	2.1	−4.1	297.9	295.9	300	297	6.00	01 LAU 048
350.0	50.0	10.0	7.8	2.1	−4.1	347.9	345.9	349	347	7.00	01 LAU 049
400.0	50.0	10.0	8.1	2.1	−4.1	397.9	395.9	399	396	8.00	01 LAU 056
450.0	50.0	10.0	8.3	2.1	−4.1	447.9	445.9	449	446	9.00	01 LAU 059
500.0	50.0	10.0	8.5	2.1	−4.1	497.9	495.9	499	496	10.00	01 LAU 066

continued

Spherical Achromats (continued)

f (mm)	ϕ (mm)	t_c (mm)	t_e (mm)	A_1H (mm)	A_2H'' (mm)	f_f (mm)	f_b (mm)	A (mm)	B (mm)	f/ϕ	PRODUCT NUMBER
550.0	50.0	10.00	8.6	2.1	-4.1	547.9	545.9	549	546	11.00	01 LAU 067
600.0	50.0	10.00	8.7	2.1	-4.1	597.9	595.9	599	596	12.00	01 LAU 068
650.0	50.0	10.00	8.8	2.1	-4.1	647.9	645.9	649	646	13.00	01 LAU 077
700.0	50.0	10.00	8.9	2.1	-4.1	697.9	695.9	699	696	14.00	01 LAU 088
750.0	50.0	10.00	9.0	2.1	-4.1	747.9	745.9	749	746	15.00	01 LAU 089
800.0	50.0	10.00	9.1	2.1	-4.1	797.9	795.9	799	796	16.00	01 LAU 099
850.0	50.0	10.00	9.1	2.1	-4.1	847.9	845.9	848	846	17.00	01 LAU 114
900.0	50.0	10.00	9.2	2.1	-4.1	897.9	895.9	898	896	18.00	01 LAU 116
950.0	50.0	10.00	9.2	2.1	-4.1	947.9	945.9	948	946	19.00	01 LAU 118
1000.0	50.0	10.00	9.2	2.1	-4.1	997.9	995.9	998	996	20.00	01 LAU 122

BEAM EXPANDER LENSES

The output beam diameter of most visible wavelength lasers is small, a millimeter or less being common. For applications such as holographic recording and playback it is useful to illuminate much larger areas. Further, diffraction will cause far-field divergence of these small beams, which is not acceptable in applications requiring the illumination of distant objects. Such applications include laser geodimeters, laser rangefinders, laser communications links, and laser alignment devices. A laser beam expander will provide larger beams for both illumination applications and reduced beam divergence.

We offer a selection of mounted beam expanders for use with visible range lasers (see page 366). The lenses described here are supplied unmounted for the convenience of those who wish to build beam expanders to their own requirements.

These beam expander lenses are available for four expansion ratios: 5X, 10X, 20X and 50X. The first three expanders are made up of two lenses each, the input and output lens respectively. The input lens is the same for all three of these beam expanders. The fourth expansion ratio, 50X is made up of a different input lens and an airspaced output doublet.

Beam expander lenses are available as individual expander pairs or in two sets. The first set includes the one common input lens and the 5X, 10X and 20X output lens. The second set contains these four lenses plus the 50X input singlet and output doublet lenses. In this second set the 50X output doublet is premounted in an anodized aluminum cell for your convenience.

A beam expander is a telescope used in reverse. The magnification of the expander, m, will increase the beam diameter by a factor m, while simultaneously reducing the beam angular divergence by a factor 1/m. The simplest beam expander consists of two elements. If the leading element is positive, and presents a real beam focus to the second element, the telescope is called Keplerian. If the leading element is negative, and presents a virtual beam focus to the second element, the telescope is called Galilean. An advantage of Galilean expanders is the minimization of package length. In Galilean expanders it is also possible to achieve better cancellation of spherical aberration than is possible in Keplerian expanders. However, spatial filtering is possible only in Keplerian expanders. Spatial filtering consists of placing a small pinhole at the common focus of the elements of a Keplerian expander. This pinhole eliminates the transmission of higher order diffracted radiation. It is generally counterproductive to use pinholes smaller than

$$d_{min} = \frac{2f\lambda}{\pi w}, \text{ where}$$

f = focal length of the leading element,

λ = wavelength of the laser light (in the same units as f and w), and

w = radius of the "$1/e^2$" irradiance contour at the

first element. Pinholes of much larger size that this are often found to be very helpful.

These expander pairs are all designed with an entrance pupil diameter of 2.0mm. Element shape and spacing have been chosen so that a ray entering the expander parallel to the optical axis, and with a ray height of 0.7mm, will ultimately emerge from the second element parallel to the optical axis. Because aberrations of the elements do not precisely cancel, rays entering the first element at distances other than 0.7mm from the optical axis will ultimately emerge from the expander with directions slightly different from that of the optical axis. By adjusting the element spacing (focusing), it is possible to direct emergent rays in the optical axis direction which originally entered the expander at distances other than 0.7mm from the optical axis. In this way a single expander can be used to collimate laser beams of various sizes.

The 5X, 10X, and 20X expanders each consist of a pair of unmounted plano-convex lenses of high-index flint glass. The first (small) element is the same for all three expansion ratios. Because of the high index, performance of plano-convex lenses closely approaches that of best-form lenses. The use of high-index glass typically reduces the far-field beam angular radius by about a factor of 2 from what is possible with BK 7. Antireflection coatings increase the transmittance.

The 50X expander is computer-optimized for all six radii, the three element thicknesses, the two high-index materials, and the element spacing. The optimization was carried out for rays entering the first element at a distance of 0.7mm from the optical axis. A variety of input laser beam sizes can be collimated by adjusting the separation of the first (small) element from the two large elements. The spacing of the large elements in this airspaced doublet is critical and is precisely set at our factory.

In the table which follows we show, as a function of the "$1/e^2$" Gaussian laser beam diameter, the fraction of total beam power which gets through the expander (neglecting reflectance losses at the coated surfaces), the fraction of power which enters the circle of 0.7mm radius for which the expanders have been designed, and the fraction of power which falls outside this circle. We assume here that spatial filters are not used.

The laser beam expanders are available in individual pairs, as shown in the following list, or in two sets.

BEAM EXPANDER THROUGHPUT. Expressed in % Input Power, as a function of Input "1/e²" Gaussian Beam Diameter.

Input $1/e^2$ Beam Diameter (mm)	% Throughput	% Inside Optimized Circle	% Outside Optimized Circle
0.5	100.00	100.00	0.00
0.6	100.00	100.00	0.00
0.7	100.00	99.97	0.03
0.8	100.00	99.78	0.22
0.9	99.99	99.21	0.79
1.0	99.97	98.02	1.98
1.1	99.87	96.08	3.92
1.2	99.61	93.43	6.57
1.3	99.12	90.17	9.83
1.4	98.31	86.47	13.53
1.5	97.14	82.49	17.51
1.6	95.61	78.37	21.63
1.7	93.72	74.24	25.76
1.8	91.53	70.18	29.82
1.9	89.10	66.24	33.76
2.0	86.47	62.47	37.53

SPECIFICATIONS: BEAM EXPANDER LENSES

Back Focal Length (f_b): ±0.5%
Optimum Performance Wavelength: 632.8nm
Useful Wavelength Range: 5X, 10X and 20X, 488 to 950nm; 50X, 600 to 950nm
Surface Accuracy: 2λ at 632.8nm
Centration: 3 arc minutes for both elements
Diameter (ϕ): ±0, -0.1mm
Maximum Entrance Pupil Diameter: 2.0mm
Edge Thickness (t_e): ±0.2mm
Material: High-index Flint Grade A Fine Annealed
Cosmetic Surface Quality: Pitch-polished, 40-20 scratch and dig
Coating: Single-layer antireflection at 632.8nm, all surfaces

We also offer a selection of mounted laser beam expanders and laser collimators which are directly compatible with the fittings on our helium neon lasers. *See page 361 for details.*

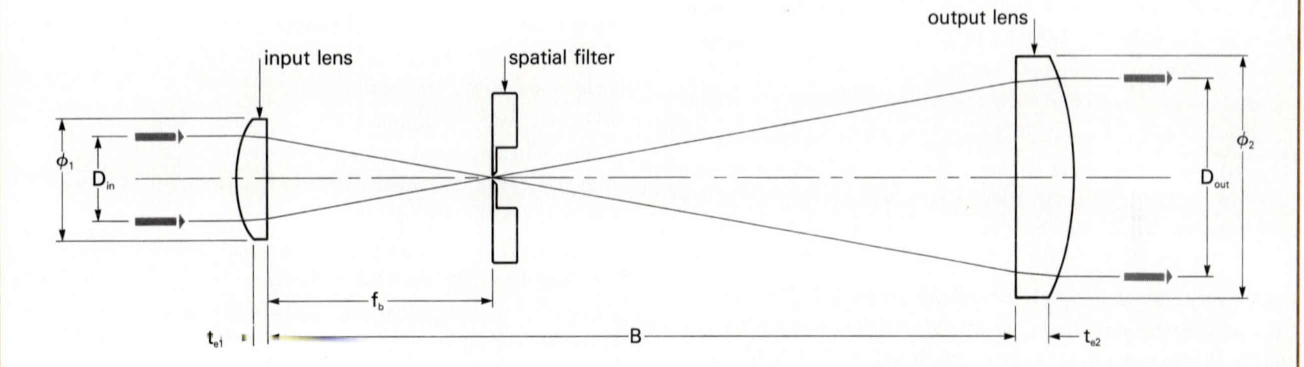

01 LBX

Beam Expander Lenses

Expansion Ratio, m	Input Lens ϕ_1 (mm)	f_b (mm)	t_1 (mm)	Output Lens ϕ_2 (mm)	t_2 (mm)	Lens Separation B (mm)	PRODUCT NUMBER
5X	5.1	9.5	1.1	17.0	2.2	59.3	01 LBX 001
10X	5.1	9.5	1.1	32.0	3.8	109.7	01 LBX 003
20X	5.1	9.5	1.1	62.2	2.7	214.5	01 LBX 005
50X	7.1	-6.6	2.9	111.8	*	274.0	01 LBX 007

*This lens is a mounted airspaced doublet.

BEAM EXPANDER LENS SETS

The expanders listed are also offered in sets for use in experimental situations where use of different expansion ratios may be desirable.

The 5X to 2X expander set contains one 9.5mm input lens and three output lenses. The 5X to 50X expander set contains both the 9.5mm and the -6.6mm input lenses plus all four output lenses.

Beam Expander Lens Sets

DESCRIPTION	PRODUCT NUMBER
5X to 20X Expander Set	01 LBX 011
5X to 50X Expander Set	01 LBX 015

LASER DIODE GLASS DOUBLETS

These cemented glass doublets have been designed to perform well throughout the entire near-infrared spectrum. Between 780 and 1550nm their operation is diffraction limited; wavefront distortion at 830mm is less than λ/5. Special antireflection coatings providing less than 0.25% reflectivity are available for three specific near-infrared wavelength bands.

These lenses are intended to be used for laser diode beam manipulation after collimation and shaping. They will be found useful for focusing, defocusing beam expansion and de-expansion. Spherical aberration, coma, astigmatism, and spherochromatism have been well corrected. Pure chromatic aberration (variation of focal length with wavelength) has been left uncorrected to provide greater freedoms in the other corrections. When used with monochromatic laser diodes this causes no aberration contribution, but will mean that systems have to be refocused for use with different wavelength laser diodes. The focal lengths over a range of 750 to 1550nm are shown graphically below.

High-efficiency antireflection coatings are available at 780, 830, 1300, and 1550nm. These provide air-glass surface reflectivities of less than 0.25% per surface.

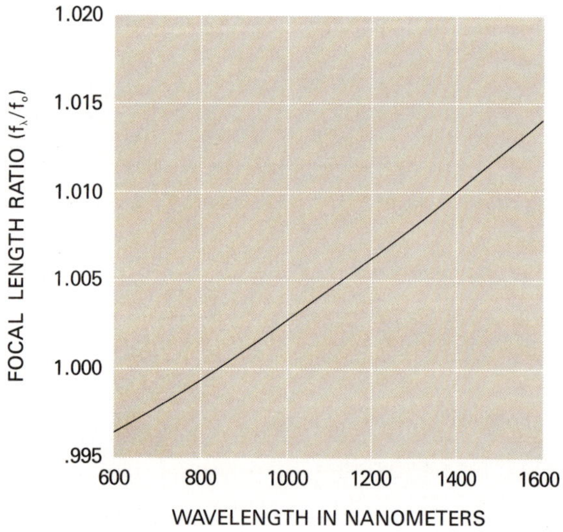

FOCAL LENGTH OF LASER DIODE GLASS DOUBLETS as a function of wavelength.

SPECIFICATIONS: LASER DIODE GLASS DOUBLETS

Paraxial Focal Length: ±2%
Design Wavelength: 830nm
Wavelength Range: 780 to 1550nm
Wavefront Distortion: <λ/5 at 830nm
Centration; 4 arc minutes
Diameter (φ): ±0.25mm
Center Thickness (t_c): ±0.25mm
Materials: SK 11 and SF 5, Grade A
Cement: Ultraviolet cured synthetic polyester
Cosmetic Surface Quality: Pitch polished, 80-50 scratch and dig
Coating: Must be specified by suffix
Circumference is ground and outside edges have a slight protective bevel.

COATING OPTIONS

The Coating Suffix system should be used to request antireflection coatings for these lenses. If no Coating Suffix is appended to the Product Number, the lens will be supplied uncoated.

Antireflection Coatings; Normal Incidence

Wavelength (mm)	Maximum Reflectance (%)	COATING SUFFIX
780-830	0.25	/076
1300	0.25	/168
1550	0.25	/169

See page 454 for a mini-lens holder.

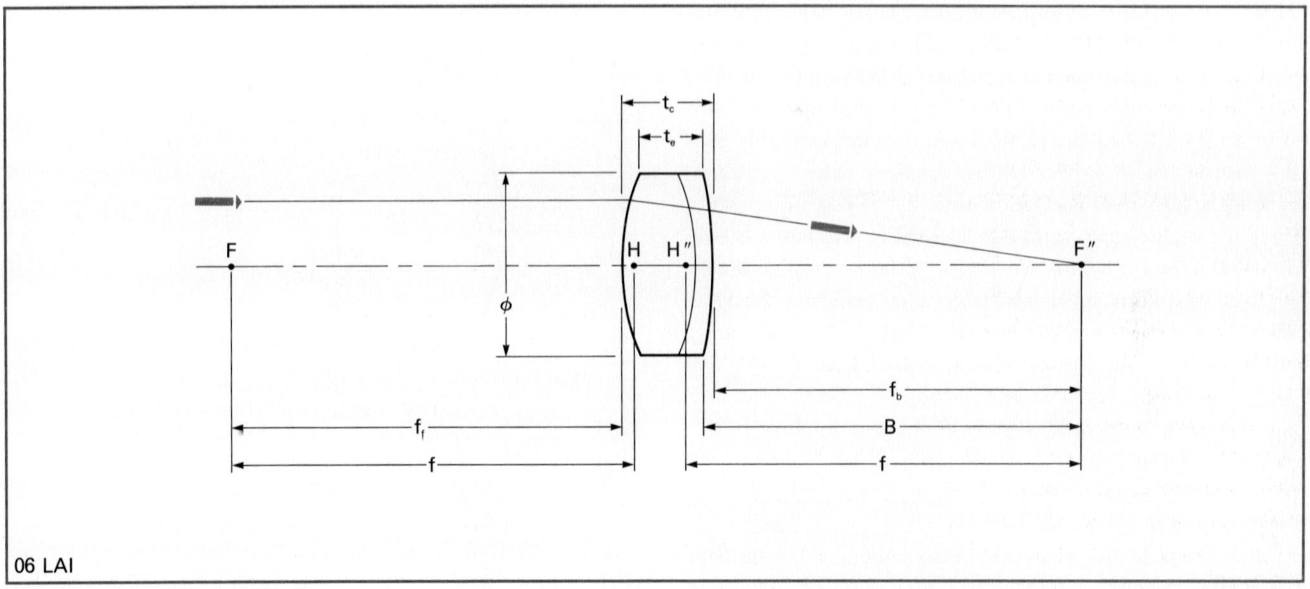

Laser Diode Glass Doublets

for λ=830nm							for λ=1300nm		for λ=1550nm		PRODUCT
f	f_b	f/ϕ	ϕ	t_c	t_e	B	f	f_b	f	f_b	NUMBER
10.0	8.2	2.0	5.0	3.5	2.8	8	10.1	8.3	10.1	8.3	06 LAI 001
25.0	22.4	2.5	10.0	5.0	3.9	23	25.2	22.6	25.3	22.8	06 LAI 003
40.0	37.2	2.7	15.0	5.5	4.0	38	40.3	37.6	40.5	37.8	06 LAI 005
60.0	56.7	3.0	20.0	6.5	4.7	57	60.5	57.1	60.8	57.4	06 LAI 007
80.0	75.6	3.2	25.0	8.5	6.4	76	80.6	76.2	81.0	76.6	06 LAI 009
100.0	94.8	3.3	30.0	10.0	7.6	95	100.8	95.6	101.3	96.1	06 LAI 011
145.0	138.5	3.6	40.0	12.5	9.5	139	146.1	139.6	146.8	140.4	06 LAI 013
190.0	183.0	3.8	50.0	13.5	10.0	184	191.4	184.5	192.4	185.5	06 LAI 015

Supplied uncoated. For antireflection coated lenses, append coating suffix from the coating options chart.

We offer the largest selection of stock optics which have been designed and manufactured specifically for use with laser diodes. See page 369.

STEINHEIL ACHROMATIC TRIPLETS

These consist of a low index symmetric convex crown glass element cemented between two identical high index meniscus flint glass elements. Because these triplets are symmetric in structure, they satisfy the "symmetrical principle" at unit conjugate ratio. These lenses will therefore perform well from unit conjugate ratio up to ratios of about 5 to 1. At higher conjugate ratios the symmetry is a handicap, and good achromatic doublets perform better. These triplets are achromatized for the red (656.27nm) and blue (486.13nm) lines of hydrogen (C & F). Because they are symmetric, these lenses are reversible; there is no preferred conjugate plane at any specific conjugate ratio. These lenses make good relay optics for both on- and off-axis applications.

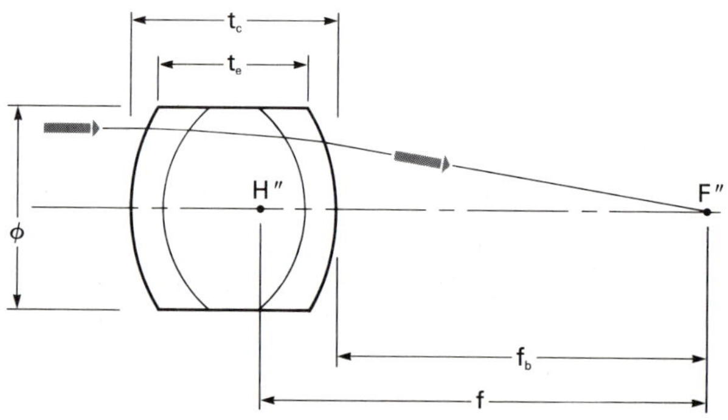

01 LAT

Steinheil Achromatic Triplets

f (mm)	f_b (mm)	ϕ (mm)	t_c (mm)	PRODUCT NUMBER
7.5	5.5	6.5	5.2	01 LAT 011
10.0	7.5	8.5	6.6	01 LAT 009
12.5	9.9	9.5	7.0	01 LAT 003
20.0	16.5	12.5	9.0	01 LAT 013
25.0	21.0	15.0	10.2	01 LAT 005
30.0	25.1	17.0	12.6	01 LAT 015
40.0	33.4	23.0	17.0	01 LAT 007

SPECIFICATIONS: STEINHEIL ACHROMATIC TRIPLETS

Focal Length (f): ±5%
Centration: <0.1mm within lens
Diameter (ϕ): ±0.25mm
Center Thickness (t_c): ±0.25mm
Materials: A Fine Annealed flint for external elements
Grade A Fine Annealed crown for central element
Cosmetic Surface Quality: Pitch-polished, 80-50 scratch and dig
Circumference is ground and outside edges are beveled.

The off-axis performance of finite conjugate achromats can often be dramatically improved by a computer design optimized to the specific performance goals. Melles Griot designers have considerable experience in multi-element designs of this type. Further, our production facilities in California and the Republic of China are equipped to provide reliable delivery at competitive prices for OEM quantitites. Contact one of our Applications Engineers for further details.

CYLINDRICAL LENSES

In this section we describe our broad selection of cylindrical lenses. These are manufactured from the best optical quality glass or synthetic fused silica.

First we list our plano-cylindrical singlets, both positive and negative. These are available in two grades. Next, we describe our unique range of cylindrical doublets, useful for those who need better performance than can be achieved with simple optical elements.

Cylindrical optics, which provide magnification in one dimension only, are very important in laser scanning and many other applications.

Contents **Page Number**

PLANO-CYLINDRICAL GLASS LENSES

These lenses are used in applications requiring magnification in one dimension only, such as in transforming a point image into a line image or changing the height of an image without changing its width or vice versa. Typical applications include slit and line detector array illumination. Earlier remarks made about plano-convex and plano-concave spherical lenses, with regard to aberrations and conjugate ratios, are also applicable to cylindrical lenses, though these image in only one (instead of two) dimensions. This is because the cylinder lens surface profiles are circular, just as the meridional profiles of the spherical lenses are. The decision parameter values previously tabulated for spherical lenses (see page 28) are directly applicable to cylindrical lenses.

A large variety of single and multilayer antireflection coatings is available for these lenses. See the preceding section on coatings (page 69). A corresponding variety of high reflectance coatings is also available through the suffix system allowing any cylindrical lens (whether positive or negative) to be transformed into a cylindrical mirror. Mirror surface radius is found from the lens design index n_0, published lens focal length f_0, and the earlier formula for plano-lens surface radius.

Cylindrical lenses are either plano-convex or plano-concave in form and rectangular in shape. To obtain circular shapes, any of the listed cylindrical lenses can be edge-ground to a specific diameter. Other special shapes are available on request.

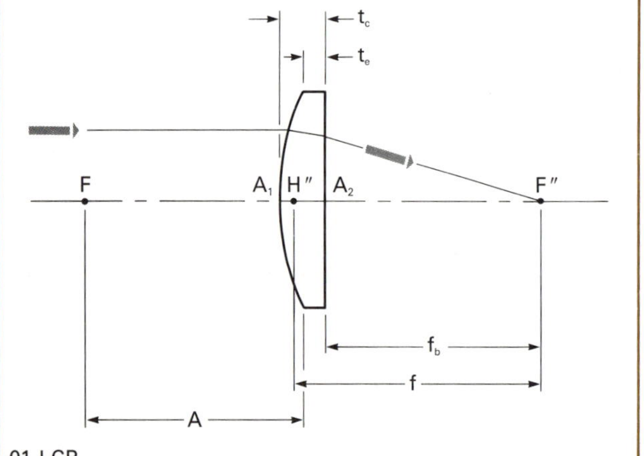

01 LCP

SPECIFICATIONS: PLANO-CYLINDRICAL GLASS LENSES

Paraxial Focal Length (f_0): ±5%
Design Wavelength: (λ_0): 587.1nm
Design Index (n_0): 1.5168 ±0.001
Wedge = $\Delta t_c/x$: ±8 milliradians
Cylinder Surface Profile: Circular
Length (x): ±0.6mm
Width (y): ±0.6mm
Flat Island Width (Δ): 0.5mm ±0.4mm
Clear Aperture Width (y_1): $y_1 = y - 2\Delta$
Material: BK 7, Grade A fine annealed
Cosmetic Surface Quality: Pitch polished, 80-50 scratch and dig
All edge surfaces and flat islands are ground and have a slight protective bevel.

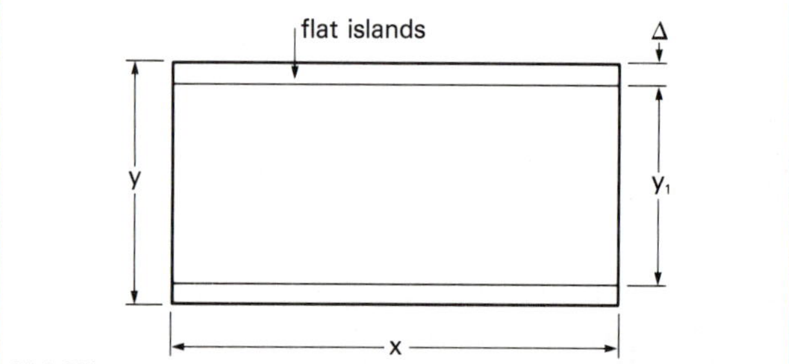

01 LCN

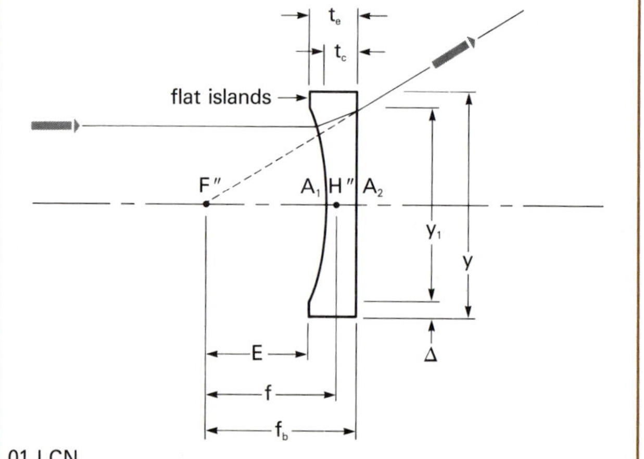

01 LCN

Plano-Cylindrical Glass Lenses, Positive

f_o (mm)	x (mm)	y (mm)	t_c (mm)	t_e (mm)	A_1H (mm)	A_2H'' (mm)	f_b (mm)	A (mm)	f_oy	PRODUCT NUMBER
6.35	25.0	6.0	3.5	1.6	0	-2.3	4.0	8.3	1.06	01 LCP 000
12.7	25.0	12.0	5.0	1.1	0	-3.3	9.4	16.6	1.06	01 LCP 002
19.0	25.0	19.0	8.5	1.2	0	-5.6	13.4	26.3	1.00	01 LCP 004
25.4	60.0	22.0	7.0	1.0	0	-4.6	20.8	31.4	1.15	01 LCP 006
40.0	60.0	15.0	2.7	1.3	0	-1.8	38.2	41.4	2.67	01 LCP 001
40.0	130.0	15.0	2.7	1.3	0	-1.8	38.2	41.4	2.67	01 LCP 003
60.0	60.0	20.0	3.0	1.3	0	-2.0	58.0	61.7	3.00	01 LCP 005
60.0	130.0	20.0	3.0	1.3	0	-2.0	58.0	61.7	3.00	01 LCP 021
80.0	60.0	30.0	4.0	1.2	0	-2.6	77.4	82.8	2.67	01 LCP 009
80.0	130.0	30.0	4.0	1.2	0	-2.6	77.4	82.8	2.67	01 LCP 007
100.0	60.0	50.0	8.0	1.6	0	-5.3	94.7	106.4	2.00	01 LCP 011
150.0	60.0	50.0	6.0	1.9	0	-4.0	146.0	154.1	3.00	01 LCP 013
200.0	60.0	50.0	5.0	1.9	0	-3.3	196.7	203.1	4.00	01 LCP 015
250.0	60.0	50.0	4.5	2.1	0	-3.0	247.0	252.4	5.00	01 LCP 017
300.0	60.0	50.0	4.0	2.0	0	-2.6	297.4	302.0	6.00	01 LCP 019

Plano-Cylindrical Glass Lenses, Negative

f_o (mm)	x (mm)	y (mm)	y_1 (mm)	t_c (mm)	t_e (mm)	A_1H (mm)	A_2H'' (mm)	f_b (mm)	E (mm)	f_o/y_1	PRODUCT NUMBER
-6.35	25.0	7.0	6.0	3.1	5.1	0	-2.0	-8.4	3.34	-1.06	01 LCN 000
-12.7	25.0	13.0	12.0	3.2	7.1	0	-2.1	-14.8	7.71	-1.06	01 LCN 001
-19.0	25.0	16.0	15.0	3.6	7.1	0	-2.4	-21.4	14.29	-1.27	01 LCN 002
-25.4	60.0	16.0	15.0	3.7	6.0	0	-2.4	-27.8	21.79	-1.69	01 LCN 003
-40.0	60.0	16.0	15.0	4.6	6.0	0	-3.0	-43.0	37.02	-2.67	01 LCN 004
-60.0	60.0	21.0	20.0	4.4	6.1	0	-2.9	-62.9	56.84	-3.00	01 LCN 005
-80.0	60.0	31.0	30.0	3.2	6.0	0	-2.1	-82.1	76.09	-2.67	01 LCN 006
-100.0	60.0	51.0	50.0	3.0	9.4	0	-2.0	-102.0	92.53	-2.00	01 LCN 007
-150.0	60.0	51.0	50.0	5.4	9.5	0	-3.6	-153.6	144.02	-3.00	01 LCN 008
-200.0	60.0	51.0	50.0	6.5	9.6	0	-4.3	-204.3	194.72	-4.00	01 LCN 009
-250.0	60.0	51.0	50.0	7.0	9.4	0	-4.6	-254.6	245.17	-5.00	01 LCN 011
-300.0	60.0	51.0	50.0	7.5	9.5	0	-4.9	-304.9	295.42	-6.00	01 LCN 012

Convenient holders for Melles Griot cylindrical lenses will be found on page 455.

PRECISION PLANO-CYLINDRICAL GLASS LENSES

These precision cylindrical lenses meet much more stringent specifications than the preceding cylindrical lenses, therefore serving a variety of more demanding applications. These include code scanning, holography, optical information processing and computing, laser projection, and the illumination of slits or linear detector arrays.

Each precision cylindrical lens is individually compared by interference methods to a master test plate to make sure that surface accuracy and irregularity specifications are met. The approach to diffraction limited performance as f-number is increased is much more rapid than for ordinary cylindrical lenses. Precision cylindrical lenses are offered only in plano-shapes.

A large variety of antireflection coatings and high reflectance coatings (for converting lenses to mirrors) is available for precision cylindrical lenses via the Coating Suffix system (see page 69). Edge grinding service is also available if change of shape or size is desired.

In the specifications which follow, optical centerline and mechanical centerline have special meanings appropriate for cylindrical lenses. The optical centerline is that axis which passes through the center of curvature of the cylinder and is orthogonal to the plano-surface. Mechanical centerline means the axis of symmetry of the lens paralleling the x-direction. The specified upper limit on the separation of these centerlines is applicable throughout the plano surface.

SPECIFICATIONS: PRECISION PLANO-CYLINDRICAL GLASS LENSES

Paraxial Focal Length (f_o): $\pm 2\%$
Design Wavelength (λ_o): 587.6nm
Design Index (n_o): 1.5168 + 0.001
Surface Accuracy: Cylindrical to within $\lambda/2$ in y direction, and $\lambda/4$ per centimeter in the x-direction, at 546.1nm.
Wedge $=\Delta t_c/x$: <1.0 milliradian within lens
Centration: ≤ 0.1mm and 3 arc minutes.
Length (x): ± 0.1mm Width (y): ± 0.14mm
Flat Island Width (Δ): 0.5mm, +0.1mm, -0.4mm
Clear Aperture Width (y_1): $y_1 = y - 2\Delta$
Center Thickness (t_c): $\pm .5$mm
Material: BK 7, Grade A fine annealed
Cosmetic Surface Quality: Pitch polished, 60-40 scratch and dig

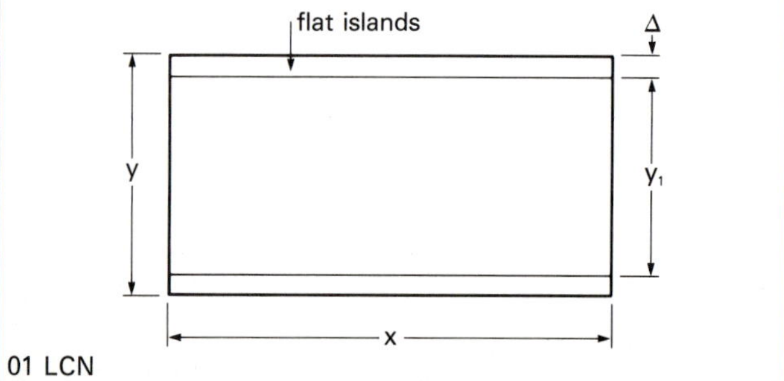

01 LCN

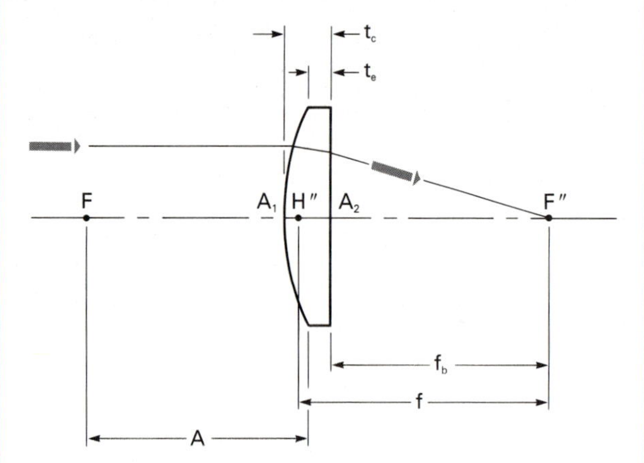

01 LCP

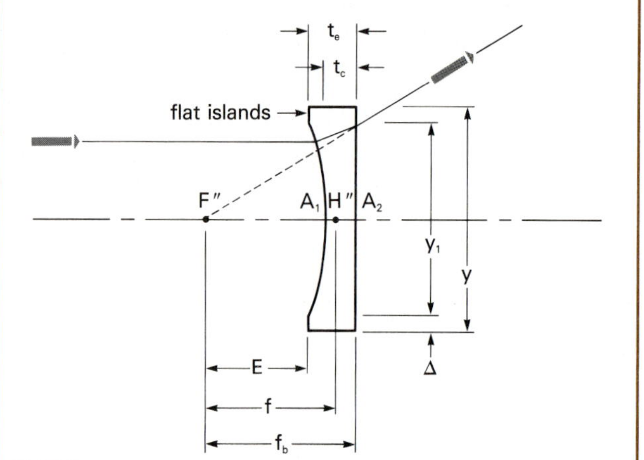

01 LCN

Precision Plano-Cylindrical Glass Lenses, Positive

f_o (mm)	x (mm)	y (mm)	t_c (mm)	t_e (mm)	A_1H (mm)	A_2H'' (mm)	f_b (mm)	A (mm)	f_o/y	PRODUCT NUMBER
6.35	25.0	6.0	2.95	1.0	0	-2.0	4.4	8.3	1.06	01 LCP 124
12.7	25.0	11.0	4.00	1.0	0	-2.6	10.1	15.7	1.15	01 LCP 125
19.0	25.0	16.0	5.10	1.0	0	-3.4	15.6	23.1	1.19	01 LCP 126
22.2	20.0	12.5	5.75	3.9	0	-3.8	18.4	24.1	1.78	01 LCP 123
25.4	60.0	22.0	7.00	1.0	0	-4.6	20.8	31.4	1.15	01 LCP 127
38.1	60.0	26.0	6.90	2.0	0	-4.6	33.5	43.0	1.47	01 LCP 129
40.0	60.0	15.0	5.75	4.3	0	-3.8	36.2	41.4	2.67	01 LCP 128
50.0	60.0	26.0	5.50	2.0	0	-3.6	46.4	53.5	1.92	01 LCP 133
60.0	60.0	20.0	5.75	4.1	0	-3.8	56.2	61.7	3.00	01 LCP 136
76.2	60.0	26.0	4.20	2.0	0	-2.8	73.4	78.4	2.93	01 LCP 134
80.0	60.0	30.0	5.75	2.9	0	-3.8	76.2	82.0	2.67	01 LCP 145
100.0	60.0	50.0	8.89	2.4	0	-5.9	94.1	106.4	2.00	01 LCP 155
150.0	60.0	50.0	5.75	1.6	0	-3.8	146.2	154.1	3.00	01 LCP 166
200.0	60.0	50.0	6.35	3.3	0	-4.2	195.8	203.1	4.00	01 LCP 178
250.0	60.0	50.0	5.40	3.0	0	-3.6	246.4	252.4	5.00	01 LCP 135
300.0	60.0	50.0	6.00	4.0	0	-4.0	296.0	302.0	6.00	01 LCP 137

Precision Plano-Cylindrical Glass Lenses, Negative

f_o (mm)	x (mm)	y (mm)	y_1 (mm)	t_c (mm)	t_e (mm)	A_1H (mm)	A_2H'' (mm)	f_b (mm)	E (mm)	f_o/y_1	PRODUCT NUMBER
-6.35	25.0	6.0	5.0	5.0	6.2	0	-3.3	-9.7	3.49	-1.27	01 LCN 123
-12.7	25.0	11.0	10.0	5.0	7.3	0	-3.3	-16.0	8.68	-1.27	01 LCN 124
-19.0	25.0	16.0	15.0	5.0	8.5	0	-3.3	-22.3	13.81	-1.27	01 LCN 125
-25.4	60.0	16.0	15.0	5.0	7.4	0	-3.3	-28.7	21.34	-1.69	01 LCN 126
-38.1	60.0	26.0	25.0	5.0	9.5	0	-3.3	-41.4	31.92	-1.52	01 LCN 127
-40.0	60.0	26.0	25.0	5.0	9.2	0	-3.3	-43.3	34.09	-1.60	01 LCN 128
-50.0	60.0	26.0	25.0	5.0	8.2	0	-3.3	-53.3	45.07	-2.00	01 LCN 129
-60.0	60.0	26.0	25.0	5.0	7.6	0	-3.3	-63.3	55.67	-2.40	01 LCN 133
-76.2	60.0	26.0	25.0	5.0	7.0	0	-3.3	-79.5	72.46	-3.05	01 LCN 134
-80.0	60.0	26.0	25.0	5.0	6.9	0	-3.3	-83.3	76.36	-3.20	01 LCN 135
-100.0	60.0	26.0	25.0	5.0	6.5	0	-3.3	-103.3	96.76	-4.00	01 LCN 136
-150.0	60.0	26.0	25.0	5.0	6.0	0	-3.3	-153.3	147.28	-6.00	01 LCN 137
-200.0	60.0	26.0	25.0	5.0	5.8	0	-3.3	-203.3	197.54	-8.00	01 LCN 138
-250.0	60.0	26.0	25.0	5.0	5.6	0	-3.3	-253.3	247.69	-10.00	01 LCN 139
-300.0	60.0	26.0	25.0	5.0	5.5	0	-3.3	-303.3	297.79	-12.00	01 LCN 144

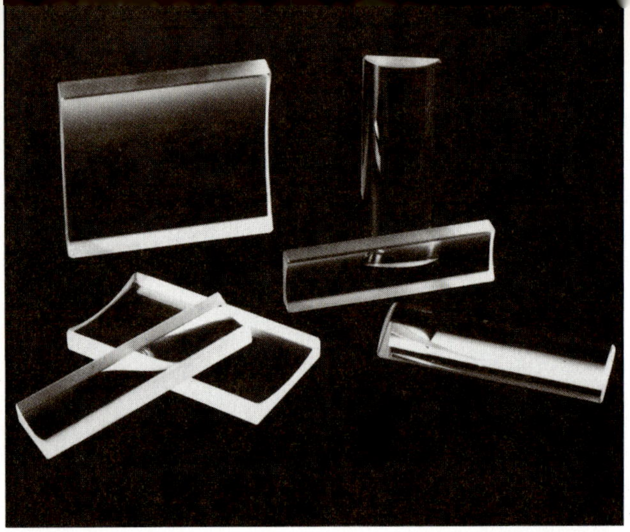

PRECISION PLANO-CYLINDRICAL SYNTHETIC FUSED SILICA LENSES

Except for material the present lenses meet essentially the same specifications, and are suited for the same variety of applications as the precision glass cylindrical lenses. See the earlier discussion of precision plano-cylindrical glass lenses.

Synthetic fused silica cylinders are superior to the corresponding glass cylinders in that they have a useful spectral range extending deep into the ultraviolet and into the near-infrared. They have low dispersion, a low coefficient of thermal expansion, are resistant to thermal shock, and are resistant to scratching. Their internal transmittances from the near-ultraviolet to the near-infrared are extremely high. (See the earlier presentation of synthetic fused silica properties in the materials section on page 55). Because the visible refractive index is slightly less than that of BK 7, spherical aberration is slightly worse than in the glass. As a specific comparison, positive cylinders of 300mm focal length have full focal linewidths, under optimum conditions at 633nm, of 25.1 microns and 26.6 microns respectively, for BK 7 and synthetic fused silica.

In the specifications which follow, optical centerline and mechanical centerline have special meanings appropriate for cylindrical lenses. The optical centerline is that axis which passes through the center of curvature of the cylinder and is orthogonal to the plano-surface. Mechanical centerline means the axis of symmetry of the lens paralleling the x-direction. The specified upper limit on the separation of these centerlines is applicable throughout the plano surface.

SPECIFICATIONS: PRECISION PLANO-CYLINDRICAL SYNTHETIC FUSED SILICA LENSES

Paraxial Focal Length (f_o): ±2%
Design Wavelength (λ_o): 587.6nm
Design Index (n_o): 1.4585 ±0.00003
Surface Accuracy: Cylindrical to within $\lambda/2$ in y direction, and $\lambda/4$ per centimeter in the x-direction, at 546.1nm
Wedge = $\Delta t_c/x$: <1.0 milliradian within lens
Centration: ≤0.1mm and 3 arc minutes.
Length (x): ±0.1mm Width (y): ±0.14mm
Flat Island Width (Δ): 0.5mm, +0.1mm, -0.3mm
Clear Aperture Width (y_1): $y_1 = y - 2\Delta$
Center Thickness (t_c): ±0.5mm
Material: UV grade synthetic fused silica
Cosmetic Surface Quality: Pitch polished, 60-40 scratch and dig

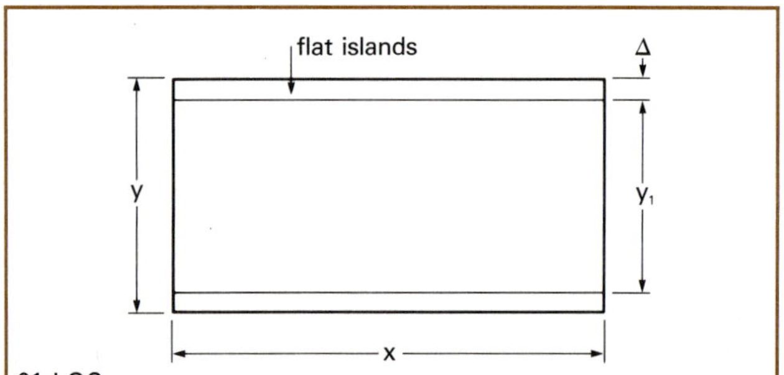

01 LQC

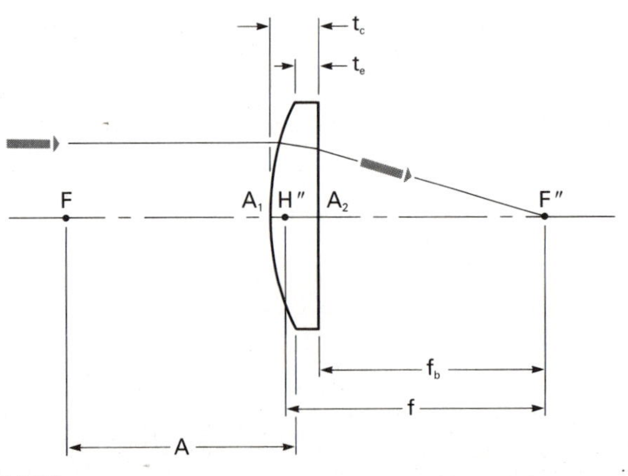

01 LQC

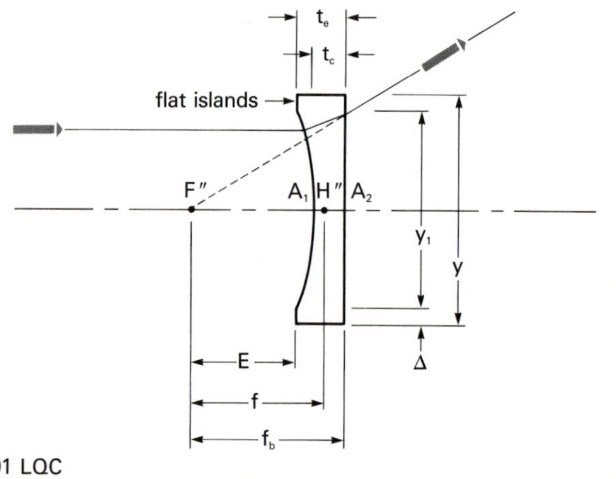

01 LQC

Precision Plano-Cylindrical Synthetic Fused Silica Lenses, Positive

f_o (mm)	x (mm)	y (mm)	t_c (mm)	t_e (mm)	A_1H (mm)	A_2H'' (mm)	f_b (mm)	A (mm)	f_o/y	PRODUCT NUMBER
12.7	25.0	10.0	3.84	1.0	0	-2.63	10.1	15.5	1.27	01 LQC 000
19.0	25.0	15.0	5.28	1.0	0	-3.62	15.4	23.3	1.27	01 LQC 001
25.0	60.0	15.0	4.79	2.0	0	-3.29	21.7	27.8	1.67	01 LQC 002
40.0	60.0	26.0	7.40	2.0	0	-5.08	34.9	45.4	1.54	01 LQC 003
50.0	60.0	26.0	6.04	2.0	0	-4.14	45.9	54.0	1.92	01 LQC 004
76.2	60.0	26.0	4.51	2.0	0	-3.09	73.1	78.7	2.93	01 LQC 005
100.0	60.0	26.0	3.88	2.0	0	-2.66	97.3	101.9	3.85	01 LQC 006
200.0	60.0	26.0	3.93	3.0	0	-2.69	197.3	200.9	7.69	01 LQC 007
300.0	60.0	26.0	3.62	3.0	0	-2.48	297.5	300.6	11.54	01 LQC 008

Precision Plano-Cylindrical Synthetic Fused Silica Lenses, Negative

f_o (mm)	x (mm)	y (mm)	y_1 (mm)	t_c (mm)	t_e (mm)	A_1H (mm)	A_2H'' (mm)	f_b (mm)	E (mm)	f_o/y	PRODUCT NUMBER
-12.7	25.0	10.0	9.0	5.0	7.1	0	-3.4	-16.1	9.0	-1.41	01 LQC 009
-19.0	25.0	15.0	14.0	5.0	8.5	0	-3.4	-22.4	13.9	-1.36	01 LQC 011
-25.4	60.0	15.0	14.0	5.0	7.3	0	-3.4	-28.8	21.5	-1.81	01 LQC 012
-40.0	60.0	26.0	25.0	5.0	9.9	0	-3.4	-43.4	33.5	-1.60	01 LQC 013
-50.0	60.0	26.0	25.0	5.0	8.7	0	-3.4	-53.4	44.7	-2.00	01 LQC 014
-76.2	60.0	26.0	25.0	5.0	7.3	0	-3.4	-79.6	72.3	-3.05	01 LQC 015
-100.0	60.0	26.0	25.0	5.0	6.7	0	-3.4	-103.4	96.7	-4.00	01 LQC 016
-200.0	60.0	26.0	25.0	5.0	5.9	0	-3.4	-203.4	197.6	-8.00	01 LQC 017
-300.0	60.0	26.0	25.0	5.0	5.6	0	-3.4	-303.4	297.9	-12.00	01 LQC 018

A large variety of antireflection coatings is available for precision synthetic fused silica cylindrical lenses via the coating suffix system (see page 69). Edge grinding service is also available if change of shape or size is desired.

Convenient holders for Melles Griot cylindrical lenses will be found on page 455.

PRECISION OPTIMIZED CYLINDRICAL AIR-SPACED ACHROMATS

Applications of cylindrical air-spaced achromatic doublets of this high quality include: boxcar and container code scanners, Fourier transform holography, acoustic imaging, holographic astigmatism correction, optical information processing, optical computing, spatial frequency spectrum analysis, multichannel coherent filtering, and processing of synthetic aperture (side-looking) radar data.

In an air-spaced doublet the shapes of four air-glass interfaces, the element separation, element thicknesses, and two material choices can all be controlled to achieve performance far superior to that of a cemented doublet. In the present series of cylindrical doublets, durable HEBBAR™ (high efficiency broadband antireflective) coatings are used on all four air-glass interfaces, thus assuring that no surface reflectance penalty is paid in exchange for the other advantages gained. Cylindrical doublets are optimized for maximum polychromatic performance at full aperture and infinite conjugate ratio when used alone.

Several physical limitations are imposed on doublet performance. Diffraction requires that cylinder doublet MTF curves must fall at or below the diffraction limit line for which we have previously presented the formula. Aberrations also limit performance. Most applications require a flat or nearly flat focal plane, and for these focal surface curvature ("curvature of field") is one of the two most important residual aberrations. Petzval's theorem implies that three or more elements must be present if this curvature is to be suppressed. We offer a third field flattener element for this purpose. Unless it is used, sharpest focus on a flat plane can be obtained only along the intersection with the true focal surface. True focal surface shape departs from the Petzval surface, and instead conforms to the tangential astigmatic surface. The sagittal astigmatic surface does not exist for cylindrical lenses. Thus astigmatism exists only to shape the true focal surface, but does not impair image quality on this surface. For this reason the off-axis performance of a cylindrical lens is superior to that of a spherical lens of the same material and meridional profile. Also for this reason, field flatteners do much more for cylindrical lenses than for spherical lenses. Astigmatic blurring in spherical lenses is caused by separation of the sagittal and tangential surfaces.

The true shape of the focal surface of a cylindrical doublet is that of a parabolic cylinder. In the discussions which follow, "curved field" always refers to MTF on the circular cylindrical surface best approximating the true focal surface. "Flat field" refers to MTF on the optimum choice of flat planes intersecting the true focal surface.

Coma is an almost insignificant aberration in the present series of optimized cylindrical doublets. It is evident only in the three lowest f-number doublets (approximately f/4.0) and only at the largest field angle (2.5 degrees). Under these conditions coma does cause the monochromatic MTF curves to fall slightly below the diffraction limit line. Coma suppression is one of the major results of the optimization. Spherical aberration is completely absent.

These cylindrical achromats are all manufactured to precision test-plate specifications, as determined by tolerance effect studies. Assembly tolerances are likewise determined. This assures that the doublets will perform as expected. No elements are plano. Cylindrical achromatic doublets are supplied in a sturdy metal mount with element positions securely set.

For the benefit of those customers who may be able to exploit the advantages of curved field performance (with curved image sensors, fiber optics faceplates, refocusing, or by other means), we have tabulated below the radii of the cylinders most closely approximating the optimum focal surface.

SPECIFICATIONS: PRECISION CYLINDRICAL AIR-SPACED ACHROMATS

Paraxial Focal Length (f) at 550 nm: $\pm 2\%$
Design Wavelengths: 450nm, 550nm, and 650nm
Transmission Range: 340nm to 2500nm
Surface Accuracy: Cylindrical to within $\lambda/4$ at 546.1nm over 90% of aperture
Field of View: ± 2.5 degrees
MTF Variation in x-direction: ≤ 0.04
Mount Diameter (ϕ): 50.0mm ± 0.25mm
Clear Aperture (y_1, x): ± 0.5mm
Endplane Separation (E_1 to E_2): 25.0mm ± 0.25mm
Cylinder Surface Profiles: Circular
Materials: Grade A fine annealed crown and flint glasses. Cell is 6061 aluminum, black anodized.
Maximum Service Temperature: 93 °C
Cosmetic Surface Quality: 60-40 scratch and dig
Coating: HEBBAR™ for normal incidence, all surfaces
Engraved arrow points to focal plane.

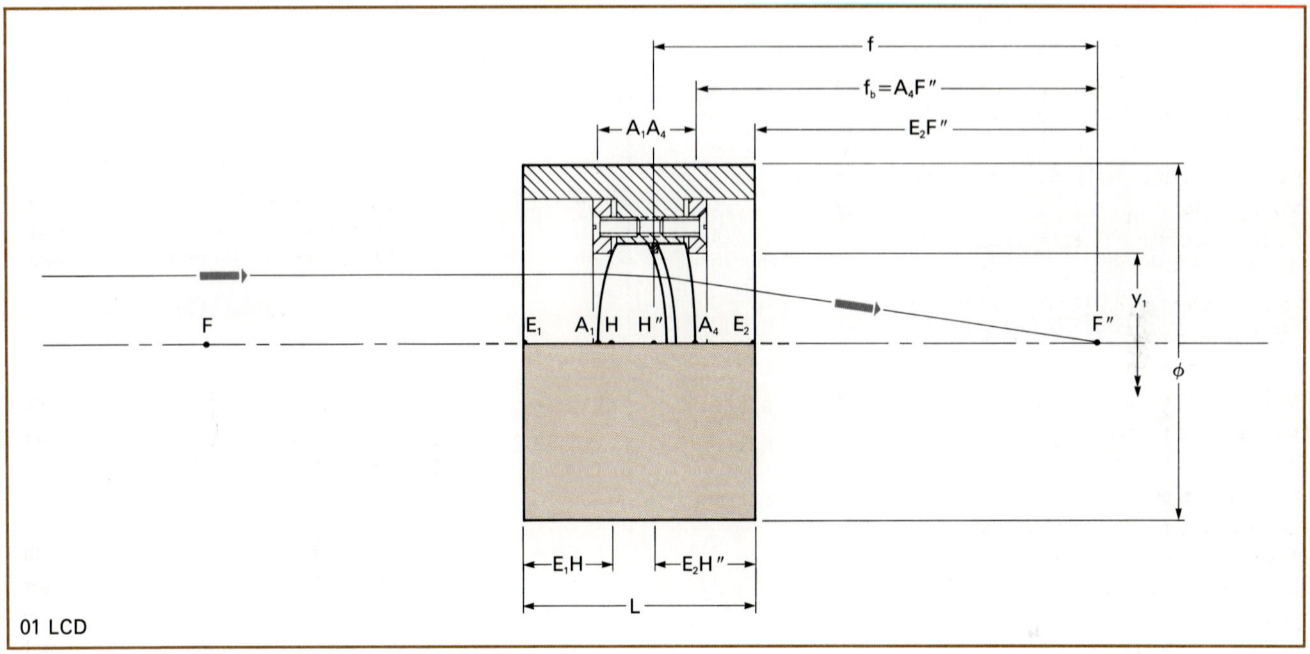

Precision Computer-Optimized Cylindrical Doublet Air-Spaced Achromats

f (mm)	Clear Aperture width y_1 (mm)	Clear Aperture length x (mm)	Combination Principal Points A_1H (mm)	Combination Principal Points A_4H'' (mm)	E_1H (mm)	E_2H'' (mm)	f_b (mm)	Rear Flange Focal Length E_2F'' (mm)	f/y_1	Focal Surface Radius (mm)	PRODUCT NUMBER
50.8	12.7	25.0	0.45	-4.5	11.4	-10.7	46.3	40.1	4.00	14.31	01 LCD 000
76.2	19.0	25.0	0.48	-4.8	10.9	-10.9	71.4	65.3	4.01	21.29	01 LCD 001
101.6	25.0	25.0	0.50	-5.2	10.4	-11.2	96.4	90.4	4.06	28.28	01 LCD 002
127.0	25.0	25.0	0.51	-4.8	10.8	-10.9	122.2	116.1	5.08	35.25	01 LCD 003
152.4	25.0	25.0	0.51	-4.6	11.1	-10.7	147.8	141.7	6.10	42.20	01 LCD 004
203.2	25.0	25.0	0.51	-4.3	11.5	-10.5	198.9	192.7	8.13	56.10	01 LCD 005
254.0	25.0	25.0	0.52	-4.2	11.7	-10.4	249.8	243.6	10.16	69.99	01 LCD 006
304.8	25.0	25.0	0.52	-4.1	11.9	-10.3	300.7	294.5	12.19	83.90	01 LCD 007
457.2	25.0	25.0	0.53	-3.9	12.1	-10.1	453.3	447.0	18.29	125.59	01 LCD 008
508.0	25.0	25.0	0.52	-3.8	12.2	-10.1	504.2	497.9	20.32	139.49	01 LCD 009

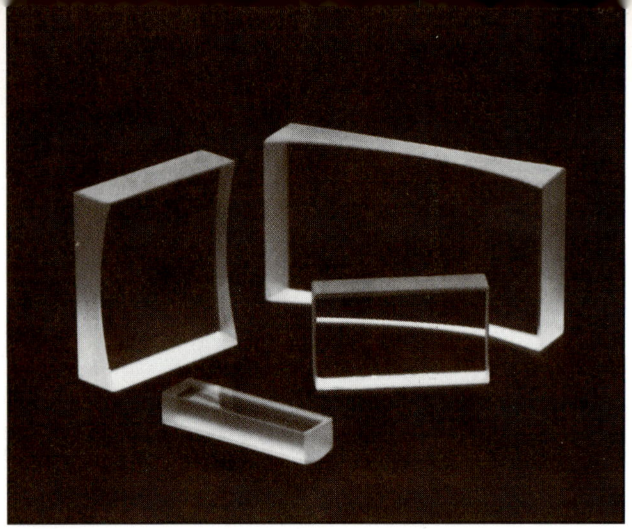

PRECISION OPTIMIZED CYLINDRICAL FIELD FLATTENERS

These are companion elements for the precision optimized cylindrical doublet achromats (LCD Series) previously presented. It is intended that these doublets be operated at full aperture with their companion flatteners. Flatteners are offered for all cylinder doublets.

Flatteners introduce the spherical aberration associated with a "plate" which slightly degrades the axial performance. However the flattened-field makes the companion doublet useful out to its full field angle of $\pm$ 2.5 degrees.

The correct position for the optimized field flattener is with its flat surface in a plane 2mm $\pm$0.5mm toward the doublet from the experimentally determined plane of best focus for the combination. Within this plane the flattener should be centered on the doublet focal line to within 2% of the flattener y-dimension for best results. The curved surface must face the doublet.

Field flatteners are supplied unmounted only. Their dimensions admit field angles up to 2.5 degrees at infinite conjugate ratio. Field angles will be more restricted, and field flattener effectiveness will be diminished, at other conjugate ratios.

In the specifications which follow, optical centerline and mechanical centerline have special meanings appropriate for cylindrical field flatteners. The optical centerline is that axis which passes through the center of curvature of the cylinder and is orthogonal to the plano-surface. Mechanical centerline means the axis of symmetry of the lens paralleling the x-direction. The specified upper limit on the separation of these centerlines is applicable throughout the plano surface. The maximum angle or centration between these centerlines is 4.0 milliradians.

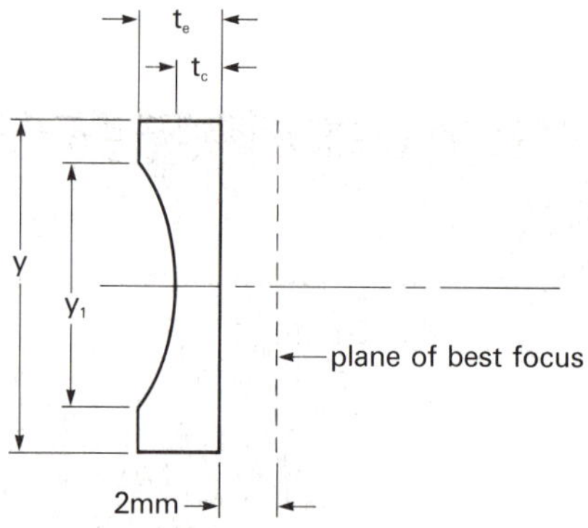

01 LCF

SPECIFICATIONS: PRECISION OPTIMIZED CYLINDRICAL FIELD FLATTENERS

Field of View: 2.5 degrees
Maximum MTF variation along combination image line (x-direction): 0.04 at any spatial frequency if flattener is correctly positioned
Centration: $<$0.05mm and 4.0 milliradian
Length (x): 25.0mm $\pm$0.1mm
Width (y): $\pm$0.1mm
Center Thickness (t_c): 5.0mm $\pm$0.5mm
Edge Thickness (t_e): $\pm$0.5mm
Material: BK 7, Grade A Fine Annealed
Cosmetic Surface Quality: Pitch-polished, 60-40 scratch and dig
Coating: HEBBAR™ for normal incidence on both surfaces
Edge surfaces and flat islands are ground and have a slight protective bevel.

See page 455 for a convenient range of holders for cylindrical lenses.

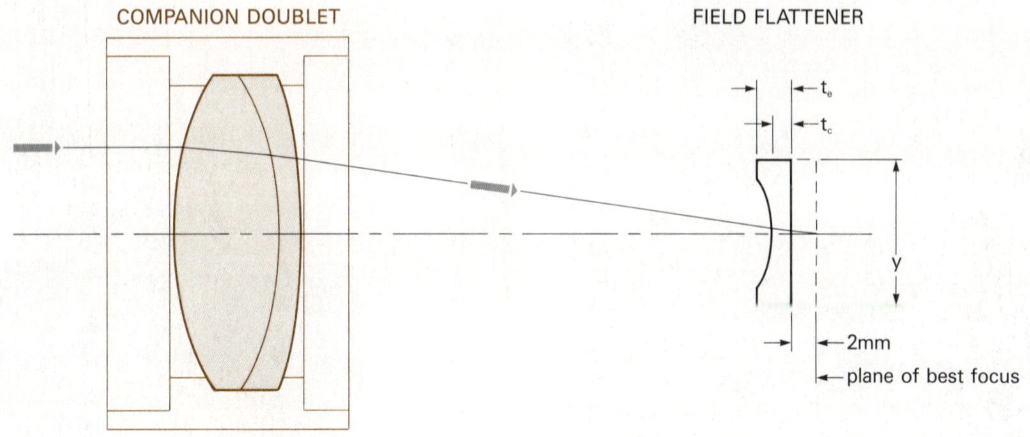

01 LCF

Precision Optimized Cylindrical Field Flatteners

x (mm)	y (mm)	t_c (mm)	t_e (mm)	COMPANION DOUBLET PRODUCT NUMBER	PRODUCT NUMBER
25.0	4.44	5.0	*4.7	01 LCD 000	01 LCF 000
25.0	6.65	5.0	*4.4	01 LCD 001	01 LCF 001
25.0	9.87	5.0	5.5	01 LCD 002	01 LCF 002
25.0	12.09	5.0	5.7	01 LCD 003	01 LCF 003
25.0	14.31	5.0	5.9	01 LCD 004	01 LCF 004
25.0	18.74	5.0	6.3	01 LCD 005	01 LCF 005
25.0	23.18	5.0	6.7	01 LCD 006	01 LCF 006
25.0	27.62	5.0	7.1	01 LCD 007	01 LCF 007
25.0	40.92	5.0	8.4	01 LCD 008	01 LCF 008
25.0	45.36	5.0	8.9	01 LCD 009	01 LCF 009

*Contrary to common expectations, these two optimized field flatteners are positive.

OPTICAL SYSTEMS

The Melles Griot Optical Systems Division in Rochester, New York, designs and builds prototypes and volume manufactures complex, multi-element lens systems for the Original Equipment Manufacturer. Many years of experience in sophisticated optical design provide you with a unique set of advantages, including the know-how to create systems which are reliable, efficient and buildable, systems which combine optimized specifications with affordable prices. Examples include:

LASER SCANNING LENSES

F-theta lenses which are designed to provide an output scan length (over a flat field) which is linearly proportional to the field angle. Scan lens systems have been designed and manufactured to cover a variety of specific user requirements. Focal lengths from 60 to 540mm with scan image lengths from 33 to 466mm and spot sizes averaging 20 microns are available as standard designs. Modified designs, for the user's specific requirements, are easily developed from these standard versions.

CATHODE RAY TUBE LENSES

A complete family of lenses designed to image the face of cathode ray tubes onto a variety of imaging mediums is available. These lenses are designed specifically to deal with the inherent illumination and field problems associated with CRT imaging systems. Existing designs with focal lengths from 20 to 300mm and nominal magnifications from 2:1 to 1:2 are available with spectral correction for a wide variety of specific phosphors. Focal ratios for this type of lens are normally in the region of f/2.0, while half-field angle converage runs from 15 to 30 degrees. Units with built-in Melles Griot mechanical or electronic shutters are available as standard.

FACSIMILE SCAN LENSES

Electronic data transmission systems require an optical system design which combines specific MTF requirements with unusual spectral and photometric demands. Our experience in designing and manufacturing lens systems of this type provides the know-how to design systems which efficiently meet the performance demands while satisfying the user's magnification requirements in what typically is a very restricted volume of space.

IMAGE SENSING SYSTEMS

Optical systems which utilize solid-state photodiode arrays require special designs to assure image recognition. We have designed and manufactured a large variety of lenses of this type for specific pattern and bar code recognition systems and we are fully aware of the specific needs and requirements for these optical systems.

Contact your nearest Melles Griot office or representative for further information.

CONDENSERS

Condensers are a special category of lenses which have been optimized for efficient collection of light. They are used in illumination or collection systems whenever energy throughput is the most important concern.

Here we describe two types of condenser. The first is an aspheric glass singlet and the second is a two element condenser of pyrex or fused silica.

Contents **Page Number**

ASPHERIC GLASS CONDENSER LENSES

The aspheric lens is designed to have a much shorter focal length than is possible with a spherical glass lens of equal diameter and equal spherical aberration. It has one aspheric surface only, the other surface being either plano, spherical convex, or spherical concave. A correctly formed aspheric lens surface exactly cancels the spherical aberration that would otherwise be present in an optical system, or reduces both spherical aberration and coma to insignificance, in which case the system is said to be aplanatic. Thus more energy can be concentrated into a small area, such as the entrance pupil of a projection lens system, or the sensitive area of a detector, than would otherwise be possible. Without the aspheric surface, marginal rays (which enter or exit the lens near its edge) may be severely aberrated and fail to contribute to the illumination of the desired target. The aspheric surface puts marginal rays on-target, and allows the potential collecting area of the system to be fully utilized at f-numbers as low as 0.6.

Aspheric lenses are ideally suited for low f-number and high throughput applications, such as elements adjacent to the source in condensing, projection, and illumination systems. Other applications include optical communications equipment, smoke and fire (turbulence) alarms, pollution monitors, intrusion alarms, and chemical vapor alarms. They should be considered whenever highly efficient use of either sources or detectors is important, or whenever space is limited and short focal lengths or low f-numbers are advantageous. Aspherics save energy.

The schematic drawing of a typical projection system shows an aspheric lens used in combination with a plano-convex spherical lens. The aspheric surface faces the long conjugate (in this case, away from the source), making ray deviations at the two surfaces more nearly equal. This minimizes reflection losses. For this orientation, the contour of the aspheric surface has been chosen to minimize the aberration of combined aspheric and plano-convex lenses, thereby concentrating maximum power into the projection lens entrance pupil.

Lens combination conjugate ratios can be adjusted, with aberration minimized, by changing the focal length of the spherical plano-convex element. For ratios near unity, the plano-convex lens should be replaced by a second aspheric element, oriented so that the aspheric surfaces face each other. For ratios much less than unity, the positions of the aspheric and plano lenses should be interchanged.

Aspheric lenses are used either in lens combinations, as shown, or alone at large or infinite conjugate ratio, as with a detector in a radiometer, or in a narrow-beam spotlamp. Where sharp focus is required the plano aspherics are especially effective, forming extremely small focal spots and sharp axial imagery, used alone or in combinations, even at full aperture.

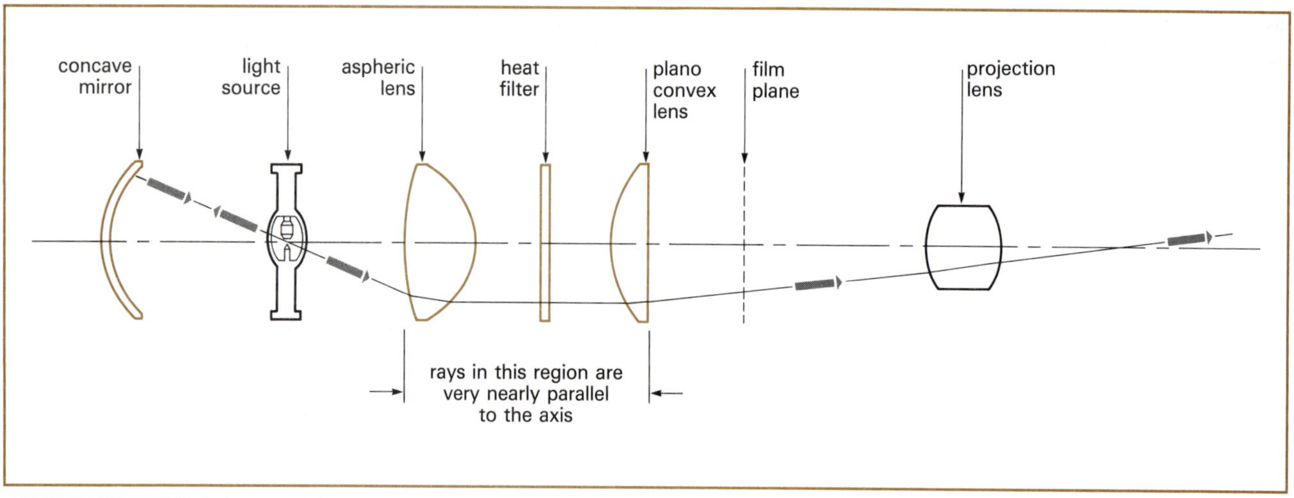

ASPHERIC LENS being used in a projection system. Both lenses and the filter in the condenser are sometimes antireflection coated.

We offer precision molded aspherics with three distinct second surfaces: plano, spherical-convex, or spherical-concave. Each type has its particular advantage. The plano back aspherics minimize aberration. The spherical-convex back aspherics have the lowest f-numbers and the highest transmissions, because equal deviation of marginal rays is most closely approached in this form. The spherical-concave back aspherics offer more clearance for the light source and forced-air cooling. Concave back aspherics are heated more uniformly than the other shapes when used in close proximity to powerful light sources.

The spherical or plano sides of these lenses are felt-polished, and the aspheric surfaces are molded. Extreme care is taken in the fabrication and polishing of the molds, resulting in the finest molded aspheric surfaces commercially available. When scattering of laser light from these surfaces is visually compared with that from felt-polished surfaces, no significant differences are seen.

Only single layer antireflection coatings (/066) are recommended for the aspheric surface. Multilayer coatings do not work well due to the steep curvature of the surface. Single layer coatings are reasonably effective however, even at the large angles of incidence encountered on the steep aspheric surfaces. Coatings provide a much larger improvement in the overall transmission of aspheric lenses than for other types of lenses because of the steep surfaces.

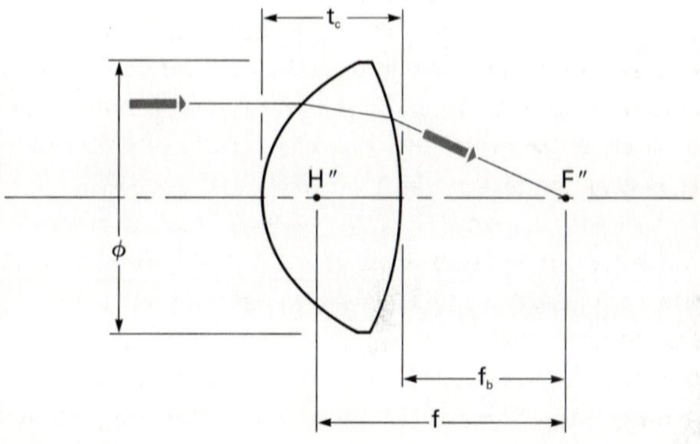

01 LAG

SPECIFICATIONS: ASPHERIC GLASS CONDENSER LENSES

Focal Length (f): ±7%
Diameter (ϕ): ±0.4mm
Center Thickness (t_c): ±7% down to ±0.5mm
Material: Optical Crown
n_d=1.523, vd=58.8, n_f - n_c=0.0089
Cosmetic Surface Quality: Aspheric side precision molded, 80-50 scratch and dig; Flat or spherical side felt-polished, 80-50 scratch and dig
Maximum Service Temp.: 177°C (350°F),

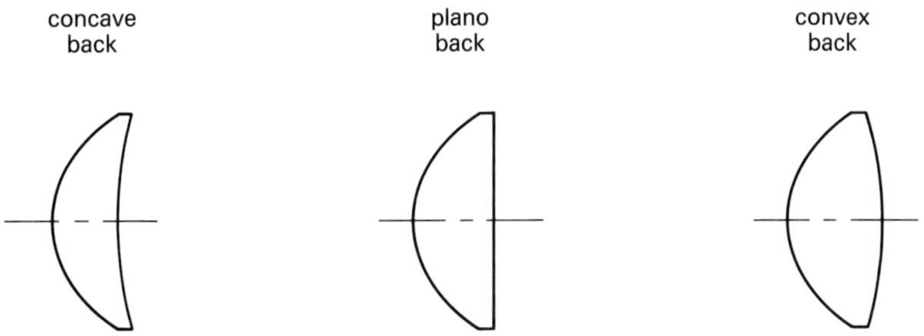

CROSS-SECTIONAL VIEWS of three standard types of aspheric lenses.

Aspheric Glass Condenser Lenses

f (mm)	ϕ (mm)	t_c (mm)	f_b (mm)	f/ϕ	NAS* Shape	PRODUCT NUMBER
8.5	12.0	5.5	5.8	0.71	CX	01 LAG 000
12.0	15.0	5.5	8.4	0.8	P	01 LAG 001
12.0	17.0	8.0	6.7	0.71	P	01 LAG 002
12.0	18.0	8.8	6.9	0.67	CX	01 LAG 111
15.5	18.0	7.0	10.9	0.86	P	01 LAG 003
17.0	19.0	7.0	12.4	0.89	P	01 LAG 004
18.0	24.0	10.0	11.4	0.75	P	01 LAG 005
23.5	32.5	13.5	15.1	0.72	CX	01 LAG 115
25.0	32.5	13.0	16.5	0.77	P	01 LAG 117
26.5	30.0	11.0	19.3	0.88	P	01 LAG 007
29.5	40.0	15.5	20.9	0.74	CX	01 LAG 009
33.0	52.0	25.0	16.6	0.63	P	01 LAG 010
34.5	38.0	12.0	26.6	0.91	P	01 LAG 012
35.0	35.0	11.5	27.4	1.00	P	01 LAG 119
35.0	50.0	20.6	22.9	0.70	P	01 LAG 013
36.0	40.0	19.4	19.7	0.90	CC	01 LAG 014
37.0	52.0	23.0	21.9	0.71	P	01 LAG 123
38.5	50.0	20.0	24.3	0.77	P	01 LAG 125
38.5	55.0	26.0	23.2	0.70	CX	01 LAG 015
39.0	60.0	27.5	25.0	0.65	CX	01 LAG 017
42.0	45.0	15.7	30.1	0.93	P	01 LAG 011
43.0	60.0	23.0	29.3	0.72	CX	01 LAG 018
50.0	75.0	29.1	31.2	0.67	P	01 LAG 023
53.0	65.0	23.5	37.6	0.82	P	01 LAG 019
53.0	68.0	24.0	36.6	0.78	CX	01 LAG 127
79.0	84.0	26.0	60.1	0.94	CC	01 LAG 025

*Shape of non-aspheric side (NAS) is plano (P), convex (CX), or concave (CC).

Edge grinding and shaping services are available to modify these lenses to meet the material constraints of your system.

TWO-ELEMENT AIR-SPACED CONDENSERS

Condensers are used in all applications where one wishes to project or display an image of an illuminated object. Examples are film and slide projectors, integrated circuit mask projectors, optical comparators, profile gages, and similar equipment. A condenser has two primary duties in projection systems. It must first transfer radiant energy efficiently from a light source to the entrance pupil of a projection lens system, so that brightness of the image finally projected is maximized. Secondly, it must provide highly uniform illumination of the object plane or film gate, which is the location of a slide, motion picture, or microfilm frame to be projected.

Our two-element, air-spaced condensers have been designed to do both of these jobs efficiently over a large range of conjugate ratios at a low f-number. They are available in two materials and will find application in most energy transfer applications requiring both efficiency and uniformity.

Efficient transfer of radiant energy from a source into the projection lens system is assured by using a condenser close to the source, and by designing the condenser for minimum spherical aberration at the resultant large conjugate ratio. Low f-number operation is also obviously advantageous. As was previously mentioned in the discussion of aspheric lenses, spherical aberration at low f-numbers can cause a significant fraction of the energy not to be collected by the projection lens system. In the present condensers, minimization of spherical aberration has been accomplished with a two-element air-spaced design.

For all of the condensers the f-number is 1.0.

Uniformity of film gate illumination is achieved by placing the film gate as close as possible to the condenser secondary principal plane. The irradiance distributions at the condenser principal planes are essentially identical. If the source were an ideal point source, the irradiance distribution at the condenser primary principal plane would clearly be non-uniform, by the inverse square law approximation. The finite size of real sources considerably improves the uniformity of the distribution at the condenser primary principal plane. There is clearly a trade-off to be made between uniformity of film gate illumination and proximity of condenser to source (large conjugate ratio). The present condensers have been designed to be short, and to allow close approach to the condenser principal planes. If necessary, a flat hot mirror and/or heat-absorbing glass filter can be interposed between condenser and film gate. Film, filter (if any), and condenser can all be forced-air cooled. The housings of our condensers are provided with six symmetrically located vent holes for compatibility with such cooling. Lenses can easily be removed from the housings, cleaned, and carefully restored to their proper positions.

In the preliminary design of projection and illumination systems it is especially helpful to be able to locate the conjugate points of the condenser. We have earlier pointed out, in the discussion of lens combinations, that the simple lens formula

$$\frac{1}{s} + \frac{1}{s''} = \frac{1}{f}$$

is applicable to combinations where f is the combination focal length, provided that the conjugate distances s and s″ are measured from the combination principal points. These remarks apply, in particular, to our two-element air-spaced condensers. To facilitate this use of the simple lens formula, we have tabulated both the condenser effective focal lengths f_1 and the positions of the condenser (combination) principal points H_1 and H''_1. These positions are referenced to the external lens vertices and the mount. We believe that you will find this information to be helpful.

SELECTING THE PROPER CONDENSER

We now present a procedure by which you can select the condenser best suited for your particular task. Frequent reference will be made to the following figure. We assume that you have a definite projection task in mind, and have already determined values for the following quantities:

$$m_2 = \frac{s_2''}{s_2} ,$$

from known film and desired image size

f_2 = Projection lens system effective focal length

D_2 = clear aperture of projection lens system, given by f_2 divided by f-number

The conjugate distance s_2, from film gate position to projection lens system primary principal plane, is given by

$$s_2 = f_2\left(1 + \frac{1}{m_2}\right).$$

Specifically allowing room for an optional heat absorbing filter, we may now set

$$s_1'' = s_2 + a.$$

Next we presume that a source has been selected, and that a source height h_1 is known. h_1 might be the diagonal of a rectangular filament coil, or the diameter of either the emitting volume in an arc lamp or the emitting button of an induction lamp. Setting $h_2 = 0.9\ D_2$, we know the source magnification (condenser conjugate ratio)

$$m_1 = \frac{h_2}{h_1} = \frac{s_1''}{s_1},$$

from which $s_1 = (s_2 + a)\dfrac{h_1}{h_2}$.

We must now make a decision based on a comparison of s_1 as just calculated with the required clearance $(r + b)$ between source center and condenser primary principal plane H_1. The distance r is the radius of the envelope of the source. Depending on the outcome of this comparison, we have one of two special cases:

Case 1

If s_1 is larger than $(r + b)$, there is no difficulty in positioning the source at the proper conjugate distance. The ideal focal length f_1 for the condenser is then calculated from

$$f_1 = \frac{(s_2 + a)h_1}{h_1 + h_2},$$

which is the simple lens formula with appropriate values for s_1 and s_1''.

Case 2

If s_1 is smaller than $(r + b)$, some changes must be made. Redefine $s_1 = (r + b)$; then

$$s_1'' = \frac{h_2}{h_1}(r + b),$$

and the simple lens formula gives

$$f_1 = \frac{(r + b)\,h_2}{h_1 + h_2}.$$

In either case one would order a condenser of focal length f_1 or larger. If one chooses a focal length much larger than f_1, significant energy may be wasted in illuminating the region surrounding the film gate, rather than just the film gate itself. Ideally the region illuminated at the plane of the film gate should have a diameter about 10% larger than the diagonal of the film frame that is to be projected.

CONDENSER CONJUGATE RATIO RANGES

We offer two-element air-spaced condensers for two specific ranges of conjugate ratio (source magnification), and it is important that you choose the right type for your needs.

Our Optimized Air-Spaced Condensers are intended for close and efficient source coupling. These are computer-optimized designs suitable for the range of conjugate ratios from 3.0 to 15.0 with peak performance at 5.0.

Remember that high source magnification (small source-to-condenser distance) implies more efficient collection by the condenser of radiation emitted by the source. Since all of our condensers are designed for f/1.0, f-number cannot serve as a basis for comparison.

Our Symmetric Air-Spaced Condensers are intended for use at or close to unit magnification. These condensers are made up of pairs of plano-convex lenses. They are well suited for conjugate ratios from 0.3 up to 3.0, with peak performance at 1.0.

SOME EXAMPLE SOURCE SIZES h_1

h_1	Source
4.0mm	1000-watt Xenon compact arc lamp, Canrad-Hanovia 976C-1.
5.8mm	Tungsten halogen FCS (24 volt, 150 watt) lamp, General Electric or Osram (No. 64640).
11.7mm	Tungsten halogen BCK lamp (120 volt, 500 watt) General Electric.

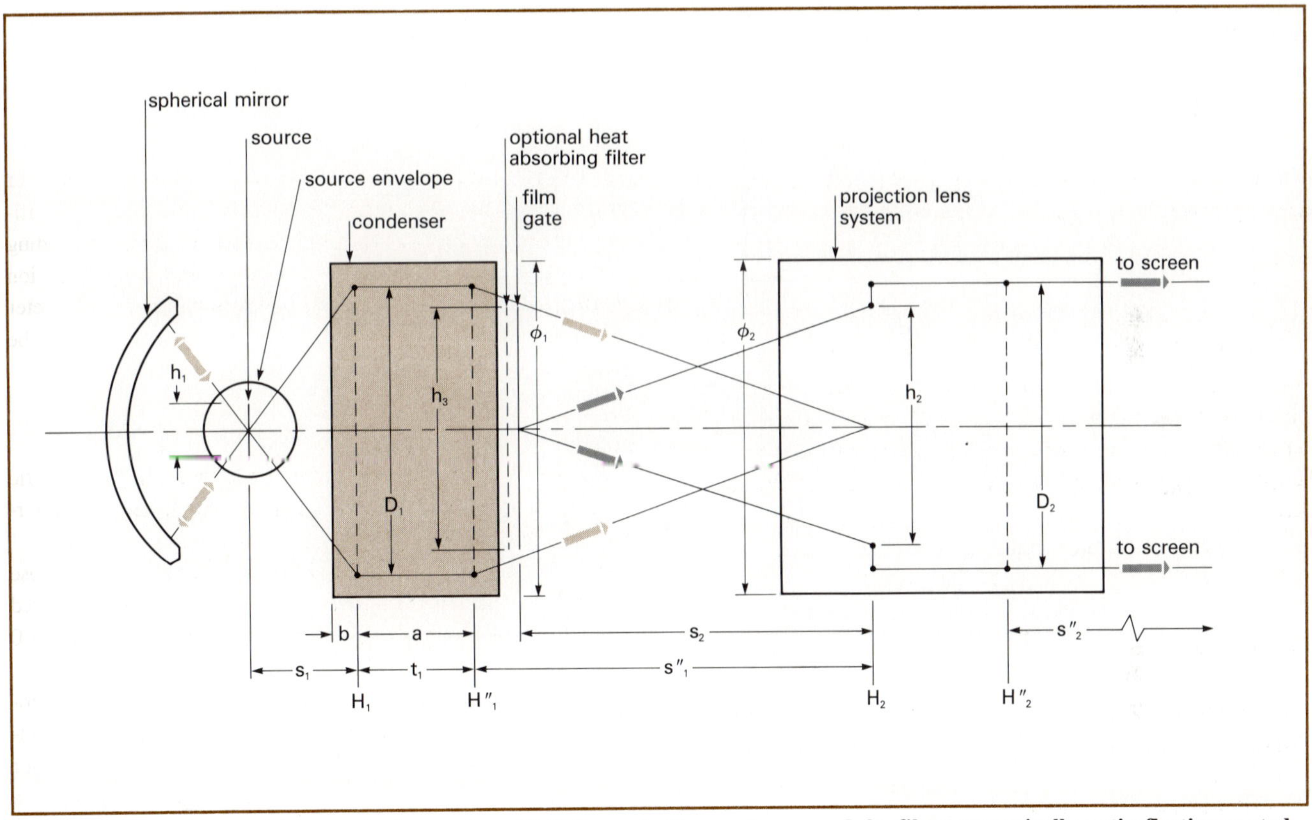

CONDENSER being used in a projection system. Both lenses in the condenser and the filter are typically antireflection coated.

CONDENSER MATERIAL CHOICES

We offer a choice of two distinct condenser materials suitable for a broad range of potential needs. Least expensive are our Pyrex condensers, suitable for normal applications and offering high transmission. Synthetic fused silica condensers combine superb transmission with extreme resistance to high temperatures and thermal shock. We suggest that synthetic fused silica condensers not be used at temperatures above 500 °C which is a limit imposed by approach to the melting point of the aluminum housing.

Fused silica condensers also offer high ultraviolet transmission, and are therefore ideally suited for use in photoresist exposure. For special applications in the very deep ultraviolet (down to 180nm), a UV grade synthetic fused silica option can be requested.

SPECIFICATIONS: TWO-ELEMENT AIR-SPACED CONDENSERS

Focal Length (f): ±2%
Focal Ratio (f/D): 1.0
Principal Plane Locations (H_1, H''_1): ±2%
Barrel Diameter (ϕ): ±0.25mm
Clear Aperture (D): ±0.13mm
Housing Length (L): ±0.25mm
Optical Materials: Pyrex (n_e = 1.477), or optical quality synthetic fused silica (n_e = 1.460)
Mount Material: 6061 Aluminum (brass retainer clamp rings)
Cosmetic Surface Quality: Pitch polished, 80-50 scratch and dig
The housing is black anodized aluminum with engraved identification.

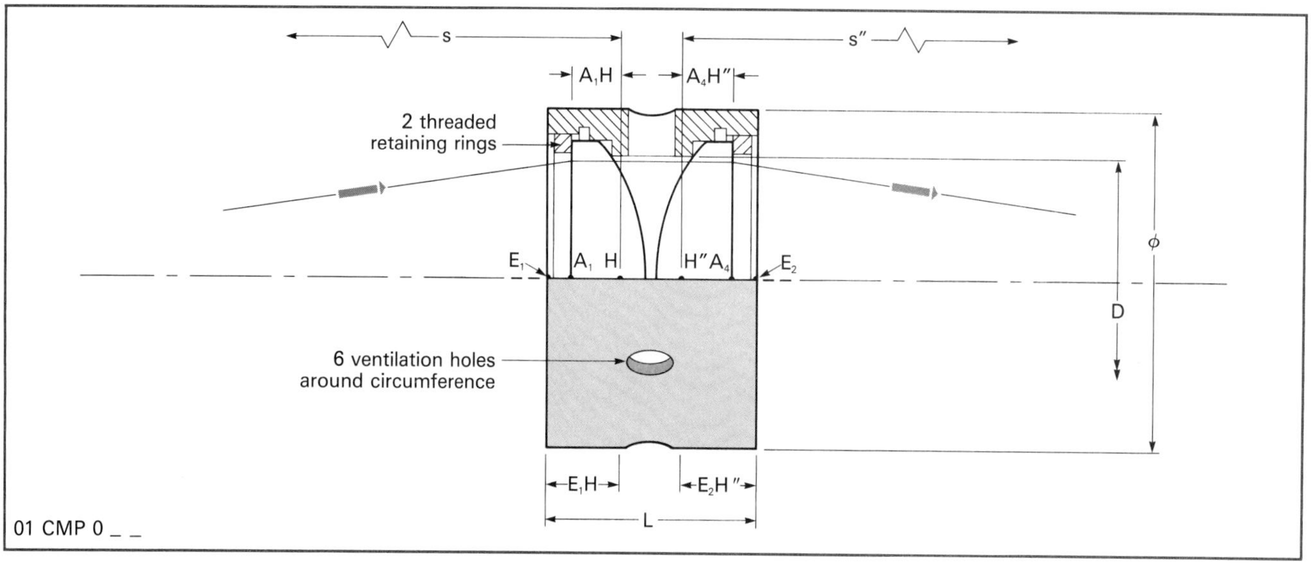

Symmetric Air-Spaced Pyrex Condensers; (0.3-3.0 Conjugate Ratio Range)

f (mm)	D (mm)	ϕ (mm)	L (mm)	A_1H (mm)	A_4H'' (mm)	E_1H (mm)	E_2H'' (mm)	PRODUCT NUMBER
10	10	16.0	14.27	2.42	-2.42	6.1	-6.1	01 CMP 001
15	15	22.0	15.88	2.87	-2.87	6.7	-6.7	01 CMP 005
25	25	33.0	19.05	3.81	-3.81	7.9	-7.9	01 CMP 009
35	35	43.0	22.23	4.76	-4.76	9.0	-9.0	01 CMP 013
50	50	58.0	26.97	6.54	-6.54	10.5	-10.5	01 CMP 017
75	75	85.0	34.93	8.93	-8.93	13.6	-13.6	01 CMP 021
100	100	110.0	42.85	11.67	-11.67	16.0	-16.0	01 CMP 025

Symmetric Air-Spaced Synthetic Fused Silica Condensers; (0.3-3.0 Conjugate Ratio Range)

f (mm)	D (mm)	ϕ (mm)	L (mm)	A_1H (mm)	A_4H'' (mm)	E_1H (mm)	E_2H'' (mm)	PRODUCT NUMBER
10	10	16.0	14.27	2.51	-2.51	6.1	-6.1	01 CMP 003
15	15	22.0	15.88	2.99	-2.99	6.7	-6.7	01 CMP 007
25	25	33.0	19.05	3.98	-3.98	7.8	-7.8	01 CMP 011
35	35	43.0	22.23	4.98	-4.98	8.9	-8.9	01 CMP 015
50	50	58.0	26.97	6.85	-6.85	10.5	-10.5	01 CMP 019
75	75	85.0	34.93	9.37	-9.37	13.3	-13.3	01 CMP 023
100	100	110.0	42.85	12.26	-12.26	15.9	-15.9	01 CMP 027

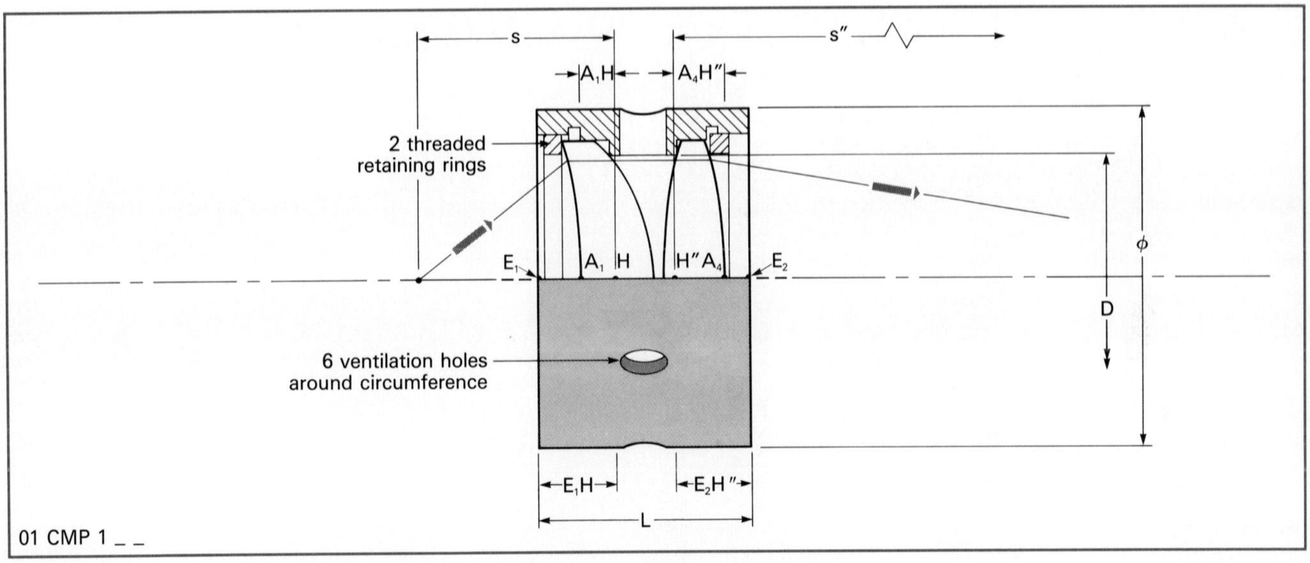

Optimized Air-Spaced Pyrex Condensers; (0.3-15.0 Conjugate Ratio Range)

f (mm)	D (mm)	ϕ (mm)	L (mm)	A_1H (mm)	A_4H'' (mm)	E_1H (mm)	E_2H'' (mm)	PRODUCT NUMBER
10	10	16.0	14.27	3.76	-1.02	8.0	-4.6	01 CMP 101
15	15	22.0	19.05	4.42	-1.85	10.2	-6.1	01 CMP 105
25	25	33.0	22.23	6.50	-2.57	12.0	-5.1	01 CMP 109
35	35	43.0	26.97	9.80	-3.58	15.4	-5.0	01 CMP 113
50	50	58.0	30.15	11.48	-4.67	16.4	-5.2	01 CMP 117
75	75	85.0	34.93	13.92	-6.38	20.2	-5.1	01 CMP 121
100	100	110.0	44.45	18.54	-8.51	25.9	-5.7	01 CMP 125

Optimized Air-Spaced Synthetic Fused Silica Condensers; (0.3-15.0 Conjugate Ratio Range)

f (mm)	D (mm)	ϕ (mm)	L (mm)	A_1H (mm)	A_4H'' (mm)	E_1H (mm)	E_2H'' (mm)	PRODUCT NUMBER
10	10	16.0	14.27	3.56	-1.55	7.5	-4.9	01 CMP 103
15	15	22.0	19.05	5.36	-2.34	10.1	-6.1	01 CMP 107
25	25	33.0	22.23	6.27	-3.15	11.7	-6.1	01 CMP 111
35	35	43.0	26.97	8.76	-4.42	—	—	01 CMP 115
50	50	58.0	30.15	10.85	-6.12	14.4	-6.6	01 CMP 119
75	75	85.0	34.93	14.94	-6.60	20.9	-3.6	01 CMP 123
100	100	110.0	44.45	19.02	-8.79	26.3	-3.8	01 CMP 127

APPLICATIONS ENGINEERING

"Service" is our password. Melles Griot, as a component supplier, recognizes that service to researchers and manufacturers is our business. That's why you'll find highly skilled optical and mechanical engineers at Melles Griot waiting to help solve your problems.

All you have to do is call us. You'll be put through to a friendly and knowledgeable Applications Engineer whose function is to help you. Backed up with sophisticated computer systems, these engineers can advise you all the way from providing current specifications and prices to designing a complex optical system.

Our service does not stop with the Optics Guide — call your nearest Melles Griot office for further assistance.

MIRRORS, WINDOWS, AND FLATS

We start this section by presenting the fundameltals of reflection, followed by a brief summary of the materials and coatings available.

Our reflective optics make up the most comprehensive selection available anywhere. Whatever your mirror requirements, we can meet them.

We also describe a selection of windows and precision optical flats.

Contents **Page Number**

REFLECTION

When light is incident at the boundary between two different media, some of that light is reflected. Many common optical components, including mirrors, beamsplitters, prisms, and polarizers make use of that reflection. A basic familiarity with the laws of reflection is helpful when specifying these components.

The first and fundamental law of reflection is that the angle of incidence is equal to the angle of reflection.

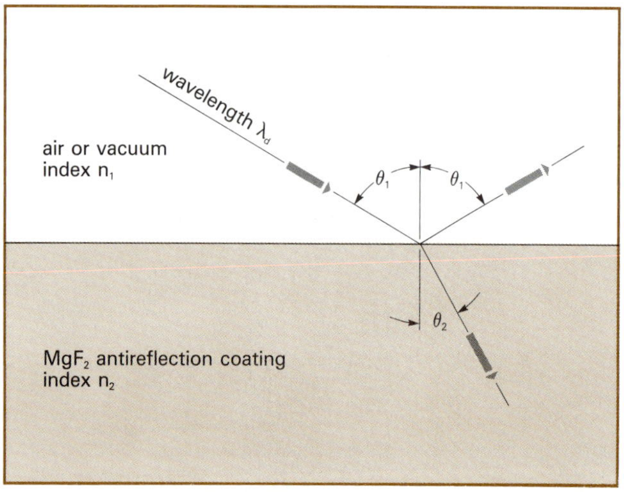

REFLECTION AND REFRACTION at a boundary between two dielectric media.

In order for useful specular reflection to occur surface irregularities must be small in relation to the wavelength of the incident light. The intensity of light reflected depends on wavelength, angle of incidence, and state of polarization. Polarization effects are always present in reflective components and may be a either great benefit or an inconvenience. It is important to think in terms of polarization when considering reflective optics.

EXTERNAL REFLECTION AT A DIELECTRIC BOUNDARY

Fresnel's laws of reflection precisely describe the amplitude and phase relationships between reflected and incident light at a boundary between two dielectric media. It is convenient to think of incident radiation as the superposition of two plane-polarized beams, one with its electric field parallel to the plane of incidence (p-polarized) and the other with its electric field perpendicular to the plane of incidence (s-polarized). Fresnel's laws can be summarized in the following two equations giving the reflectance of the s- and p-polarized components:

$$r_s = \left(\frac{\sin(\theta_1 - \theta_2)}{\sin(\theta_1 + \theta_2)} \right)^2$$

$$r_p = \left(\frac{\tan(\theta_1 - \theta_2)}{\tan(\theta_1 + \theta_2)} \right)^2$$

In the limit of normal incidence Fresnel's laws reduce to the following simple equation:

$$r = \left(\frac{n-1}{n+1} \right)^2$$

It can easily be seen that, for a refractive index of 1.5 this gives a reflectance of 4%. This important result shows that about 4% of all illumination incident normally on an air-glass surface will be reflected. In multi-element lens systems reflection losses would be very high if antireflection coatings were not used.

The variation of reflectance with angle of incidence for both the s- and p-polarized components can be seen in the figure. It can be seen that the reflectance remains close to 4% over about 30° incidence, and that it rises rapidly to 100% at grazing incidence. Furthermore, note that the p-component vanishes at 56°39′. this angle is called Brewster's angle. It is the angle at which the reflected light is completely polarized. This situation occurs when

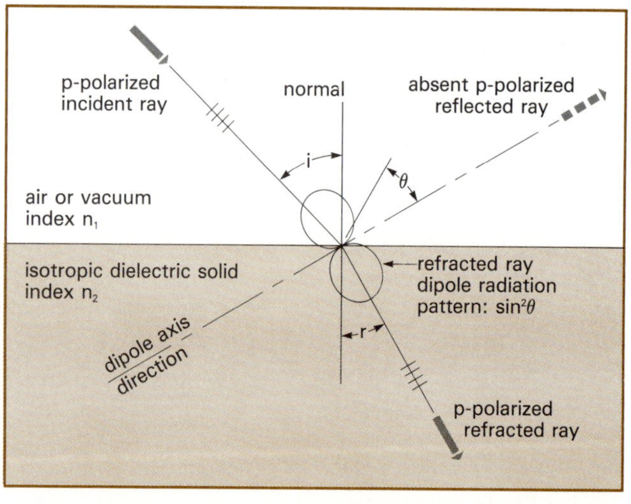

BREWSTER ANGLE. At this angle the p-polarized component is completely absent in the reflected ray.

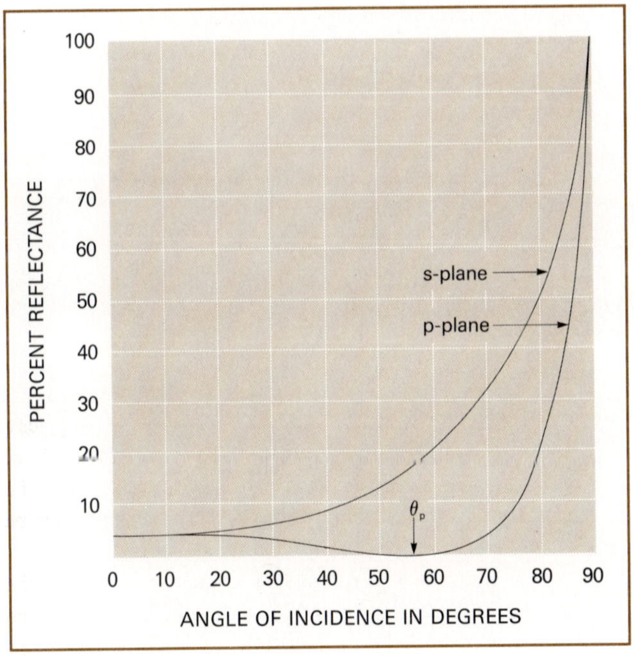

EXTERNAL REFLECTION at a glass surface (n = 1.54) showing s- and p-polarized components.

the reflected and refracted rays are perpendicular to each other ($\theta_1 + \theta_2 = 90°$). Under these conditions electric dipole oscillations of the p-component would be along the direction of propagation of, and could therefore not contribute to, the reflected ray. At the polarizing angle, reflectance of the s-component is about 15%.

INTERNAL REFLECTION AT A DIELECTRIC BOUNDARY

For light incident from the higher to the lower refractive index medium we can apply the results of Fresnel's laws in exactly the same way. The angle in the high index material at which polarization occurs is smaller by the ratio of the refractive indices in accordance with Snell's law. The internal polarizing angle is 33°21′ corresponding to the Brewster angle (56°39′) in the external medium.

The angle at which the emerging refracted ray is at grazing incidence is called the critical angle. For an external medium of air or

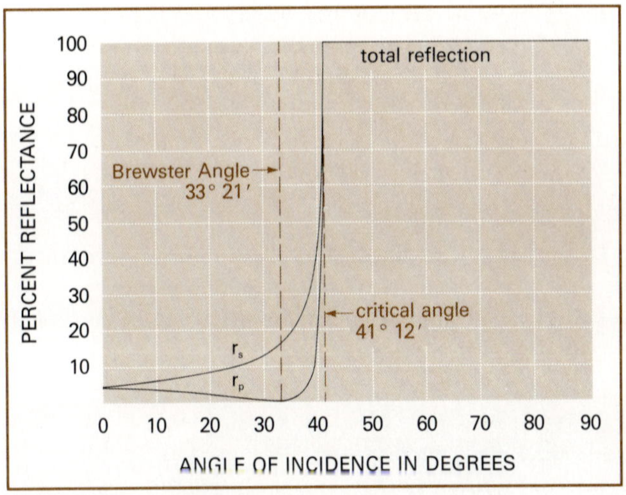

INTERNAL REFLECTION at a glass surface (n = 1.54) showing s- and p-polarized components.

vacuum (n = 1), the critical angle is given by

$$\theta_c(\lambda) = \arc\sin\left(\frac{1}{n(\lambda)}\right)$$

and depends on the refractive index n(λ) which is a function of wavelength. For all angles of incidence higher than the critical angle, total internal reflection occurs.

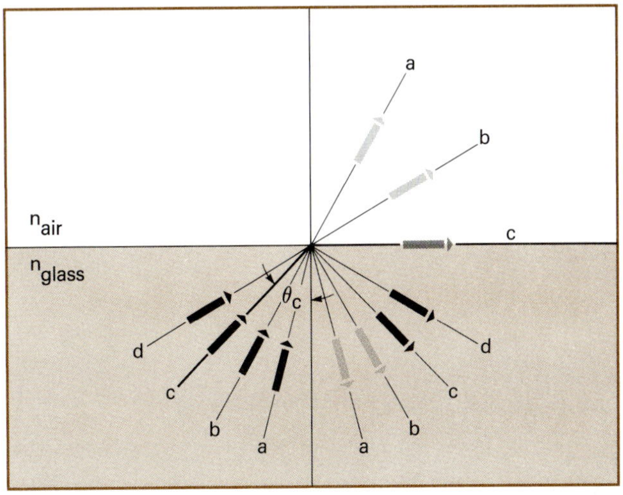

CRITICAL ANGLE. At this angle the emerging ray is at grazing incidence.

REFLECTION AT METAL SURFACES

Metal surfaces are much better reflectors than dielectrics. Their reflectance at normal incidence is higher but they too, exhibit polarization selective reflection properties in the same way as dielectrics. However the p-reflectance of metals does not drop to zero. The angle at which the p-reflectance reaches a minimum is called the principal angle.

The accompanying table shows the reflectance, at normal incidence, of a selection of commonly used metal reflectors. (From the American Institute of Physics Handbook, by permission of McGraw-Hill Book Company.)

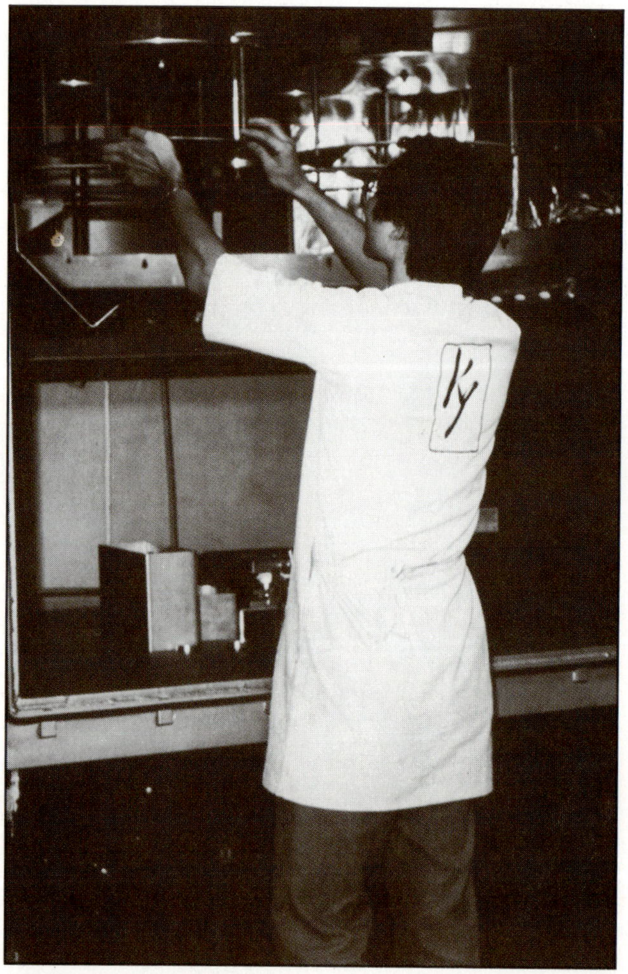

Percent Reflectance of Freshly Evaporated Mirror Coatings; Normal Incidence

Wavelength λ	Reflectance (%)				
(microns)	Al	Ag	Au	Cu	Rh
0.220	91.5	28.0	27.5	40.4	57.8
0.240	91.9	29.5	31.6	39.0	63.2
0.260	92.2	29.2	35.6	35.5	67.7
0.280	92.3	25.2	37.8	33.0	70.7
0.300	92.3	17.6	37.7	33.6	73.4
0.315	92.4	5.5	37.3	35.5	75.0
0.320	92.4	8.9	37.1	36.3	75.5
0.340	92.5	72.9	36.1	38.5	76.9
0.360	92.5	88.2	36.3	41.5	78.0
0.380	92.5	92.8	37.8	44.5	78.1
0.400	92.4	95.6	38.7	47.5	77.4
0.450	92.2	97.1	38.7	55.2	76.0
0.500	91.8	97.9	47.7	60.0	76.6
0.550	91.5	98.3	81.7	66.9	78.2
0.600	91.1	98.6	91.9	93.3	79.7
0.650	90.5	98.8	95.5	96.6	81.1
0.700	89.7	98.9	97.0	97.5	82.0
0.750	88.6	99.1	97.4	97.9	82.6
0.800	86.7	99.2	98.0	98.1	83.1
0.850	86.7	99.2	98.2	98.3	83.4
0.900	89.1	99.3	98.4	98.4	83.6
0.950	92.4	99.3	98.5	98.4	83.9
1.000	94.0	99.4	98.6	98.5	84.2
1.500	97.4	99.4	99.0	98.5	87.7
2.000	97.8	99.4	99.1	98.6	91.4
3.000	98.0	99.4	99.3	98.6	95.0
4.000	98.2	99.4	99.4	98.7	95.8
5.000	98.4	99.5	99.4	98.7	96.4
6.000	98.5	99.5	99.4	98.7	96.8
7.000	98.6	99.5	99.4	98.7	97.0
8.000	98.7	99.5	99.4	98.8	97.2
9.000	98.7	99.5	99.4	98.8	97.4
10.000	98.7	99.5	99.4	98.9	97.6
15.000	98.9	99.6	99.4	99.0	98.1
20.000	99.0	99.6	99.4	—	—

SUBSTRATE MATERIALS

We offer a variety of substrate materials, spanning a range of thermal expansion coefficients and hence costs. For applications in which thermal shock is absent and thermal stability is not critical, optical crown glass is a suitable and inexpensive material. For applications involving a moderate risk of thermal shock, pyrex is appropriate. For applications requiring high thermal stability or involving severe thermal shock, synthetic fused silica is a good choice. The respective room temperature coefficients of linear thermal expansion of optical crown, pyrex, and synthetic fused silica are $9 \times 10^{-6}/°C$, and $3.2 \times 10^{-6}/°C$ and $0.5 \times 10^{-6}/°C$. For applications demanding the ultimate in thermal stability and thermal shock resistance, a "zero expansion" glass-ceramic such as Schott ZERODUR® is the best choice.

LOW EXPANSION SUBSTRATE MATERIALS

Some materials may exhibit zero or even negative thermal expansion coefficients over limited temperature intervals. These materials include both crystalline and amorphous substances. The reason for this surprising behavior, for substances of either kind, is that as temperature varies so also does the ratio of the energies in vibrational modes which are parallel to or orthogonal to chemical bond directions. This in turn is because of the quantization of vibrational energy levels, and the fact that allowed levels for orthogonal modes need not fall at the same energies. Thus zero and negative expansion coefficients are large scale manifestations of quantum theory.

Low expansion glass-ceramics combine a negative thermal expansion coefficient crystalline phase with a positive coefficient amorphous phase, in such proportions as to result in an expansion coefficient of almost zero over a large temperature interval. Schott ZERODUR® is such a material. By design these materials typically exhibit a change in the sign of the coefficient of linear thermal expansion near room temperature.

A reasonable upper bound on the absolute value of the expansion coefficient of low expansion ZERODUR® glass-ceramic at room temperature is $0.15 \times 10^{-6}/°C$. The coefficient can be much smaller, and can have either sign. The value of the bound is significantly smaller than the room temperature coefficient for synthetic fused silica. ZERODUR® makes ideal mirror substrates for such stringent applications as multiple-exposure holography, holographic interferometry, interferometry in general, manipulation of moderately powerful laser beams, and spaceborne imaging systems.

For more information about the materials, see the section on optical materials starting on page 51.

HIGHREFLECTION COATINGS

It was previously shown that about 4% of incident light is reflected from any uncoated glass surface at normal incidence. As the angle of incidence is increased, so also is the reflectance, although polarization effects start to occur (see the prior discussion on page 174 for a full explanation). In order to increase the reflectivity of a surface, high reflection coatings may be applied. Multilayer stacks of 15 to 35 layers are common but even greater numbers of layers may be needed for same applications. These coatings may be applied to the outside of a component, such as a flat piece of glass, to produce a "first-surface" mirror. They may also be applied to an internal surface, of a prism for example, to produce a "second-surface" mirror.

The coatings we offer may be applied to many of our components to increase the reflectivity. Prism faces, cylindrical lenses and plano concave lenses are examples of common applications for high reflection coatings in addition to the more usual flat mirrors.

We offer a variety of high reflection coatings for use at any specific angle of incidence or range of angles you may require. Our coatings fall into two categories, metallic and dielectric. Coatings in either category will meet or exceed all applicable military specifications.

METALLIC COATINGS

These are used primarily for mirrors and for prism surfaces. Metallic coatings can in turn be overcoated with dielectric films to increase the reflectance over a desired range of wavelengths or angles of incidence. In this case the metallic coating is said to be enhanced.

By overcoating metallic coatings with hard dielectric layers of ½ wavelength optical thickness, abrasion and tarnish resistance can be increased. The optical properties are only slightly affected, assuming that the dielectric has high internal transmittance at the wavelengths of interest. Depending on the dielectric used, such overcoated metals are sometimes referred to as durable, protected or hard coatings.

Metals which we routinely supply as high reflection coatings include:

Aluminum	Ultraviolet-enhanced aluminum
Protected aluminum	Gold
Enhanced aluminum	Silver (second surface only)

DIELECTRIC COATINGS

If the index of a dielectric quarter wave optical thickness film is lower than the index of the substrate, reflectance of the coated surface will be reduced at normal incidence. The reflectance of this film will, in fact, be quite low over a wide range of angles of incidence.

If the index of the quarter wave film is instead higher than the index of the substrate, reflectance of the coated surface will be substantially increased.

If both types of quarter wave films are applied, alternating between high and low index, very high levels of reflectance may be obtained. These stacks achieve their very high reflectance in a narrow wavelength interval surrounding the wavelength for which the thickness of the high index plus the low index layers is a halfwave. The reflectance achieved is very sensitive to the film index ratio, and to the number of periods or pairs of high and low index layers. Increase either and you increase the reflectance.

The width of the high reflectance wavelength interval is also determined by the film index ratio. The higher the ratio the wider the interval. There are two ways in which the wavelength interval can be made larger than the limit set by index ratio. First, it is possible to deposit stacks for slightly different wavelengths on top of each other. Second, the interval can be made wider by systematically perturbing the layer thicknesses in a single stack to be different from precisely quarterwave optical thickness values at one wavelength. Computer optimization programs exist for arriving at efficient multilayer designs. The reflectances of such films can easily be made to exceed the highest metallic reflectances over very large wavelength intervals, spanning almost the entire visible spectrum for example. Such films are effective for both the s- and p-polarizations, and over a wide range of angles of incidence.

Depending on the materials chosen for the multilayer, it is also possible for the durability and abrasion-resistance of such films to be superior to that of metallic films. It is a natural characteristic of multilayer dielectric broadband reflecting coatings that their reflectance values diminish at oblique incidence.

METALLIC HIGH-REFLECTION COATINGS

We offer six forms of standard metallic high reflection coating. All of these coatings are formed by vacuum deposition. The following is a brief summary of the properties of these metals, giving some indication of where and why they are used. They are convenient in that they can be used at any angle of incidence. These coatings may be applied to most of the optical components listed. To do so simply append the coating suffix number to the component product number.

Metallic reflective coatings are delicate and require care during cleaning. Dielectric overcoats substantially improve abrasion resistance but they are not impervious to abrasive cleaning techniques. If it is necessary to clean these coatings, care is required. Clean, dry pressurized gas can be used for blowing off loose particles, then clean deionized water, mild detergent and alcohol may be used. Gentle cleaning with a swab is recommended.

ALUMINUM

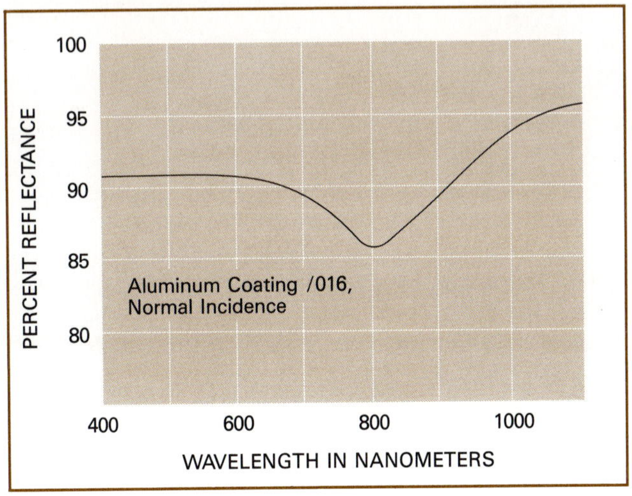

Aluminum is the most widely used metal for reflecting films, offering consistently high reflectance throughout the visible, near-infrared, and near-ultraviolet regions of the spectrum. While silver exhibits slightly higher reflectance than aluminum through most of the visible spectrum, the advantage is temporary because of tarnishing due to oxidation. Aluminum also oxidizes, though more slowly, and its oxide is tough and corrosion-resistant. Oxidation significantly reduces aluminum reflectance in the ultraviolet, and the oxide causes slight scattering throughout the spectrum.

PROTECTED ALUMINUM

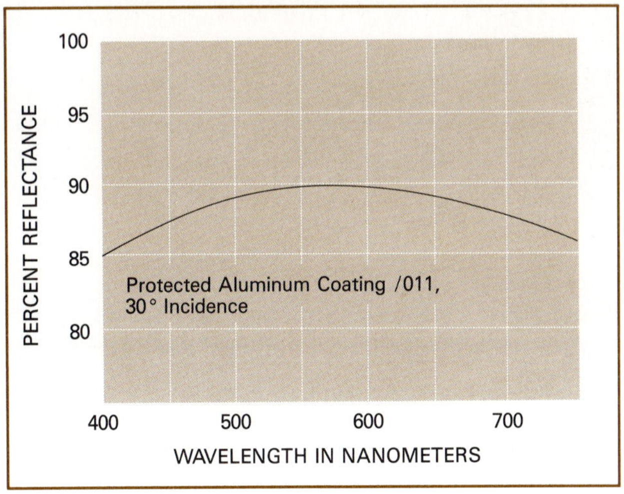

Protected aluminum is the very best general purpose metallic coating for use as an external reflector in the visible and near-infrared spectrum. All of our mirrors are coated with protected aluminum unless we specify otherwise or you specifically request a different coating. The Coating Suffix need not be appended. Protected aluminum is aluminum overcoated with a dielectric film of silicon monoxide (Si0) of 1/2-wavelength optical thickness at 550nm. The protective film arrests oxidation and helps to maintain the initial high reflectance. It is hard enough to protect the aluminum from minor abrasions.

ENHANCED ALUMINUM

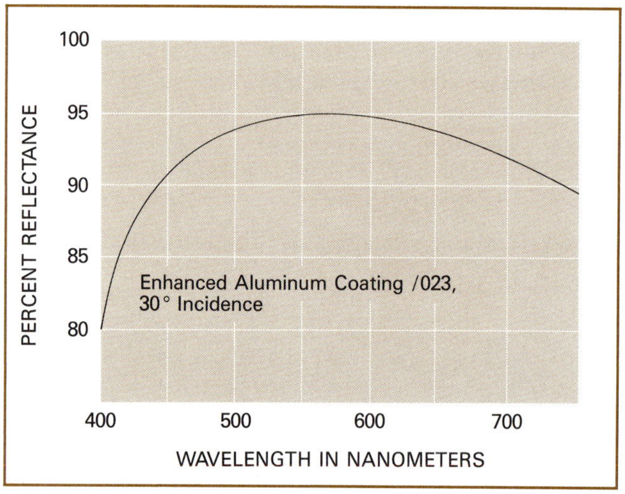

By overcoating the aluminum with a multilayer dielectric film, reflectance is increased over a wide range of wavelengths. The durable enhancing multilayer produces a peak reflectance of 95% with an average across the visible spectrum of 93%. This coating is well suited to applications requiring the durability and reliability of protected aluminum, but with higher reflectance in the mid-visible regions. The reflectance of enhanced aluminum peaks between 530 and 580nm, and is high from 400 to 800nm.

ULTRAVIOLET-ENHANCED ALUMINUM

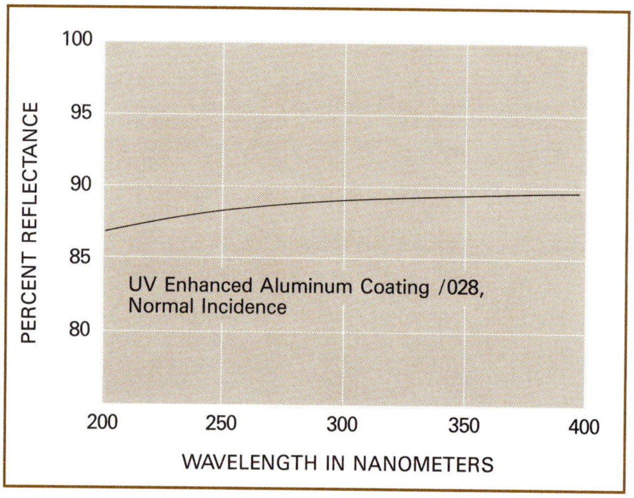

By applying a film of an ultraviolet-transmitting dielectric (usually MgF_2), the reflectance of pure bare aluminum can be preserved in the ultraviolet. The dielectric layer prevents oxidation of the aluminum surface and provides abrasion resistance. While the resulting surface is not as abrasion-resistant as our protected aluminum, this coating may be cleaned with care. Reflectance averages over 88% from 200nm to 400nm and over 85% throughout the visible spectrum. This coating can be applied to all of our mirrors with the exception of the paraboloidal reflectors and ellipsoidal reflectors.

SILVER

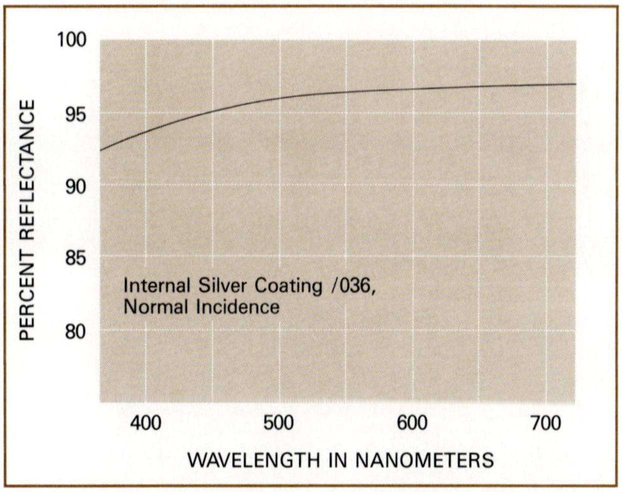

Through most of the visible and in the near-infrared spectrum, silver has higher reflectance than aluminum, at least for a short time followinng deposition. Because the advantage is temporary, silver is rarely used in external reflection. Instead, silver is used almost exclusively in internal reflection. Oxidation and tarnish are prevented by over-coating the external surface with an additional layer of either Inconel or copper. The Inconel or copper layers are subsequently painted to increase abrasion resistance. In this way, the high initial reflectance of silver is indefinitely preserved.

Silver is frequently used in the near-infrared (the interval containing the neodymium and gallium arsenide laser lines) because it avoids the small dip in reflectance which aluminum exhibits in this interval. In the near-ultraviolet, silver has very low reflectance, and aluminum is a preferable choice. From the visible into the middle-infrared spectrum, silver offers the highest internal reflectance available from a metallic coating. Silver has less effect than aluminum on the polarization state in these regions of the spectrum.

Call us with your OEM requirements for metallized mirrors.

GOLD

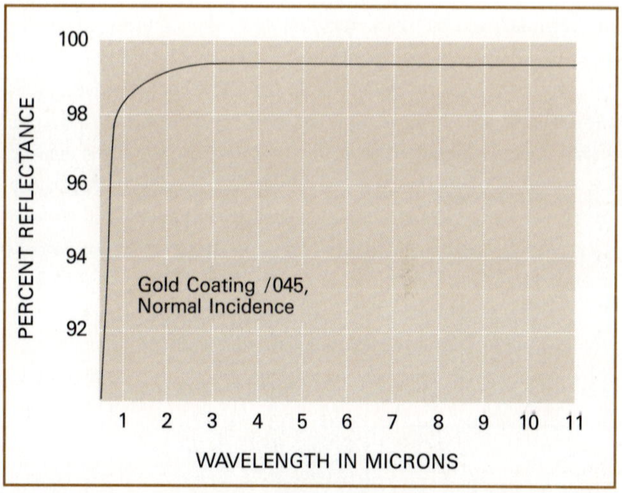

Combining good tarnish resistance with consistently high reflectance throughout the near-, middle-, and far-infrared, gold is by far the most widely used reflecting material in these regions. While it is possible to construct multilayer films which may surpass the reflectance of gold at specific wavelengths, the useful range of gold is unequaled. Gold is especially effective in the control of thermal radiation. Bare gold is soft and scratches easily, so bare gold mirrors should be cleaned only by flow-washing with solvents and clean water or by blowing the surface clean with a low-pressure stream of dry air.

Metallic High Reflection Coatings

Coating Type	Wavelength Range (nm)	Average Reflectance (%)	COATING SUFFIX
Aluminum	200-IR	90	/016
Protected Aluminum	400-800	87	/011
Enhanced Aluminum	450-750	93	/023
UV-Enhanced Aluminum	200-400	88	/028
Silver (internal)	400-IR	98	/036
Gold	700-IR	99	/045

To order append coating suffix to product number.

LASERLINE MAX-R COATINGS

These multilayer coatings are stacks intended to achieve the highest possible reflectances at specific laser line wavelengths and at particular angles of incidence. At these wavelengths and angles, laser line MAX-R coatings outperform MAXBRIte™. For limited ranges of angle of incidence (centered about the intended angles) the reflectances of these coatings will meet or exceed our published specifications at the design wavelength. We offer coatings for angles of incidence of 0 and 45 degrees. The layer thicknesses differ for these two angles so the coatings cannot be used interchangeably. Laser line MAX-R coatings are intended for external beam manipulation applications. Specified coating reflectances apply to the p-polarization (except at normal incidence, where the p- and s-polarization states are indistinguishable). Reflectances for the s-polarization generally exceed those for the p-polarization.

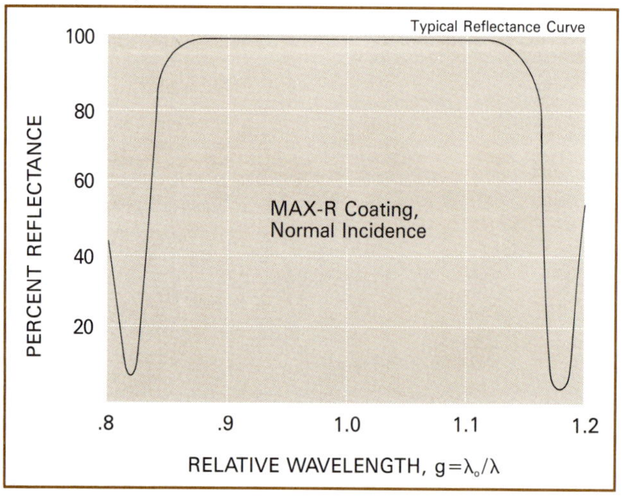

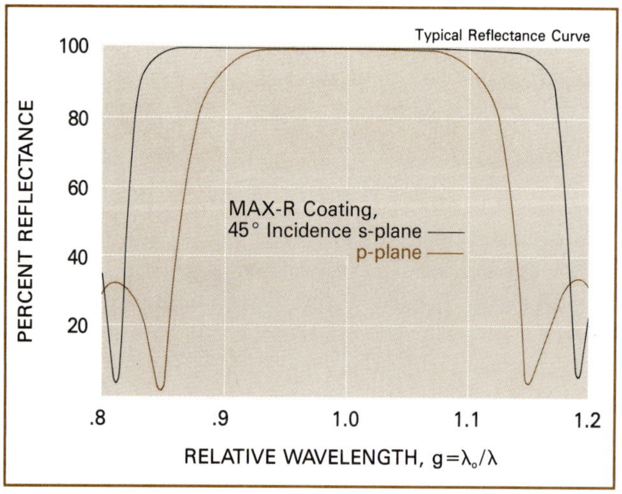

These coatings are intended for low to medium power applications. We also offer a range of guaranteed HEL Mirror Coatings for use with high energy lasers (see pages 206-209).

Laser Line MAX-R Coatings; Normal Incidence

Wavelength (nm)	Laser Type	Minimum Reflectance R_p (%) 0° Incidence	Minimum Reflectance R_p (%) 0°±15° Incidence	COATING SUFFIX
351	Ar ion	98.0	95.0	/205
364	Ar ion	98.0	95.0	/207
442	HeCd	99.3	99.0	/209
458	Ar ion	99.5	99.3	/211
466	Ar ion	99.5	99.3	/213
473	Ar ion	99.5	99.3	/215
476	Ar ion	99.5	99.3	/217
488	Ar ion	99.5	99.3	/219
496	Ar ion	99.5	99.3	/221
502	Ar ion	99.5	99.3	/222
514	Ar ion	99.5	99.3	/223
532	YAG (2nd harm.)	99.5	99.3	/225
543	HeNe	99.5	99.3	/226
633	HeNe	99.5	99.3	/229
694	Ruby	99.3	99.0	/231
700	GaAlAs	99.3	99.0	/233
830	GaAlAs	99.3	99.0	/237
850	GaAlAs	99.3	99.0	/238
904	GaAs	99.3	99.0	/239
1064	YAG	99.2	99.0	/241
1300	InGaAsP	99.2	99.0	/245
1523	HeNe	99.2	99.0	/247
1550	InGaAsP	99.2	99.0	/247

To order append coating suffix to product number.

Laser Line MAX-R Coatings; 45° Incidence

Wavelength (nm)	Laser Type	Minimum Reflectance R_p (%) 45° Incidence	Minimum Reflectance R_p (%) 45°±10° Incidence	COATING SUFFIX
351	Ar ion	97.0	95.0	/255
364	Ar ion	97.0	95.0	/257
442	HeCd	99.0	98.0	/259
458	Ar ion	99.3	98.0	/261
466	Ar ion	99.3	98.5	/263
473	Ar ion	99.3	98.5	/265
476	Ar ion	99.3	98.5	/267
488	Ar ion	99.3	98.5	/269
496	Ar ion	99.5	98.5	/271
502	Ar ion	99.5	98.5	/272
514	Ar ion	99.5	98.5	/273
532	YAG (2nd harm.)	99.5	98.5	/275
543	HeNe	99.5	98.5	/276
633	HeNe	99.5	98.5	/279
694	Ruby	99.0	98.5	/281
780	GaAlAs	99.0	98.5	/283
830	GaAlAs	99.0	98.5	/287
850	GaAlAs	99.0	98.5	/288
904	GaAs	99.0	98.5	/289
1064	YAG	99.0	98.0	/291
1300	InGaAsP	99.0	98.5	/295
1523	HeNe	99.2	9.0	/297
1550	InGaAsP	99.0	98.5	/297

To order append coating suffix to product number.

We can supply MAX-R Coatings for other laserlines not listed here. Call your Melles Griot office with your special requests.

MAXBRite™ COATING

The MAXBRIte™ (Multilayer All-dielectic Xerophilous Broadband Reflecting Interference) coating is, without a doubt, the very best broadband mirror coating available anywhere. It reflects well over 99% of any incident visible laser radiation. Furthermore, it is durable and can withstand high levels of illumination without damage.

This coating exhibits exceptionally high reflectance for both the s- and p-polarizations throughout most of the visible spectrum. For most applications it is superior to any metallic or enhanced metallic coatings.

At the most important visible laser wavelengths and for angles of incidence as high as 45 degrees the average of s- and p-reflectances exceeds 99%.

This coating is suitable for instrumental and external laser beam manipulation tasks. It is the ideal choice for use with tunable dye and parametric oscillator systems. The structural design of this coating is such that flatness specifications as tight as 1/20 wavelength can be maintained using low expansion substrate materials.

Because of the wide general applications of this superior coating, we carry a number of our better mirrors in stock already coated with MAXBRIte™.

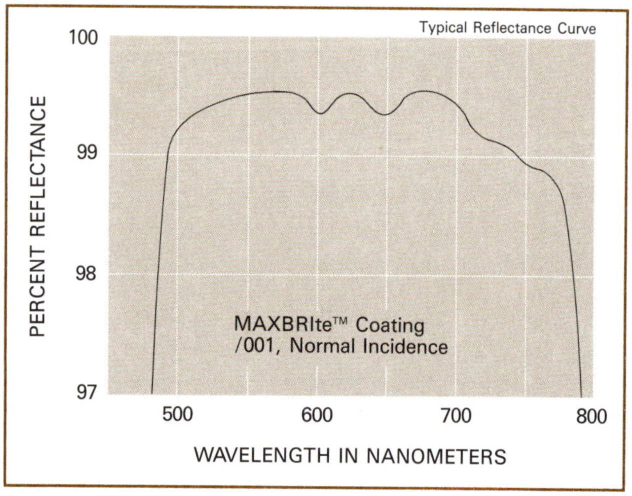

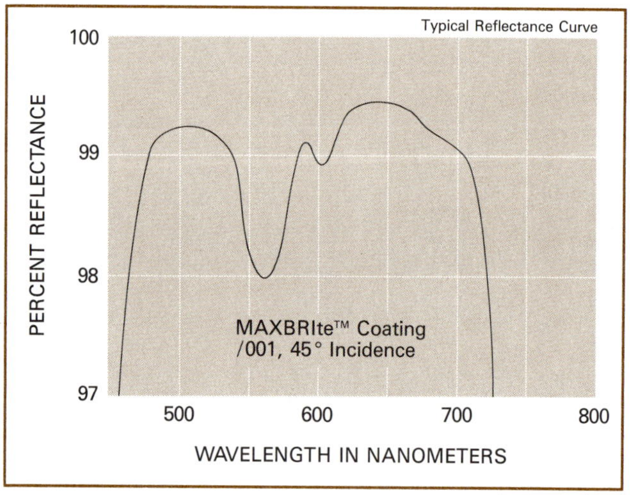

MAXBRIte™ Coating

Wavelength Range (nm)	Minimum Reflectance (%)	Angle of Incidence (deg)	COATING SUFFIX
480-700	99	0 ± 45	/001

To order append coating suffix to product number.

FLAT MIRRORS

First-surface flat mirrors, sometimes called plane mirrors, are one of the most commonly used optical components. Whatever the optical arrangement, there is often a need to fold or bend the optical path. Ideally a flat mirror should reflect 100% of the light incident on it, without altering the wavefront shape. As we have noted earlier (p174), reflective properties are dependent on wavelength, angle of incidence and polarization, so it is important to consider the application when selecting a mirror. The flatness of the mirror substrate and its mechanical and thermal stability are also important. We offer a selection of first-surface mirrors covering a range of materials and flatnesses. They may each be coated with any of the high reflection coatings we have previously described simply by appending the appropriate Coating Suffix to the Product Number. If you do not append a Coating Suffix you will be supplied with a mirror having a protected aluminum (/011) coating. Should you require an uncoated mirror substrate, for any reason, be sure to add a /000 Coating Suffix. A summary list of available coatings is presented in the accompanying table for your convenience. Full details and specifications for these coatings are given on pages 179-184.

When selecting a mirror be sure to take note of the flatness specification, explained in detail on page 43. Flatness specifications apply to the central 90% of the aperture (edge dimension or diameter). The outer 5% can be expected to roll downward by one to two times the tolerance. Remember that the reflected wavefront distortion will be twice the surface accuracy of the mirror.

LASER MIRRORS

The mirrors specified here are highly suitable for use with lasers. Our MAXBRIte™ Laser Mirrors are the very best type of mirror available for use with lasers. They provide greater than 99% reflectance in both s- and p-planes of polarization for all common laser wavelengths in the visible spectrum. These MAXBRIte™ Laser Mirrors are held in stock for your convenience.

If your requirement is restricted to one specific laser wavelength, you may prefer to select a MAX-R Coating which has been optimized for that wavelength.

In addition to the range of mirrors listed here we offer a selection of high energy laser mirrors which are guaranteed against damage at their specified energy levels. If you are working with high energy lasers, please see these specific products on pages 206-209.

High Reflection Coatings

Coating Type	Wavelength (nm)	COATING SUFFIX Normal Incidence	COATING SUFFIX 45 degrees Incidence
Uncoated		/000	
MAXBRIte™	480-700	/001	/001
Protected Aluminum	400-800	/011	/011
Aluminum	200-IR	/016	/016
Enhanced Aluminum	450-750	/023	/023
UV-Enhanced Aluminum	200-400	/028	/028
Silver (internal)	400-IR	/036	/036
Gold	700-IR	/045	/045
MAX-R	351	/205	/255
MAX-R	364	/207	/257
MAX-R	442	/209	/259
MAX-R	458	/211	/261
MAX-R	466	/213	/263
MAX-R	473	/215	/265
MAX-R	476	/217	/267
MAX-R	488	/219	/269
MAX-R	496	/221	/271
MAX-R	502	/222	/272
MAX-R	514	/223	/273
MAX-R	532	/225	/275
MAX-R	543	/226	/276
MAX-R	633	/229	/279
MAX-R	694	/231	/281
MAX-R	700	/233	/283
MAX-R	830	/237	/287
MAX-R	850	/238	/288
MAX-R	904	/239	/289
MAX-R	1064	/241	/291
MAX-R	1300	/245	/295
MAX-R	1523	/247	/297
MAX-R	1550	/247	/297

For full specifications of these coatings see pages 179-184.

SQUARE FLAT MIRRORS

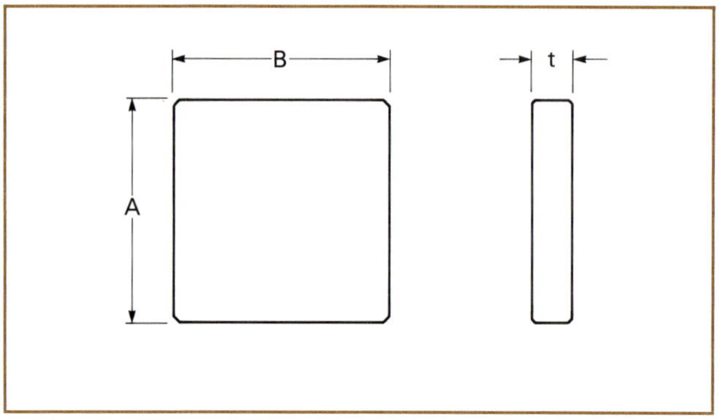

SPECIFICATIONS: SQUARE FLAT MIRRORS

Surface Flatness: at 546nm over central 90% of aperture
Dimensions: ±0.25mm
Cosmetic Surface Quality
- Crown: 80-50 scratch and dig
- Pyrex: 60-40 scratch and dig
- Fused Silica: 60-40 scratch and dig
- ZERODUR: 60-40 scratch and dig

Coating: Any of the high reflection coatings previously described may be requested by appending the appropriate Coating Suffix to the Product Number. Protected Aluminum will be supplied if no option is indicated.

When using mirrors, the precision and stability of the mount can be just as important as the mirror itself. That's why we took great care in design and testing of our range of mirror mounts. These mounts, described on pages 472-479 are available in sizes to suit our selection of mirrors.

Square Flat Mirrors

Flatness	A=B (mm)	t (mm)	Substrate Material	PRODUCT NUMBER
3λ per 25mm	5.0	1	Crown	01 MFG 001
	10.0	1	Crown	01 MFG 003
	12.5	1	Crown	01 MFG 005
	25.0	1	Crown	01 MFG 007
	50.0	3	Crown	01 MFG 011
	75.0	3	Crown	01 MFG 013
	100.0	3	Crown	01 MFG 015
	150.0	6	Crown	01 MFG 017
	200.0	6	Crown	01 MFG 019
λ/4	12.5	6	Pyrex	02 MFG 000
	25.0	6	Pyrex	02 MFG 001
	38.0	10	Pyrex	02 MFG 003
	50.0	10	Pyrex	02 MFG 005
	75.0	14	Pyrex	02 MFG 007
	100.0	14	Pyrex	02 MFG 009
	150.0	25	Pyrex	02 MFG 011
	200.0	38	Pyrex	02 MFG 013
λ/10	12.5	6	Pyrex	02 MPG 002
	12.5	6	Silica	02 MPQ 002
	25.0	6	Pyrex	02 MPG 001
	25.0	6	Silica	02 MPQ 001
	38.0	10	Pyrex	02 MPG 003
	50.0	10	Pyrex	02 MPG 005
	50.0	10	Silica	02 MPQ 005
	12.5	6	Silica	02 MLQ 001
λ/20	12.5	6	ZERODUR®	02 MLE 001
	25.0	6	Silica	02 MLQ 003
	25.0	6	ZERODUR®	02 MLE 003
	50.0	10	Silica	02 MLQ 007
	50.0	10	ZERODUR®	02 MLE 007

All mirrors on this page are supplied with a protected aluminum coating unless otherwise specified.

To request a different coating simply append the appropriate Coating Suffix.

ROUND FLAT MIRRORS

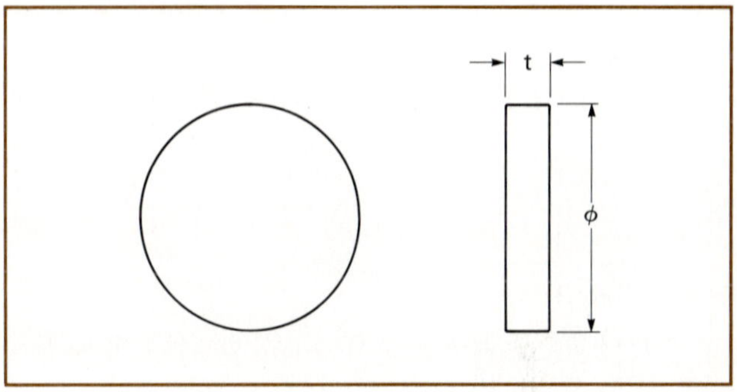

SPECIFICATIONS: ROUND FLAT MIRRORS

Surface Flatness: at 546nm over central 90% of aperture
Dimensions: ±0.25mm
Cosmetic Surface Quality
- Crown: 80-50 scratch and dig
- Pyrex: 60-40 scratch and dig
- Fused Silica: 60-40 scratch and dig
- ZERODUR: 60-40 scratch and dig

Coating: Any of the high reflection coatings previously described may be requested by appending the appropriate Coating Suffix to the Product Number. Protected Aluminum will be supplied if no option is indicated.

Round Flat Mirrors

Flatness	ϕ (mm)	t (mm)	Substrate Material	PRODUCT NUMBER
3λ per 25mm	6.4	1	Crown	01 MFG 021
	12.5	1	Crown	01 MFG 023
	25.0	1	Crown	01 MFG 033
	50.0	3	Crown	01 MFG 027
	75.0	3	Crown	01 MFG 028
	100.0	3	Crown	01 MFG 030
λ/4	12.5	6	Pyrex	02 MFG 014
	25.0	6	Pyrex	02 MFG 015
	30.0	6	Pyrex	02 MFG 016
	38.0	10	Pyrex	02 MFG 017
	50.0	10	Pyrex	02 MFG 019
	60.0	10	Pyrex	02 MFG 018
	75.0	14	Pyrex	02 MFG 021
	100.0	14	Pyrex	02 MFG 023
	150.0	25	Pyrex	02 MFG 025
	200.0	38	Pyrex	02 MFG 027
	250.0	50	Pyrex	02 MFG 029

continued

All mirrors on this page are supplied with a protected aluminum coating unless otherwise specified.

To request a different coating simply append the appropriate Coating Suffix.

Round Flat Mirrors (continued)

Flatness	ϕ (mm)	t (mm)	Substrate Material	PRODUCT NUMBER
λ/10	12.5	6	Pyrex	02 MPG 006
	12.5	6	Silica	02 MPQ 013
	25.0	6	Pyrex	02 MPG 007
	25.0	6	Silica	02 MPQ 007
	30.0	6	Pyrex	02 MPG 012
	38.0	10	Pyrex	02 MPG 009
	50.0	10	Pyrex	02 MPG 011
	50.0	10	Silica	02 MPQ 011
	60.0	10	Pyrex	02 MPG 016
	100.0	19	Pyrex	02 MPG 017
	100.0	19	Silica	02 MPQ 017
λ/20	12.5	6	Silica	02 MLQ 009
	12.5	6	ZERODUR®	02 MLE 009
	25.0	6	Silica	02 MLQ 011
	25.0	6	ZERODUR®	02 MLE 011
	30.0	6	Silica	02 MLQ 012
	50.0	10	Silica	02 MLQ 015
	50.0	10	ZERODUR®	02 MLE 015
	60.0	10	Silica	02 MLQ 016
	100.0	19	Silica	02 MLQ 017
	100.0	19	ZERODUR®	02 MLE 017

SPHERICAL CONCAVE MIRRORS

These accurate spherical concave mirrors are suitable for a variety of applications, from laser beam manipulation to instrument construction and quality imaging. They are offered in a variety of sizes and focal lengths, and can be coated with any of the reflecting coatings previously described. The longer focal lengths are especially convenient for such tasks as laser beam collimation.

In the accompanying figure, all conjugate distances are positive. A and B are conjugate points, C is conjugate to itself and F is conjugate to infinity. The radius curvature is negative because the mirror is concave to the left. The formula which relates the conjugate distances in the paraxial limit is

$$\frac{1}{s} + \frac{1}{s''} = -\frac{2}{r} .$$

In case the object distance s is infinite, the image distance s″ becomes the mirror focal length f, and the relationship between radius of curvature and focal length is

$$f = -\frac{r}{2} .$$

In the mirror list which follows we neglect the radius sign convention and publish absolute value of radius.

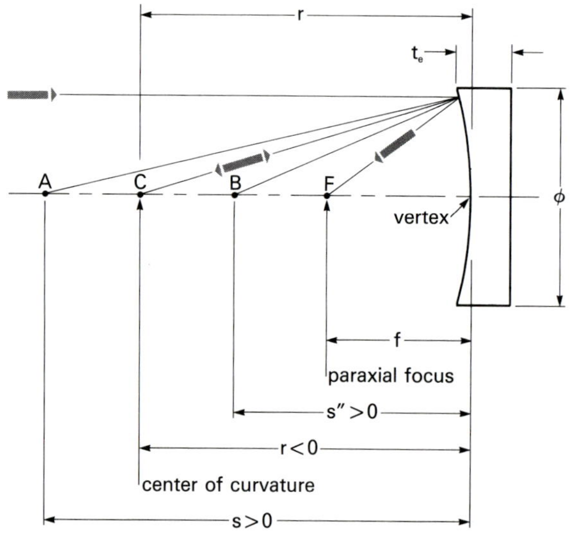

02 SCM

SPECIFICATIONS: SPHERICAL CONCAVE MIRRORS

Surface Accuracy: λ/10 at 546nm over central 90% of diameter
Diameter (ϕ): ±0.5mm
Material: Pyrex
Cosmetic Surface Quality: 60-40 scratch and dig
Coating Options: Any of the high reflection coatings previously described may be requested by appending the appropriate Coating Suffix to the product number. Protected aluminum will be supplied if no option is indicated.

CIRCULAR CYLINDRICAL MIRRORS, BOTH CONCAVE AND CONVEX

Any of the high reflectance coatings previously described can be applied to the curved surface of any of our simple cylinder or precision simple cylinder lenses, thereby forming a cylindrical mirror. To obtain such mirrors, simply append the coating suffix to the cylinder lens product number. Unless you make a special request to the contrary, only the curved surface will be coated. Surface radii for cylindrical lenses can be found via earlier formulas.

Spherical Concave Mirrors

Radius of Curvature r (meters)	Diameter ϕ (mm)	Focal Length f=r/2 (meters)	Edge Thickness t_e (mm)	f/ϕ	PRODUCT NUMBER
1± 1%	25	0.5	9.5	20	02 SCM 001
1± 1%	50	0.5	15	10	02 SCM 002
1± 1%	100	0.5	20	5	02 SCM 003
2± 2%	25	1	9	40	02 SCM 004
2± 2%	50	1	15	20	02 SCM 005
2± 2%	100	1	20	10	02 SCM 006
4± 3%	25	2	9.5	80	02 SCM 007
4± 2%	50	2	15	40	02 SCM 008
4± 2%	100	2	20	20	02 SCM 009
10± 3%	25	5	9.5	200	02 SCM 011
10± 3%	50	5	15	100	02 SCM 012
10± 3%	100	5	20	50	02 SCM 013
20±10%	25	10	9.5	400	02 SCM 014
20± 5%	50	10	15	200	02 SCM 015
20± 3%	100	10	20	100	02 SCM 016
40±10%	25	20	9.5	800	02 SCM 017
40±10%	50	20	15	400	02 SCM 018
40± 5%	100	20	20	200	02 SCM 019
60±20%	25	30	9.5	1200	02 SCM 022
60±15%	50	30	15	600	02 SCM 023
60± 5%	100	30	20	300	02 SCM 024

Supplied with a protected aluminum coating unless otherwise specified.
To request a different coating simply append the appropriate Coating Suffix.

Much of our business is composed of special, non-catalog items. Standard mirrors may be modified to special diameters, have cordal sections removed or circular perforations made on- or off-axis. Contact an Application Engineer for further details.

CONCAVE SPHERICAL REFLECTORS

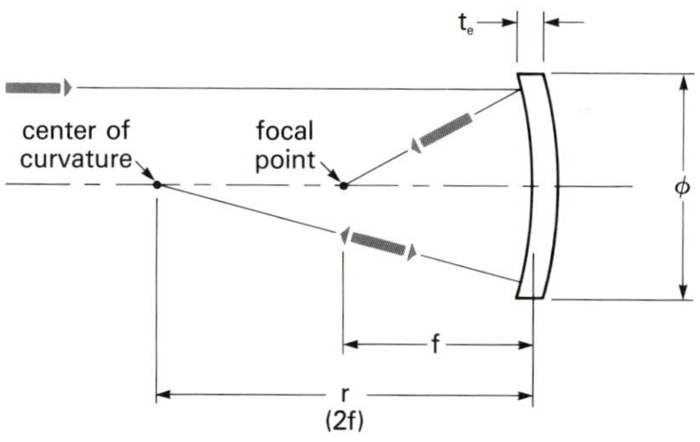

01 MCG

Concave spherical reflectors are often used to increase the light output of illumination or projection systems, by increasing the efficiency with which light sources are utilized. For this purpose the reflector is located with its center of curvature coinciding with the source and with the reflector itself opposite the condenser from the source. This has been illustrated earlier in schematic drawings (see condensor section page 163).

These are first-surface reflectors, for the reason that the substrates absorb much less heat than they would if they were second-surface reflectors. Also well suited for this application are the concave cold mirrors presented later (see page 247).

We list spherical reflectors by their paraxial focal lengths, which are equal to half of their radii of curvature. The substrate material for our concave spherical reflectors is optical crown glass. Reflecting surfaces are ground and polished. Those who need reflectors with surface radii different from those of the standard reflectors may wish to consider coating a plano-concave lens for this purpose. All necessary information has been presented earlier.

Standard reflectors may be modified to special diameters. Contact an Applications Engineer for further details.

All of the reflective options in the coating suffix system are also available for concave spherical reflectors. Protected aluminum is automatically supplied at the listed price (no suffix necessary). The other coating options may be selected at an additional cost by specifying the Coating Suffix.

SPECIFICATIONS: CONCAVE SPHERICAL REFLECTORS

Dimensions: ±0.50mm
Focal Length (f): ±5%
Substrate Material: Optical crown (coefficient of thermal expansion = $9 \times 10^{-6}/°C$)
Cosmetic Surface Quality: 80-50 scratch and dig
Coating: Protected aluminum
Circumference is ground and both edges are beveled.

Concave Spherical Reflectors

r (mm)	ϕ (mm)	f (mm)	t_e (mm)	PRODUCT NUMBER
25.0	30.0	12.5	2.5	01 MCG 007
20.0	36.0	10.0	3.0	01 MCG 009
35.0	51.0	17.5	3.0	01 MCG 013
45.0	80.0	22.5	4.0	01 MCG 021
50.0	60.0	25.0	3.0	01 MCG 017
60.0	70.0	30.0	4.0	01 MCG 019
60.0	90.0	30.0	4.0	01 MCG 023
80.0	100.0	40.0	4.0	01 MCG 025
137.0	140.0	68.5	4.0	01 MCG 027
150.0	200.0	75.0	4.0	01 MCG 033

PARABOLOIDAL AND ELLIPSOIDAL REFLECTORS

Concave paraboloidal and ellipsoidal reflectors are the basis for a variety of high-efficiency illumination, light collection, and light concentration systems. High efficiency is achieved because collection or concentration can take place throughout very large solid angles. This in turn follows from fundamental geometrical properties of the parabola and ellipse. A tangent to a parabola makes equal angles with the axis of the parabola and a line from the focus to the point of tangency. From the law of reflection applied at the point of tangency it then follows that rays passing through the focus are reflected parallel to the axis, and conversely. By similar reasoning, rays which pass through (or originate from) one focal point of an ellipse must, after reflection, pass through the other. In these situations, in which the geometric foci are optical conjugate points, both paraboloidal and ellipsoidal surfaces are free of spherical aberration. These reflectors will not, in general, be satisfactory for imaging purposes; they have been designed specifically for energy collection and concentration applications.

Access hole sizes, in both paraboloidal and ellipsoidal replicated reflectors, have been chosen for compatibility with the most widely available of the modern mercury and xenon short arc lamps. Because of lamp-to-lamp variability, the best position for the lamp should be determined experimentally, and lamp mountings should provide some adjustability for this purpose. Note that reflector focal lengths are referenced from reflector vertices which do not lie in the planes of the access holes. This is another reason why it is best to plan on determining optimum lamp position experimentally. In connection with such determination for ellipsoids, it is especially helpful to know in advance the focal point separation $s''-s$. With source and target separated by this fixed distance, all that remains to be found is the optimum relative position of lamp and reflector. It is for this reason that we have tabulated the difference $s''-s$.

In addition to their dimensions, we have tabulated below the solid angles subtended by the reflectors at their foci (primary foci only, for the ellipsoids). These values are given by the integral

$$\Omega = \frac{1}{4\pi} \int\int_R \frac{\vec{r} \cdot \vec{n}}{r^3} \, dA \ ,$$

where Ω is the solid angle in parts of a sphere, n is the outer unit vector normal to the reflector surface at the area element dA, r is the radius vector from the focus to the area element dA, the dot symbolizes the scalar product, and the domain of integration R is the entire reflector. The vectors are shown in the following drawing.

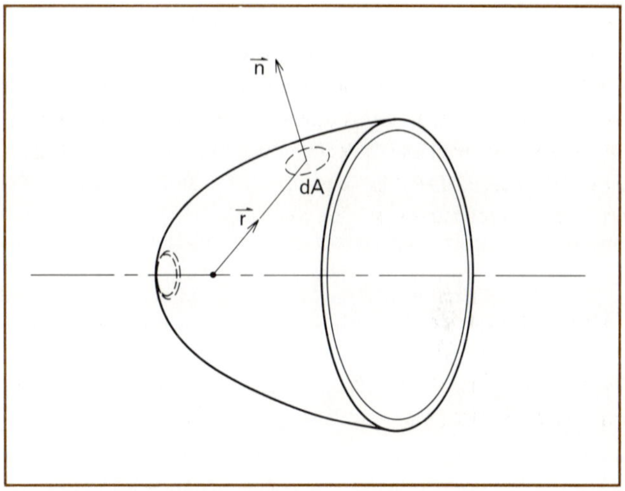

VECTOR DIAGRAM showing elements used in calculating the solid angle (Ω) subtended by these reflectors.

The tabulated solid angles are convenient measures of the efficiency with which sources (or isotropic receivers) at the foci can be utilized. Multiplied by surface reflectance, these solid angles, in our units, give reflector efficiency directly. As our unit of solid angle we have chosen here to use the intuitive "sphere," instead of the steradian. To convert from spheres to steradians, simply multiply by 4π.

Unless your volume requirements are very considerable, you will find the cost of special focal length paraboloidal and ellipsoidal reflectors to be prohibitive. This is because the initial tooling and jigging costs for these components are very high. While we will always be pleased to discuss special requirements, we would recommend that every effort be made to utilize standard designs for production requirements.

PARABOLOIDAL REFLECTORS

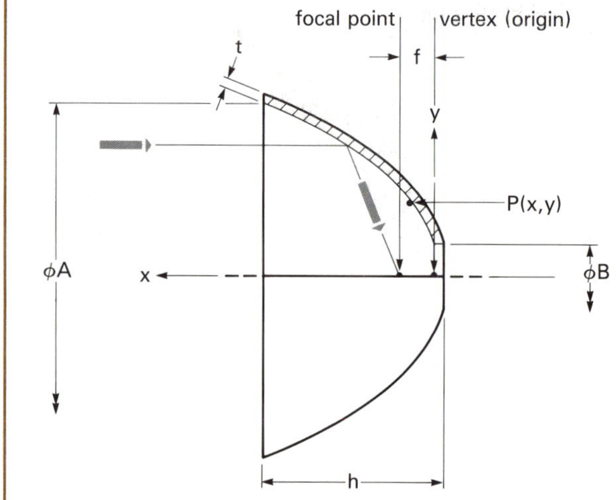

These reflectors are produced by an electrolytic replication process which very accurately reproduces the geometry and finish of a carefully generated master surface. Replication begins with electrolytic deposition of a nickel substrate. Upon separation from the master, the nickel substrate is electrolytically coated with a highly reflective yet extremely hard and durable rhodium film. Rhodium reflects well throughout the ultraviolet, visible, and infrared, as is evident from the representative rhodium reflectance graph.

Rhodium adheres well to nickel throughout large temperature excursions, and is resistant to tarnish and oxidation, even at high temperatures and in hostile ozone-rich atmospheres. These reflectors are therefore ideally suited for use with modern high-intensity high-power arc sources. These reflectors ordinarily perform well at temperatures up to 260° C (500° F), and are sufficiently tough to typically withstand several arc lamp explosions before sustaining enough damage to require replacement. Glass reflectors rarely survive arc lamp explosions.

Electrolytically replicated reflectors achieve a surface accuracy intermediate between that achieved by conventional grinding and polishing and that achieved by such processes as molding. There is little difference between the reflective properties of replicated reflectors and rhodium-coated mirrors which are conventionally ground and polished. By conventional methods it is, of course, impossible to produce the very deep paraboloids and ellipsoids which are essential for high-efficiency applications. We normally specify that a ray, after reflection from the replicated surface, shall not deviate from its ideal direction by more than 6 arc minutes.

Such surface accuracy makes possible a wide variety of applications, including:

- Portable narrow-beam spotlamps of very high brightness and efficiency, based on modern short arc lamps.
- Collimation of high-power sources
- Solar energy collection
- Intensity interferometry
- Efficient area illumination, achieved by deliberate defocusing.
- Laser cavity pumping

SPECIFICATIONS: PARABOLOIDAL REFLECTORS

Maximum angular deviation of reflected ray from ideal path: 6 arc minutes (does not apply within 7mm of either edge)
Nominal equation of the paraboloidal profile: $y^2 = 4fx$
Reflectance: See graph of nominal rhodium reflectance values.
f, t, and ϕB: $\pm$ 0.13mm
ϕA: $\pm$ 0.25mm (except 02 RPM 033: see below)
h, (ϕ+2t): $\pm$ 0.40mm
Coating: Rhodium, electrolytically deposited
Maximum Operating Temperature: 260°C (500°F)
Coating Durability: Bright rhodium meets or exceeds all applicable military and DIN specifications.

For 02 RMP 033 only,
t: $\pm$ 0.25
ϕA: $\pm$ 2.5

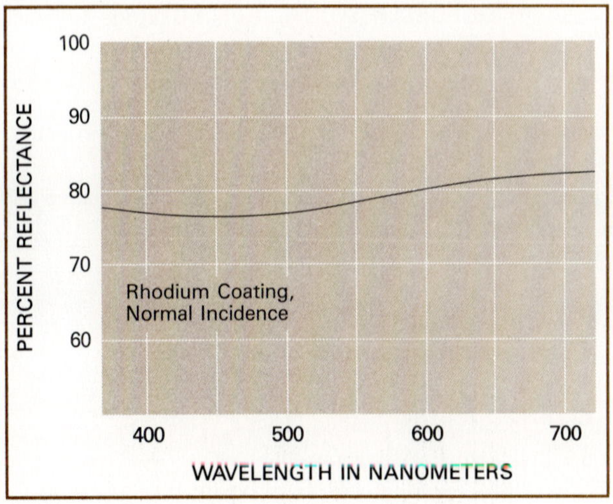

RECOMMENDED MAXIMUM REFLECTOR WATTAGES

Compatibility between lamp and reflector depends upon comaprative geometry, lamp wattage, the cooling system, and several other factors. One should check before ordering, from lamp specifications, that the reflector geometry is acceptable.

Experience has shown that is possible to associate meaningful maximum power ratings with reflectors almost independent of lamp type, i.e., the same ratings apply whether lamps are of the compact filament, short-arc xenon, or short-arc mercury types. Our recommended maximum reflector wattages (published in the next-to-last columns of the reflector product tables) are intended as rough guidelines only, and not as absolute limits.

Recommended maximum reflector wattages above 250 presume the presence of forced air cooling.

Paraboloidal Reflectors

f (mm)	ϕA (mm)	ϕB (mm)	h (mm)	t (mm)	Ω (spheres)	Max. Power (watts)	PRODUCT NUMBER
10.2	88	22	45	0.6	0.59	100	02 RPM 006
10.2	88	16	46	0.6	0.69	100	02 RPM 001
10.2	123	22	91	0.6	0.67	150	02 RPM 007
10.2	123	16	92	0.6	0.77	100	02 RPM 005
19.1	123	22	49	0.8	0.64	350*	02 RPM 011
19.1	152	40	72	0.8	0.59	500*	02 RPM 008
19.1	170	47	88	0.8	0.56	1000*	02 RPM 016
19.1	236	47	177	0.8	0.63	1000*	02 RPM 009
19.1	236	40	179	0.8	0.69	500*	02 RPM 023
33.0	241	57	105	1.0	0.61	1600*	02 RPM 017
33.0	241	44	107	1.0	0.67	1000*	02 RPM 018
59.7	356	89	125	1.2	0.57	2500*	02 RPM 036
59.7	356	89	125	1.2	0.57	2500*	02 RPM 028
59.7	356	57	130	1.2	0.64	1000*	02 RPM 029
176.1	762	159	200	2.3	.49	5000*	02 RPM 033

*NOTE: Forced air cooling required on these models.

ELLIPSOIDAL REFLECTORS

Ellipsoidal reflectors find various applications in condensing and illumination systems. A single ellipsoidal reflector can function as an essentially complete condenser. In pairs they can almost completely surround both source and target, offering the ultimate in efficiency of radiative transfer. Like aspheric lenses (as discussed earlier), ellipsoidal reflectors are used to illuminate the film gates of projection systems (especially microfilm systems and systems for large outdoor theaters). Effective f-numbers can be extremely low (some of our reflectors have s/ϕ ratios of less than 0.13). Other applications include flash photolysis, laser pumping, solar simulation, and fiber optic illuminator input.

These reflectors are produced by an electrolytic replication process which very accurately reproduces the geometry and finish of a carefully generated master surface. Replication begins with electrolytic deposition of a nickel substrate. Upon separation from the master, the nickel substrate is electrolytically coated with a highly reflective yet extremely hard and durable rhodium film. Rhodium reflects well throughout the ultraviolet, visible, and infrared, as is evident from the representative rhodium reflectance graph.

SPECIFICATIONS: ELLIPSOIDAL REFLECTORS

s, t, and ϕB: ±0.13mm ϕA: ±0.25mm
h: ±0.40mm s″: ±0.80mm
Maximum separation of optical and mechanical axes at vertex: ±0.13mm
Maximum angular deviation of reflected ray from ideal path: 6 arc minutes (does not apply within 7mm of either edge)
Nominal equation of the ellipsoidal profile:

$$\frac{x^2}{a^2} + \frac{y^2}{b^2} = 1$$, origin midway between foci

Maximum Operating Temperature: 260° C (500° F)
Coating: Rhodium electrolytically deposited
Reflectance: See graph of nominal rhodium reflectance no prior page
Coating durability: Bright rhodium meets or exceeds all applicable military and DIN specifications.

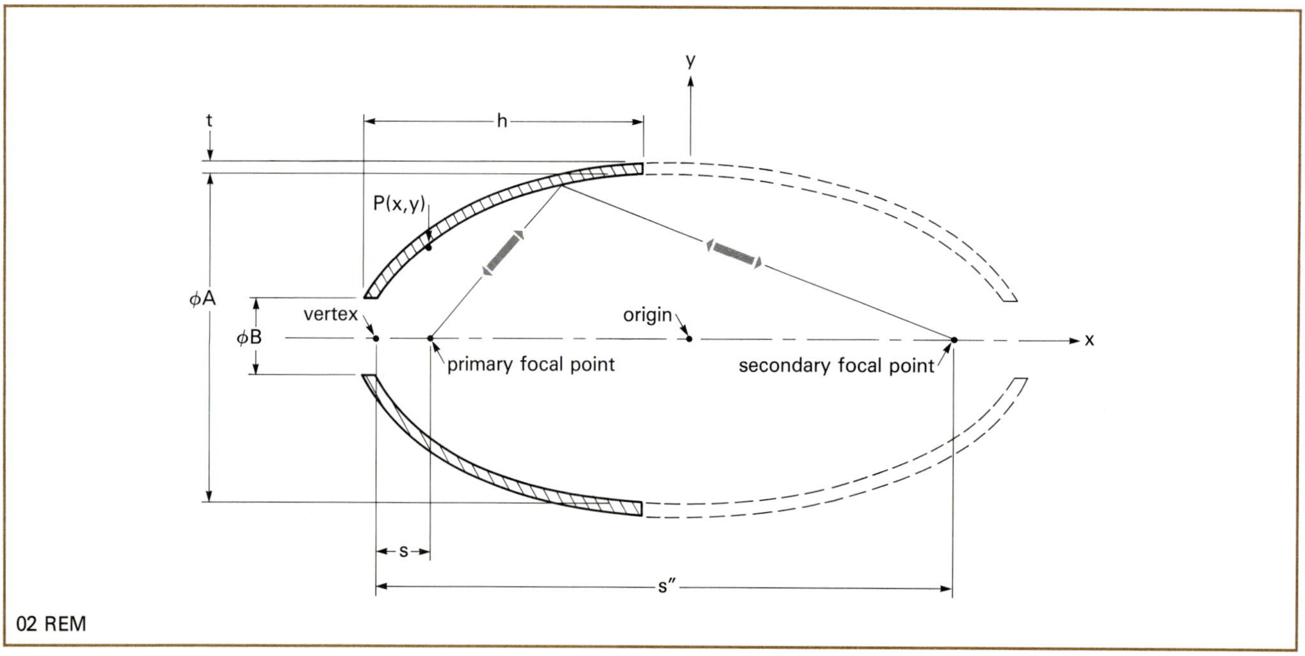

Ellipsoidal Reflectors

s (mm)	ϕA (mm)	h (mm)	s″ (mm)	s″−s (mm)	ϕB (mm)	t (mm)	Ω (spheres)	Max. Power (watts)	PRODUCT NUMBER
12.1	74.4	56.4	115.0	102.8	25.4	0.76	0.67	350*	02 REM 005
12.1	74.4	59.3	115.0	102.8	12.7	0.76	0.84	100	02 REM 001
12.8	86.4	59.2	155.0	142.3	31.8	0.76	0.59	350*	02 REM 003
12.8	74.7	38.1	155.0	142.3	12.7	0.76	0.72	100	02 REM 004
12.8	86.4	63.8	155.0	142.3	12.7	0.76	0.82	100	02 REM 006
17.2	142.7	93.5	414.6	397.4	44.5	0.76	0.58	350*	02 REM 007
17.2	123.2	65.0	414.6	397.4	31.8	0.76	0.64	500*	02 REM 008
17.2	142.7	97.3	414.6	397.4	31.8	0.76	0.70	350*	02 REM 009
20.6	123.2	58.2	261.9	241.3	44.5	0.76	0.56	500*	02 REM 012
20.6	142.9	101.3	261.9	241.3	47.0	0.76	0.63	500*	02 REM 013
20.6	123.2	62.5	261.9	241.3	25.4	0.76	0.70	350*	02 REM 014
20.6	142.9	105.4	261.9	241.3	31.8	0.76	0.76	500*	02 REM 011
30.0	169.7	73.9	420.0	390.0	44.5	0.76	0.62	1000*	02 REM 015
30.0	203.0	125.2	420.0	390.0	44.5	0.76	0.73	1000*	02 REM 017
30.0	215.9	159.8	420.0	390.0	44.5	0.76	0.77	1000	02 REM 018
30.0	203.0	127.5	420.0	390.0	31.8	0.76	0.78	500*	02 REM 019
33.9	254.0	177.0	541.9	508.0	76.2	0.89	0.64	3000*	02 REM 022
33.9	236.2	140.0	541.9	508.0	63.5	0.89	0.66	2500*	02 REM 024
33.9	254.0	179.1	541.9	508.0	69.9	0.89	0.67	3000*	02 REM 025
33.9	236.2	141.5	541.9	508.0	57.2	0.89	0.69	1600*	02 REM 026
33.9	254.0	180.6	541.9	508.0	63.5	0.89	0.70	2500*	02 REM 028
33.9	254.0	182.1	541.9	508.0	57.2	0.89	0.73	1600*	02 REM 023
33.9	254.0	183.6	541.9	508.0	50.8	0.89	0.76	1600*	02 REM 016
33.9	254.0	184.4	541.9	508.0	47.0	0.89	0.78	1000*	02 REM 027
43.2	123.2	36.1	129.5	86.4	31.8	0.76	0.52	500*	02 REM 029
43.2	147.3	66.0	129.5	86.4	50.8	0.76	0.59	350*	02 REM 033
60.3	368.3	181.4	873.1	812.8	86.4	1.52	0.67	5000*	02 REM 034
60.3	393.7	221.7	873.1	812.8	76.2	1.52	0.73	5000*	02 REM 035

*Forced air cooling required on these models.

Standard reflectors may be modified to special diameters. Contact an Applications Engineer for further details.

OFF-AXIS PARABOLOIDAL REFLECTORS

These mounted reflectors are convenient for a variety of applications in which diffraction limited performance is not essential. The optical or symmetry axis of the reflector segment does not pass through the reflector segment itself, but instead passes by in nearby space. The user has unrestricted access to the focal point and a sizable region surrounding it, without fear of shadowing. Imagery is achromatic and free of spherical aberration. The useful field of view for imaging purposes is extremely limited, as the off-axis aberrations are severe. For illumination and collection applications the field of view is quite adequate. Mounted off-axis paraboloidal reflectors are made from segments cut from our regular paraboloidal reflectors, and are hence subject to the same surface accuracy specification. Namely, with the off-axis reflector correctly positioned and oriented, reflected rays do not deviate from their ideal directions by more than 6 arc minutes. The substrates are electroformed nickel, and the reflecting surface is of electrolytically deposited bright rhodium. The mounting is black anodized aluminum. Direction to the optical axis is identified by the mounting screw nearest which the mount departs least from the mounting surface.

All of the paraboloidal reflectors present a circular cross section when viewed along the mechanical axis; there is no projection of the mechanical axis; there is no projection of the mechanical holder beyond the edge of the reflector itself. This maximizes the access to the focal region. It is important to note that not all of these units reflect at 90 degrees to the incident light. While this is the most commonly used configuration for off-axis paraboloids, we have included other options to facilitate more complex geometries. The angle, θ, between the incident and reflected beams is shown in the drawing, and its values are tabulated in the product table. All other dimensions needed for accurate determination of the location of the focal point are given. Threaded holes are provided on the base of these reflectors to allow for mounting.

While electroformed off-axis paraboloidal reflectors cannot approach diffraction limited performance in the visible spectrum, they do very well in the infrared and have found many uses there. The surface accuracy is about 5 microns, or about 1/2 wave at a wavelength of 10 microns. This presumes that the reflectors are not subjected to undue stress or thermal gradient.

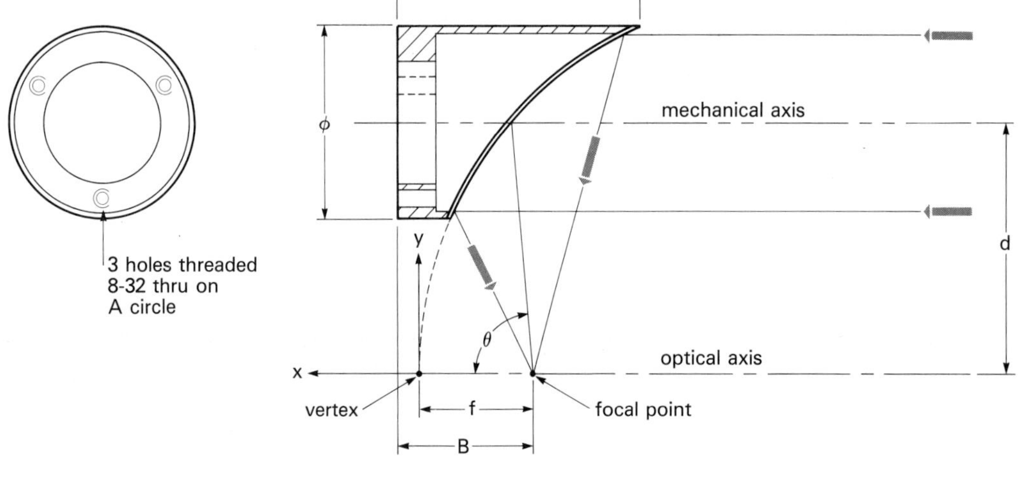

02 POA

SPECIFICATIONS: MOUNTED ELECTROFORMED OFF-AXIS PARABOLOIDAL REFLECTORS

Maximum angular deviation of reflected ray from ideal path: 6 arc minutes.

Reflectance: See graph of nominal rhodium reflection.

f, ϕ: ± 0.13mm

d, B, L: ± 0.8mm

A: ± 0.25mm

θ: for reference only

Clear aperture ϕ is concentric with mechanical axis to within 0.25mm.

Coating: Rhodium, electrolytically deposited

Coating Durability: Bright rhodium meets or exceeds all applicable military and DIN specifications.

Maximum Operating Temperature: 204° C (400° F), because of limitations of mounting material.

Mounting Screws: Three 8-32 UNF Allen Socket screws, 0.50 inch long, evenly spaced on bolt circle diameter.

SOME EXAMPLE SOURCE SIZES h_1

h_1	Source
4.0mm	100-watt Xenon compact arc lamp, Conrad-Hanovia 976C-1.
5.8mm	Tungsten halogen FCS (24 volt, 150 watt) lamp, General Electric or Osram (No. 64640).
11.7mm	Tungsten halogen BCK lamp (120 volt, 500 watt) General Electric.

Off-Axis Paraboloidal Reflectors, Mounted

f (mm)	ϕ (mm)	d (mm)	B (mm)	θ (deg)	L (mm)	A (mm)	PRODUCT NUMBER
10.2	30	26.4	20.3	105	52.8	21.6	02 POA 011
19.1	44	38.1	31.0	90	57.9	34.0	02 POA 013
33.0	50	66.0	39.9	90	64.8	40.1	02 POA 015
59.7	75	119.0	45.5	90	89.4	65.5	02 POA 017
59.7	95	83.6	68.4	70	80.8	83.8	02 POA 019
176.0	95	203.2	156.7	60	70.1	83.8	02 POA 027

A considerable portion of the cost of manufacturing electroformed reflectors is involved in tooling. While we will be pleased to manufacture items to your specific requirements, we would forewarn you that volume needs must be substantial before this becomes economically attractive.

WINDOWS

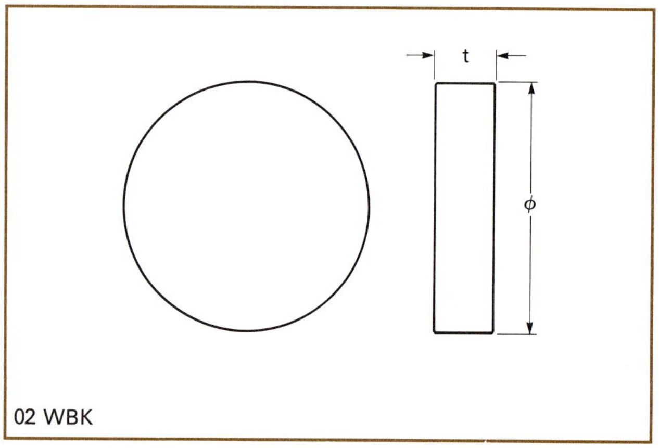

Windows are used to allow optical radiation to pass from one environment to another without allowing other elements of these environments to mix. Environments may vary widely in temperature, pressure, chemical composition, phase, or cleanliness. Considerations in selecting windows may include resistance to environments, transmission, scattering, and wavefront distortion. We offer windows of three distinct materials, from which you may choose in accordance with the properties you need: sapphire, synthetic fused silica, and BK 7 optical glass.

Tilted at Brewster's angle, windows of any of these materials offer very nearly 100% transmission for light linearly polarized parallel to the plane of incidence. (Refer to the earlier text on reflection. Page 174.)

SAPPHIRE WINDOWS

Sapphire is in many ways a superior window material. Because of its extreme surface hardness, sapphire can only be scratched by a few substances (such as diamond and boron nitride) other than itself. Chemically inert and insoluble in almost everything except at highly elevated temperatures, sapphire can be cleaned with impunity. For example, even hydrogen fluoride fails to attack sapphire at temperatures below 300° C. Sapphire exhibits high internal transmittance all the way from 150 nanometers in the vacuum ultraviolet out to 6000 nanometers in the middle-infrared, as can be seen in the graph below.

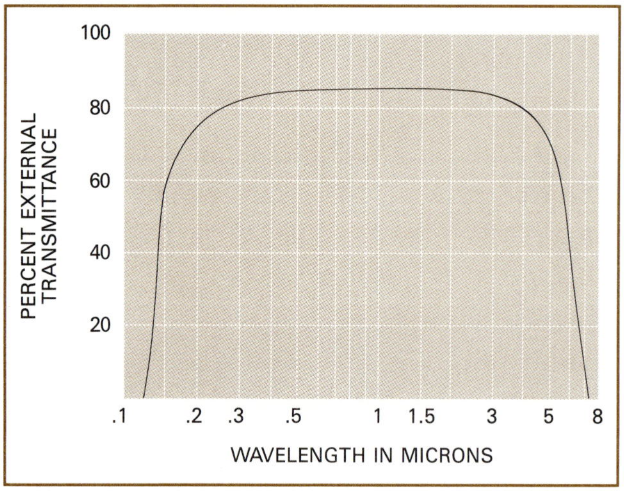

EXTERNAL TRANSMITTANCE OF SAPPHIRE: 1mm thick sample.

SPECIFICATIONS: SAPPHIRE WINDOWS

Crystalline Optic Axis Orientation: Uncontrolled
Birefringence: $(n_o - n_e) = 0.008$ for visible light in directions orthogonal to the optic axis.
Stress Birefringence Coefficient: 1 fringe per cm thickness at 632.8nm and 2.43 kilograms per cm^2.
Flatness: 1λ per 25mm at 632.8nm over 90% of aperture
Parallelism: 3 arc minutes
Diameter: ±0.25mm
Thickness: ±0.1mm
Material: Anisotropic crystalline synthetic sapphire (Al_2O_3)
Cosmetic Surface Quality: 80-50 scratch and dig
Circumference is ground and both edges are beveled.

Synthetic Sapphire Windows

ϕ (mm)	t (mm)	PRODUCT NUMBER
5.0	1	02 WSA 002
10.0	1	02 WSA 004
20.0	2	02 WSA 006
25.0	2	02 WSA 008
40.0	2	02 WSA 012

SPECIFICATIONS: BK 7 WINDOWS

Wavefront Distortion: at 632.8nm over 80% of aperture
Parallelism: 1 arc minute
Diameter (ϕ): $\pm$ 0.25mm
Thickness: $\pm$ 0.2mm
Material: BK 7 Grade A Fine Annealed
Wavefront Distortion: at 632.8nm over 80% of ϕ
Cosmetic Surface Quality: Pitch-polished, 80-50 scratch and dig
Circumference is ground and both edges are beveled.

BK 7 Windows; λ/10 Wavefront Distortion

ϕ (mm)	t (mm)	PRODUCT NUMBER
12.5	6	02 WBK 222
25	6	02 WBK 224
50	10	02 WBK 226
75	14	02 WBK 228
100	19	02 WBK 233

BK 7 Windows; 1 λ per 25mm Wavefront Distortion

ϕ (mm)	t (mm)	PRODUCT NUMBER
12.5	3	02 WBK 001
25	3	02 WBK 003
50	3	02 WBK 005
75	6	02 WBK 007
100	6	02 WBK 009
150	6	02 WBK 011

See also our high energy laser windows on page 210.

SPECIFICATIONS: SYNTHETIC FUSED SILICA WINDOWS

Wavefront Distortion: λ/10 per ϕ at 632.8mm over 80% of aperture
Parallelism: 30 arc seconds
Dimensions: $\pm$ 0.1mm
Material: Optical Quality synthetic fused silica
Cosmetic Surface Quality: 60-40 scratch and dig
Circumference is ground and both edges are beveled.

Synthetic Fused Silica Windows

ϕ (mm)	t (mm)	PRODUCT NUMBER
12.5	6	02 WLQ 003
25.0	6	02 WLQ 005
50.0	10	02 WLQ 007

We can supply interferograms of these windows on special request.

Call an Applications Engineer for further information.

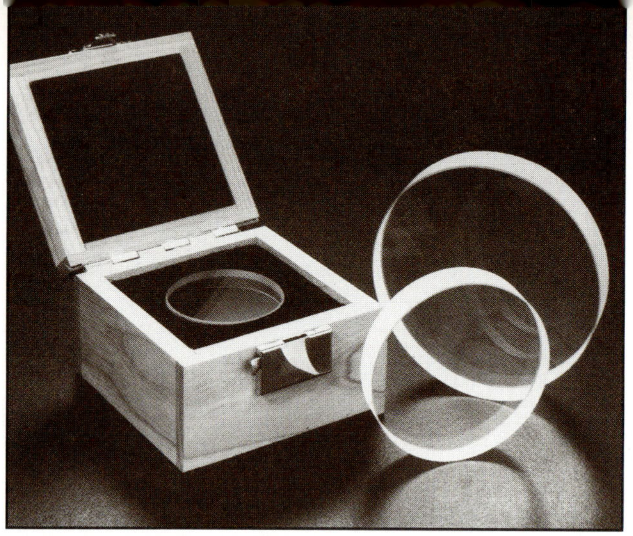

OPTICAL FLATS

Optical flats are used in the testing and evaluation of the surfaces of other optical elements. An interference pattern is formed in the air between the flat and the object being evaluated, and this pattern is usually more easily seen through the flat than through the object. The pattern consists of alternating bright and dark bands or fringes which are a contour map of the thickness of the air film. If the surface of the optical flat is significantly flatter than the surface being evaluated, it is correct to interpret the interference pattern directly as a contour map of the surface being evaluated. The contour interval of the map (the distance between the fringes of the same kind, whether light or dark) corresponds to an optical path length difference interval of 1/2-wavelength. If the surface under test is spherical, the radius of curvature can easily be derived from the fringe pattern.

If the flat is used on top of the object, and the interference pattern viewed through the flat, it is advantageous to have an antireflection coating on the top surface of the flat *(the surface which does not touch the object being evaluated).* Apart from its reflectance, quality and flatness of the top surface are not a matter of concern. It is for this reason that the second surface of our pyrex flats meets a flatness specification of only 2 waves across the surface. When two sufaces of our flats are finished to different flatness specifications, their circumferences are marked with arrows pointing to the first surface (the side meeting the tightest specifications).

The reason for offering the fused silica flats with both surfaces finished to equally stringent flatness specifications is that flats tend to become shopworn through years of intensive service, and a flat with two standard surfaces has twice the useful life expectancy of a flat with only one. If you want to take advantage of the antireflection coated topside, we suggest that you choose a flat with only one standard surface.

The thickness of these flats exceeds the radiation coherence lengths of the more commonplace monochromatic light sources, such as sodium lamps, so surface parallelism is not normally a matter of concern. We usually control only flatness, not parallelism. If you wish to use laser illumination, in which case coherence length may greatly exceed flat thickness, interference will not necessarily be confined to the air film. Fringes can also form between the test object and the topside of the flat, and between the surfaces of the flat. Wedged flats, which minimize these problems, are available on special request.

We offer optical flats of two materials, namely optical quality pyrex and optical quality synthetic fused silica. Both materials are highly transparent, for maximum fringe visibility.

We supply each flat in a hardwood box for permanent protection. Larger size flats, and flats without boxes, are available on special request. To request an antireflection coating for the second surface of a flat with only one standard surface, append the appropriate suffix to the product number of the flat. For example, "02 FPS 015/066, one side only" would be a 100mm diameter pyrex flat with a single $\lambda/20$ standard surface, and having the topside single-layer MgF_2 AR-coated for 550nm at normal incidence. The circumference is ground, and the edges are lightly beveled.

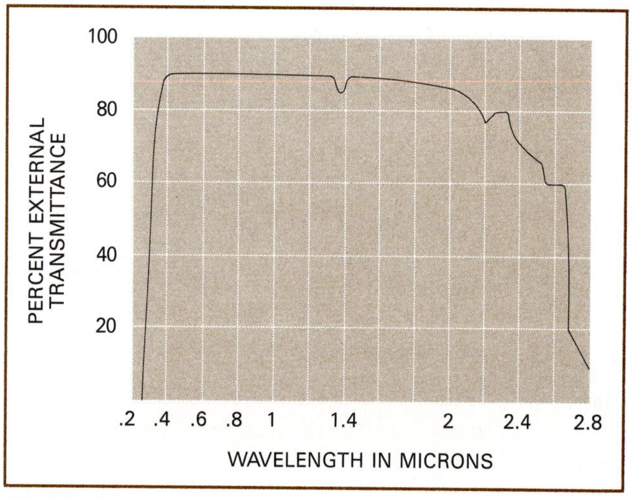

EXTERNAL TRANSMITTANCE OF PYREX, 8mm thick sample.

Also available on a special request basis are optical flats with a hard high-reflectance quarterwave optical thickness layer on the standard surface only. This has the effect of significantly increasing the finesse and visibility of the interference fringes. Fringe sharpness is a function of the geometric mean of the reflectances of the surfaces being compared. A single quarterwave layer of titanium dioxide on a synthetic fused silica flat has a reflectance of more than 35% throughout the range of angles of incidence from 60 degrees to zero. The geometric mean and fringe sharpness are correspondingly increased. Titanium dioxide is an extremely durable film.

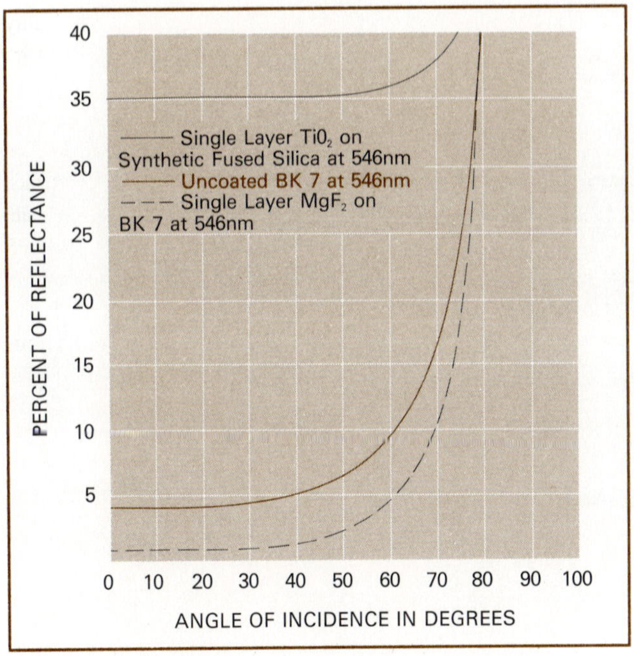

EFFECT OF COATING on reflectance of optical flats.

SPECIFICATIONS: OPTICAL FLATS

Flatness: at 546nm over central 90% of aperture

Dimensions: ± 1.0mm

Materials: Corning 7740 Pyrex ($\alpha = 3.2 \times 10^{-6}/°C$)
Synthetic Fused Silica ($\alpha = 0.5 \times 10^{-6}/°C$)

Surface Quality: 60-40 scratch and dig

Cosmetic Surface Quality: 60-40 scratch and dig

Optical Flats

Flatness 1st Surface	Flatness 2nd Surface	ϕ (mm)	t (mm)	Material	PRODUCT NUMBER
λ/20	2λ	25.0	6.0	Pyrex	02 FPS 003
λ/20	2λ	50.0	10.0	Pyrex	02 FPS 007
λ/20	2λ	75.0	14.0	Pyrex	02 FPS 011
λ/20	2λ	100.0	19.0	Pyrex	02 FPS 015
λ/20	λ/20	25.0	6.0	Fused Silica	02 FQD 003
λ/20	λ/20	50.0	10.0	Fused Silica	02 FQD 007
λ/20	λ/20	75.0	14.0	Fused Silica	02 FQD 011
λ/20	λ/20	100.0	19.0	Fused Silica	02 FQD 015

DESIGN AND ENGINEERING

Our in-house optical and mechanical design, engineering, and prototyping facilities enable us to undertake the production of special systems and assemblies for OEM customers.

HIGH ENERGY LASER OPTICS

In order to withstand the effects of irradiation from high energy lasers, optics must be made from special materials and the whole manufacturing and coating process needs to be carefully controlled. We offer here a selection of optical components with very high damage thresholds suitable for use with high energy lasers.

Contents **Page Number**

LASER INDUCED DAMAGE AND HOW WE MINIMIZE ITS PROBABILITY

It is well known that high energy lasers can destroy optics in a single pulse. Many details of the damage process are still not well understood, and are the subject of active investigation. We cannot begin to do justice to the subject here, but we would refer the interested reader to the proceedings of the annual conference on laser damage:

Laser Induced Damage in Optical Materials.
1983 — NBS Special Publication 688;
1982 — NBS Special Publication 669;
prior years back to 1970 also available.

Three processes by which laser induced damage can occur are acoustic damage caused by stimulated Brillouin scattering, simple heating by absorption and electron avalanche. The acoustic process, similar to the shattering of a wine glass by an opera singer's note, is regarded as rare and unimportant. Heating, either by bulk absorption or localized absorption at inclusions, is unimportant in high purity materials, but may follow from cleanliness problems associated with the production of multilayer dielectric films.

Electron avalanche is the most important damage process, whether in substrates or in multilayers. Electron avalanche appears as an optical frequency spark, associated with shock waves and explosive vaporization. Most often the spark begins at a dielectric surface or interface. Avalanche is initiated by ionization of a single atom or contaminant molecule, such as water, adsorbed on a surface or trapped at an interface. Contaminant sites are localized, mimicing inclusions. If the ionization occurs at an interior point, we may describe it as excitation of an electron from the valence band to the conduction band of a pure dielectric. Such ionization cannot normally occur if the energy of the incident photon is less than the photoelectric threshold energy of the dielectric or contaminant. In the very high photon fluxes of some laser beams, especially if focused, pairs (or larger groups) of photons in coincidence may accomplish otherwise impossile ionizations. This is multiphoton ionization. An equivalent description is that the optical electric fields in some laser beams are so large as to polarize atoms until they come apart.

Multiphoton ionization is a quantum mechanical tunneling phenomenon. Its most outstanding characteristic is its probabilistic nature. There is no threshold pulse energy or pulse peak irradiance for multiphoton ionization. Whether breakdown and damage will inevitably follow an initial multiphoton ionization depends upon laser pulse shape and duration. A variety of dissipative processes compete with avalanche growth. It is usually easier to maintain an avalanche than to start it, so the statistics of laser induced damage usually follow the statistics of multiphoton ionization. Because there are no thresholds for multiphoton ionization, it is fundamentally wrong to attempt to identify a damage threshold with any single damage event.

DAMAGE THRESHOLD

Order emerges from the chaos when a large number of damage events is viewed statistically. For any specific laser pulse shape, *damage probability* is a highly repeatable function of peak electric field, while damage itself is not. See, for example, *Bass and Barrett, "Avalanche Breakdown and the Probabilistic Nature of Laser-Induced Damage," IEEE Journal of Quantum Electronics, VQE8, N3, March 1972, pp. 338-343,* wherein damage probability is represented by a simple formula. "Damage threshold" has come to be identified with damage probability rather than with certainty.

Damage threshold figures in this catalog correspond to probabilities sufficiently low that you are unlikely to see damage. By "damage" we mean permanent damage, ranging from visible cosmetic effects to complete destruction of coating or element. Thus our thresholds should be regarded as recommended safe maximum pulse irradiances. While conservative, these levels are high enough to be extremely useful. We have expended great effort to make them so.

Fortunately, there is much that can be done to minimize the probability of laser induced damage. One point which cannot be overstressed is that *preparation of the substrate is equally as important as design and control of the coating deposition.* For this reason, we restrict our high energy coatings to only the substrates listed with them (except on special order).

SUBSTRATE MATERIALS

We only use the very best materials in our substrate fabrication. A choice of either BK 7 optical glass or synthetic fused silica is offered.

The BK 7 which we use for these components is platinum-free. When glass is made it can retain some platinum ions from the manufacturing process. These platinum inclusions have been shown to correlate with damage sites on test samples. We eliminate this possibility by using only BK 7 which has been made in ceramic vessels.

Our fused silica is the highest purity form of ultraviolet grade synthetic fused silica (UVGSFS). It is substantially free from defects and striae.

Choice of material will depend on your application. Fused silica has the advantage of a low thermal expansion coefficient which helps to retain the surface figure if the optic is heated or cooled. It also has superior transmission at low wavelengths and is therefore recommended for most transmission optics. BK 7 is satisfactory for most reflective optics unless they are subject to thermal variations. It also has slightly superior transmission characteristics above 1 micron wavelength, and may be advantageous for transmission optics in the near infrared.

SUBSTRATE FABRICATION

All high energy laser substrates are specially made in a way which avoids the presence of surface and subsurface defects. Only diamond tools are used in the rough grinding and shaping stage to minimize surface microfracture. Both pre-polishing with alumina compounds and final polishing with special fine-grain compounds are carried out very slowly. The polishing process takes two to three times as long as conventional polishing and the polishing compounds are gradually diluted, resulting in a zero-defect polished surface. Continuous polishing techniques are employed to assure uniform optical figure.

CLEANING

Very careful hand cleaning is carried out prior to coating. Special detergent is used to remove gross contamination. Then a deionized waterwash, rinse and vapor dry cycle is carried out. No abrasive cleaners are used. Final inspection and jigging is done in a Class 100 clean cabinet. Only highly skilled technicians are employed in this process. Even the time-cycle between fabrication, cleaning and coating is carefully controlled.

COATING DESIGN

Computer optimized coating designs are used for these optics. As well as producing the desired spectral transmission, great care is taken to ensure the highest laser damage threshold characteristics. Undercoats are employed to reduce stress at the substrate-film boundary. Overcoating and special optimization procedures are used to reduce the electric field strength in critical areas. Frequently, graded index layers will be used to reduce the probability of damage within the multilayer stack.

Only by knowledgeable combination of these techniques in conjunction with powerful optimization programs is it possible to design coatings which will withstand high energy laser radiation.

COATING TECHNIQUES

Control of the coating process is a most critical factor in the production of high energy laser optics. Deposition parameters must be closely maintained for each layer of the coating. A quartz crystal, used to determine film thickness and deposition rate, is coupled to a microprocessor controller. All significant variables are monitored and nothing is left to chance. Only in this way are coating designs repeatably realized in practice.

Long heating and cooling cycles, together with careful control of substrate temperatures and deposition rates, help to minimize stress and improve coating purity.

TESTING

All optics are carefully tested prior to shipment. Spectrophotometer transmittance curves are produced after deposition and compared with the specifications. These are included with the optics at no extra charge.

Periodic batch testing for laser damage is carried out on a continuing basis in an attempt to improve production techniques for high damage threshold optics. These tests are done by independent laboratories throughout the world. Notably, we cite specific corroboration of our claims by the Lawrence Livermore Laboratory in California and the Rutherford Laboratory in England.

Laser damage threshold has been found to vary with wavelength, angle of incidence, polarization and pulse length. The figures we quote are based on actual test results and represent the fair average threshold which can be expected. In many cases these figures may be greatly exceeded in practice.

HIGH ENERGY LASER MIRRORS, NORMAL INCIDENCE

These high energy laser (HEL) mirrors are suitable for use as cavity reflectors in a selection of excimer, gas, solid-state and dye lasers. They may also be used in other applications requiring high reflectivity at normal incidence.

They are available on platinum-free BK 7 or ultraviolet grade synthetic fused silica substrates, in three sizes. We strongly recommend the fused silica substrates for reasons of thermal stability.

All coatings are computer optimized and use overcoats and undercoats to minimize the risk of laser induced damage. They are resistant to ether, acetone, trichlorethylene, methanol and dilute hydrochloric (acid except where indicated).

Every mirror is supplied with a spectrophotometer transmittance trace.

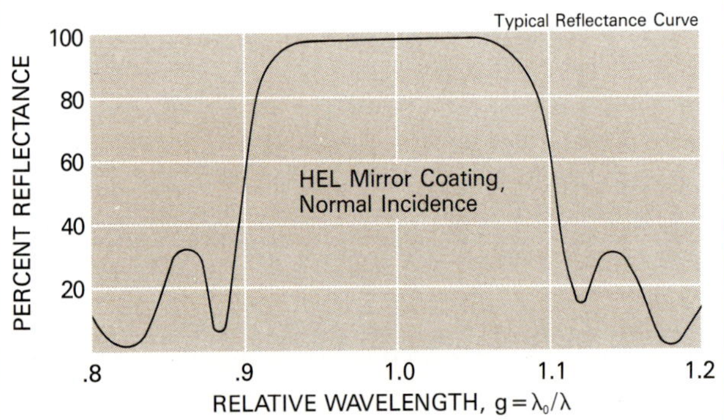

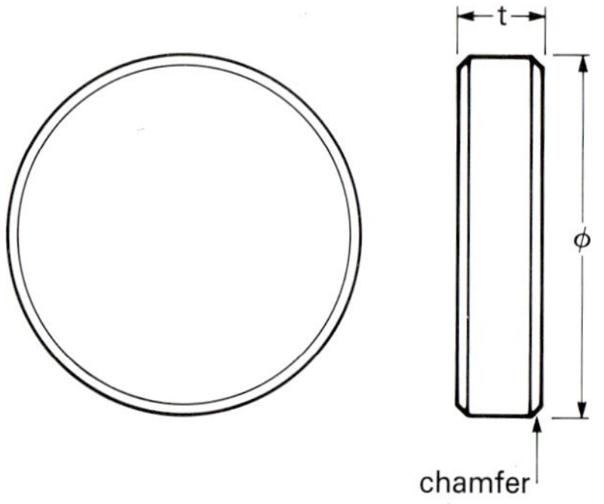

PRODUCT NUMBER	ϕ	t	Material
08 MLB 001	12.5	6	BK 7
08 MLB 003	25	6	BK 7
08 MLB 005	50	10	BK 7
08 MLQ 001	12.5	6	UVGSFS
08 MLQ 003	25	6	UVGSFS
08 MLQ 005	50	10	UVGSFS

A high energy laser mirror at 1064nm of the type 08MLB005/326 was tested by Lawrence Livermore Laboratory and found to have a laser damage threshold between 6.9 and 8.5 joules/cm² at 1.2 n.sec pulse duration.

SPECIFICATIONS: HEL MIRRORS, NORMAL INCIDENCE

Material: UV grade synthetic fused silica, or platinum-free BK 7 optical glass

Diameter: + 0.00, − 0.15mm

Thickness: ± 0.15mm

Surface Flatness: Front surface; $<\lambda/10$ at 546nm over central 90% of diameter

Rear surface; $<\lambda/5$ at 546nm

Parallelism: <2 arc minutes

Cosmetic Surface Quality: Laser grade polish, 10-5 scratch and dig or better

Complies With: MIL-C-675 and MIL-M-13508 (except /302)

HEL Mirror Substrates

Material	Diameter ϕ (mm)	Thickness t (mm)	PRODUCT NUMBER
BK 7	12.5	6	08 MLB 001
BK 7	25	6	08 MLB 003
BK 7	50	10	08 MLB 005
UVGSFS	12.5	6	08 MLQ 001
UVGSFS	25	6	08 MLQ 003
UVGSFS	50	10	08 MLQ 005

To order, select substrate product number above and append coating suffix from table on following page.

HEL Mirror Coatings; Normal Incidence

Wavelength (nm)	Laser Type	Minimum Reflectance (%)	Angular Tolerance (± deg)	Coating Type	Laser Damage Threshold at Normal Incidence	COATING SUFFIX
248	KrF	99	15	*Durable, low stress with graded index layers; gas compatible all flouride.	2 joules/cm^2 in 10 n.sec pulses	/302
308	XeCl	99.4	30	Ultra hard, high energy, low stress, oxide	2 joules/cm^2 in 10 n. sec pulses	/305
337	N_2	99.4	30	Ultra hard, high energy, low stress, oxide	2 joules/cm^2 in 1 n.sec pulses	/307
351/2	XeF	99.5	30	Ultra hard, high energy, low stress, oxide	2 joules/cm^2 in 1 n.sec pulses	/312
694	Ruby	99.5	30	Ultra hard, high energy, medium stress, oxide	3.5 G watts/cm^2 in 30 n.sec pulses	/322
730-790	Alexandrite	99.5	10	Ultra hard, high energy, medium stress, oxide	2 G watts/cm^2 in 100 n.sec pulses	/323
1053/60	Nd	99.5	30	Ultra hard, high energy, medium stress, oxide	10 joules/cm^2 in 1 n.sec pulses	/326
1064	YAG	99.5	30	Ultra hard, high energy, medium stress, oxide	10 joules/cm^2 in 1 n.sec pulses	/326
1315	I_2	99.5	30	Ultra hard, high energy, medium stress, oxide	10 joules/cm^2 in 1 n.sec pulses	/328
1320	YAG (CW)	99.5	30	Ultra hard, high energy, medium stress, oxide	5 M watts/cm^2 CW	/328

*The KrF coating is not resistant to organic solvents.

HEL Mirror Coatings, Broad Band; Normal Incidence.

Wavelength Range (nm)	Laser Type	Minimum Reflectance (%)	Angular Tolerance (± deg)	Coating Type	Laser Damage Threshold at Normal Incidence	COATING SUFFIX
320-450	Dye	99	10	Ultra hard, high energy, medium stress, oxide	2 joules/cm^2 in 10 n.sec pulses	/341
440-660	Dye	99	10	Ultra hard, high energy, medium stress, oxide; electric field optimization	2 joules/cm^2 in 10 n.sec pulses	/343
610-860	Dye	99	10	Ultra hard, high energy, medium stress, oxide; electric field optimization	2 joules/cm^2 in 10 n.sec pulses	/345

HIGH ENERGY LASER MIRRORS, 45° INCIDENCE

These high energy laser mirrors are suitable for beam deviating tasks with a wide selection of excimer, gas and solid state lasers.

They are available on platinum-free BK 7 or ultraviolet grade synthetic fused silica substrates, in three sizes. We strongly recommend the fused silica substrates for reasons of thermal stability.

All coatings are computer optimized and use overcoats and undercoats to minimize the risk of laser induced damage. They are resistant to ether, acetone, trichlorethylene, methanol and dilute hydrochloric acid (except for the krypton fluoride coating).

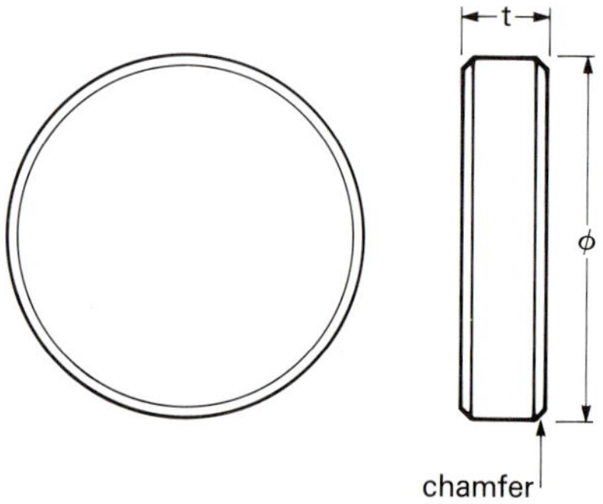

PRODUCT NUMBER	φ	t	Material
08 MLB 001	12.5	6	BK 7
08 MLB 003	25	6	BK 7
08 MLB 005	50	10	BK 7
08 MLQ 001	12.5	6	UVGSFS
08 MLQ 003	25	6	UVGSFS
08 MLQ 005	50	10	UVGSFS

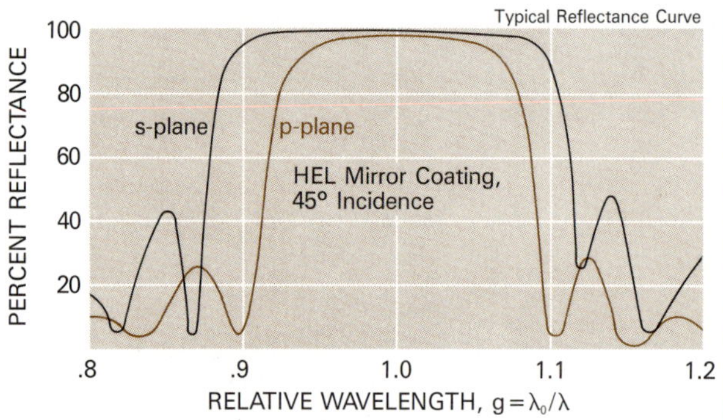

A 45° high energy laser mirror at 1064nm of the type 08 MLB 005/426 was tested by the Rutherford Laboratory using 1 to 2 n.sec pulses. It was found to have a laser damage threshold as high as 15 joules/cm² for the s-plane of polarization.

SPECIFICATIONS: HEL MIRRORS, 45 DEGREE INCIDENCE

Materials: UV grade synthetic fused silica, or platinum-free BK 7 optical glass

Diameter: + 0.00, − 0.15mm

Thickness: ± 0.15mm

Surface Flatness: Front surface; <l/10 at 546nm over central 90% of diameter

Rear surface; <l/5 at 546nm

Parallelism: <2 arc minutes

Cosmetic Surface Quality: Laser grade polish, 10-5 scratch and dig or better

Complies With: MIL-C-675 and MIL-M-13508 (except /402)

HEL Mirror Substrates

Material	Diameter φ (mm)	Thickness t (mm)	PRODUCT NUMBER
BK 7	12.5	6	08 MLB 001
BK 7	25	6	08 MLB 003
BK 7	50	10	08 MLB 005
UVGSFS	12.5	6	08 MLQ 001
UVGSFS	25	6	08 MLQ 003
UVGSFS	50	10	08 MLQ 005

To order, select substrate product number above and append coating suffix from table on following page.

HEL Mirror Coatings; 45° Incidence

Wavelength (nm)	Laser Type	Reflectance $R_{av}(\%)$	$R_s(\%)$	$R_p(\%)$	Coating Type	Laser Damage Threshold at 45° Incidence in s-plane of Polarization**	COATING SUFFIX
248	KrF	99.0	99.2	98.0	*Durable, low stress coating, graded index layers	1 joule/cm² in 10 n.sec pulses	/402
264	Nd (4th Harmonic)	99.2	99.5	98.8	Ultra hard, high energy, low stress, oxide	0.5 joules/cm² in 1 n.sec pulses	/404
266	YAG (4th Harmonic)	99.2	99.5	98.8	Ultra hard, high energy, low stress, oxide	0.5 joules/cm² in 1 n.sec pulses	/404
308	XeCl	99.5	99.7	99.0	Ultra hard, high energy, low stress, oxide	1 joule/cm² in 10 n.sec pulses	/405
337	N_2	99.5	99.7	99.0	Ultra hard, high energy, low stress, oxide	1 joule/cm² in 10 n.sec pulses	/407
347	Ruby (2nd Harmonic)	99.6	99.8	99.0	Ultra hard, high energy, low stress, oxide	0.5 G watt/cm² in 30 n.sec pulses	/412
351/3	XeF	99.6	99.8	99.0	Ultra hard, high energy, low stress, oxide	0.5 joule/cm² in 10 n.sec pulses	/412
355	YAG (3rd Harmonic)	99.6	99.8	99.0	Ultra hard, high energy, low stress, oxide	1 joule/cm² in 1 n.sec pulses	/412
527/30	Nd (2nd Harmonic)	99.6	99.8	99.0	Ultra hard, high energy, medium stress, oxide	1.5 joules/cm² in 1 n.sec pulses	/418
532	YAG (2nd Harmonic)	99.6	99.8	99.0	Ultra hard, high energy, medium stress, oxide	1.5 joules/cm² in 1 n.sec pulses	/418
694	Ruby	99.6	99.8	99.0	Ultra hard, high energy, medium stress, oxide	2 G watts/cm² in 30 n.sec pulses	/422
730-790	Alexandrite	99.4	99.6	98.6	Ultra hard, high energy, medium stress, oxide	1 G watt/cm² in 100 n.sec pulses	/423
1053/60	Nd	99.6	99.8	99.0	Ultra hard, high energy, medium stress, oxide	5 joules/cm² in 1 n.sec pulses	/426
1064	YAG	99.6	99.8	99.0	Ultra hard, high energy, medium stress, oxide	5 joules/cm² in 1 n.sec pulses	/426
1315	I_2	99.5	99.8	99.0	Ultra hard, high energy, medium stress, oxide	5 joules/cm² in 1 n.sec pulses	/428
1320	YAG(CW)	99.5	99.8	99.0	Ultra hard, high energy, medium stress, oxide	2.5 M watts/cm² CW	/428

*The KrF coating is not to MIL specification and is not resistant to organic solvents.

**Note that laser damage thresholds for the p-plane of polarization are two times the value stated for the s-plane.

HIGH ENERGY LASER WINDOWS

These laser polished, parallel sided windows are suitable for isolating equipment which uses a wide range of excimer, gas, solid-state and dye lasers.

They are made from platinum-free BK 7 or ultraviolet grade synthetic fused silica and can be antireflection coated both sides by selection of the appropriate coating suffix. Two sizes are offered.

Only fused silica windows should be selected for wavelengths lower than 500nm, for reasons of higher transmission. Above 500nm the choice is optional but fused silica has better thermal properties. Above 1000nm the transmittance of BK 7 is slightly higher than fused silica, so BK 7 may be a better choice for applications in which energy loss considerations are paramount.

All coatings are fully computer optimized and the single wavelength coatings are all electric field optimized for minimum laser induced damage. They are resistant to ether, acetone, trichlorethylene, methanol and dilute hydrochloric acid (except for the krypton fluoride coating).

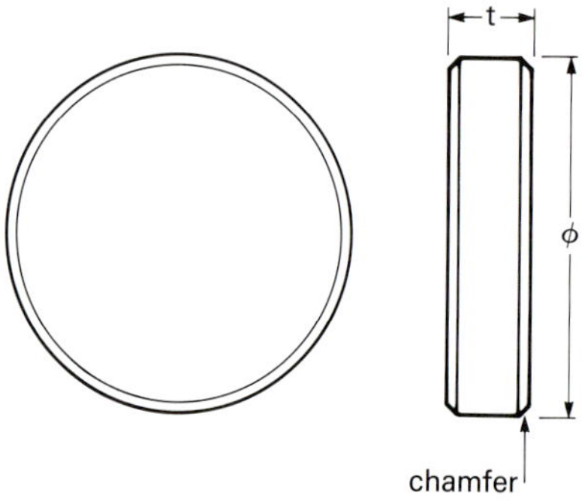

PRODUCT NUMBER	ϕ	t	Material
08 WLB 003	25	6	BK 7
08 WLB 005	50	10	BK 7
08 WLQ 003	25	6	UVGSFS
08 WLQ 005	50	10	UVGSFS

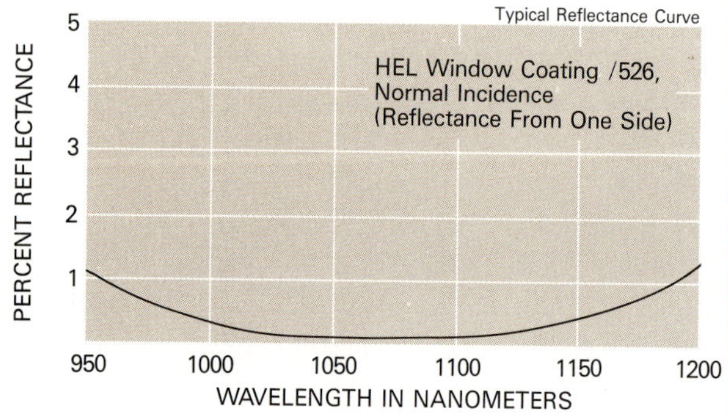

A laser window for 1064nm of the type 08 MLB 003/526 was tested by Lawrence Livermore Laboratory and found to have a laser damage threshold between 5.9 and 7.2 joules/cm² at 1.2 n.sec pulse duration.

SPECIFICATIONS: HEL WINDOWS

Materials: UV grade synthetic fused silica, or platinum-free BK 7 optical glass

Diameter: + 0.00, - 0.15mm

Thickness: ± 0.15mm

Surface Flatness: <λ/10 at 546nm over central 90% of diameter.

Parallelism: <λ10 arc seconds

Cosmetic Surface Quality: Laser grade polish, 10-5 scratch and dig or better

Complies With: MIL-C-675 and MIL-M-13508 (except /502)

HEL Window Substrates

Material	Diameter ϕ (mm)	Thickness t (mm)	PRODUCT NUMBER
BK 7	25	6	08 WLB 003
BK 7	50	10	08 WLB 005
UVGSFS	25	6	08 WLQ 003
UVGSFS	50	10	08 WLQ 005

To order, select substrate product number above and append coating suffix from table on following pages.

HEL Window Coatings; Normal Incidence Antireflection Coating Both Sides

Wavelength (nm)	Laser Type	Minimum Transmittance (%)	Coating Type	Laser Damage Threshold at Normal Incidence	COATING SUFFIX
248	KrF	98.0	*Durable, low stress	1.5 joules/cm^2 in 10 n.sec pulses	/502
264	Nd (4th Harmonic)	98.5	Ultra hard, high energy, low stress, oxide	0.5 joules/cm^2 in 1 n.sec pulses	/504
266	YAG (4th Harmonic)	98.5	Ultra hard, high energy, low stress, oxide	0.5 joules/cm^2 in 1 n.sec pulses	/504
308	XeCl	99.0	Ultra hard, high energy, low stress, oxide	1.5 joules/cm^2 in 10 n.sec pulses	/505
337	N_2	99.0	Ultra hard, high energy, low stress, oxide	1.5 joules/cm^2 in 10 n.sec pulses	/507
347	Ruby (2nd Harmonic)	99.0	Ultra hard, high energy, low stress, oxide	0.6 G watt/cm^2 in 30 n.sec pulses	/512
351/3	XeF	99.0	Ultra hard, high energy, low stress, oxide	0.5 joule/cm^2 in 10 n.sec pulses	/512
355	YAG (3rd Harmonic)	99.0	Ultra hard, high energy, low stress, oxide	1 joule/cm^2 in 1 n.sec pulses	/512
527/30	Nd (2nd Harmonic)	99.0	Ultra hard, high energy, medium stress, oxide	2 joules/cm^2 in 1 n.sec pulses	/518
532	YAG (2nd Harmonic)	99.0	Ultra hard, medium stress, oxide	2 joules/cm^2 in 1 n.sec pulses	/518

*The KrF coating is not to MIL specification and will not withstand organic solvents.

HEL Window Coatings; Normal Incidence Antireflection Coating Both Sides (continued)

Wavelength (nm)	Laser Type	Minimum Transmittance (%)	Coating Type	Laser Damage Threshold at normal incidence	COATING SUFFIX
694	Ruby	99.0	Ultra hard, high energy, medium stress, oxide	2.5 G watts/cm^2 in 30 n.sec pulses	/522
730-790	Alexandrite	99.0	Ultra hard, high energy, medium stress, oxide	1.5 G watts/cm^2 in 100 n.sec pulses	/523
1053/60	Nd	99.0	Ultra hard, high energy, medium stress, oxide	6 joules/cm^2 in 1 n.sec pulses	/526
1064	YAG	99.0	Ultra hard, high energy, medium stress, oxide	6 joules/cm^2 in 1 n.sec pulses	/526
1315	I_2	99.0	Ultra hard, high energy, medium stress, oxide	6 joules/cm^2 in 1 n.sec pulses	/528
1320	YAG (CW)	99.0	Ultra hard, high energy, medium stress, oxide	3 M watts/cm^2 CW	/528

Specials

We will be happy to discuss your requirements for high energy laser optics not listed in this catalog.

As the substrate quality is integral to the performance of these optics, we regret we cannot apply HEL coatings to other substrates unless these are specially made to order.

This may take a little longer but it enables us to assure you of the performance you require.

HEL Window Coatings, Dual Wavelength; Normal Incidence Antireflection Coating Both Sides

Wavelengths (nm)	Laser Type	Minimum Transmittance (%)	Coating Type	Laser Damage Threshold at Normal Incidence	COATING SUFFIX
532/1064	YAG	99.0	Ultra hard, high energy, low stress, oxide	2 joules/cm^2 in 1 n.sec pulses	/531

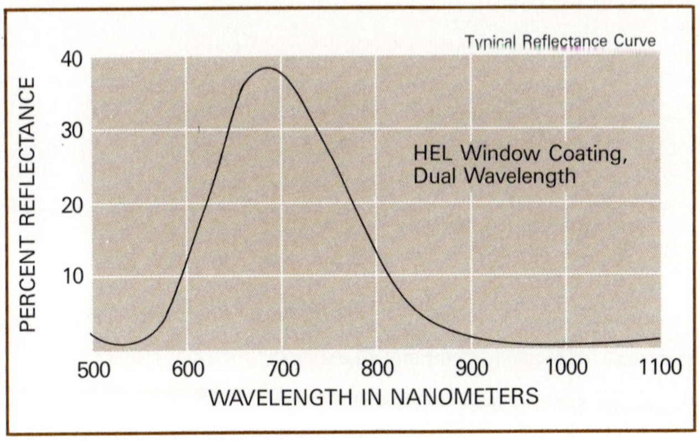

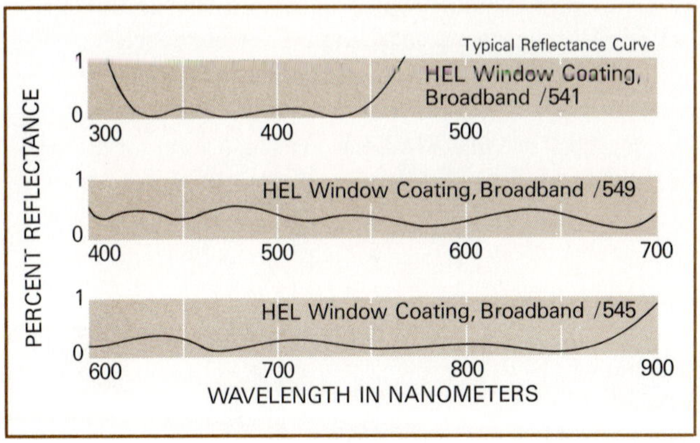

HEL Window Coatings, Broadband; Normal Incidence Antireflection Coating Both Sides

Wavelength range (nm)	Laser Type	Average Transmittance (%)	Coating Type	Laser Damage Threshold at normal incidence	COATING SUFFIX
320-450	Dye	99.0	Ultra hard, high energy, low stress, oxide	2 joules/cm^2 in 10 n.sec pulses	/541
400-700	Dye	99.0	Ultra hard, high energy, low stress, oxide	2 joules/cm^2 in 10 n.sec pulses	/549
610-860	Dye	99.0	Ultra hard, high energy, low stress, oxide	2 joules/cm^2 in 10 n.sec pulses	/545

HIGH ENERGY LASER BEAMSAMPLERS

These laser polished, parallel side, plate type beamsamplers are useful for sampling of high energy laser beams from a range of excimer, gas and solid state lasers.

They are made from platinum-free BK 7 glass or ultraviolet grade synthetic fused silica in two sizes. BK 7 should not be used below 500nm because of its lower transmittance. We recommend the use of fused silica for reasons of thermal stability.

These samplers are simply uncoated on one face and antireflection coated on the other. The natural reflectivity of the substrate at 45° is used to split off a sampling beam. Reflectance is stated for both s- and p-planes of polarization.

All coatings are ultra-hard, high-energy oxide coatings, except for the krypton fluoride coating which is a durable, low-stress coating. All coatings are under and over coated as well as being electric field optimized for minimization of the risk of laser induced damage. They are resistant to ether, acetone, trichlorethylene, methanol and dilute hydrochloric acid (except for the krypton fluoride coating).

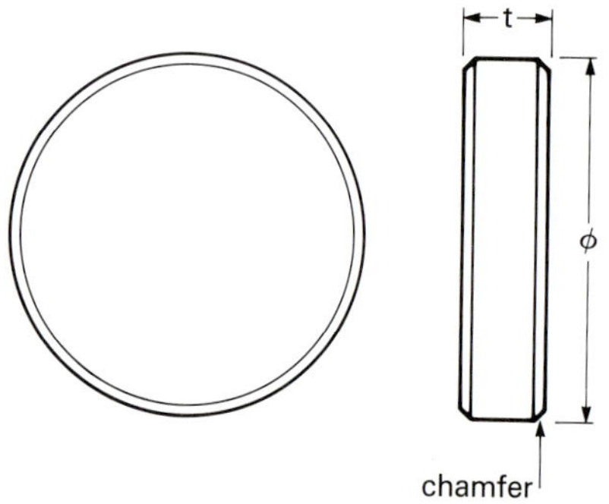

PRODUCT NUMBER	φ	t	Material
08 BSB 003	25	6	BK 7
08 BSB 005	50	10	BK 7
08 BSQ 003	25	6	UVGSFS
08 BSQ 005	50	10	UVGSFS

We offer a wide range of optical coating services. Ask for your free copy of our brochure.

SPECIFICATIONS: HEL BEAMSAMPLERS

Materials: UV grade synthetic fused silica, or platinum-free BK 7 optical glass

Diameter: + 0.00, − 0.15mm

Thickness: ± 0.15mm

Surface Flatness: $<\lambda/10$ at 546nm over central 90% of diameter.

Parallelism: < 10 arc seconds

Cosmetic Surface Quality: Laser grade polish, 10-5 scratch and dig or better

Complies With: MIL-C-675 and MIL-M-13508 (except /602)

HEL Beamsampler Substrates

Material	Diameter φ (mm)	Thickness t (mm)	PRODUCT NUMBER
BK 7	25	6	08 BSB 003
BK 7	50	10	08 BSB 005
UVGSFS	25	6	08 BSQ 003
UVGSFS	50	10	08 BSQ 005

To order select substrate product number above and append coating suffix from table on the following page.

Beamsampler Coatings, 4%, 45°; No Coating One Side, Antireflection Coating Other Side

Wavelength (nm)	Laser Type	Reflectance on BK 7 $R_s(\%)$	Reflectance on BK 7 $R_p(\%)$	Reflectance on F.S. $R_s(\%)$	Reflectance on F.S. $R_p(\%)$	Laser Damage Threshold at 45° in s-plane of Polarization**	COATING SUFFIX
248	KrF			9.30	0.87	*1 joule/cm^2 in 10 n.sec pulses	/602
264	Nd (4th harmonic)			9.17	0.84	0.5 joule/cm^2 in 1 n.sec pulses	/604
266	YAG (4th harmonic)			9.17	0.84	0.5 joule/cm^2 in 1 n.sec pulses	/604
308	XeCl			8.70	0.79	1 joule/cm^2 in 10 n.sec pulses	/605
337	N_2			8.65	0.76	1 joule/cm^2 in 10 n.sec pulses	/607
347	Ruby (2nd harmonic)			8.60	0.75	0.5 G watt/cm^2 in 30 n.sec pulses	/612
351/3	XeF			8.60	0.75	0.5 joule/cm^2 in 10 n.sec pulses	/612
355	YAG (3rd harmonic)			8.60	0.75	1 joule/cm^2 in 1 n.sec pulses	/612
527/30	Nd (2nd harmonic)	9.67	0.93	8.28	0.69	1.5 joules/cm^2 in 1 n.sec pulses	/618
532	YAG (2nd harmonic)	9.67	0.93	8.28	0.69	1.5 joules/cm^2 in 1 n.sec pulses	/618
694	Ruby	9.51	0.91	8.15	0.67	2 G watts/cm^2 in 30 n.sec pulses	/622
760	Alexandrite	9.47	0.89	8.11	0.66	1 G watt/cm^2 in 100 n.sec pulses	/623
1053/60	Nd	9.36	0.88	8.02	0.64	5 joules/cm^2 in 1 n.sec pulses	/626
1064	YAG	9.36	0.88	8.02	0.64	5 joules/cm^2 in 1 n.sec pulses	/626
1315	I_2	9.28	0.86	7.94	0.63	5 joules/cm^2 in 1 n.sec pulses	/628
1320	YAG (CW)	9.28	0.86	7.94	0.63	2 M watts/cm^2 CW	/628

*The KrF coating is not to MIL specification and is not resistant to organic solvents.

**Note that the laser damage thresholds for the p-plane of polarization are two times the value stated for the s-plane.

HIGH ENERGY LASER BEAMSPLITTERS

These laser polished, parallel side, plate type beamsplitters are useful for equal splitting of high energy laser beams from a range of excimer, gas and solid state lasers.

They are made from platinum-free BK 7 glass or ultraviolet grade synthetic fused silica in two sizes. BK 7 should not be used below 500nm because of its lower transmittance. We recommend the use of fused silica for reasons of thermal stability.

These beamsplitters are used for beam separation and merging. They are coated for 50% (±2%) average reflectance at 45° on one side and are antireflection coated on the other. Reflectance is stated for both s- and p-planes of polarization.

All coatings are ultra-hard, high-energy oxide coatings, except for the krypton fluoride coating which is a durable, low-stress coating. All coatings are under and over coated as well as being electric field optimized for minimization of the risk of laser induced damage. They are resistant to ether, acetone, trichlorethylene, methanol and dilute hydrochloric acid (except for the krypton fluoride coating).

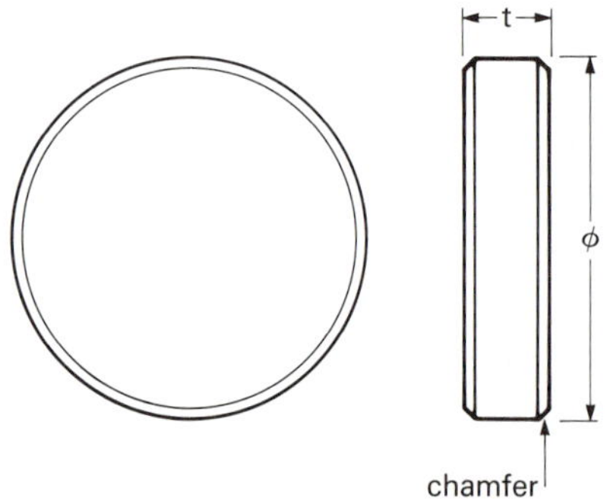

PRODUCT NUMBER	φ	t	Material
08 BSB 003	25	6	BK 7
08 BSB 005	50	10	BK 7
08 BSQ 003	25	6	UVGSFS
08 BSQ 005	50	10	UVGSFS

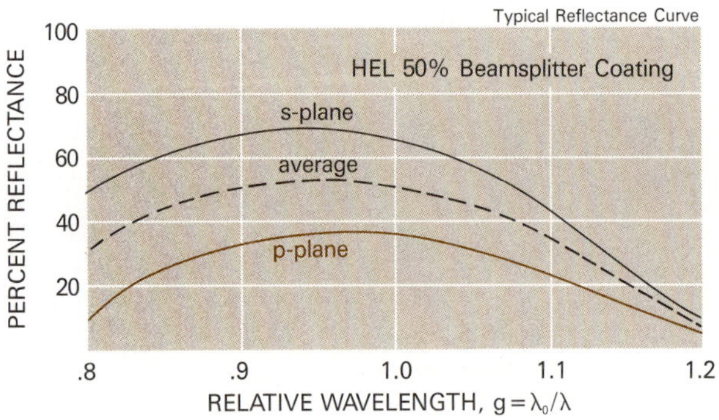

See page 475 for a beamsplitter mount which will accept these beamsplitters with minimum vignetting at 45°.

SPECIFICATIONS: HEL BEAMSPLITTERS

Materials: UV grade synthetic fused silica, or platinum-free BK 7 optical glass

Diameter: + 0.00, - 0.15mm

Thickness: ± 0.15mm

Surface Flatness: $<\lambda/10$ at 546nm over central 90% of diameter.

Parallelism: < 10 arc seconds

Cosmetic Surface Quality: Laser grade polish, 10-5 scratch and dig or better

Complies With: MIL-C-675 and MIL-M-13508 (except /602)

HEL Beamsplitter Substrates

Material	Diameter φ (mm)	Thickness t (mm)	PRODUCT NUMBER
BK 7	25	6	08 BSB 003
BK 7	50	10	08 BSB 005
UVGSFS	25	6	08 BSQ 003
UVGSFS	50	10	08 BSQ 005

To order select substrate product number above and append coating suffix from table on the following page.

Beamsplitter Coatings, 50%, 45°; Partial Reflector One Side, Antireflection Coating Other Side

Wavelength (nm)	Laser Type	Reflectance R_s(%)	R_p(%)	Laser Damage Threshold at 45° in s-plane of Polarization**	COATING SUFFIX
248	KrF	71.0	29.0	*1 joule/cm² in 10 n.sec pulses	/802
264	Nd (4th harmonic)	57.0	43.0	0.5 joule/cm² in 1 n.sec pulses	/804
266	YAG (4th harmonic)	62.0	38.0	0.5 joule/cm² in 1 n.sec pulses	/804
308	XeCl	63.0	37.0	1 joule/cm² in 10 n.sec pulses	/805
337	N_2	63.0	37.0	1 joule/cm² in 10 n.sec pulses	/807
347	Ruby (2nd harmonic)	63.0	37.0	0.5 G watt/cm² in 30 n.sec pulses	/812
351/3	XeF	63.0	37.0	0.5 joule/cm² in 10 n.sec pulses	/812
355	YAG (3rd harmonic)	63.0	37.0	1 joule/cm² in 1 n.sec pulses	/812
527/30	Nd (2nd harmonic)	65.0	35.0	1.5 joules/cm² in 1 n.sec pulses	/818
532	YAG (2nd harmonic)	66.0	34.0	1.5 joules/cm² in 1 n.sec pulses	/818
694	Ruby	66.0	34.0	2 G watts/cm² in 30 n.sec pulses	/822
760	Alexandrite	66.0	35.0	1 G watt/cm² in 100 n.sec pulses	/823
1053/60	Nd	67.0	33.0	5 joules/cm² in 1 n.sec pulses	/826
1064	YAG	67.0	33.0	5 joules/cm² in 1 n.sec pulses	/826
1315	I_2	67.0	33.0	5 joules/cm² in 1 n.sec pulses	/828
1320	YAG (CW)	67.0	33.0	2 M watts/cm² CW	/828

*The KrF coating is not to MIL specification and is not resistant to organic solvents.

**Note that the laser damage thresholds for the p-plane of polarization are two times the value stated for the s-plane.

HIGH ENERGY LASER OUTPUT COUPLERS

These output couplers are normal incidence partial reflectors. They are suitable for providing feedback in ruby, alexandrite, YAG and iodine lasers. In general, the amount of feedback needed depends on the gain of the laser. We offer a range of reflectances but we make the following recommendations:

- For high gain lasers — 30%R
- For moderate gain lasers — 70%R
- For CW lasers — 90%R

Output couplers are manufactured from platinum-free BK 7 optical glass and are 25mm in diameter.

The coatings are ultra-hard, high-energy, medium-stress, all-oxide coatings. They are overcoated and undercoated as well as being electric field optimized to reduce the probability of laser induced damage. These coatings are resistant to ether, acetone, trichlorethylene, methanol and dilute hydrochloric acid.

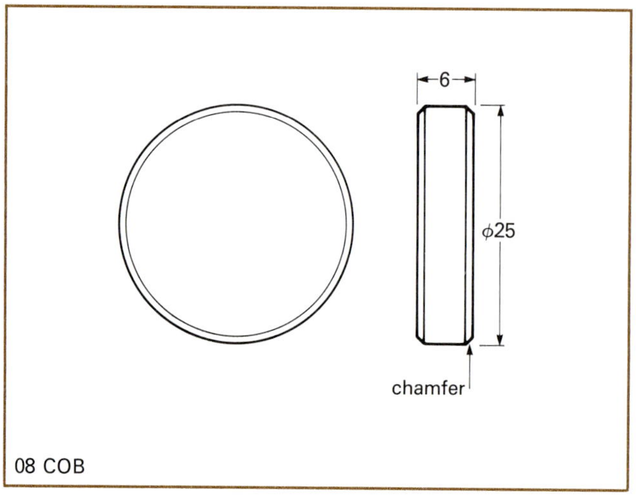

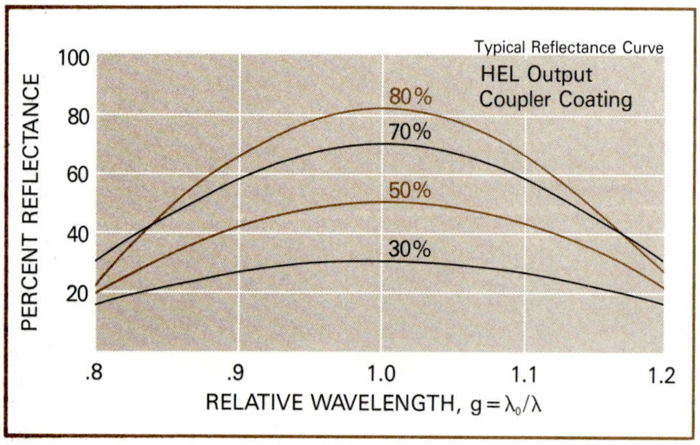

To order see table on following page.

SPECIFICATIONS: HEL OUTPUT COUPLERS

Material: Platinum-free BK 7 optical glass
Diameter: 25mm +0.00, - 0.15mm
Thickness: 6mm ± 0.15mm
Surface Flatness: <λ/10 at 546nm over central 90% of diameter
Parallelism: <10 arc seconds
Cosmetic Surface Quality: Laser grade polish, 10-5 scratch and dig or better
Complies With: MIL-C-675 and MIL-M-13508

Partial reflectors of other reflectivities can be supplied to special order. Spherical or non-standard substrates can also be produced. However, please allow more time for delivery if your requirements differ from our stock item specifications.

HEL Output Couplers; Normal Incidence Partial Reflectors

Wavelength (nm)	Laser Type	Reflectance (%)	Reflectance (± tolerance %)	Laser Damage Threshold at Normal Incidence	PRODUCT NUMBER
694	Ruby	30	3	2.5 G watts in 30 n.sec pulses	08 COB 003
		50	2	2.5 G watts in 30 n.sec pulses	08 COB 005
		70	2	2.5 G watts in 30 n.sec pulses	08 COB 007
		80	2	2.5 G watts in 30 n.sec pulses	08 COB 008
760	Alexandrite	30	3	1.5 G watts in 100 n.sec pulses	08 COB 013
		50	2	1.5 G watts in 100 n.sec pulses	08 COB 015
		70	2	1.5 G watts in 100 n.sec pulses	08 COB 017
		80	2	1.5 G watts in 100 n.sec pulses	08 COB 018
1064	YAG	30	3	6 joules/cm² in 1 n.sec pulses	08 COB 023
		50	2	6 joules/cm² in 1 n.sec pulses	08 COB 025
		70	2	6 joules/cm² in 1 n.sec pulses	08 COB 027
		80	2	6 joules/cm² in 1 n.sec pulses	08 COB 028
		90	2	6 joules/cm² in 1 n.sec pulses	08 COB 029
1315	I_2	30	3	6 joules/cm² in 1 n.sec pulses	08 COB 033
		50	2	6 joules/cm² in 1 n.sec pulses	08 COB 035
		70	2	6 joules/cm² in 1 n.sec pulses	08 COB 037
		80	2	6 joules/cm² in 1 n.sec pulses	08 COB 038
		90	2	6 joules/cm² in 1 n.sec pulses	08 COB 039
1320	YAG (CW)	30	3	5 M watts/cm² CW	08 COB 033
		50	2	5 M watts/cm² CW	08 COB 035
		70	2	5 M watts/cm² CW	08 COB 037
		90	2	5 M watts/cm² CW	08 COB 039

HIGH ENERGY LASER HARMONIC SEPARATORS

These harmonic separators are specially coated substrates which reflect the fundamental and transmit the harmonic in order to separate the two wavelengths. (In the case of the YAG 3rd harmonic separator, the reverse is true). They are provided for both YAG and ruby lasers.

These optics are manufactured from platinum free BK 7 optical glass or ultraviolet grade synthetic fused silica as appropriate in two sizes.

The coatings are ultra-hard, high-energy, medium-stress all-oxide coatings. They are fully computer optimized and are over-coated and undercoated to reduce the probability of laser induced damage. These coatings are resistant to ether, acetone, trichlorethylene, methanol and dilute hydrochloric acid.

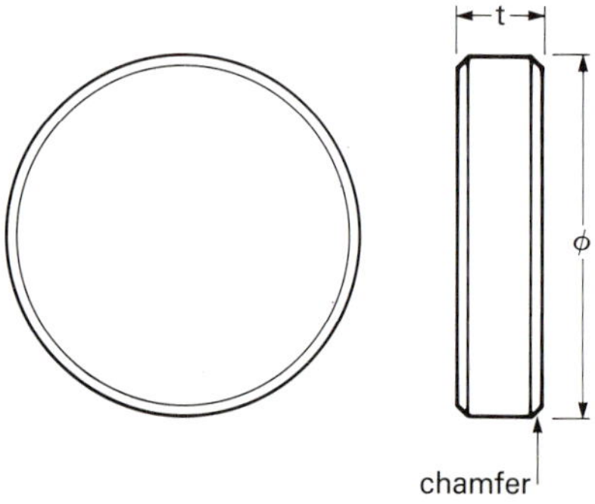

PRODUCT NUMBER	ϕ	t	Material
08 HSB 001	25	6	BK 7
08 HSB 002	50	10	BK 7
08 HSB 005	25	6	BK 7
08 HSB 006	50	10	BK 7
08 HSQ 003	25	6	UVGSFS
08 HSQ 004	50	10	UVGSFS

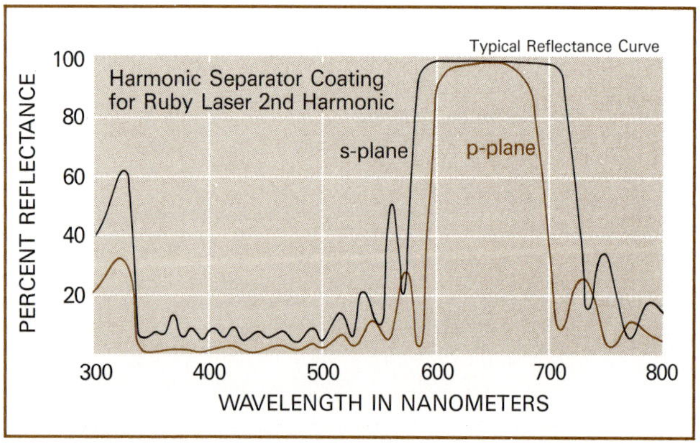

To order see table on following page.

Whatever your production optics needs, we can help. Ask for your free copy of our OEM Optics brochure.

SPECIFICATIONS: HEL HARMONIC SEPARATORS

Material: UV grade synthetic fused silica, or platinum free BK 7 optical glass

Diameter: +0.00, - 0.15mm

Thickness: ± 0.15 mm

Surface Flatness: $<\lambda/10$ at 546nm over central 90% of diameter

Parallelism: < 10 arc seconds

Cosmetic Surface Quality: Laser grade polish, 10-5 scratch and dig or better

Complies With: MIL-C-675 and MIL-M-13508

HEL Harmonic Separators, Ruby

Type	Substrate Material	Diameter ϕ (mm)	Thickness t (mm)	Minimum Transmittance or Reflectance 694nm R_s (%)	347nm T_p (%)	Laser Damage Threshold at 45° in s-plane of Polarization	PRODUCT NUMBER
Ruby 2nd harmonic	UVGSFS	25	6	99.7	90.0	1 joule/cm² in 1 n.sec pulses	08 HSQ 003
Ruby 2nd harmonic	UVGSFS	50	10	99.7	90.0	1 joule/cm² in 1 n.sec pulses	08 HSQ 004

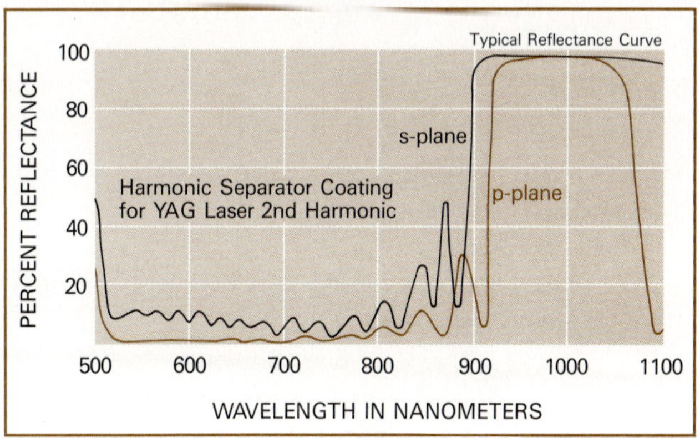

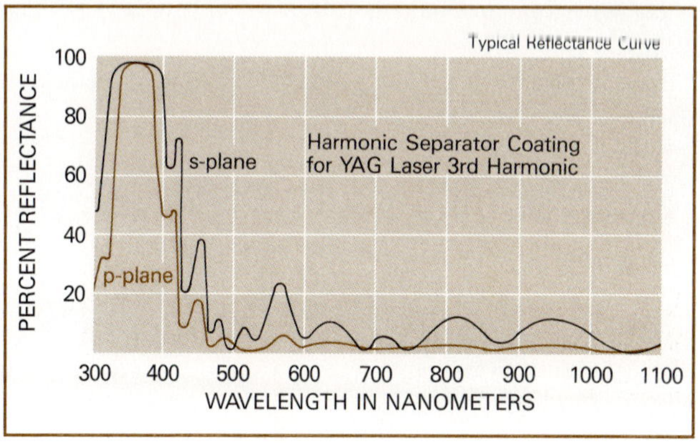

HEL Harmonic Separators, YAG

Type	Substrate Material	Diameter ϕ (mm)	Thickness t (mm)	Minimum Reflectance or Transmittance 1064nm R_s (%)	1064nm T_s (%)	532nm T_p (%)	355nm R_s (%)	Laser Damage Threshold at 45° in s-plane of Polarization	PRODUCT NUMBER
YAG 2nd harmonic	BK 7	25	6	99.7	—	90.0	—	3 joules/cm² in 1 n.sec pulses	08 HSB 005
YAG 2nd harmonic	BK 7	50	10	99.7	—	90.0	—	3 joules/cm² in 1 n.sec pulses	08 HSB 006
YAG 3rd harmonic	BK 7	25	6	—	90.0	90.0	99.5	1 joule/cm² in 1 n.sec pulses	08 HSB 001
YAG 3rd harmonic	BK 7	50	10	—	90.0	90.0	99.5	1 joule/cm² in 1 n.sec pulses	08 HSB 002

HIGH ENERGY LASER POLARIZERS

These polarizers are thin film plate polarizers. They are used with radiation incident at the Brewster Angle (56°) and utilize polarization sensitive reflectivity to separate the s- and p-planes of polarization. At each successive film layer boundary radiation in the s-plane of polarization is reflected while radiation in the p-plane is transmitted. After passage through the multilayer stack, the transmitted beam is highly p-polarized and the reflected beam is highly s-polarized.

These optics are manufactured from platinum-free BK 7 optical glass and are 50mm in diameter. They are provided for both YAG and Ruby lasers.

The coatings are ultra-hard, high-energy, medium-stress, all-oxide coatings. They are fully computer optimized and are electric field optimized to minimize the probability of laser induced damage.

These coatings are resistant to ether, acetone, trichlorethylene, methanol and dilute hydrochloric acid.

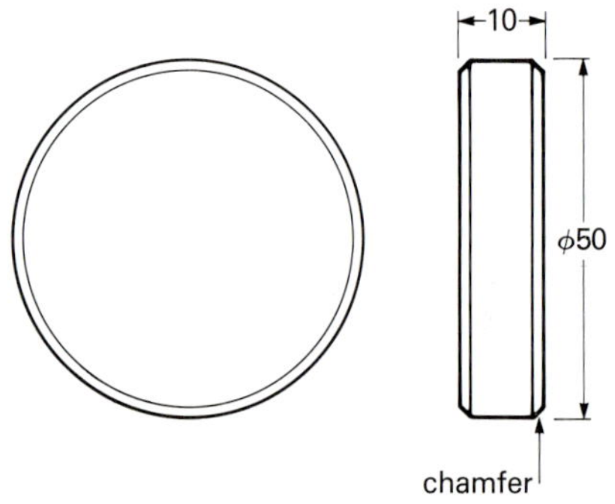

08 PFB

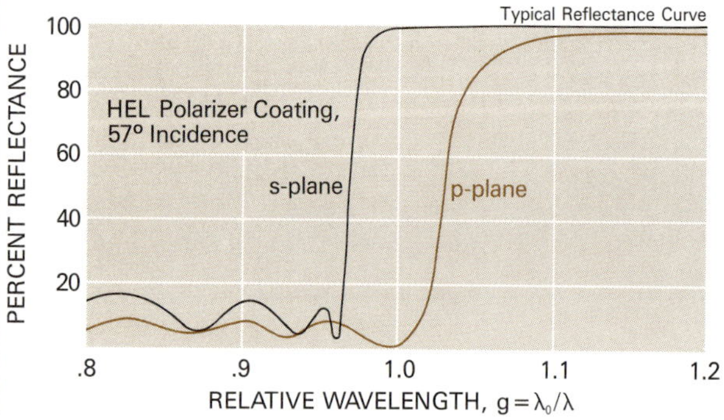

SPECIFICATIONS: HEL POLARIZERS

Material: Platinum-free BK 7 optical glass
Diameter: 50mm, +0.00, − 0.15mm
Thickness: ± 0.15mm
Surface flatness: <λ/10 at 546nm over central 90% of diameter
Parallelism: < 10 arc seconds
Cosmetic Surface Quality: laser grade polish, 10-5 scratch and dig or better
Complies With: MIL-C-675 and MIL-M-13508

A polarizer for 1064nm of the type 08 PFB 026 was measured by the Rutherford Laboratory. It was found to have a laser damage threshold of 6 joules/cm² for the s-plane and 15 joules/cm² for the p-plane.

HEL Polarizers

Wavelength (mm)	Laser Type	Minimum Reflectance R_s (%)	Minimum Transmittance T_p (%)	Angle of Incidence (deg)	Laser Damage Threshold at 56° in s-plane of Polarization**	PRODUCT NUMBER
694	Ruby	99%	95%	56° ±2°	1.5 G Watts/cm² in 30 n.sec pulses	08 PFB 022
1064	YAG	99%	95%	56° ±2°	4 joules/cm² in 1 n.sec pulses	08 PFB 026

**Note that the laser damage threshold for the p-plane of polarization is two times the value stated for the s-plane.

PRISMS AND BEAMSPLITTERS

In this section we introduce a broad selection of prisms and beamsplitters. We show their usefulness and limitations, presenting their function with the aid of clear three dimensional drawings.

Contents **Page Number**

PRISMS

Prisms may be used in an optical system to deflect or deviate a beam of light. They can invert or rotate an image, they can disperse light into its component wavelengths and they can be used to isolate separate states of polarization. The numerous applications for prisms make them an important component of any optical system.

Prisms are, essentially, blocks of optical material with flat polished sides which are arranged to be at precisely controlled angles to each other. The angular specifications of prisms are of prime importance since small angular face errors can lead to incorrect function of the prism.

All the prisms we offer are designed for specific optical application. They are carefully specified and controlled in manufacture to ensure correct operation.

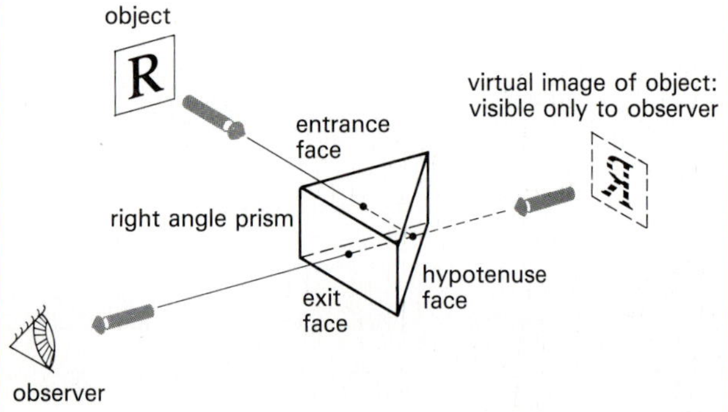

EXPLANATION OF PRISM DRAWINGS

Prisms by themselves are incapable of forming real images. The meanings of the real images which we show in many of the prism drawings which follow, and in earlier drawings of plate and cube-type beamsplitters, require explanation. Using the right-angle prism as an example, we illustrate all of the detail necessary to visualize the effect of the prism on image orientation. In the drawing above, we show the situation as it actually occurs. A viewer looks through the prism at an object and sees a virtual image. This image may be displaced from the original object, such as in the example, or it may coincide with the object, as in the case of the dove prism. Furthermore, the orientation of the image may differ

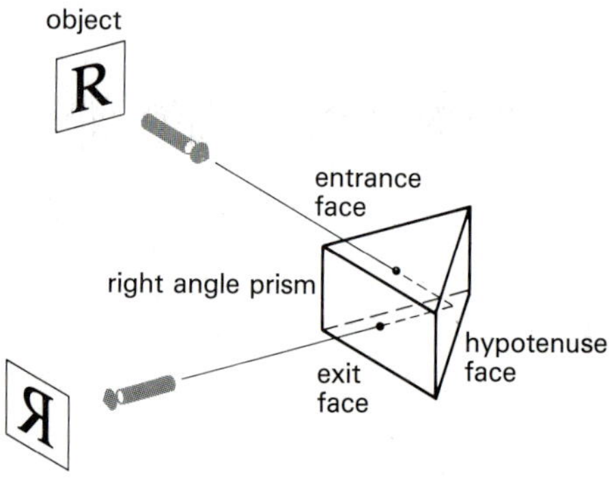

from that of the object; the right angle prism shown above has reversed left and right in this situation. The drawing above shows the same situation using the convention used in all subsequent and prior prism and beamsplitter drawings. A real image is shown; its orientation is a result of the action of the prism. It must be kept in mind, however, that a real image would only be formed if there were some sort of imaging optics present in the system. If no imaging optics are present, the image is virtual. The virtual image will have the same orientation as the real image we show, but it can only be viewed by an observer looking back through the prism system. Also, its location is not where the real image is shown in the drawing. Image orientation information is of vital importance to the designer of optical systems using prisms. This illustration should be kept in mind while interpreting all subsequent prism drawings.

ABERRATIONS INTRODUCED BY PRISMS

The arrangement by means of which we have explained the prism drawings corresponds closely to the way in which prisms should actually be used in optical systems. If used in an optical system where the beam is either convergent or divergent, prisms will introduce aberrations which would not otherwise be present. These can be avoided or minimized by making sure that prisms are used only in colliamated or nearly-collimated beams. Conjugate distances which include prisms should be long. Alternatively, other elements of the system can be designed to compensate for aberrations introduced by the prism in the presence of convergence or divergence. The system as a whole (specifically including prisms) must be designed for an acceptable level of aberration.

TOTAL INTERNAL REFLECTION

Many prisms operate by total internal reflection (TIR). Rays internally incident upon an air/glass boundary, at angles greater than the critical angle are reflected with 100% efficiency regardless of their initial polarization state.

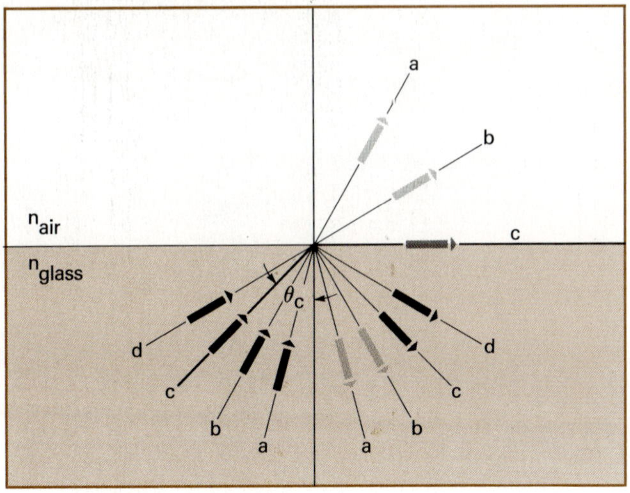

The critical angle is given by

$$\theta_c(\lambda) = \text{arc sin}\left(\frac{1}{n_\lambda}\right) ,$$

and depends on the refractive index which is a function of wavelength. See the discussion of Internal Reflection on page 174. If, at some wavelength, the refractive index should fall to less than $\sqrt{2} \approx 1.414$, the critical angle will exceed 45 degrees, and TIR will fail for a collimated beam internally incident at 45 degrees on the hypotenuse face of a right-angle prism. Reflectance decreases rapidly at angles of incidence less than the critical angle. (See graph on page 175.)

As can be seen from the dispersion formula and table of BK 7 indices presented earlier, the index of BK 7 is sufficiently high to guarantee TIR of a collimated beam at 45 degrees internal incidence over the visible and near-infrared wavelength region. The posibility of significant TIR failure with convergent or divergent beams should, however, be kept in mind — especially if polarization is important.

It is often useful to know how large the field of view can be without crossing the boundary from total to partial reflection. The graph shows field angular radii which are calculated from material dispersion formulae.

TIR is possible for very large fields in synthetic fused silica in the ultraviolet, but the TIR field is severely restricted in the infrared. At a wavelength of about 3.2 microns the boundary between total and partial reflection passes across the center of the field. This does not happen for BK 7.

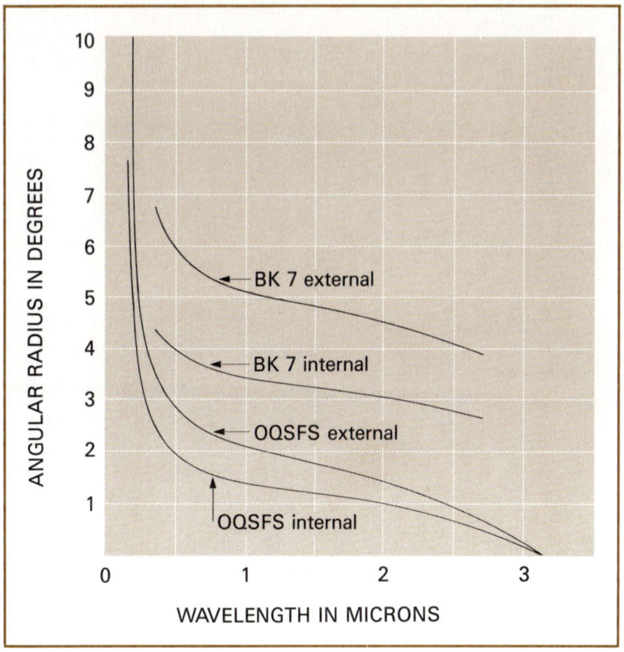

TIR FIELD LIMITATION for right angle prisms.

Total internal reflection will also fail if the hypotenuse face is not kept scrupulously clean. An almost invisible fingerprint, for example, can lead to TIR failure. In applications where a right-angle TIR prism must be frequently handled, cannot be kept clean, or where strongly convergent (or divergent) beams are used, an aluminized or silvered hypotenuse face is recommended. There is then a few percent reflectance loss at all internal angles of incidence, and no critical angle exists. For many applications this loss is negligible.

RIGHT-ANGLE GLASS PRISMS

The right-angle prism is often preferable to an inclined mirror in applications involving severe acoustic or inertial loads, because it is easier to mount, and deforms much less than a mirror in response to external mechanical stress. Very high transmission can be achieved by using the hypotenuse face in total internal reflection (TIR), and by antireflection coating the entrance and exit faces. Antireflection coatings must be requested by using a Coating Suffix. Alternatively, the hypotenuse face can be used in external reflection, and coated with any one of the metallic reflective coatings listed earlier. (See page 179.)

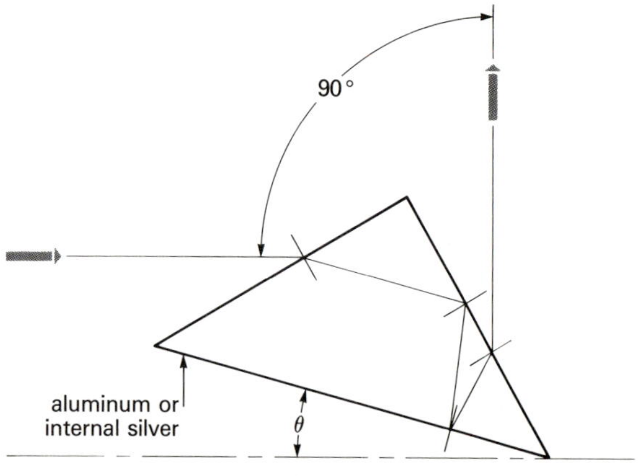

RIGHT-ANGLE PRISM used for constant deviation.

If it desired to use the hypotenuse face in internal reflection, but field-angle requirements exceed TIR acceptance limits, or the environment does not permit the hypotenuse face to be kept as clean as TIR requires, a right-angle prism with hypotenuse face aluminized will solve the problem. It is for this reason that we carry in stock both unaluminized (TIR) and aluminized right-angle prisms. The aluminized hypotenuse is protected by a coating of black paint which is acetone-soluble, and which you can easily remove should you wish to do so. Internal silver coating is also available for the hypotenuse.

As long as acceptance-angle limitations for TIR from the minor (roof) faces are not exceeded, the right-angle prism can serve as a retroreflector. This is illustrated in the drawing below. If the hypotenuse face is antireflection coated, reflective efficiency is very high, and is insensitive to orientation about axes normal to the plane of the page. Reflective efficiency is extremely sensitive to orientation about an axis parallel to the hypotenuse face and in the plane of the page.

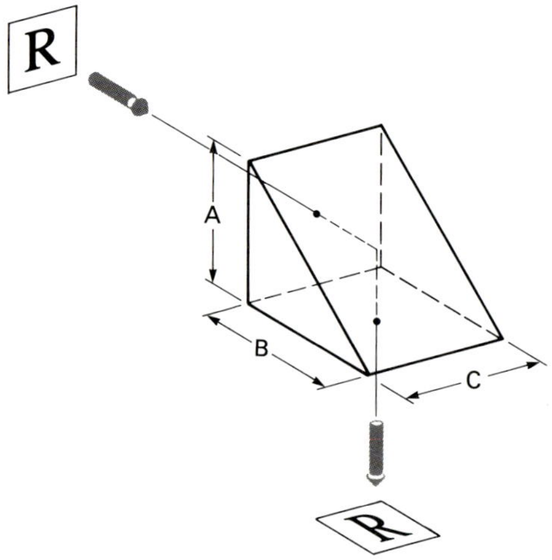

01 PRA/PRS

If the hypotenuse face is aluminized or silvered, there is a beam path through the prism which guarantees precise right angle deflection while allowing the prism a single degree of rotational freedom. See Taylor & Hafner, "Turning a perfect corner," *American Journal of Physics,* Jan. 1979, pp. 113-114.

All right-angle prisms listed here have entrance and exit faces of identical size and shape: A = B = C in the figure. Triangular faces are fine ground, and all edges are slightly beveled.

COATINGS FOR RIGHT-ANGLE PRISMS

Any of the coatings described in the coatings section on pages 69-85 can be applied to prism faces. The commonly specified antireflection coatings and high reflection coatings are listed here for your convenience. When specifying a coating option, please be sure to indicate which surfaces you want coated.

Coatings; for Right-Angle Prisms

Type	COATING SUFFIX
MgF_2	/066
HEBBAR™	/078
"V" at 632.8nm	/123
Protected Aluminum	/011
MAXBRite™	/001
MAX-R at 632.8nm, 45 degrees	/279

Glass Right Angle Glass Prisms; Aluminized Hypotenuse

A=B=C (mm)	PRODUCT NUMBER ±30 arc sec	PRODUCT NUMBER ±3 arc min
3.2	—	01 PRA 001
5.0	01 PRT 003	01 PRA 003
6.4	—	01 PRA 005
8.0	—	01 PRA 007
10.0	01 PRT 009	01 PRA 009
12.7	01 PRT 011	01 PRA 011
15.0	—	01 PRA 013
16.5	—	01 PRA 015
18.0	—	01 PRA 017
20.0	01 PRT 019	01 PRA 019
25.0	—	01 PRA 020
25.4	01 PRT 021	01 PRA 021
30.0	01 PRT 023	01 PRA 023
35.0	—	01 PRA 025
38.1	—	01 PRA 027
40.0	01 PRT 029	01 PRA 029
45.0	—	01 PRA 033
50.0	—	01 PRA 034
50.8	01 PRT 035	01 PRA 035
55.0	—	01 PRA 037
63.5	—	01 PRA 041
76.2	—	01 PRA 045
100.0	—	01 PRA 055

For antireflection coatings append Coating Suffix and specify faces to be coated.

SPECIFICATIONS: RIGHT-ANGLE GLASS PRISMS

Dimensions: ± 0.15mm
Flatness: 2λ at 546.1nm over central 90% of edge dimension
Angles: see table
Material: BK 7 Grade A Fine Annealed.
Cosmetic Surface Quality: 80-50 scratch and dig
Apex Bevel Width: 0.25mm

Glass Right-Angle Glass Prisms; Unaluminized (TIR)

A=B=C (mm)	PRODUCT NUMBER ± 30 arc sec	PRODUCT NUMBER ± 3 arc min
0.7	—	01 PRS 401
1.0	—	01 PRS 403
1.3	—	01 PRS 405
2.0	—	01 PRS 407
2.7	—	01 PRS 409
3.2	—	01 PRS 001
4.0	—	01 PRS 411
4.8	—	01 PRS 413
5.0	01 PRP 003	01 PRS 003
6.4	—	01 PRS 005
8.0	—	01 PRS 007
10.0	01 PRP 009	01 PRS 009
12.7	01 PRP 011	01 PRS 011
15.0	—	01 PRS 013
16.5	—	01 PRS 015
18.0	—	01 PRS 017
20.0	01 PRP 019	01 PRS 019
25.0	—	01 PRS 020
25.4	01 PRP 021	01 PRS 021
30.0	01 PRP 023	01 PRS 023
35.0	—	01 PRS 025
38.1	—	01 PRS 027
40.0	01 PRP 029	01 PRS 029
45.0	—	01 PRS 033
50.0	—	01 PRS 034
50.8	01 PRP 035	01 PRS 035
55.0	—	01 PRS 037
63.5	—	01 PRS 041
76.2	—	01 PRS 045
100.0	—	01 PRS 055

For antireflection or reflective coatings append Coating Suffix and specify faces to be coated.

RIGHT-ANGLE FUSED SILICA PRISMS

These right-angle prisms are made from UV grade synthetic fused silica. They combine the mounting and acoustic isolation advantages of a prism with the excellent transmission of this material.

Very high prism transmission can be achieved by using the hypotenuse face in total internal reflection (TIR), and by antireflection coating the entrance and exit faces. Antireflection coatings were discussed earlier.

Because the index of synthetic fused silica is somewhat lower than that of BK 7, the critical angle for TIR is greater (thereby reducing the TIR field acceptance limits). For larger field angles or non-collimated beams, the hypotenuse face can be coated with any of the metallic reflective coatings. (Silver is not recommended for UV applications.) Prisms both with and without aluminized hypotenuse faces are listed below. The aluminized hypotenuse is protected by a coating of black paint which is acetone-soluble, and which you can easily remove should you wish to do so.

All right-angle prisms listed have entrance and exit faces of identical size and shape: A=B=C in the figure. Triangular faces are fine ground, and all edges are slightly beveled.

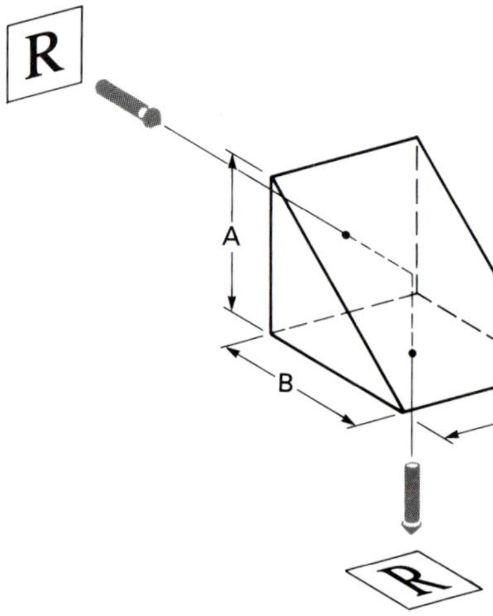

01 PRA/PRS

COATINGS FOR RIGHT-ANGLE PRISMS

Most coatings described in the coatings section on pages 69-85 can be applied to prism faces. The commonly specified antireflection coatings and high reflection coatings are listed below for your convenience. When specifying a coating option please be sure to indicate which surface you want coated.

Coatings; for Right-Angle Prisms

Type	COATING SUFFIX
MgF_2	/066
HEBBAR™	/078
"V" at 632.8nm	/123
Protected Aluminum	/011
MAXBRite™	/001
MAX-R at 632.8nm, 45 degrees	/279

SPECIFICATIONS: RIGHT-ANGLE FUSED SILICA PRISM

Dimensions: ± 0.15mm
Flatness: λ/4 at 546.1nm over central 90% of edge dimension
Angle: ± 30 arc seconds
Material: UV grade synthetic fused silica
Cosmetic Surface Quality; 60-40 scratch and dig
Bevel Width: 0.15mm to 1.0mm

Right-Angle Fused Silica Prisms

A=B=C (mm)	PRODUCT NUMBER Unaluminized	PRODUCT NUMBER Aluminized Hypotenuse
10.0	01 PQB 000	01 PQA 000
12.7	01 PQB 001	01 PQA 001
25.4	01 PQB 002	01 PQA 002

For coatings append Coating Suffix and specify faces to be coated.

DOVE PRISMS

Dove prisms are used as image rotators in a variety of optomechanical systems. As the prism is rotated, the image passing through will rotate at twice the angular rate of the prism. Light entering the dove prism must be parallel or collimated, because the length of the prism is typically four or five times the dimension A. The hypotenuse face of a dove prism is the face of the largest area. Normally the dove prism is used in the total internal reflection (TIR) mode, with the hypotenuse face unaluminized. Cleanliness of this face is important.

The useful aperture of a dove prism can be conveniently doubled by aluminizing or silvering the hypotenuse face of two identical prisms, and subsequently cementing these faces together. The length of the combination is the same as for a single prism. Overall transmission is slightly reduced by the substitution of the metallic reflecting film for the usual TIR surface.

Either bare aluminum or internal silver may be coated on these prisms on a special request basis. When this is done, the coatings will be protected by an overcoating of black lacquer.

The clear aperture of these prisms is square and is given by the dimension A in the figure. Single-layer AR coatings and broadband HEBBAR™ antireflection coatings are available for the entrance/exit faces on special request.

We supply substantial numbers of modified standard prisms to original equipment manufacturers. Please contact an Applications Engineer for further details.

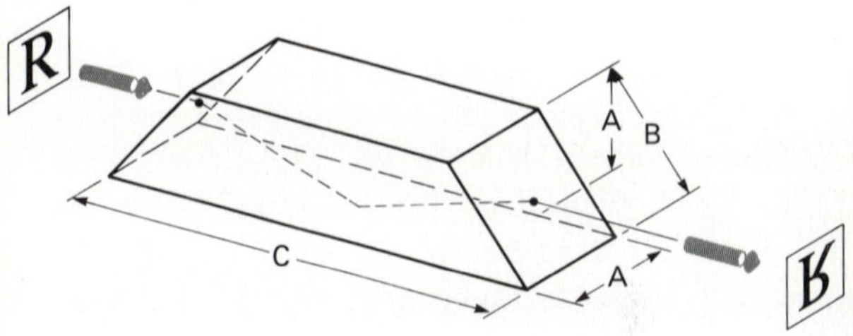

01 PDE

SPECIFICATIONS: DOVE PRISMS

Dimensions: ± 0.15mm
Design Index: 1.5168
Design Wavelength: 587.6nm
Flatness: 2 λ at 587.6nm over central 90% of edge dimension
Angles: ± 3 arc min
Material: BK 7 Grade A fine annealed
Cosmetic Surface Quality: 80-50 scratch and dig
Coatings: See text.
Sides are fine ground. All edges are lightly beveled.

Unaluminized Dove Prisms (TIR)

A (mm)	B (mm)	C* (mm)	PRODUCT NUMBER
1.2	1.7	5.5	01 PDE 401
2.0	2.8	8.2	01 PDE 403
2.5	3.5	10.3	01 PDE 405
3.6	5.4	15.0	01 PDE 407
6.0	8.5	23.0	01 PDE 409
10.0	14.1	42.3	01 PDE 001
15.0	21.2	63.4	01 PDE 003
20.0	28.3	84.5	01 PDE 005

*To theoretical sharp edges (unbeveled).

ROOF PRISMS

The roof or Amici prism deviates or deflects the image through an angle of 90 degrees. The figure shows its effect upon the image orientation, where the precise meanings of object and image were explained earlier. Basically, the roof prism is a right-angle prism whose hypotenuse has been replaced by a 90-degree TIR roof. Its shape is further complicated by the fact that glass which does not contribute to useful aperture has been trimmed away to reduce size and weight. Cleanliness and quality of the TIR roof surfaces is important. The 90-degree deviation applies only to rays which are precisely normal to the entrance and exit faces, other rays being deflected by other amounts, exactly as in the case of the ordinary right-angle prism. The roof prism finds application in situations which demand both right-angle deflection and image erection (a combination of left-to-right reversion and top-to-bottom inversion, equivalent to a 180-degree rotation about the optical axis), as in some instrument viewfinders.

These prisms are normally sold uncoated, for use in TIR mode. For wide field applications, beyond TIR limits, protected aluminum or internal silver coatings should be requested for the roof surfaces. Antireflection coated entrance/exit faces are also available. These options can be requested by using the approrpiate Coating Suffix introduced earlier.

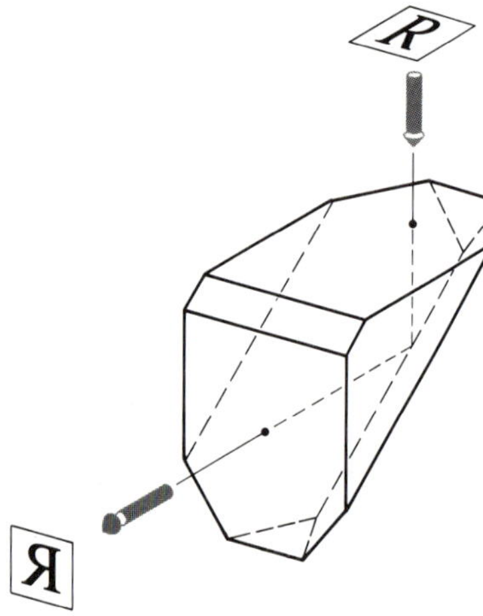

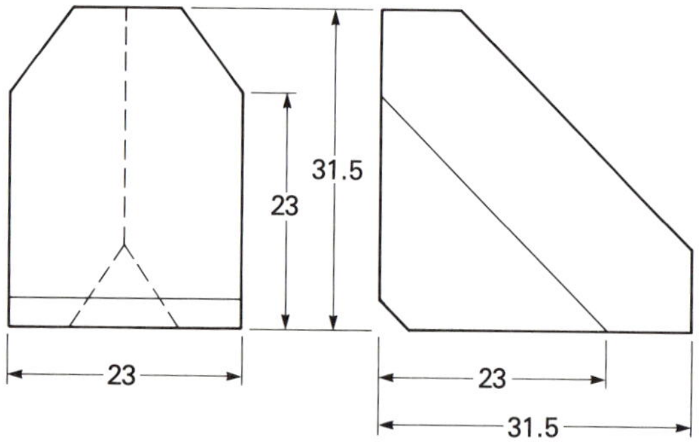

01 PRF 001

SPECIFICATIONS: ROOF PRISMS

Dimensions: ± 0.15mm
Flatness: 1 λ at 546.1nm over central 90% of edge dimension
Roof Angles: ± 5 arc seconds
Other Angles: ± 30 arc seconds
Material: BK 7 Grade A Fine Annealed
Cosmetic Surface Quality: 80-50 scratch and dig
Coatings: See text.
Sides are fine ground. All edges are lightly beveled.

Roof Prisms; Unaluminized

A (mm)	B (mm)	C (mm)	PRODUCT NUMBER
23.0	23.0	31.5	01 PRF 001

If you require volumes of special prisms not shown here, please contact one of our Applications Engineers.

PENTA-PRISMS

Penta-prisms have two important properties. The first is that the image is neither inverted nor reversed as it is deviated by 90 degrees. Secondly, the penta-prism is a constant deviation prism, meaning that the same exact 90-degree deviation angle applies to all rays transmitted by the useful aperture — regardless of the angles between these rays and the optical axis (or entrance and exit face normals). The deviation is thus independent of the orientation of the prism, making the penta-prism especially important in applications in which prism orientation cannot be precisely controlled (because of vibration, for example). This prism is commonly used in optical tooling, alignment, rangefinding, surveying, and other instrumental applications. For example, reflection normal to the surface of a calm pool of mercury, seen through a penta-prism, establishes an accurate horizontal reference.

The internal geometry of these penta-prisms is such that the reflecting surfaces must be coated with a metallic or dielectric reflective coating. TIR cannot be used. Standard penta-prism reflecting surfaces are coated with aluminum. HEBBAR™ or "V" coatings for entrance/exit faces can be requested via the Coating Suffix. Reflecting multilayer dielectric coatings are available for specific wavelengths, such as 488.0nm.

We supply penta-prisms in two levels of deviation tolerance which are listed on this page.

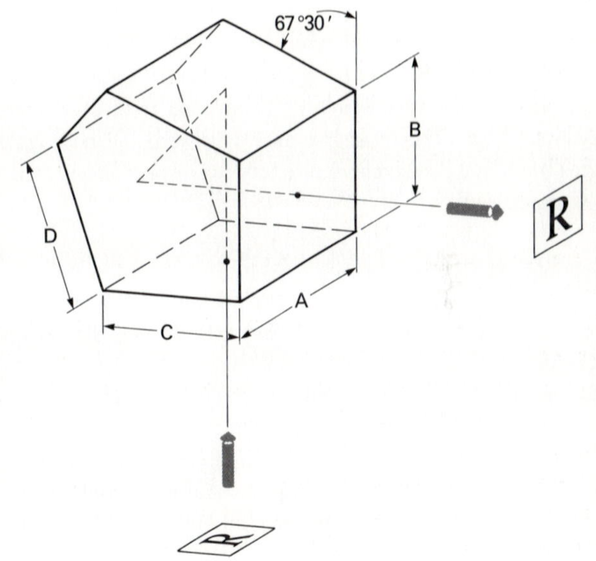

01 PPA/PPS

SPECIFICATIONS: PENTA-PRISMS

Dimensions: ± 0.15mm
90 Degree Deviation Tolerance
Standard Series: < 3 arc minutes
Precision Series: < 10 arc seconds
Flatness Standard Series: 2λ at 546.1nm over central 90% of edge dimension
Precision Series: λ/4 at 546.1nm
Reflectivity: >90% per face from 400 to 800nm.
Material: BK 7 Grade A Fine Annealed
Cosmetic Surface Quality Standard: 80-50 scratch and dig
Precision: 60-40 scratch and dig
Coating: Bare aluminum with black lacquer overcoat.
Pentagonal surfaces are fine ground. All edges are beveled.

Aluminized Penta-Prisms; Standard

A (mm)	B=C (mm)	D (mm)	PRODUCT NUMBER
10.0	10.0	10.8	01 PPA 001
8.5	12.0	13.0	01 PPA 003
12.0	13.0	14.0	01 PPA 005
10.0	15.0	16.0	01 PPA 007
15.0	15.0	16.0	01 PPA 009
20.0	20.0	22.0	01 PPA 011
25.0	25.0	27.1	01 PPA 013

Aluminized Penta-Prisms; Precision

A=B=C (mm)	D (mm)	PRODUCT NUMBER
10.0	10.8	01 PPS 003
12.5	13.7	01 PPS 005
20.0	21.6	01 PPS 009
25.0	27.1	01 PPS 011

EQUILATERAL DISPERSING PRISMS

Equilateral prisms are used as dispersing elements in instruments and experiments requiring wavelength separation. While prism dispersion (output ray direction versus wavelength) is nonlinear, spectral brightness is superior to that of diffraction gratings, and there is no possibility of confusion from overlapping spectral orders. The 60-degree angles between the rectangular faces result in a good compromise between dispersion and reflection losses.

A light ray is twice refracted in passing through the prism with total deviation denoted by D in the figure. Deviation is a function of refractive index, and hence of wavelength. Angular dispersion (ϕD) is the difference in deviation for light rays having different wavelengths. This difference varies with prism orientation. The F (blue) and C (red) lines of the atomic hydrogen spectrum are frequently used to define angular dispersion. It is desirable to have large angular dispersions, as this implies high instrumental spectral resolution at fixed slit width.

Reflection losses are, for an unpolarized ray of some particular wavelength, minimized for the condition known as minimum deviation, in which the ray travels parallel to the base of the prism while passing through it. This is also the condition in which the angular dispersion of the prism for nearby wavelengths is most nearly linear. In this condition wavelengths of unknown spectrum lines can be approximated by linear interpolation between closely spaced lines of known wavelength. Externally, the minimum deviation condition is identified by the fact that, for the ray and wavelength of concern, the angle of incidence at the entrance surface is equal to the angle of emergence (both angles measured with respect to the surface normals) at the exit surface.

We offer equilateral dispersing prism of two materials, crown (BK 7) and flint (F 2). BK 7 is the more durable material, and has superior transmission (both internally and at the surfaces). F 2 flint offers superior dispersion, but is somewhat more fragile and sensitive to staining. Antireflection coatings can be applied to all surfaces, serving both to protect from staining and improve transmission. Such coatings also reduce polarization at the prism surfaces by increasing the total transmittance.

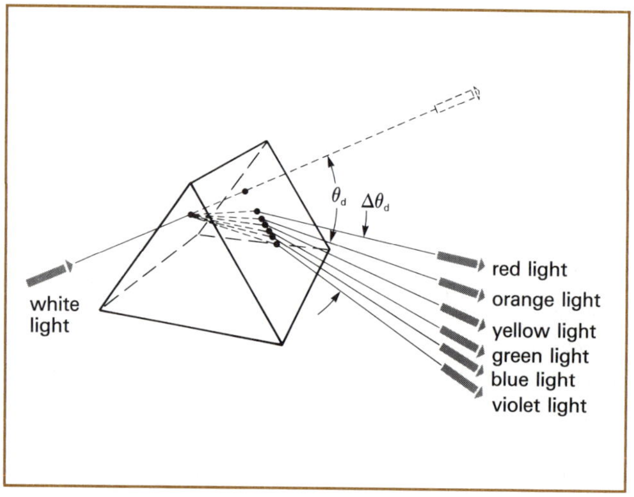

SPECIFICATIONS: EQUILATERAL DISPERSING PRISMS

Dimensions: $\pm$ 0.2mm
Surface Flatness: 3 λ at 546.1nm over 90% of edge dimension
Materials:
  Crown: BK 7 Grade A fine annealed
  Flint: F 2 Grade A fine annealed
Cosmetic Surface Quality: 80-50 scratch and dig
Coatings: Single-layer MgF_2 (/066) or HEBBAR™ (/078), by appending the appropriate Coating Suffix.

Equilateral Dispersing Prisms

Edge Dimensions (mm)	PRODUCT NUMBER BK 7 (Crown)	PRODUCT NUMBER F 2 (Flint)
25.0	01 PES 001	01 PEH 010
30.0	01 PES 003	01 PEH 011
40.0	01 PES 005	01 PEH 013
45.0	01 PES 007	01 PEH 015
50.0	01 PES 009	01 PEH 017
60.0	01 PES 011	01 PEH 019

BREWSTER AND LITTROW DISPERSING PRISMS

The apex angles of Brewster prisms are chosen so that, for one particular wavelength and linear polarization, rays at minimum deviation (which travel parallel to the base of the prism while inside it) both enter and exit the prism at the Brewster angle. For light within the prism linearly polarized with electric field vector orthogonal to the prism base, the entrance/exit faces are perfectly transmitting. These prisms, like Brewster-angle windows, can be used inside the cavities of lasers operating on very low gain laser transitions, where even slight reflection losses may be intolerable.

Because reflectance for the p-polarization is extremely low, and not very sensitive to either angle or wavelength at angles of incidence near the Brewster angle, we offer only a Brewster prism designed for 488nm, slightly short of the middle of the visible spectrum. At wavelengths of 350.7 and 676.5nm, and at minimum deviation for these wavelengths, this prism has single-surface p-reflectances of only 3×10^{-5} and 7×10^{-6} respectively. Thus this prism is satisfactory throughout this wavelength interval.

When a Brewster prism is cut in half along the plane which is the bisector between the two Brewster entrance/exit faces, two Littrow prisms are formed. The surfaces resulting from bisection may be coated with a high-reflectance multilayer dielectric film designed for near-normal incidence at and near the wavelength desired. The Littrow prism is used as the planar element of a hemispherical laser resonator. Tuning is again accomplished by tilting. We offer two Littrow prisms, one for general purpose use designed for 488nm, and the other designed for 633nm (to prevent the undesirable infrared oscillation in helium-neon). With suitable reflective coatings, either of these prisms can be used over wide wavelength ranges. We can also supply prisms of either type made to different dimensions or tolerances on special request. Please contact our Applications Engineers for further information regarding alternative coatings.

SPECIFICATIONS: ISOSCELES BREWSTER AND LITTROW DISPERSING PRISMS

Dimensions: ± 0.15mm
Wavelength Range: 350-700nm
Flatness: λ/20 at 546.1nm over central 90% of edge dimension
Angles: ± 2 arc minutes
Material: UV grade synthetic fused silica
Cosmetic Surface Quality: Pitch-polished for minimum scatter, 10-5 or better scratch and dig

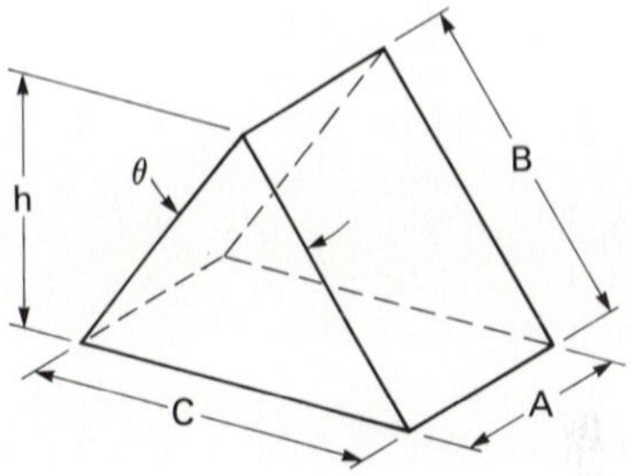

02 IBD 001

Isosceles Brewster Dispersing Prisms

A (mm)	B (mm)	C (mm)	h (mm)	φ (degrees)	PRODUCT NUMBER
12.7	23.0	26.0	19.0	68°42″	02 IBD 001

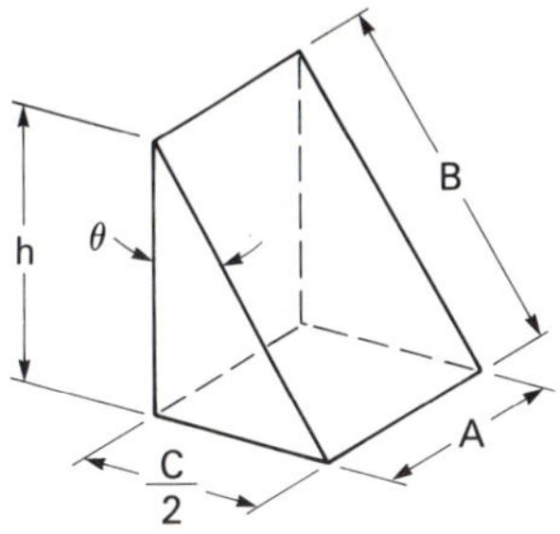

02 LDP

Littrow Dispersing Prisms

A (mm)	B (mm)	C/2 (mm)	h (mm)	φ/2 (degrees)	PRODUCT NUMBER
12.7	23.0	13.0	19.0	34°21′	02 LDP 001
12.7	22.4	12.7	18.5	34°27′	02 LDP 002

WEDGE PRISMS

LASER BEAM STEERING

Wedge prisms are used as beam steering elements in optical systems, playing a role in optics analogous to that of the wobble plate in mechanics. The minimum deviation or deflection D experienced by a ray in passing through a thin wedge of apex angle A is approximately given by

$$D = (n - 1) A,$$

where n is the refractive index. The "power" (Δ) of a prism is measured in prism diopters, a prism diopter being defined as a deflection of 1 cm at a distance of 1 meter from the prism. Thus

$$\Delta = 100 \tan(D).$$

By combining two wedges of equal power (equal deviation), in near contact, and independently rotating them about an axis roughly parallel to the normals of their adjacent faces, a ray passing through the combination can be steered in any direction, within a narrow cone, about the path of the undeviated ray. The angular radius of this cone is approximately 2D. Wedge prisms are most frequently used for laser beam steering and Schlieren system calibration. Deviations for other wavelengths can be estimated from the above formula and appropriate nominal refractive index values. Apex angle is controlled to within very tight tolerances in the manufacturing process. As a result of the melt-to-melt index tolerance, deviation angles (functions of wavelength) are only nominally specified.

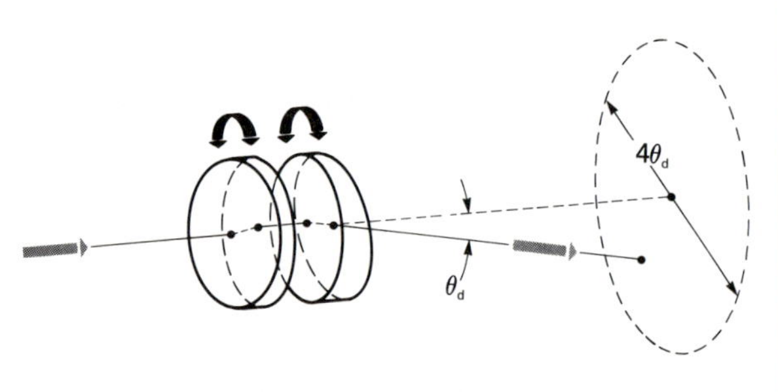

LASER BEAMSTEERING using wedge prisms.

The deviation angles are specified with the assumption that the input beam is *normal* to the perpendicular face. At other input angles the deviation will, of course, be different. To determine the deviation angle for the same input direction but other wavelengths, the equation is:

$$D = \arcsin(n \sin A) - A,$$

where D is the deviation angle, A is the wedge angle, and n is the nominal index at the appropriate wavelength for BK 7 (see earlier dispersion formula and index table).

Wedge prisms are available in alternative materials, such as synthetic fused silica, and in different shapes and sizes, or with various antireflection coatings.

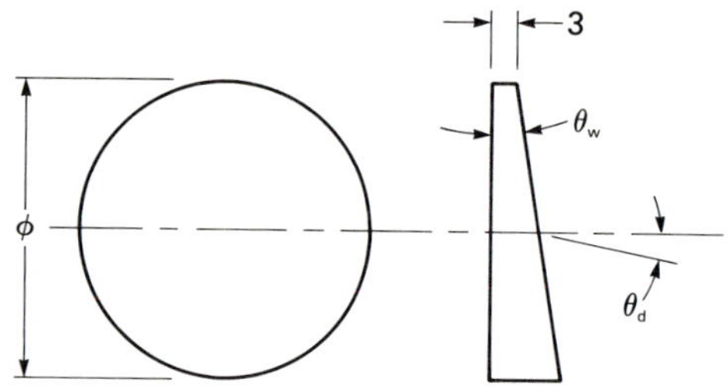

02 PRW

SPECIFICATIONS: WEDGE PRISMS

Dimensions: ± 0.2mm
Flatness: $\lambda/4$ at 546.1nm over central 90% of edge dimensions
Wedge Angle: ± 30 arc seconds
Material: BK 7 Grade A Fine Annealed
Cosmetic Surface Quality: Pitch polished. 60-40 scratch and dig
Coatings: Single-layer MgF_2 (/066), HEBBAR™ (/078), or various "V" coatings, by appending the Coating Suffix

Wedge Prisms For Laser Beam Steering

ϕ (mm)	Nominal Deviation for 632.8nm D (degrees)	Wedge Angle A (degrees)	"Power" (prism) diopters) (Δ)	PRODUCT NUMBER
25.0	1.00	1.94	1.75	02 PRW 001
25.0	2.00	3.87	3.50	02 PRW 003
25.0	4.00	7.67	7.00	02 PRW 005
25.0	6.00	11.36	10.50	02 PRW 007
25.0	8.00	14.85	14.00	02 PRW 009
25.0	10.00	18.13	17.60	02 PRW 011

RETROREFLECTORS

The trihedral retroreflector or corner cube reflector has the property that any ray entering the effective aperture will be reflected and emerge from the entrance/exit face parallel to itself, but with opposite direction of propagation. An incident beam, exactly filling the effective aperture, is reflected exactly back on itself. These properties are, within acceptance angle limits, independent of the orientation of the retroreflector. Retroreflectors therefore find frequent application in situations where orientation is difficult or impossible to control, and where a mirror would therefore be unsatisfactory.

The retroreflector, like the right angle prism, depends upon the principle of total internal reflection (TIR) for its operation. Like TIR right-angle prisms, retroreflectors have acceptance angle limitations. Because the geometry of the retroreflector is complicated, the acceptance angle limitations cannot be simply specified.

It is usually possible to choose a retroreflector orientation such that acceptance angle limitations are, for all practical purposes, nonexistent. It is always possible to guarantee, by appropriate orientation of neighboring reflectors in a retroreflector array (consisting of at least two retroreflectors), that part of the array will be functional even at very large angles of incidence. All acceptance angle limitations can be removed, at the cost of several percentage points reduction in reflective efficiency, by applying reflective coatings to the rear reflecting surfaces and thereby avoiding TIR failure.

Success of the TIR process at the rear surfaces of the retroreflctor demands that these surfaces be kept scrupulously clean. When retroreflectors are used as elements of high-power laser cavities, survival of the retroreflector may depend upon the cleanliness of these surfacees. In applications where a retroreflector must be frequently handled, or otherwise cannot be kept scrupulously clean, and where powers are not so high as to assure coating destruction, we suggest that the three rear reflecting surfaces be coated with protected aluminum, or with internal silver overcoated with copper or Inconel and paint. This assures reliable reflectivity of these surfaces, and removes acceptance angle limitations. Where this slight reduction in reflective efficiency is unacceptable, multilayer reflective coatings can be used. Antireflection coatings are available for the entrance/exit face.

We maintain tight control over angles and surface accuracy in manufacture, assuring a parallelism between incident and returned beams of 2 arc seconds or better. Roof and other edges are lightly beveled to minimize the risk of chipping.

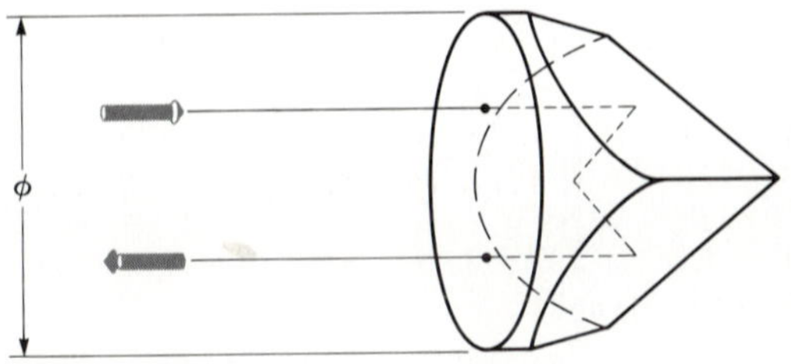

02 CCG

SPECIFICATIONS: SOLID GLASS RETROREFLECTORS

Dimensions: $\pm$ 0.15mm
Deviation: 180 degrees $\pm$ 2 arc seconds
Wavefront Distortion: $<\lambda/4$ at 632.8nm
Material: BK 7 Grade A Fine Annealed
Coatings: MgF_2 (/066), HEBBAR™ (/068) or "V"-coatings by appending appropriate Coating Suffix

Solid Glass Retroreflectors; Uncoated

Aperture ϕ (mm)	PRODUCT NUMBER
15.0	02 CCG 001
25.0	02 CCG 003
38.0	02 CCG 005
50.0	02 CCG 007
70.0	02 CCG 008

These retroreflectors may be conveniently mounted using the retroreflector holders described on page 457.

PLATE BEAMSPLITTERS

These plate beamsplitters consist of a thin (1mm) plate of optical crown glass on each side of which is deposited a different type of coating. The first-surface is coated with an all-dielectric film having partial reflection properties over either the visible or the near-infraredspectrum. The benefit of this type of coating is that it has low absorption; typically 0.5% for a 50/50 splitter at 45°. The second-surface has a HEBBAR™ broad band antireflection coating optimized for 45°.

The transmittance and reflectance graphs are for unpolarized inputs at an angle of incidence of 45°. Total transmittance or reflectance are shown without regard for polarization state.

These beamsplitters are offered in two sizes, 50mm square and 50mm diameter. The circular beamsplitter can be conveniently mounted in our beamsplitter mounts described on page 475.

If you have a need for plate beamsplitters having different characteristics please contact one of our Applications Engineers.

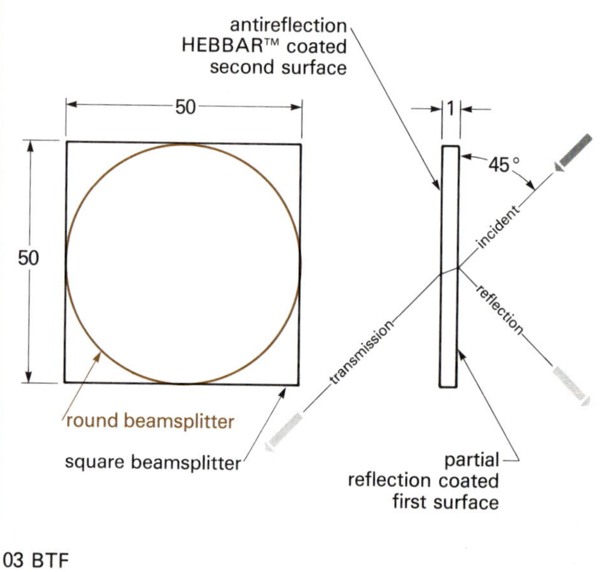

SPECIFICATIONS: PLATE BEAMSPLITTERS

Dimensions: ± 0.2mm
Substrate Material: Optical Crown glass
Flatness: 3λ per 25mm over central 90% of edge dimension
Parallelism: 5 arc minutes
Coatings
  1st surface: all-dielectric partial reflection coating for 45°
  2nd surface: HEBBAR™ for 45°
R/T Ratio: ± 5% at 550nm (VIS)
  ± 5% at 850nm (I-R)
Cosmetic Surface Quality: 80-50 scratch and dig
Edges are ground

Plate Beamsplitters

Shape	Side or Diameter (mm)	Nominal R/T Ratio	PRODUCT NUMBER Visible	Near I-R
Square	50	30/70	03 BTF 001	03 BTF 013
	50	50/50	03 BTF 007	03 BTF 015
	50	70/30	03 BTF 011	03 BTF 017
Round	50	30/70	03 BTF 021	03 BTF 027
	50	50/50	03 BTF 023	03 BTF 029
	50	70/30	03 BTF 025	03 BTF 033

The round beamsplitters may be conveniently mounted in our beamsplitter mount which provides 45° access from both sides. (see page 475.)

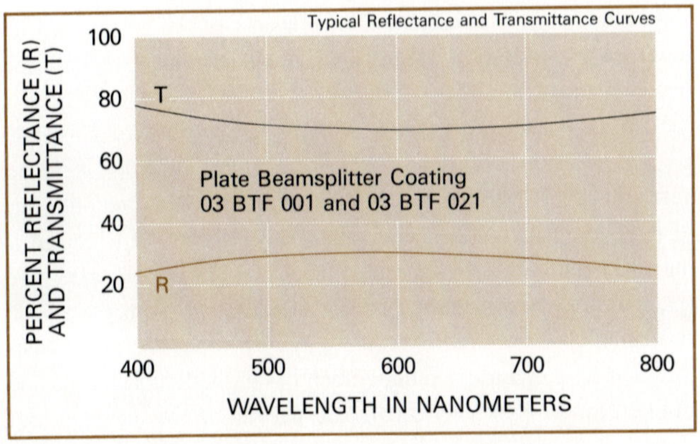
Typical Reflectance and Transmittance Curves
Plate Beamsplitter Coating
03 BTF 001 and 03 BTF 021
PERCENT REFLECTANCE (R) AND TRANSMITTANCE (T)
WAVELENGTH IN NANOMETERS
T
R

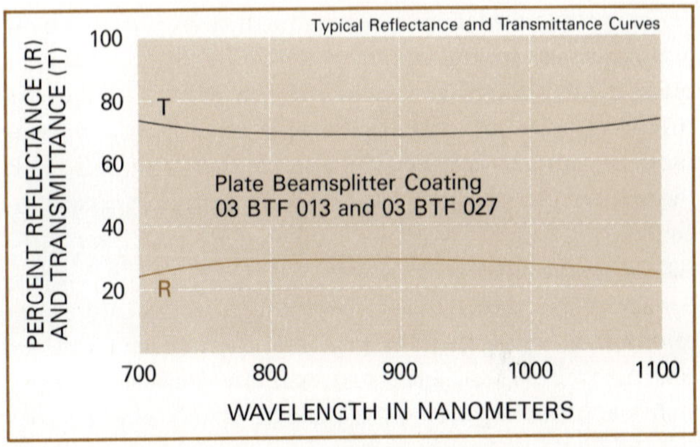
Typical Reflectance and Transmittance Curves
Plate Beamsplitter Coating
03 BTF 013 and 03 BTF 027
PERCENT REFLECTANCE (R) AND TRANSMITTANCE (T)
WAVELENGTH IN NANOMETERS
T
R

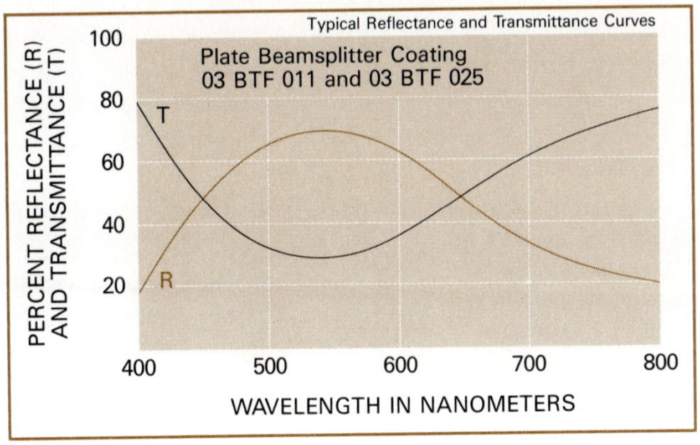
Typical Reflectance and Transmittance Curves
Plate Beamsplitter Coating
03 BTF 011 and 03 BTF 025
PERCENT REFLECTANCE (R) AND TRANSMITTANCE (T)
WAVELENGTH IN NANOMETERS
T
R

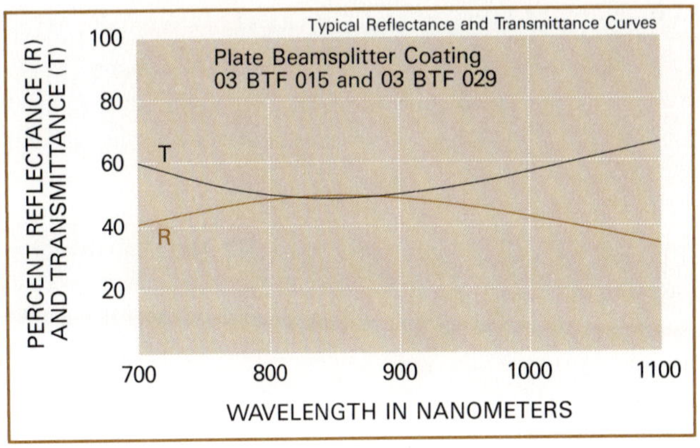
Typical Reflectance and Transmittance Curves
Plate Beamsplitter Coating
03 BTF 015 and 03 BTF 029
PERCENT REFLECTANCE (R) AND TRANSMITTANCE (T)
WAVELENGTH IN NANOMETERS
T
R

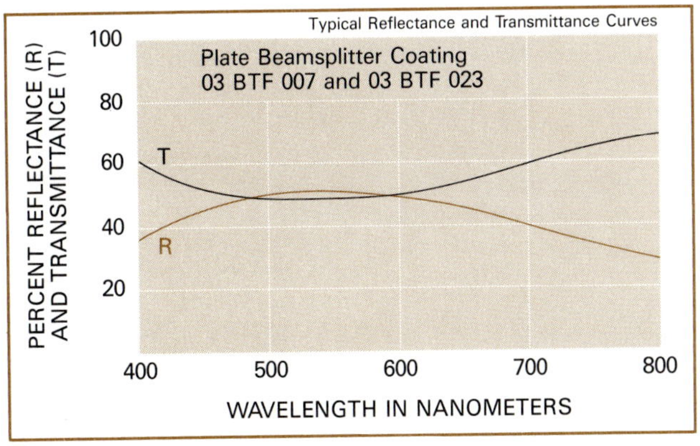
Typical Reflectance and Transmittance Curves
Plate Beamsplitter Coating
03 BTF 007 and 03 BTF 023
PERCENT REFLECTANCE (R) AND TRANSMITTANCE (T)
WAVELENGTH IN NANOMETERS
T
R

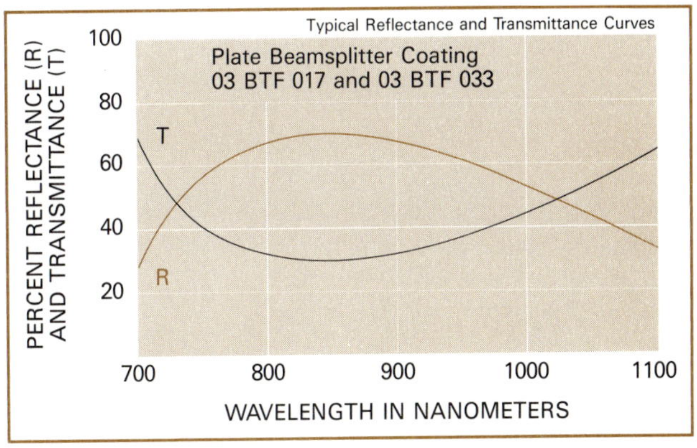
Typical Reflectance and Transmittance Curves
Plate Beamsplitter Coating
03 BTF 017 and 03 BTF 033
PERCENT REFLECTANCE (R) AND TRANSMITTANCE (T)
WAVELENGTH IN NANOMETERS
T
R

CUBE BEAMSPLITTERS

Cube beamsplitters have several advantages over plate beamsplitters and are widely used for the following reasons. A cube beamsplitter deforms less in response to external mechanical stress and, most importantly, is free of ghost images. It gives excellent performance over wide ranges of angles of incidence, is rugged, easy to mount and ideal for beam superposition applications.

If cube beamsplitters are used in convergent or divergent portions of an optical beam, they will contribute substantial amounts of unwanted aberration. This can be avoided or minimized by making sure these components are used in collimated, or nearly collimated, beams. Conjugate distances which include cubes should be long. Alternately other elements of the system can be designed to compensate for any aberrations introduced by the cube in a convergent beam.

Cube beamsplitters consist of pairs of identical right-angle prisms, with their hypotenuse faces cemented together. Prior to cementing, a partial reflection film is vacuum deposited onto one of the hypotenuse faces. (The particular prism which is coated is marked with a small dot on one of the ground faces.) We offer two types of cube beamsplitter, classified by the nature of the partial reflection coatings used. These are hybrid, metal-dielectric coatings and all-dielectric coatings. Each has its particular merits and limitations, so your selection should depend on the intended use.

HYBRID PARTIAL REFLECTION COATINGS

Hybrid, metal-dielectric, coatings combine the benefits of both metals and dielectrics to produce a moderate absorption beamsplitter with little polarization sensitivity. Typically, reflection and transmission will each be about 40%, with s- and p-polarized components within 10% of each other.

These coatings exibit reasonable spectral flatness enabling them to be used over a broad range of wavelengths. Another advantage of the hybrid coating is that it stays chromatically neutral in convergent or divergent beams. Our hybrid coatings cover the visible spectrum (400-700nm) or the very near-infrared for laser diode use (700-1100nm). These coatings are not offered further into the infrared due to severe material limitations.

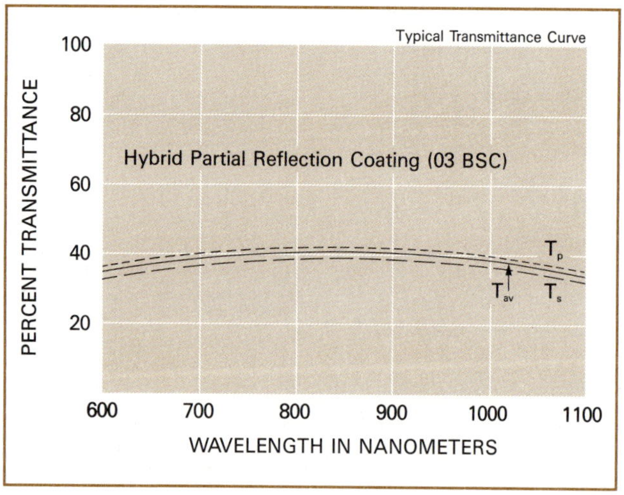

ALL-DIELECTRIC PARTIAL REFLECTION COATINGS

All-dielectric partial reflection coatings provide a high degree of efficiency, having negligible absorption and giving very close to 50% reflection and transmission. However, they are extremely polarization sensitive, having s- and p-components which can be as much as 70% apart. Care should therefore be taken, when using all-dielectric beamsplitters, to consider polarization implications in the system.

We offer all-dielectric partial reflection coatings for the visible spectrum (450-650nm), the laser diode short-waveband (650-900nm), the neodymium region (900-1300nm) and the laser diode telecommunications waveband (1300-1600nm).

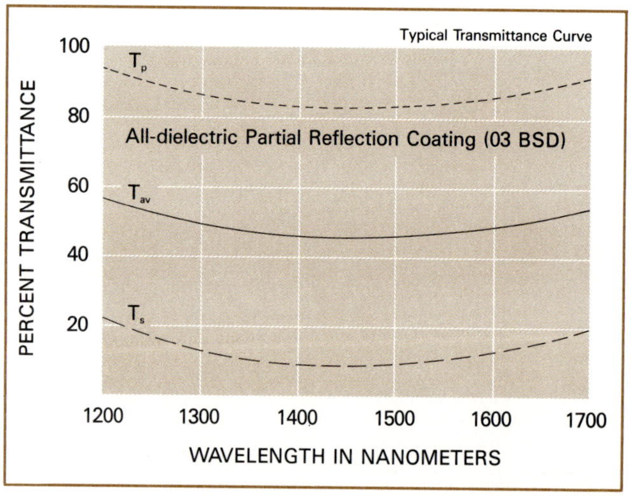

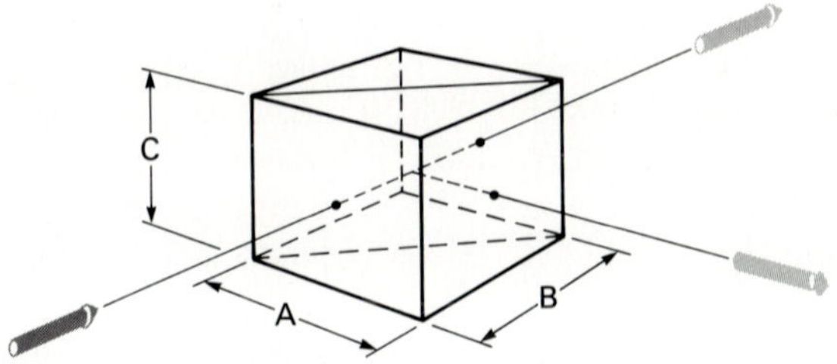

03 BSC

SPECIFICATIONS: CUBE BEAMSPLITTERS

Dimensions: ± 0.3mm
Material: BK 7 grade A fine annealed
Angles: 3 arc minutes
Face Flatness: 2λ at 546nm
Transmission
  Hybrid: 41 ±6% (±5% variation with wavelength)
  All-dielectric: 50 ±5% (±3% variation with wavelength)
Absorption
  Hybrid: < 27%
  All-dielectric: < 5%
Coatings: All 4 faces HEBBAR™ antireflection coated
Cosmetic Surface Quality: 80-50 scratch and dig

ANTIREFLECTION COATINGS

All our cube beamsplitters are antireflection coated on all four faces to minimize ghost images and reflection losses. They are normally supplied with a multilayer HEBBAR™ antireflection coating. There is no need to append a Coating Suffix.

Cube Beamsplitters; Visible Wavelength Range

A=B=C (mm)	PRODUCT NUMBER Hybrid Coating 400-700nm	All-Dielectric Coating 450-650nm
5.0	03 BSC 001	03 BSD 002
10.0	03 BSC 003	03 BSD 004
12.7	03 BSC 005	03 BSD 006
20.0	03 BSC 007	03 BSD 008
25.4	03 BSC 009	03 BSD 012
30.0	03 BSC 011	03 BSD 014
40.0	03 BSC 013	03 BSD 016
50.8	03 BSC 015	03 BSD 018

Cube Beamsplitters; Near-IR Wavelength Range

A=B=C (mm)	PRODUCT NUMBER Hybrid Coating 700-1100nm	All-Dielectric Coating 650-900nm	900-1300nm	1300-1600nm
5.0	03 BSC 023	03 BSD 042	03 BSD 062	03 BSD 024
10.0	03 BSC 025	03 BSD 044	03 BSD 064	03 BSD 026
20.0	03 BSC 027	03 BSD 048	03 BSD 068	03 BSD 028
25.4	03 BSC 029	03 BSD 052	03 BSD 072	03 BSD 032
50.8	03 BSC 035	03 BSD 058	03 BSD 078	03 BSD 038

PELLICLE BEAMSPLITTERS

A pellicle beamsplitter is made of a high tensile strength elastic membrane, stretched like a drumhead over a black anodized flat metal frame, and bonded to the precision lapped edge of this frame. The membrane has a thickness of 5 microns or less, so that the separation of the primary and ghost images at 45° angle of incidence is only about 7 microns, making them indistinguishable. Because the membrane is so very thin, pellicle beamsplitters have some advantages over plate beamsplitters: chromatic and spherical aberration in converging beams is negligible, absorption is very low, and ghost image problems are virtually eliminated.

Uncoated pellicles transmit about 92% of incident radiation, and reflect about 8%, throughout the visible and near-infrared spectral regions. The membrane material, nitrocellulose, does exhibit absorption in the ultraviolet, and in the infrared beyond two microns. In beamsplitting applications, coatings are deposited on one side of the membrane. Graphs showing the transmission and reflectance properties of these coatings follow. In convergent light, low-contrast interference fringes appear because of interference between reflections from the front and rear surfaces of the membrane. Because the reflectance of the metallic coated surface far exceeds that of the uncoated surface, the visibility or contrast of these fringes is very low. Pellicles are often applied to situations in which the light is uncollimated.

Pellicles are not affected by mechanical shock transmitted through the frame, or by climatic variations. They are, however, acoustically coupled to the atmosphere, and have low resonant frequencies. To avoid surface vibration they must be isolated from mechanical and acoustical disturbances.

For convenience of mounting, each frame is provided with two threaded holes placed diametrically opposite each other. Since the flatness of a pellicle is determined by the frame itself, pellicles should not be stressed or distorted by excessively firm mounting techniques.

Pellicle surfaces must not be touched! Clean only with a gentle flow of clean, dry air.

Each pellicle is supplied in a functional, protective plastic container.

Several different beamsplitting coatings may be ordered for these pellicles, and these are illustrated in the graphs. To order these coatings, append the indicated coating suffix to the product number. If you append no suffix, the pellicle will be shipped uncoated.

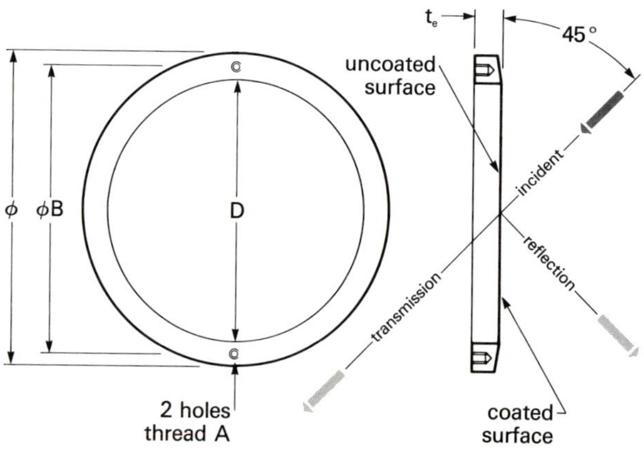

PRODUCT NUMBER	φ	D	t_e	thread A	φB
03 BPL 001	34.9	25.4	4.8	2-56	30.2 BCD
03 BPL 003	63.5	50.8	6.4	2-56	57.2 BCD
03 BPL 005	88.9	76.2	6.4	2-56	82.6 BCD
03 BPL 007	114.3	101.6	6.4	2-56	108 BCD
03 BPL 009	139.7	127	7.9	6-32	133.4 BCD
03 BPL 011	165.1	152.4	9.5	6-32	158.6 BCD

Pellicles are the best type of beamsplitters if you are concerned about ghost images.

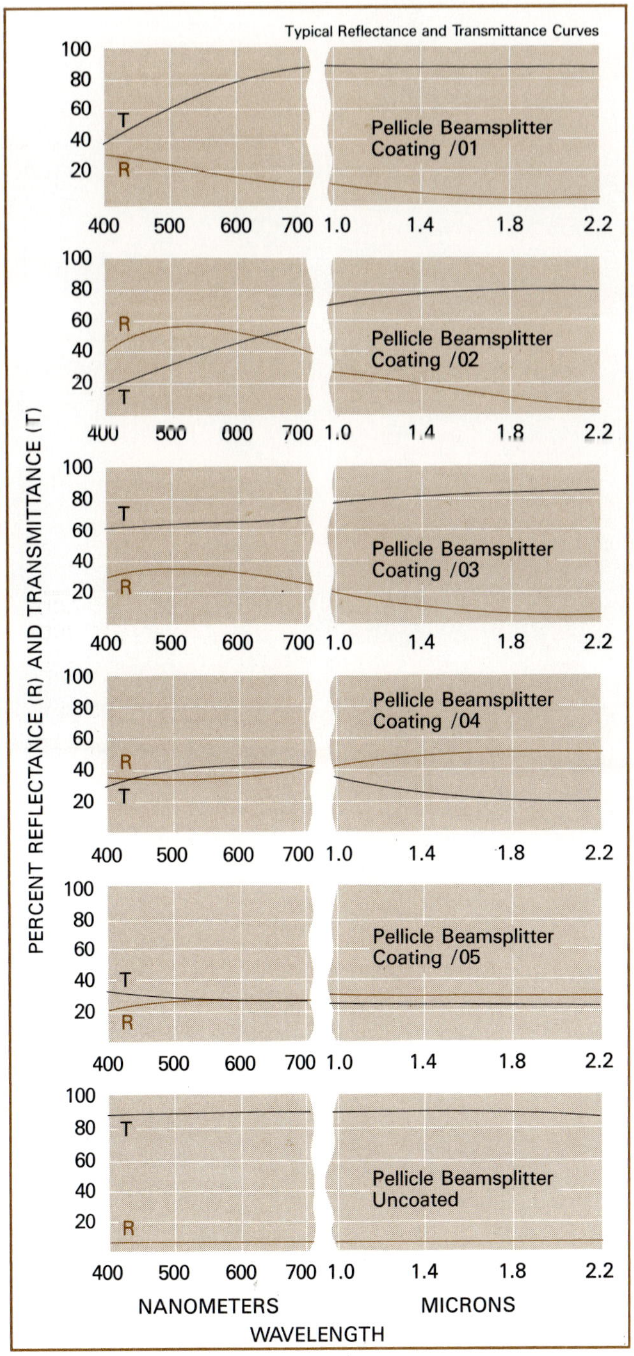

SPECIFICATIONS: PELLICLE BEAMSPLITTERS

Pellicle Thickness: ≤5 microns
Material: Select optical grade nitrocellulose
Index of Refraction (n_d): 1.5
Thickness Uniformity: 2λ per 25mm
Cosmetic Surface Quality: 40-20 scratch and dig
Temperature Range: –40°C to + 125°C
Humidity Range: ≤ 95% relative humidity
Frame: Hard aluminum alloy, black anodized

Pellicle Beamsplitters

D (mm)	ϕ (mm)	t_e (mm)	PRODUCT NUMBER
25.4	34.9	4.8	03 BPL 001
50.8	63.5	6.4	03 BPL 003
76.2	88.9	6.4	03 BPL 005
101.6	114.3	6.4	03 BPL 007
152.4	165.1	9.5	03 BPL 011

To order select pellicle and apped Coating Suffix from table.

Pellicle Beamsplitter Coatings

Nominal R/T Ratio 546nm	1.06μm	COATING SUFFIX
20/70	10/90	/01
60/40	30/70	/02
35/65	20/80	/03
40/40	40/40	/04
30/30	30/30	/05

These coatings are specific to pellicles. They can not be applied to other substrates.

OEM OPTICS

Much of our business is involved in the supply of optical components for use by original equipment manufacturers. Some of these components are the same as those listed in this Optics Guide; others are made to the manufacturer's specifications.

We are ready to meet your volume optics requirements, so please send us your requests for quotation. Here's how we can help the OEM:

- We offer highly competitive volume pricing.
- We accept yearly blanket orders with monthly or biweekly shipping schedules.
- We can analyze your design and present ways to reduce your component costs.
- We will adopt your QA procedures, thereby reducing your need for inspection.
- We carry back-up inventory to meet your "just-in-time" inventory requirements.

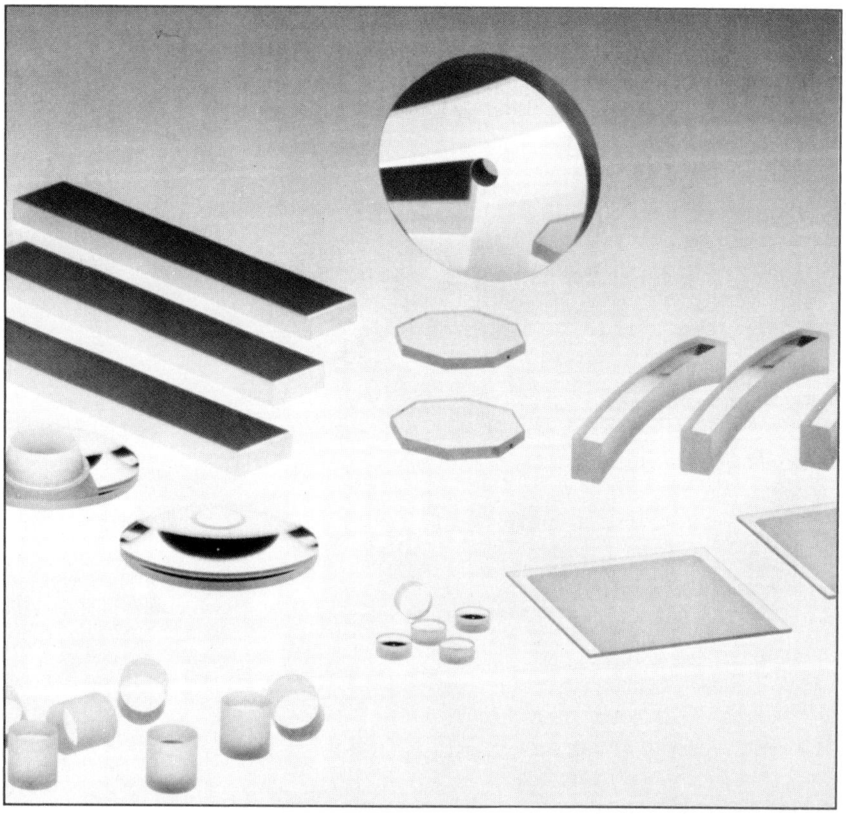

FILTERS

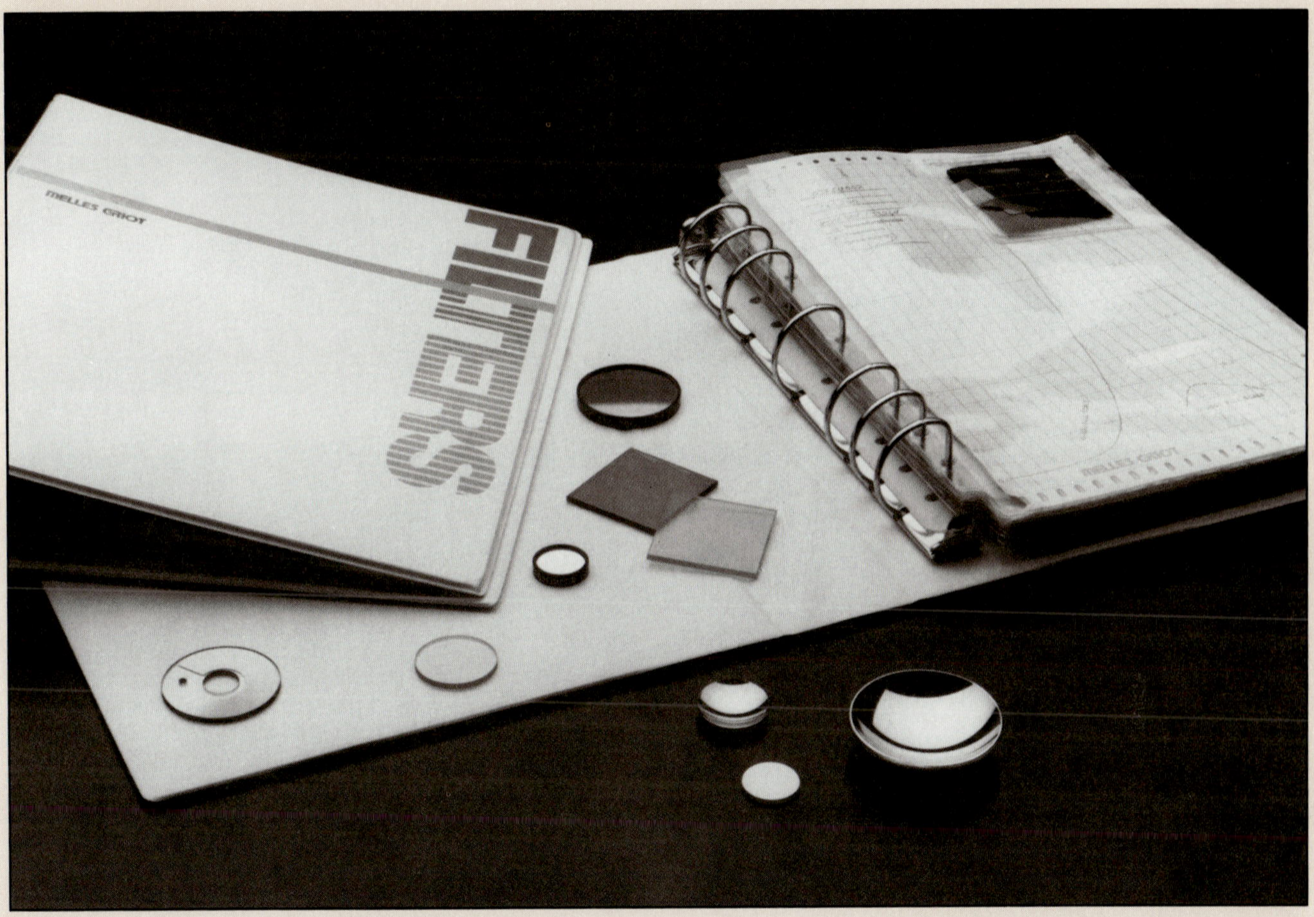

Here we describe a selection of colored glass, neutral density, and passband interference filters. All our filters are individually calibrated using a spectrophotometer. The calibration curves are supplied at no extra charge. We also offer sets of filters, together with their calibration curves, in a convenient storage binder.

Contents **Page Number**

HEAT-ABSORBING FILTERS

HEAT-CONTROLLING ELEMENTS

These elements are used in illumination systems to reduce local operating temperatures to acceptable values. This is done by separating infrared from visible radiation and either blocking infrared from places where it is unwanted, or diverting it to places from which the resulting heat may be dissipated. These elements are of two distinct types. One type is made of heat-absorbing or infrared opaque glass, while the other is a spectrally selective reflector (hot or cold mirror) based upon multilayer thin-film principles. These principles are explained later in connection with interference filters.

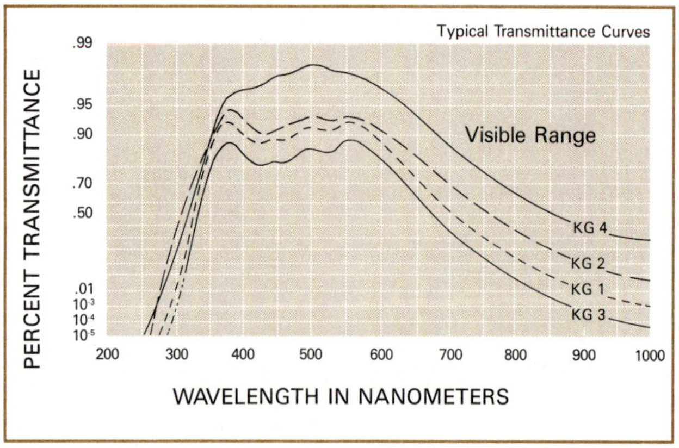

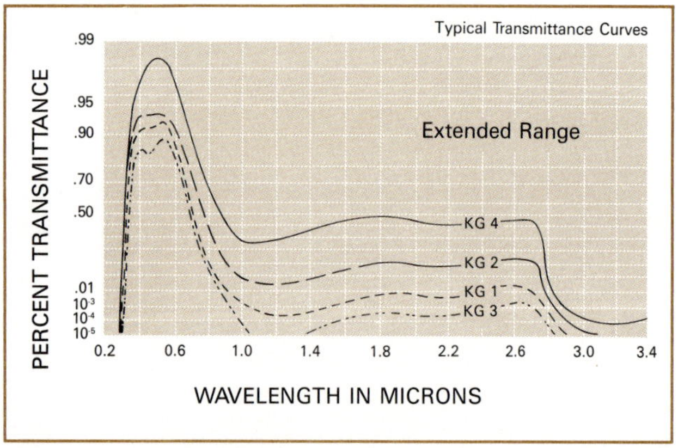

HEAT-ABSORBING FILTERS

The graphs show the transmittance of Schott heat-absorbing glass types KG1, KG2, KG3, and KG4. The curves are for thicknesses of 3mm, with reflectance losses at the surfaces not considered, i.e., these are internal transmittance curves for 3mm thickness. The correction factor, when multiplied by an internal transmittance curve, yields an external transmittance curve which does account for reflectance losses. The correction factor applies throughout the visible spectrum, within which the refractive indices vary only slightly, and is valid for arbitrary thicknesses. Such correction factors will be offered in connection with other filters, made of Schott glasses. The symbolism t_1t_2 for the correction factor is explained later in connection with colored glass filters, and has previously appeared in the discussion of lens material properties.

These filters are made to precise specifications of parallelism and their surface quality is superior to common practice for heat-absorption elements.

SPECIFICATIONS: HEAT-ABSORBING FILTERS

Correction Factor (t_1t_2): 0.92
Dimensions: 50 x 50 x 3 (± 0.25) mm
Parallelism: 2 arc minutes
Material: Schott KG glass
Cosmetic Surface Quality: Pitch-polished, 80-50 scratch and dig
Suggested Maximum Operating Temperature: 250 °C (482 °F)

Heat-Absorbing Filters

Schott Glass Type	PRODUCT NUMBER
KG 1	03 FCG 163
KG 2	03 FCG 165
KG 3	03 FCG 167
KG 4	03 FCG 169

CONCAVE COLD MIRRORS

Concave cold mirrors are used in illumination and projection systems, where they simultaneously increase the illumination of, and reduce the heat load on, the subject of illumination (often a film gate). They are ground and polished first-surface spherical mirrors, intended for use behind a source with the source at the center of curvature (see illustrations of projection systems with aspherics and mounted condensers — pages 164 and 169). The first surface is coated with a special multilayer dielectric coating which has high transmittance for near-infrared radiation, but high reflectance for visible light (see curve). The rear surface is also polished, to maximize the efficiency with which infrared radiation is transmitted. Average reflectance through the visible is in excess of ninety-seven percent. Heat loads on heat-absorbing glass filters elsewhere in the system are dramatically reduced. Concave cold mirrors are listed below by their paraxial focal lengths (f), which are half of their radii of curvature. Cold mirrors with other diameters and focal lengths are available on request.

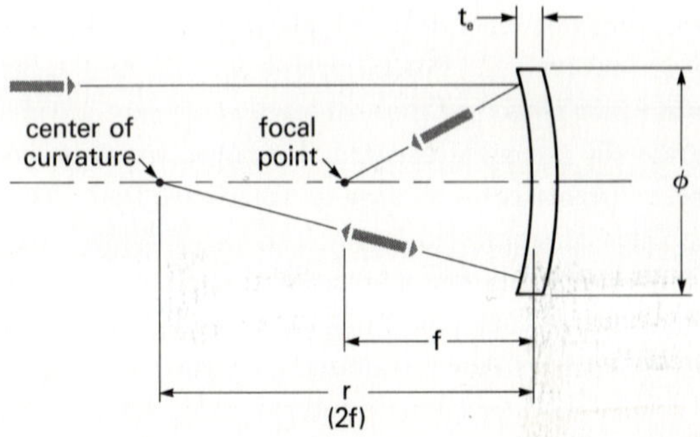

01 MCC

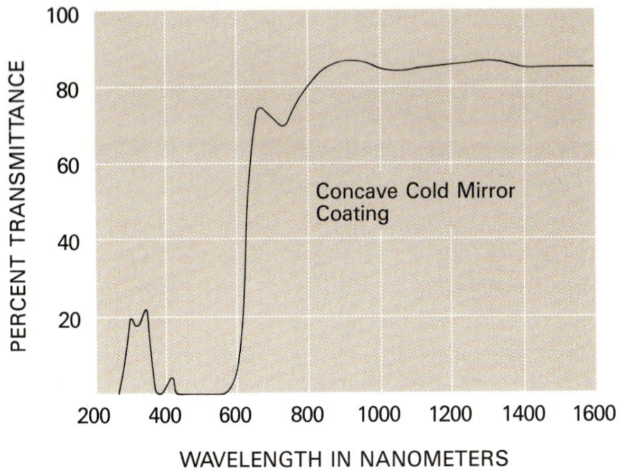

SPECIFICATIONS: CONCAVE COLD MIRRORS

Focal Length: ±1mm
Dimensions: ±0.2mm
Substrate: Optical crown
Cosmetic Surface Quality: 80-50 scratch and dig
Coating: Multilayer dielectric

Concave Cold Mirrors

f (mm)	ϕ (mm)	t_e (mm)	PRODUCT NUMBER
12.5	30.0	2.1	03 MCC 001
15.0	30.0	2.5	03 MCC 003
15.0	40.0	2.5	03 MCC 005
21.0	46.0	2.5	03 MCC 007
21.0	50.0	3.0	03 MCC 009
21.5	65.0	3.0	03 MCC 011

Cold mirrors with other diameters and focal lengths are readily produced. We will be pleaed to quote for your specific requirements in both large and small quantities. For further information, please contact one of our Applications Engineers.

PRECISION METALLIC NEUTRAL-DENSITY FILTER SETS

Neutral-density filter sets make possible the attenuation, comparison, splitting, or combining of beams in a wide range of irradiance ratios without significant change in chromaticity. These carefully prepared sets find wide application in the precise attenuation or control of light. For example, beams can be attenuated to levels at which photometers or radiometers are most accurate and linear, thereby extending the useful range of such instruments. Initial beam irradiances can then be calculated from accurately known filter densities.

Each filter set is supplied in a special binder containing both the filters and the individual spectrophotometer curves recorded from the filters. Each set includes a blank (uncoated) substrate, of the same material and thickness as used for other filters in the set, which is often helpful in the alignment and focusing of an optical system. Circular neutral density wedges are supplied with individual microdensitometer curves instead of spectrophotometer curves.

All of our neutral-density filters pass stringent optical and mechanical tests using the most modern test equipment. Our sets and replacement filters are standard and in inventory. Our Applications Engineers will be pleased to assist you in the selection and application of either standard or custom filters.

Optical density (D) is defined as the base 10 logarithm of the reciprocal of transmittance (T):

$$D = \log_{10}(1/T), \text{ or } T = 10^{-D}.$$

The definition of optical density is analogous to the definition of decibel as used in electronics. That is, densities of neutral-density filters used in combinations are *additive* provided multiple reflections between filters do not occur in the direction of interest. The reciprocal of the transmittance, 1/T, is called the opacity. Also in widespread use is the relative optical density D_r, defined as the difference between the density D of a coated substrate and the density D_o of an uncoated region of the same substrate:

$$D_r = D - D_o, \text{ or } D = D_r + D_o$$

In terms of the substrate refractive index n,

$$D_o = 2 \log_{10} \frac{(n+1)^2}{4n},$$

which at 550nm is typically about 0.0376 for BK 7, and about 0.0309 for synthetic fused silica. Relative Denisty D_r, and not the absolute density D, is the quantity which appears on the individual microdensitometer traces supplied with our circular neutral density wedge filters. The reason for this is that in many instrumental applications of wedge filters there is a requirement for constancy of focal plane position or constancy of plate aberrations, and this requirement prohibits removal or perforation of the substrate as would be necessary to achieve accurately the value D = 0.0. In such instruments the density differences, and not the densities themselves, are important.

Transmittance and density values may, like reflectance values, refer to either small angular fields (specular or undeviated values) or very large angular fields (diffuse or hemispherical values). Measurements which determine hemispherical values include both specular and scattered contributions. Density and relative density values which we publish for neutral-density filters are specular values based on external transmittance.

Because of Beer's Law, and because of Fechner's Law (sensation proportional to logarithm of stimulus, applicable to vision as a special case), it has been historically convenient to use the logarithmic density scale, instead of a transmittance scale. While optical density is dimensionless, the notation 0.50D is sometimes used to mean 0.50 "density units," or simply a density of 0.50.

All of our neutral-density filters are made by vacuum deposition of a thin film of several special metallic alloys on a substrate of BK 7 grade A fine annealed glass or optical quality synthetic fused silica. The compositions of these alloys have been chosen to result in a spectral density curve which is a flatter over a wide spectral range than the curves of most pure metals. The spectral density curves shown illustrate this flatness. Substrate materials are chosen for their homogeneity, uniformity of transmittance, finishing characteristics, and (in the case of synthetic fused silica) ultraviolet transmittance.

Our substrates are pitch-polished to minimize light scattering. Neutral-density filters can be used at any wavelength between 200nm and 2500nm. In their operation they depend both on absorption in, and reflection from, the thin metallic film.

Neutral-density filters can be used in combinations of two or more at a time to achieve values of transmittance or density not otherwise conveniently available. If they are arranged so that multiple reflections between them do not occur in the direction of interest, transmittances are multiplicative, and optical densities are *additive*. By combining the various filters in the set, fifty-five separate density values may be achieved.

When used in high-intensity beams, neutral-density filters should be oriented with the metallic film facing toward the source, so as to minimize absorption and heating in the substrate. Our alloy films are corrosion-resistant and do not age at normal temperatures. Adhesion of alloy films to their substrates is tenacious, and unaffected by moisture and most solvents throughout the temperature range from -73°C (-100°F) to + 150°C (+302°F). Exposure to higher temperatures should be avoided, as film oxidation and increased transmittance will result.

FILTER SET CONTENTS

Accurate spectrophotometer recordings of optical density (D) accompany every neutral density filter in each neutral-density filter set. Each individual filter curve is the original curve. Wavelength ranges are from 200 to 700nm for sets on synthetic fused silica, and from 350 to 700nm for sets on BK 7 substrates. Curves extending beyond 700nm into the infrared as far as 2.5 microns are available on special request.

Each set includes a blank (uncoated) substrate of the same material and thickness as used for other filters in the set. This blank filter is often very helpful in the alignment and focusing of an optical system, before inserting the required neutral-density filter.

Each neutral-density filter set is packaged in a convenient book-like binder. Plastic pages of these binders display and protect the individual spectrophotometer curves recorded for the filters in the set. There is one such page for each filter. Each plastic page also features a special transparent pocket which securely holds the actual filter from which the spectrophotometer curve of that page was recorded. These pages keep both the filter and its curve conveniently together. Filters can easily be removed from their plastic pages, if necessary, so that detailed measurements, calculations, and photocopies may be made. Each filter, and the curve recorded from it, is permanently marked with the nominal density value at 550nm. This identification serves to preserve the correspondence between curve and filter within the set.

Spectrophotometer curves are recorded at room temperature on a Varian-Cary 14 or similar instrument, calibrated with standards directly traceable to the U.S. National Bureau of Standards.

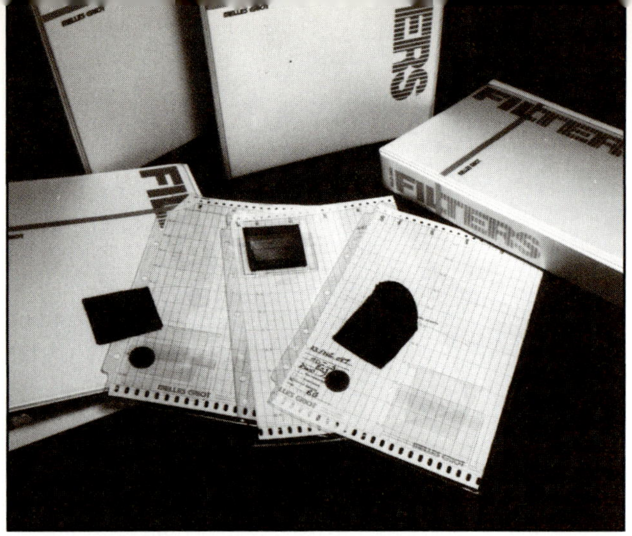

NEUTRAL-DENSITY FILTER SETS

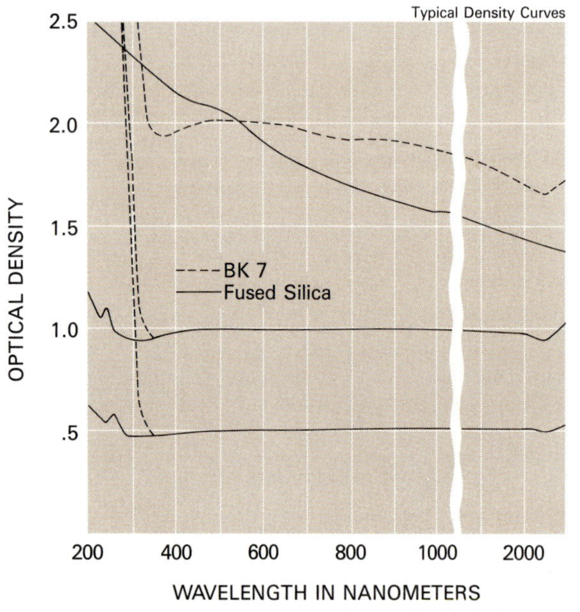

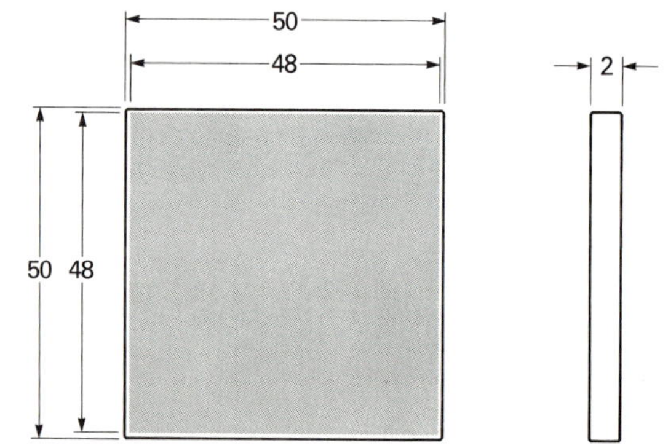

03 FSG/FSQ

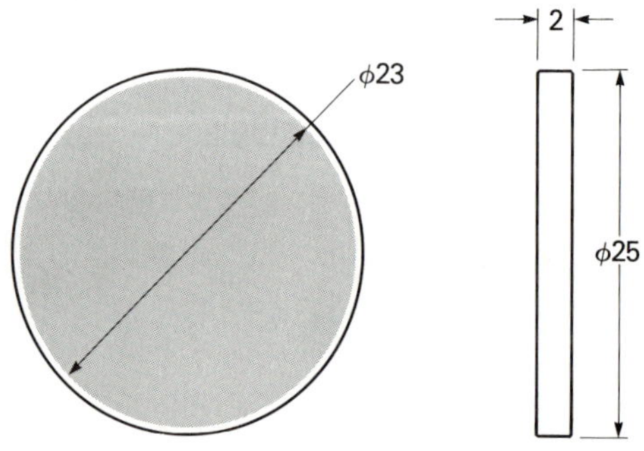

03 FSG/FSQ

SPECIFICATIONS: PRECISION METALLIC NEUTRAL-DENSITY FILTER SETS

Surface Flatness: 2λ per 25mm

Deviation allowed from catalog nominal density at 550nm: ± 0.02 density for densities of less than 0.5; ± 5% of density value at or above density 0.5

Accuracy of spectrophotometer curves between 200nm and 700nm: ± 1% of full scale

Clear Apertures: 23mm ϕ and 48x48mm

Dimensions: ±0.2mm

Thickness: 2mm

Parallelism: 3 arc minutes

Substate Material: BK 7 grade A fine annealed or optical quality synthetic fused silica

Cosmetic Surface Quality: Pitch-polished, 80-50 scratch and dig

Coatings: Vacuum-deposited metallic alloy

See page 458 for a convenient filter holder which permits stacking of these filters.

Precision Metallic Neutral-Density Filter Sets

Included Nominal Densities (D)	Number of Filters*	Size Shape	Sub-strate Material	PRODUCT NUMBER
0.04* 0.1 0.3 0.5 1.0 2.0 3.0	7	25mm Round	BK 7	03 FSG 013
0.04* 0.1 0.3 0.5 1.0 2.0 3.0	7	50mm Square	BK 7	03 FSG 017
0.03* 0.1 0.3 0.5 1.0 2.0 3.0	7	25mm Round	Fused Silica	03 FSQ 011
0.03* 0.1 0.3 0.5 1.0 2.0 3.0	7	50mm Square	Fused Silica	03 FSQ 015

*This includes a blank substrate of BK 7 (D_o = 0.0376) or fused silica (D_o = 0.0309).

Neutral density filters with different sizes, shapes and/or substrate material can be readily supplied in both large and small quantities. Please request a quotation from one of our Applications Engineers.

FILTER BINDERS AND STORAGE PAGES

You will find our filter binder system to be both protective and convenient. Double-pocketed vinyl pages hold one 50mm x 50mm (or smaller) filter and the corresponding spectrophotometer or microdensitometer curve. Individual pages, complete with filters and curves, may be assembled into our filter binders for convenient storage. The filter pocket location is varied on each page to facilitate page stacking. Filter storage pages are supplied in packages of twelve.

Filter Binder

	PRODUCT NUMBER
	03 FBB 001

Filter Storage Pages

	PRODUCT NUMBER
Set of 12	03 FBP 001

INDIVIDUAL NEUTRAL-DENSITY FILTERS

For the benefit of those who may wish to replace precision metallic neutral-density filters which have been lost, and for those who may wish to augment their sets, we offer single filters having the nominal densities at 550nm as indicated. Each filter is permanently marked with the nominal density at 550nm, and is supplied with a spectrophotometer curve recorded between 700nm and either 200nm or 350nm (depending on the substrate material). Each filter and its curve is supplied in a protective plastic page which may be bound into any of our filter sets. Filter binders and further individual plastic pages may be purchased separately. They will provide you with a convenient way to store existing neutral-density filters, colored glass filters, interference filters, and their curves, for easy use and lasting protection.

Replacement and Supplemental Neutral-Density Filters

Nominal Density at 550nm (D)	Percent Transmittance (%T)	PRODUCT NUMBER: BK 7 25mm Round	BK 7 50mm Square	Fused Silica 25mm Round	Fused Silica 50mm Square
0.04	91.20	03 FNG 040*	03 FNG 002*	03 FNQ 040*	03 FNQ 002*
0.1	79.43	03 FNG 041	03 FNG 003	03 FNQ 041	03 FNQ 003
0.2	63.10	03 FNG 043	03 FNG 005	03 FNQ 043	03 FNQ 005
0.3	50.12	03 FNG 045	03 FNG 007	03 FNQ 045	03 FNQ 007
0.4	39.81	03 FNG 089	03 FNG 083	03 FNQ 089	03 FNQ 083
0.5	31.62	03 FNG 099	03 FNG 085	03 FNQ 099	03 FNQ 085
0.6	25.12	03 FNG 047	03 FNG 009	03 FNQ 047	03 FNQ 009
0.7	19.95	03 FNG 049	03 FNG 011	03 FNQ 049	03 FNQ 011
0.8	15.85	03 FNG 055	03 FNG 013	03 FNQ 055	03 FNQ 013
0.9	12.59	03 FNG 111	03 FNG 087	03 FNQ 111	03 FNQ 087
1.0	10.00	03 FNG 057	03 FNG 015	03 FNQ 057	03 FNQ 015
1.3	5.012	03 FNG 061	03 FNG 019	03 FNQ 061	03 FNQ 019
1.5	3.162	03 FNG 113	03 FNG 115	03 FNQ 113	03 FNQ 115
2.0	1.000	03 FNG 065	03 FNG 023	03 FNQ 065	03 FNQ 023
2.5	0.3162	03 FNG 067	03 FNG 025	03 FNQ 067	03 FNQ 025
3.0	0.1000	03 FNG 069	03 FNG 027	03 FNQ 069	03 FNQ 027
4.0	0.0100	03 FNG 077	03 FNG 029	03 FNQ 077	03 FNQ 029

*These filters are uncoated substrates. Absolute density (D) is 0.0376 for BK 7 and 0.0309 for Fused Silica. These units will be found to be useful for system adjustment prior to inserting required neutral-density filters.

A convenient filter holder is described on page 458.

CIRCULAR LINEAR-WEDGE NEUTRAL-DENSITY FILTERS

Circular linear-wedge neutral-density filters are neutral-density filters, circular in shape, with relative density (D_r) an accurate linear function of angle as measured from a reference radius. Any desired density between the limits for the filter can be coated in a 300° sector adjoining the reference radius. Nominal relative density ranges are specified by appending one of five suffixes (B through F) to the wedge filter product number. Each wedge filter is packaged with a microdensitometer recording of its relative density profile, is of the metallic-film type, and is of protective laminated construction for durability and permanence. From the microdensitometer tracing, relative density upon any radius line within the coated sector can be accurately determined.

Relative density can be converted to absolute density by adding $D_0 = 0.038$.

Densities of unknown samples can immediately be determined from the position angle at which filter density and unknown sample density are photoelectrically equal.

Circular linear-wedge neutral-density filters are widely used in spectrophotometers and other instruments, where they are typically in continuous rotation about their centers. Photoelectric equality of filter and unknown densities triggers readout of a shaft-angle encoder. From this readout the unknown density is automatically computed in each rotation. For such applications, beams are often focused down to narrow radial lines at the plane of the filter. The cylindrical lenses listed previously are very effective in transforming light from a point source or beam into a line image.

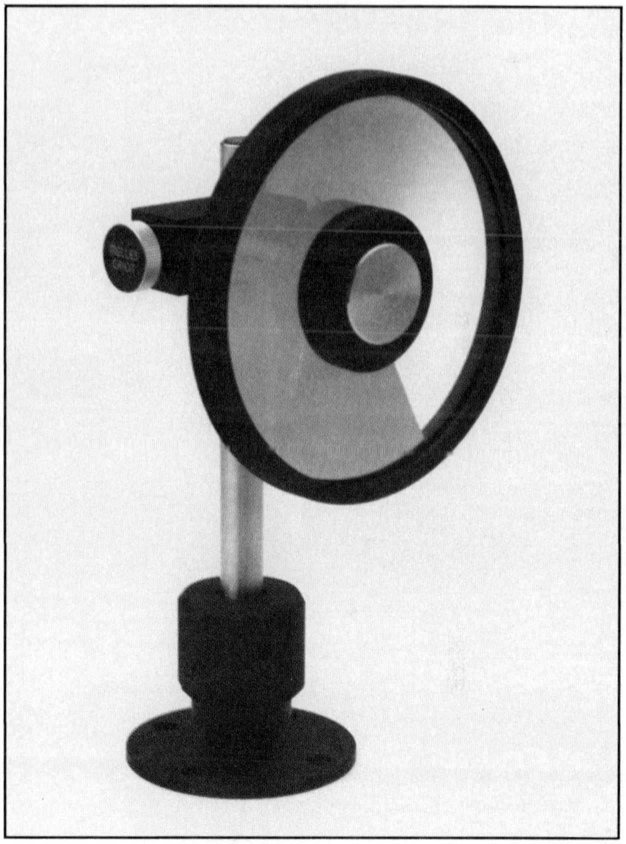

A convenient mount which permits rotation of the filters to any desired density position is offered on page 459.

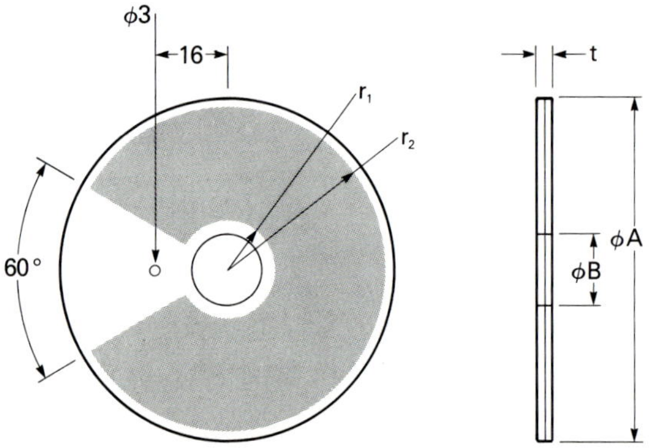

PRODUCT NUMBER	ϕA	ϕB	r_1	r_2	t
03 FDC 009	25	6	5	12	3
03 FDC 001	50	15	10	24	3
03 FDC 003	100	15	13	48	2
03 FDC 005	150	15	23	73	2
03 FDC 007	200	15	23	98	2

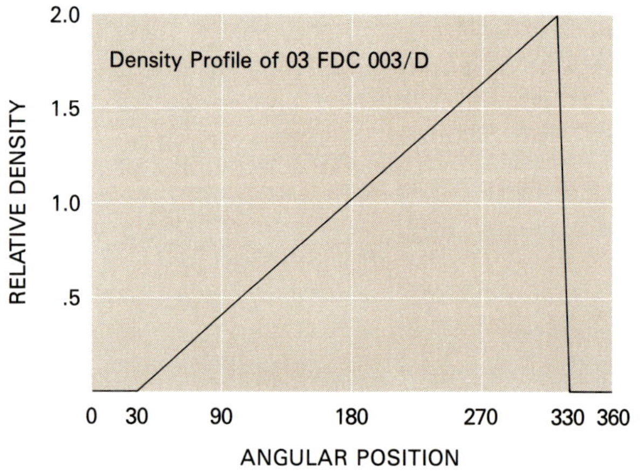

Circular Linear-Wedge Neutral-Density Filters

Diameter (mm)	t (mm)	PRODUCT NUMBER
25	3	03 FDC 009
50	3	03 FDC 001
100	2	03 FDC 003
150	2	03 FDC 005
200	2	03 FDC 007

To order select filter Product Number and append Coating Suffix for desired relative density range.

Relative Density Range Coatings

Relative Density (D_r) Range	COATING SUFFIX
0.0 to 1.0	/B
0.0 to 1.5	/C
0.0 to 2.0	/D
0.0 to 3.0	/E
0.0 to 4.0	/F

SPECIFICATIONS: CIRCULAR LINEAR-WEDGE NEUTRAL-DENSITY FILTERS

Accuracy of Recorded Density: ±2% of full scale (microdensitometer curve supplied with each filter)

Linearity of relative density wedge profile (relative density vs. angle): ±7% of maximum value

Deviations allowed from catalog nominal relative density at range endpoints: ±5%

Dimensions: ± 0.2mm

Coated Sector Angular Width: 300°

Parallelism: 3 arc minutes

Cosmetic Surface Quality: Pitch-polished, 80-50 scratch and dig

Coating: Vacuum-deposited metallic alloy

Construction: Vacuum-deposited film is laminated between two pieces of polished optical crown.

See page 252 for individual neutral density filters.

COLORED GLASS FILTERS

The colored glass filters we keep in inventory are made from the highest quality Schott colored filter glass. The spectral properties of these filters are uniform over their entire aperture, and are invariant with time at temperatures below 250 °C (482 °F). Spectral properties of these filters are insensitive to angle of incidence. Schott glass types are indicated below along with our product numbers.

Standard filters are 50mm square by 3mm thick (other sizes and shapes upon request), with both surfaces ground and pitch-polished. The parallelism is held to within 2 arc minutes. The 3mm thickness assures ample breakage resistance and guarantees adequate short-wavelength blockage for the sharp cutoff filters.

TRANSMITTANCE CURVE INTERPRETATION

At any specific wavelength the external transmittance T of a colored glass filter is given by the approximate formula

$$T = t_1 t_2 e^{-\beta cx}$$

where t_1 is the transmittance of the first air-glass interface, t_2 is the transmittance of the second air-glass interface, e is the base of the natural system of logarithms, β is a positive constant determined by wavelength, c is the concentration of pigmentation in the glass, and x is the filter thickness. This formula neglects possible multiple reflections which may occur between the interfaces. The exponential

$$e^{-\beta cx}$$

is known as the internal transmittance. The interface transmittances t_1 and t_2 are equal, and their product t_1t_2 is called the correction factor. The correction factor is related to the refractive index n of the glass by

$$t_1 t_2 = 1 - 2\left(\frac{n-1}{n+1}\right)^2 + \left(\frac{n-1}{n+1}\right)^4$$

The nominal filter transmittance curves shown here are graphs of internal transmittance versus wavelength for a thickness, x, of 3mm. This is the thickness of our standard filters. Thus the curves neglect all effects of reflections at the air-glass interfaces. Values of the correction factor also appear below, and can be used to determine nominal external transmittances. Nominal internal transmittance graphs have been included in this catalog primarily for the convenience of those of you who may need to calculate nominal external transmittances at non-standard thicknesses.

All colored glasses exhibit melt-to-melt variations of transmittance. For the sharp-cutoff glasses, curves may deviate ±7.0nm from our published nominal curves. Cary 14 spectrophotometer external transmittance curves are supplied with each filter.

TEMPERATURE DEPENDENCE

Sharp-cutoff glasses exhibit a reversible shift of the absorption edge position (of which the half-power point is a convenient indicator) as the temperature is increased. The shift is roughly constant over the range from 10° to 90°C, and is tabulated (for sharp-cutoff filters only) in mm/°C under the heading $\Delta\lambda/\Delta T$. The shift is in the direction of longer wavelengths as the temperature is increased.

ADVANTAGES OF MELLES GRIOT COLORED GLASS FILTERS

1. All colored glass filters are supplied with individual Varian-Cary 14 or equivalent spectrophotometer curves. These provide precise calibration for serious measurement and future reference.
2. Curves and filters are supplied in our specially designed protective vinyl binder page. Binders are available separately to allow you to assemble convenient sets.
3. Each filter is edge-marked with the last three digits of our product number for identification. The corresponding number is marked on each curve.

SPECIFICATIONS: COLORED GLASS FILTERS

Spectral Curves: Varian-Cary 14 or equivalent spectrophotometer curves are supplied with each filter. Accuracy of transmittance curves is ±2% of full scale.

Dimensions: 50 x 50 x 3 (±0.25)mm

Parallelism: 2 arc minutes

Material: Schott colored glass

Cosmetic Surface Quality: Pitch-polished, 80-50 scratch and dig

Suggested Maximum Service Temperature: 250 °C (482 °F). Irreversible transmittance changes may occur at higher temperatures.

Identification: Each filter is marked with the last three digits of its product number.

Packaging: Each filter and its curve are supplied in a protective vinyl binder page. Binders are available separately.

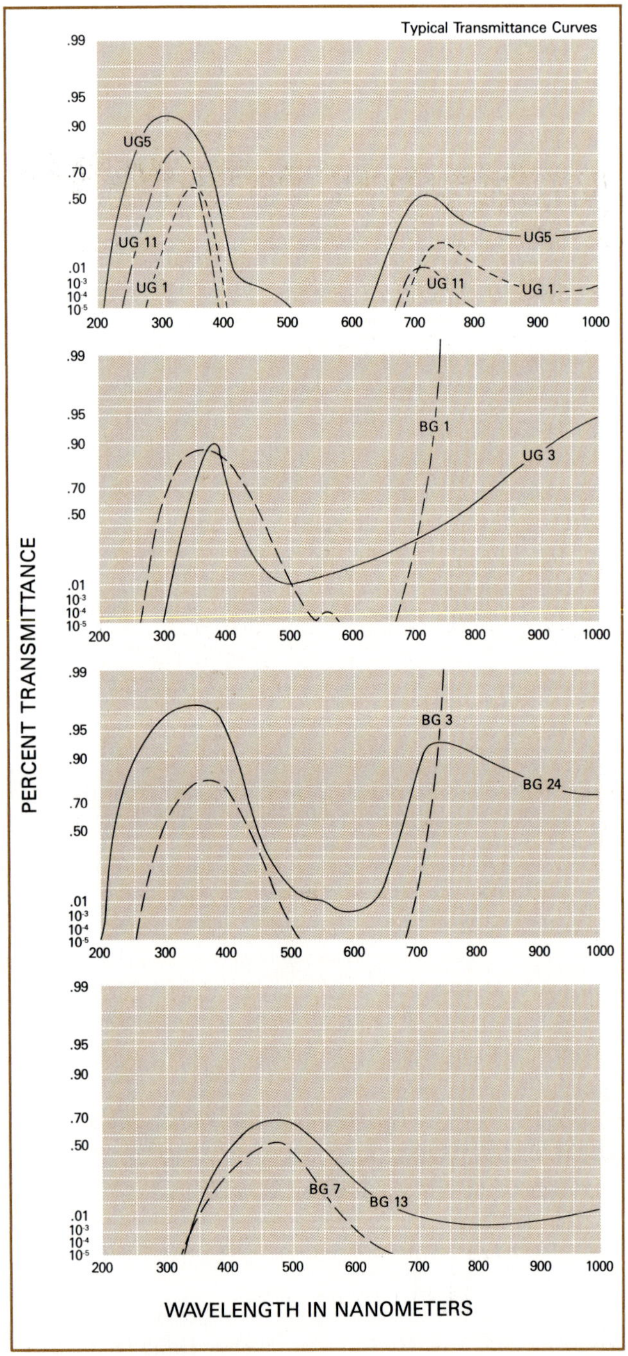

UV Transmitting Black Glass Filters

Schott Glass Type	Correction Factor (t_1t_2)	PRODUCT NUMBER
UG 5	0.91	03 FCG 177
UG 11	0.91	03 FCG 179
UG 1	0.91	03 FCG 001

UV and Red Transmitting Blue-Violet Glass Filters

Schott Glass Type	Correction Factor (t_1t_2)	PRODUCT NUMBER
BG 1	0.92	03 FCG 007
UG 3	0.90	03 FCG 003

UV and Red Transmitting Blue-Violet Glass Filters

Schott Glass Type	Correction Factor (t_1t_2)	PRODUCT NUMBER
BG 24	0.915	03 FCG 187
BG 3	0.92	03 FCG 009

Red Absorbing Blue Glass Filters

Schott Glass Type	Correction Factor (t_1t_2)	PRODUCT NUMBER
BG 13	0.915	03 FCG 189
BG 7	0.92	03 FCG 011

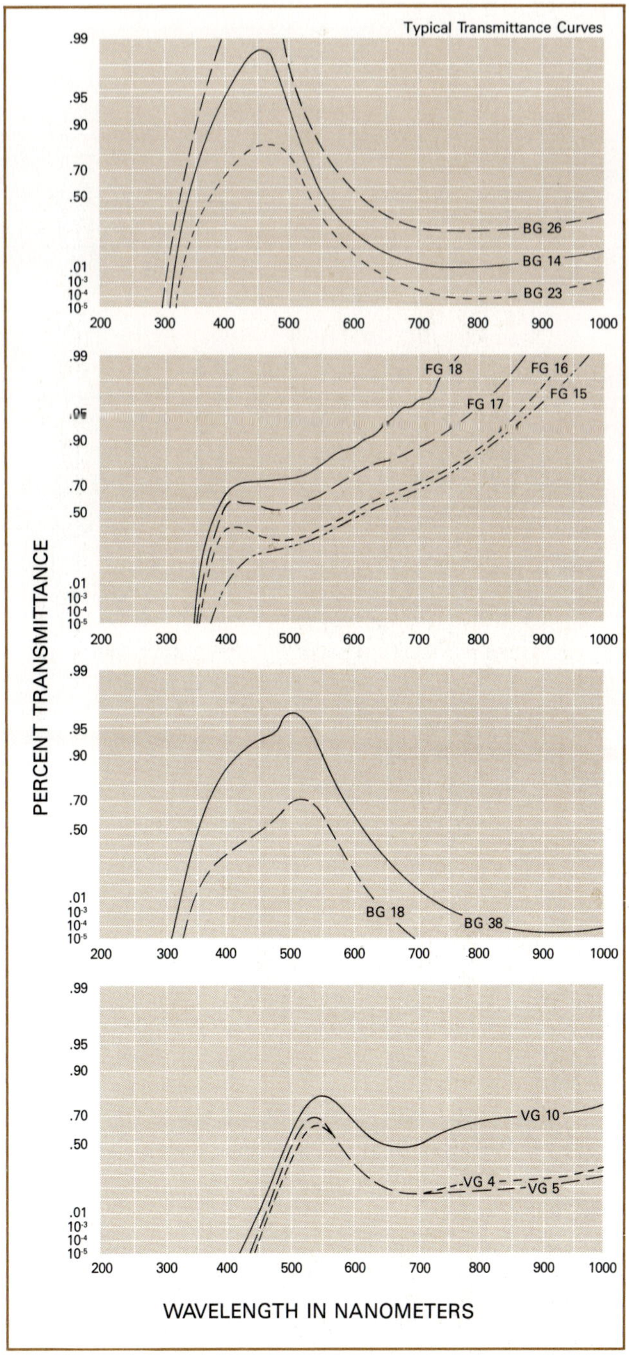

Red Absorbing Blue Glass Filters

Schott Glass Type	Correction Factor (t_1t_2)	PRODUCT NUMBER
BG 26	0.92	03 FCG 023
BG 14	0.92	03 FCG 015
BG 23	0.92	03 FCG 019

Amber Tinted Glass Filters; for Color Conversion

Schott Glass Type	Correction Factor (t_1t_2)	PRODUCT NUMBER
FG 15	0.896	03 FCG 144
FG 16	0.914	03 FCG 146
FG 17	0.925	03 FCG 148
FG 18	0.915	03 FCG 158

Red Absorbing Blue-Green Glass Filters

Schott Glass Type	Correction Factor (t_1t_2)	PRODUCT NUMBER
BG 38	0.915	03 FCG 033
BG 18	0.91	03 FCG 014

Green Glass Filters

Schott Glass Type	Correction Factor (t_1t_2)	PRODUCT NUMBER
VG 10	0.91	03 FCG 043
VG 5	0.91	03 FCG 037
VG 4	0.915	03 FCG 035

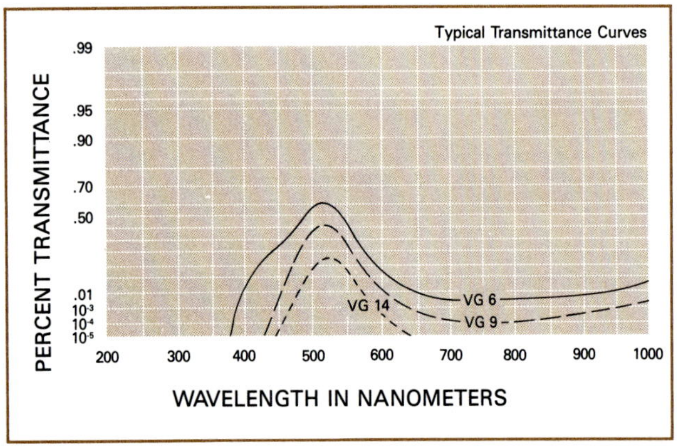

Green Glass Filters

Schott Glass Type	Correction Factor (t_1t_2)	PRODUCT NUMBER
VG 6	0.91	03 FCG 039
VG 9	0.91	03 FCG 041
VG 14	0.91	03 FCG 045

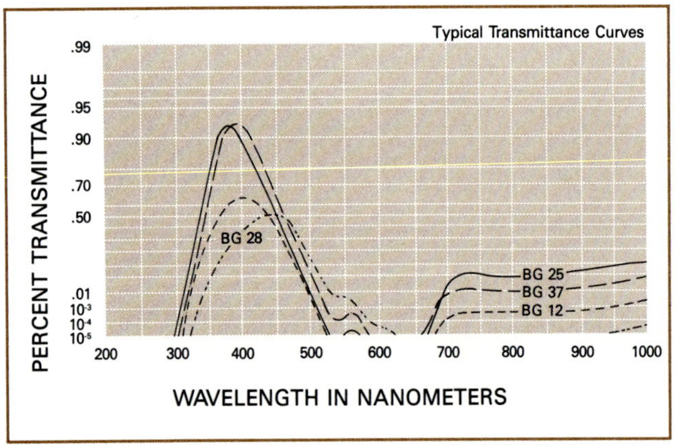

Red Absorbing Blue-Violet Glass Filters

Schott Glass Type	Correction Factor (t_1t_2)	PRODUCT NUMBER
BG 25	0.92	03 FCG 021
BG 37	0.92	03 FCG 029
BG 12	0.92	03 FCG 013
BG 28	0.92	03 FCG 025

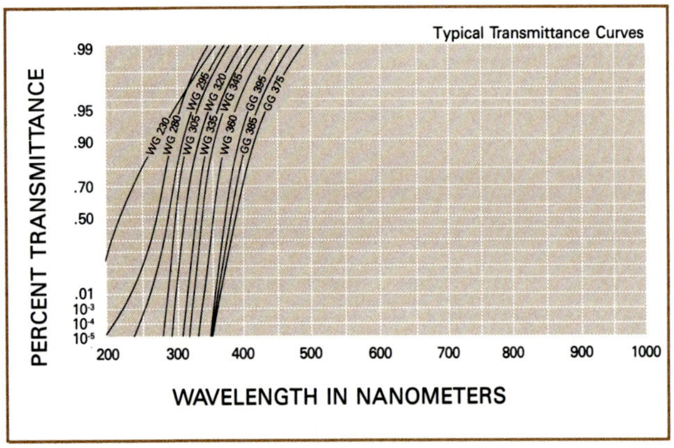

Colorless Sharp Cut-Off Glass Filters

Schott Glass Type	$\frac{\Delta\lambda}{\Delta T}$	Correction Factor (t_1t_2)	PRODUCT NUMBER
WG 230	0.02	0.915	03 FCG 198
WG 280	0.02	0.925	03 FCG 117
WG 295	0.03	0.92	03 FCG 119
WG 305	0.03	0.92	03 FCG 121
WG 320	0.05	0.91	03 FCG 123
WG 335	0.06	0.89	03 FCG 125
WG 345	0.07	0.885	03 FCG 127
WG 360	0.08	0.885	03 FCG 129
GG 375	0.06	0.92	03 FCG 047
GG 385	0.07	0.905	03 FCG 049
GG 395	0.08	0.895	03 FCG 055

$\frac{\Delta\lambda}{\Delta T}$ is the temperature coefficient of half-power point shift, in nm/ °C. See text.

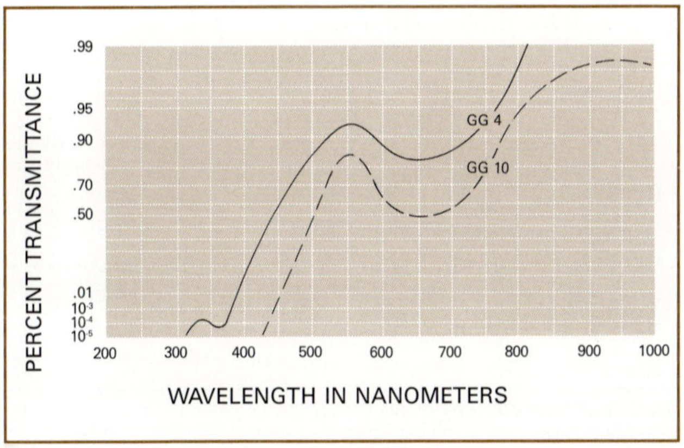

Greenish-Yellow Glass Filters

Schott Glass Type	Correction Factor (t_1t_2)	PRODUCT NUMBER
GG 4	0.915	03 FCG 069
GG 10	0.92	03 FCG 079

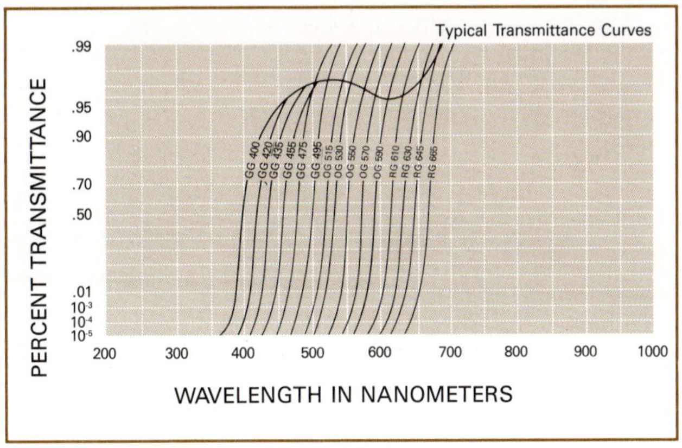

Yellow, Orange and Red Sharp Cut-Off Glass Filters

Schott Glass Type	$\frac{\Delta\lambda}{\Delta T}$	Correction Factor (t_1t_2)	PRODUCT NUMBER
GG 400	0.07	0.91	03 FCG 057
GG 420	0.07	0.91	03 FCG 059
GG 435	0.07	0.91	03 FCG 061
GG 455	0.08	0.915	03 FCG 063
GG 475	0.09	0.915	03 FCG 065
GG 495	0.10	0.915	03 FCG 067
OG 515	0.11	0.915	03 FCG 083
OG 530	0.12	0.915	03 FCG 085
OG 550	0.13	0.915	03 FCG 087
OG 570	0.14	0.915	03 FCG 089
OG 590	0.15	0.915	03 FCG 098
RG 610	0.16	0.915	03 FCG 101
RG 630	0.17	0.915	03 FCG 103
RG 645	0.17	0.915	03 FCG 105
RG 665	0.17	0.915	03 FCG 107

$\frac{\Delta\lambda}{\Delta T}$ is the temperature coefficient of half-power point position shift, in nm/°C. See text.

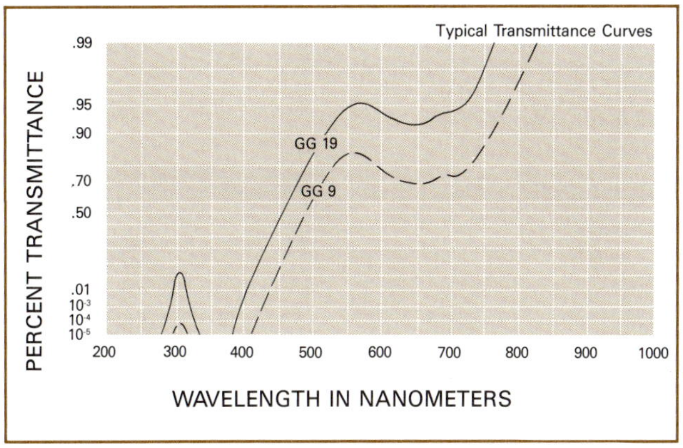

UV Transmitting Greenish Yellow Glass Filters

Schott Glass Type	Correction Factor (t_1t_2)	PRODUCT NUMBER
GG 19	0.915	03 FCG 201
GG 9	0.915	03 FCG 077

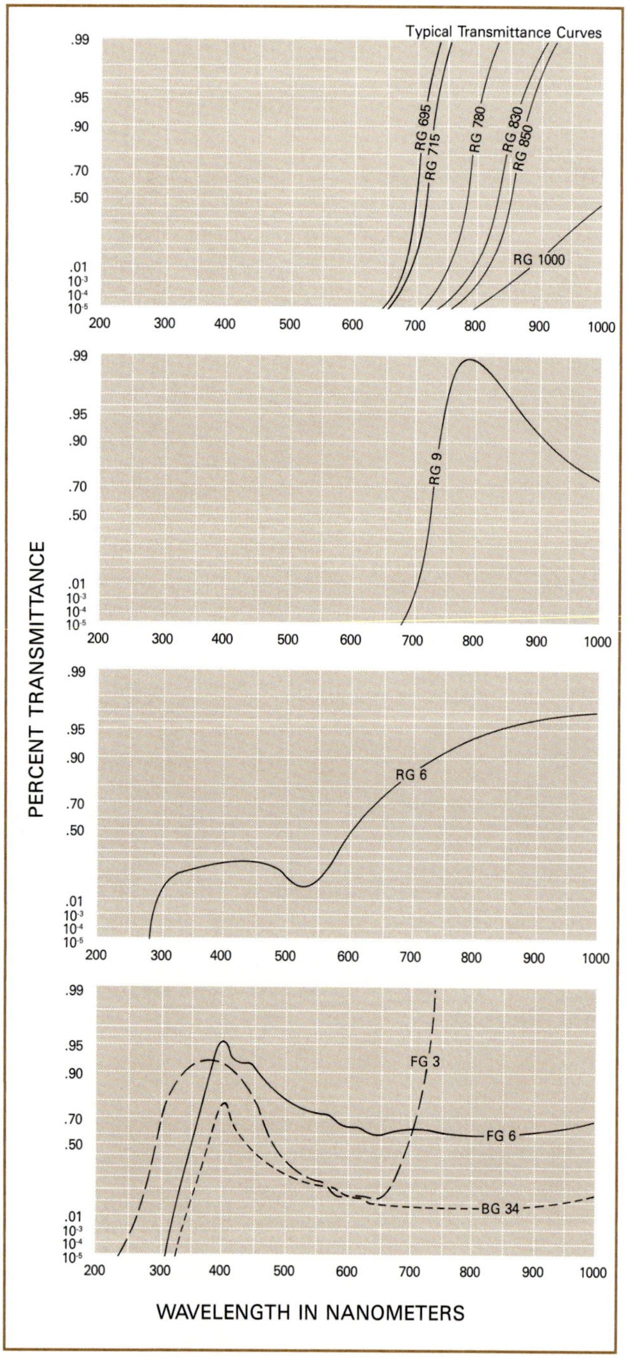

IR Transmitting Black Glass Filters

Schott Glass Type	$\frac{\Delta\lambda}{\Delta T}$	Correction Factor (t_1t_2)	PRODUCT NUMBER
RG 695	0.18	0.915	03 FCG 109
RG 715	0.18	0.915	03 FCG 111
RG 780	0.22	0.915	03 FCG 112
RG 1000	0.18	0.915	03 FCG 113
RG 830	0.28	0.915	03 FCG 114
RG 850	0.30	0.915	03 FCG 118

$\frac{\Delta\lambda}{\Delta T}$ is the temperature coefficient of half-power point position shift, in nm/°C. See text.

Near-IR Transmitting Black Glass Filters

Schott Glass Type	$\frac{\Delta\lambda}{\Delta T}$	Correction Factor (t_1t_2)	PRODUCT NUMBER
RG 9	0.07	0.915	03 FCG 115

$\frac{\Delta\lambda}{\Delta T}$ is the temperature coefficient of half-power point position shift, in nm/°C. See text.

Rose Tinted Glass Filters; for Color Conversion

Schott Glass Type	Correction Factor (t_1t_2)	PRODUCT NUMBER
RG 6	0.92	03 FCG 205

Blue Tinted Glass Filters; for Color Conversion

Schott Glass Type	Correction Factor (t_1t_2)	PRODUCT NUMBER
FG 6	0.90	03 FCG 145
FG 3	0.925	03 FCG 139
BG 34	0.90	03 FCG 027

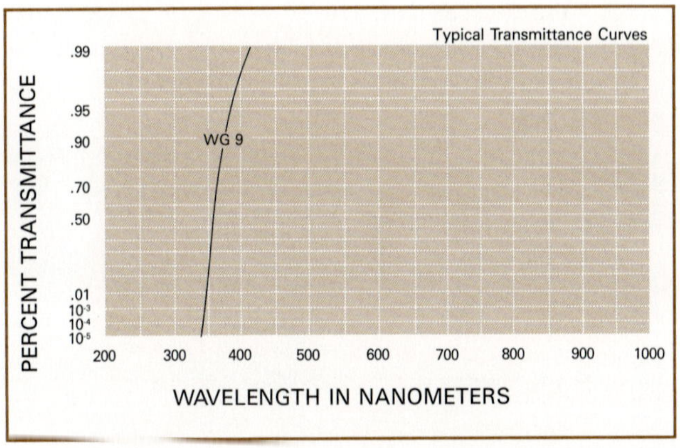

Light Blue and Greenish-Yellow Fluorescent Glass Filters

Schott Glass Type	Correction Factor (t_1t_2)	PRODUCT NUMBER
WG 9	0.92	03 FCG 133

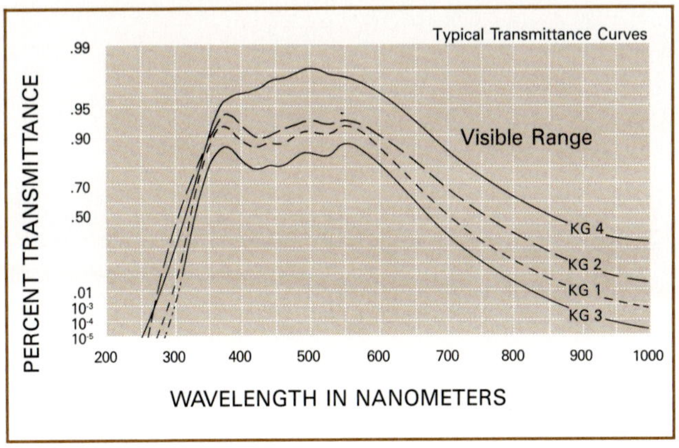

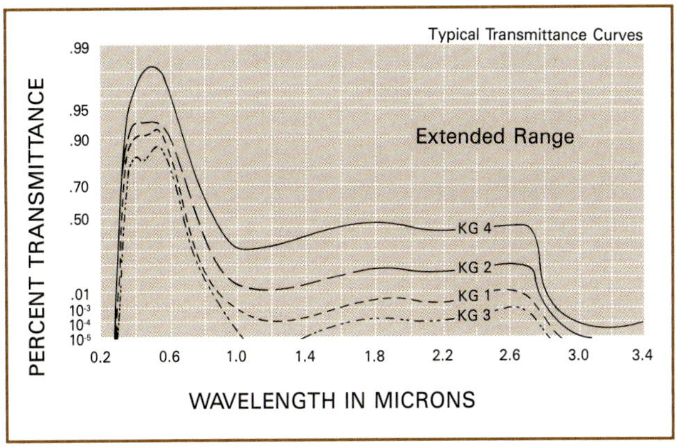

Heat-Absorbing Filters

Schott Glass Type	PRODUCT NUMBER
KG 1	03 FCG 163
KG 2	03 FCG 165
KG 3	03 FCG 167
KG 4	03 FCG 169

The nominal filter transmittance curves shown here are graphs of internal transmittance versus wavelength for a thickness of 3mm.

COLOR TEMPERATURE CONVERSION FILTERS

Color temperature conversion filters allow source color temperatures to be either elevated or depressed by predetermined amounts. In psychophysics and colorimetry it is customary to specify color temperature not in °K but in mireds (acronym for micro reciprocal degrees). To convert an absolute temperature (T) in °K to mireds, take the reciprocal (1/T) and multiply by 10^6.

Color Temperature Conversion Table

Absolute Temperature (°K)	Mireds
infinity	0
1,000,000	1
10,000	100
5000	200
3333	300
2500	400
2000	500

Color temperature conversion filters are rated in accordance with the mired shift, s, which they produce. By definition,

$$s = (10^6)\,[(1/T_2) - (1/T_1)],$$

where T_1 is the initial source color temperature and T_2 is the color temperature after filtration. The mired shift is positive for the yellowish filters which depress color temperature, and negative for the bluish filters which elevate color temperature. For example if T_1 is 2500 °K, a +100 mired shift yields 2000 °K, a +60 mired shift yields 2137 °K, a –100 mired shift yields 3333 °K and a -60 mired shift yields a source temperature T_2 of 2941 °K.

Each color process set contains six color temperature conversion filters of +20, +60, +120, -20, -60, and –120 mireds. These filters can be used individually and in combination to achieve other shift values. Mired shifts, like optical densities, are additive when multiple reflections between filters do not propagate in the direction of interest. The following nominal shift values can be achieved through the use of various individual filters and combinations of filters from our set: –200, –180, –160, –140, –120, –80, –60, –20, +20, +60, +80, +120, +140, +160, -180, and +200. This enables you to vary a 2500 °K source in steps from 1667 °K to 5000 °K.

Following are nominal external transmittance curves for the six color temperature conversion filters which are contained in each color process set. Curves are identified by the mired shift they produce.

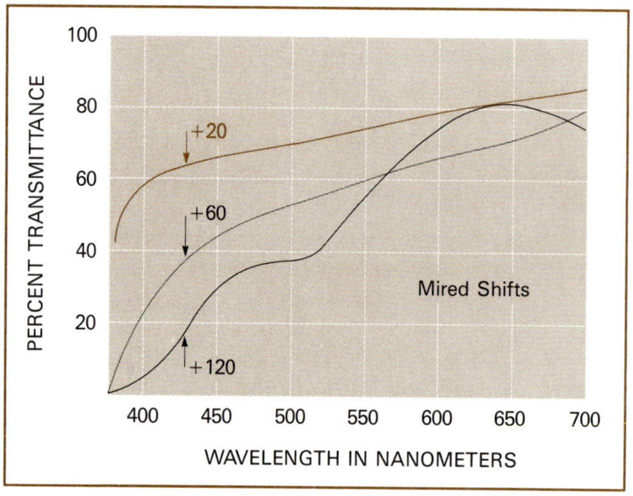

COLOR TEMPERATURE REDUCING filter transmittance graphs.

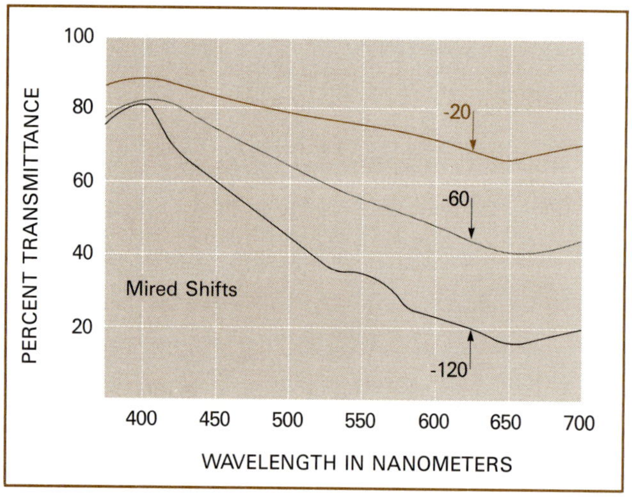

COLOR TEMPERATURE INCREASING filter transmittance graphs.

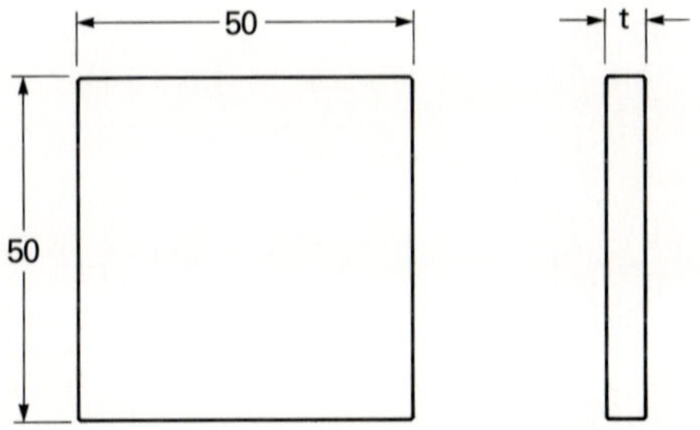

03 FCG

All filters are 50mm square. Varian-Cary 14 or equivalent spectrophotometer curves are supplied with each filter. Filter and curve are contained in individual transparent vinyl storage pages.

Color Temperature Conversion Filters

Curve	Thickness (t) (mm)	PRODUCT NUMBER
+20	2.5	03 FCG 251
+60	2.5	03 FCG 253
+120	2.5	03 FCG 255
-20	2.5	03 FCG 257
-60	2.5	03 FCG 259
-120	2.5	03 FCG 261

Color Process Filter Set

	PRODUCT NUMBER
6 elements as listed above	03 CPS 005

COLOR TEMPERATURE CONVERSION NOMOGRAM

The following nomogram covers the color temperature range of 2000 °K to 10,000 °K. Any three points which can be joined by a single straight line will satisfy the formula relating T_1, T_2, and s. In this way you can readily estimate the color temperature change that any filter or filter combination will produce.

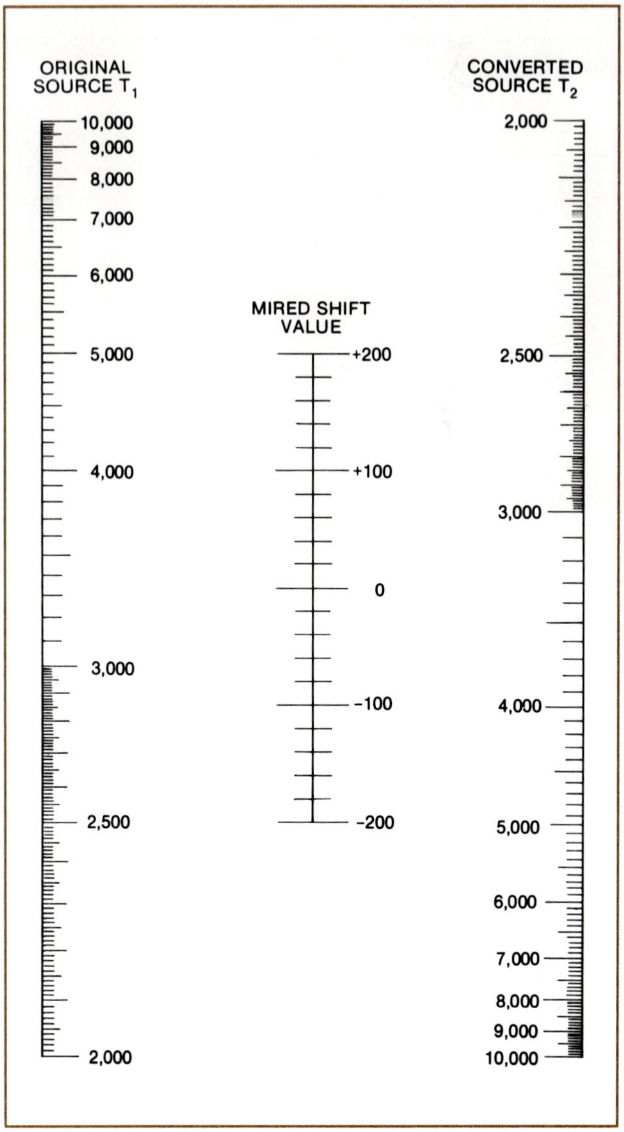

COLORED GLASS FILTER SET

Accurate spectrophotometer recordings of external transmittance (T) accompany each of the twelve colored glass filters in a set. One of the filters (Schott WG 305), highly transparent except at the extreme violet end of the visible spectrum, is suitably used as a blank, and helpful in the focusing and alignment of optical systems which employ other filters. The set has been prepared to yield the most effective coverage possible of the UV, visible, and IR spectral regions.

Each set is packaged in a convenient book-like binder. Clear vinyl pages of these binders display and protect the individual spectrophotometer curves recorded for each filter. There is one page for each filter. Each page also features a special transparent pocket which securely holds the actual filter from which their pockets for use. Curves can be removed from their vinyl pages, if ever necessary, so that detailed measurements, calculations, and photocopies may be made. Every filter, and the curve recorded from it, is permanently marked with the filter product number, which serves to preserve the correspondence between curve and filter within the set.

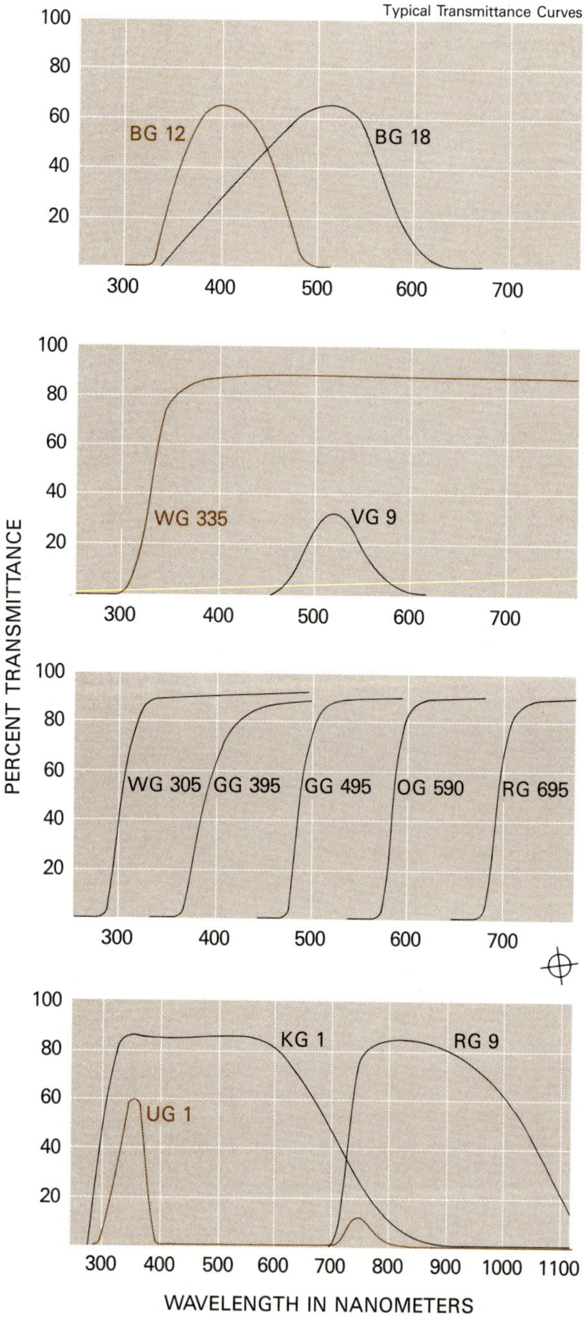

SPECIFICATIONS: COLORED GLASS FILTER SET

Consists of 12 colored glass filters all with original Varian-Cary 14 or equivalent spectrophotometer curves.

Dimensions: 50 x 50 x 3 ($\pm$0.25)mm

Parallelism: 2 arc minutes

Cosmetic Surface Quality: 80-50 scratch and dig

Suggested Maximum Service Temperature: 250 °C

Packaging: Each filter and its associated curve are contained in an individual transparent vinyl filter storage page. Both filter and original curve may be removed from the apge. Pages comprising a filter set are held in a special binder as shown in the photograph above.

Colored Glass Filter Set

	PRODUCT NUMBER
12 elements	03 CFS 007

INTERFERENCE FILTERS

INTERFERENCE FILTERS: WHAT THEY ARE AND HOW TO USE THEM

Applications of interference filters and interference filter sets are extremely diverse, including diagnosis of disease (by tracing of fluorescent antibodies), spectral radiometry (which is the basis for all remote sensing schemes), colorimetry, and color separation in television cameras. Used in connection with even the least expensive broadband photometers or radiometers, our interference filter sets enable you to do rapid and accurate measurements of spectral distributions. While such a combination is the very simplest of spectrophotometers, it has an enormous throughput advantage. This is because the collecting area of the filters is very large by comparison with that of instrumental slits. Additionally, interference filters make possible the viewing and near-instantaneous recording of complete yet very spectrally selective images, making available a wealth of information which spatially and spectrally scanning instruments can provide only in much longer times.

Narrowband interference filters permit isolation of wavelength intervals down to a few nanometers or less in width without the use of dispersion elements such as prims or gratings. For example, the strength of a single line in the emission spectrum of a flame can be monitored without confusion from other nearby lines, or the signal from a laser communications transmitter can be received without interference from a brightly sunlit landscape. Colored glass and gelatin filters are incapable of such discrimination.

Interference filters are multilayer thin-film devices. While many interference filters can be correctly described as "all-dielectric" in construction, metallic layers are often present in the auxiliary blocking structures. Broadband interference filters almost always contain a metallic layer (in their spacer; not in their stacks). Interference filters come in two basic kinds, edge filters and bandpass filters which transmit a desired wavelength interval while simultaneously rejecting both longer and shorter wavelengths.

The transmittance or reflectance curve of an interference filter is the result of interference among the reflections which occur between the many thin-film boundaries, a greatly refined version of the same process which gives rise to the colors seen in soap bubbles and oil films. This interference is constructive for some wavelengths, and destructive for others, sometimes being strongly resonant, depending on filter structure and whether the filter is viewed in transmission or reflection. Absorption in unblocked filters is usually made negligible in the region of interest by suitable choice of construction materials, so that energy which is not transmitted is specularly reflected. Orientation of this type of filter is not important.

In reflection a blocked bandpass filter has, from at least one side, a colorless shiny "metallic" appearance, even though it may contain no metal. The reflectance spectra of the two sides of an interference filter may differ considerably, because of the position and influence of absorbing materials used for blocking. As a general rule, the "metallic" side of a filter should face toward the light source to reduce the thermal load. Most of the depression of peak transmittance values below the 100% level for complete filters can be attributed to absorption in the auxiliary blocking structures. Bandpass interference filters can correctly be thought of as thin solid Fabry-Perot interferometers having spacing which is fixed.

BANDPASS FILTERS

By analogy with interferometers, the simplest bandpass interference filters are sometimes called cavities. Two or more such filters can be used in tandem, or deposited one on top of the other (separated by "absentee" layers of 1/2-wavelength optical thickness) in the same vacuum deposition process to form a multiple-cavity filter. The resulting final transmittance passband is approximately given by the product of the passbands of the cavities of which the multiple-cavity filter is band slopes, improved nearband rejection, and "square" (not Gaussian or Lorentzian) passband tops. This last result is especially desirable in intermediate bandwidth filters, and is achieved in part by reducing stack reflectance and thereby broadening the individual cavity passbands.

In the accompanying figures we show a cross-sectional view of a typical two-cavity interference filter, and show graphically the effect of a number of cavities on final passband shape with final FWHM (*f*ull *w*idth at *h*alf-*m*aximum) held fixed at 10nm. Notice that the optical thicknesses of all layers in the reflective stacks are precisely 1/4-wavelength; other stacks, in which layer optical thicknesses depart considerably from 1/4-wave values, are essential in the construction of edge filters.

MOISTURE RESISTANCE

Melles Griot interference filters are mounted in a black anodized aluminum edge ring with a special epoxy cement (see cross section drawing). The metal ring protects the filter edges from chipping while, at the same time, greatly increasing the in-epoxy path length that water molecules must diffuse through to reach the multilayers. This greatly increases the moisture resistance of these filters (by comparison with the more usual edge-seal filter). Even with the aluminum ring, however, it is advisable to avoid prolonged exposure to environments in which high humidity and large rapid temperature fluctuations are simultaneously present.

TABLE OF NORMALIZED INTERFERENCE FILTER PASSBAND SHAPES

Applicable to ZnS/Na_3AlF_6 interference filters of any FWHM value, the present table is of broader applicability than the following graph. Filters made of other materials have similar characteristics. The table shows the functional dependence of normalized passband shape on the number of cavities used in filter construction, with FWHM arbitrary but held fixed. Transmittance is normalized to peak value, so the table is applicable to both blocked and unblocked filters. To apply the table to a specific filter, simply multiply by peak transmittance. Both minimum and maximum full bandwidths are shown at various normalized transmittance levels. The difference between minimum and maximum full bandwidths allows for both possible choices of spacer material and for filter-

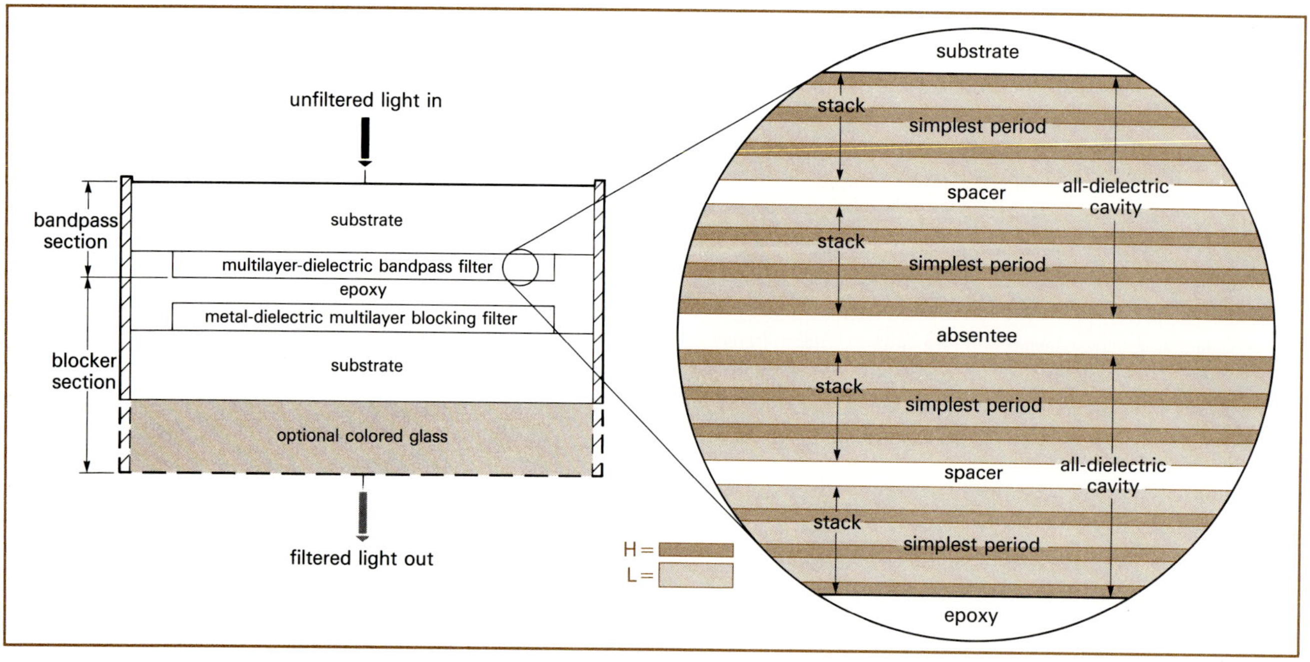

CROSS-SECTION OF A TYPICAL TWO-CAVITY PASS INTERFERENCE FILTER, showing detailed structure of the all-dielectric multilayer bandpass filter film. H symbolizes a precisely quarter-wavelength optical thickness layer of high index material (typically ZnS), while L symbolizes a precisely quarter-wavelength optical thickness layer of low-index material (typically cryolite, Na_3AlF_6). By wavelength we mean wavelength of peak transmittance. Layers are formed by vacuum deposition. The aluminum rings protect the edges while the epoxy cement serves both to protect the films from moisture and to laminate the bandpass and blocker sections together. Depending on the center wavelength and FWHM, other filters may have many more layers than those illustrated here.

to-filter variation. Normal incidence is assumed. Beyond the spectral range displayed here, our filters of 2-, 3-, and 4- cavity construction are supplied with blocking structures which limit absolute transmittance to less than 10^{-4}. More stringent blocking is possible, but increases cost and compromises maximum transmission. This table is strictly applicable only from 400 to 1100nm, but our ultraviolet filters (of different composition) have very similar spectral characteristics.

Bandwidth at various normalized transmittances

Number of Cavities	Normalized Trans-mittance Level (% of peak)	Multiplier for Full Bandwidth at Indicated Transmittance	
		Minimum (in FWHM units)	Maximum
	90	0.3	0.35
1	10	2.5	3.0
	1	8.0	10.0
	90	0.5	0.6
	10	1.6	2.0
2	1	2.8	3.5
	0.1	5.5	6.3
	0.01	10.0	15.0
	90	0.7	0.8
	10	1.2	1.5
3	1	1.9	2.2
	0.1	2.9	3.2
	0.01	4.9	5.4
	90	0.85	0.9
	10	1.1	1.25
4	1	1.5	1.65
	0.1	2.0	2.25
	0.01	3.5	4.25

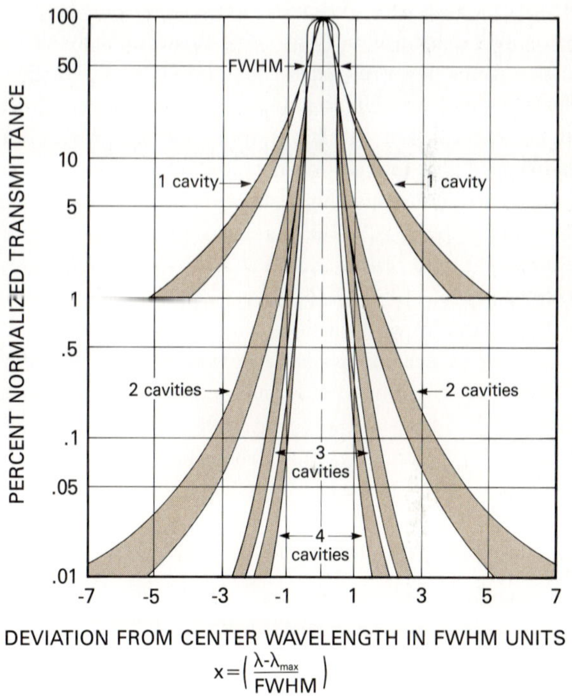

NORMALIZED LOG SCALE TRANSMITTANCE CURVES FOR ZnS/Na_3AlF_6 BANDPASS INTERFERENCE FILTERS OF 10nm FWHM (full width at half-maximum), as functions of the number of cavities used in filter construction. Filters made of other materials have similar characteristics. The curves are applicable from 400 to 1100nm. Normalization is to peak transmittance. To apply to any specific filter of this type, simply multiply by peak transmittance. The shaded regions indicate the nominal limits of filter characteristics at normal incidence, accounting for both possible choices of spacer layer material and for filter-to-filter variations.

WAVELENGTH DEPENDENCE ON ANGLE OF INCIDENCE

It is a common characteristic of single and multilayer dielectric coatings, and interference filters, that both their transmittance and reflectance spectra shift to shorter wavelengths as they are tilted from normal to oblique incidence. This applied to both edge and bandpass filters. For small angles the shift is without distortion or reduction of peak transmittance. In narrowband filters constructed with metallic layers in their stacks, there is also a splitting of the transmittance peak into two separate and orthogonally polarized transmittance peaks which shift to shorter wavelengths at different rates as tilt is increased. This splitting is not seen in any of our narrowband filters, whose stacks are of all-dielectric multilayer construction.

The shift to shorter wavelengths at oblique incidence is very useful in tuning bandpass filters from one wavelength to another, or adjusting the half-power-point wavelengths of edge filters, in collimated light. If the peak transmittance wavelength is slightly too long, you can correct it by tilting. This adds considerably to the usefulness of interference filter sets. Any filter in the variable bandpass sets can be used, by tilt-tuning, at wavelengths down to that at which the next filter in the set functions at normal incidence.

The wavelengths of the transmittance peaks or cavity resonances for both Fabry-Perot interferometers and bandpass interference filters are approximately governed, for observers within the cavity or spacer, by the equation

$$2n_e t \cos\theta = m\lambda,$$

where n_e is the spacer refractive index, t is the spacer thickness, θ is the internal angle of incidence (measured within the cavity or spacer), m is the order number of the interference (a nonzero positive integer), and λ is the wavelength of a particular resonance transmittance peak. This equation is often called the monolayer approximation. The formula can be satisfied simultaneously for many different combinations of order number and wavelength. Corresponding to each such combination there is, in principle, a different resonance transmittance peak for an unblocked filter. For an all-dielectric filter there are, between cavity resonance transmittance peaks, additional broader peaks corresponding to wavelengths at which the dielectric stacks are ineffective as resonant reflectors. Only a single resonance transmittance peak is selected for use, and allowed to appear in the output spectrum of a complete (blocked) interference filter. Blocking techniques are highly effective.

EXAMPLE:

A 3-cavity filter at the 1% normalized transmittance level (1% of peak) would have a nominal full bandwidth (FW1%M) between the limits of 1.9 and 2.2 FWHM units. If the FWHM were 5.0nm, the FW1%M would be between 9.5 and 11.0nm.

In terms of the external angle of incidence ϕ, it can be shown that the wavelength of peak transmittance at small angles of incidence is given by

$$\lambda = \lambda_{max} \sqrt{1 - \left(\frac{n_o}{n_e}\right)^2 \sin^2\phi} \; ,$$

where n_o is the refractive index of the external medium ($n_o = 1$ in air) and n_e is the "effective refractive index" of the spacer. The difference $\lambda_{max} - \lambda$ is the "angle shift." The effective index of the spacer is dependent on wavelength, film material and order number. Because of effects of the multilayer reflecting stacks, the effective index and actual refractive index of the spacer material are not equivalent, although we have used the same symbol n_e for both. By curve-fitting the second formula above (from which t is absent) to measured angle shifts at small angles, both the effective index and the angle at which blocker displacement of the peak becomes significant can in principle be found.

In the absence of actual measurements, the formula probably should not be trusted much beyond 5 or 10 degrees. With suitable interpretation the formula can be applied to prominent landmarks in the transmittance and reflectance spectra of edge filters, multilayer coatings, and single-layer coatings, and is applicable to all interference filters.

Later, in the context of individual interference filters, we will illustrate an advanced method for effectively eliminating angle shift and its effects even in wide-field very-narrow-band radiometry or spectral imaging. In many applications, angle shifts can safely be ignored. Advanced radiometer designs are necessary only if you simultaneously need wide fields and narrow bandwidth. For example, if you were to use primitive methods, a filter with a very narrow 1.0nm FWHM, and ask that its transmittance for a desired monochromatic signal be at least 90% of T_{peak}

throughout some field, you would find that this field could have an angular radius of only about 2.5°.

In most of our filters we use a high-index (usually zinc sulfide) spacer to minimize angle shift. In a few filters, which we will specifically identify later, we have used low-index (usually cryolite) spacers to achieve higher transmittances or narrower bandwidths than would otherwise be possible without a significant increase in cost.

CORRECT INTERFERENCE FILTER ORIENTATION

A good rule of thumb in interference filter use, which is especially important if there is risk of overheating or solarization, is that interference filters should always be oriented with the shiniest "metallic-looking" and most nearly colorless side facing toward the source in the radiant flux. This orientation will minimize the thermal load on the absorbing-glass blocking components. Reversing the orientation of the filter will have no effect on filter transmittance near or within the passband.

TEMPERATURE EFFECTS, LIMITS, AND THERMAL SHOCK

The transmittance spectrum of an interference filter is slightly temperature dependent. As temperature increases, for example, all layer thicknesses increase. At the same time, all layer indices change. These effects combine in such a way that the transmittance spectrum shifts slightly to longer wavelengths with increasing temperature. The thermal coefficient of wavelength shift is a function of wavelength, as is shown in the following table of typical values:

Our interference filters are designed for use at 20°C. Unless bandpass filters with extremely narrow FWHM's are used at very different temperatures, the transmittance shifts indicated in the table are negligible. Our standard interference filters can be used at temperatures down to -50°C, below which differential thermal contraction will result in permanent filter damage. High temperature limits depend on filter design, +70°C being a safe and conservative limit for all filters. Some of our standard filters can accommodate temperatures up to +125°C (check with us for specific applications). As a general rule it is unwise to subject interference filters to serve thermal shock. Especially as the lower limit of -50°C is approached, temperature change rates should not exceed 5°C per minute.

Temperature Dependence of Peak Transmittance

Wavelength (nm)	Temperature Coefficient of Shift (nm per °C)
400	0.016
476	0.019
508	0.020
530	0.021
557	0.021
608	0.023
630	0.023
643	0.024
710	0.026
820	0.027

Special size and shape interference filters can be supplied in both large and small volumes. Please contact one of our Applications Engineers for a quotation.

INTERFERENCE FILTER SETS

Interference filter sets are useful in the rapid measurement of spectral irradiance distributions, measurement of spectral responses, measurement of tristimulus values in colorimetry, and in the construction and evaluation of prototype optical systems. Our filter sets are indispensable companions of modern electronic radiometer/photometer systems. Each interference filter set contains filters designed to do a specific job quickly and conveniently.

The bandpass filter sets are of the interference type and may be used to isolate and identify specific spectral regions. These sets cover the ultraviolet, visible, and infrared spectrum at equally spaced wavelengths and with FWHM values of 10nm, 40nm, and 80nm. The color additive and subtractive sets also contain absorbing glass filters which change the apparent color temperature of a source.

INTERFERENCE FILTER SET CONTENTS

Accurate spectrophotometer recordings of external transmittance (T) versus wavelength (λ) accompany every filter of every interference filter set. Each individual filter curve is an original curve, assuring you of accurate data for use in calibration and measurement. Wavelength ranges spanned by individual interference filter curves include all wavelengths of interest for the particular filter, and are recorded on a Varian-Cary 14 or equivalent quality spectrophotometer. These instruments have been calibrated with standards which are directly traceable to the U.S. National Bureau of Standards. Absence of unwanted energy transmission beyond the intervals spanned by individual filter curves is assured by statistical quality control procedures involving transmittance measurements on extended wavelength intervals.

Each interference filter set is packaged in a convenient book-like binder. Plastic pages of these binders display and protect the individual spectrophotometer curves recorded for the filters in the set. There is one such page for each filter. Each plastic page also features a special transparent pocket which securely holds the filter from which the spectrophotometer curve of that page was recorded. Filters can easily be removed from their pockets for use. Curves can be removed from their plastic pages, if necessary, so that detailed measurements, calculations, and photocopies may be made.

Each filter is mounted in a black anodized aluminum ring with a special moisture resistant epoxy. The ring serves to protect the edges of the filter from chipping and greatly increases the resistance to moisture. Product numbers are individually marked on each filter ring. Each bandpass filter, and the curve recorded from it, is permanently marked with the nominal center wavelength.

Every set also includes a blank substrate of uncoated BK 7 or fused silica, having the same thickness and dimensions as the interference filters in the set. This blank is often very helpful in the alignment and focusing of optical systems which employ interference filters.

We carry a large variety of individual interference filters in inventory. We also are able to supply special filters, in small and large quantities, peaked at your specified wavelength and with a choice of bandpass, size and shape.

Volume filter users frequently do not require an individual spectrophotometer curve with each individual filter. Discounts are available to OEM users who purchase filters without curves.

UV-10 FILTER SET

This ten element ultraviolet filter set consists of seven narrow-band interference filters, two special-purpose glass filters and a fused silica blank. The interference filters are 10 nanometers in bandwidth, and cover the range from 260nm to 380nm in equally spaced steps of 20nm. This filter set is useful in spectral distribution measurements in the near-UV, and for isolating particular UV wavelength regions of interest. It can be used with sources such as hydrogen and deuterium.

The GG 395 filter is a sharp-cutoff visible-transmitting glass for which a nominal internal transmittance curve was presented earlier. It is especially helpful in the detection of fluorescence, by eliminating the ultraviolet excitation energy at the detector. The WG 9 filter is strongly fluorescent and helpful in locating ultraviolet beams which may be invisible and yet dangerous to the unprotected eye.

UV-10 Filter Set

	PRODUCT NUMBER
10 elements	03 IFS 002

SPECIFICATIONS UV-10 FILTER SET

Comprising seven interference filters at 260, 280, 300, 320, 340, 360, and 380nm; two special glass filters (glass types WG 9 and GG 395); all with original Varian-Cary 14 or equivalent spectrophotometer curves, plus one fused silica blank (without curve).

FWHM: 10 $\pm$2nm

Filter-to-Filter Center Wavelength Interval: 20 $\pm$2nm

Peak Transmittance: 13 to 25%

Blocking: 0.01% (X-ray to 3.5 (μm)

Clear Aperture: 21mm diameter

Dimensions: 25ϕ $\pm$0.5mm

Thickness: 6.9mm maximum

Cosmetic Surface Quality: 80-50 scratch and dig

Temperature Limits: -50°C to +70°C

Construction: Laminated and sealed with moisture resistant epoxy in a protective aluminum ring.

Packaging: Each filter and its associated curve are contained in an individual transparent vinyl filter storage page. Pages comprising a filter set are held in a special binder.

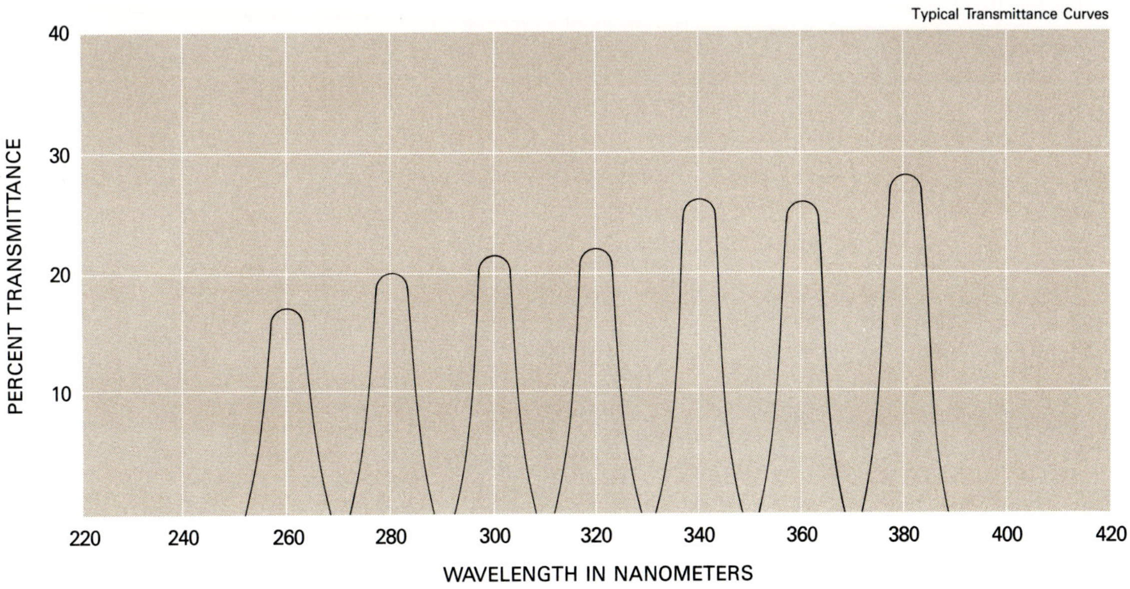

UV MERCURY LINE FILTER SET

This seven element ultraviolet filter set consists of six narrowband interference filters and one fused silica blank. The interference filters are 10 nanometers in bandwidth and cover the mercury vapor discharge line spectrum from 254nm to 365nm.

This set is designed specifically for isolation of the mercury ultraviolet lines, and will prove useful in most spectrographic, analytical, and photochemical laboratories. The center wavelengths of these filters have been toleranced to be plus three minus zero nanometers from the nominal, to allow for finite field angular radius and tilt tuning. The fused silica blank will be found useful for alignment and focusing before insertion of the desired filter.

UV-10 Filter Set

	PRODUCT NUMBER
7 elements	03 IFS 004

SPECIFICATIONS: UV MERCURY LINE FILTER SET

Comprising six interference filters, 25mm diameter, centered at 254, 280, 289, 313, 334, and 365 nanometers; all with original Cary 14 or equivalent spectrophotometer curves, plus one fused silica blank (without curve).

FWHM: 10 ±2nm

Center Wavelength Tolerance: +3, -0nm

Minimum Peak Transmittance: 13 to 25%

Blocking: 0.01% (X-ray to 3.5 μm)

Clear Aperture: 21mm diameter

Dimensions: 25ϕ ±0.5mm

Outside Diameter: 25mm

Thickness: 6.9mm maximum

Cosmetic Surface Quality: 80-50 scratch and dig

Temperature Limits: -50°C to +70°C

Construction: Laminated and sealed with moisture resistant epoxy in a protective aluminum ring.

Packaging: Each filter and its associated curve are contained in an individual transparent vinyl filter storage page. Pages comprising a filter set are held in a special binder.

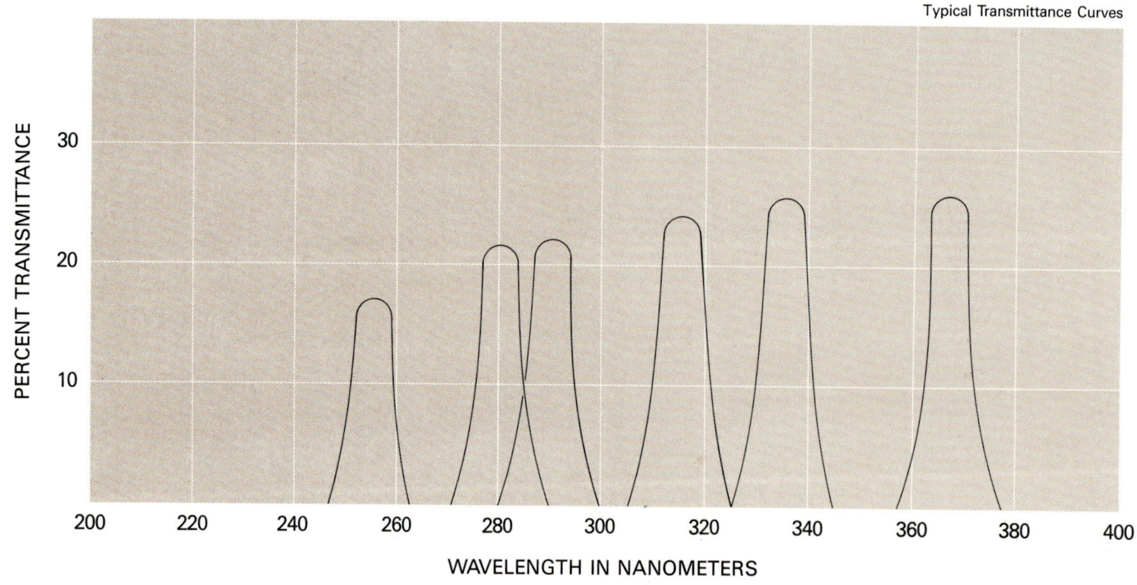

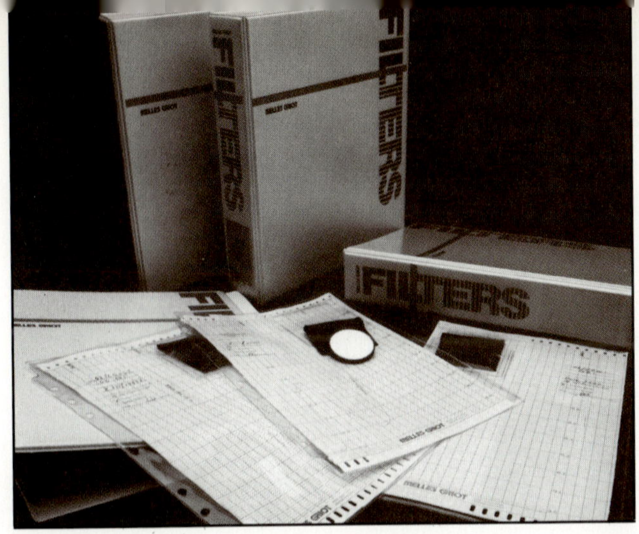

VISIBLE-10 FILTER SET

This eight element visible filter set consists of seven narrowband interference filters and one BK 7 blank. The interference filters are 10 nanometers in bandwidth and cover the visible spectral region of 400nm to 700nm in 50nm equally spaced steps.

Principal applications of this filter set will include the measurement of spectral radiance and irradiance distributions and contours, photodetector spectral response, photochemical response, measurement of colorimetric tristimulus values, and the construction and evaluation of prototype optical systems. The BK 7 blank will be found useful in alignment and focusing of systems prior to insertion of the desired filters.

Visible-10 Filter Set

	PRODUCT NUMBER
8 elements	03 IFS 006

SPECIFICATIONS: VISIBLE-10 FILTER SET

Comprising seven interference filters at 400, 450, 500, 550, 600, 650, and 700 nanometers; all with original Varian-Cary 14 or equivalent spectrophotometer curves, plus one BK 7 blank (without curve).

FWHM: 10 ±2nm

Filter-to-Filter Center Wavelength Interval: 50 ±5nm

Peak Transmittance: 40 to 55%

Blocking: 0.01% (X-ray to 3.5 μm)

Clear Aperture: 46mm diameter

Dimensions: 50ϕ ±0.5mm

Outside Diameter: 50mm

Thickness: 6.9mm maximum

Cosmetic Surface Quality: 80 50 scratch and dig

Temperature Limits: -50°C to +70°C

Construction: Laminated and sealed with moisture resistant epoxy in a protective aluminum ring.

Packaging: Each filter and its associated curve are contained in an individual transparent vinyl filter storage page. Pages comprising a filter set are held in a special binder

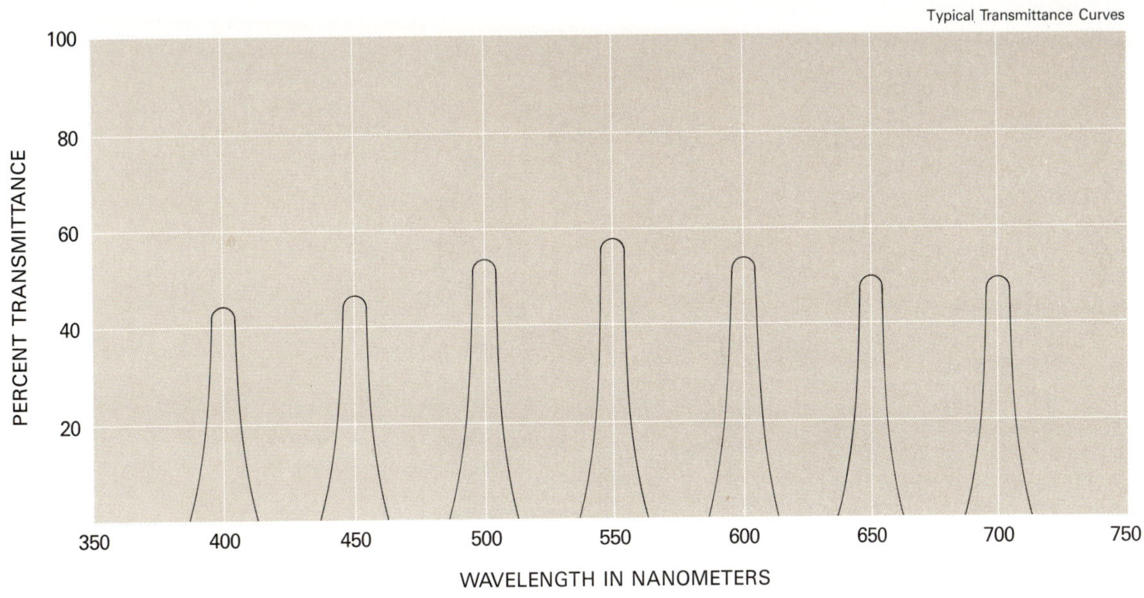

VISIBLE-40 FILTER SET

This eight element visible filter set is made up of seven medium-band interference filters and one BK 7 blank. The interference filters are 40 nanometers in bandwidth and cover the visible spectrum from 400nm to 700nm in 50nm equally spaced steps.

This set is similar to our Visible-10 Filter set, described previously, and will have corresponding applications. Because of its increased bandwidth, this set will smooth the results of spectral distribution measurements. The increased bandwidth will pass more energy (increase throughput) and reduce angleshift sensitivity in weak signal and wide field applications.

Visible-40 Filter Set

	PRODUCT NUMBER
8 elements	03 IFS 008

SPECIFICATIONS: VISIBLE-40 FILTER SET

Comprising seven interference filters, 50mm diameter, at 400, 450, 500, 550, 600, 650, and 700 nanometers; all with original Varian-Cary 14 or equivalent spectrophotometer curves, plus one BK 7 blank (without curve).

FWHM: 40 ±8nm

Filter-to-Filter Center Wavelength Interval: 50 ±5nm

Minimum Peak Transmittance: 50 to 55%

Blocking: 0.01% (X-ray to 3.5 μm)

Clear Aperture: 46mm diameter

Dimensions: 50ϕ ±0.5mm

Outside Diameter: 50mm

Thickness: 6.9mm maximum

Cosmetic Surface Quality: 80-50 sratch and dig

Temperature Limits: -50°C to +70°C

Construction: Laminated and sealed with moisture resistant epoxy in a protective aluminum ring.

Packaging: Each filter and its associated curve are contained in an individual transparent vinyl filter storage page. Pages comprising a filter set are held in a special binder

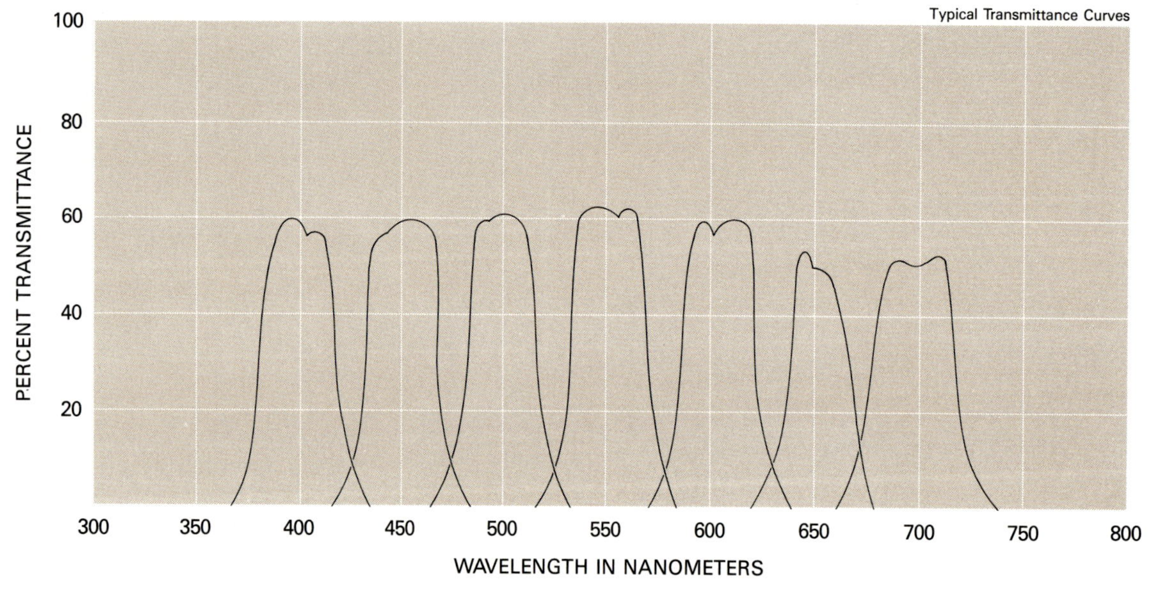

VISIBLE-80 FILTER SET

This eight element visible filter set consists of seven wideband interference filters and one BK 7 blank. The interference filters are 80 nanometers in bandwidth and cover the visible spectral region of 400nm to 700nm in 50nm equally spaced steps.

This set has many of the same applications as our Visible-10 and Visible-40 sets. The increased bandwidth of this set for large-field spectral distribution studies at low signal levels. This set will be found useful for all detector response measurements where both energy input levels are restricted and wide angular fields of view are required. The blank filter is useful in alignment and focusing. Because the passbands overlap, this set offers continuous coverage of the visible spectrum.

Visible-80 Filter Set

	PRODUCT NUMBER
8 elements	03 IFS 014

SPECIFICATIONS: VISIBLE-80 FILTER SET

Comprising seven induced-transmission type (metal spacer & dielectric stacks) at 400, 450, 500, 550, 600, 650, and 700 nanometers; all with original Varian-Cary 14 or equivalent spectrophotometer curves, plus one BK 7 blank (without curve).

FWHM: 80 +16-20mm

Filter-to-Filter Center Wavelength Interval: 50 ±5nm

Minimum Peak Transmittance: 60 to 75%

Blocking: 0.01% (X-ray to 3.5 μm)

Clear Aperture: 46mm diameter

Dimensions: 50ϕ ±0.5mm

Outside Diameter: 50mm

Thickness: 6.9mm maximum

Cosmetic Surface Quality: 80-50 scratch and dig

Temperature Limits: -50°C to +70°C

Construction: Laminated and sealed with moisture resistant epoxy in a protective aluminum ring.

Packaging: Each filter and its associated curve are contained in an individual transparent vinyl filter storage page. Pages comprising a filter set are held in a special binder.

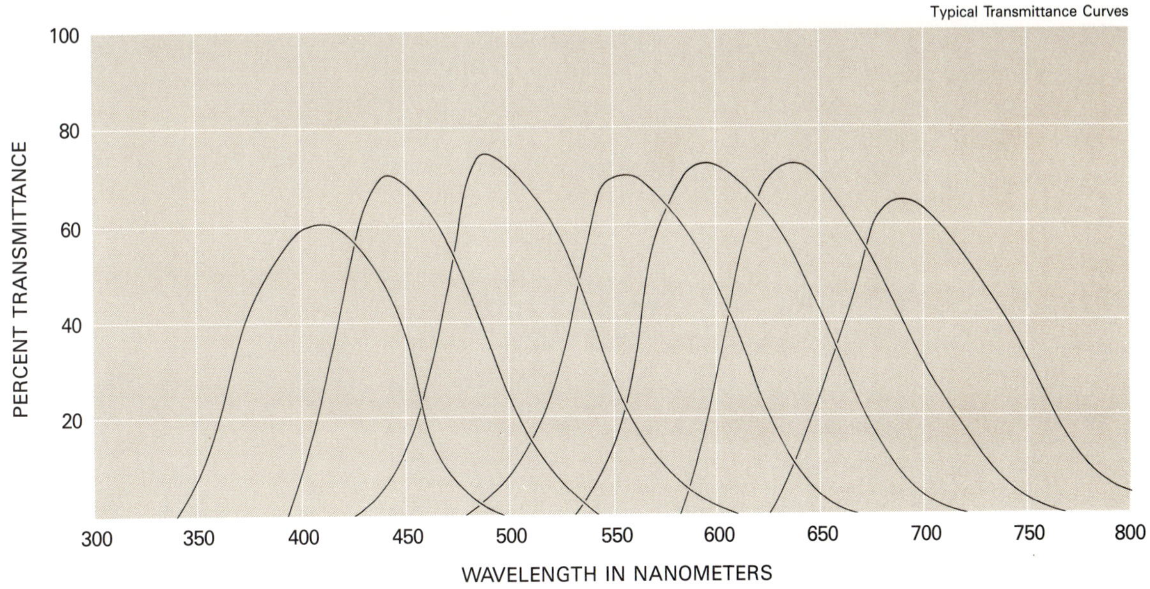

SPECTRAL LINE FILTER SET

This eight element spectral line filter set is made up of seven narrowband interference filters and one BK 7 blank. The interference filters are 10 nanometers FWHM, and cover the most frequently used visible spectral lines from low pressure discharge lamps.

The set is designed primarily for isolation of the most prominent mercury discharge lines, the sodium yellow doublet, and the two principal hydrogen lines. This set will prove useful in most educational physics and chemistry laboratories. The center wavelengths of these filters have been toleranced to be plus two, minus zero nanometers from the nominal, to allow for finite field angular radius and tilt tuning. The BK 7 blank will be found useful in alignment and focusing of systems prior to insertion of the desired filters.

Spectral Line Filter Set

	PRODUCT NUMBER
8 elements	03 IFS 016

SPECIFICATIONS: SPECTRAL LINE FILTER SET

Comprising seven interference filters, 25mm diameter, with transmittance peaks at 405, 436, 486, 546, 577, 589, and 656 nanometers; plus one BK 7 blank (without curve). Original Varian-Cary 14 or equivalent spectrophotometer curves accompany each filter.

FWHM: 10 ±2nm

Center Wavelength Tolerance: +2, -0nm

Minimum Peak Transmittance: 40 to 50%

Blocking: 0.01% (X-ray to 3.5 μm)

Clear Aperture: 21mm diameter

Dimensions: 25ϕ ±0.5mm

Outside Diameter: 25mm

Thickness: 6.9mm maximum

Cosmetic Surface Quality: 80-50 scratch and dig

Temperature Limits: -50°C to +70°C

Construction: Laminated and sealed with moisture resistant epoxy in a protective aluminum ring.

Packaging: Each filter and its associated curve are confined in an individual transparent vinyl filter storage page. Pages comprising a filter set are held in a special binder.

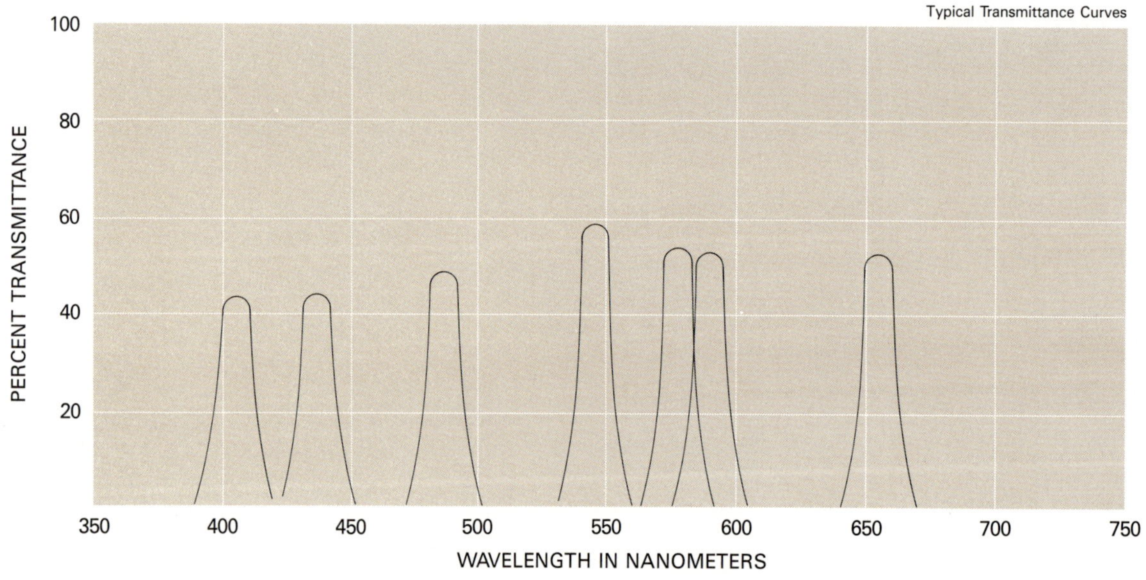

IR-10 FILTER SET

This nine element infrared filter set is made up of eight narrow-band interference filters and one BK 7 blank. The interference filters are 10 nanometers in bandwidth, and cover the range of 750nm to 1100nm in 50nm equally spaced steps.

This filter set is a continuation of our UV-10 and Visible-10 Filter Sets, and allows you to perform spectral distribution, detector response studies, and evaluation of prototype optical systems in the near infrared. With higher throughput and greater convenience, these filter sets will do many of the routine and exploratory studies previously reserved for spectrophotometers. The BK 7 blank will be found useful for alignment and adjustment by visible light before insertion of the required infrared filters.

IR-10 Filter Set

	PRODUCT NUMBER
9 elements	03 IFS 018

SPECIFICATIONS: IR-10 FILTER SET

Comprising eight interference filters, 50mm diameter, at 750, 800, 850, 900, 950, 1000, 1050, and 1100 nanometers; all with original Varian-Cary 14 or equivalent spectrophotometer curves, plus one BK 7 blank (without curve).

FWHM: 10 ±2nm

Filter-to-Filter Center Wavelength Interval: 50 ±5nm

Minimum Peak Transmittance: 40 to 45%

Blocking: 0.01% (X-ray to 3.5 μm)

Clear Aperture: 46mm diameter

Dimensions: 50ϕ ±0.5mm

Thickness: 6.9nm maximum

Cosmetic Surface Quality: 80-50 scratch and dig

Temperature Limits: -50°C to +70°C

Construction: Laminated and sealed with moisture resistant epoxy in a protective aluminum ring.

Packaging: Each filter and its associated curve are contained in an individual transparent vinyl filter storage page. Pages comprising a filter set are held in a special binder.

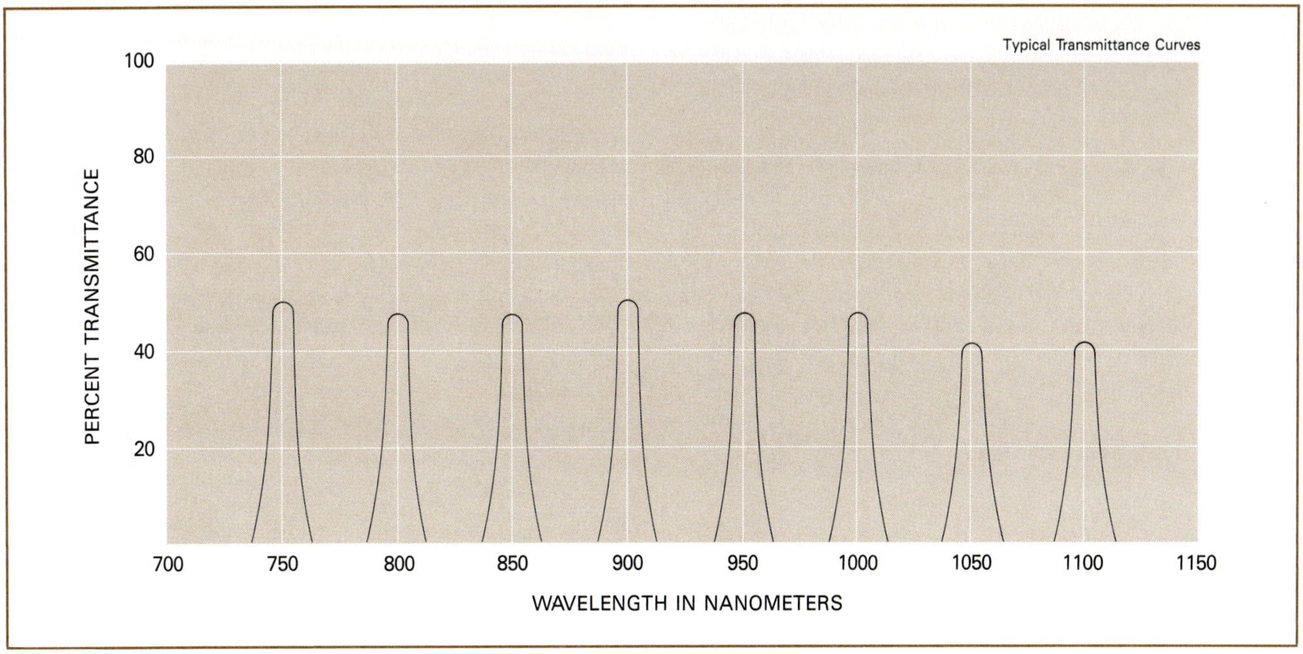

INDIVIDUAL INTERFERENCE FILTERS

Individual interference filters are available at all important wavelengths from the ultraviolet to the near infrared. They are listed in three sizes and in three bandwidths. Some filters have FWHM's down to 1.0nm. All of these filters have the same characteristics as those which have previously appeared in our sets. These filters may serve to replace damaged or lost filters.

All individual filters are round and are mounted in black anodized aluminum rings with a special epoxy cement. The rings protect the filter edges from chipping, and greatly increase the filter's resistance to moisture. Outside diameters are 50mm; clear apertures are 48mm. Maximum thickness is 7mm.

All filters are supplied in plastic binder pages of the same kind used in our filter sets. Each page holds the filter in a protective pocket from which the filter can be removed for use. Each page also holds the individual spectrophotometer curve for the filter. Curves can be removed from the pages for photocopying or calculations. Spectrophotometer curves are originals, and are made on a Varian-Cary 14 or equivalent instrument.

Each filter and curve is permanently marked with its catalog product number. Accurate bandpass filter center wavelengths, bandwidths, and edge filter 50%-points, can all be read from the individual filter spectrophotometer curves. You can easily incorporate filters into existing sets. On a volume special-request basis, filters can be supplied without spectrophotometer curves or binder pages, at a reduced price.

In the list of interference filters which follows we have tabulated the number of cavities used in construction. With this information you can apply the normalized log scale transmittance curves and table of normalized interference filter passband shapes (both presented earlier) to the various filters in our inventory. For those who are concerned about angle shift in wide-field and very narrowband applications, we have also tabulated either the spacer material or the nominal effective index n_e.

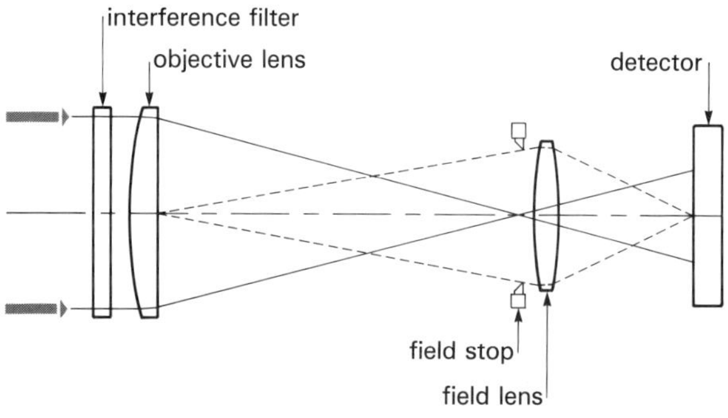

INTERFERENCE FILTER USAGE. As we have pointed out before, narrowband interference filters are extremely angle-sensitive. The transmittance of a 1.0nm FWHM filter will fall by 10%, at the desired wavelength, for field angles of only 2.5 degrees. For field angles of 5 degrees the transmittance will have fallen by over 90%. It is important, therefore, that narrowband interference filters be used in locations that assure maximum collimation.

The illustration shows the design of a narrow-field spectral radiometer for infinite conjugate ratio use, and indicates proper interference filter location. The radiometer consists of an interference filter, objective lens, field lens, field stop and detector.

The field lens, which images the objective lens onto the detector's sensitive area, assures uniform illumination of the detector. The field of view is limited by a field stop placed close to the field lens.

See pages 458-459 for a convenient selection of filter holders.

Visible Bandpass Filters (See previous set descriptions for specifications.)

	Wavelength (nm)		FWHM (nm)	Peak Trans- mittance Minimum	Size (mm)	Effective Index (n_e)	Number of Cavities	PRODUCT NUMBER
	400	+ 3, -0	10 ± 2	40%	50φ	1.45	3	03 FIV 002
	400	+10, -0	40 ± 8	50%	50φ	*	*	03 FIV 026
	400	+25, -0	80 + 16, -20	60%	50φ	*	*	03 FIB 002
Hg	404.7	+ 2, -0	10 ± 2	40%	25φ	1.45	3	03 FIM 002
Hg	435.8	+ 2, -0	10 ± 2	40%	25φ	1.45	3	03 FIM 004
	450	+ 3, -0	10 ± 2	40%	50φ	1.45	3	03 FIV 004
	450	+10, -0	40 ± 8	50%	50φ	*	*	03 FIV 028
	450	+25, -0	80 + 16, -20	65%	50φ	*	*	03 FIB 004
H	486.1	+ 2, -0	10 ± 2	45%	25φ	2.10	3	03 FIR 002
	500	+ 3, -0	10 ± 2	50%	50φ	2.10	3	03 FIV 006
	500	+10, -0	40 ± 8	55%	50φ	*	*	03 FIV 038
	500	+25, -0	80 + 16, -20	70%	50φ	*	*	03 FIB 006
Hg	546.1	+ 2, -0	10 ± 2	55%	25φ	2.10	3	03 FIM 006
	550	+ 3, -0	10 ± 2	55%	50φ	2.10	3	03 FIV 008
	550	+10, -0	40 ± 8	55%	50φ	*	*	03 FIV 044
	550	+25, -0	80 + 16, -20	70%	50φ	*	*	03 FIB 008
Hg	577	+ 2, -0	10 ± 2	50%	25φ	2.10	3	03 FIM 008
Na	589.3	+ 2, -0	10 ± 2	50%	25φ	2.10	3	03 FIR 004
	600	+ 3, -0	10 ± 2	50%	50φ	2.10	3	03 FIV 018
	600	+10, -0	40 ± 8	50%	50φ	*	*	03 FIV 046
	600	+25, -0	80 + 16, -20	70%	50φ	*	*	03 FIB 012
	650	+ 3, -0	10 ± 2	45%	50φ	2.10	3	03 FIV 022
	650	+10, -0	40 ± 8	45%	50φ	*	*	03 FIV 048
	650	+25, -0	80 + 16, -20	70%	50φ	*	*	03 FIB 014
H	656	+ 2, -0	10 ± 2	45%	25φ	2.10	3	03 FIR 006
	700	+ 3, -0	10 ± 2	45%	50φ	2.10	3	03 FIV 024
	700	+10, -0	40 ± 8	45%	50φ	*	*	03 FIV 058
	700	+25, -0	80 + 16, -20	70%	50φ	*	*	03 FIB 016

*These filters do not have spacers in the normal sense, and the index information does not apply. In fact these are "induced transmission" filters based on principles very different from those previously discussed in this text. A metallic film has been substituted for the former spacer. The presence of the dielectric stacks profoundly modifies the transmittance of this film, giving rise to a transmittance peak at the wavelength for which the stacks are quarter-wave. This peak is unique (there are no other peaks corresponding to different order numbers), by nature slightly asymmetric, and both broader and higher than can conveniently be achieved in all-dielectric filters. Filters of the induced transmission type are frequently incorporated into the blocking structures of narrow-band filters.

Ultraviolet and Infrared Bandpass Filters (See previous set descriptions for specifications.)

	Wavelength (nm)		FWHM (nm)	Peak Trans-mittance Minimum	Size (mm)	Effective Index (n_e)	Number of Cavities	PRODUCT NUMBER
Hg	253.7	+ 2, −0	10 ± 2	13%	25φ	*	*	03 FIM 016
	260	+ 2, −0	10 ± 2	13%	25φ	*	*	03 FIU 002
Hg	280	+ 2, −0	10 ± 2	15%	25φ	*	*	03 FIM 018
Hg	289	+ 2, −0	10 ± 2	15%	25φ	*	*	03 FIM 022
	300	+ 2, −0	10 ± 2	15%	25φ	1.45	3	03 FIU 004
Hg	313	+ 2, −0	10 ± 2	15%	25φ	1.45	3	03 FIM 024
	320	+ 2, −0	10 ± 2	25%	25φ	1.45	3	03 FIU 006
Hg	334	+ 2, −0	10 ± 2	25%	25φ	1.45	3	03 FIM 026
	340	+ 2, −0	10 ± 2	25%	25φ	1.45	3	03 FIU 008
	360	+ 2, −0	10 ± 2	25%	25φ	1.45	3	03 FIU 012
Hg	365	+ 2, −0	10 ± 2	25%	25φ	1.45	3	03 FIM 028
	380	+ 2, −0	10 ± 2	25%	25φ	1.45	3	03 FIU 014
	750	+ 3, −0	10 ± 2	45%	50φ	2.1	3	03 FII 002
	800	+ 3, −0	10 ± 2	45%	50φ	2.1	3	03 FII 004
	850	+ 3, −0	10 ± 2	45%	50φ	2.1	3	03 FII 006
	900	+ 3, −0	10 ± 2	45%	50φ	2.1	2	03 FII 008
	950	+ 3, −0	10 ± 2	45%	50φ	2.1	2	03 FII 016
	1000	+ 3, −0	10 ± 2	45%	50φ	2.1	2	03 FII 018
	1050	+ 3, −0	10 ± 2	40%	50φ	2.1	2	03 FII 022
	1100	+ 3, −0	10 ± 2	40%	50φ	2.1	2	03 FII 024

*These filters are of a complicated structure involving repeated metallic and dielectric layers. They exhibit polarization splitting. Their behavior cannot be adequately described by a single effective index value.

We can supply a vast selection of interference filters, not listed here, to special order. Pleae give us full specifications when requesting a quotation.

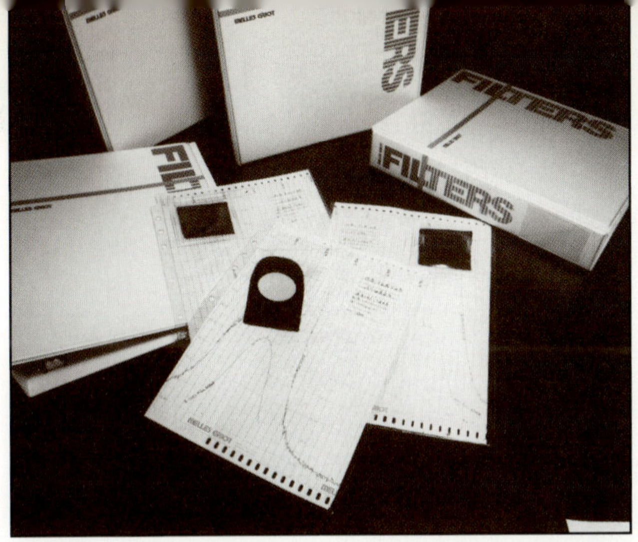

LASERLINE INTERFERENCE FILTERS

We offer filters of various bandwidths for the most important laser lines, suitable for discrimination against unwanted background radiation. Such filters permit laser range finders, laser communications equipment, and fringe counting interferometer systems to be used over large distances in direct sunlight.

The 50mm round laser line interference filters are companion elements for 01 LAO 238 and for our laser line singlet focusing lenses (01 LFS series), intended for radiometer construction. The optimized area of the focusing singlets is of 47mm diameter, 1mm more than the coated aperture specification for these corresponding filters. We recommend that these lenses and filters be used in conjunction with 46mm diameter stops. It is important that all three elements be accurately centered for optimum radiometer performance.

All laser line filters are mounted in a black anodized aluminum ring with a special epoxy cement. Maximum thickness is 7mm.

FILTER BINDERS AND STORAGE PAGES

If you have many loose filters, you will find our filter binder system to be both protective and convenient. Double-pocketed vinyl pages hold one 50mm x 50mm (or smaller) filter and the corresponding spectrophotometer or microdensitometer curve. Individual pages, complete with filters and curves, may be assembled into our filter binders for convenient storage and ready access. The filter pocket location is varied on each page to facilitate page stacking. Filter pages are supplied in sets of twelve.

Filter Binder

PRODUCT NUMBER
03 FBB 001

Filter Storage Pages

	PRODUCT NUMBER
Set of 12	03 FBP 001

Laser Line Interference Filters

Wave-Length (nm)	Tolerance (nm)	FWHM (nm)	Peak Trans-mittance Minimum	Size (mm)	Effective Index n_e	Number of Cavities	PRODUCT NUMBER
Argon-Ion							
488	+1.5, -0	10 ± 2	50	25ϕ	2.1	3	03 FIL 002
514.5	+1.5, -0	10 ± 2	50	25ϕ	2.1	3	03 FIL 004
Helium Neon							
543.5	+1.5, -0	10 ± 2	50	25ϕ	2.1	3	03 FIL 048
543.5	+1.5, -0	10 ± 2	50	50ϕ	2.1	3	03 FIL 049
632.8	+0.2, -0	1 ± 0.2	42	25ϕ	2.1	2	03 FIL 006
632.8	+0.2, -0	1 ± 0.2	42	50ϕ	2.1	2	03 FIL 008
632.8	+0.5, -0	3 ± 0.5	50	25ϕ	2.1	2	03 FIL 018
632.8	+0.5, -0	3 ± 0.5	50	50ϕ	2.1	2	03 FIL 022
632.8	+1.5, -0	10 ± 2	50	25ϕ	2.1	3	03 FIL 024
632.8	+1.5, -0	10 ± 2	50	50ϕ	2.1	3	03 FIL 026
Ruby							
694.3	+1.5, -0	10 ± 2	45	25ϕ	2.1	3	03 FIL 028
GaAlAs Diode							
830.0	+1.5, -0	10 ± 2	45	25ϕ	2.1	3	03 FIL 038
Neodymium Glass/Neodymium YAG, Fundamental							
1064	+2.0, -0	10 ± 2	45	25ϕ	2.1	3	03 FIL 036

RAPID RESPONSE

We know that when you need a component, you usually need it immediately. That's why we've developed sophisticated systems to respond to your needs.

We carry extensive inventories in Irvine, California; Zevennar, The Netherlands; and Tokyo, Japan. In fact, at any time we have at least 90% of the items listed in this Optics Guide in stock.

Our computer based inventory control system enables us to ship products from stock within a few hours of the receipt of your order.

POLARIZATION COMPONENTS

Polarization is an important, and often misunderstood, phenomenon in optics. Here we discuss polarization and how it can affect you. We also present a broad selection of components which modify the polarization state of incident illumination.

Contents **Page Number**

POLARIZATION COMPONENTS

This section describes a group of components which modify the state of polarization of optical radiation in various useful ways. Polarized light carries valuable information about where the light has been and the various physical parameters which have been acting upon it. Magnetic fields, chemical interactions, molecular structures, and mechanical stress all impose changes in the polarization state of an optical beam (and electromagnetic radiation in general). Applications relying on the study of these changes of polarization cover a vast range and include astrophysics, agricultural production, electric power generation, molecular biology and various industrial processes.

Polarization states are customarily classified as linear, circular or elliptical according to the path traced by the tip of the electric field vector in a passing wave train. Unpolarized light, such as that radiated from a thermal source like an incandescent light or from an electrical discharge, can be thought of as a superposition of all possible linear, circular and elliptical states — with all having an equal probability. Randomly polarized light, when spoken of as that emitted from many types of laser, is composed of two orthogonally linearly polarized co-linear beams whose power randomly (and inversely) varies over time with respect to each other. Random, when used in this way, can be a misleading word; the radiation is clearly linearly polarized at any one moment in time. In the same way, de-polarized light (such as from the Cornu pseudo-depolarizer described on page 300) is usually linearly polarized light which has been randomized by either temporal or spatial variations of retardation across or along the beam. If the integration of these various retardations is great enough, the beam will appear to be de-polarized. However, care should be taken with depolarizers; the randomization process usually varies the linear polarization in a fairly smooth and predictable fashion. Considerable beam integration must occur before the beam appears again to be essentially unpolarized.

While analysis of a polarized beam will provide information of an optical beam's past history, the failure to anticipate and deal with the polarization state of an optical beam frequently can cause great problems. Radiometers, monochromators, and spectrometers all are particularly polarization sensitive and if care is not exercised substantial errors can be introduced if the polarization state of the radiation being measured is not known and dealt with.

The components described here utilize birefringence (double refraction), dichroism (a special case of birefringence), optical activity, polarization by reflection (at or near Brewster's angle), and polarization by a linear array (wire grid). Birefringence is the most common basis for the operation of many optical components, and we therefore discuss these principles first.

BIREFRINGENCE

A birefringent crystal, such as calcite, will divide an entering beam of monochromatic light into two beams having opposite (orthogonal) polarizations. The beams will usually propagate in different directions and both have different propagation speeds. Depending on whether the birefringent crystal is uniaxial or biaxial, there will be only one or two directions (the optic axis directions) within the crystal along which the beam will remain collinear and continue to propagate at the same speed. If the crystal is a plane-parallel plate, and the optic axis directions are not collinear with the beam, the radiation will emerge as two separate, orthogonally polarized beams. Where the beams overlap upon emergence, that portion of the beam will remain unpolarized.

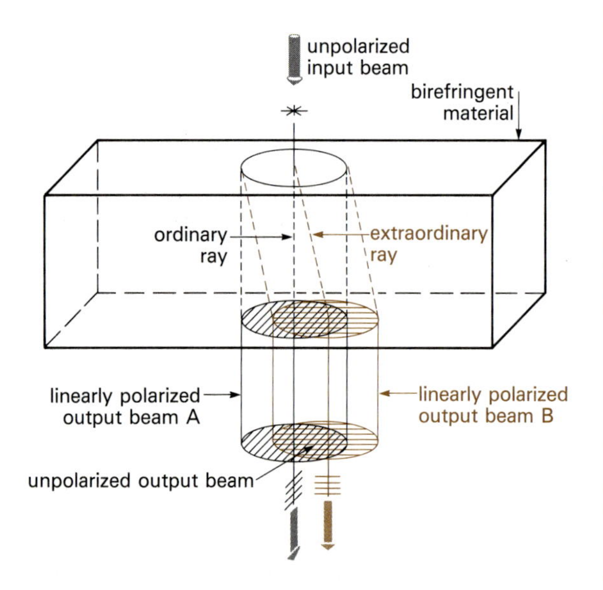

DOUBLE REFRACTION in a birefringent crystal.

The two new beams within the birefringent material are distinguished from each other by more than just polarization and velocity. The rays within one beam are referred to as extraordinary, in that they violate both Snell's Law and the Law of Reflection. These rays need not be confined to the plane of incidence. Furthermore, the velocity of these rays changes with direction. Thus the index of refraction for the extraordinary rays is also a continuous function of direction. The index of refraction for the other ray (the ordinary ray) is a constant and is independent of direction. The two indices of refraction are equal only in the direction of an optic axis within the crystal. The dispersion curve for the ordinary ray is described as a single unique curve when the ordinary index of refraction is plotted against wavelength. The dispersion curve for the extraordinary ray is a family of curves with different curves for different directions. Unless in a particular special polarization state, or unless the crystalline surface is perpendicular to an optic axis, a ray normally incident on a birefringent crystalline surface will be divided in two at the boundary. The extraordinary ray will be deviated while the ordinary ray will not. The ordinary ray index n_0, and the most extreme (whether greater or smaller) extraordinary ray index n_E, are together known as the principal indices of refraction of the material.

In the same way, if a beam of linearly polarized monochromatic light enters a birefringent crystal along a direction not parallel to the optical axis of the crystal, the beam will be divided into two separate beams. Each will be polarized at right angles to the other, and they will travel in different direction. The energy of the original beam will be divided between the two new beams in a manner which depends on the original orientation of the electric field vector to the crystal. The ratio between the two orthogonally polarized beams can be of any value, including the possibility that all the energy will go into one or the other of the new beams. If the crystal is cut as a plane-parallel plate these beams will recombine upon emergence to form an elliptically polarized beam.

As we shall see later, both aspects of the difference between the ordinary and the extraordinary ray may be used to create birefringent crystal polarization devices. In some cases the difference in refractive index is used primarily to separate the rays (as for example in the various Glan type polarizers). In other cases the changes in direction of propagation between the two rays is optimized to separate the incoming beam into two separately polarized beams (as in the Wollaston prisms).

DICHROISM

Dichroism is the selective absorption of one plane of polarization in preference to the other orthogonal polarization during transmission through a material. Sheet type polarizers, such as those described later, are manufactured from organic materials which have been imbedded into a plastic sheet. The sheet is then stretched, thereby aligning the molecules and causing them to be birefringent, and then dyed with a pigment. The dye molecules selectively attach themselves to the aligned polymer molecules, with the result that absorption is very high in one plane and relatively weak in the other. The transmitted beam is linearly polarized. Polarizers made of such material are very useful for low power and visual applications. The usable field-of-view is large (up to grazing incidence) and diameters in excess of 100mm are available.

OPTICAL ACTIVITY

The earlier description of birefringence is applicable to non-active crystals, such as calcite. These are crystals in which there is a specific direction in which the index of refraction is exactly equal for both the ordinary (O) and the extraordinary (E) ray. In "active" crystals, such as quartz, there is no such axis; there is no direction within the crystal in which the indices of refraction are equal for both the ordinary and the extraordinary ray. There is, however, a specific direction in which the difference in index between the O and E indices is a minimum. This direction is the optical axis for active crystals. If a quartz plate is cut such that its optic axis is normal to the surfaces of the plate, and a ray of linearly polarized light is incident parallel to the optical axis, the ray will be separated into two collinear, circularly polarized rays. The ordinary (O) and the extraordinary (E) rays will have opposite senses of circular polarization and will travel at different speeds. The difference in speeds in the O and E rays has the effect of causing the resultant plane of polarization to rotate about the optic axis as the ray penetrates the plate. The amount of rotation is directly proportional to the depth of penetration, and ultimately to the thickness of the plate. The result of the superposition of the counter-rotating circular polarizations is always to produce linear polarization without any intermediate elliptical polarization states. This is the distinction between optical activity or circular retardation and the more commonly known linear retardation. For crystalline quartz at the yellow sodium line wavelength (589.3nm) the difference in the refractive indices for opposite circular polarizations propagating along the optic axis is only about 1 part in 10,000. Correspondingly the plane of resultant linear polarization rotates at a rate of about 21.7 degrees per millimeter travel. Crystalline quartz exists in two distinct crystalline forms. Natural crystals of these forms, and their respective molecules, have shapes which are mirror images of each other. The plane of resultant linear polarization rotates in opposite senses in these forms, known as left-handed and right-handed quartz.

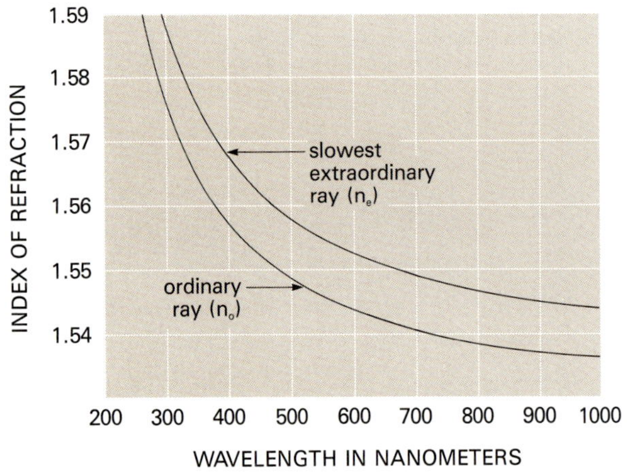

REFRACTIVE INDICES OF CRYSTALLINE QUARTZ (AT 18°C) VERSUS WAVELENGTH. The extraordinary index may have any value between n_0 and n_e depending on the direction of propagation in relation to the crystalline optic axis direction.

The polarization of the O and E rays in quartz rapidly changes from circular to elliptical even for directions which depart only slightly from the optic axis direction. For sodium light at an angle of only 5 degrees from the axis, the polarization ellipse has a major to minor axis ratio of 2.37. For the ellipse to be even approximately circular, much smaller angles are required. For this reason devices which depend on circular polarization (such as the Cornu pseudo-depolarizer) are effective only in highly collimated light propagating quite accurately parallel to the optic axis direction.

POLARIZATION BY REFLECTION

We have discussed polarization effects occurring at Brewster's angle elsewhere (see page 174). When a beam of ordinary light is incident at the polarizing angle on a transmissive dielectric such as glass, the emerging refracted ray is partially linearly polarized. For a single surface of glass (with $n = 1.50$) at Brewster's angle 100% of the light vibrating parallel to the plane of incidence is transmitted, whereas for the perpendicular vibration only 85% is transmitted (the other 15% being reflected). Clearly, the degree of polarization resulting from a single surface reflection is small.

If a number of plates are stacked parallel to one another and the entire pile is orientated at the polarizing angle, some of the vibrations perpendicular to the plane of incidence will be reflected at each surface and all those parallel to it will be refracted. The net result will be that the refracted beam is partially plane polarized due to having lost more and more of its perpendicular components at each reflection interface. By making the number of plates within the stack large (greater than 25) quite high degrees of linear polarization may be achieved. This is the polarization method utilized in the Melles Griot polarizing beamsplitters described later in this section. The pile-of-plates is made up of many layers of quarter-wave dielectric thin films laid onto the interior prism angle of the beamsplitter. The result is a beamsplitter which separates the incident laser beam into two mutually perpendicular beams each of which is polarized in the opposite sense.

WIRE GRID POLARIZATION

When radiation is incident on an array of parallel reflective stripes whose spacing is of the order of or less than the wavelength of the radiation, the radiation whose electric vector is perpendicular to the direction of the array is reflected. The result is that the transmitted radiation is largely linearly polarized. The difficulty in this method of polarization is manufacturing a sizeable reflective array with uniform spacing of the grid at short enough wavelengths to be useful optically. Wire grid arrays for polarization of radar and radio waves are fairly straightforward, but only recently have techniques been developed which enable this method to be used in the near-infrared.

The Melles Griot infrared polarizers described later in this section utilize this newly developed method to make possible useful and effective polarizers in the near-IR. These polarizers are considerably more effective than the dichroic type polarizers previously available for this wavelength range.

BIREFRINGENT POLARIZERS

Calcite is an anisotropic, uniaxial crystal which exhibits birefringence. Since the refractive index for the extraordinary ray is less than that for the ordinary ray the material is considered optically negative. Numerous other materials exhibit birefringence, both positive and negative and many have a greater birefringence value ($n_o - n_E$). Yet no other material is as readily available in adequate size, clarity and freedom from problems (such as being poisonous, hygroscopic, overly fragile or optically unworkable) as calcite. The material used for Melles Griot calcite polarizers is mined in Mexico and carefully selected to meet the requirements of MIL-G-174A. These requirements assure that the material is free of visible striae, veils, inclusions or defects.

Calcite has a wide range of optical transmission with an absorption coefficient for both the E-ray and the O-ray of less than 0.1 from 260.0nm to 1.7μ. The O-ray has a somewhat greater transmission range (especially in the infrared), but as the O-ray is typically the rejected ray in a calcite polarizer this greater transmission range is not advantageous. The dichroic nature of the transmission is a further result of the anisotropic nature of the material. The index of refraction of calcite is given for both the O-ray and the E-ray in the accompanying table. The dispersion formulae for the O- and E-rays are as follows:

$$n_o = (2.863919 - 0.010849\lambda^2 + 0.033682\lambda^{-2} - 0.006746\lambda^{-4} + 0.001518\lambda^{-6} - 0.000112\lambda^{-8})^{1/2}$$

$$n_E = (2.173670 + 0.000814\lambda^2 + 0.021472\lambda^{-2} - 0.006434\lambda^{-4} + 0.001422\lambda^{-6} - 0.000107\lambda^{-8})^{1/2}$$

Calcite crystals may be formed into very efficient polarizers by causing the ordinary and extraordinary rays to confront an angle within the material that exceeds the critical angle for one of the rays, but not for the other. In this way one of the polarizations is completely separated from the other by total internal reflection. Many forms of calcite polarizer have been designed, but in practice only three, the Glan-Thompson, Glan-Taylor and the Wollaston, styles are of great benefit. While all three styles are offered by Melles Griot, and all are similar, the differences between these styles tend to trade the interdependent relationships between power handling, useful acceptance angle, and beam separation.

Both Glan-type prisms have been designed such that both the entrance and exit faces are normal to the intended direction of use and the prism angle has been cut such that the O-ray is totally internally reflected at the first face. In the case of the Glan-Thompson prism, the two halves are cemented together with an optical cement. In the case of the Glan-Taylor prism, the two halves of the polarizer are separated by an air-space. While the air-space separation does allow the prism to handle substantially higher power levels than the cemented prisms, the useful angular field is reduced and the losses due to internal Fresnel reflections are slightly increased.

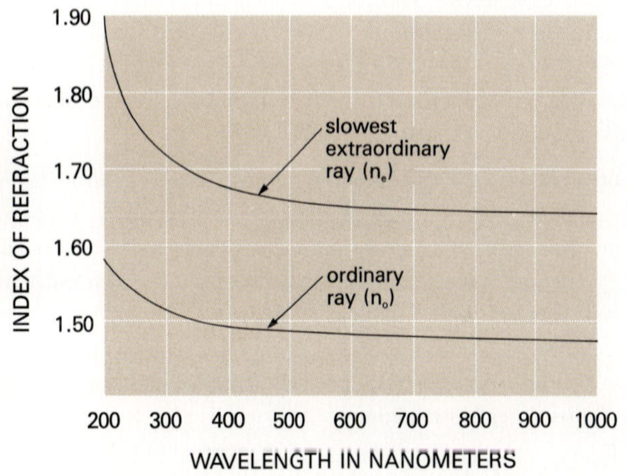

REFRACTIVE INDEX OF CALCITE for the ordinary (n_o) and extraordinary (n_e) rays.

Wollaston prisms transmit both linearly polarized beams. The beams are polarized in mutually perpendicular planes and both emerge from the prism traveling in different directions. In this way both beams may be worked with independently further down the optical path.

Whatever form of Glan polarizer is chosen, the prism angle is selected such that it lies mid-range within the boundaries of acceptable polarizing angles. One side is bound by the angle at which total internal reflection occurs for both the O- and the E-ray; the other is bound by the angle at which transmission begins for both rays. This angle can be optimized only for one wavelength in any particular polarizer and will vary depending on the type of polarizer and whether or not the prism is cemented or air-spaced. In general terms, the acceptance angle for a Glan-Taylor polarizer is typically 8.5° and for a Glan-Thompson prism is 18.5°. Specific details of the acceptance angle as a function of wavelength are given under each specific product.

GLAN-TAYLOR POLARIZING PRISMS

These Glan-Taylor polarizers are manufactured from the very finest optical grade natural calcite and are free from defects and blemishes. They are particularly useful in applications requiring a high degree of polarization purity, high total transmission and low- to medium-power requirements.

Both entrance and exit faces are cut and polished to be normal to the mechanical axis and the transmitted, linearly polarized extraordinary beam is not deviated, upon emergence, from its initial direction. Because these prisms are air-spaced, their transmission is excellent from 215nm to 2.3 microns. Overall transmission is optimized since the transmitted beam is incident on the air-space interface at approximately Brewster's angle, thereby reducing reflection losses to a minimum.

The length-to-aperture ratio of these prisms is 0.85. The usable, full angular field-of-view is approximately 8.5° and is symmetrical about the mechanical axis. The field-angle varies as a function of wavelength as shown on the accompanying graph. However, it should be remembered that the useful amount of flux which can be transmitted through a prism or other optical component is dependent on both the angular aperture and the cross-sectional area of the aperture. The maximum throughput (defined as the product of the solid angle and the clear aperture area perpendicular to the prism axis) of a Glan-Taylor prism with a relatively small field-angle may be comparable to a small Glan-Thompson prism with a larger field-of-view.

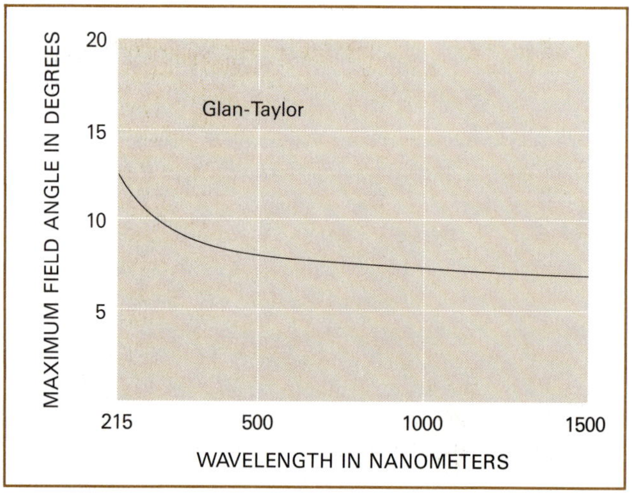

Please call your Applications Engineer for help with your polarization questions.

These polarizing prisms will be found especially useful in conjunction with low to medium power lasers. The rejected (ordinary) beam is absorbed by a black polymeric mounting material within the prism housing and it is the energy level of this beam which sets the upper power levels the polarizer can safely handle. Power levels above 2 watts continuous wave are not recommended for extended times as unacceptably rapid outgassing of the polymer will occur.

Single layer, magnesium fluoride, antireflection coatings are available for these polarizers. These coatings, which are effective over a broad range of wavelengths may be centered on 550nm for visible range use or an 830nm for near-infrared range applications.

Melles Griot Glan-Taylor polarizing prisms are available in three clear aperture sizes: 10 x 10mm, 15 x 15mm, and 20 x 20mm. All are mounted in black anodized aluminum cylindrical housings for ease of use and mounting. The Melles Griot Product Number, as well as the orientation of the plane of emergent polarization, are permanently engraved on the barrel.

SPECIFICATIONS: GLAN-TAYLOR POLARIZING PRISMS

Wavelength Range: 215nm to 2.3 microns
Transmission (ratio of total output to total unpolarized input) for a Single Prism Acting Alone: ½ $(k_1+k_2)=38\%$
Open Transmission for Pair of Prism Polarizers (H_o): >18%
Extinction Ration (H_{90}): $<1x10^{-5}$
Useful Field Angle: see graph
Length/Aperture: 0.85
Dimensions: ±0.25mm
Mechanical to Optical Axis: ±5 arc minutes
Cosmetic Surface Quality: 80-50 scratch and dig
Mounting: Cylindrical black anodized aluminum housing with direction of polarization and Product Number permanently engraved on side.

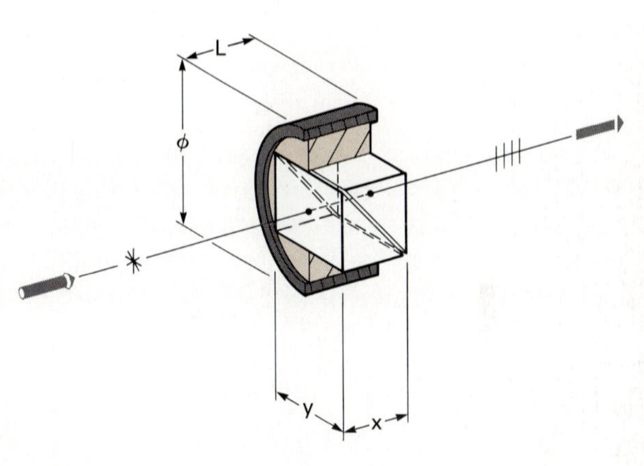

03 PTA

Glan-Taylor Polarizing Prisms

Outside Diameter (mm)	Housing Length (mm)	Clear Aperture (mm)	PRODUCT NUMBER
25.0	10.5	10x10	03 PTA 001
30.0	14.8	15x15	03 PTA 003
40.0	19.0	20x20	03 PTA 005

For antireflection coated prisms please append Coating Suffix from table.

Antireflection coatings for Polarizing Prisms

Center Wavelength (nm)	Wavelength Range (nm)	Maximum Reflectance (%)	COATING SUFFIX
550	400-700	2.0	/A
830	650-1100	2.0	/C

Adapters are available to allow mounting of these prisms into a Melles Griot polarizer holder (see page 461).

GLAN-THOMPSON POLARIZING PRISMS

Melles Griot Glan-Thompson calcite polarizing prisms combine a large field-of-view with high polarization purity and ease of use. For low-power polarization applications, these polarizers are preferred over all others.

These polarizing prisms are offered in two length to aperture ratios, 2.5 and 3.0. The useable angular field-of-view of each is symmetrical about the prism axis. While the specific angle varies as a function of wavelength, a usable field of greater than 12° for the 2.5 ratio prism and 24° for the 3.0 ratio prism may be expected. For specific wavelength requirements see the accompanying chart.

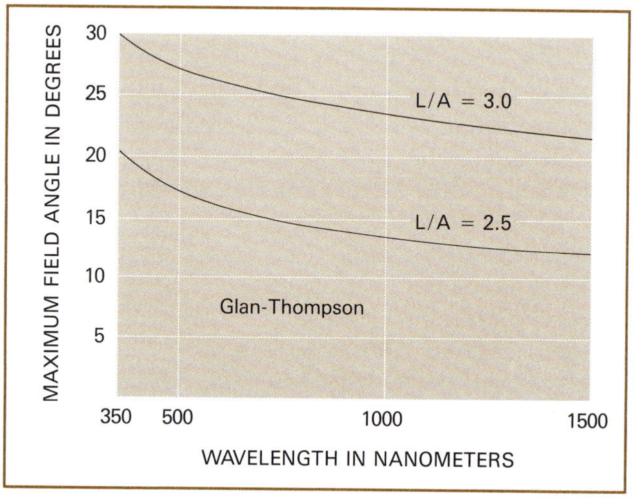

The entrance and exit faces and the optic axis direction are all perpendicular to the mechanical axis (and entrance beam), thereby providing convenience and certainty in mounting and prealignment. The interface between the two prism halves is cemented. This results in a somewhat reduced ultraviolet transmission. The useful transmission range is from 350nm to 2.3 microns.

Glan-Thompson prisms will be found to be most useful in low-power application where their increased angular field-of-view allows the user to increase the system throughput by maximizing the cone angles through the system. The prisms themselves are mounted within a black anodized aluminum cylinder for ease of handling. A black polymeric sealant holds the prisms in place within the cylinder and absorbs the non-transmitted ordinary ray. Failure of this polymer through thermal absorption is the upper boundary of power handling capacity for these prisms. Power levels greater than 2 watts continuous wave are not recommended.

Single layer, magnesium fluoride, antireflection coatings are available for these polarizers. These coatings, which are effective over a broad range of wavelengths may be centered on 550nm for visible range use or an 830nm for near-infrared range applications.

Melles Griot Glan-Thompson polarizing prisms are available in three clear apterure sizes: 10x10mm, 15x15mm, and 20x20mm. All are mounted in black anodized aluminum cylindrcal housings for ease of use and mounting. The Melles Griot Product Number, as well as the orientation of the plane of emergent polarization, are permanently engraved on the barrel.

Melles Griot HeNe lasers are now available for three wavelengths; 543nm, 633nm and 1523nm (see page 345).

SPECIFICATIONS: GLAN-THOMPSON POLARIZING PRISMS

Wavelength Range: 350nm to 2.3 microns

Transmission (ratio of total output to total unpolarized input) for a Single Prism Acting Alone: $\frac{1}{2}(k_1+k_2)=36\%$

Open Transmission for Pair of Prism Polarizers (H_o): >17%

Extinction Ratio (H_{90}): $<1 \times 10^{-5}$

Useful Field Angle: see graph

Dimensions: ±0.25mm

Mechanical to Optical Axis: ±5 arc minutes

Cosmetic Surface Quality: 80-50 scratch and dig

Mounting: Cylindrical black anodized aluminum housing with direction of polarization and Product Number permanently engraved on side.

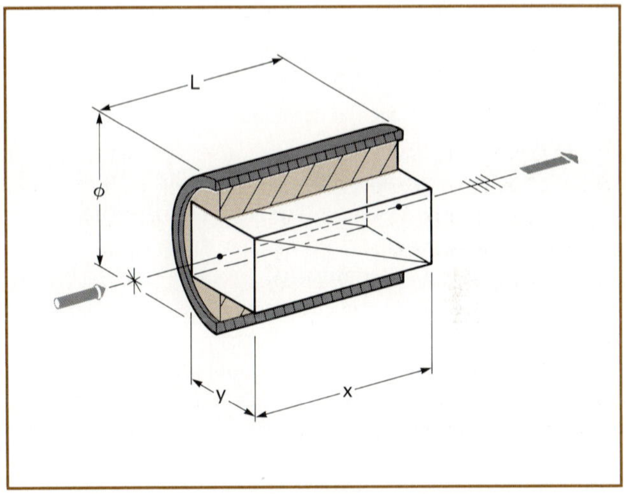

Glan-Thompson Polarizing Prisms

Length/ Aperture	Outside Diameter (mm)	Housing Length (mm)	Clear Aperture (mm)	PRODUCT NUMBER
2.5	25.0	27.0	10x10	03 PTH 001
2.5	30.0	39.5	15x15	03 PTH 003
2.5	40.0	52.0	20x20	03 PTH 005
3.0	25.0	32.0	10x10	03 PTH 012
3.0	30.0	47.0	15x15	03 PTH 014
3.0	40.0	62.0	20x20	03 PTH 016

Antireflection Coatings for Polarizing Prisms

Center Wavelength (nm)	Wavelength Range (nm)	Maximum Reflectance (%)	COATING SUFFIX
550	400-700	2.0	/A
830	650-1100	2.0	/C

Adapters are available to allow mounting of these prisms into a Melles Griot polarizer holder (see page 461).

WOLLASTON PRISMS

Melles Griot Wollaston prisms provide a simple and convenient way to split a beam of light into two mutually orthogonal linearly polarized beams. Two angular separation versions are available. After passing through the prisms, the beams diverge from one another by an angle of approximately 15° or 20°.

Wollaston prisms provide a convenient method of providing double images of single sources, such as the Young two-slit experiment and scatter plate interferometry.

Beam separation angle between the orthogonally plane polarized output is wavelength dependent. Each beam is deviated approximately half of the divergence from the input beam. There is some asymmetery of the deviation between the two beams. The specific separation angle (for both separation values), as a function of wavelength, is best seen by inspecting the graph below. The two halves of these prisms are cemented together. Useful transmission is from 350nm to 2.3 microns.

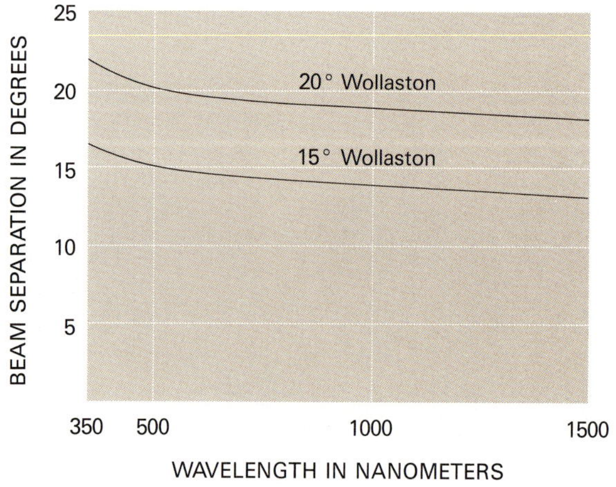

Single layer, magnesium fluoride, antireflection coatings are availalbe for these prisms. These coatings which are effective over a broad range of wavelengths may be centered on 550nm for visible range use or on 830nm for near-infrared range applications.

Melles Griot Wollaston prisms are mounted in a black anodized cylindrical aluminum housing. This housing mounts easily and conveniently in adapters for the Melles Griot polarizer holder

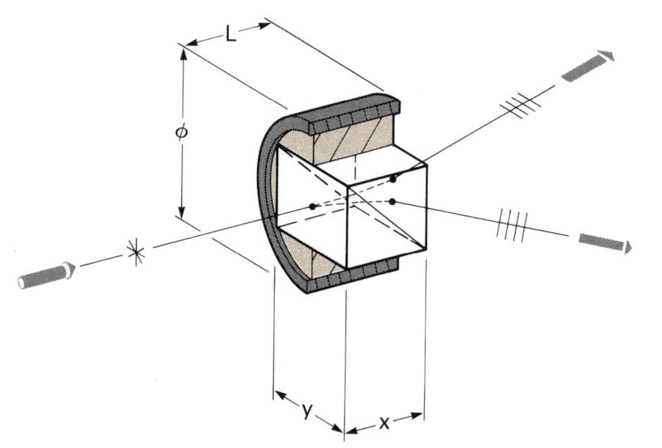

03 PPW

SPECIFICATIONS: WOLLASTON PRISMS

Wavelength Range: 350nm to 2.3 microns

Transmission (ratio of total output to total unpolarized input): $(k_1+k_2)=84\%$

Extinction Ratio (H_{90}): $<1x10^{-5}$

Length/Aperture: 1.0

Separation Angle: see chart

Dimensions: ±0.25mm

Mechanical to Optical Axis: ±5 arc minutes

Cosmetic Surface Quality: 80-50 scratch and dig

Mounting: Cylindrical black anodized aluminum housing with Product Number permanently engraved on side.

Wollaston Prisms

Outside Diameter (mm)	Length (mm)	Beam Deviation (degrees)	Clear Aperture (mm)	PRODUCT NUMBER
25.0	12.0	15	10x10	03 PPW 001
25.0	17.0	15	15x15	03 PPW 003
25.0	12.0	20	10x10	03 PPW 012
25.0	17.0	20	15x15	03 PPW 014

Antireflection Coatings for Polarizing Prisms

Center Wavelength (nm)	Wavelength Range (nm)	Maximum Reflectance (%)	COATING SUFFIX
550	400-700	2.0	/A
830	650-1100	2.0	/C

CALCITE

Calcite ($CaCO_3$) is a commonly occurring mineral which takes on many different forms including limestone and marble. Calcite which is used in the manufacture of polarizing components is a pure, colorless, rhombohedral crystal. Occurrence of this form in sufficiently large and perfect crystals is fairly rare.

The calcite used by Melles Griot is mined in northern Mexico. It is specified as best optical quality calcite. When a laser beam is passed through one of these crystals, no significant scatter can be detected from imperfections and inclusions in the crystal.

The strong birefringence of calcite makes it one of the best-known materials for the manufacture of polarizing components.

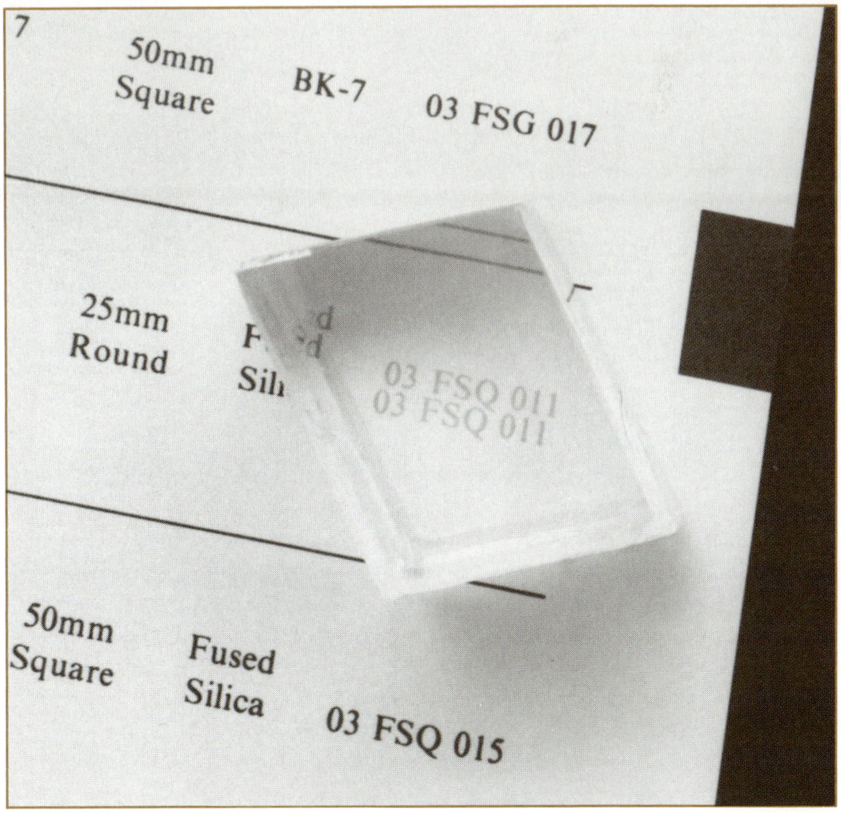

POLARIZING BEAMSPLITTERS

These prisms function as polarizers, beamsplitters, or both. Pairs of right angle prisms are cemented together, hypotenuse-face-to-hypotenuse-face, with a special multilayer dielectric film in between. Monochromatic unpolarized light which is normally incident upon the external faces of the resulting cube (internally incident at 45° upon the multilayer film) is separated into two polarized beams which emerge from the cube through adjacent faces and in directions which are accurately 90° apart. The beam which passes straight through the cube emerges linearly polarized to a purity of 98% or better, with the plane of the electric field vector parallel to the plane of incidence defined for the multilayer film (p-polarized). The beam which emerges from the cube at right angles to the incident beam (having been reflected by the multilayer film) is also linearly polarized to a purity of 98% or better, this time with the electric field vector orthogonal to the plane of incidence defined for the multilayer film (s-polarized). An incident beam of linearly polarized monochromatic light is similarly divided, with the ratio of the emergent beam irradiances depending on the orientation of the electric field vector of the incident beam.

If one of these polarizing cubes is preceded by a half-wave retardation plate, the result is a variable ratio beamsplitter for linearly polarized monochromatic input. The output beam irradiance ratio can be continuously varied from below 1:49 to above 49:1, or fixed at any value in between, by suitably rotating the half-wave retarder within its plane. A ratio of 1:1 is easily achieved. For unpolarized monochromatic input at the intended wavelength, and without the use of any retarder, these beamsplitters always achieve a very accurate 1:1 ratio regardless of beamsplitter orientation.

The operation of these polarizing beamsplitters is best understood by thinking of them as pile-of-plates polarizers. The pile of plates is in this case a quarter-wave resonant reflective multilayer dielectric stack designed for 45° incidence. Because the total thickness of the multilayer film is only a few wavelengths, there are no ghost image problems associated with this film. At every dielectric interface within the stack, the radiation is incident at or near Brewster's angle (also called the polarizing angle), and the p-polarized component is therefore transmitted with almost no reflection. The s-polarized component, in contrast, is partially reflected at each interface. By resonance among these many partial reflections, the total reflectance of the stack for s-polarized radiation is raised to 98% or better.

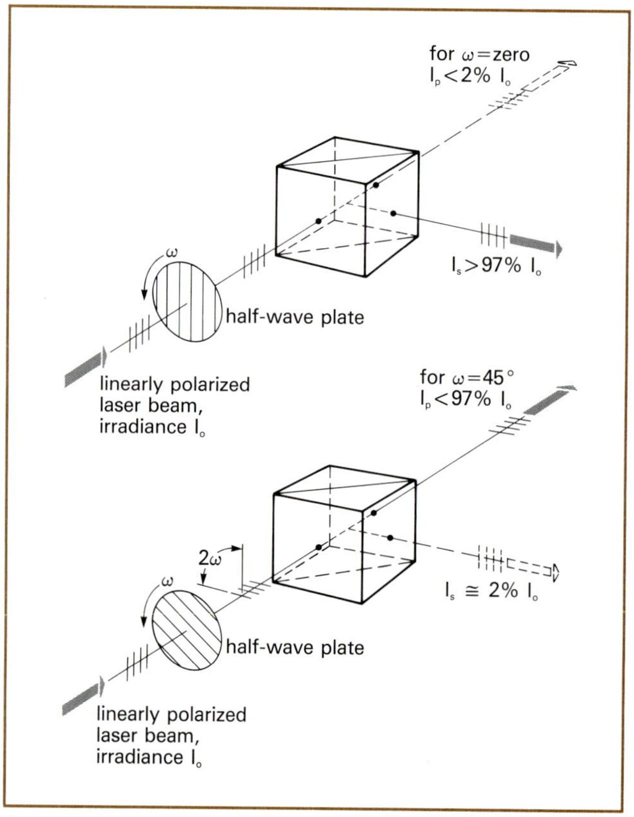

VARIABLE RATIO BEAMSPLITTER: The half-wave plate rotates with angular velocity ω while the permitted plane of output polarization (here suggested by lines on the highly transparent retarder) rotates at angular velocity 2ω. The polarized beams which finally emerge from the cube have the irradiances I_s I_p indicated. Intermediate irradiances are seen at intermediate times.

These polarizing cubes are suitable for most of the monochromatic applications to which other polarizers (such as calcite polarizers) might be applied, such as isolator construction, or the construction of electro-optic or magnetic-optic modulators. They can also be used in laser Q-switches, though the use of cement and the multilayer film, prohibit use at high powers.

These polarizing beamsplitters may be conveniently mounted on the appropriate Melles Griot prism table. The prism tables provide the user with smooth ±3° tilt in two orthogonal axes and

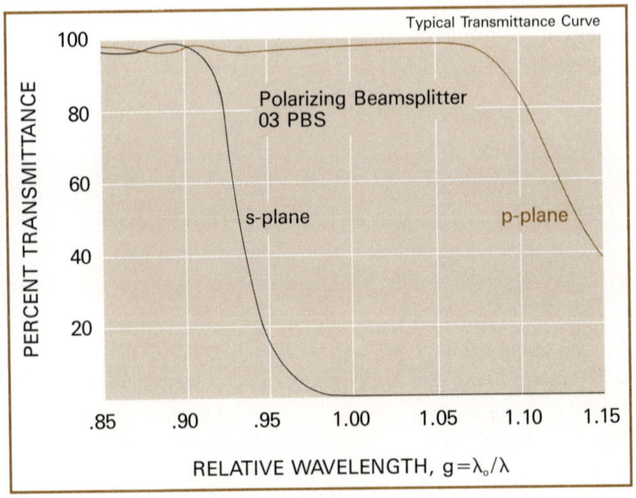

±3° rotational control. All motions are controlled by thumb screws to provide assured and controlled beamsteering functions in all axes. (See page 456.)

Polarizing beamsplitters are available coated for eight common laser wavelengths. They provide a polarization purity of 98% or better at their design wavelength. Where greater extinction is needed than crossed pairs of the present polarizing cubes can provide, each cube can be replaced by a pair of cubes in identical orientation. The resulting extinction will be the square of that of a single pair. Principal transmittance, extinction ratios, etc., are defined for these polarizer in exactly the same way as for sheet dichroic) polarizers.

All four entrance and exit faces of the beamsplitter are multilayer antireflection coated with high-efficiency coatings which reflect less than 0.25% per surface at the specified wavelengths.

SPECIFICATIONS: POLARIZING BEAMSPLITTERS

Dimensions: ±0.2mm

Principal Transmittance: >98%T for p-polarization; <2%T for s-polarization

Principal Reflectance: >98%R for s-polarization; <2%R for p-polarization.

Beam Polarization Purity (with unpolarized input): 98% pure p-polarization in straight-through beam, 98% pure s-polarization in 90° deviated beam

Transmission (ratio of straight-through output to total unpolarized input) for single polarizer acting alone $\frac{1}{2}(k_1+k_2)=49\%$

Open Straight-Through Transmission of pair assuming unpolarized input: $H_o=\frac{1}{2}k_2^2+k_2^2)=48\%$

Extinction Ratio or closed straight-through transmission of pair, assuming unpolarized input: $H_{90}=k_1k_2=0.02$

Beam Deviation: <3 arc minutes

Useful Angular Diameter: 10°

Entrance/Exit Surface Flatness: <λ/10 at 633nm

Wavefront Distortion: <λ/2at 633nm

Cosmetic Surface Quality: 40-20 scratch and dig

Material: BK 7 Grade A Fine Annealed

AR Coating: ±0.25%R — all four faces

Unused faces are fine ground, and all edges are lightly beveled.

Polarizing Beamsplitters

A=B=C (mm)	PRODUCT NUMBER Wavelength (nm) 488	514	543	633	780	830	1300	1550
3.2	—	—	—	—	03 PBS 051	03 PBS 061	03 PBS 071	03 PBS 081
5.0	—	—	—	—	03 PBS 052	03 PBS 062	03 PBS 072	03 PBS 082
10.0	03 PBS 013	03 PBS 023	03 PBS 033	03 PBS 043	03 PBS 053	03 PBS 063	03 PBS 073	03 PBS 083
20.0	03 PBS 015	03 PBS 025	—	03 PBS 045	03 PBS 055	03 PBS 065	03 PBS 075	03 PBS 085
25.4	03 PBS 017	03 PBS 027	03 PBS 037	03 PBS 047	03 PBS 057	03 PBS 067	—	—
38.1	03 PBS 019	03 PBS 029	—	03 PBS 049	—	—	—	—

DICHROIC SHEET POLARIZERS

Sheet-type or dichroic polarizers are made of dichroic materials, which are birefringent materials in which one of the two orthogonal polarizations (either ordinary or extraordinary) is subject to strong absorption, while the other polarization is not. Historically, the name dichroic has its origin in the fact that certain crystalline materials of this type, when subject to white linearly polarized illumination, characteristically exhibit two distinct colors. The word literally means two-color, and is applied in this literal sense to our color additive and color subtractive interference filters. The earliest sheet-type polarizers were made with tiny dichroic crystals, all oriented in a special way, embedded in a plastic film. Modern sheet polarizer material is instead made of polymeric plastic, the molecules of which are oriented by stretching (thereby achieving birefringence), and to which pigment molecules are selectively attached. The preferential absorption of one polarization is related to the orientation of chemical bonds at the attachment sites.

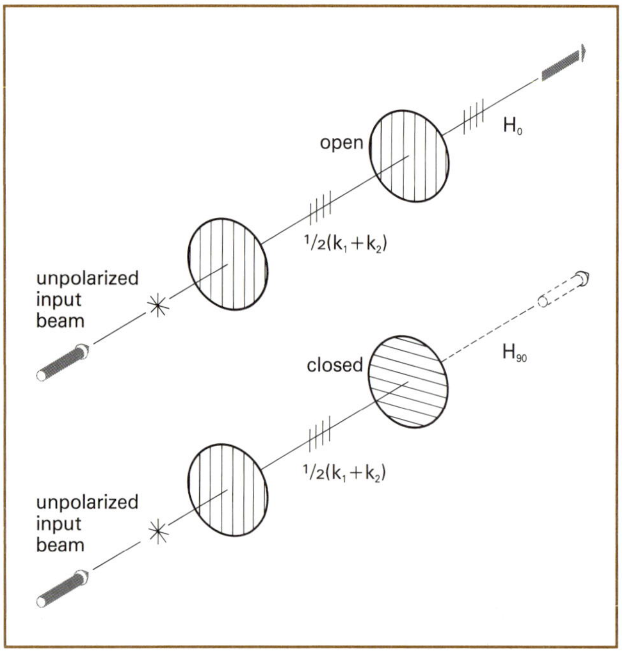

Our polarizers are made of a plastic dichroic polarizing sheet sandwiched between selected strain-free glass plates. Because dichroic materials function by selective absorption, they cannot be used in high-power laser beams. Advantages of these polarizers include large apertures and acceptance angles, excellent extinction ratios and simplicity in mounting.

The transmittance T of a single-sheet polarizer in a beam of linearly polarized incident light is given by

$$T = k_1 \cos^2(\theta) + k_2 \sin^2(\theta)$$

where θ is the angle between the plane of polarization of the incident beam (more accurately, the plane of the electric field vector of the incident beam) and the plane of preferred transmission of the polarizer. The orientation of the latter plane is clearly marked by engravings on the ring in which the polarizer is mounted. k_1 and k_2 are the principal transmittances of the polarizer, and both are functions of wavelength. Ideally, $k_1 = 1$, and $k_2 = 0$. In reality, k_1 is always somewhat less than unity, and k_2 always has some small but nonzero value.

If the incident beam is unpolarized, and the angle θ is redefined to be the angle between the planes of preferred transmission (planes of polarization) of two sheet polarizers in near contact, it can easily be shown that the transmittance of the pair is given by

$$T_{pair} = k_1 k_2 \sin^2\theta + \tfrac{1}{2}(k_1^2 + k_2^2)\cos^2\theta.$$

If we define

$$H_{90} = T_{pair}(90°) = k_1 k_2 \text{ and}$$
$$H_0 = T_{pair}(0°) = \tfrac{1}{2}(k_1^2 + k_2^2),$$

the above formula can be simplified to

$$T_{pair} = H_{90}\sin^2\theta + H_0 \cos^2\theta$$
$$= H_{90} + (H_0 - H_{90})\cos^2\theta.$$

The quantity H_{90} is called the closed transmittance or extinction ratio, while the quantity H_0 is called the open transmittance. Both quantities are wavelength dependent, as is shown in the graphs. Because of the large ranges of open and closed transmission, it is convenient to graph the optical densities corresponding to these transmissions, rather than the transmissions themselves. The open and closed optical densities are defined as follows:

$$\Delta_{90} = \log_{10}\left(\frac{1}{H_{90}}\right),$$

and

$$\Delta_0 = \log_{10}\left(\frac{1}{H_0}\right)$$

Applications for these polarizers include production and/or analysis of light in various polarization states, glare reduction, isolator construction (see discussion of quarter-wave plate applications), and the construction of electro-optic and magneto-optic modulators.

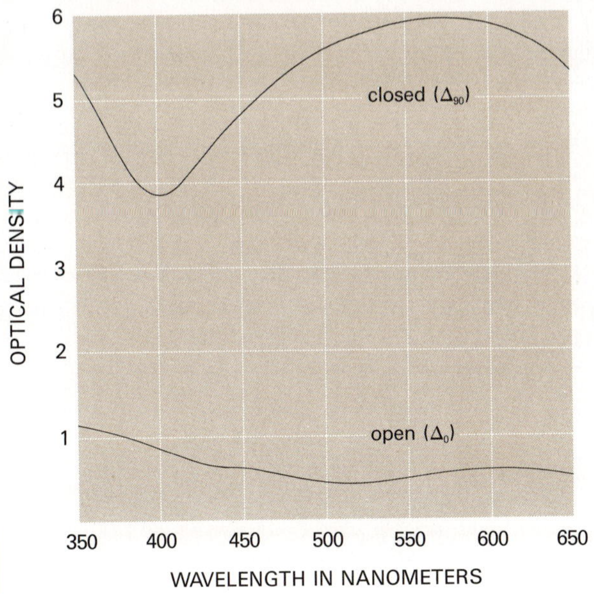

Mounted Sheet Polarizers

Outside Diameter (mm)	Clear Aperture (mm)	Thickness t (mm)	PRODUCT NUMBER
20.6	16.9	2.5	03 FPG 001
30.2	26.5	2.5	03 FPG 003
41.3	38.4	2.5	03 FPG 005
50.8	45.4	2.5	03 FPG 007
63.5	58.0	3.5	03 FPG 009
82.6	76.0	3.5	03 FPG 011
93.0	85.0	3.5	03 FPG 013
103.0	94.0	3.5	03 FPG 015
119.0	110.0	3.5	03 FPG 017

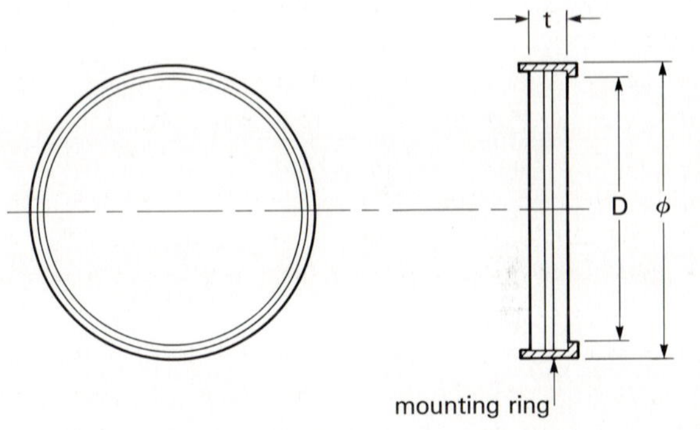

03 FPG

SPECIFICATIONS: SHEET-TYPE POLARIZERS

Transmission (ratio of total output to total unpolarized input) for a single dichroic polarizer acting alone: $\frac{1}{2}(k_1+k_2)=33\%$

Open transmission for pair of polarizers (H_o): $\geq 20\%$

Extinction Ratio (H_{90}): 10^{-4} for visible white light, closed pair of polarizers

Useful Field Angle: 90° (can be used at grazing incidence)

Diameter: ± 0.25mm

Thickness: +0.0,-0.5

Mounting: Between selected strain-free glass discs, held in a black metal ring. Polarization axis plane of electric field vector in preferred transmission) is indicated by engravings on the mount.

The 30.2mm outside diameter Melles Griot dichroic sheet polarizer may be conveniently mounted in a Melles Griot polarizer holder. This holder features large, firm brass on stainless steel conical bearings for smooth, wobble-free rotation (see page 460).

NEAR-INFRARED POLARIZERS

These newly developed near-infrared polarizers are metallic thin film devices which operate on the same principle as wire grid polarizers. Two types are available to cover the wavelength ranges of 780nm to 1250nm and 1250nm to 1550nm. Because they do not absorb radiation in the polarization process, they will be found far more useful and effective than the dichroic polarizers previously available for use within these wavelength ranges.

The polarizer itself consists of an array of thin, parallel metallic whiskers all aligned in the same direction. The whiskers cause incident radiation to be polarized by much the same mechanism used by an array of parallel metal wires. Historically this type of polarizer was used by Hertz prior to 1900 to polarize radio waves. The array of parallel metal elements acts as a polarizer when the spacing between wires is less than the wavelength of the incident radiation and the grid material is reflective to the radiation at the wavelength to be polarized. Incident unpolarized light is resolved into components parallel and perpendicular to the direction of the metallic wires. The polarization component parallel to the wires is reflected and the polarization component perpendicular to the wires is transmitted. It is usually the transmitted (polarized) beam which is used.

High vacuum evaporative techniques are used to deposit a thin, parallel array of metallic whiskers onto a smooth glass substrate. When radiation is incident normal to the film surface, the unwanted polarization component is reflected by the wire-array surface while the desired polarization component is transmitted. The thin-film surface is too delicate to survive touching or cleaning, so it is hermetically sealed by another piece of smooth glass into a black anodized aluminum cell. A slight positive pressure purging with nitrogen during assembly further protects the active surface from degradation. The finally assembled component is rugged and capable of surviving a wide range of thermal, vibrational, and pressure environments. To optimize transmission all glass surfaces are antireflection coated with a single layer of magnesium fluoride.

The direction of the polarization axis is permanently marked on the edge of the aluminum mounting ring. The Melles Griot Product Number is also marked in the mounting ring for permanent ease of identification. The outside diameters of these polarizers have been chosen so that they can be easily mounted in a Melles Griot polarizer holder. (See page 460.)

These polarizers have been designed to be used in a wide range of applications and are therefore thin and intended to work best with normally incident radiation. Divergent or convergent input cone angles up to 20 degrees can be accepted. The advantages of using these polarizers include the absence of dead-zones within the active area, a broad wavelength region, large apertures, stability in normal temperature and humidity ranges, and the ability to reflect the unwanted polarization component. As with any thin-film device, long term exposure to high humidity combined with cyclical temperature changes will cause degradation and eventual failure.

Definitions of the transmittance (T), open transmittance (H_o), closed transmittance (H_{90}), and optical density are equivalent to the definitions we have used for dichroic sheet polarizers (see page 296). Curves for both wavelength ranges showing opened and closed densities as a function of wavelength are shown on this page. Notice that the polarization effectiveness is largely wavelength independent.

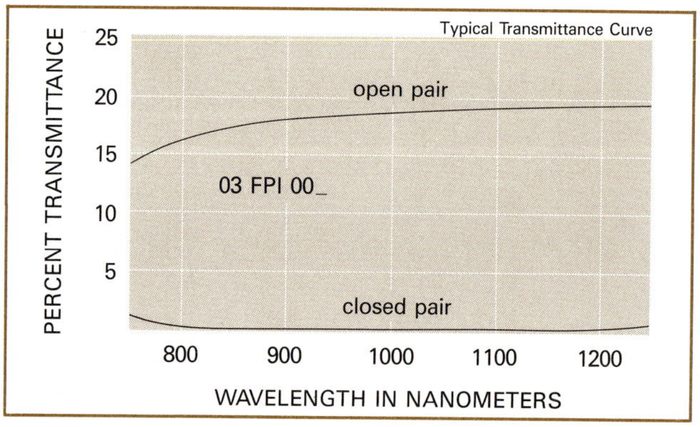

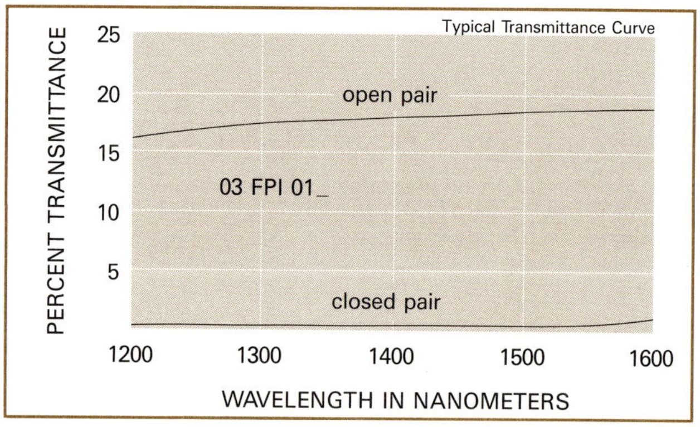

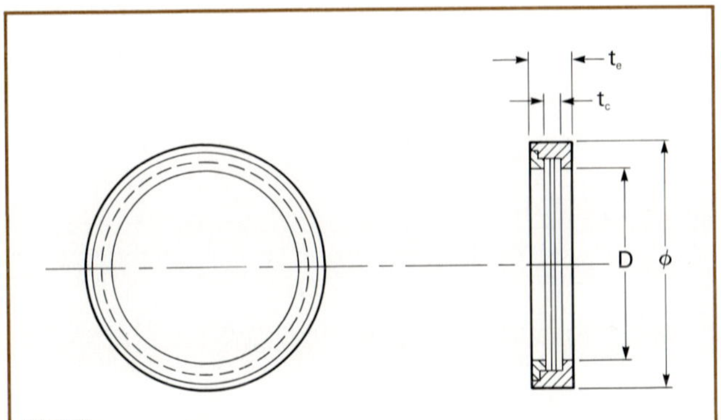

03 FPI

SPECIFICATIONS: NEAR-INFRARED POLARIZERS

Wavelength Regions: 780 to 1250nm
1250 to 1550nm

Transmission (ratio of total output to total unpolarized input) for a single IR polarizer acting alone: $\frac{1}{2}(k_1+k_2) = >20\%$

Open transmission for pair of polarizers (H_o): > 16%

Extinction Ratio (H_{90}): $<10^{-2}$ within the specified wavelength region

Useful Field Angle: ±20° about normal

Dimensions: ±0.25mm

Thickness: 2.75 ±0.2mm

Cosmetic Surface Quality: Float surface, 80-50 scratch and dig

Mounting: Between two glass discs, hermetically mounted in a black anodized aluminum ring. Polarization plane of maximum transmission indicated by engravings on the edges.

Near-Infrared Polarizers

Wavelength Range (nm)	Outside Diameter (mm)	Clear Aperture (mm)	PRODUCT NUMBER
750-1250	20.0	15.0	03 FPI 001
	30.0	25.0	03 FPI 003
	40.0	35.0	03 FPI 005
1250-1550	20.0	15.0	03 FPI 012
	30.0	25.0	03 FPI 014
	40.0	35.0	03 FPI 016

These polarizers are made by the patented "Slocum" process. They have improved specifications compared with dichroic sheet polarizers and withstand infrared radiation without bleaching.

CORNU PSEUDODEPOLARIZER

In laser power measurement, radiometry, and spectral radiometry, polarization properties of the source often interact with detection apparatus to produce significant measurement error. This happens because oblique reflection from either dielectric or metallic surfaces, oblique transmission by dielectric surfaces, dispersion by either gratings or prisms, and passage through slits may modify the polarization state.

True depolarizers do not exist. Pseudodepolarizers transform one polarization state into a continuum of states which under suitable circumstances may mimic the behavior of unpolarized radiation. Properly employed in radiometric measurements, pseudodepolarizers can dramatically improve both the repeatability and absolute accuracy of measurement.

The Cornu pseuddepolarizer may be conveniently mounted on the smallest Melles Griot prism table. This will provide the user with smooth ±3° tilt in two orthogonal axes and ±3° rotational control. All motions are controlled by thumb screws to provide assured and controlled alignment in all axes. (See page 456.)

The Cornu pseudodepolarizer (CPD) employes optical activity in crystalline quartz (see earlier description of optical activity). It operates on collimated beams of light, transforming linear polarization states into a complicated and spatially changing continua of linear polarization states. The CPD transmits circularly polarized light without change, and CPD effectiveness for elliptically polarized light is greatly diminished. In general it can be said that the Cornu pseudodepolarizer is more effective in polychromatic than in monochromatic light, since the mix of output polarizations is more complex in the polychromatic case. The discussion which follows concentrates on the more difficult monochromatic case.

Assume that linearly polarized monochromatic light propagates from left to right through the CPD. It does not matter which end of the CPD is made of left or right-handed quartz. What is important is that the two halves be made of the opposite kinds of quartz, as is the case in the CPD. In both halves the optic axis is normal to the XY plane of the figure (and hence parallels the direction of propagation). A well collimated axial beam is presumed. This beam experiences circular retardation of opposite sign in the two halves, and hence the plane of polarization rotates about the direction of propagation. The amount of rotation is a continuous function of the y-coordinate on the exit face, but is independent of the x-coordinate. This structure of the CPD assures that the collimated beam is transmitted without deviation, and further has the result that the output polarization plane is twice as sensitive to the y-coordinate as it would have been if ordinary glass had been used for one half of the CPD.

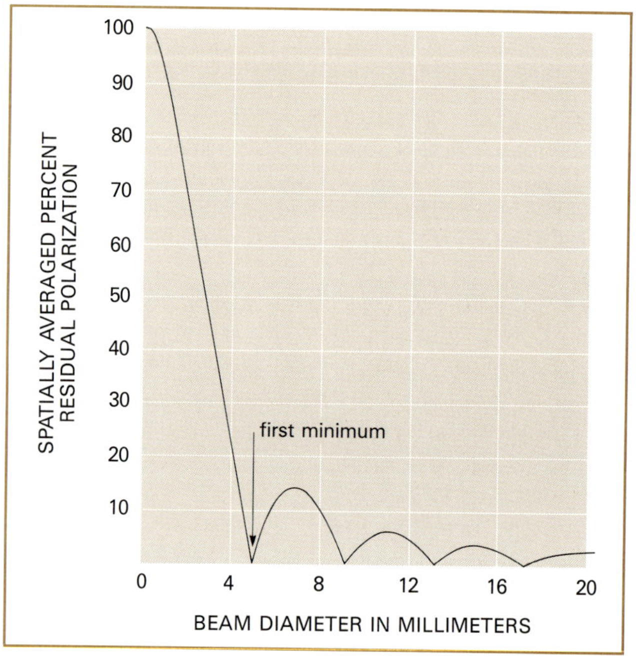

The residual "degree of polarization" at each point (x,y) in the output beam has a strict mathematical definition in terms of the Stokes parameters (see Shurcliff & Ballard, *Polarized Light*, p. 78). The effectiveness of the CPD is conveniently measured by weighted averaging of the residual degree of polarization over the output beam irradiance distribution at the exit face. For monochromatic input, CPD effectiveness demands well-collimated axial input, and is additionally very sensitive to input polarization state, beam size, beam position, and beam irradiance distribution. Beam position is subject to mechanical control. Beam size can be controlled with iris diaphragms and/or beam expanders. For any beam shape or irradiance distribution, beam position and beam size can be experimentally optimized to achieve minimum spatially averaged residual polarization. In special cases, assuming optimum positioning, it is possible to calculate optimum beam size as a guide to such experimentation.

The graph is the numerical result for the special case of a circular beam of constant irradiance. The calculation presumed linear input polarization, yellow light (589.5nm), and that the beam was optimally positioned. The graph is of spatially averaged residual polarization as a function of beam diameter. The calculation was continued up to beam diameters twice as large as the CPD can accommodate. It is apparent that there are several different beam diameters for which the spatially averaged residual polarization vanishes. The first minimum, at a diameter of 5.05mm for this wavelength, is of spatial interest because it is attainable with a real CPD. The beam diameter at first minimum is a function of wavelength, and for many wavelengths only the first minimum is attainable with a CPD of 10mm aperture. For a beam of constant irradiance the diameter at first minimum bears a special relationship to a characteristic distance ΔY_{180} on the exit face. Specifically, the beam diameter at first minimum is approximately 3/4 of this characteristic distance. For a beam of constant irradiance and rectangular shape the optimum beamwidth in the y-direction is approximately 2/3 of this same characteristic distance ΔY_{180}.

The significance of the characteristic distance Y_{180} is that, for linearly polarized input to the CPD, points on the exit face whose y-coordinates differ by this distance have identical states of linear polarization. The plane of output polarization rotates smoothly by 180 degrees as the y-coordinate is changed by this distance. This characteristic distance can easily be calculated for any wavelength. Specifically,

$$\Delta Y_{180} = \frac{90^\circ}{\alpha} + \frac{w}{4},$$

where w is the 10mm clear aperture width of the CPD, and α is the specific rotation or optical rotatory power of crystalline quartz in degrees per millimeter. T. Martin Lowry, in his book *Optical Rotary Power* (Dover, 1964, page 258), publishes the following dispersion formula for specific rotation:

$$\alpha = \frac{9.5639}{\lambda^2 - 0.0127493} - \frac{2.3113}{\lambda^2 - 0.000974} - 0.1905 .$$

In this formula the wavelength λ must be expressed in microns. With the preceding formulas it is possible to estimate the optimum beam diameter for any wavelength.

For any beam diameter, whether optimum or not, the beam may be focused to a spot which more closely mimics the ideal of unpolarized radiation.

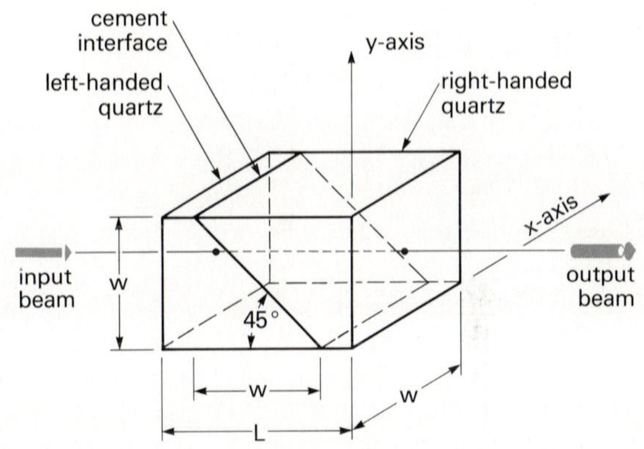

04 LDP 001

SPECIFICATIONS: CORNU PSEUDODEPOLARIZER

Clear Aperture (w): 10mm square
Length (L): 15mm
Dimensional Tolerance: $\pm$0.5mm
Material: Optical grade crystalline quartz
Cement: Optical grade epoxy
Useful Spectral Range: 0.35 to 2.5 microns

Cornu Pseudopolarizer

	PRODUCT NUMBER
	04 LDP 001

Call us with your special polarization component needs.

RETARDATION PLATES

Retardation plates, or phase shifters, including quarter-wave, half-wave, and full-wave plates, are elements primarily used in the synthesis and analysis of light in various states of polarization. The quarter-wave plate is especially useful, finding application in the construction of isolators used in laser-interferometry, in multi-stage traveling-wave laser amplifiers (to prevent the stages from behaving as oscillators), and in electro-optic modulators.

The simplest retardation plate is a slice cut out of a uniaxial crystal such that the slice contains the crystalline optic axis direction. The velocity difference between the ordinary (O) and extraordinary (E) beams within the plate, resulting from an unpolarized beam of light which is normally incident upon it, is therefore maximized. As the O and E beams traverse the plate, a phase difference accumulates between these beams which is proportional to the distance traveled within the plate. At emergence the O and E beams recombine to form a second unpolarized beam.

If the thickness of the plate is such that the phase difference (retardation of the slow ray by comparison with the fast ray at emergence) is 1/4-wavelength, the plate is called a first-order quarter-wave plate. If the phase difference at emergence is 1/2-wavelength, the plate is called a first-order half-wave plate. If the phase difference at emergence is some multiple of 1/4-or 1/2-wavelength, the plate is called a multi-order or higher-order plate. Notice that it is the phase difference, and not the physical thickness of the plate, that these names refer to. Mica and quartz are usually used to make retardation plates, and the actual thickness of even a first-order 1/4-wave is a large number of wavelengths. The usual form is a flat thin disc. Our mica discs are sandwiched between layers of glass for added strength. Mica happens to be biaxial, and quartz exhibits optical activity (rotation of the plane of linear polarization as a function of distance through the medium), facts which do not significantly affect the end results.

Since both the O and E ray refractive indices of most materials are strongly wavelength dependent, the retardation which accumulates within a plate of specified thickness is also wavelength dependent. If a particular value of retardation is desired, this value can be exactly achieved for normal incidence at only one wavelength, and the wavelength itself must also be specified. Mica is exceptional in that its principal indices of refraction vary slowly across the visible spectrum. Thus a retarder made for 550nm and normal incidence will produce approximately the same retardation at other visible wavelengths. Our mica retarders are made for 550nm, and the retardation tolerance is $\pm 10\%$. Retardation plates with a retardation tolerance of 2% are available on special order.

Within the plane of the retarder, the crystalline optic axis and another axis normal to it are often called the "fast" or "slow" axes (whichever is appropriate, depending on whether the uniaxial crystal is positive or negative). By rotating the retarder slightly about one of these axes or the other, it is possible to adjust the value of retardation achieved. Rotation around the crystalline optic axis increases the effective thickness of the plate, but does not affect the velocity difference between the O and E rays, thus increasing the accumulated retardation. Rotation around the other axis both increases the effective thickness of the plate and reduces the velocity difference between the O and E rays. The latter effect dominates for small rotations, reducing the accumulated retardation. In these ways a narrowband retarder (or combination of retarders) may be tuned over a limited range of retardations at fixed wavelength, or over a limited range of wavelengths at fixed retardation.

The effect of a retarder on a beam of polarized light depends on three things; the initial state of polarization, the orientation of the retarder about an axis perpendicular to the disc (measured with respect to the axes of polarization of the incident beam), and the net value of the retardation. The principal plane of the retarder is the plane passing through both the symmetry axis of the disc and the crystalline optic axis within the disc. In the examples of the retarder applications which follow we assume that the monochromatic incident light is collimated and normally incident upon the disc.

HALF-WAVE PLATE APPLICATIONS

If the retarder is half-wave plate (either first- or multiple-order), and the angle between the electric field vector in a plane or linearly polarized incident beam and the retarder principal plane (defined in the preceding paragraph) is θ (which we assume to be acute), the electric field vector in the emergent beam (also plane or linearly polarized) will make an acute angle ϕ with the retarder principal plane. Thus the effect of the half-wave plate is to rotate the plane of polarization through an angle 2θ. The angle θ is continuously adjustable (leave the plane of incident, polarization fixed, and rotate the half-wave plate about its normal), and thus the plane of polarization of the emergent beam is continuously adjustable. Half-wave plates are used in "laser-line rotators" which allow the plane of polarization of a laser beam, as determined by its Brewster's angle windows, to be rotated without having to physically rotate the laser itself. This same result can be accomplished by placing a pair of quarter-wave plates, identically oriented, back-to-back. A half-wave plate followed by a polarizing cube (see page 294) will make a variable ratio beamsplitter for monochromatic linearly polarized inputs. The half-wave plate will convert left circularly polarized light into right circularly polarized light, or vice versa (by reversing the direction of propagation). In the same way it will convert left elliptically polarized light into right elliptically polarized light, or vice versa, effectively reflecting the major axis of the ellipse in the retarder principal plane. A pair of half-wave plates, identically oriented and back-to-back, make up a full-wave or "tint" plate as is used in photoelastic stress analysis. Similarly, half-wave plates can be assembled from quarter-wave plates.

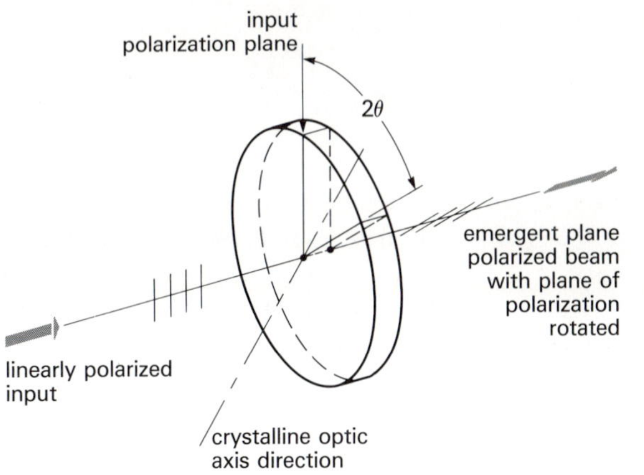

HALF-WAVE PLATE EFFECT ON LINEARLY POLARIZED LASER BEAM. The plane of polarization of the beam can be rotated at will without rotating the laser. The plane of polarization is essentially reflected in a plane containing the crystalline optic axis, and the output plane rotates at twice the angle rate of the retarder.

QUARTER-WAVE PLATE APPLICATIONS, INCLUDING ISOLATOR CONSTRUCTION

If the retarder is a quarter-wave plate, and the angle θ (between the electric field vector of the incident linearly polarized beam and the retarder principal plane) is +45 degrees, the emergent beam is circularly polarized. Reversing θ to –45 degrees reverses the sense of the emergent circular polarization. Reversing the direction of propagation, a quarter-wave plate will transform circularly polarized light into linearly polarized light.

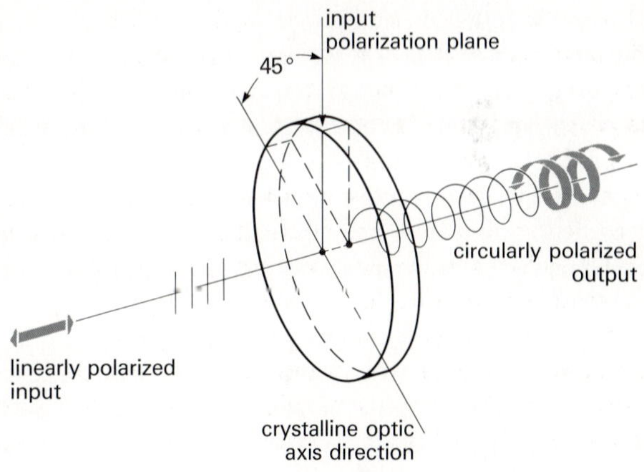

QUARTER-WAVE RETARDATION PLATE: The transformation from linearly to circularly polarized light (or vice versa), which occurs when the angle between the plane of linear polarization and the crystalline optic axis is exactly 45°, is shown. For other angles the transformation is from linear to elliptical (or vice versa).

Especially important is the case in which θ is ± 45 degrees and the direction of propagation of the circularly polarized beam is reversed by reflection in a mirror or other surface at normal incidence, for in this case the sense of the circular polarization is reversed before the second transformation into linear polarization. The result is that the ultimate linear polarization of the reflected beam, as it finally emerges from its second trip through the retarder, is orthogonal to the incident linear polarization with which we began. Suppose that the polarization of the incident beam has been determined by a polarizer, or that a suitably oriented polarizer were present. Then this polarizer would extinguish the reflected beam. The combination of a polarizer and a quarter-wave plate, with θ equal to either +45 or –45 degrees, traversed in the order stated, is opaque to specularly reflected radiation. Such a combination is called an isolator.

Isolators are used to prevent feedback from interferometers into lasers, with the resulting troublesome mode-pulling and de-tuning of the lasers.

Similarly, inside a laser-illuminated Twyman-Green interferometer, quarter-wave plates are often used to suppress unwanted reflections.

If θ has some value other than 45 degrees (+ or –), a quarter-wave plate transforms linearly polarized light into (right or left) elliptically polarized light. Again the transformation is reversed by reversing the direction of propagation. Quarter-wave plates also find application in electro-optic modulators for laser beams. A typical amplitude modulator appears as follows.

A continuous input beam is linearly polarized by a polarizer (A) which is optional if the beam originates in a laser with Brewster's angle windows within the laser cavity. A block of electro-optic material, such as KDP, (B) is either rendered birefringent or has its state of birefringence altered by an externally impressed electric field. In either case, the electro-optic block can be made equivalent, by appropriate choice of block length and a D.C., bias voltage, to a quarter-wave plate. The angle θ (between radiation electric field vector and retarder principal plane) can then be modulated about a value of 45 degrees. The beam which emerges from it is (approximately) elliptically and variably polarized. The elliptically polarized beam is transformed into one of linear and variable polarization by a suitably oriented quarter-wave plate (C). The linearly polarized beam must pass through a second polarizer (D) oriented to achieve about 50% transmission of the incident linearly polarized beam with zero modulation voltage at element (B). Small variations in modulation voltage are then transformed into amplitude modulation of the final output beam.

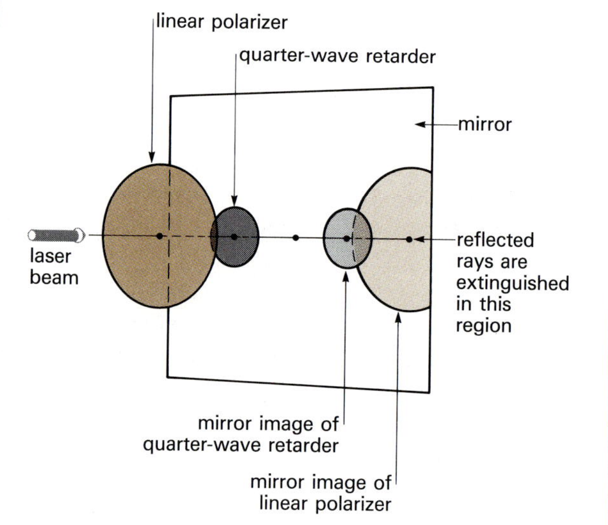

QUARTER-WAVELENGTH RETARDER ISOLATOR CONSTRUCTION. The incident laser beam encounters first the polarizer, next the quarter-wave retarder, and finally the mirror. Specular reflections (not diffuse reflections) are extinguished upon their return to the polarizer. The crystalline optic axis of the quarter-wave plate must be oriented at precisely 45° to the plane of polarization (plane of allowed electric field vector orientations) of the polarizer.

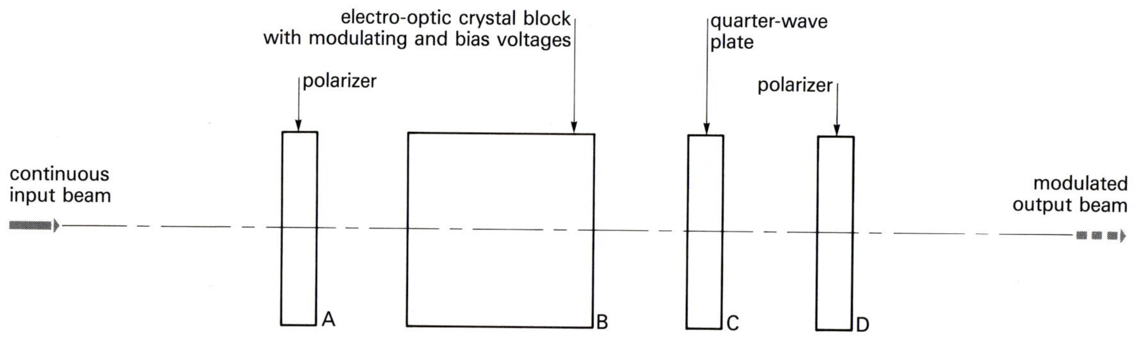

MICA RETARDATION PLATES

Mica retardation plates, because of their relatively high absorption coefficient and occasional inhomogeneities, are normally used for relatively low power applications such as tint plates in microscopes, visual stress analyzers, and in conjunction with helium neon lasers. For higher power applications quartz retardation plates are recommended. These retardation plates are offered in two types.

Type 1 mica retarders are broadband devices suitable for use in the visible spectrum (400-700nm). They are first order (of single order thickness) retarders for 550nm, made of a mica sheet between glass plates to increase strength.

Type 2 mica retarders made for any specific wavelength between 400 and 2500nm. They are only exactly 1/4 wave or 1/2 wave at the specific wavelength for which the retardation tolerance will be $\lambda/50$.

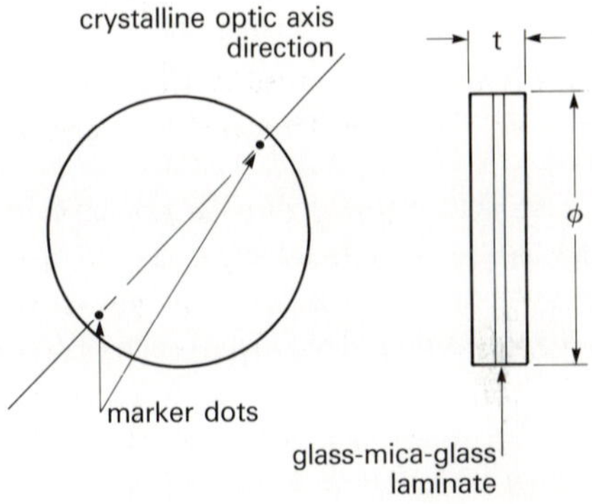

All of the $\lambda/2$ and the $\lambda/4$ 10mm and 20mm diameter mica retardation plates may be mounted in the Melles Griot polarizer holder for ease of use. *(See page 460.)*

SPECIFICATIONS: MICA RETARDATION PLATES

Wavelength Range, Type 1: 400 to 700nm
Type 2: 400 to 2500nm
Wavefront Distortion, Type 1: 2λ at 550nm
Type 2: $\lambda/2$ at 550nm
Material: Selected Mica Sheet
Mounting: Cemented between protective glass discs.
Retardation Tolerance, Type 1: $\lambda/20$
Type 2: $\lambda/50$
Diameter Tolerance: ± 0.25mm
Optic Axis: Crystalline axis direction indicated by pair of diametrically opposed dots.

Mica Retarders; Type 1

Diameter, ϕ (mm)	t	PRODUCT NUMBER Retardation $\lambda/2$	$\lambda/4$
10	2	02 WRM 021	02 WRM 001
20	2.5	02 WRM 023	02 WRM 003
30	2.5	02 WRM 025	02 WRM 005
40	3	02 WRM 027	02 WRM 007
50	3	02 WRM 029	02 WRM 009

Mica Retarders; Type 2

Diameter, ϕ (mm)	t	PRODUCT NUMBER Retardation $\lambda/2$	$\lambda/4$
10	2	02 WRM 031	02 WRM 011
20	2.5	02 WRM 033	02 WRM 013
30	2.5	02 WRM 035	02 WRM 015
40	3	02 WRM 037	02 WRM 017
50	3	02 WRM 039	02 WRM 019

To order Type 2 retarders you must give Product Number and *specify wavelength.*

QUARTZ RETARDATION PLATES

These are 1/4 or 1/2-wave first-order or multiple-order plates, suitable for high- and low-power laser applications. Multiple-order quartz retarders are made from a single crystalline plate, while first order retarders are assembled from pairs of optically contacted crystalline quartz plates whose effects partially counterbalance each other. These component plates have orthogonal optic axis directions, so that the roles of the ordinary and extraordinary rays are interchanged in passing from one plate to the other. The retardation of first-order plates is essentially invariant with temperature, while the retardation of multiple-order plates is slightly temperature dependent. The temperature coefficient of retardance (phase difference between O and E rays at emergence) in multiple-order quartz retarders is approximately 0.07 mm/°C. The net retardation in first-order plates is a function of the thickness difference between the two plates. The net retardations are almost temperature invariant because both plates are of very nearly equal thickness. Both plates expand and contract essentially the same amount, and so almost cancel each other's effect.

SPECIFICATIONS: QUARTZ RETARDATION PLATES

Wavelength: Specified by customer

Wavelength Range: 200nm to 2300nm

Retardation Tolerance: λ/500

Wavefront Distortion: λ/10 peak-to-peak

Parallelism: 0.5 arc seconds

Material: Crystal Quartz

Optic Axis: Normal to facet on circumference of retarder

Cosmetic Surface Quality: 20-10 scratch and dig

AR Coatings: For Multiple Order Plates, see suffix system.

For First Order Plate, coatings can be supplied to special order only.

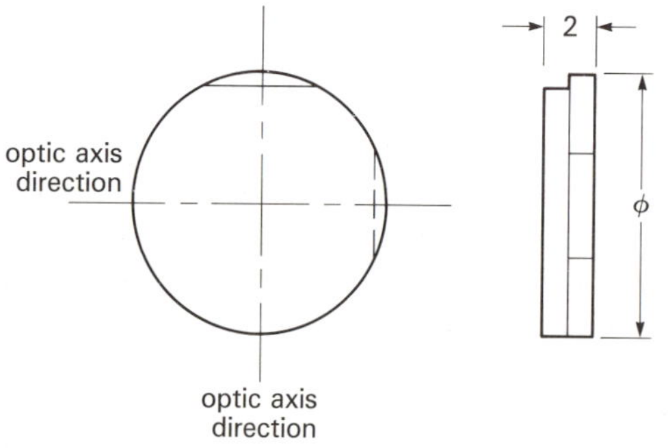

02 WRQ/First Order

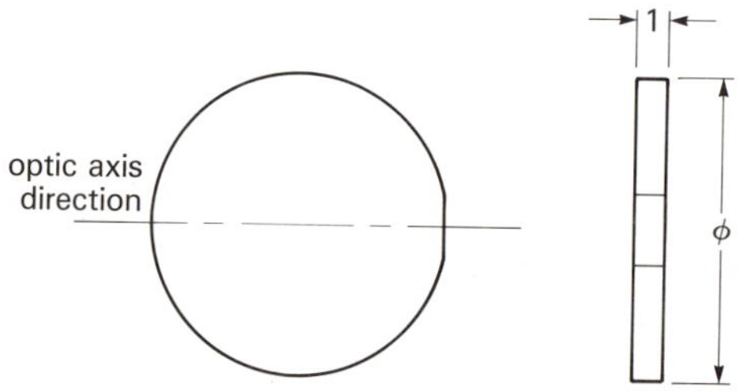

02 WRQ/Multiple Order

Quartz Retardation Plates; First-Order

ϕ (mm)	PRODUCT NUMBER Retardation λ/2	λ/4
10	02 WRQ 023	02 WRQ 003
20	02 WRQ 027	02 WRQ 007
30	02 WRQ 031	02 WRQ 011

Quartz Retardation Plates; Multiple-Order

ϕ (mm)	PRODUCT NUMBER Retardation λ/2	λ/4
10	02 WRQ 021	02 WRQ 001
20	02 WRQ 025	02 WRQ 005
30	02 WRQ 029	02 WRQ 009

Please specify the wavelength when ordering these retardation plates. A limited number of commonly requested wavelengths are held in inventory for your convenience.

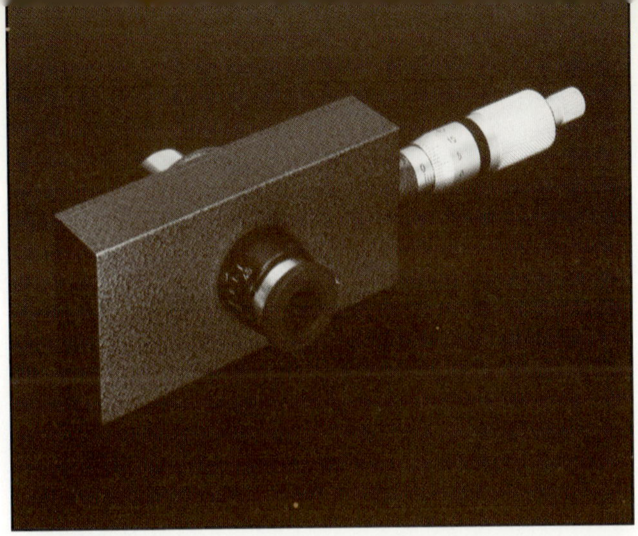

SOLEIL-BABINET COMPENSATOR

The Soleil-Babinet compensator can be thought of as a continuously adjustable retardation plate. At any wavelength from 250nm to 3500nm it is possible to achieve any desired relative retardation by simply turning a micrometer screw. The selected retardation value is constant over the entire working aperture of the compensator, as has been assured by using only non-magnetic metal in its construction (thus minimizing apparent retardation variations due to the Faraday effect). Available retardations depend on wavelength (see specifications), and range from zero up to two full wavelengths (at 546.1nm). The micrometer screw is calibrated to read in units of 0.001 wavelengths (at 546.1nm), and retardation differences can be preselected to this resolution. The micrometer reading at which zero retardation results (the "null" point) varies slightly from compensator to compensator, and the first thing the user must do is establish this reading. If retardation difference accuracies of better than the equivalent of $\lambda/100$ at 546nm are required (ten of the finest divisions), strong magnetic fields should be avoided or appropriate magnetic shielding should be provided.

The principle of the Soleil-Babinet compensator is easily understood. Consider two crystalline quartz plates of equal thickness, each cut so that the crystalline optic axes parallel the surfaces of the plates. If these are stacked one on top of the other, so that the crystalline optic axes of the plates are perpendicular to each other, the roles of the ordinary and extraordinary rays are interchanged as light passes from one plate into the other. Because of this interchange, and because the plates are of equal thickness, any phase difference or retardation that may have accumulated in the first plate is exactly canceled out in passing through the second plate. By replacing one of the fixed-thickness plates with an effective plate of variable thickness (a pair of wedges which move relative to one another), any desired net retardation can be obtained, and a Soleil-Babinet compensator results. Refer to the following illustration.

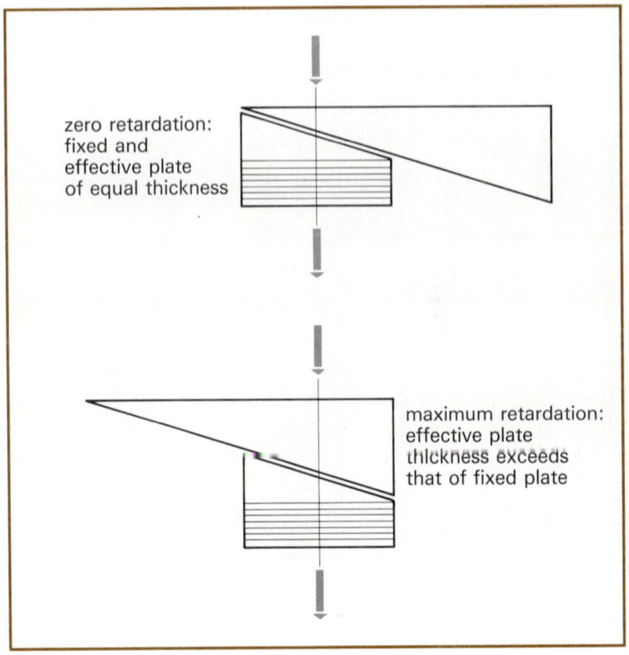

One wedge is optically contacted to the lower plate, while the other wedge, separated from the first by a small air space, is moved by the micrometer screw. The retardation exhibited by the emergent beam is proportional to the difference in thickness between the fixed and effective plates. Unknown retardations can be measured by adjusting the compensator to produce an exactly canceling retardation.

SOLEIL-BABINET COMPENSATOR APPLICATIONS

As we have mentioned, quarter-wave plates can be used to determine the approximate composition of light which may be a mixture of several unknown polarization states and intensities. These procedures can be made extremely accurate if a Soleil-Babinet compensator is used. With the compensator one can accurately determine the planes of polarization, phase difference, and intensity ratio of the linear polarization states of which any elliptically polarized beam is composed. Compensators can also be used to synthesize radiation in any desired polarization state. These abilities are the basis for a multitude of compensator applications.

Compensators make possible the science of ellipsometry, by which the properties of surfaces and the thicknesses of extremely thin films (thicknesses down to an extremely tiny fraction of a wavelength) can be determined. The information revealed is not available by other methods. Compensators are important elements in polarization process-inspection instrumentation used in mineralogical analysis, organic chemical analysis, molecular structure determination, textile fiber analysis, medical diagnosis, phase equilibria studies in ceramics, and pollution monitoring. Compensators, because they permit accurate measurement of strain-

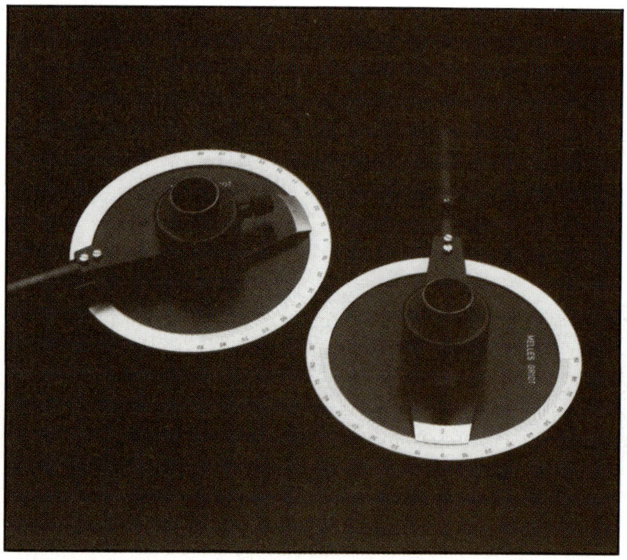

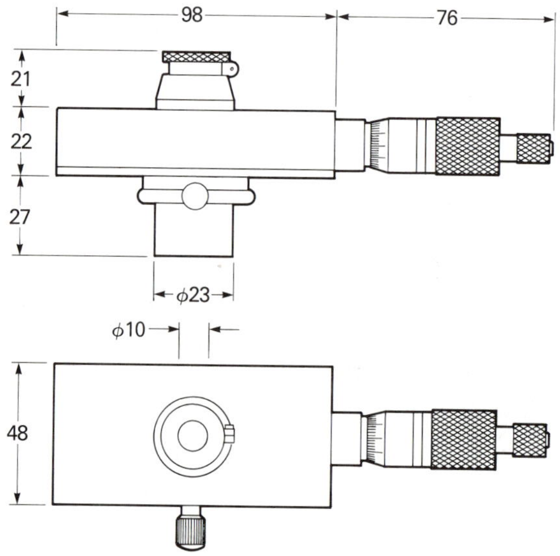

04 SBC 001

A divided circle mount for holding and adjusting the soleil-babinet compensator is available and is described in detail in the hardware section of this catalog. *(See page 461.)*

induced birefringence, are valuable in photoelastic stress analysis. Compensators are also useful in the testing of optical surfaces — especially aspheric surfaces. See references cited by De Vany in *Applied Optics,* V18, N24, 15 De. 1979, page 4066.

SPECIFICATIONS: SOLEIL-BABINET COMPENSATOR

Wavelength Range: 250.0nm to 3.5 micrometers
Retardation Range:
- 0 to $4\lambda \pm 2\%$ at $\lambda = 300$nm
- 0 to $2\lambda \pm 1\%$ at $\lambda = 546.0$nm
- 0 to $1\lambda \pm 0.5\%$ at $\lambda = 1000.0$nm
- 0 to $0.5\lambda \pm$ at $\lambda = 2000.0$nm

C.A.: 10mm ϕTemperature Limits: -20 °C to +80 °C
Retardation Resolution: 0.001λ at $\lambda = 546.1$nm
Materials: Schlieren-free crystalline quartz and non-magnetic metals
Wavefront Distortion: 1/4-wave or less at 546.1nm

Soleil-Babinet Compensator

PRODUCT NUMBER
04 SBC 001

OPTO-MECHANICAL PRODUCTS

In this section we describe a range of microscope components, apertures, test targets and shutters.

Contents **Page Number**

MICROSCOPE COMPONENTS

Microscope components are useful both in imaging and laser-beam handling. These components have been designed in accordance with an international system of specifications, and are therefore freely interchangeable within that system. When used with the mechanical components which we offer, these microscope components will allow you to accurately position, aim, and control the location of object and image, They will also facilitate spatial filtering or controling an optical beam with precision.

Since all our components conform to the international standards established by the Royal Microscopy Society, they may be used with most existing research microscopes.

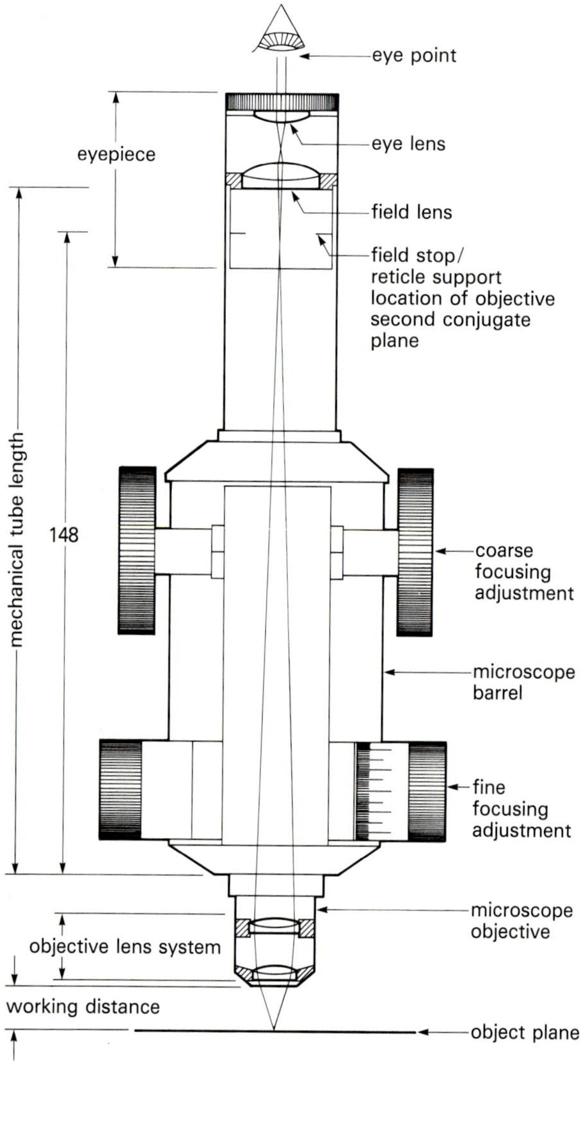

MICROSCOPE BARREL ASSEMBLY with eyepiece, objective, and focusing mechanism

MICROSCOPE OPTICAL TERMS

Magnification

Microscope optics have been designed to work at a specific conjugate ratio. Both objectives and eyepieces are mechanically designed such that tube length and objective shoulder location will be approximately correct when mounted in a standard microscope housing. The term mechanical "tube length" is defined as the distance from the rear shoulder of the microscope objective mounting flange to the rear shoulder of the eyepiece field stop. This distance is usually 160mm, although 170mm is sometimes used.

Objective and eyepiece magnifications are stated with the assumption that conjugate ratios are fixed. Variations from these distances will produce slightly different magnifications, and large variations will result in a degradation of image quality. If the tube length is 160mm, the magnification of a microscope system is the product of the magnification of the objective and the magnification of the eyepiece.

Resolution

The resolving power of a microscope objective is limited by its numerical aperture. The greater the numerical aperture, the better the resolution. Numerical aperture (N.A.) was defined earlier to be

$$\text{N.A.} = n \sin(\theta),$$

where in the present context n is the index of refraction of the medium between the object and the objective, and θ is the angular radius of the cone of light entering the objective. Some microscope objectives with unusually high resolving powers are designed for use with an immersion liquid between the object and the objective front element. Liquid immersion objectives may have numerical apertures greater than unity. Immersion diminishes the wavelength of light relative to the object.

In practice the theoretical resolution limit is often not achieved because of partial coherence effects in the illumination of the object being viewed. In back-lighted microscopes a practical limit of resolution is

$$\text{Practical Limiting Resolution} = \frac{\lambda}{\text{N.A.}}$$

The practical upper limit of resolution, without resorting to oil immersion, is approximately 500nm for visible light.

Objective Second Conjugate Plane

As explained previously, the magnification of microscope objectives is calculated with an assumed fixed location of the secondary conjugate plane. The secondary conjugate plane for all our microscope objectives is located 148mm behind the rear shoulder of the objective (i.e., the seat of the mounting barrel thread). The secondary conjugate plane is the correct location of the eyepiece field stop.

Depth of Focus

The depth of focus for a microscope objective of high numerical aperture can be extremely shallow, and the actual thickness of the object should be kept in mind when designing a system. Depth of focus in microns is approximately equal to $\lambda/4\,(\text{N.A.})^2$ in visible light.

Cover Glass Correction

Many objectives, especially those with numerical apertures approaching and greater than one, have been designed with biological applications in mind. In these applications a thin (0.17mm) cover glass is often placed over the object being viewed. To compensate for this layer of glass, a slight negative spherical aberration term has been included in the objective design. When such higher numerical aperture objectives are used without a cover glass between the object and objective, a slight deterioration of image quality should be expected.

MICROSCOPE OBJECTIVES

All of our microscope objectives are achromatic (corrected at two colors, 656.3nm and 486.1nm) and are supplied with standard microscope body threads. All objectives may be mounted in our focusing tubes and turret nose pieces, which are described later. Flat field objectives with powers above 20X magnification have a spring-loaded front element, to reduce possible damage from over-focusing. Objective bodies are made of free-turning brass with a chromium protective plating. Objective magnification and numerical aperture are engraved on each objective for ease of identification.

We offer two basic types of microscope objectives, standard and flat field. Standard objectives always exhibit some curvature of the object plane. Our specially designed line of six-element flat field objectives corrects this problem and produces a wider field of fully-corrected view.

Standard Microscope Objectives

Power	N.A.	Focal Length (mm)	Working Distance (mm)	PRODUCT NUMBER
1.0X	0.05	80.0	160	04 OAS 001
2.0X	0.07	46.8	60	04 OAS 003
2.5X	0.10	45.0	45	04 OAS 005
3.0X	0.10	36.0	15	04 OAS 007
3.5X	0.10	31.0	24	04 OAS 009
5.0X	0.16	25.0	24	04 OAS 011
7.0X	0.20	18.0	22	04 OAS 013
10.0X	0.30	16.0	5.8	04 OAS 015
18.0X	0.45	10.0	4.2	04 OAS 017
20.0X	0.50	9.0	1.6	04 OAS 019
30.0X	0.65	6.0	1.0	04 OAS 021*
40.0X	0.65	5.0	0.42	04 OAS 023
45.0X	0.82	4.0	0.30	04 OAS 027
50.0X	0.85	4.0	0.65	04 OAS 029
60.0X	0.85	3.0	0.45	04 OAS 033

*These objectives are designed for 170mm mechanical tube lengths.

SPECIFICATIONS: MICROSCOPE OBJECTIVES

Barrel Material: Chrome-plated brass
Thread Pitch: 36 TPI (1.42 threads per mm)
Pitch Angle: 55 degrees
Pitch Diameter: 20.1mm
Thread Form: Whitworth
Coating: All optical surfaces are single layer, magnesium fluoride, antireflection coated.
All internal surfaces are blackened.

Flat Field Objectives

Power	N.A.	Focal Length (mm)	Working Distance (mm)	PRODUCT NUMBER
5.0X	0.18	26.9	19	04 OAP 001
10.0X	0.25	16.0	4.6	04 OAP 003
20.0X	0.45	8.4	0.70	04 OAP 005
40.0X	0.65	4.5	0.32	04 OAP 007

EYEPIECES

An eyepiece magnifies the image from a microscope or telescope objective, and presents this image to the observer's eye for viewing.

Eyepieces are available in two major categories, the negative or Huygenian type and the positive or Ramsden type. With positive eyepieces the focal plane falls outside the optical system on the side toward the objective. A reticle placed within the tube of an eyepiece, at the eyepiece focal plane, will be magnified by the nominal magnification of the eyepiece.

The focal plane of a negative eyepiece lies between the eyelens (the lens closest to the eye) and the field lens (the lens furthest from the eye). The reticle is therefore magnified only by the power of the eyelens and a correction factor must be determined to establish the exact magnification.

In general, we recommend that reticles be used with positive or Ramsden-type eyepieces only. With the exception of the Huygenian and Periplan eyepieces, all of our eyepieces are positive.

All eyepieces have the internationally standard 23mm barrel diameter and are antireflection coated for maximum transmission and image clarity.

HUYGENIAN EYEPIECES

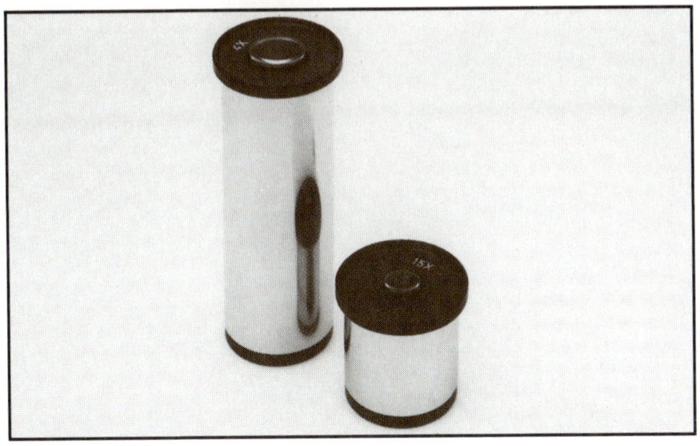

Huygenian (negative) eyepieces consist of two simple plano-convex lenses with the plano sides oriented toward the eye. The focal plane falls between these two lenses. The ratio of powers of the two lenses is chosen to minimize coma and reduce spherical aberration. As previously mentioned in the presentation of lens combination formulas, Huygenian eyepieces are achromatic even though both elements may be made of the same material.

Huygenian eyepieces may be used with any of the microscope objectives listed earlier. All eyepieces are AR coated, for maximum transmission and image clarity. All internal housing surfaces are blackened.

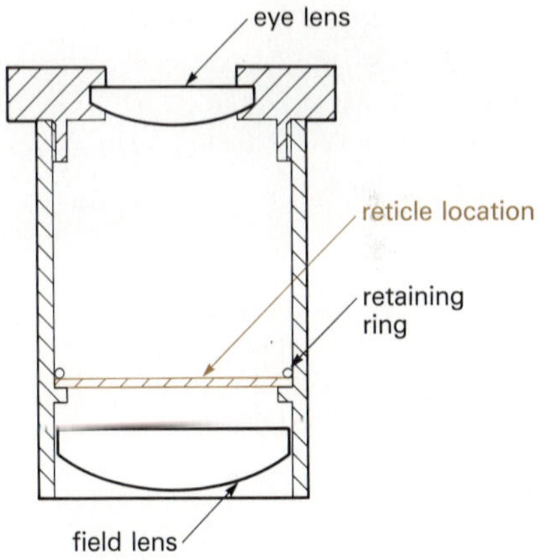

04 EHY

Standard Huygenian Eyepieces

Magnification	Focal Length (mm)	PRODUCT NUMBER
4X	60	04 EHY 001
5X	50	04 EHY 003
6X	40	04 EHY 005
8X	30	04 EHY 007
10X	25	04 EHY 009
12X	20	04 EHY 011
15X	17	04 EHY 013

See page 317 for microscope bodies.

PERIPLAN EYEPIECES

The Periplan eyepiece is a Huygenian type eyepiece with an additional correcting element. This element allows the eyepiece to cover a larger field of view, and improves the correction of both field curvature and lateral chromatic aberration. These eyepieces provide superior performance when used with any of the microscope objectives listed earlier. All optical surfaces are antireflection coated, for maximum transmission and image clarity. All internal surfaces are blackened.

Periplan Eyepieces

Magnification	Focal Length (mm)	PRODUCT NUMBER
5X	50.0	04 EPP 001
10X	25.0	04 EPP 007
15X	17.0	04 EPP 011
20X	12.5	04 EPP 013
25X	10.0	04 EPP 015*

*This eyepiece will not accept a reticle.

RAMSDEN EYEPIECES

Ramsden (positive) eyepieces consist of two plano-convex lenses placed with the convex surfaces facing each other. The spacing between the lenses is approximately two-thirds of the average of the focal length of the two lenses.

The focal plane of all Ramsden eyepieces lies outside of the eyepiece on the side toward the objective. A reticle placed within the eyepiece tube, at the focal plane of the eyepiece, will be magnified by exactly the specified eyepiece magnification. Any of the reticles listed later may be used with these eyepieces.

All lens surfaces are antireflection coated. All internal surfaces are blackened.

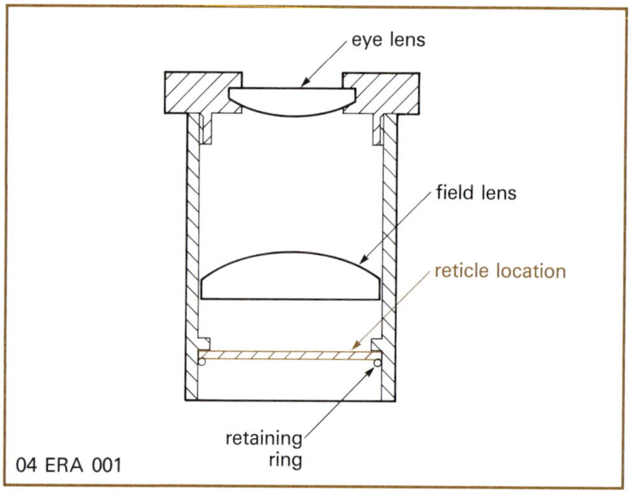

Standard Ramsden Eyepiece

Magnification	Focal Length (mm)	PRODUCT NUMBER
10X	25	04 ERA 001

WIDE-FIELD EYEPIECES

These eyepieces are available in three varieties; the widefield eyepieces, compensating wide-field eyepieces, and focusing wide-field eyepieces.

Our standard wide-field eyepieces are modified Ramsden eyepieces made up of an achromatic doublet and a plano-convex singlet. Coma and lateral chromatic aberration have been reduced to a minimum. These eyepieces cover a field of view approximately twice as wide as our standard Huygenian and Ramsden eyepieces, and are virtually free from field distortion.

All optical surfaces are antireflection coated, for maximum transmission and image clarity. All internal surfaces are blackened.

Wide-Field Eyepieces

Magnification	Focal Length (mm)	PRODUCT NUMBER
10X	25	04 EWA 001
10X*	25	04 EWR 001
15X	17	04 EWA 003
20X**	12	04 EWP 001

*This eyepiece is supplied with a built-in plain cross-hair reticle.
**This eyepiece utilizes two achromatic doublets for an especially flat field and may be used with either our Standard or Flat Field objectives.

COMPENSATING WIDE-FIELD EYEPIECE

The wide-field compensating eyepieces have built-in specific negative aberration which exactly compensates for the slight residual aberrations (especially lateral color) in the flat field objectives. All lens surfaces are AR coated. All internal surfaces are blackened.

Compensating Wide-Field Eyepieces

Magnification	Focal Length (mm)	PRODUCT NUMBER
10X	25	04 ECW 001

FOCUSING WIDE-FIELD EYEPIECES

These positive eyepieces are designed such that the eyelens may be adjusted to sharply focus a reticle. All surfaces are antireflection coated.

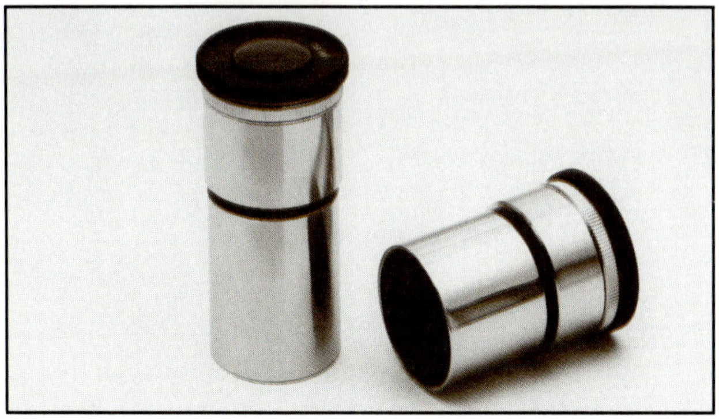

Focusing Wide-Field Eyepieces

Magnification	Focal Length (mm)	PRODUCT NUMBER
10X	25	04 EWF 001
15X	17	04 EWF 003

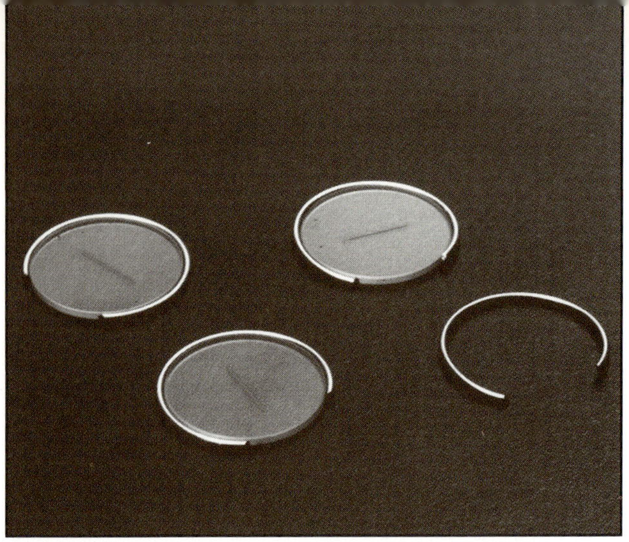

RETICLES

Our reticles are precision photochrome reproductions mounted on ground and polished plate glass. All line widths are 0.01mm.

These reticles fit all standard eyepieces. Negative eyepieces must be disassembled to mount the reticle on the first shoulder below the eyelens. A reticle locking ring, pushed into the barrel, will hold the reticle firmly in place. For convenience, a locking ring is supplied with each reticle. Spares are available below.

All reticles are 21.0±0.25 mm in diameter and 1.0±0.2mm in thickness.

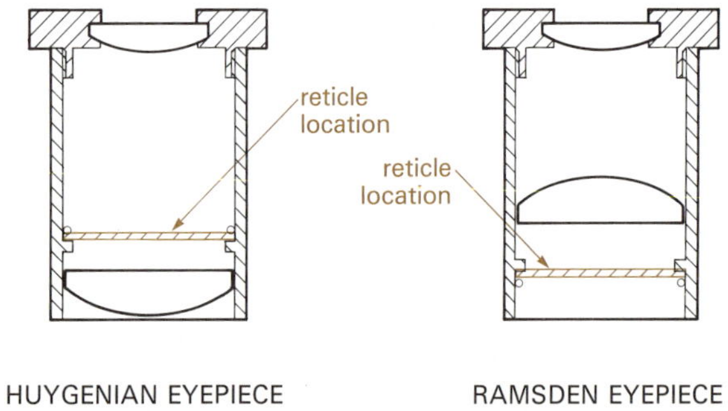

Standard Reticles

Pattern		PRODUCT NUMBER
Plain Cross Hair	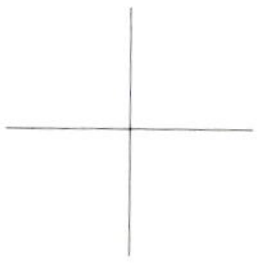	04 RET 001
Cross Hair with 10mm scale in 100 divisions	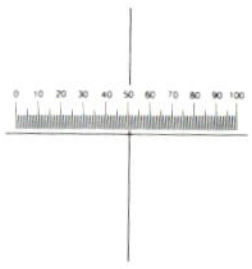	04 RET 011
Crossed 10mm scales in 100 divisions	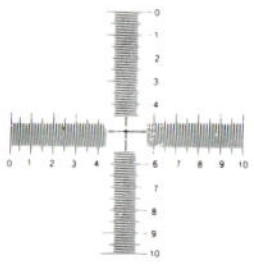	04 RET 013
Cross Hair with concentric circles. Outside circle 10mm ϕ	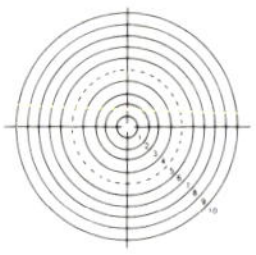	04 RET 003
10mm scale in 100 divisions	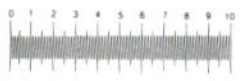	04 RET 005
5mm scale in 100 divisions		04 RET 007

Special sizes, shapes and patterns are available for volume users of reticles. If you have specific reticle requirements, please contact one of our Applications Engineers for a quotation.

RETICLE LOCKING RING

Spare spring-wire locking ring to hold reticles firmly in place.

Reticle Locking Ring

	PRODUCT NUMBER
	04 SWR 021

FOCUSING MICROSCOPE BODIES

Focusing microscope bodies will immediately accept all of our standard microscope objectives, eyepieces, and accessories. They are useful in scientific, engineering, and production equipment. For example, they are widely used in lens and optical component testing, metrology, image evaluation, and semiconductor test equipment.

By using focusing microscope bodies, one may conveniently interchange eyepieces or objectives of different powers. This is especially useful in micro/telephotometry applications.

The eyepiece tube and objective threads are made to international standards, and will accept all quality eyepieces and objectives. The dovetails are hand-lapped, and antibacklash gears are used throughout. Two drilled and countersunk mounting holes are provided on the bottom of the base for your convenience. Mounting other accessories (such as our angular eyepiece holders, adjustable eyepiece holders, or nosepieces) is easily accomplished by simply unscrewing and removing the standard attachments.

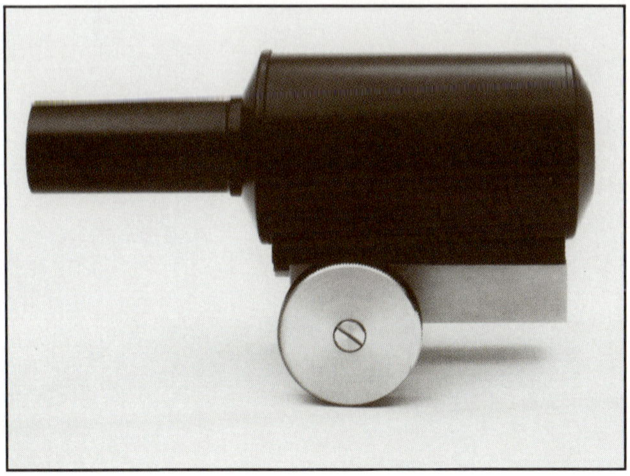

SPECIFICATIONS: COARSE FOCUSING MICROSCOPE BODY

Barrel Diameter: 50mm
Mechanical Tube Length: 153mm
Travel Range: 63mm
Travel per 360° Knob Rotation: 18mm
Material: Black anodized aluminum tube and housing with brass slideways and steel drive mechanism.

Coarse Focusing Microscope Body

	PRODUCT NUMBER
	04 TCF 001

SPECIFICATIONS: FINE FOCUSING MICROSCOPE BODY

Barrel Diameter: 50mm
Mechanical Tube Length: 153mm
Travel Range: Coarse 63mm; Fine 3mm
Fine Focus Accuracy: 4 microns per engraved division
Material: Black anodized aluminum tube and housing with brass slideways and steel drive mechanism.

Fine Focusing Microscope Body

	PRODUCT NUMBER
	04 TFF 001

FOCUSING MECHANISMS

These focusing mechanisms are similar to those designed for the focusing microscope bodies listed previously, and utilize the same high-quality hand-lapped ways and antibacklash gearing throughout. These mechanisms are especially useful in applications where precise, smooth, and controlled motions of optical components and subsystems are required.

The top plates may be easily modified or adapted to special applications. Two mounting holes are drilled and countersunk on the bottom of the base. Additionally, the top plate is drilled and tapped for your convenience. Unlimited-life nylon bearings, as in the focusing microscope body assemblies, are utilized in each of the fine focusing mechanisms.

Four models are offered: two with extended top plates for greater travel and increased capacity, and two with standard top plates.

SPECIFICATIONS: COARSE FOCUSING MECHANISM

Dimensions: ±0.25mm
Travel Range: 04 MFS 001 — 63mm
04 MFS 003 — 125mm
Travel per 360° Knob Rotation: 18mm
Material: Chromium-plated brass with steel drive mechanism.

Coarse Focusing Mechanism

	PRODUCT NUMBER
Short-Top Coarse Focusing Mechanism	04 MFS 001
Long-Top Coarse Focusing Mechanism	04 MFS 003

SPECIFICATIONS: FINE FOCUSING MECHANISM

Dimensions: ±0.25mm
Coarse Travel Range: 04 MFF 001 — 63mm
04 MFF 003 — 125mm
Fine Travel Range: 3mm
Fine Focus Accuracy: 4 microns per engraved division
Material: Chromium-plated brass with steel drive mechanism.

Fine Focusing Mechanism

	PRODUCT NUMBER
Short-Top Fine Focusing Mechanism	04 MFF 001
Long-Top Fine Focusing Mechanism	04 MFF 003

MICROSCOPE BODY ACCESSORIES

ANGULAR EYEPIECE HOLDER

The angular eyepiece holder has a built-in prismatic system which bends light 45°. The eyelens can then be rotated to more accessible or comfortable locations in difficult or prolonged viewing situations. This holder screws into our standard microscope bodies after the straight eyepiece holder is removed.

SPECIFICATIONS: ANGULAR EYEPIECE HOLDER

Angle of Offset: 45°
Material: Black enameled brass tube with black anodized aluminum mount.

Angular Eyepiece Holder

PRODUCT NUMBER
04 AEH 003

ADJUSTABLE EYEPIECE HOLDER

In compound microscope assemblies, the adjustable eyepiece holder allows adjustment from the standard 160mm to a nearby magnification. This is useful when setting a reticle scale to an arbitrary length. Not recommended for use with objective numerical apertures greater than 0.80.

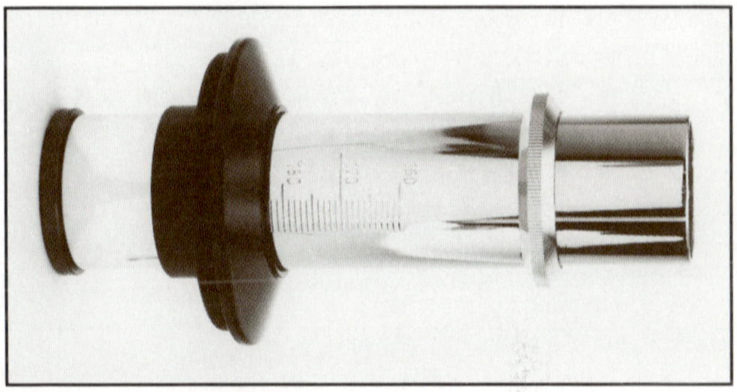

SPECIFICATIONS: ADJUSTABLE EYEPIECE HOLDER

Maximum/Minimum Mechanical Tube Length: 140 to 200mm
Barrel Diameter: 25.4mm
Material: Chromium-plated brass tube with black anodized aluminum mount

Adjustable Eyepiece Holder

PRODUCT NUMBER
04 AEH 001

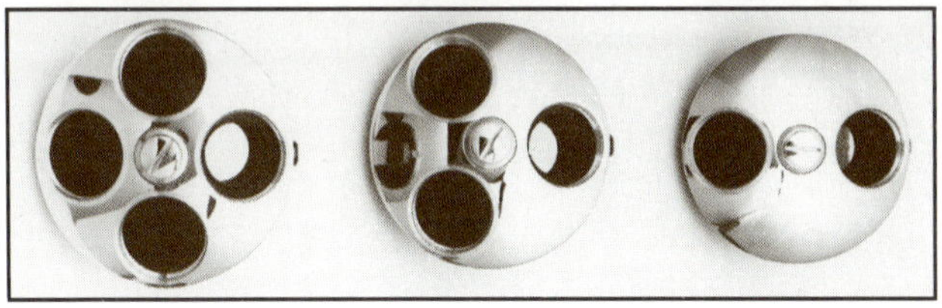

NOSE PIECES

Our chrome-plated brass nosepieces make objective changing fast and convenient. Available with capacities for two, three or four objectives.

Nose Pieces

OBJECTIVE CAPACITY	PRODUCT NUMBER
2	04 NOP 001
3	04 NOP 003
4	04 NOP 005

PRECISION PINHOLES

The following components will be found especially useful in physical optics, laser technology, spatial filtering, image analysis, and image processing.

Since these pinholes are precision laser drilled in solid stainless steel stock, they are completely free of the defects normally seen in pinholes prepared by vacuum deposition on dielectric substrates. Containing no dielectrics, they may be used at any wavelength in the electromagnetic spectrum.

Pinholes, of known size and precisely round in shape, are useful for many applications. When used with a condensing system (such as a microscope objective), they provide excellent spatial filtering. Laser beam diameters may be reduced to a specific size, and any number of alignment and mapping/scanning functions may be performed.

We offer individual precision pinholes and a precision pinhole set. All pinholes are individually laser-drilled in the center (±0.13mm) of a 9.5mm diameter stainless steel disc. The thickness of the disc depends on the pinhole diameter and is listed. Each pinhole is exceptionally round and free of edge distortions.

Given enough time, even with great care, micron size pinholes will pick up dust particles. We advise that our pinholes be kept in their dust-free envelopes (or similarly protected) when not in use. If dust particles are picked up, the pinholes may be cleaned with normal ultrasonic cleaning techniques.

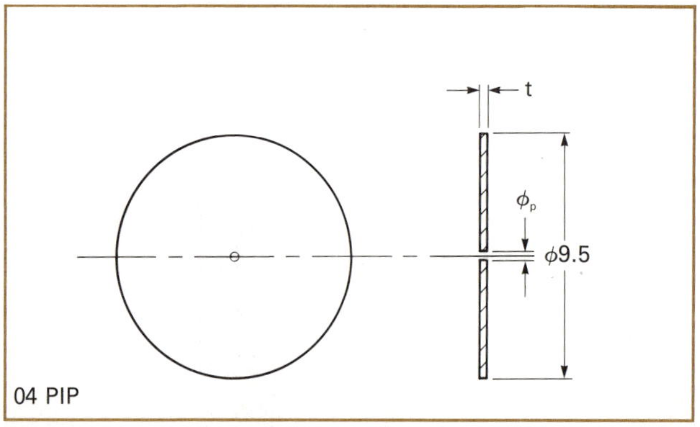

Precision Individual Pinholes

Pinhole Diameter (microns)	Diameter Tolerance (microns)	Substrate Thickness (mm)	PRODUCT NUMBER
1	+0.5,–0.0	*	04 PIP 002
2	±0.5	*	04 PIP 001
3	±0.5	0.013	04 PIP 027
4	±0.5	0.013	04 PIP 029
5	±0.8	0.013	04 PIP 003
6	±0.8	0.013	04 PIP 033
8	±1.0	0.013	04 PIP 035
10	±1.0	0.013	04 PIP 005
12	±1.0	0.013	04 PIP 037
15	±1.5	0.013	04 PIP 007
20	±2.0	0.013	04 PIP 009
25	±2.0	0.013	04 PIP 011
30	±2.0	0.013	04 PIP 039
40	±3.0	0.013	04 PIP 045
50	±3.0	0.013	04 PIP 013
75	±3.0	0.013	04 PIP 014
100	±4.0	0.013	04 PIP 015
150	±6.0	0.013	04 PIP 016
200	±6.0	0.013	04 PIP 017
400	±8.0	0.013	04 PIP 019
600	±8.0	0.013	04 PIP 021
800	±8.0	0.013	04 PIP 023
1000	±10.0	0.013	04 PIP 025

*0.0025mm stainless steel foil bonded to 0.076mm thick disc.

PINHOLE MOUNTING PLATE

A 50mm x 50mm x 2mm thick aluminum plate, drilled and countersunk at the center to accept the 9.5mm diameter pinhole substrates. An acetone soluble cement may be used for mounting.

Pinhole Mounting Plate

PRODUCT NUMBER
04 PMP 001

HIGH ENERGY LASER PRECISION PINHOLES

These pinholes are of the same high quality as the precision pinholes previously described. However the materials used are capable of withstanding irradiation from high energy lasers.

The pinholes are laser-drilled in thin copper sheet which provides rapid dissipation of heat. They are also gold plated on one side for high reflectivity at YAG (1.06 microns) and CO_2 (10.6 microns) wavelengths. The reverse side is matte black coated to reduce spurious reflections. The result is a pinhole capable of withstanding high power densities. (Densities as high as 50 megawatts per square millimeter have been used without damage.)

In use, care must be taken that energy is incident on the gold plated side.

The diameter of the copper substrate is 9.5mm making these pinholes directly interchangeable with the standard precision pinholes. They may be mounted in the pinhole mounting plate previously described.

HEL Precision Pinholes

Pinhole Diameter (microns)	Diameter Tolerance (microns)	Substrate Thickness (mm)	PRODUCT NUMBER
5	±0.75	0.026	04 PHP 003
10	±1.00	0.026	04 PHP 005
15	±1.50	0.051	04 PHP 007
25	±2.00	0.051	04 PHP 011
50	±3.00	0.051	04 PHP 013
100	±4.00	0.051	04 PHP 015
200	±6.00	0.051	04 PHP 017

PRECISION PINHOLE SETS

These sets include seven precision pinholes or HEL precision pinholes in a protective hardwood box. Each pinhole is individually mounted in a 50mm square aluminum plate 2mm thick, and marked with pinhole diameter and product number. Diameter tolerance and substrate thickness are as listed for the individual pinholes.

Precision Pinhole Set

	Pinhole Diameters (microns)	PRODUCT NUMBER
7 elements	2 5 10 25 50 100 200	04 PPS 001

HEL Precision Pinhole Set

	Pinhole Diameters (microns)	PRODUCT NUMBER
7 elements	5 10 15 25 50 100 200	04 PHS 001

If these components need to be cleaned, they should be puffed with clean freon gas, immersed in alcohol or put in an ultrasonic bath. No acetone should be used as this will dissolve the mounting cement.

PRECISION AIR-SLITS

Air slits of precisely known width and constant parallelism are useful in a number of applications. When scanned across the focal point of an optical system they yield information from which the point spread function and MTF can be calculated. They may be used as the entrance or exit slits of spectrophotometers. Slits form the basis for many image scanning or dissecting systems, and may be used in a large variety of alignment and mapping functions.

Because these are air slits (rather than scribed through a vacuum-deposited layer on a dielectric substrate) they may be used at any wavelength, including the vacuum ultraviolet. Being laser machined from opaque solid material assures that the slits are absolutely free of pinhole defects. These slits are individually laser-cut in the center (±0.3mm) of a 9.5mm diameter stainless steel disc. Each slit is extremely straight, with parallel sides and constant width. Slit edges are free of local distortions.

As with pinholes, air slits will pick up dust particles over a period of time. We therefore advise that our slits be kept in their dust-free envelopes, or be similarly protected, when not in use. If dust particles are picked up, they may be removed with normal ultrasonic cleaning techniques.

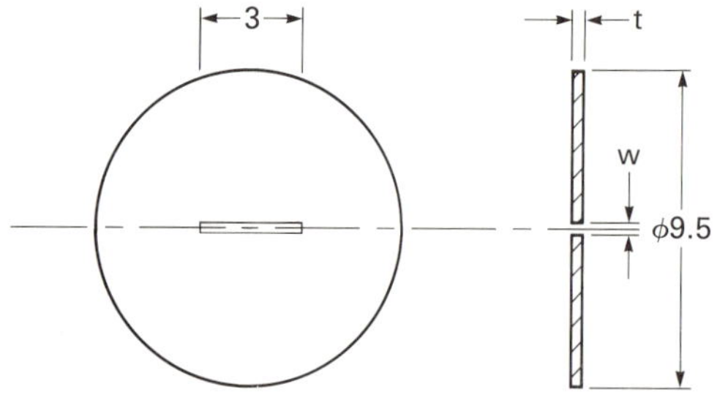

04 PAS

Precision Air Slits

Slit Width (microns)	Slit Length (mm)	PRODUCT NUMBER
5.0 ±1	3.0 ±0.25	04 PAS 001
10.0 ±1	3.0 ±0.25	04 PAS 002
25.0 ±2	3.0 ±0.25	04 PAS 004
50.0 ±3	3.0 ±0.25	04 PAS 005
100.0 ±4	3.0 ±0.25	04 PAS 006
150.0 ±4	3.0 ±0.25	04 PAS 007
200.0 ±4	3.0 ±0.25	04 PAS 008

SLIT MOUNTING PLATE

A 50mm x 50mm x 2mm thick aluminum plate, drilled and counter sunk at the center to accept the 9.5mm diameter air slit and pinhole substrates. An acetone soluble cement may be used for mounting.

Slit Mounting Plate

	PRODUCT NUMBER
	04 PMP 001

PRECISION AIR SLIT SET

This set includes all seven precision air slits in a protective hardwood box. Each slit is individually mounted in a 50mm square by 2mm thick aluminum plate. The slits are very accurately aligned to the edges of the mounting plate for ease of use and set-up. Each slit is individually marked with its width and product number. The set contains all widths as listed above.

Precision Air Slit Set

	PRODUCT NUMBER
7 Elements	04 PSS 001

MOUNTED PRECISION PINHOLES AND SLITS

For ease of handling we offer this range of mounted pinholes and slits. These are the same precision laser drilled components which have been previously listed, unmounted. For full specifications please refer back to the preceding page (pages 320-322).

Each is supplied in a 16mm diameter black anodized aluminum disc. These discs are engraved with the pinhole diameter or slit width in microns for ease of identification. The pinholes or slits are centered in the disc to ±0.25mm accuracy.

These mounted components may be directly attached to the pinhole holder described on page 467. This holder permits lateral (x-y) positioning for ease of alignment.

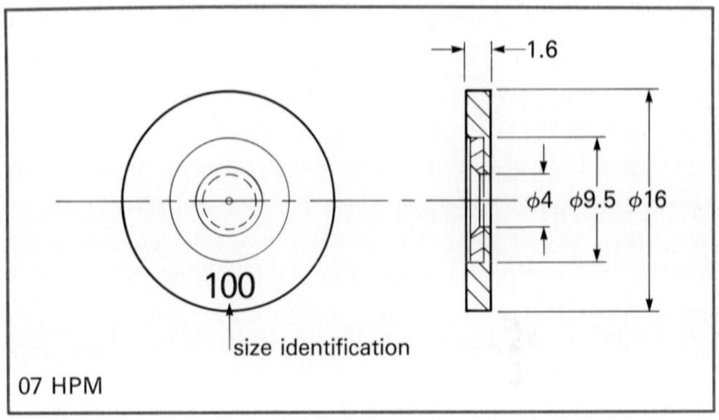

Mounted Precision Pinholes

Pinhole Diameter (microns)	PRODUCT NUMBER
1	04 PPM 002
2	04 PPM 001
3	04 PPM 027
4	04 PPM 029
5	04 PPM 003
6	04 PPM 033
8	04 PPM 035
10	04 PPM 005
12	04 PPM 037
15	04 PPM 007
20	04 PPM 009
25	04 PPM 011
30	04 PPM 039
40	04 PPM 045
50	04 PPM 013
75	04 PPM 014
100	04 PPM 015
150	04 PPM 016
200	04 PPM 017
400	04 PPM 019
600	04 PPM 021
800	04 PPM 023
1000	04 PPM 025

Mounted HEL Precision Pinholes

Pinhole Diameter (microns)	PRODUCT NUMBER
5	04 PHM 003
10	04 PHM 005
15	04 PHM 007
25	04 PHM 011
50	04 PHM 013
100	04 PHM 015
200	04 PHM 017

Mounted Precision Slits

Slit Width (microns)	PRODUCT NUMBER
5	04 PSM 001
10	04 PSM 002
25	04 PSM 004
50	04 PSM 005
100	04 PSM 006
150	04 PSM 007
200	04 PSM 008

IRIS DIAPHRAGMS

Iris diaphragms are useful for smoothly controlling the f-number of an optical system, for reducing or expanding a beam diameter while keeping the energy per unit area constant, or for use as a variable field stop.

Each black spring-steel leaf of these diaphragms is mounted on a low-friction steel bearing pin which assures a smooth action and long life. These diaphragms have been tested and shown to withstand in excess of one million openings and closings without failure. Diaphragms are factory-lubricated, and will require no service throughout their life. The leaf profile has been carefully calculated to closely approximate a circular aperture throughout the operational diameter range of each iris.

TAB ACTUATED IRIS DIAPHRAGMS

Actuation is by a side-mounted plastic tab with an excursion of approximately 90°. The housing is of black anodized aluminum.

PIN ACTUATED IRIS DIAPHRAGMS

These diaphragms are pin-operated. A 90° motion is required to fully open or close the iris, and the motion is smooth and consistent throughout the travel. Pin lengths and iris outside diameters vary with the size of the diaphragm, and are listed below. Except the largest pin, which is black anodized aluminum, all pins are chrome plated brass.

Housings of 80mm outside diameter and smaller are made of black enameled brass. Larger size housings are black anodized aluminum. All leaves are made of blackened, corrosion-resistant spring steel.

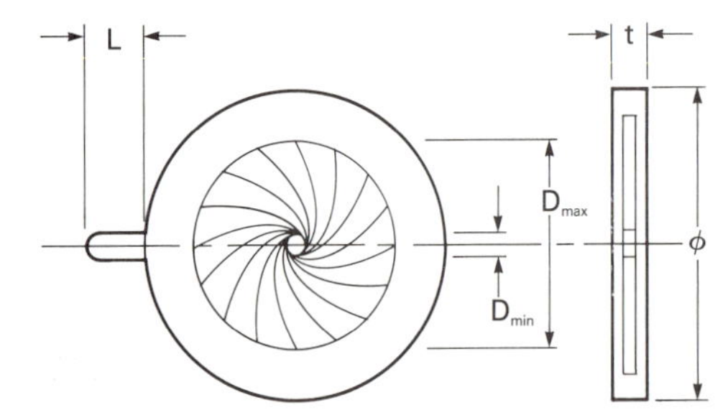

04 IDM

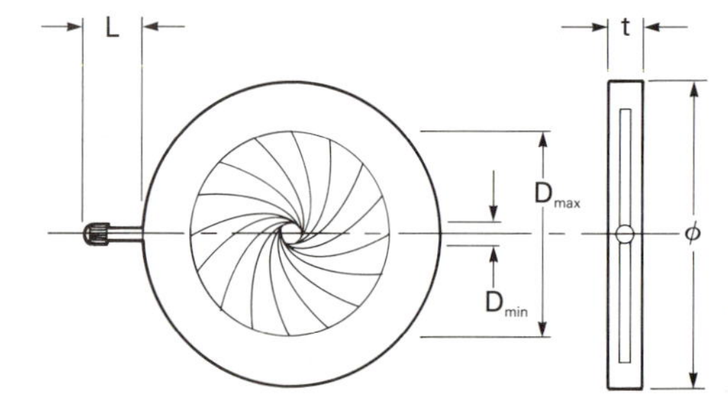

04 IDC

Tab Actuated Iris Diaphragms

Opening		Diameter	Thickness	Tab	
D_{max} (mm)	D_{min} (mm)	ϕ (mm)	t (mm)	Length L	PRODUCT NUMBER
8.5	1.0	15.0	5.0	8.2	04 IDM 001
14.0	1.0	22.0	5.5	11.5	04 IDM 005
18.0	1.2	31.0	7.0	9.5	04 IDM 011
27.0	1.2	40.0	6.0	11.7	04 IDM 015
30.0	1.2	48.5	7.5	10.2	04 IDM 017
40.0	2.5	60.0	7.8	13.0	04 IDM 021
60.0	4.0	82.0	10.0	10.4	04 IDM 033

Pin Actuated Iris Diaphragms

Opening		Diameter	Thickness	Number	Pin Length	
D_{max} (mm)	D_{min} (mm)	ϕ (mm)	t (mm)	of Leaves	L (mm)	PRODUCT NUMBER
8	0.7	14.8	4.5	8	6	04 IDC 000
12	0.8	19.8	5.0	10	11	04 IDC 001
15	0.8	24.0	5.0	12	11	04 IDC 003
18	0.8	28.0	5.0	12	13	04 IDC 005
20	0.8	30.0	5.5	12	13	04 IDC 007
22	0.8	33.0	5.5	14	13	04 IDC 009
25	0.8	37.0	5.5	14	13	04 IDC 011
28	1.2	40.0	5.5	16	13	04 IDC 013
34	1.2	49.0	6.5	14	13	04 IDC 015
36	1.2	50.0	6.0	16	13	04 IDC 017
37	1.2	53.0	6.0	16	13	04 IDC 019
41	1.2	55.0	6.5	18	13	04 IDC 021
42	1.2	58.0	6.5	18	13	04 IDC 023
50	2.0	70.0	7.5	16	16	04 IDC 025
59	7.0	77.0	8.0	20	16	04 IDC 027
58	5.0	80.0	8.0	18	16	04 IDC 029
58	3.5	81.5	8.0	18	15	04 IDC 031
70	4.5	92.0	9.0	20	15	04 IDC 033
75	4.0	100.0	9.0	20	15	04 IDC 035
81	4.5	110.0	10.0	16	15	04 IDC 037
90	5.0	120.0	12.0	20	15	04 IDC 039
98	5.0	130.0	12.0	20	20	04 IDC 041
110	5.0	150.0	13.0	20	20	04 IDC 043
120	6.0	165.0	15.0	18	20	04 IDC 045
138	10.0	180.0	13.0	20	20	04 IDC 047
158	11.0	208.0	16.0	20	20	04 IDC 049
225	12.0	300.0	19.0	18	20	04 IDC 051

On a special request basis we are able to supply a number of additional sizes in either pin, tab, or ring actuated iris diaphragms. Please contact one of our Applications Engineers for further information.

TEST TARGETS

The need to assign numerical tolerances to the performance of optical systems and photographic processes has resulted in the general acceptance of the USAF 1951 resolution test target. In this target the number of line pairs per millimeter doubles with every seventh target element (a "line pair" being a dark bar plus an equally spaced clear bar). An element consists of two target patterns of three lines each, at right angles to each other. These six elements are known as a group.

USAF TEST TARGET

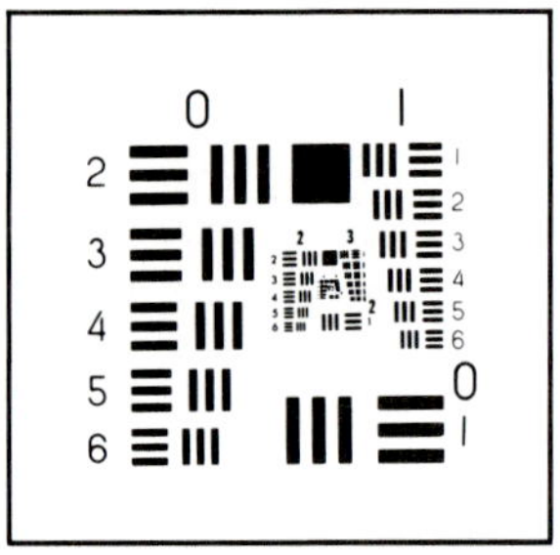

Positive

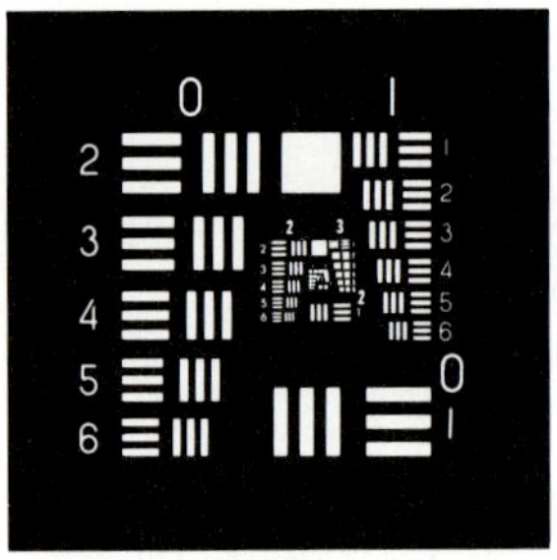

Negative

Our chromium test targets cover the range from group 0 to group 7, while the emulsion targets cover 0 to 6. USAF test targets, of either chromium or emulsion type, are available in either positive or negative form. The corresponding numbers of line pairs per millimeter appear in the following table.

Line Pairs Per Millimeter

Element Number	Group Number 0	1	2	3	4	5	6	7*
1	1.00	2.00	4.00	8.00	16.0	32.0	64.0	128
2	1.12	2.24	4.49	8.98	17.95	36.0	71.8	144
3	1.26	2.52	5.04	10.1	20.16	40.3	80.6	161
4	1.41	2.83	5.66	11.3	22.62	45.3	90.5	181
5	1.59	3.17	6.35	12.7	25.39	50.8	102	203
6	1.78	3.56	7.13	14.3	28.51	57.0	114	228

*Chromium only.

Also in widespread use is the NBS 1963A test target. In appearance it is self-explanatory, the number nearest each pattern element being the number of line pairs per millimeter for that element. NBS test targets are chromium positive only.

NBS TEST TARGET

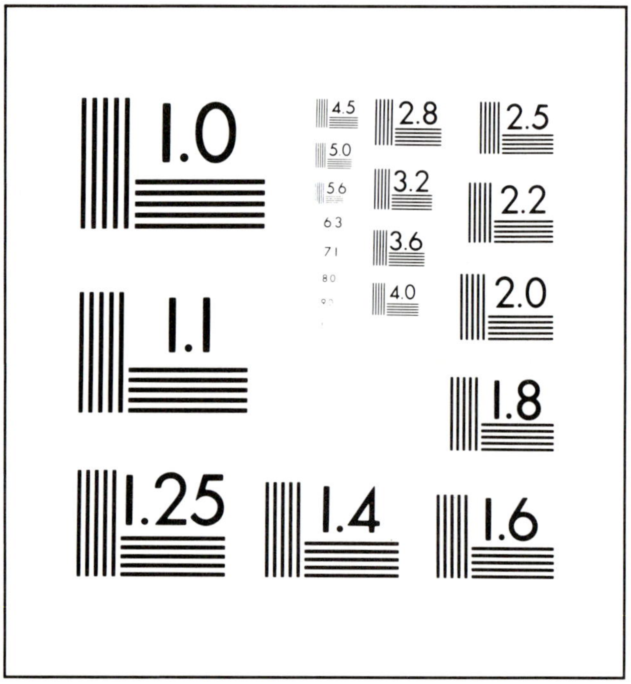

SPECIFICATIONS: TEST TARGETS

Substrate: 50 x 50 x 1.5mm for USAF pattern
63.5 x 63.5 x 1.5mm for NBS pattern
Type: Evaporated chromium or photographic emulsion
Pattern: USAF 1951 or NBS 1963A

Range:		
	USAF chromium type	1-228 lp/mm
	USAF emulsion type	1-114 lp/mm
	NBS chromium type	1-18 lp/mm

Test Targets

Description	PRODUCT NUMBER
USAF 1951 emulsion positive	04 TRP 001
USAF 1951 chromium positive	04 TRP 003
USAF 1951 emulsion negative	04 TRN 001
ISAF 1951 chromium negative	04 TRN 003
NBS 1963A chromium positive	04 TRP 005

SHUTTERS

Shutters are used to turn on and off electromagnetic radiation. They accomplish this by mechanically removing and imposing a completely opaque barrier across the optical path. Two basic forms of shutter are normally used. "Between-the-lens" shutters open and close a symmetric aperture (usually consisting of five thin metal blades) for a predetermined length of time. These shutters are usually placed between the forward and rear sets of optical elements in a photographic objective, at a location where rays are neither strongly convergent nor strongly divergent and equal time history of transmission for all elements of the field is preserved. The second common form of shutter is a "focal-plane-shutter" which transports a variable width slit across the focal plane at a fixed (or variable) speed. The width and speed of the motion determines the amount of light transmitted.

"Between-the-lens" shutters were originally designed for specific camera applications, but their general usefulness in the optical laboratory is so great that a clear understanding of their operation is essential.

BETWEEN-THE-LENS SHUTTERS

We manufacture two types of between-the-lens shutters: electronic shutters and mechanical shutters. Both operate by swinging thin metallic shutter blades from the center outward. When fully open, they form a circular aperture for a preset period of time. Both types also incorporate a 10 or 12 bladed iris diaphragm which may be independently operated to vary the diameter of the clear aperture as seen through the shutter.

Electronic shutters are activated by an electromagnetic solenoid. When the solenoid is energized, a lever arrangement simultaneously draws all the shutter blades open. De-energizing the solenoid allows the blades to be redrawn closed by a spring. Duration of the open time is specified by an electronic control unit known as a Speedcomputer® . The Speedcomputer® is housed separately and may be located at some distance from the shutter itself. Facilities for remote electrical operation of the Speedcomputer® (and shutter) are provided.

Mechanical shutters are activated by a spring driven gear wheel assembly. The spring is wound by a cocking lever on the shutter housing. Upon release, the spring drives a high-to-low ratio gear train which engages a pinion-ratchet drive. This arrangement produces a constant energy dissipation system which opens and then closes the leaves of the shutter. The timing for the duration of opening is provided by an escapement-type timing mechanism similar to that found in a watch. Remote operation of mechanical shutters is possible with a push cable release system. Remote distance is restricted by the mechanical nature of the cable.

While the actuation method varies between the electronic and the mechanical shutter versions, both rely upon a mechanical opening and closing mechanism. Both types therefore are subject to mechanical wear and eventual failure. Electronic shutters are inherently more accurate, stable, and longer-lived than mechanical shutters because they contain so few mechanical parts and linkages.

Both types of shutters have been designed to assure as long a trouble-free and accurate life as possible. Under normal operating conditions, a life of 200,000 operations should be expected from mechanical shutters and 1,000,000 operations from electronic shutters.

There are two causes for time aberrations which occur in between-the-lens shutters. When the opening time is long (about one second), there is little concern about the time required for opening and closing, but when the opening times are in milliseconds, both the reaction time and the changing size of the opening in transition determine the actual open time. Opening and closing times vary with the shutter size, as well as with ambient operating conditions, cleanliness of the mechanism and wear history. A small aperture is uncovered and covered much more quickly than is a large aperture. Consequently, with a fast shutter speed, the exposure will be shorter for small apertures than for large apertures.

Opening and closing times of both mechanical and electronic shutters are of the order of 3 milliseconds. With electronic shutters, the time between command and opening is both constant and repeatable. Furthermore, variations in opening and closing times are markedly less for electronic shutters than for mechanical shutters.

These shutters are manufactured by Melles Griot Optical Systems in Rochester, New York, and were previously known as Ilex shutters.

MECHANICAL SHUTTERS

These mechanical shutters are self-cocking automatic shutters. They are operated simply by pressing the finger release lever, i.e., cocking and tripping are accomplished in one operation. Operation of mechanical shutters can be accomplished using the finger release lever, or by using a cable release attached to the cable guide. The cable guides are made to accept the standard ASA type cable release thread, No. 5-44.

Each size has available an independently operated "press-focus release" lever which permits the shutter blades to be opened without operating the shutter mechanism or the necessity of changing the shutter speed setting to "T" for focusing.

Mechanical shutters have built-in ten-blade iris diaphragms. Diaphragm opening is controlled by a lever on the shutter housing. An f-number scale is provided to indicate the opening. A separately moveable mechanical stop may be set on the f-number scale to repeat a frequently used diaphragm opening position.

Each shutter has built-in "X" synchronization with a bi-post connector designed to fire flash equipment of any kind. The "X" synchronization contacts are silver plated. The contacts can be utilized to signal exactly that point in time at which the shutter blades have become fully opened.

These mechanical shutters are equipped with "B" (bulb) and "T" (time) speed settings. The "B" setting is the setting at which the shutter leaves will open and remain open until the activating finger release lever is released. The "T" setting is the setting at which the shutter leaves will open and remain open until the activating finger release lever is depressed a second time.

SPECIFICATIONS: MECHANICAL SHUTTERS

Dimensions: ± 0.2mm
Shutter speed settings: T, B, 1, 1/2, 1/5, 1/10, 1/25, 1/50, 1/100 seconds
Flash synchronization: "X"
Cable release socket: No. 5-44

Mechanical Shutters

Aperture Max. (mm)	Aperture Min. (mm)	Number of Shutter Blades	Number of Iris Leaves	PRODUCT NUMBER
25.4	1.6	5	10	04 IMS 001
34.9	2.5	5	10	04 IMS 003
42.7	3.4	5	10	04 IMS 004
63.5	4.5	7	12	04 IMS 005

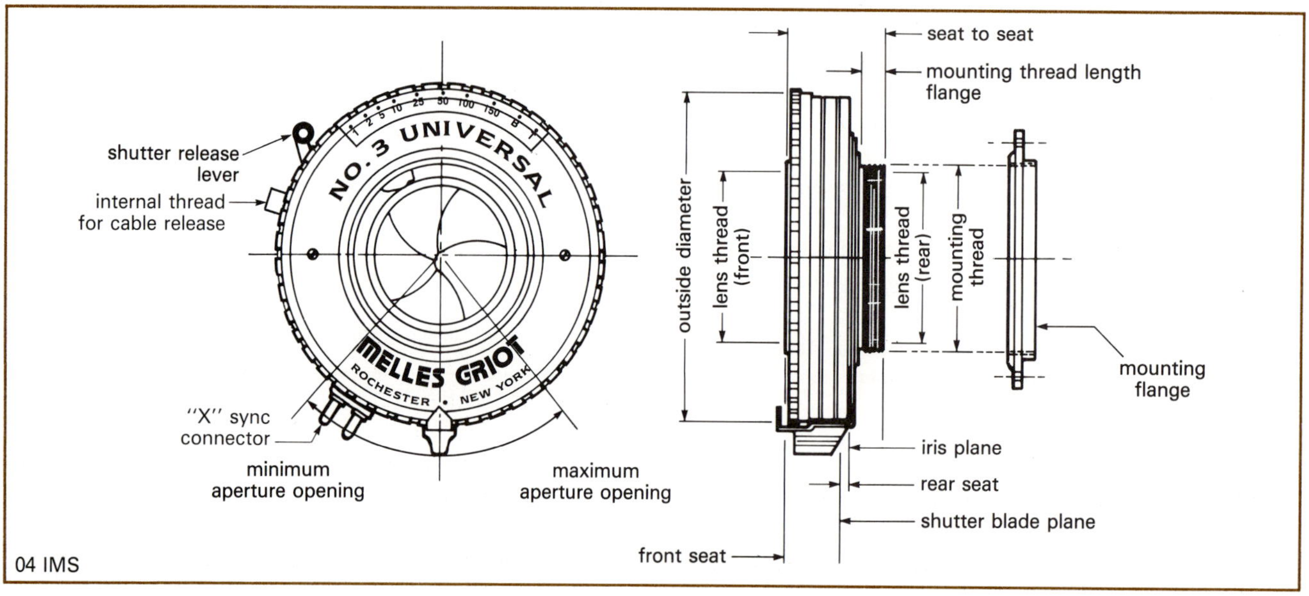

ELECTRONIC SHUTTERS

The simplicity, reliability and accuracy of these shutters make them well suited to situations demanding precise operation and long trouble-free life. These shutters may be operated with the Speedcomputer® timing electronics described below or with customer supplied DC open and hold signals.

The Melles Griot electronic shutter is the only completely electronic shutter of its kind. While most electronic shutters are electronically controlled only during long exposures, and revert to mechanical control at the high speed, this electronic shutter is electronically controlled for all exposure times, resulting in greater accuracy, consistency, and fewer control mechanism parts to become defective.

Electronic shutters are available with four maximum aperture sizes: 25.4mm, 34.9mm, 42.7mm and 63.5mm. All but the largest are five-bladed shutters; the largest has seven blades. All are equipped with built-in iris diaphragms. Diaphragms may be adjusted from full aperture to approximately 10% of full aperture diameter. Diaphragm opening is controlled by a lever on the shutter housing.

A microswitch in the shutter provides electronic flash synchronization. Alternatively, this normally open switch can be used as an event trigger.

Electronic shutters may be used with or without the Speedcomputer® timing and actuation system. The Speedcomputer® provides accurate pre-calibrated timing signals to operate the shutter at speeds from 4 seconds to 1/125 second (1/60 second maximum speed in the two larger aperture shutters). "Bulb" and "Time" operation controls are also provided.

Unlike other electrical shutters whose power source is built into their housing, these electronic shutters may be operated by other circuitry. To open the shutter a DC signal 4 times the rated solenoid voltage should be applied to the bi-post connector for 3 to 5 milliseconds. To keep the shutter open, a holding DC voltage of one-half the solenoid rating must then be maintained after opening. Removal of this signal will quickly close the shutter.

Standard solenoids are rated at 48VDC. (Optional solenoid ratings of 3, 6, 12 and 24VDC are available on special order.) Thus the nominal opening signal should be 192VDC for 3 to 5 milliseconds with a standard solenoid, and the holding signal should be 24VDC.

As a standard feature, these electronic shutters have an independently operated "press focus" lever which permits the shutter blades to be opened without operating the shutter mechanism. In addition, the mounting dimensions of all Melles Griot electronic shutters are the same as those of the corresponding number size of Melles Griot mechanical shutters. This permits the added flexibility of interchanging lenses between mechanical shutters and electronic shutters.

All these shutters, mechanical and electronic, conform to applicable ASA and DIN standards. Housings are made of aluminum and lacquer finished for long durable life. Shutter blades are made of steel; iris blades are made of phosphore bronze. Both are finished matte black. All electrical shutter contacts are silver plated.

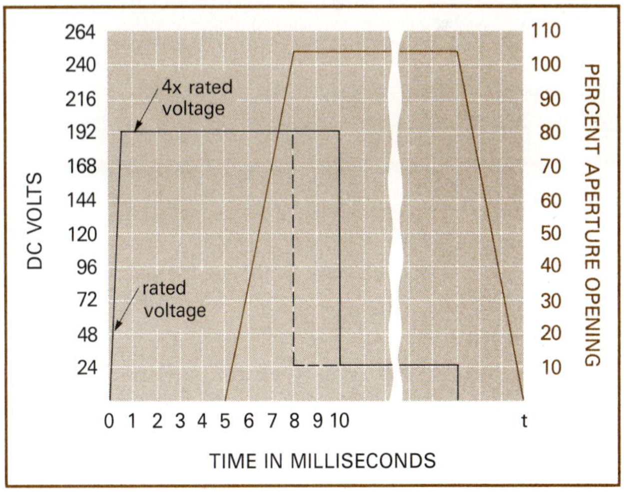

TYPICAL OPENING AND CLOSING TIMES for Melles Griot Electronic Shutters. The 3 millisecond reaction time is a function of both aperture and shutter age, but is nevertheless typical. The lower diagram illustrates the relation between the initial DC pulse and the shutter reaction. The 5 millisecond delay is both repeatable and constant. The voltages are typical for the standard 48VDC solenoid. Non-standard voltage solenoids operate at the same 8:1 ratio of DC voltages.

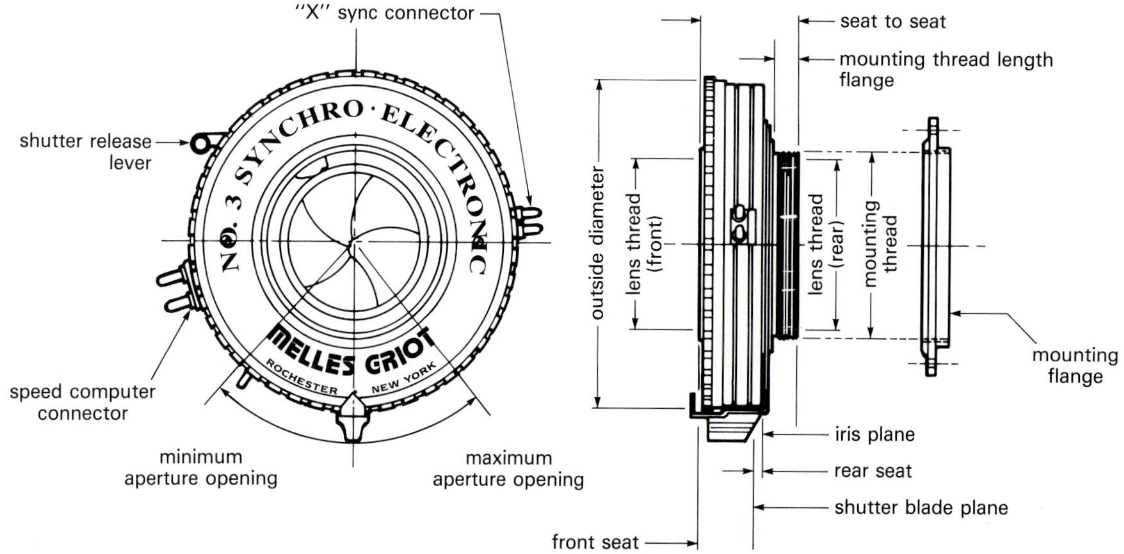

04 IES

SPECIFICATIONS: ELECTRONIC SHUTTERS

Dimensions: ±0.2mm

Shutter Speed settings: B, 4, 2, 1, 1/2, 1/4, 1/8, 1/15, 1/30, 1/60 seconds

04 IES 001 and 003 only: 1/125 seconds

04 IES 001, 004, 004 only: T

04 IES 005 only: 5

Flash synchronization: "X"

Solenoid voltage: 48VDC (other voltages available on special request)

Maximum current: 4mA

Electronic Shutters

Aperture Max. (mm)	Aperture Min. (mm)	Number of Shutter Blades	Number of Iris Leaves	Power Used (watts)	PRODUCT NUMBER
25.4	1.6	5	10	2 to 5	04 IES 001
34.9	2.2	5	10	4 to 7	04 IES 003
42.7	3.4	5	10	4 to 8	04 IES 004
63.5	5.1	7	12	4 to 9	04 IES 005

Shutters modified and/or designed for specific applications are supplied routinely by Melles Griot. A substantial number of special OEM shutters are currently being manufactured for specific customers. Typical applications include medical x-ray devices, thermographs, oscilloscope photography, census information processing, laser research and products, military instrumentation, microfilm printers, surveillance cameras, commercial photography, and various research applications.

ELECTRONIC SHUTTER CONTROLLER

The Speedcomputer® is an electronic control unit for use with Melles Griot electronic shutters. The Speedcomputer® provides precise drive voltage to open the electronic shutters for a predetermined length of time. The shutter may be triggered by the front panel push button.

The Speedcomputer® provides the following calibrated shutter speeds: 4, 2, 1, 1/2, 1/8, 1/15, 1/30, 1/60 and 1/125 seconds. (Maximum of 1/60 second on the two larger Melles Griot Shutters.) A three-way front panel switch provides a "Bulb" setting, which will hold the shutter open until it is released, and a "Time" setting, which will hold the shutter open until the switch is turned off.

A front panel light indicates when the shutter is open. The Speedcomputer® may be easily hand held or mounted in a convenient location. A 2.4 meter cable connects the Speedcomputer® to the Melles Griot electronic shutter. A further 2.4 meter cable is supplied for connection of the Speedcomputer® to a 115VAC (or 220VAC) power source.

Speedcomputer® Electronic Shutter Controller; 115v operation

For Use With	Supply Voltage (VAC)	Supply Frequency (Hz)	PRODUCT NUMBER
04 IES 001/04 IES 003	115	60	04 ISC 001
04 IES 004/04 IES 005	115	60	04 ISC 004

Speedcomputer® Electronic Shutter Controller; 230v operation

For Use With	Supply Voltage (VAC)	Supply Frequency (Hz)	PRODUCT NUMBER
04 IES 001/04 IES 003	230	50	04 ISC 501
04 IES 004/04 IES 005	230	50	04 ISC 504

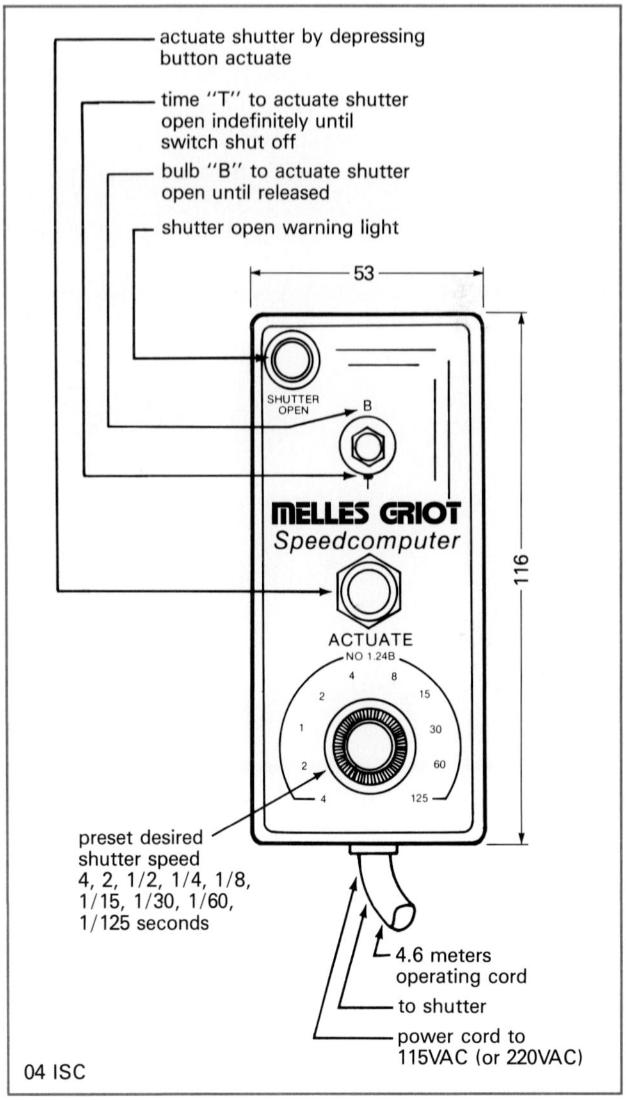

SPEEDCOMPUTER®

The Speedcomputer® is a control device used to actuate the shutter. With its solid state circuitry it controls all shutter speeds with great accuracy.

LASER PRODUCTS

Recent in-house research by Melles Griot Laser Products in San Marcos, California, has resulted in several startling new products. The GreNe™ is a HeNe Laser which provides an intense collimated, coherent beam of green light. An infrared version of this same laser, known as the IReNe™ produces a near infrared output beam. Both offer dramatic new application possibilities for the helium neon laser. A completely new self-contained 8-watt CO_2 laser has been introduced which offers reliability, portability and a dependable source of CW output power.

These new products are further indications of Melles Griot's innovative design abilities and market leadership. The great majority of Melles Griot's laser business is to OEM's — people who use lasers as a component in the equipment which they in turn manufacture. The volume of HeNe lasers used by OEM's today is very great and, as the preferred supplier to most of these companies, Melles Griot is now the largest manufacturer of helium neon laser plasma tubes in the world.

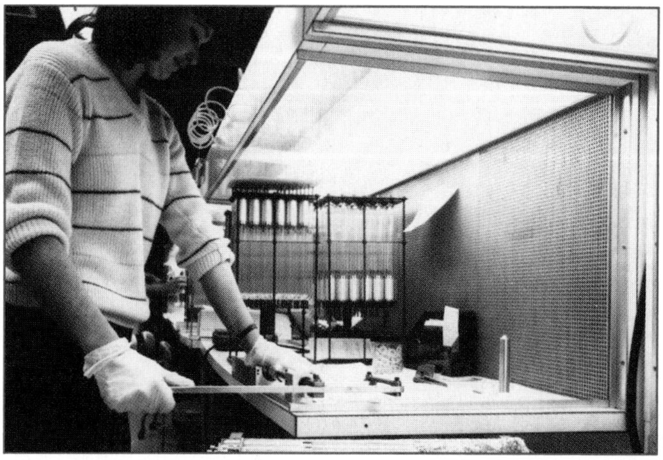

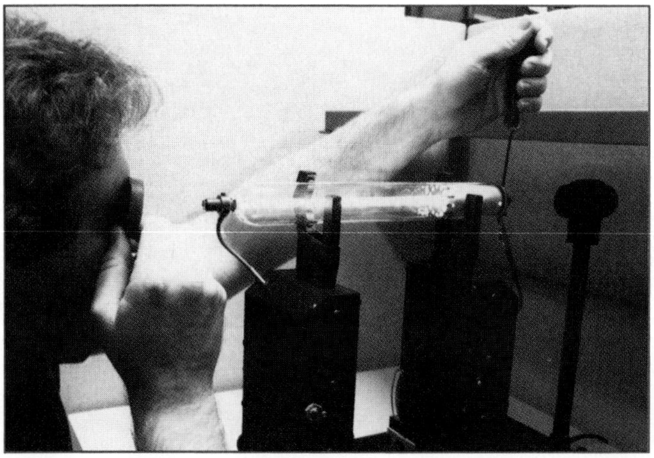

Literally, we make more gas lasers than anyone. Melles Griot has achieved this market leadership by making a better product. Our laser tubes are reliable, stable, dependable and rugged; we provide technical support and prompt delivery; we work with customers to solve problems and produce results; we manufacture only lasers (not end-user equipment which competes with yours).

If you are an Original Equipment Manufacturer and expect a volume requirement for HeNe or CO_2 Lasers, please call your Melles Griot office or representative for further information.

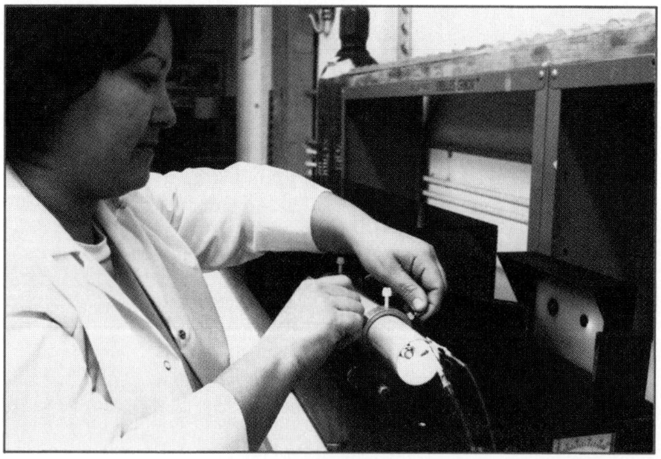

HELIUM NEON LASERS AND ACCESSORIES

Here we describe the basic principles of Helium Neon lasers and Gaussian Beam Optics. We then list our superior range of Helium Neon lasers (currently available at three wavelengths) and laser accessories.

Contents **Page Number**

HOW DOES A HELIUM NEON LASER WORK?

All lasers depend for their operation on the phenomenon of stimulated emission of radiation. For this to occur, it is necessary for some atoms to exist in an excited energy state. In an excited atom an electron may decay to a lower energy level, emitting a photon of energy $E = h\nu$, where h is Planck's constant and ν is the photon frequency. This is spontaneous emission of radiation, or luminescence. If the emitted photon passes close to another atom in a similar excited state, it may induce an identical transition. In this case the emitted and stimulating photons will be identical in every respect, having the same frequency, phase, direction, and polarization — they will be spatially and temporally coherent. This is stimulated emission of radiation.

To achieve laser oscillation it is necessary to make the probability of stimulated emission exceed that of spontaneous emission and absorption. This is done by creating a population inversion, in which more atoms exist in an excited energy state than in some lower state to which direct radiative transition is possible. This situation must be maintained or laser action will cease.

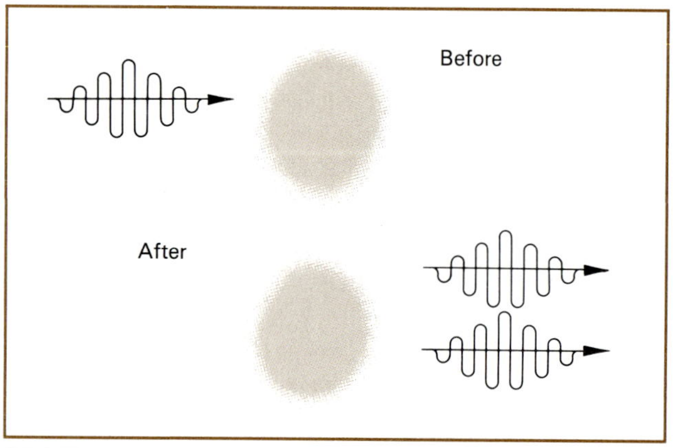

STIMULATED EMISSION, an artist's conception.

In the helium neon system, a mixture of helium and neon gases is contained in a narrow bore tube. These gases are replenished as needed from a reservoir. The mixture is predominantly helium, with about 10% neon. Melles Griot helium neon lasers are filled with a proprietary mixture which has the effect of maximizing the ratio of beam power to plasma volume. An electrical discharge ionizes the gas inside the narrow bore. The plasma so formed contains many helium atoms in metastable states. In these states one of the valence electrons has been raised from the 1s and 2s level, where it will remain for a long time, the downward transition being forbidden by the selection rules of quantum mechanics. Neon has an atomic number of 10, and has 10 electrons normally residing in the $1s^2 2s^2 2p^6$ configuration. Collisions between metastable helium atoms and neon atoms cause one of the neon electrons in the 2p level to be raised to the 4s or 5s levels. The energy match between the 2s helium levels and the 4s and 5s neon levels is very close. It is therefore much more probable that this excitation will occur than one to the 3s or 3p levels. On exciting a neon atom in this way, the helium atom drops back to the ground state. However, because there are many more helium atoms than neon, the metastable collision probability remains high. For neon laser action to occur, it is necessary to maintain a greater population of neon atoms with electrons in the 4s and 5s levels than in the 3p level. This condition is satisfied by the selective excitation process just described, and by the low natural population of the 3p levels. A number of alternative downward transitions are available to the excited neon electrons. The major neon laser line transitions are indicated on the energy level diagram.

Some of the radiation emitted will be in a direction approximately parallel to the tube bore. The rest will be emitted through the tube walls and does not contribute to the laser operation. For a photon traveling along the bore, there is a chance of passage close to a second excited neon atom and stimulation of the emission of another identical photon. The probability of this occurrence depends on the gas density and length of the bore. For a single pass along the bore the probability is on the order of 10%.

Mirrors are placed at either end of the tube, reflecting photons many times along the length of the bore. In this way the likelihood of stimulated emission is further increased. Each stimulated photon is capable of stimulating another, and each is in perfect coherence with those already present. In this way, light amplification is achieved.

The two mirrors form a resonant optical cavity, and it is this cavity which is responsible for the selective gain characteristics of the laser. Cavity mirrors are coated to have a very high reflectance at the laser wavelength of interest, but to transmit unwanted laser lines. By retaining only the desired wavelength photons in the bore, amplification is restricted to only this wavelength. Similarly, misdirected radiation of the desired wavelength escapes from the cavity and is not amplified, giving rise to the highly directional nature of the laser output. In order to make use of the desired laser radiation, some of it must be allowed to escape from the cavity as a beam. In practice, one of the mirrors is coated to permit about 1% transmission at the laser wavelength. It is this 1% of light which escapes that forms the usable output of the laser. It follows that the internal beam is 100 times more powerful than the output beam.

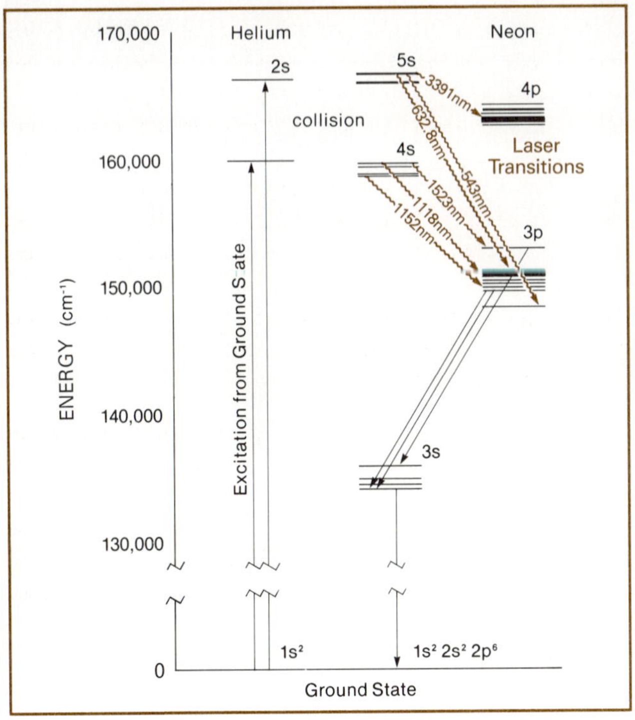

ENERGY LEVEL DIAGRAMS for helium and neon.

Various shapes of resonant cavity have been used. The plane parallel resonator is never used. It is inherently unstable and has large diffraction losses.

The confocal resonator has two equal radius concave spherical mirrors, each placed at the center of curvature of the other. The cavity utilizes a large fraction of the plasma volume and produces high power. It is less sensitive to alignment, but is slightly sensitive to mirror separation. The confocal resonator is rarely used when TEM_{00} output is required.

The hemispherical resonator has a plane mirror at the center of curvature of a concave spherical mirror. The resonator becomes quasi-hemispherical if this mirror separation is slightly reduced. This quasi-hemispherical configuration is easy to adjust and pro-

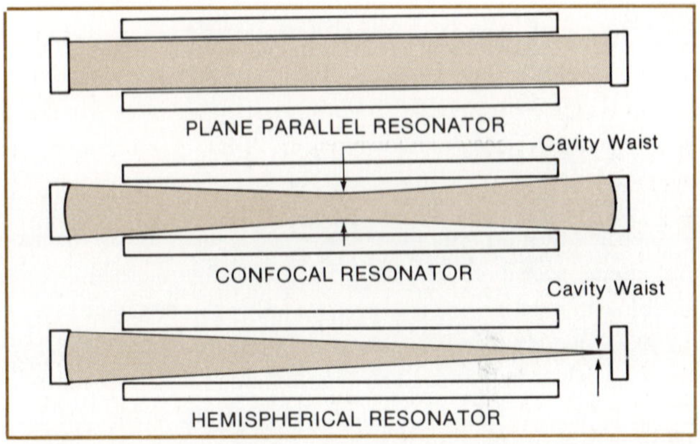

LASER CAVITY TYPES, showing the shape and size of the active plasma volume.

duces highly coherent output. The disadvantage of this cavity is that it only utilizes about one-third the volume of available plasma due to the focusing effect of the concave mirror. However, its ease of adjustment, and stability once aligned more than outweigh the disadvantage of power limitation. Quasi-hemispherical cavities are, therefore, the most commonly used in commercial lasers.

The presence of an optical cavity leads to longitudinal modes of oscillation. Only the light wavelengths which correspond to standing waves are amplified. Out of phase reflections are quickly lost through destructive interference. In order for a mode or standing wave to exist, the round trip cavity length must be exactly equal to an integer number of wavelengths:

$$2L = m\lambda \text{ (where m is a large integer).}$$

This corresponds to the series of resonant frequencies

$$\nu = \frac{mc}{2L} \text{ (where c = velocity of light).}$$

These frequencies each define an extremely narrow band within which laser oscillation can occur. The frequency separation of these longitudinal modes or resonances is

$$\Delta\nu = \frac{c}{2L},$$

and is inversely proportional to the length of the cavity. Thus, for a 50cm cavity length, the mode spacing will be:

$$\Delta\nu = \frac{3 \times 10^{10}}{2 \times 50}\ \text{Hz}$$

$$= 300\,\text{MHz}$$

In order for these modes to exhibit optical gain, their individual frequencies must lie within the finite spectral line width of the natural atomic transition. The net effect of random thermal atomic motion within the plasma is to produce a Doppler broadened spectral line. Modes which fall within the Doppler broadened spectral envelope will be amplified. The linewidth of a single laser mode is extremely narrow compared with the Doppler linewidths ordinarily encountered in atomic-spectroscopy.

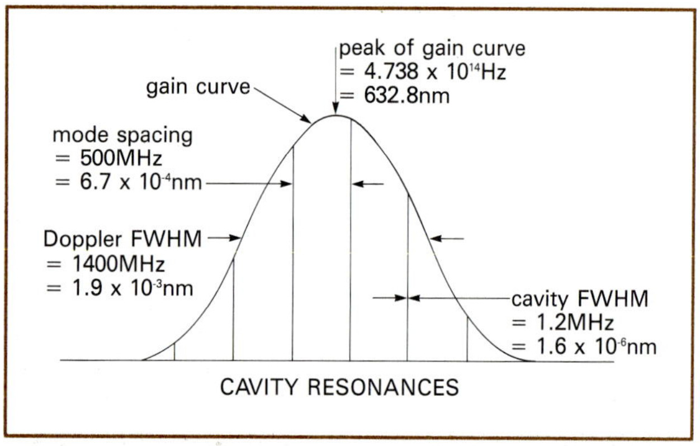

DOPPLER BROADENED GAIN CURVE supporting 6 longitudinal modes or cavity resonances.

In addition to longitudinal modes, the cavity can support a number of Transverse Electric and Magnetic modes (TEM modes). These are modes in which the irradiance distribution along the electric and magnetic field vectors exhibits one or more nulls. Transverse modes are identified by their irradiance distributions and are designated TEM_{pq}, where the subscripts p and q refer to the number of nulls along two orthogonal axes which are perpendicular to the cavity axis. The lowest order TEM_{00} mode has a Gaussian irradiance distribution. This mode experiences the minimum possible diffraction loss, has minimum divergence, and can be focused to the smallest possible spot. For these reasons, it is often desirable that the laser should be restricted to operation in this fundamental mode. Higher order modes have a larger spread and usually suffer high diffraction losses.

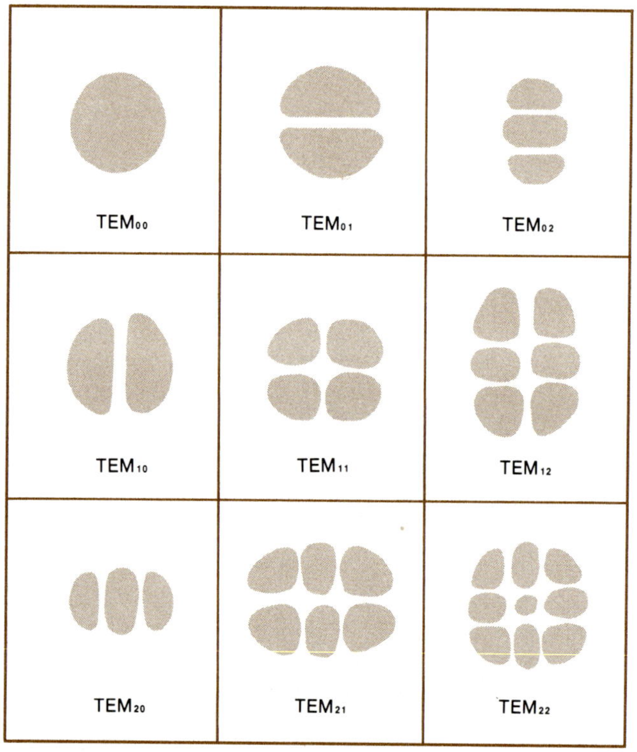

TRANSVERSE ELECTRIC AND MAGNETIC MODES of a laser cavity.

The polarization of laser radiation is an important effect which occurs as a direct result of the stimulated emission process. Just as the stimulated photon is in phase, frequency, and directional agreement with the stimulating photon, so it also carries the same state of polarization. The resonant cavity sets up standing waves of constant linear polarization. This results in alternate longitudinal modes having orthogonal states of polarization. Likewise, adjacent transverse modes have orthogonal states of polarization. The resultant polarization depends upon the precise mix of modes present, and this mix may change rapidly with time. So called "randomly polarized" lasers have an indeterminate combination of orthogonally polarized radiation. To eliminate this uncertainty, laser outputs are often deliberately linearly polarized. A single Brewster angle window is placed in the cavity, effectively eliminating one state of polarization, and preventing it from being amplified. The output is more stable and has a known state of polarization, namely the p-polarization defined by the window.

CHARACTERISTICS OF LASER LIGHT

Melles Griot helium neon lasers are available with three different output wavelengths; 543.5nm, 632.8nm, and 1523.5nm. All deliver relatively intense coherent electromagnetic output radiation. They are able to do this reliably, for many hundreds of hours, without variation of output power or beam direction. The temporal coherence of a laser is a measure of the ability of the beam to produce interference effects as a result of differences in path lengths, and is important in interferometry and holography. The spatial coherence is of importance in applications where it is necessary to focus all of the laser's output into an extremely small spot.

Both temporal and spatial coherence have long been sought for industrial and laboratory optical applications. Before the laser was invented, monochromatic sources with various degrees of coherence were in use. The relatively low output levels available, and the Doppler broadening of most atomic sources, were such that only limited applications were ever achieved.

TEMPORAL COHERENCE

If the phase at a particular instant in time along a traveling wavefront is the same as its phase after the wave has traveled a distance x in x/c seconds, for all values of x, the wave is defined to be perfectly temporally coherent. Michelson interference fringes are a direct result of temporal coherence, and they are visible only over the distance in which the temporal coherence is maintained. For conventional sources this distance is rarely more than a few millimeters, whereas for a multimode HeNe laser it is typically 20 to 30 centimeters. That this distance is not infinite demonstrates that lasers are not perfectly monochromatic or perfectly coherent. Off-the-shelf lasers are good enough, however, to permit precise interferometric measurement of distances hundreds of times longer than was formerly possible.

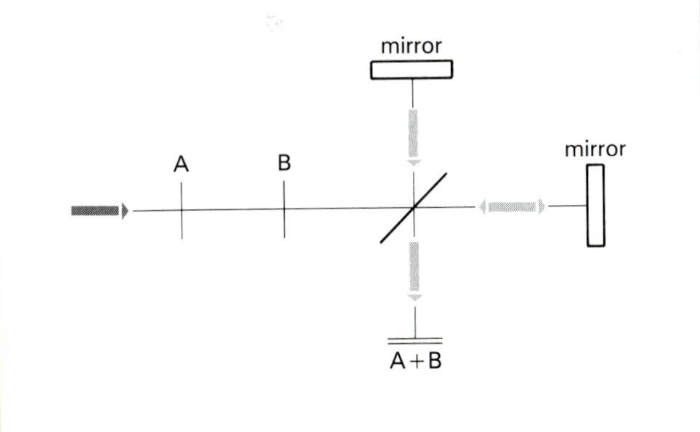

MICHELSON INTERFEROMETER, with path difference adjusted so as to superimpose wavefronts A and B in the output beam. The interferometer measures coherence length, and coherence length is the measure of temporal coherence.

Visibility of interference fringes is defined as

$$V = \frac{I_{max} - I_{min}}{I_{max} + I_{min}} ,$$

where I_{max} and I_{min} are the maximum and minimum observed fringe intensities. Unity visibility is associated with full temporal coherence. Zero visibility (absence of fringes) corresponds to $I_{max} = I_{min}$, and is associated with complete incoherence.

The minimum path length difference for which V is zero (i.e., fringes vanish) is called the coherence length (S_c). The time delay corresponding to this length is the coherence time (Δt). These are related by $S_c = c\Delta t$. The uncertainty relation between time and frequency leads to $\Delta\nu\Delta t = 1$, where $\Delta\nu$ is the line width (frequency spread) of the non-monochromatic laser output line. Lasers operating in a single longitudinal mode have linewidths of approximately 1 MHz. A typical FWHM for a commercial multimode HeNe laser is $\Delta\nu \cong 1500$ MHz, because the Doppler broadened bandwidth provides an envelope for the gain of a number of longitudinal modes.

SPATIAL COHERENCE

Correlation of phase in light beams transverse to the direction of travel is best illustrated by Young-type interference fringes. The light from two transverse portions of the same beam is allowed to interfere, producing fringes. Visibility is defined as above. Visibility of unity will again correspond to complete coherence, whereas visibility of less than unity implies partial coherence.

Spatial and temporal coherence are independent of one another; a source may be spatially coherent while temporally incoherent and vice versa. The mode discrimination properties of the laser cavity produce spatial coherence, and the Doppler broadened bandwidth limits temporal coherence.

DIRECTIONALITY AND ENERGY DENSITY

The extreme directionality of laser radiation follows directly from the coherence of the stimulated emission process. Because all of the stimulated photons are alike in every way, they carry no information about the location of the excited neon atoms they come from. Thus all of the excited neon atoms are functionally equivalent, and the laser output radiation behaves as if it all had originated in a tiny imaginary volume, having dimensions of the order of a wavelength, situated precisely on-axis at the center of the cavity waist. This remains true regardless of the size or shape of the active plasma volume.

When an off-axis neon atom is stimulated to emit a photon, the appearance presented to the outside world is that this photon originates on-axis. This behavior is attributable to transverse coherence. When a neon atom far from the waist plane is stimulated to emit, the photon still appears to come from the waist plane. This behavior is attributable to temporal coherence far in excess of the monochromaticity requirements of the beam optics formulas. Because all the coherent energy output of the active medium is effectively funneled by stimulated emission through that tiny imaginary volume, energy densities in focused laser beams can far exceed the highest energy densities actually occurring in the source plasma, a situation impossible with spontaneous emission.

When we look at an extended incoherent source, such as the surface of an incandescent solid or the spontaneous emission of a gaseous discharge, we can plainly see that it is extended. This is because the photons carry directional information telling us which parts of the source they come from. Laser radiation carries no such information.

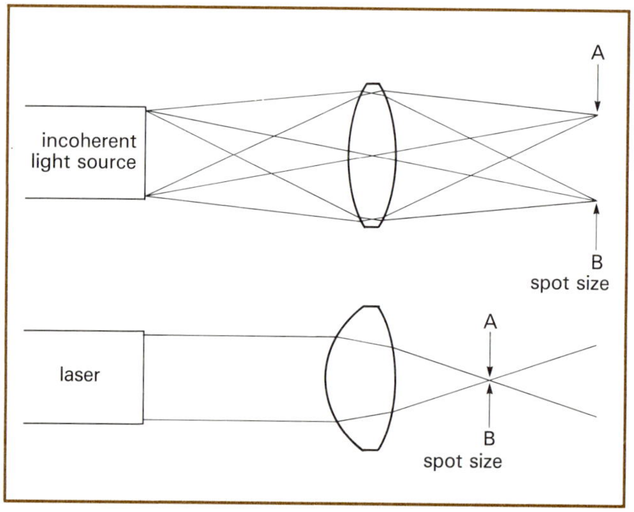

FOCUSABILITY COMPARISON for light from laser and conventional sources. Extended sources have extended images. The laser is the functional equivalent of a true point source situated at the corner of the cavity waist.

SPECKLE

A smooth but diffuse reflector has a surprising appearance under HeNe laser illumination. Though the beam irradiance distribution may be very smooth, the apparent distribution at the surface is granular. The granule size is approximately the size of the resolution element of the eye. Neighboring granules vary enormously in brightness. A bright granule looks bright because of cooperative in-phase reflections from two or more area elements within the granule, directed at the eye. Such interference is possible because coherence length exceeds the scale of surface roughness. Speckle is a great nuisance in laser imaging and holography, but carries information valuable in non-contacting metrology. Radar "glint" is the same effect on a larger scale, and hampers radar imagery.

Now you can have the convenience of HeNe lasers at 543nm and 1523nm, too.

GAUSSIAN BEAM OPTICS

Making use of a laser often involves the use of lenses and other optical elements to modify or shape the laser beam.

In some respects laser component design and tolerancing are more critical than for conventional optical components. The designs tend to be simpler, however, because many of the constraints associated with imaging systems are not present. Laser beams are nearly always used on axis, eliminating the need to design for asymmetric aberration correction. Chromatic aberrations are of no concern in helium neon lasers, although they are critical for some dye laser applications. In fact the only significant aberration in most helium neon applications is primary (3rd order) spherical aberration. Focal ratios are often very high, making even this aberration insignificant.

Scatter from surface defects, inclusions, dust, or damaged coating gives more concern than in incoherent systems. Speckle content arising from object texture and radiation coherence can limit system performance.

Because laser light is generated coherently over a relatively large volume, it is not subject to some of the limitations normally associated with incoherent sources. All parts of the wavefront act as if they originated from the same point, and consequently the emergent wavefront may be precisely defined. Starting out with a well-defined wavefront permits more precise focusing and control of the beam than would otherwise be possible. In laser component design both aberration and diffraction effects must be considered.

In order to gain an appreciation of the principles and limitations of beam optics it is necessary to understand the nature of the laser output beam. In the TEM_{00} mode the beam emitted from a laser has a perfect plane wavefront and a Gaussian transverse irradiance profile. The Gaussian shape is truncated at some diameter either by the laser bore or some limiting aperture in the optical train. To specify and discuss the propagation characteristics of a laser beam we must define its diameter in some way. The commonly adopted definition is the diameter at which the transverse field amplitude has fallen to a fraction 1/e of its peak axial value. At this same diameter the corresponding beam irradiance will have fallen to $1/e^2$ (13.5%) of its peak or axial value.

A Gaussian beam can be collected and focused to a small area even at large distances from the laser, but all optics must approach diffraction limited performance.

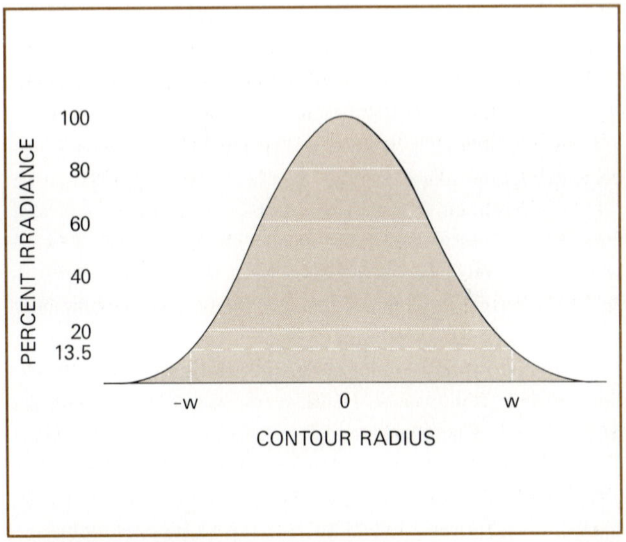

GAUSSIAN IRRADIANCE PROFILE for TEM_{00} mode, showing definition of beam radius w.

BEAM DIVERGENCE

Physical law dictates that waves must spread transversely as they propagate, and it is therefore impossible to have a perfectly collimated beam. The spreading of HeNe laser beams is in precise accord with the predictions of pure diffraction theory; aberration is totally insignificant in the present context. Under quite ordinary circumstances the spreading can be so slight as to be very difficult to see, and many people do not realize that the spreading occurs at all. The formulas which follow accurately describe beam spreading. Through them it is easy to see the capabilities and limitations of laser beams. The notation is consistent with much of the laser literature, and in particular with Siegman's excellent *Introduction to Lasers And Masers* (McGraw-Hill).

Even if a Gaussian TEM_{00} laser beam wavefront were made perfectly flat at some plane, with all elements moving in precisely parallel directions, it would quickly acquire curvature and begin spreading in accordance with

$$R(z) = z\left[1 + \left(\frac{\pi w_o^2}{\lambda z}\right)^2\right]$$

and

$$w(z) = w_o\left[1 + \left(\frac{\lambda z}{\pi w_o^2}\right)^2\right]^{1/2},$$

where z is the distance propagated from the plane where the wavefront was flat, λ is the wavelength of light, w_o is the radius of the $1/e^2$ irradiance contour at the plane where the wavefront was flat, w(z) is the radius of the $1/e^2$ contour after the wave has propagated a distance z, and R(z) is the wavefront radius of curvature after propagating a distance z. R(z) is infinite at z = zero, passes through a minimum at some finite z, and rises again toward infinity as z is further increased, asymptotically approaching the value of z itself. The wavefront shape and irradiance distribution changes are precisely analogous to the changes between the near-field and far-field patterns of a dipole antenna. These changes occur without violation of the speed of light constancy postulate of Special Relativity, because of the distinction between group and phase velocity. The plane z = zero marks the location of a Gaussian "waist," or a place where the wavefront is flat, and w_o is called the waist radius. A waist occurs naturally at the mid-plane of the symmetric confocal cavity. Another waist occurs at the surface of the planar mirror of the quasi-hemispherical cavity used in Melles Griot lasers.

The irradiance distribution of the Gaussian TEM_{oo} beam, namely

$$I(r) = I_o e^{-2r^2/w^2} = \frac{2}{\pi w^2} e^{-2r^2/w^2} ,$$

is the same at all cross sections of the beam, where w = w(z) and P is the total power in the beam. The invariance of the form of the distribution is a special consequence of the presumed Gaussian distribution at z = zero. If instead a uniform irradiance distribution had been presumed at z = zero, the pattern at z = ∞ would have been the familiar Airy disc pattern given by a Bessel function, while the pattern at intermediate z-values would have been enormously complicated. See Born & Wolf, *Principles of Optics*, 2nd edition, Pergamon/Macmillan.

Simultaneously as R(z) asymptotically approaches z for large z, w(z) asymptotically approaches the value

$$w(z) \cong \frac{\lambda z}{\pi w_o} ,$$

where z is presumed much larger than $\pi w^2/\lambda$, so that the $1/e^2$ irradiance contours asymptotically approach a cone of angular radius

$$\theta = \frac{w(z)}{z} = \frac{\lambda}{\pi w_o} .$$

This value is the far field angular radius of the Gaussian TEM_{oo} beam. The vertex of the cone lies at the center of the waist.

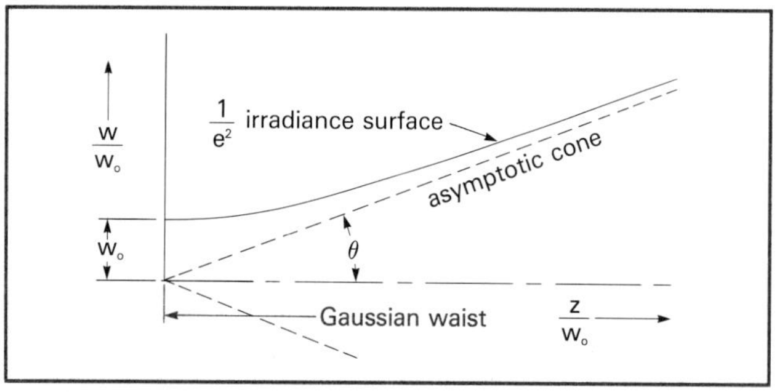

GROWTH IN $1/e^2$ CONTOUR RADIUS with distance propagated away from Gaussian waist. The effect is due to pure diffraction.

At every z-value, including ∞, an irradiance contour of radius 1.5 times larger than w(z) is, to an excellent approximation, the 99% throughput contour (meaning that 99% of the beam power passes through it). Corresponding to the cone of angular radius θ mentioned above is a larger cone of angular radius 1.5θ and sharing the same vertex. Unless beam energy losses of 1% or more (along with other undesirable effects) are acceptable, all optical elements through which it is intended that the beam should pass must be large enough to accept this larger cone. It is a value 1.5θ or larger, and not θ itself, which should be identified with f-number requirements for these elements.

The spectrally selective coating of the spherical output mirror is actually supported on the concave surface of a weak spherical positive meniscus lens. This lens operates on-axis at very high f-number, and so introduces no spherical or other aberration. Its power has been chosen to focus the output beam very slightly, just enough to form a new and much larger waist (flat wavefront) near the lens exit pupil. This improves (reduces) the beam divergence θ by a large factor, and correspondingly increases beam usefulness. Our published $1/e^2$ beam diameters are for the $1/e^2$ irradiance contour of this large output waist.

As an example, the Melles Griot red 4 milliwatt laser 05 LHR 141 (see page 357) has a published beam diameter of $2w_0$ = 0.8mm. Accordingly,

$$\theta = \frac{\lambda}{\pi w_o} = \frac{6328 \times 10^{-7}}{(\pi)(0.4)} = 5.04 \times 10^{-4} \text{ radians.}$$

At a distance of z of 100 meters

$$w(z) = z\theta = (10^5)(5.04 \times 10^{-4}) = 50.4\text{mm},$$

which is about 126 times larger than w_o.

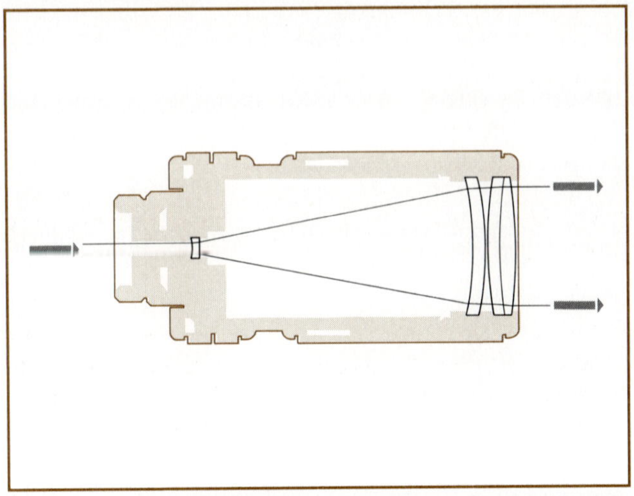

LASER BEAM EXPANDER 09 LBM 003 (reversed telescope).

Suppose instead that this same laser is directed into a laser beam expander (reversed telescope) of angular magnification m = 10, such as Melles Griot 09 LBM 005 (see page 366), and that this expander is focused so as to form a waist of radius $10w_o = 4.0$mm at the expander output lens. By definition θ is reduced by a factor 10, and so for z = 100 meters

$$w(z) = \frac{(10^5)(5.04 \times 10^{-4})}{10} = 5.04\text{mm}.$$

For the expanded beam the ratio $w(z)/w_o$ is only a factor 1.26 for a distance z of 100 meters, while it was a factor 126 for the same distance when the laser was used alone.

In the preceding example we specified that the expander was so focused as to produce a waist at its output end. This is not necessarily the optimum condition either for minimizing w(z) at specified z or for minimizing the variation of w(z) on some z-interval. Siegman shows a condition in which w(z) does not vary by more than a factor $\sqrt{2}$ on an interval of length

$$2z_R = \frac{2\pi w_o^2}{\lambda}.$$

where w_o is a waist radius at the mid-plane of the interval, z_R is called the Rayleigh range, and $w(\pm z_R) = w_o\sqrt{2}$ at either end of the range. Applied to the expander example, we may set

$$w(z) = w_o\sqrt{2} = 2.0\text{mm}$$

at the output end of the expander. Then

$$w_o = \frac{4}{\sqrt{2}} = 2\sqrt{2}\text{ mm},$$

and

$$2z_R = \frac{2\pi w_o^2}{\lambda} = \frac{16\pi}{6328 \times 10^{-7}}$$

$$= 7.944 \times 10^4\text{ mm}$$

$$\cong 80\text{ meters}.$$

The change in beam size over this distance would be almost imperceptible to the eye. As Siegman also points out on his page 314, the beam centerline can be photoelectrically located to a precision of $w_o/1000 = \sqrt{2} \times 10^{-3}$mm $= \sqrt{2}$ microns anywhere in this 20 meter interval. This remarkable precision would be difficult to achieve by any non-laser method. Careful focusing would be required to achieve these conditions.

BEAM CONCENTRATION

To find the size of the focal spot formed when a perfectly collimated laser beam is focused with an aberration free lens (such as a Melles Griot LAO, see page 134), we may again apply the asymptotic formula

$$w(z) = \frac{\lambda z}{\pi w_o}.$$

In the present problem the distance z becomes the lens focal length, f, $w(z) = w(f) = w$ is the $1/e^2$ contour radius at the lens, and w_o is the contour radius of the new and very small waist formed near the lens focal point. Thus

$$w_o = \frac{\lambda f}{\pi w}.$$

For a combination of laser and laser line focusing singlet, such as Melles Griot 05 LHR 141 (page 357) and 01 LFS 033 (page 120), the formula should be modified to become:

$$w_0 \cong \frac{4\lambda f}{3\pi w} = \frac{4(6328 \times 10^{-7})(7)}{30.4\pi}$$

$$= 4.70 \times 10^{-3}\,\text{mm}$$

$$= 4.7 \text{ microns.}$$

The factor 4/3 arises because of the careful balance of spherical aberration and diffraction designed into the singlet. The ratio f/w is proportional to lens f-number, but does not equal it.

In the diffraction limit the minimum spot size is determined by the ratio f/w. To minimize spot size this ratio and hence f/ϕ must be minimized. There is sometimes advantage to expanding a beam before attempting to concentrate it, to achieve a better match of beam diameter (99% or higher throughput contour) to the entrance pupil diameter of a highly corrected focusing system. It is essential that any such expansion be aberrationless. The angular radii of the $1/e^2$ and 99% throughput convergence cones are respectively

$$\theta = \arctan(w/f)$$

and

$$\theta_{99} = \arctan(1.5w/f).$$

The lens clear aperture diameter ϕ, f-number (f/ϕ), and numerical aperture must respectively satisfy

$$\phi \geq 3w,$$
$$f/\phi \leq f/3w, \text{ and}$$
$$(N.A.) \geq \sin\theta_{99}$$
$$\geq 1.5\, w/f.$$

If an especially small spot is desired, there is an advantage to using a well-corrected high numerical aperture microscope objective (see page 312) to concentrate the laser beam. The objective selected for this purpose should not be an oil immersion objective unless it is intended to operate under oil-immersed conditions. Thus the rated numerical aperture of the objective must be less than or equal to unity. The principal advantage of the microscope objective over a simple lens is the diminished level of spherical aberration. While microscope objectives are often used for this purpose, microscope objectives are never designed for use at infinite conjugate ratio. Suitably optimized lens systems, which Melles Griot has designed and built on special request, are more effective in beam concentration tasks.

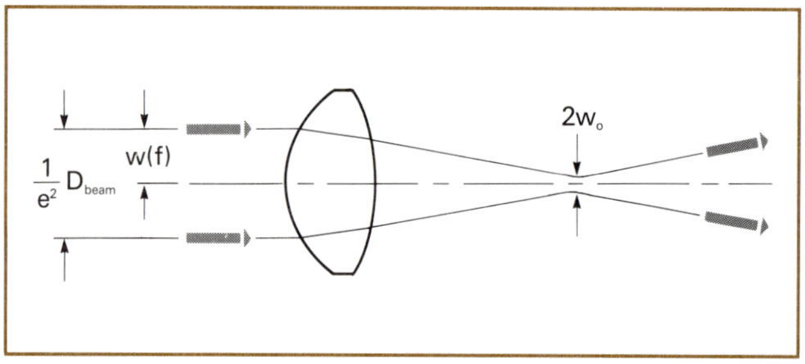

CONCENTRATION OF LASER BEAM by a laser line focusing singlet. Size of the focal waist has been greatly exaggerated for illustrative purposes.

Depth of focus $\pm\Delta z$ can be derived from the complete formula

$$w(z) = w_0\left[1 + \left(\frac{\lambda z}{\pi w_0^2}\right)^2\right]^{1/2}$$

by setting $w(z) = 1.05\, w_0$ (to allow for 5% increase) and solving for $z = \Delta z$. The result is

$$\Delta z \approx \pm\frac{0.32\,\pi w_0^2}{\lambda}.$$

Applying this result to the combination of the laser 05 LHR 141 and laser line focusing singlet 01 LFS 033, we find

$$\Delta z = \pm\frac{0.32\,\pi(0.75 \times 4.70 \times 10^{-3})^2}{6328 \times 10^{-7}}$$

$$= \pm\, 19.7 \times 10^{-3}\,\text{mm}$$

$$= \pm\, 19.7 \text{ microns.}$$

The factor 4/3 has been taken out of w_0 to prevent overestimation of Δz.

SPATIAL FILTERING

In consideration of holographic principles it is obvious that laser light scattered from dust particles residing on optical surfaces may interfere with the direct laser light to produce interference patterns resembling zone plates. Such patterns are very bothersome in interferometric and holographic applications, where they constitute a highly detailed, contrasty, and confusing background from which desired information must be salvaged. Spatial filtering is a simple way of suppressing this background and maintaining very smooth beam irradiance distributions. Spatial filtering depends on the fact that the scattered light propagates in directions different from the direct laser light, and hence arrives at a lens focal plane at places different from that where the direct beam is focused. By centering a small aperture around the focal spot of the direct beam, it is possible to block the scattered light while allowing the direct beam to pass unscathed. The result is a fan or cone of light with a very smooth irradiance distribution, and which may be refocused to form a collimated beam that is almost equally smooth.

At the plane of the aperture the scattered light is identified with low spatial frequencies, while the direct beam carries the highest spatial frequencies (the opposite is true at the lens surfaces), and so the aperture is frequently called a high-pass spatial filter.

For ease of alignment it is best that the aperture diameter be about 2 times the $1/e^2$ beam contour diameter at the focus, or about 1.33 times the 99% throughput contour diameter. A Melles Griot spatial filter may be used, consisting of a focusing optics and a pinhole, with relative position controlled by a fine xyz motion.

Suppose that the laser 05 LHR 141 is to produce a broad and smooth fan of light for hologram illumination, and that for this fan the ratio $z/w(z)$ is to be near 8.0. What combination of focusing optics and pinhole should be used? The numerical aperture corresponding to the illumination cone is 0.12.

The focal length of the appropriate focusing optics is determined by reference to the graph on page 363.

The objective focal length is found by matching the numerical aperture to the input laser beam diameter. Thus the focusing optics focal length should be 4mm so use Melles Griot 09 LSL 001 (page 363) as the focusing optics. The pinhole diameter is given by

$$D_p = 4w_o = 4(\lambda f/\pi w)$$

$$= 4\,\frac{(6328 \times 10^{-7})4}{0.4\pi}$$

$$= 8.057 \times 10^{-3}\text{mm}$$

$$\cong 8 \text{ microns}$$

so use Melles Griot 8 micron pinhole 09 LSP 011 (page 363).

Thus to accomplish this task use Melles Griot Spatial Filter 09 LSL 001 with Focusing Optics 09 LSL 001 and Pinhole Assembly 09 LSP 011. (See pages 362-363 for details of these products.)

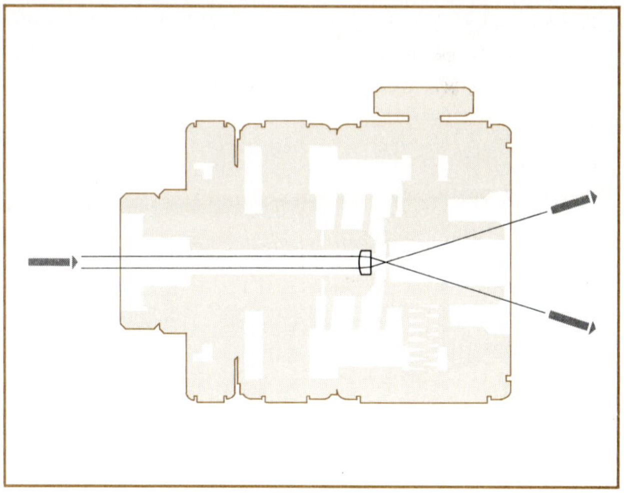

Laser accessories which provide for beam expansion, collimation and spatial filtering are described on pages 361-365.

HELIUM NEON LABORATORY LASERS

Melles Griot helium neon lasers are reliable, trouble-free laboratory tools which provide the user with stable, directional, monochromatic beams of coherent light. Three output wavelength versions are available, green (543.5nm), red (632.8nm), and infrared (1523.5nm). Both linearly polarized and randomly polarized beam outputs are available at most wavelengths. Built-in power supplies facilitate operation by simply plugging the laser into a normal grounded AC outlet.

The amplification and directionality of the output beam depends on a stable set of mechanical, optical, environmental and electrical conditions. Producing and maintaining these stable conditions requires not only controlled and repeatable manufacturing conditions, but also a skillful understanding of the whole system design.

The following pages will provide you with insight into the design of a laser system.

WARRANTY

Melles Griot helium neon lasers and their power supplies are guaranteed to operate to specification for 18 months after the date of your purchase. If for any reason, other than breakage or abuse, they should fail to perform to this level, we will replace them free of charge. This is, we believe, the best warranty in the business. It reflects our confidence in these lasers and our belief that they will have a long and trouble free life. If you should ever have reason to use this warranty provision, we request that you contact one of our offices for shipping instructions prior to returning the laser.

A complete copy of our warranty is packed with every laser that we ship.

LASER SAFETY

Melles Griot helium neon lasers are low output power devices which emit only non-ionizing radiation at 543.5, 632.8 or 1523.5 nanometers. They emit no X-rays or other harmful radiation. They have been designed and built to comply with the relevant United States federal regulations contained in 21 CFR, Chapter 1, Subchapter J. Warning labels are affixed to each laser for identification and safety.

A number of specific safety features are built into Melles Griot lasers for your protection. A green pilot light on the rear panel indicates that the power supply is on and warns that laser radiation may be present. A delay is built into the power supply so that power is not passed to the plasma tube until 3 to 5 seconds after the pilot light goes on. Also, a mechanical shutter, which can be closed when the laser is not in use, is built into the front of our lasers to prevent inadvertent exposure to the beam. In this way, the laser beam may be interrupted without forcing the laser to cool down from its stable operating condition.

As with any extremely bright light source, common sense is most important. DO NOT look directly into the laser beam or at any specular reflections of the laser beam.

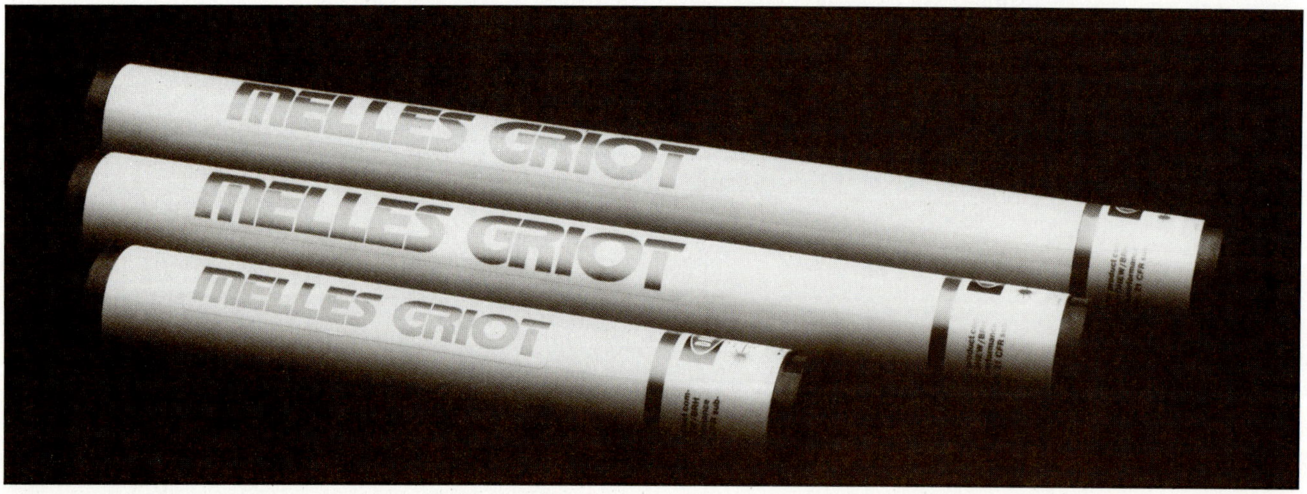

SELF-CONTAINED RED, GREEN AND INFRARED HENE LASERS

The self-contained lasers described here are exciting for two reasons. First, they represent the first time that compact, reliable helium neon lasers are available in alternate wavelengths other than red. Both green and infrared helium neon lasers are available and are completely described. Second, these lasers are truly superior products. They are convenient to use in the laboratory, classroom, and industrial research environment. they are good-looking, easy to use and compatible with a wide range of accessories and applications. They are also reliable — so reliable they are backed by the finest warranty in the laser business.

All these lasers have been designed, developed and are manufactured by Melles Griot Laser Products in San Marcos, California. The development of HeNe lasers at new wavelengths is a further indication of Melles Griot's technical competence. The majority of Laser Products' business is manufacturing HeNe plasma tubes for original equipment manufacturers. OEM customers demand reliability, stability, and a long and trouble-free life when buying lasers. The know-how and experience we have gained by serving these customers assures you that these are the very finest lasers available anywhere.

A number of features will be immediately apparent:

- Completely self-contained power supply. Just plug it in.
- Convenient operating controls. On the back panel, out of the way.
- Three "T" shaped mounting slots on both sides and the bottom.
- Safety features. Indicator light, mechanical shutter, keyswitch (on Class IIIb versions).
- An 18-month "no-nonsense" warranty. Proof of performance.
- A set of useful accessories. Fully described in this catalog.

The features that are not so apparent are the ones inside. The features providing you better stability, longer life, pointing accuracy, quiet output, fast warm-up, easy starting, great reliability.

POWER

Output powers of 1/2, 1, 3 and 5 milliwatts are available with these lasers. The 1/2mW lasers are classified by the US Bureau of Radiological Health as a Class II laser. Both the 3 and 5mW versions are classified as Class IIIb. The safety precautions, attached labels and various safety devices for the Class IIIb lasers are somewhat more stringent than for Class II.

The 1/2 milliwatt version is well-suited for most classroom and laboratory applications where the need for bright, high density beams is not so great.

MOUNTING CONVENIENCE

These lasers have been designed to be convenient to use in the laboratory and classroom. The black case has T-slots along the length of both sides and the bottom. These slots will accept the head of a standard 1/4-20 hex-head bolt and will capture it to prevent rotation. In this way one may easily create special set-ups and mounting jigs to accommodate unusual situations. A set of special clamps is available which utilize these slots when fixing the laser to an optical table or breadboard. These slots are identical to the side slots in our range of new optical breadboards (see pages 415-419). This combination of accessory mounting techniques gives you more options for your experiments and prototypes than any other laser.

The front end of these lasers has a one-inch diameter, 32 TPI threaded hole centered on the laser output beam. This may be used to mount the spatial filter, beam expanders, and collimators listed later in this catalog (see page 361). Accessories manufactured by other companies may be mounted into this standard interface as well. The front end also has four 4-40 tapped holes in a square pattern for attaching special devices.

The center of the laser beam is 45mm above the mounting surface making these lasers easily interchangeable with earlier laser designs from other manufacturers.

BORE DESIGN

The design of the capillary bore in helium neon lasers is a tradeoff. If the bore diameter is too narrow, diffraction losses in the laser beam will limit output. If the diameter is too large, the plasma is inefficiently utilized (current density is sub-optimal) and the output mode purity will be poor. Design of Melles Griot lasers utilizes an approach which maximizes the power output while controlling the transverse modes and holding diffraction losses to a minimum. Careful design of the bore diameter in relation to the cavity results in better utilization of plasma than has been available before. The TEM_{00} mode is more completely filled while higher order modes are suppressed. The result is a shorter plasma tube with greater power output. The package size is therefore smaller (as a function of output power) than usual making the compact laser more convenient to use.

The bore outside diameter has been kept large to increase stability. The hard borosilicate glass we use is exceptionally rigid and results in a more stable cavity design. The precision capillary bore is firmly attached to the plasma tube at the anode end, and a unique metallic spring spider supports the cantilevered capillary tube at the cathode end. This results in a greatly improved rotational stability of the laser because gravitational sag of the bore is eliminated. In a more conventional capillary design the bore is not supported and the slight bending of the capillary results in obscuration of a portion of the internal beam. These slight motions can give rise to very great variations in the output power.

HARD SEALS

Melles Griot lasers are sealed with truly, zero-leak hard-seals. A special pre-formed ring of the frit sealing material is placed around the parts during initial assembly. Both the cathode retaining end-cap and the anode tube are manufactured of Kovar (a nickel/iron/cobalt alloy which is thermally compatible with the hard borosilicate glasses we use). Using radio-frequency induction heating techiniques, the Kovar cathode and end-cap are welded to the plasma tube envelope. Both cavity mirrors are bonded directly to their alloy cells. The pre-formed frit material flows fully onto both parts creating a proven and reliable seal. This use of glass-metal seals throughout the tube and the complete absence of epoxy bonds results in the elimination of tube leakage. The gas mixture in the plasma tube is invariant with time and is precisely known. The result is reliable, consistent operation and a long shelf life.

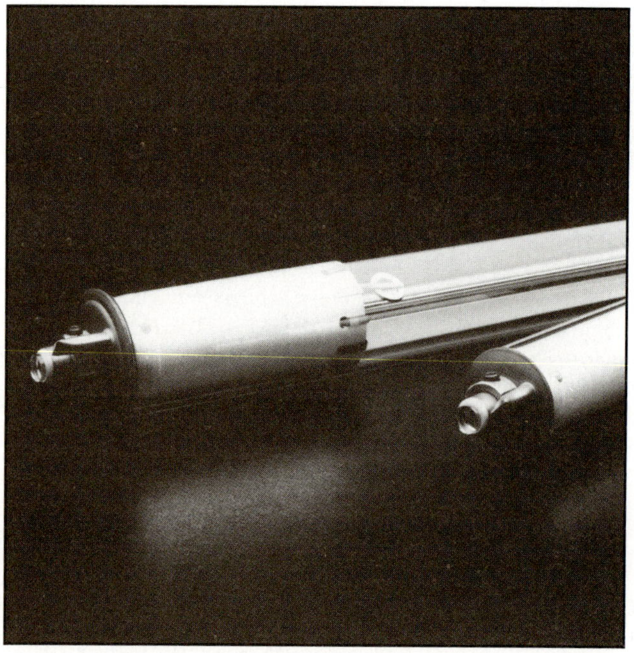

CATHODE CONSIDERATIONS

Bore design is only one part of the reason why our lasers are rotationally stable and retain great pointing accuracy. The unique patented cathode design contributes significantly to this stability as well. The deep coaxial cathode assures a symmetrical discharge. The hemispherical end-cap protects the nickel-iron end plate from bombardment by the plasma discharge. Without this protection, material from the end plate tends to sputter onto the optical surfaces, thereby shortening tube life. Further, this hemispherical end-cap design assures a more uniform current density distribution between the open end of the bore and the cathode.

POLARIZATION

Both randomly polarized and linearly polarized output models are available at most wavelengths and power levels. The polarized models produce a polarization ratio of greater than 500:1 between two orthogonal planes. The polarization plane (plane of the electric field vector) is oriented horizontally with respect to the base of the laser. The plane of this polarization can be easily confirmed by observing the reflection from a flat dielectric oriented at approximately Brewster's angle. DO NOT, of course, look into the beam; sample the beam with a piece of paper or similar material.

The randomly polarized output beams contain both orthogonal output planes of polarization, but the ratio and phases of these two components vary rapidly in time. The output, therefore, appears to be non-polarized when integrated over a fairly short period of time. Even so, care should be taken using these lasers in situations where polarization effects may occur further down the optical path.

In such cases, when polarization noise becomes a problem due to interaction of the laser output beam with polarizing or reflective surfaces, a Melles Griot polarized laser is recommended.

Polarized lasers have an internal Brewster angle window built into the plasma tube. This window selects only one phase of polarization for amplification. The resulting output polarization purity is 500:1, thereby effectively eliminating polarization noise.

POWER SUPPLY

Power supplies for these lasers are completely built-in to the housing making the lasers self-contained. The supplied power cord is detachable at the laser for convenience. Operation from either 115VAC or 230VAC is switch controlled on the rear panel of the laser. Line frequencies of either 50 or 60Hz are acceptable. The power supply is a true state-of-the-art design; output is feedback regulated and high-frequency switching circuits provide a rock-stable and reliable output. Full power laser output is available after a 3 to 5 second delay with rapid start and quick warm-up. The power supply module is mounted internally on the main circuit board together with the plasma tube.

A rear panel ON/OFF switch controls the laser. A green indicator lamp glows to show when the power supply is turned ON. The power supply (as well as the plasma tube) is covered by our standard 18-month "no-nonsense" warranty.

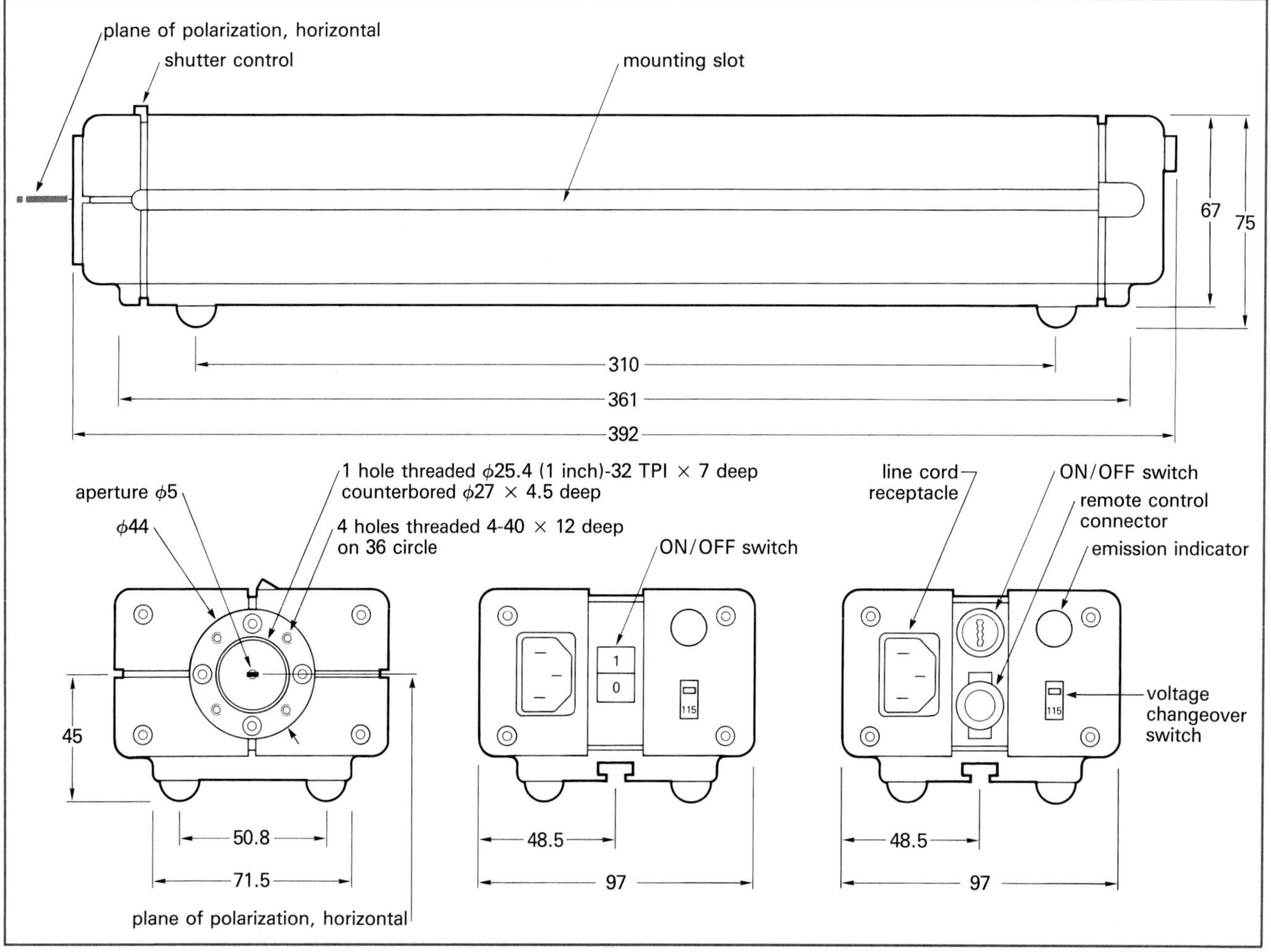

DHHS/CDRH WARNING LOGOTYPES

05 LLR 805 05 LLP 805

05 LLR 831 05 LLP 831
05 SIR 851 05 SIP 851

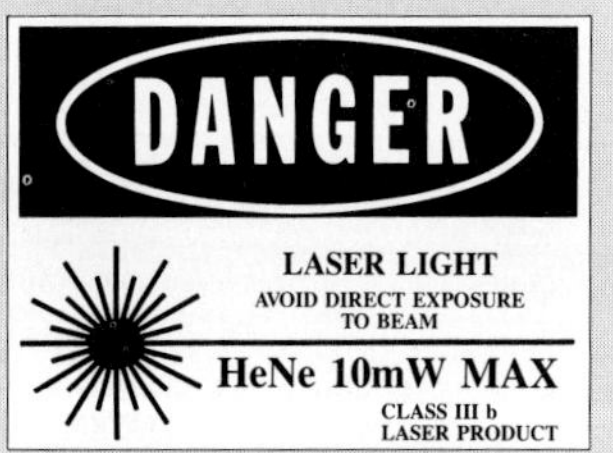

05 LLR 851 05 LLP 851

Self-Contained Red HeNe Lasers

SPECIFICATIONS	½mW 05 LLR 805	½mW 05 LLP 805	3mW 05 LLR 831	3mW 05 LLP 831	5mW 05 LLR 851	5mW 05 LLP 851
Minimum CW Power Output at 632.8nm TEM_{00}: (mW)	0.5	0.5	3.0	3.0	5.0	5.0
Beam Diameter $1/e^2$: (mm)	0.59	0.59	0.59	0.59	0.77	0.77
Beam Divergence: (mrad, full)	1.3	1.3	1.3	1.3	1.0	1.0
Polarization Ratio:	Random	>500:1	Random	>500:1	Random	>500:1
Longitudinal Mode Spacing: (MHz)	685	685	685	685	454	454
Noise Amplitude (30Hz to 10MHz, rms):	<1%	<1%	<1%	<1%	<1%	<1%
Output Power Stability:	±2%	±2%	±2%	±2%	±2%	±2%
Power Input: (VAC, 50 or 60Hz)	115/230	115/230	115/230	115/230	115/230	115/230
Weight: (kg)	1.9	1.9	1.9	1.9	1.9	1.9

PRODUCT NUMBERS

Common To All Product Numbers	
Static Alignment:	Centered within ±0.2mm and 2.0mrad
Angular Drift:	<0.03 mrad after 15 minutes
Amplitude Noise:	<1% (30Hz to 10MHz rms)
Dimensions:	±0.5mm

Environmental Specifications	Operating	Non-Operating
Temperature:	-20 to +50°C	-40 to +80°C
Altitude:	0 to 3000 meters	0 to 5800 meters
Relative Humidity:	0 to 100%	0 to 100%
Shock:	15g for 11msec	15g for 11msec

LASER TABLE CLAMPS

These extruded aluminum clamps utilize the slots in the sides of the self-contained lasers for secure mounting to optical tables and breadboards. Supplied in sets of four clamps complete with ¼-20 or M6 attaching thumb screws. Clamps are designed such that the laser may be either centered over a row of mounting holes in a breadboard or mounted between rows. Clamps are placed over the feet and when tightened, slightly compress the rubber assuring secure mounting.

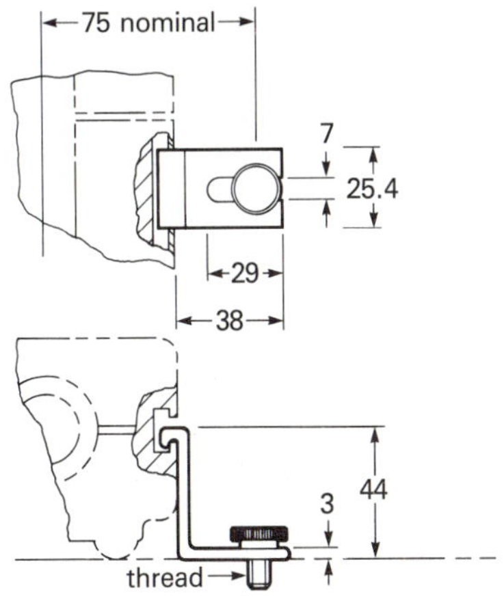

Table Clamps

	PRODUCT NUMBER
Set of four with ¼-20 thumb screws	07 CLL 001
Set of four with M6 thumb screws	07 CLL 501

A convenient tiltable table mount for these lasers is described on page 463.

SELF-CONTAINED GREEN HENE LASERS

GREEN HELIUM NEON LASERS

These GreNe'sTM are newly developed self-contained helium neon lasers which emit a collimated, coherent beam of green radiation at 543.5nm. Only randomly polarized output versions are available. In all other respects these lasers are equivalent to the widely used red helium neon lasers (emitting at 632.8nm) described on the previous pages of this section. The plasma tube in these new lasers are approximately 30cm long. It utilizes the standard Melles Griot cold-cathode HeNe discharge and is completely hard-sealed with internal mirrors.

Like other Melles Griot HeNe products, these lasers are guaranteed to operate as specified for 18 months after the date of purchase.

These lasers are available only with 0.5mW output. They are classified as Class IIIb lasers by the US Bureau of Radiogical Health; standard helium neon safety precautions should be exercised. Mechanical packaging details, safety devices, mounting details and environmental specifications are identical to those described for the red self-contained helium neon lasers.

The green output wavelength of these lasers is especially useful in a number of new applications. These applications all benefit from the low cost of these lasers (roughly equivalent wavelengths are available from argon-ion lasers costing much more) and their ease of use over a long trouble-free life. While the available power output levels are somewhat lower than equivalent red HeNe lasers, the wavelength falls exactly in the middle of the peak of the eye-response curve. The result is that the laser beams are extremely visible in even adverse conditions. A number of applications are anticipated for these lasers and are listed below; we also realize that there will be a number of other applications for these lasers which have not been anticipated.

Applications

- In conjunction with a red HeNe laser as a pointer or marker, the individual marks can be identified.
- Avionic and training displays are more visible and may be color coded.
- Contrast enhancement when compared to blood provides surgical pointing and illumination.
- Markedly improved transmission in water provides marine communication and illumination.
- When used as a pointer with high-power lasers, the green color is more visible against a hot substrate.
- Illumination for endoscopic examination and identification.
- Sensitivities in particle counting applications are greatly improved at reduced wavelengths.
- Advertising, light shows, demonstrations, two-color holography all become practical and affordable in two colors.

HOW IS IT DONE?

The green line (543.5nm) transition has long been realized to be a possible helium neon output wavelength. Many unsuccessful attempts have been made to lase this line. In November 1970, Darwin L. Perry succeeded in lasing this line (see D.L. Perry, CW Laser Oscillation at 5433A in Neon, *IEE JQE-7,* p107, 1971) utilizing a physically long laboratory HeNe device with Brewster windows or prisms and external mirrors and a hot cathode. The transition for the green line is $3s_2$-$2p_0$ as illustrated below. The Melles Griot lasers cause this transition to lase by optimizing the gas mixture, processing techniques and cavity mirror specifications. Because the entire process relies on previously optimized mechanical, optical and electrical design features of the Melles Griot 632.8nm HeNe laser, operation is both reliable and long-lived.

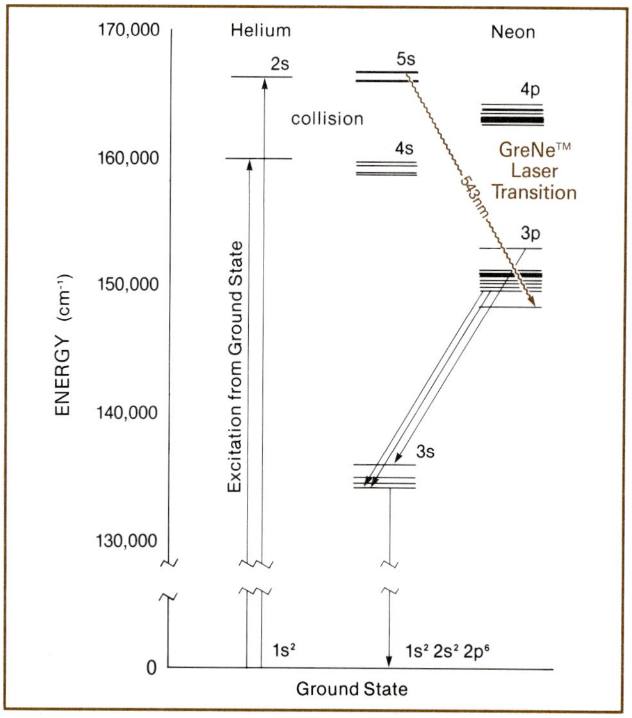

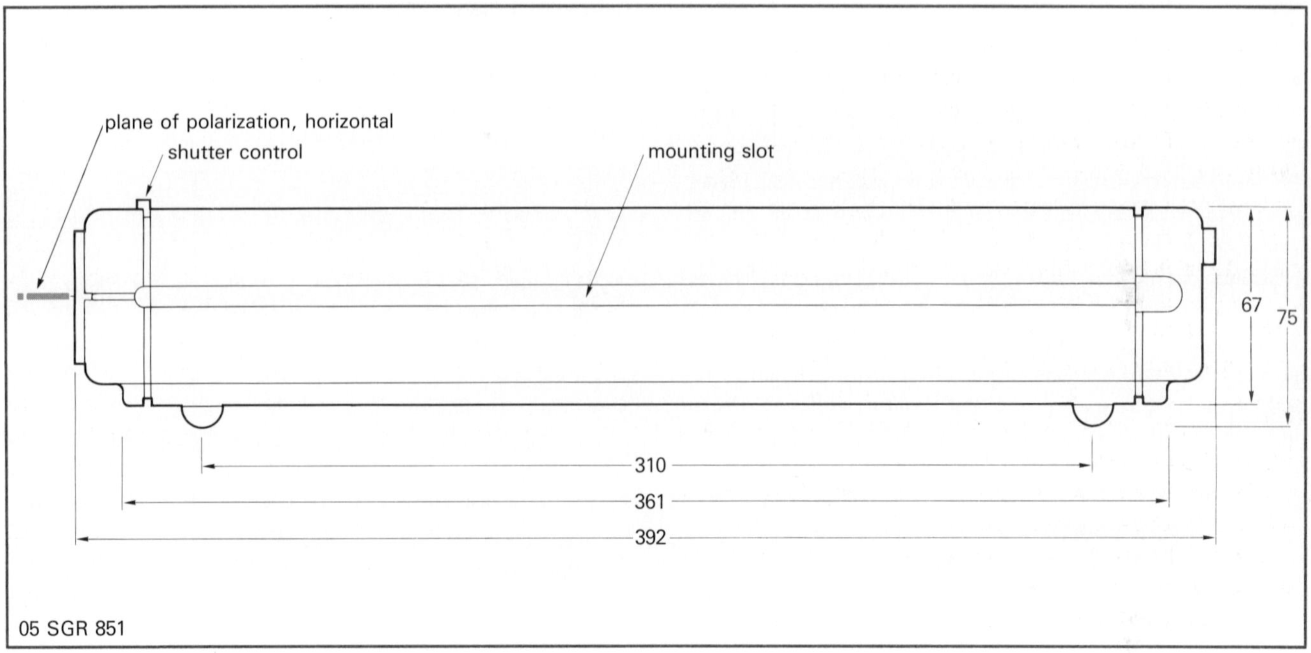

Self-Contained Green HeNe Lasers

SPECIFICATIONS	PRODUCT NUMBER 05 SGR 851
Minimum CW Power Output at 543.5nm TEM_{00}: (mW)	0.50
Beam Diameter $1/e^2$: (mm)	0.44
Beam Intensity: (W/cm^2)	260
Beam Divergence: (mrad, full)	1.57
Polarization Ratio:	Random
Longitudinal Mode Spacing: (MHz)	441
Noise Amplitude (30Hz to 10MHz, rms)	<5%
Output Power Stability:	±2%
Weight: (kg)	1.9

Common To All Product Numbers

Static Alignment: Centered within 0.20mm and 2.0mrad
Angular Drift: <0.03mrad after 15 minutes
Amplitude Noise: <1% (30Hz to 10MHz rms)
Dimensions: ±0.5mm

Environmental Specifications	Operating	Non-Operating
Temperature:	-20 to +50°C	-40 to +80°C
Altitude:	0 to 3000 meters	0 to 5800 meters
Relative Humidity:	0 to 100%	0 to 100%
Shock:	15g for 11msec	15g for 11msec

A mount for this laser is described on page 463.

SELF-CONTAINED INFRARED HENE LASERS

INFRARED HELIUM NEON LASERS

These Completely new self-contained helium neon lasers lase in the near-infrared at 1.523 microns wavelength. The monomode output beam is collimated, coherent and invisible. Both randomly and linearly polarized versions are available. An output power of 1.0 milliwatt is currently available; it is expected that higher powers for OEM applications will be available in the future. Melles Griot calls these new lasers IReNe's™.

Because these lasers utilize the same hard-seal, cold-cathode design as our other helium neon laser plasma tubes, life times of 20,000 hours are anticipated. Experience, to date, with these newly introduced lasers indicates that longevity, reliability and ease of use will be equivalent to our standard red helium neon lasers. So certain are we of this performance, that these new lasers are also covered by our standard 18-month helium neon laser warranty.

The operation of these HeNe lasers is somewhat less stable than the conventional red laser. Thermal expansion of the plasma tube causes enough movement to result in single mode power sweeping of up to nearly 50%. Active control of an external cavity mirror can limit this value to less than 1% with suitable feedback. Applications requiring stable output frequencies, such as communications applications, will require the customer to supply adequate external stabilization. Most other applications will not be affected at all by this factor.

HOW IS IT DONE?

The infrared line (1523.5nm) transition has long been realized to be a possible helium neon output wavelength. Many unsuccessful attempts have been made to lase this line. Operation of a HeNe laser at this wavelength in neon was first reported by McFarlane, Patel, Bennet and Faust in 1962 (see McFarlane et al, "New Helium Neon Optical Maser Transitions," *Proc. IRE, 50,* p2111, October 1962) using 1 and 2.5 meter long discharge tubes. No particular reason existed at the time to continue the research and little work has been done on this transition since then. The transition for the near-infrared line $2s_2$-$2p_1$ is as illustrated. The Melles Griot lasers cause this transition to lase by optimizing the gas mixture, processing techniques and cavity mirror specifications. Because the entire process relies on previously optimized mechanical, optical and electrical design features of the Melles Griot 632.8nm HeNe laser, operation is both reliable and long-lived.

APPLICATIONS

Infrared helium neon lasers offer a number of unique and intriguing application possibilities. The combination of a standard, long-life plasma tube and an invisible output beam matched to the wavelength of many laser diodes offers interesting possibilities. Applications for and features of these new lasers include:

- Eye-safety is markedly superior to other wavelengths where people may be accidentally exposed to laser radiation.
- Wavelength corresponds to new, minimum absorption fiber-optics for communications.
- Collimated output far more appealing for all applications now using laser diodes.
- Monomode output frequency with KHz widths available for long distance transmission applications (when stabilized).
- Invisibility and directionality makes ideal for intrusion and security detection.
- Heat removal problems associated with laser diodes is absent.
- Output wavelength corresponds to the peak sensitivity of germanium detectors for optimum signal detection.

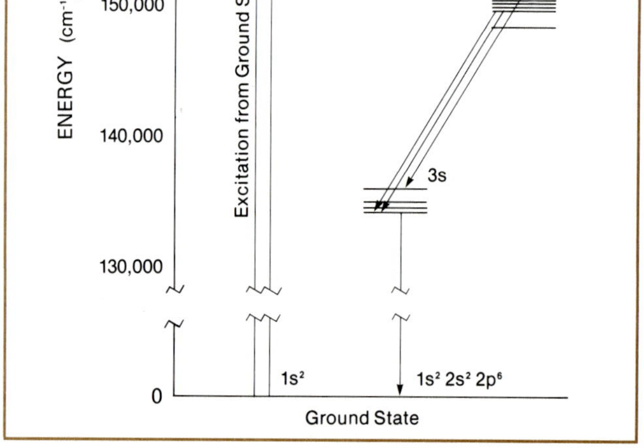

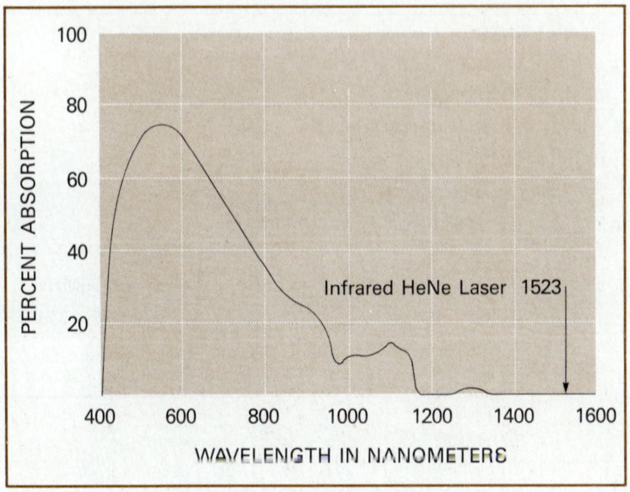

RELATIVE SPECTRAL EFFECTIVENESS CURVE for retinal thermal injury.

AN ALTERNATE TO THE LASER DIODE

Heat-removal problems usually associated with high-power laser diodes are absent in the Melles Griot HeNe laser. The greatly larger volume and surface area of a gas laser allows the heat to be dissipated easily. Also there is no themal migration of active centers and detrimental dislocations which so frequently affect semiconductors. As a result diode life is strongly affected by operating temperature. Diode operating life may be defined as:

$$\text{Diode life} = K_e^{(-8124/T)}$$

where K is a constant and T is the temperature in degrees Celsius. In other words, a 50 °C temperature rise from ambient will shorten the diode life by a factor of 48. A diode with a typical anticpated life of 2000 hours will last only 40 hours when it is operated at 70 °C. The Melles Griot infrared HeNe laser will operate at this temperature for an anticipated 20,000 hours.

Furthermore, the center frequency ν_o of laser diode gain curve shifts with temperature by 0.2nm per °C (or about 88,000MHz per °C). A helium neon laser exhibits no such change in its gain curve.

A third temperature effect that which applies to both the gas laser and the laser diode is the differential change in physical size caused by thermal expansion. For either type of laser the change in cavity length due to a temperature change ΔT is given by:

$$\Delta L = \alpha L \Delta T.$$

where α is the coefficient of linear expansion of the material separating the end mirrors from each other. Since the resonator-allowed lasing frequency is

$$\nu_R = q(c/2L),$$

where q is a large constant and c is the speed of light, a change in length of the resonator cavity, ΔL will cause a change in the resonator frequency $\Delta\nu_R$ given by:

$$\Delta\nu_R = -(qc/2L) \cdot \Delta L/L = -\nu_R(\Delta L/L).$$

Thus for a temperature change, ΔT,

$$\frac{\Delta\nu_R}{\nu_R} = -\alpha\Delta T$$

The important point here is that the relative frequency change is independent of the cavity length and depends only on the coefficient of linear expansion and the actual temperature change. Since the coefficient of linear expansion for typical laser diodes (such as InGaAsP) is essentially equal to glass, the value of $\Delta\nu_R$ is about the same for both laser diodes and helium neon lasers.

Active feedback control techniques can however be used with helium neon lasers to make the resonator frequency independent of temperature. Various techniques have been developed to achieve these feedback controls and we believe the Melles Griot infrared HeNe laser will be able to produce a stable 1.523nm output with a width of 1kHz or less when these techniques are applied.

Melles Griot does not supply these lasers in a thermally stabilized configuration at this time. However, a version of the laser with a Brewster angle window on the cathode end and a separate, unmounted external cavity mirror can be supplied to special order to allow the user to perform experiments and developments along these lines.

See pages 380-394 for a range of optics and coatings designed for 1.5 microns wavelength.

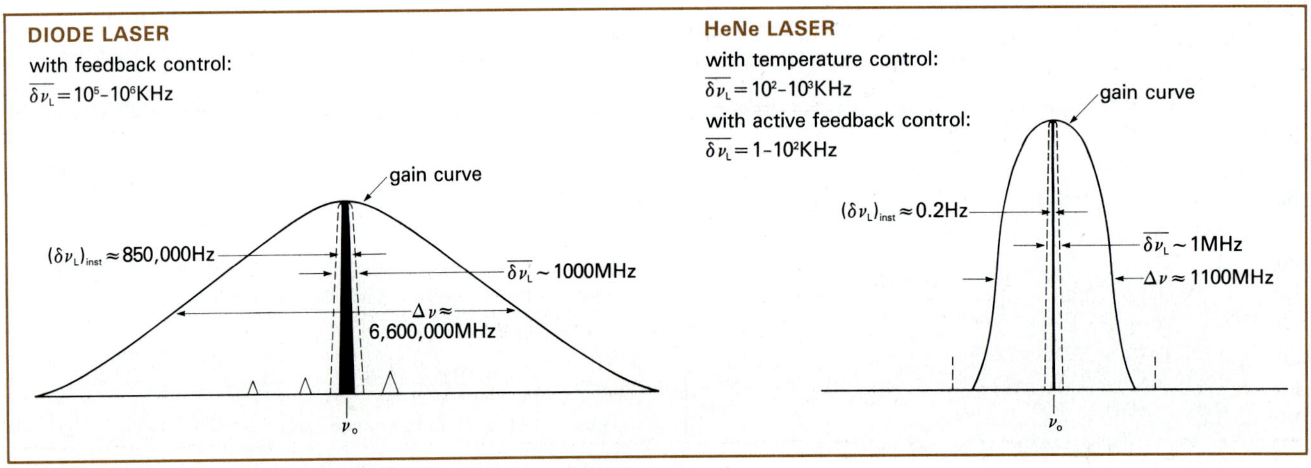

COMPARISON OF SPECTRAL PURITY of Diode and HeNe Lasers

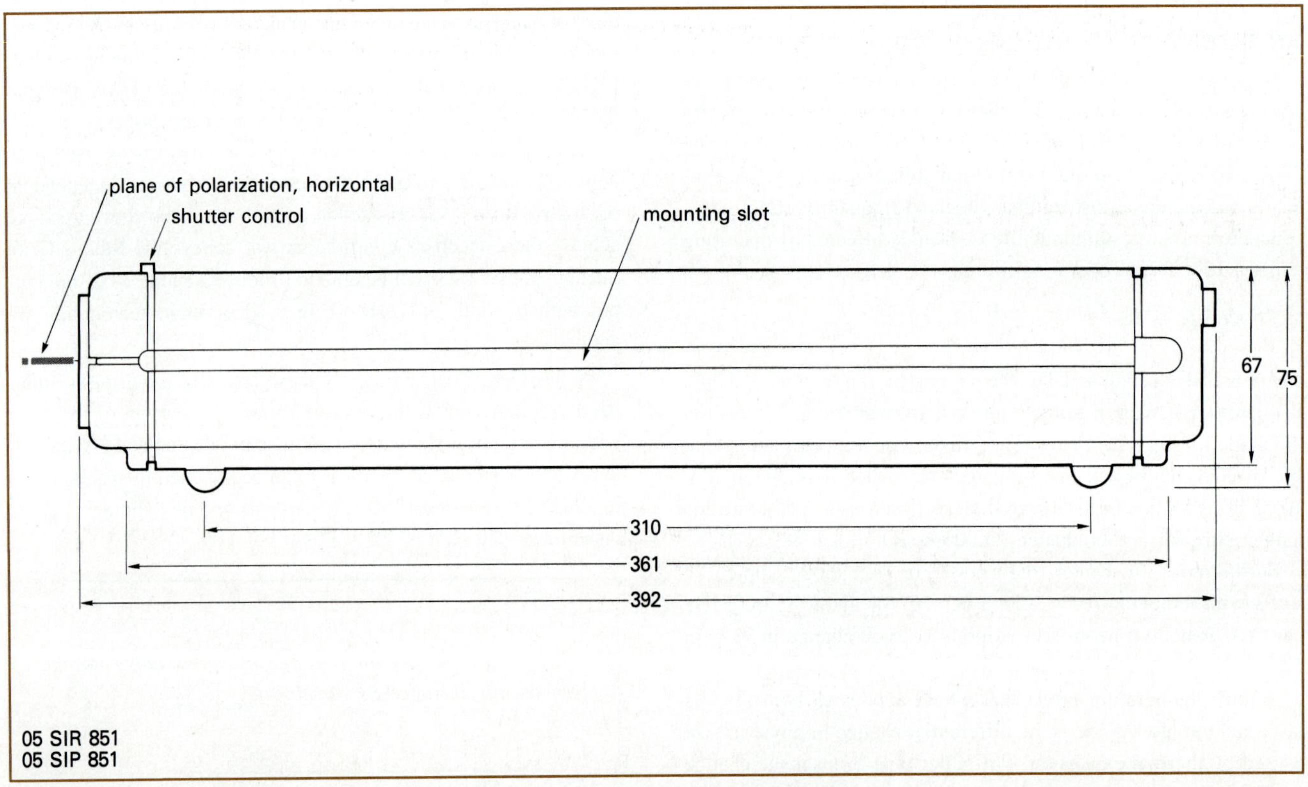

Self-Contained IR HeNe Lasers

SPECIFICATIONS	PRODUCT NUMBERS	
	05 SIP 851	05 SIR 851
Minimum CW Power Output at 1.523μ TEM_{00}: (mW)	0.8	1.00
Beam Diameter $1/e^2$: (mm)	1.25	1.25
Beam Divergence: (mrad, full)	1.55	1.55
Polarization Ratio:	>500:1	Random
Longitudinal Mode Spacing: (MHz)	441	441
Single Mode Power Sweep	<20%	<20%
Power Input: (VAC, 50 or 60Hz)	115/230	115/230
Weight: (kg)	1.9	1.9

Common To All Product Numbers

Static Alignment: Centered within 0.20mm and 2.0mrad
Angular Drift: <0.03mrad after 15 minutes
Amplitude Noise: <1% (30Hz to 10MHz rms)
Dimensions: ±0.5mm

Environmental Specifications	Operating	Non-Operating
Temperature:	-20 to +50°C	-40 to +80°C
Altitude:	0 to 3000 meters	0 to 5800 meters
Relative Humidity:	0 to 100%	0 to 100%
Shock:	15g for 11msec	15g for 11msec

We offer a wide variety of optical components and special coatings for use at 1.5 microns wavelength. See the laser diode optics section starting on page 369 and the coatings section starting on page 69.

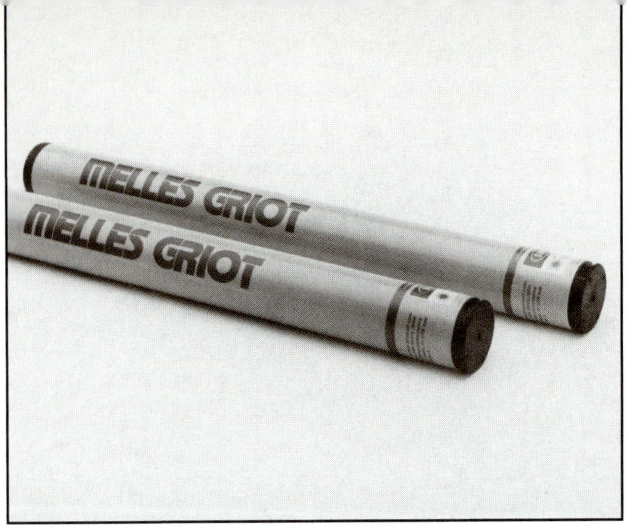

RED HENE CYLINDRICAL LASER HEADS

These Cylindrical Laser Heads are packaged red helium neon lasers. The power supply is separated from the laser head and the two are interconnected by a 1.8m cable. This allows certain mounting freedoms in restricted situations. The cylindrical housing is compatible with our 07 HLH 001, 07 HLH 002, 07 HLH 003 or 07 HLA 001 laser holders and other mechanical hardware components (see page 462).

The plasma tubes used in these lasers are the same as described earlier in this catalog and feature all the same advantages of stability, reliability and long-life. All BRH safety features are incorporated and the lasers are fully certified and labeled by Melles Griot.

Output powers of 1, 2, 4, 5 and 7 milliwatts are offered in both randomly and linearly polarized versions. The lasers are permanently and concentrically mounted within metal tubes for protection and ease of use. The connected cable is for the external power supply which should be ordered separately (see page 360).

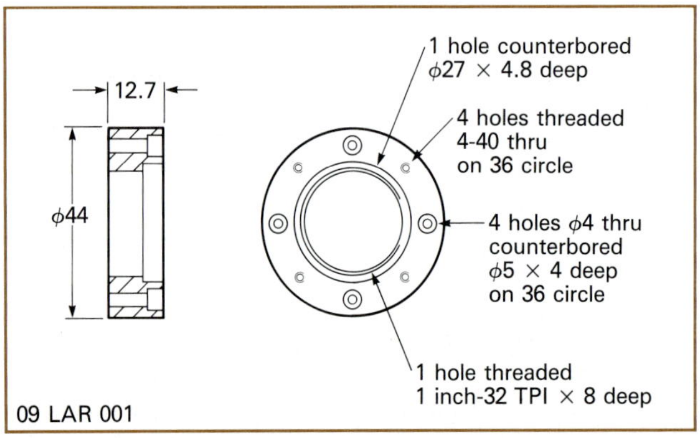

Accessory Adapter Ring

	PRODUCT NUMBER
Complete with screws	09 LAR 001

Red HeNe Cylindrical Laser Heads

SPECIFICATIONS	PRODUCT NUMBERS 05 LHR 111	05 LHP 111	05 LHR 121	05 LHP 121
Minimum CW Power Output at 632.8nm TEM_{oo}: (mW)	1	1	2	2
Beam Diameter $1/e^2$: (mm)	0.59	0.59	0.59	0.59
Beam Divergence: (mrad, full)	1.3	1.3	1.3	1.3
Polarization Ratio:	Random	500:1	Random	500:1
Longitudinal Mode Spacing: (MHz)	685	685	685	685
Operating Current, nominal: (mA)	6.5	6.5	6.5	6.5
Operating Voltage: (VDC) ±100	1800	1800	1800	1800
Recommended Minimum Power Supply Resistance: (kΩ)	50	50	50	50
Weight: (kg)	0.40	0.40	0.40	0.40
Long Term Drift:	±2%	±2%	±2%	±2%
Starting Voltage: (kVDC)	>10	>10	>10	>10
Ballast:	Included	Included	Included	Included

Common To All Product Numbers

Static Alignment:	Centered to outer cylinder within 0.25mm and 1.0 milliradian
Angular Drift:	<0.03 milliradians after 15 minutes.
Amplitude Noise:	<1% (30Hz to 10MHz rms)
Dimensions:	±0.5mm

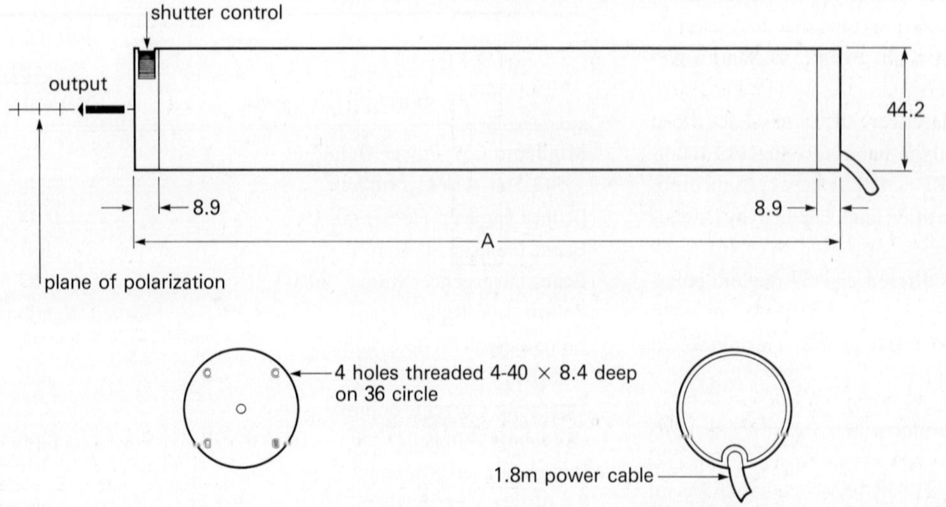

PRODUCT NUMBER	A
05 LHR 111	271.8
05 LHP 111	271.8
05 LHR 121	271.8
05 LHP 121	271.8
05 LHR 141	396.2
05 LHP 141	396.2
05 LHR 151	396.2
05 LHP 151	396.2
05 LHR 171	455.9
05 LHP 171	455.9
05 LHR 321	309.9
05 LHP 321	309.9

PRODUCT NUMBERS

05 LHR 321	05 LHP 321	05 LHR 141	05 LHP 141	05 LHR 151	05 LHP 151	05 LHR 171	05 LHP 171
2	2	4	4	5	5	7	7
0.80	0.80	0.80	0.80	0.80	0.80	1.00	1.00
1.0	1.0	1.0	1.0	1.0	1.0	0.8	0.8
Random	500:1	Random	500:1	Random	500:1	Random	500:1
570	570	435	435	435	435	380	380
6.5	6.5	6.5	6.5	6.5	6.5	7.0	7.0
1900	1900	2350	2350	2350	2350	2600	2600
50	50	65	65	65	65	75	75
0.50	0.50	0.56	0.56	0.56	0.56	0.68	0.68
±2%	±2%	±2%	±2%	±2%	±2%	±2%	±2%
>10	>10	>10	>10	>10	>10	>10	>10
Included	Included	Included	Included	Included	Included	Included	Included

Environmental Specifications	**Operating**	**Non-Operating**
Temperature:	-20 to +50°C	-40 to +80°C
Altitude:	0 to 3000 meters	0 to 5800 meters
Relative Humidity:	0 to 100%	0 to 100%
Shock:	15g for 11msec	15g for 11msec

GREEN HENE CYLINDRICAL LASER HEADS

These GreNe™ cylindrical laser heads are identical in all respects to those previously described except that they contain plasma tubes which emit laser light at the green (543.5nm) wavelength.

The plasma tubes used in these lasers are the same as described earlier in this catalog and feature all the same advantages of stability, reliability and long-life. All BRH safety features are incorporated and the lasers are fully certified and labelled by Melles Griot.

Output power of 0.5 milliwatt is offered, having random polarization. The lasers are permanently and concentrically mounted within metal tubes for protection and ease of use. The connected cable is for the external power supply.

See page 360 for a power supply for this laser.

Environmental Specifications	Operating	Non-Operating
Temperature:	-20 to +50 °C	-40 to +80 °C
Altitude:	0 to 3000 meters	0 to 5800 meters

Green HeNe Cylindrical Laser Heads

SPECIFICATIONS	PRODUCT NUMBER 05 LGR 151
Minimum CW Power Output at 543.5nm TEM_{oo}: (mW)	0.50
Beam Diameter $1/e^2$: (nm)	0.44
Beam Intensity: (W/cm^2)	260
Beam Divergence: (mrad, full)	1.57
Polarization Ratio:	Random
Longitudinal Mode Spacing: (MHz)	441
Operating Current, nominal (mA):	3.7
Operating Voltage: (VDC) ±100	1800
Recommended Minimum Power Supply Resistance: (kΩ)	65
Weight: (kg)	0.56
Starting Voltage: (kVDC)	10

Common To All Product Numbers

Static Alignment: Centered to outer cylinder within 0.25mm and 1.0mrad
Angular Drift: <0.03mrad after 15 minutes
Amplitude Noise: <1% (30Hz to 10MHz rms)
Dimensions: ±0.5mm

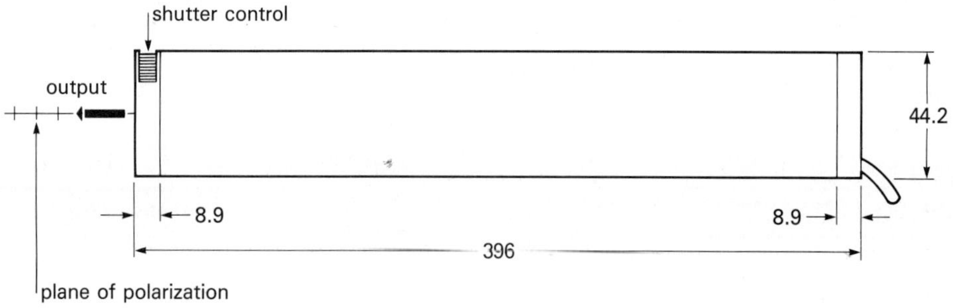

05 LGR 151

LASER ACCESSORIES

This line of accessories for helium neon lasers has been designed with the user in mind. We have incorporated top-level performance with both convenience and versatility. All these devices mount quickly and easily onto the front flange of our self-contained HeNe lasers. An adapter ring (09 LAR 001) provides an equivalent mounting interface for the cylindrical laser heads. These devices will mount on most other packaged lasers as well. For this reason they have been designed to work over a far wider wavelength region than most other laser accessories.

With these accessories one is able to expand the diameter of the laser beam up to 50mm, reduce the divergence of the beam very considerably and smooth the output beam profile to a more precise Gaussian shape; all at diffraction limited performance levels.

These various spatial filters, collimators and beam expanders are all made of quality materials. All surfaces are finished flat black to reduce stray glare and reflections. The focusing control rings have wide deep knurling on their circumference for a firm grasp. Motions are precise and smooth. All these components attach to any laser with a male 32 thread per inch (TPI), one-inch diameter threaded boss. (This interface is a common interface in the optics industry — it is used by most laboratory laser manufacturers.)

A post mounted holder for these accessories can be found on page 470.

SPATIAL FILTER

A spatial filter allows one to filter out the higher-order diffraction patterns from a laser beam. By matching the size of the pinhole to the central maximum a truly Gaussian output distribution can be achieved.

A pinhole size should be chosen to suit the $1/e^2$ diameter of the laser beam, thereby passing only the central maximum of the focal diffraction pattern and excluding the extraneous side-band spatial noise. This will produce a spherical wavefront with a smooth Gaussian power distribution. Five interchangeable focusing optics and thirteen interchangeable pinholes are available.

Two knurled knobs 90 degrees apart on the spatial filter permit easy and precise positioning of the pinhole at the center of the focused laser beam. The large knurled ring in the middle of the spatial filter is used to focus the laser beam exactly onto the plane of the pinhole for maximum transmission. A locknut permits convenient orientation.

Spatial filters may be used without a pinhole to produce a divergent cone of laser light. Be certain to choose focusing optics which will accept a larger beam than the output of your laser (preferably 30% to 50% larger) to avoid truncation of the Gaussian distribution.

Spatial filters are supplied in foam-lined, hardwood boxes for convenient and protective storage. Spaces are provided for six pinholes and five focusing optics within the box. A special insertion tool is included for mounting both the pinholes and the focusing optics.

SPECIFICATIONS: SPATIAL FILTER

Dimensions: ± 0.5mm
Pinhole Adjustment in both X and Y: 6mm total
Thermal Stability: ±1 micron in 4 °C
Material: Aluminum
Finish: Black anodized

Spatial Filter

	PRODUCT NUMBER
Spatial Filter (in hardwood box, complete with special insertion tool but without Pinholes and Focusing Optics)	09 LSF 001

Please select focusing optics and pinhole assemblies from the next page in addition to the spatial filter.

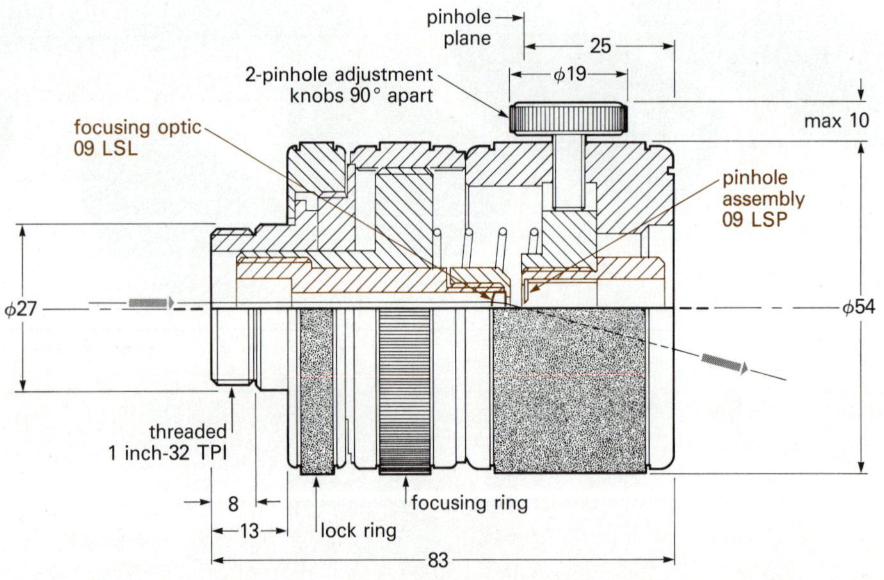

09 LSF 001

FOCUSING OPTICS AND PINHOLE ASSEMBLIES

FOCUSING OPTICS

These focusing optics are for use with the Melles Griot spatial filter to provide a focused beam at the pinhole plane. Alternatively they may be used without the pinhole to create an unfiltered spherical wavefront of varying numerical aperture. The mechanical constraints of the spatial filter are such that the maximum focal ratio which may be accepted by the system without obscuration is f/2.5.

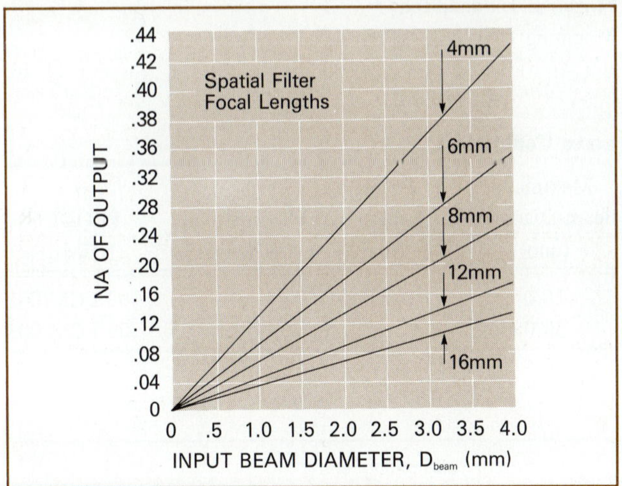

FOCUSING OPTICS selection graph.

Focusing optics are chosen by reference to the graph above and the $1/e^2$ focal spot radius (w_o) may be calculated from the following formula:

$w_o = \lambda f/\pi w$, where λ = wavelength
f = focal length of Focusing Optics
w = $1/e^2$ radius of input beam

Focusing Optics

Focal Length f_o (mm)	Maximum Aperture (mm)	Wavelength Range (nm)	PRODUCT NUMBER
4.0	1.6	450-750	09 LSL 001
6.0	2.4	450-750	09 LSL 003
8.0	3.2	450-750	09 LSL 005
12.0	4.8	450-750	09 LSL 007
16.0	5.0	450-750	09 LSL 009

PINHOLE ASSEMBLIES

Thirteeen different size pinholes are available for matching the aperture to the $1/e^2$ diameter of the focused laser beam. Each pinhole is mounted in an aluminum assembly which screws into a special mount in the spatial filter. The pinholes themselves are precisely laser-drilled in 0.013mm stainless steel (0.0025mm stainless foil in the small sizes). All pinholes are exceptionally round and free from edge defects.

Pinhole Selection Chart; for λ = 632.8nm

Focusing Optics	Pinhole Diameter (microns)		
	Input Beam Diameter		
f_o (mm)	0.6mm	0.8mm	1.0mm
4.0	10	8	6
6.0	15	12	10
8.0	20	15	12
12.0	30	25	20
16.0	40	30	25

Pinhole selection is directly related to the spot size. We recommend the use of a pinhole of diameter about 2 times the $1/e^2$ beam contour diameter at focus.

To select an appropriate pinhole for use with the most red helium neon lasers, refer to the chart above.

Pinhole Assemblies

Diameter (microns)	PRODUCT NUMBER
2.0	09 LSP 001
3.0	09 LSP 003
4.0	09 LSP 005
5.0	09 LSP 007
6.0	09 LSP 009
8.0	09 LSP 011
10.0	09 LSP 013
12.0	09 LSP 015
15.0	09 LSP 017
20.0	09 LSP 019
25.0	09 LSP 023
30.0	09 LSP 025
40.0	09 LSP 027

LASER COLLIMATORS

These laser collimators will provide a highly collimated output beam when used in conjunction with the Melles Griot spatial filter. Both 25mm and 50mm output apertures are available. Divergence of the collimated beam is inversely proportional to the beam diameter. When the output aperture is filled, the output beam divergence from the 50mm collimator is only 0.015 milliradians.

These three-element Collimators are diffraction limited in their performance. Wavefront distortion is less than 1/8 wave over the entire aperture. They are designed to provide excellent performance at 488nm and 514nm, as well as 633nm. All air/glass surfaces are coated with our HEBBAR™ high-efficiency, broad-band antireflection coating for high transmission.

Focus is adjustable from 2 meters to infinity with the large knurled control ring on the front of the housing. The rear of the collimator attaches directly to a spatial filter (09 LSF 001) using a standard 1 inch-32 TPI threaded boss. Collimators are mounted in black anodized aluminum housings. They are supplied in a fitted, foam-filled woodbox for storage and protection.

SPECIFICATIONS: LASER COLLIMATORS

Dimensions: ±0.5mm
Focusing Range: 2m to infinity
Wavefront Distortion: $\lambda/8$ at 632.8nm
Number of Elements: 3
Coatings: HEBBAR™
Wavelength Range: 450 to 750nm
Material: Aluminum
Finish: Black anodized
Mounting: 1 inch-32 TPI

Laser Collimators

Maximum Beam Diameter (mm)	Focal Length f_o (mm)	Maximum Focal Ratio	PRODUCT NUMBER
25.0	101.5	4.1	09 LCM 001
50.0	227.2	4.5	09 LCM 003

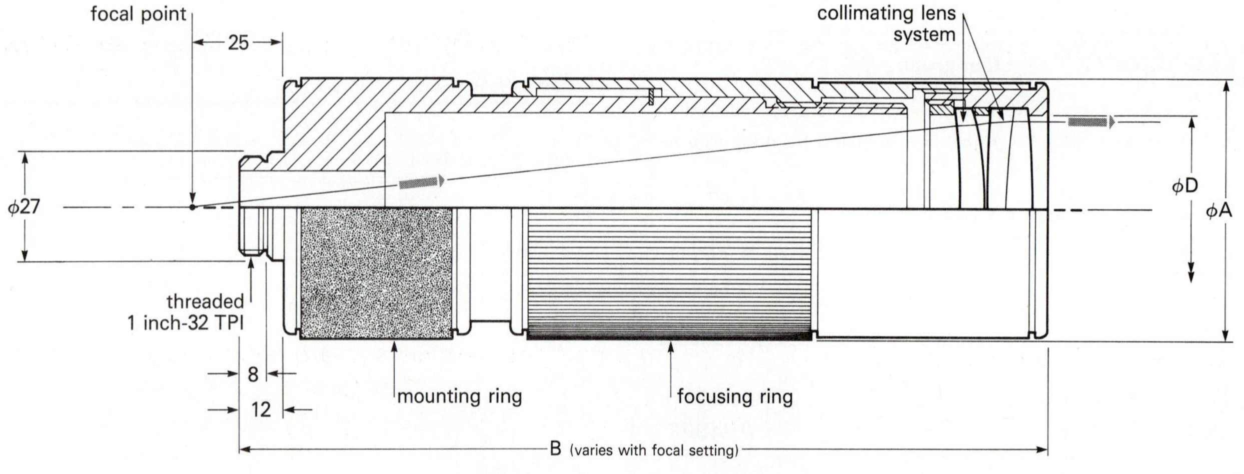

PRODUCT NUMBER	ϕA	B max	B min	ϕD
09 LCM 001	54	106.7	81.3	25
09 LCM 003	68	261.6	218.4	50

COLLIMATOR SELECTION

These two precision laser collimators will provide a wide variety of beam diameters when used in combination with different focusing optics mounted in the Melles Griot spatial filter.

Collimators and focusing optics may be selected to provide whatever output diameter is required by referring to the graphs below.

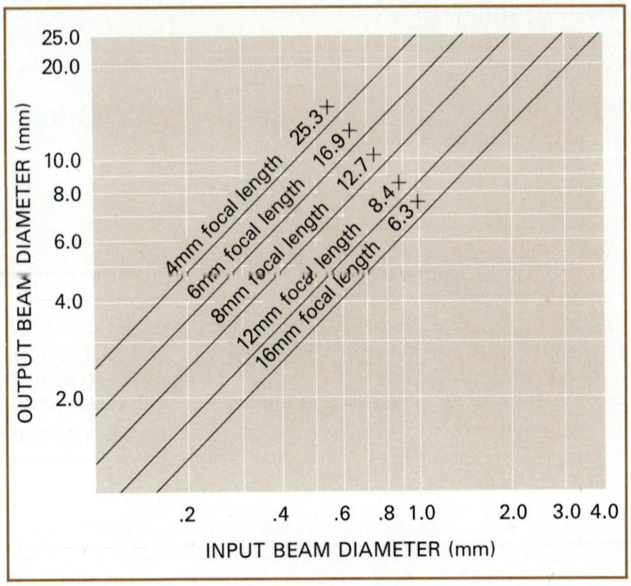

BEAM DIAMETER selection for 25mm Laser Collimator (09 LCM 001).

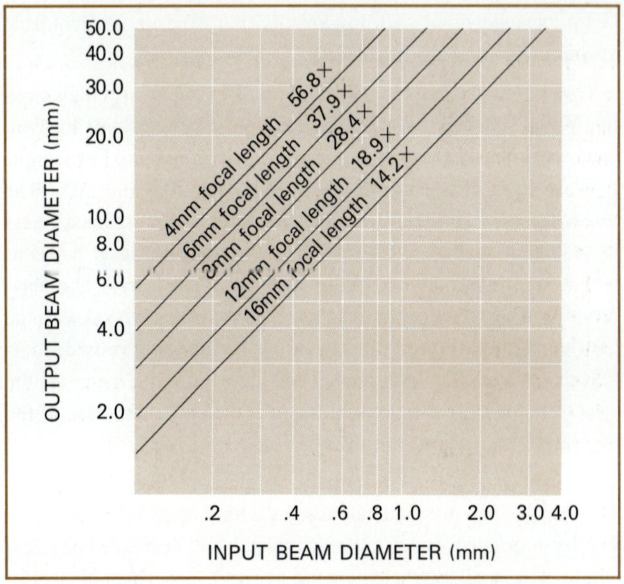

BEAM DIAMETER selection for 50mm Laser Collimator (09 LCM 003).

Did you know Melles Griot manufactures laser diode collimators, too? If you are working with laser diodes, see pages 369-394 for further details.

LASER BEAM EXPANDERS

When using a helium neon laser to illuminate an extended or distant object a beam expander becomes particularly useful. The larger output beam facilitates holographic experiments and other illumination applications. The reduced beam divergence in the far-field is useful in applications requiring target illumination of distant objects. Frequently however, the output wavefront from a beam expander is of such low quality that experiments requiring spatial coherence are impossible.

These mounted beam expanders are of exceptionally high quality. Wavefront distortion is less than one-quarter wave. Further, each expander is short and compact for its function. These units provide beam expansion ratios of 3X, 10X, 20X and 30X. The threaded rear flange screws directly into the self-contained lasers or (using an adapter, 09 LAR 001) the cylindrical laser heads as well as most other commercially available HeNe lasers. The separation of the two front and rear lens sets may be adjusted with the knurled front portion of the barrel. This adjustment provides far-field focusing of the laser beam from a position of 1 meter (in the case of the two smaller expanders) or 2 meters (in the case of the two larger expanders) to infinity.

A beam expander is essentially a telescope used backwards. These expanders are Galilean designs with a multi-element objective for improved wavefront performance and decreased physical length. The small input element is a negative lens which presents a virtual image to the positive output lenses. Because no real image is formed internally, spatial filters may not be used with these expanders. The Galilean design offers two advantages over the Keplerian design: namely, improved performance in the cancellation of spherical aberration between the two lens groups and a markedly shorter barrel length.

The design of these expanders has been optimized for a maximum input beam diameter of 1.0mm, thus matching the output beam diameters of our self-contained lasers and our cylindrical laser heads.

The beam expanders are mounted in a black anodized aluminum housing with a one-inch diameter, 32 TPI threaded rear boss. The front and rear portions of the housing are "captive" so that they may not be inadvertently disassembled when adjusting the focus. This further assures that the inner lens surfaces remain clean and free from fingerprints.

Beam expanders are supplied in hardwood boxes with preformed foam liners for convenient storage and protection.

SPECIFICATIONS: LASER BEAM EXPANDERS

Dimensions: ±0.5mm
Maximum Entrance Aperture: 1.0mm
Transmission: >95%
Wavelength Range: 450 to 750nm
Wavefront Distortion: $\lambda/4$ at 632.8nm
Coatings: HEBBAR™
Material: Aluminum
Finish: Black anodized
Mounting: 1 inch-32 TPI

Laser Beam Expanders

Expansion Ratio	Exit Aperture (mm)	Minimum Near Focus (m)	PRODUCT NUMBER
3	3.0	1	09 LBM 001
10	10.0	1	09 LBM 003
20	20.0	2	09 LBM 005
30	30.0	2	09 LBM 007

For those who wish to build their own laser beam expander system we offer a range of unmounted expander optics also.

Details of these products may be found on page 145-147.

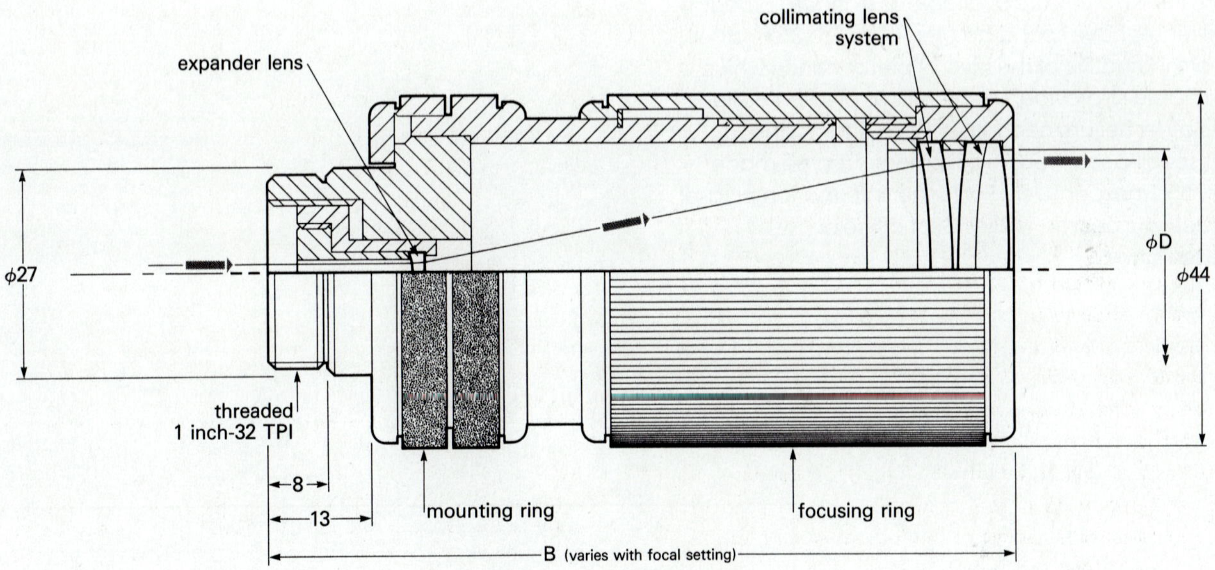

PRODUCT NUMBER	B max	B min	ϕD
09 LBM 001	95.3	87.6	3
09 LBM 003	110.5	100.3	10
09 LBM 005	135.9	128.3	20
09 LBM 007	186.7	171.5	30

ACCESSORY ADAPTER RING

A black, anodized aluminum adapter ring for the Melles Griot cylindrical lasers to allow mounting of spatial filter and beam expanders.

Accessory Adapter Ring

	PRODUCT NUMBER
Complete with screws	09 LAR 001

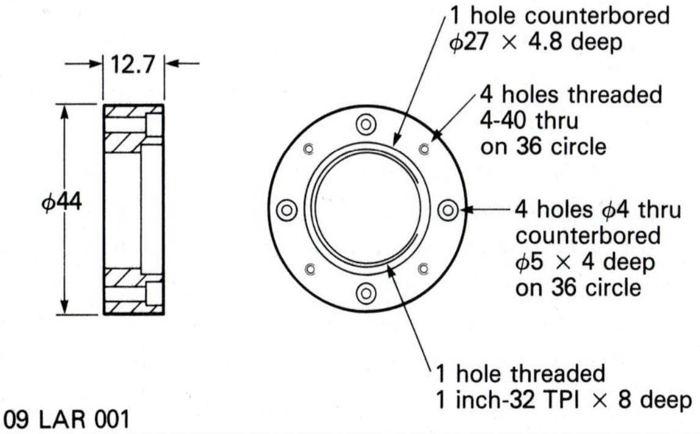

09 LAR 001

LASER SCANNING LENSES

Laser scanning lenses cause an angularly scanned laser input beam to be imaged onto a flat plane in a linear manner. Known often as f-theta lenses, these systems are highly efficient when used to read characters, print characters, plot data, read film, display graphic functions or other similar applications. Individual applications are invariably different from one another and require a specific lens system design.

Scan lenses with image lengths from 33 to 466mm, manufactured by Melles Griot Optical Systems in Rochester, New York are available as standard products. These lenses have spot diameters from 7 to 36 microns and are indicative of the level of performance which we are able to provide. Modifications of these designs to suit other helium neon laser wavelengths and to meet specific scan lens requirements can be made easily and inexpensively. For further information please contact your nearest Melles Griot office or representative.

LASER DIODE OPTICS

Laser Diodes provide a compact, reliable source of laser radiation. The components described in the following pages will help you collect and use the output of laser diodes.

Contents **Page Number**

LASER DIODES

Semiconductor laser diode technology is moving at a dramatic pace. Devices fabricated from the III-V compounds (such as gallium arsenide, indium phosphide, indium arsenide, and more complex compounds) emit in the near-infrared and are now reliable, rugged devices capable of continuous output at room temperatures. Laser diodes are being used to feed fiber optic links between telephone switching offices. Others read optical disks used in the latest audio recording systems. New types are being introduced at a rapid rate and promise increased speed of transmission, improved information density, and greater capacity. Other applications such as data storage, computer output printers, or data display interfaces are pushing the technology towards greater reliability, shorter wavelengths and higher output levels.

Laser diodes share the advantages of semiconductor devices in general. They are very small, rugged, and inexpensive to manufacture. The output beam is emitted from a spot only ten or twenty microns wide by a fraction of a micron high making them the right size for interfacing with fiber optical cores. Laser diodes require very little power to operate efficiently and the output may be modulated to carry information simply by modulating the drive current.

Yet, for all their advantages, using these devices in newly developed hardware is not as simple as one might wish. The output from a laser diode is a sharply divergent cone and is typically asymmetrical. This elliptical output cone is difficult to couple efficiently to an optical system. The near-infrared output region is not well matched to the transmission of many glasses. Further, the output from single-mode laser diodes is highly coherent, thus dictating the need for diffraction limited collection, focusing and coupling optics.

LASER DIODE OPERATION

Laser diodes are, essentially, Light Emitting Diodes (LED's) with several structural and operational reflective differences. Both are manufactured from a block of semiconductor material containing a p-n junction. Like other semiconducting devices, current is carried by the free electrons within the crystal when an electric field is applied, and by the holes or vacancies for valence electrons within the crystal lattice. Laser diodes are structured such that a large proportion of the energy released by a single recombination takes the form of a photon. Recombination of the excited carriers is stimulated before the excited energy state is able to drop back to the original, lower energy level through spontaneous emission. High drive currents produce a population inversion of excited carriers so that the light is amplified as it travels through the material. To sustain the amplification, opposite ends of the crystal are cleaved and reflective surfaces are created to provide the optical feedback needed to overcome cavity losses. In this way self-sustained laser oscillation is created within the crystal.

The stimulated emission occurs at what would be the peak wavelength of a corresponding LED, but the laser diode spectral bandwidth is very much narrower. In a laser diode, much higher current density produces a sufficiently high density of electron-hole pairs to result in population inversion. While current densities in laser diodes are high, the devices are small, and so net current is low. Laser diodes typically require only a few volts to operate, while HeNe lasers need a kilovolt or two.

The variety of laser diode structures presently available is changing rapidly. Efforts to create symmetrical output geometry, reduce the operating wavelength, increase the power output, and improve the lifetime all lead to different compromises in building techniques and choice of materials. In fiber optic systems pressure towards longer wavelengths and single mode operation provide other compromises. These varying designs produce different optical output geometries and complicate the optical problems.

In virtually every system application there is a need to collect, collimate and focus the output light from the laser diode. The conflicts in designing a system with a reasonable number of components, manufactured from acceptable glass types, and not over-critical to alignment are considerable. Readily available components are designed for other applications and other wavelengths. Special designs are expensive and time-consuming. Other related optical problems include coupling the diode output efficiently to fiber optic cores and finding correctly matched high-efficiency coatings. This section describes a line of optical components which go a long way towards solving those problems. If you require more specialized or larger volume specific components, please do not hesitate to contact us.

BEAM DIVERGENCE

The output beam emitted from a laser diode is both highly divergent and elliptical in cross-section. Beam divergence typically runs between 40 and 60 degrees; divergences up to 90 degrees are not unusual. Beam divergence in the direction perpendicular to the active layer is typically two to six times the divergence in the plane parallel to the layer.

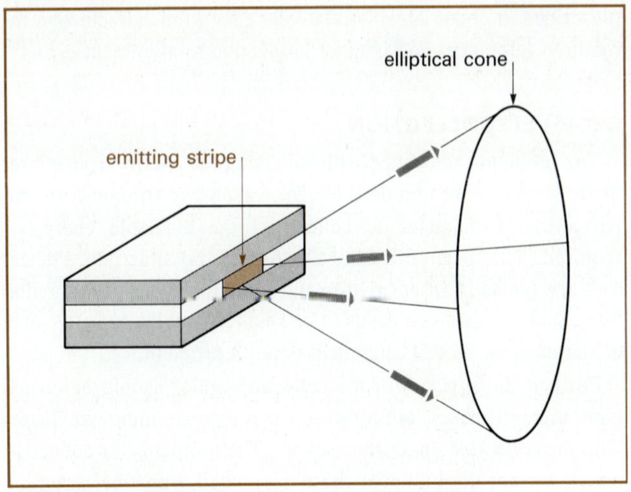

OUTPUT GEOMETRY of a typical laser diode.

Failure to correct the out-of-roundness of the beam is sometimes unimportant, but when the beam is collimated the far-field beam divergence can be different in each direction. When corrected by expanding the minor axis of the output, the far-field beam divergence will be minimized.

Index-guided laser diodes appear to be gaining in both popularity and use because they exhibit little or no astigmatism. In this case the beam is constrained in both directions by internal reflections and the beam appears to originate from the same location in both orthogonal planes. Index-guided laser diodes are considerably easier to use because of this feature and in most research applications their use will be preferred.

In gain-guided laser diodes the output beam is astigmatic. In the plane of the active layer, the beam will appear to diverge from a source which is located some 10 to 50 microns behind the surface of the window. In the opposite plane, the source will appear to diverge from the surface of the window itself. This astigmatism is caused in gain-guided devices by a difference in the constraining mechanism in the plane of the active layer compared with the perpendicular plane. In the perpendicular direction the light beam is constrained by reflections from the boundary surfaces, whereas in the opposite direction the beam is constrained by the electric current profile (the cause of the existing population inversion). This astigmatism can be corrected by the addition of an appropriate weak negative cylindrical lens.

WINDOW ABERRATIONS

Laser diodes are usually sealed on the output end of the laser with a plano glass window. The window is typically between 0.10 and 0.30mm thick. When the divergent output beam from the laser diode passes through this window, aberrations are introduced. The greater the divergence angle, the greater the aberration contributions and the more serious this problem becomes.

At numerical apertures of 0.30 or less, the aberration contributions become negligible. However, with numerical apertures of 0.50 and above, the aberration contributions quickly rise above 0.1 wave and become limiting to the system performance. These problems become even more critical when the optics are being used as an objective in conjunction with thick optical recording disks. The graph illustrates the relationship between numerical aperture and aberration contribution for window thicknesses between zero and 1.0mm. Index of refraction is nominally 1.5 and the wavelength is 830nm.

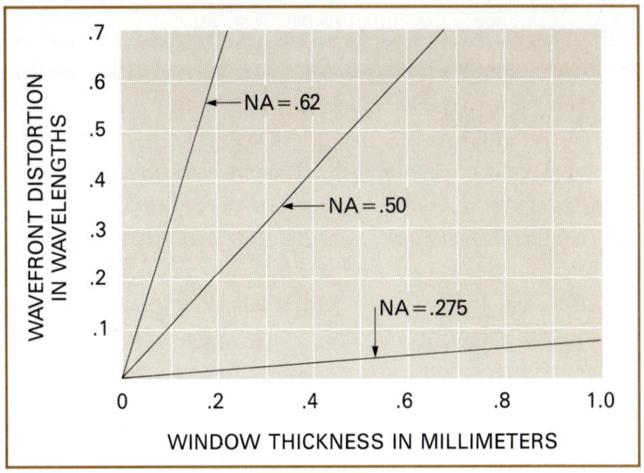

EFFECT OF A WINDOW on wavefront distortion at 830nm.

Compensation for these window aberration contributions must be provided for in the design of any high numerical aperture lenses if system performance is to be optimized. Thickness adjustment of the lens itself is one straight-forward method of providing this adjustment.

FIBER COUPLING

Coupling the output of a laser diode to an optical fiber efficiently is difficult. Fiber communications systems typically utilize a variety of devices such as isolators, attenuators, wavelength multiplexers and de-multiplexers to process and manipulate the light. Space (often several millimeters) is required between the fiber and the laser diode to perform these actions. The separation of fiber and diode would lead to considerable insertion losses if some form of coupling optics were not included.

Typically a 1:1 image of the output of the laser diode is formed on the input of the fiber core. A number of optical systems have been proposed, and in all of them the limit of performance is now the aberration contribution of the lens to the system.

Of the various types presently being used, both the gradient index rod lenses and the spherical ball lenses appear to be the easiest to use. Both types have similar limiting coupling efficiencies of 95% — an insertion loss of approximately 0.2dB. Nevertheless, both types are extremely small to handle and mount properly, and if efficient antireflection coatings are not properly applied, losses will occur. Mismatch and misalignment tolerances are essentially equivalent to butt joint tolerances, since in both cases 1:1 images are used to couple the output to the fiber.

BEAM SHAPING

In applications where either minimum beam divergence or a small focal spot is required, it may be necessary to do beam shaping. Collimating the asymmetric output cone from laser diode yields a collimated beam with an elliptical cross-section. Because of diffraction, the beam divergence will be different in the two axes of the ellipse; namely, there is a larger divergence in the minor axis of the ellipse. Where minimum beam divergence in the far-field is required, it is usually necessary to exand the beam in one dimension only, so that the cross-section is transformed from elliptical to circular. In applications where a small focal spot is desired, it may also be necessary to do this type of beam shaping. Focusing a beam with an elliptical cross-section results in an elliptical focal spot; diffraction causes the minor axis of the beam to be focused to a larger size than the major axis. This beam shaping can be accomplished with a telescope system using cylindrical lenses; however, this type of system is both expensive to manufacture and difficult to design with the necessary diffraction limited performance. Of course, each separate design would be matched to only one specific type of laser diode. Instead, Melles Griot has designed an anamorphic prism pair which performs the required beam reshaping for a wide variety of laser diodes. The system (described in detail later) is relatively easy to manufacture to the required tolerances and is quite insensitive to alignment errors.

RECEIVER COLLECTION

The problems associated with collecting the energy from a laser diode for exposure to a detector are reasonably straight-forward. An adequate diameter to collect all the available energy is required. The focal ratio can be as high as possible (since detectors are typically insensitive to the direction of origin of radiation), and the spot-size needs to be smaller than the sensitive area of the detector so that alignment is not a problem.

Further, the surfaces of the collection optics should be coated with high-efficiency antireflection coatings to minimize losses with the collection optics themselves. The problems for collection optics are not dissimilar to those associated with fiber coupling discussed above.

THE MELLES GRIOT SOLUTION

Melles Griot has designed and manufactured a complete array of optics specifically for use with laser diodes. Each component has been conceived as a solution to a set of very special problems associated with laser diodes. The quality required to perform these chores calls for stringent controls and standards, and yet the designs have been optimized to provide this level of performance and to provide the possibility of really large volume production without problems. All these components are available from stock from any of our offices worldwide.

The number of components required to accomplish these tasks has been reduced to a manageable number by choosing the simplest lens form which would provide the needed levels of correction over the entire wavelength range from 780nm to 1.55 microns. Objective lenses were designed in various numerical apertures, all having the same clear aperture so they could be used together in pairs to focus the laser energy to a sharp spot with different diameters and distances from the laser. The shorter focal lengths are intended for beam collection, while the longer focal lengths are considered more appropriate for refocusing the beam. A large diameter, moderate numerical aperture is included for producing a larger collimated beam for long distance application. For higher throughput, a beam-expanding telescope has been designed to be used in conjunction with one of the high numerical aperture collimating objectives for other long distance uses.

All the lenses have been combined with a series of mounted beam expanders, collimators and a mounted anamorphic prism pair to correct the asymmetric output of a wide variety of laser diodes. All designs have been optimized for their application and all surfaces have been coated with high efficiency antireflection coatings for increased transmission and performance.

Furthermore, all these components may be ordered in volume quantities at reduced price for Original Equipment Manufacturers. For further information, please contact any of our offices.

For further technical information regarding the design and use of these collimator lenses we refer you to a paper by John F. Forkner and David W. Kuntz entitled "Characteristics of Efficient Laser Diode Collimators" and a further paper by David W. Kuntz entitled "Specifying Laser Diode Optics." Copies of both papers are available from Melles Griot upon request.

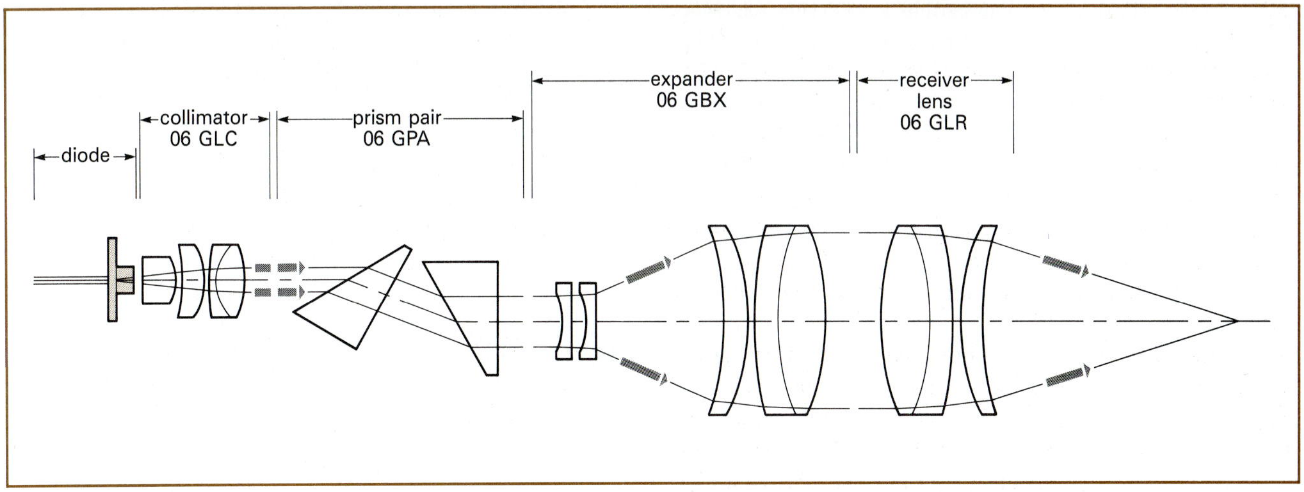

INDIVIDUAL COMPONENTS of the Melles Griot Laser Diode Optics system, showing how they may be used in relation to each other.

LASER DIODE COLLIMATING AND FOCUSING LENSES

This set of six lenses has been designed to operate as either collimating or focusing lenses for laser diodes. Five of these lenses have the same clear aperture (8.0mm) so that they may be used in interchangeable pairs to both collimate the output and then refocus the beam to a diffraction limited spot. A larger diameter, medium numerical aperture lens is also available for long distance applications where the increased diameter will reduce beam divergence. The two shorter focal length (larger numerical aperture) lenses have thick rear elements with a plano final surface. This allows the thickness of these elements to be modified to compensate for the thickness of a cover plate or window on the output of the laser diode. When ordering these lenses, please contact us if you plan to use a window.

These lenses are all diffraction limited. Their design assures the user of optimum performance with minimum application difficulty. The number of elements in each system was chosen to balance cost and performance. Spherical aberration, coma, astigmatism, and sphero-chromatism have been corrected by the design. Both chromatic aberration and field curvature were left uncorrected to allow optimization of the other parameters, since neither has any significant effect on the performance of these systems when used with laser diodes. The correction of sphero-chromatism is important in this application because it permits use of these lenses at a wide variety of (monochromatic) wavelengths by simply adjusting the back focal position to compensate. Correction of the off-axis aberrations permits a reasonable tolerance in setup to compensate for misalignment and decentration in the laser diode.

For collimating applications, the numerical aperture of the lens should be selected to match the angular distribution of radiation from the laser diode. These lenses, being diffraction limited, produce an angular divergence of as little as 0.127 milliradians in the output beam. This relates to the meridian for which the collimator is filled. To obtain a uniform aspect ratio of the output, it is necessary to convert the beam shape to near circular. The anamorphic prism pair is ideal for this purpose.

All air-to-glass surfaces are coated with a single layer magnesium fluoride coating to improve overall transmission. Because high index glasses are used throughout these systems, this results in a very effective coating.

Lenses are all mounted in black anodized aluminum housings for convenience and utility. The focal length is clearly marked on each lens.

SPECIFICATIONS: LASER DIODE COLLIMATING AND FOCUSING LENSES

Paraxial Focal Length: ±2% at 830nm
Wavefront Distortion: $<\lambda/4$ at 830nm
Design Wavelength: 830nm
Wavelength Range: 780 to 1550nm
Clear Aperture: ±0.5mm
Dimensions: ±0.25mm
Coating: Single layer MgF_2 for 839nm
Housing: Aluminum
Finish: Black anodized

WINDOW ABERRATIONS

Window thicknesses of typical laser diodes range between 0.1 and 0.3mm. If the numerical aperture of the collimating lens system is greater than 0.5, this window contributes an undesirable amount of spherical aberration and will degrade the overall performance of the system. To allow for compensation in these cases we have designed the higher numerical aperture collimators with a thick, flat rear element. To correct for the diode window thickness, this rear element is adjusted to compensate for the additional material (06 GLC 001 and 06 GLC 002 only).

The following equation predicts the lens thickness adjustment which must be made, based on third order aberration correction:

$$t_L' = t_L - t_W \frac{n_L^3(n_W^2 - 1)}{n_W^3(n_L^2 - 1)}$$

The lenses, as listed, are optimized for use without windows. For your convenience when using 06 GLC 001 or 06 GLC 002 with diodes having windows you may append a suffix for the appropriate diode window thickness.

Window Thickness Options; for Laser Diode Collimators.

For Use With	Nominal Window Thickness (mm)	Useful Range of Window Thickness (mm)	OPTION SUFFIX
06 GLC 001	0.00	0.00 to 0.15	no suffix
	0.15	0.00 to 0.30	/B
	0.30	0.15 to 0.45	/D
06 GLC 002	0.00	0.00 to 0.25	no suffix
	0.30	0.10 to 0.50	/D

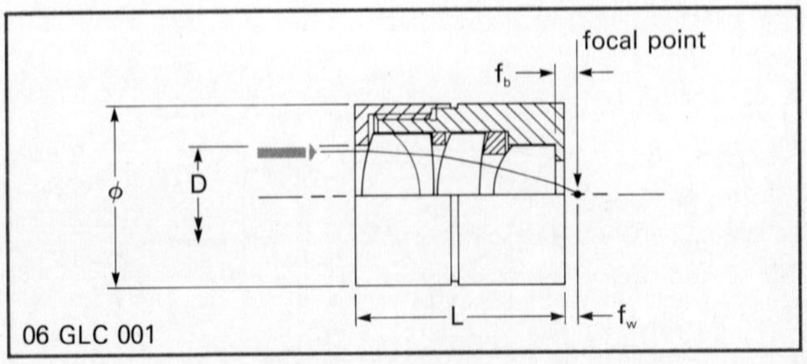

06 GLC 001

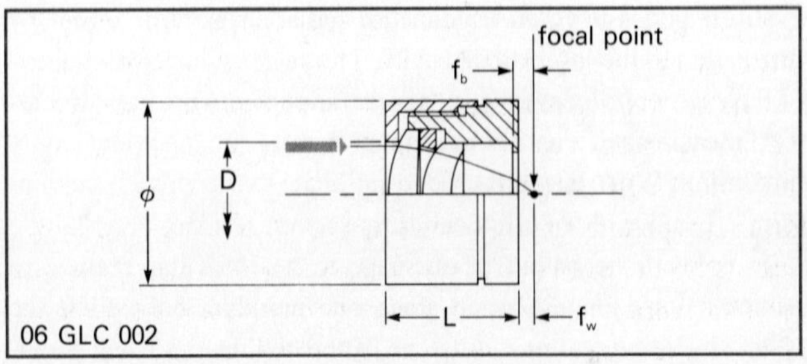

06 GLC 002

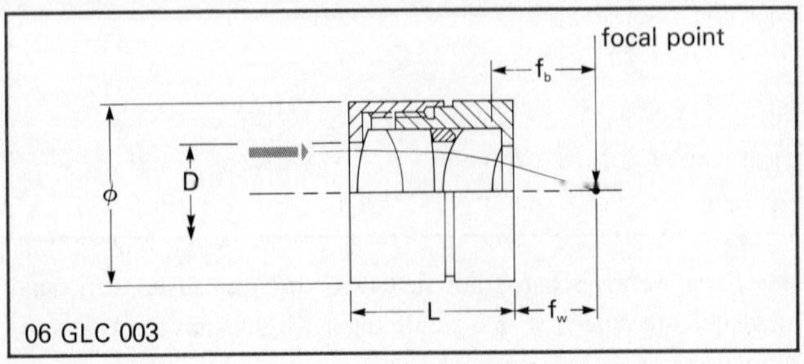

06 GLC 003

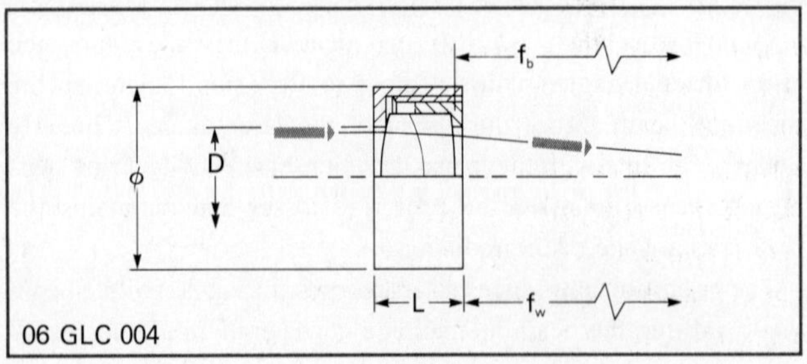

06 GLC 004

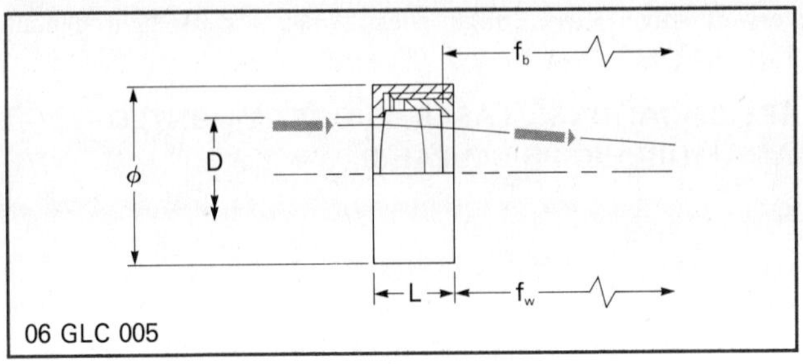

06 GLC 005

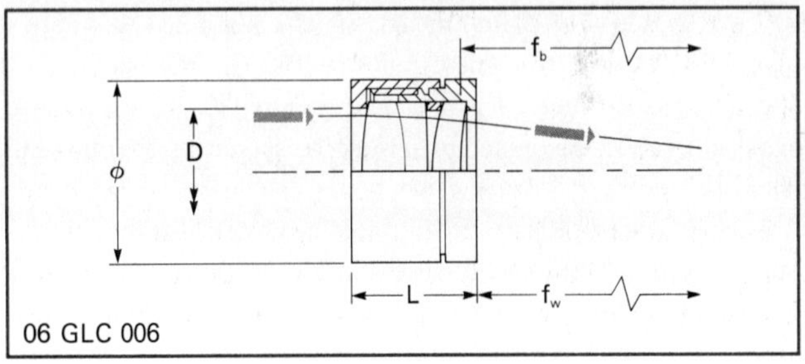

06 GLC 006

Laser Diode Collimating and Focusing Lenses

f_o (mm)	Clear Aperture D (mm)	N.A.	Field Angular Radius (Deg.)	f_b (mm)	Working Distance f_w (mm)	Spot Size (microns)	Diameter ϕ (mm)	Length L (mm)	PRODUCT NUMBER
6.5	8.0	0.615	±0.7	1.21	.78	0.82	14.29	16.48*	06 GLC 001
8.0	8.0	0.500	±1.0	1.56	1.13	1.02	14.29	10.79*	06 GLC 002
14.5	8.0	0.276	±1.0	8.48	6.72	1.85	14.29	12.95	06 GLC 003
25.6	8.0	0.156	±1.0	22.60	19.96	3.26	14.29	7.21	06 GLC 004
48.0	8.0	0.083	±1.0	45.10	44.40	6.11	14.29	6.40	06 GLC 005
50.0	20.0	0.200	±0.7	43.50	41.10	2.55	28.58	20.07	06 GLC 006

* Length varies if window options are specified.
Request special coatings if used at 1300 or 1550nm.

LASER DIODE ANAMORPHIC PRISM PAIRS

Laser diodes have an asymmetric radiation pattern caused by diffraction at the laser diode cavity. The rectangular cross-section of the cavity produces a wide angular output from the narrow cavity dimension and a narrower angular output from the broad cavity dimension. When a lens is used to collimate this asymmetrical pattern, a beam with an elliptical cross-section results.

To convert the output beam shape to near-circular requires a device having magnification along one meridian only. First the beam should be collimated by one of the collimators listed previously.

In an anamorphic prism pair, two identical prisms are mounted at an incident angle, which is close to Brewster's angle, to the incoming beam. Since the output of the laser diode is linearly polarized in line with the broad direction of the diode stripe, this arrangement minimizes the reflection losses and maintains the polarization state of the radiation.

The prism pair introduces an anamorphic expansion of the beam while shifting the beam a small amount laterally. Any degree of ellipticity can be corrected with this type of device. Nevertheless, we have chosen to offer four models which cover fixed ellipticities of 2x, 3x, 4x, and 6x. These ratios will satisfy the majority of applications. However, if you should require some other precise expansion ratio, unmounted prisms can be supplied (see following page).

Anamorphic prism pairs are mounted in a black anodized aluminum housing. The beam expansion ratio is engraved into the side of the housing for easy identification. Any of the four shorter focal length collimator systems (Product Numbers 06 GLC 001, 06 GLC 002, 06 GLC 003, and 06 GLC 004) will fit into the front of the anamorphic prism pair housing, where they may be held in place with a nylon set-screw.

Surfaces of the anamorphic prism pair which are used at Brewster's angle are uncoated but surfaces which are used at normal incidence are coated with a single layer magnesium fluoride antireflection coating for enhanced transmission. These coatings are very effective because high-index glasses are used throughout these systems.

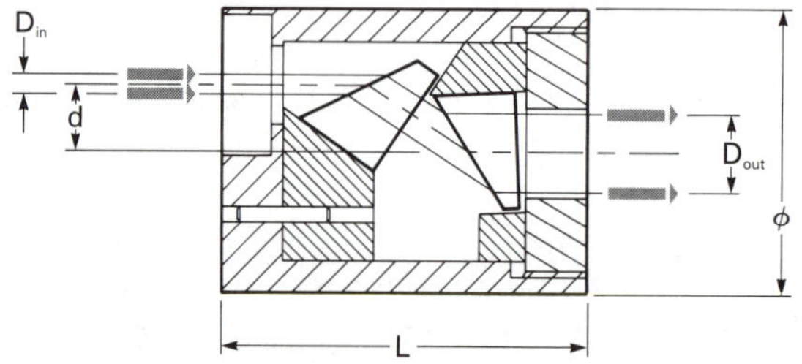

06 GPA

SPECIFICATIONS: LASER DIODE MOUNTED ANAMORPHIC PRISM PAIRS

Design Wavelength: 830nm
Wavelength Range: 780 to 1550nm
Wavefront Distortion: $<\lambda/4$ at 830nm
Clear Aperture: $\pm$0.5mm
Dimensions: $\pm$0.25mm
Coating: Single layer MgF_2 for 830nm
Housing: Aluminum
Finish: Black anodized

Laser Diode Mounted Anamorphic Prism Pairs

Anamorphic Magnification	Beam Displacement, d (mm)	Beam Size D_{in} (mm)	Beam Size D_{out} (mm)	Diameter ϕ (mm)	Length L (mm)	PRODUCT NUMBER
2x	5.7	8.0 x 4.0	8.0	30.2	38.1	06 GPA 001
3x	7.7	8.0 x 2.7	8.0	30.2	38.1	06 GPA 004
4x	6.8	8.0 x 2.0	8.0	30.2	38.1	06 GPA 002
6x	7.4	8.0 x 1.3	8.0	30.2	38.1	06 GPA 003

Request special coatings if used at 1300 or 1550nm.

UNMOUNTED ANAMORPHIC PRISMS

For those who wish to incorporate anamorphic prisms into their own mechanical housings, we offer, unmounted, the same prisms we use in our anamorphic prism pairs.

These prisms are supplied with a single layer magnesium fluoride coating on the exit (perpendicular) face and with no coating on the entrance (Brewster angle) face. They should be mounted so that the angle of incidence at the entrance face of each prism is the same.

An anamorphic expansion or reduction ranging from 2 to 6 times can be achieved by varying the angle of incidence as shown in the accompanying graph and table.

The table gives values for the orientation and position of the anamorphic prism pairs over a range of magnifications. Angles, α, measured clockwise from the reference line are listed as positive; angles measured counterclockwise are negative. The 1.5mm spacing between the prisms is arbitrary; this separation does not affect the magnification. However, if a different spacing is used, both the e and d displacement values will be altered.

Note that these prisms are supplied in pairs.

SPECIFICATIONS: UNMOUNTED ANAMORPHIC PRISMS

Flatness: $\lambda/8$ at 830nm
Angles: $\leq$30 arc seconds
Dimensions: $\pm$0.25mm
Material: SF 11 grade A fine annealed
Cosmetic Surface Quality: 60-40 scratch and dig
Coatings: Single layer MgF_2 for 830nm on perpendicular faces
Sides are fine ground and edges are lightly bevelled.

Unmounted Anamorphic Prisms

	A (mm)	B (mm)	C (mm)	PRODUCT NUMBER
Pair	8.5	12.0	12.0	06 GPU 001

Melles Griot also offers 1.5 micron self-contained helium neon lasers. See page 352 for further information.

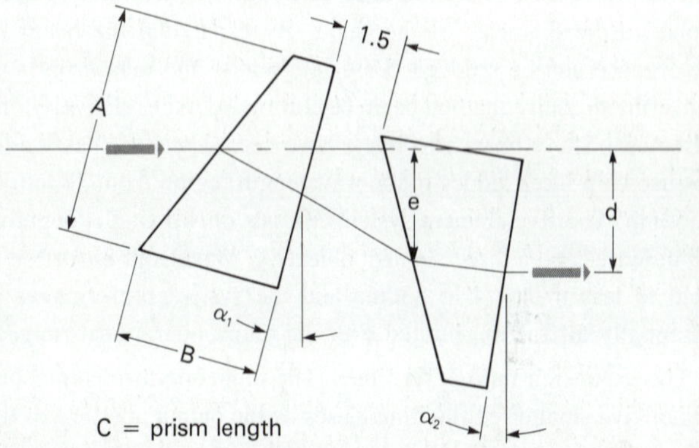

06 GPU 001

Anamorphic Expansion of Prism Pairs

Anamorphic Magnification	Prism Angles α_1 (deg)	Prism Angles α_2 (deg)	Displacement e (mm)	Displacement d (mm)
2.0	21.2	6.0	5.1	5.3
2.2	23.8	4.3	5.4	5.6
2.4	25.8	2.9	5.7	5.8
2.6	27.6	1.8	6.0	6.0
2.8	29.1	0.9	6.2	6.2
3.0	30.4	0.1	6.4	6.4
3.2	31.6	-0.6	6.6	6.5
3.4	32.7	-1.1	6.7	6.7
3.6	33.6	-1.6	6.9	6.8
3.8	34.4	-2.1	7.0	6.9
4.0	35.2	-2.5	7.1	7.0
4.2	35.9	-2.8	7.2	7.1
4.4	36.6	-3.1	7.3	7.2
4.6	37.2	-3.4	7.4	7.3
4.8	37.7	-3.7	7.5	7.3
5.0	38.2	-3.9	7.6	7.4
5.2	38.7	-4.1	7.6	7.5
5.4	39.2	-4.3	7.7	7.5
5.6	39.6	-4.5	7.8	7.6
5.8	40.0	-4.7	7.8	7.6
6.0	40.4	-4.8	7.9	7.7

LASER DIODE BEAM EXPANDERS

The divergence of a collimated laser beam is inversely proportional to the diameter of the beam. By expanding the beam its divergence can be reduced. Laser alignment and measuring systems often require that the beam be collimated over relatively long distances. These beam expanders have been designed specifically for use with laser diodes in the wavelength region from 780nm to 1550nm. The five element system consists of two singlet negative input lenses and a three element objective. Wavefront distortion is held to less than $\lambda/4$ at 830nm and the system performance is essentially diffraction limited over the entire operational range.

The expansion ratio is six times. The input aperture is matched to our five smaller collimating lenses or the output aperture of the anamorphic prism pair. Either a collimator or a prism pair may be attached to the beam expander by sliding the housings into each other and securing them with a nylon set screw.

All air-to-glass surfaces of the beam expanders are coated with a single layer magnesium fluoride antireflection coating for enhanced transmission. Because high-index glasses are used throughout these systems, this results in a very effective coating. The beam expander is mounted in a black anodized aluminum housing.

SPECIFICATIONS: LASER DIODE BEAM EXPANDERS

Paraxial Focal Length: ±2% at 830nm
Wavefront Distortion: $<\lambda/4$ at 830nm
Design Wavelength: 830nm
Clear Aperture: ±0.5mm
Dimensions: ±0.25mm
Coating: Single layer MgF_2 for 830nm
Housing: Aluminum
Finish: Black anodized

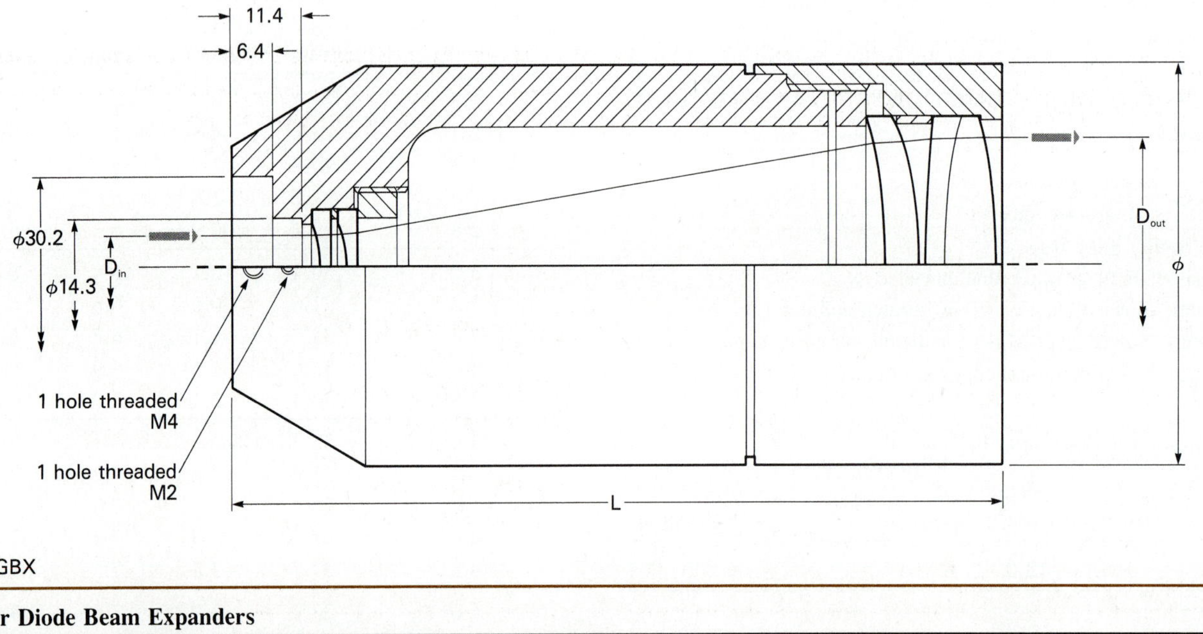

06 GBX

Laser Diode Beam Expanders

Expansion	Wavelength Range (nm)	Beam Diameter D_{in} (mm)	D_{out} (mm)	Field Of View (degrees)	Divergence (mrad)	Diameter ϕ (mm)	Length L (mm)	PRODUCT NUMBER
6x	780-870	8.0	48.0	±0.25	0.022	63.5	127	06 GBX 001
6x	1300-1550	8.0	48.0	±0.25	0.057	63.5	127	06 GBX 003

LASER DIODE RECEIVER LENSES

These lenses have been designed to collect efficiently a full 48mm diameter beam for focusing onto a detector. The input diameter was chosen to match the beam expander telescope output. When collecting laser diode radiation in the far-field it is desirable to use large diameter lenses to collect as much energy as possible.

Both lenses have been designed to optimize the input to a detector by assuring that the maximum amount of energy is concentrated into the central core of the focal spot. Focal spot sizes of 30 and 60 microns are well matched to detectors normally used with laser diodes. Both lenses utilize a color-corrected cemented doublet which has been designed to operate effectively from 750 to 1550nm. The shorter focal length receiver lens also utilizes a high-index meniscus, air-spaced behind the achromat.

All glass-to-air surfaces are coated with a single layer magnesium fluoride antireflection coating. Because high index glasses are used throughout these systems, this produces a very efficient antireflection coating.

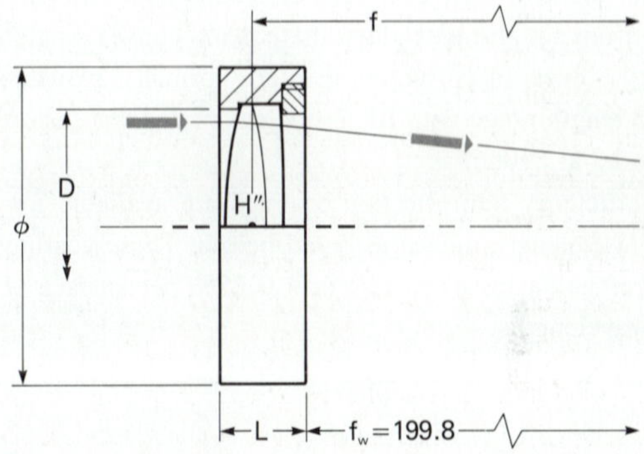

06 GLR 002

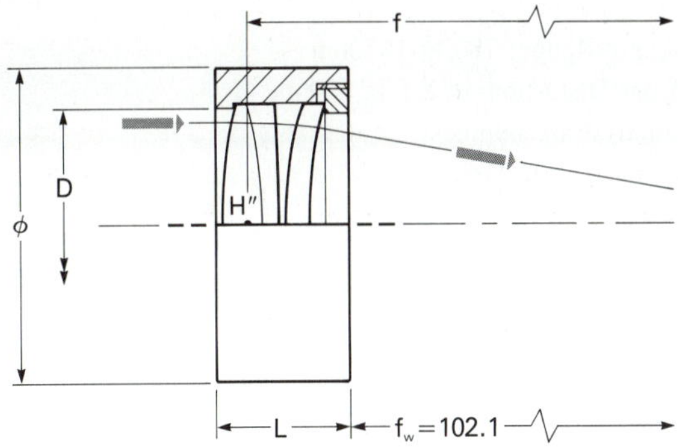

06 GLR 001

Both receiver lenses are mounted in black anodized aluminum housings. The focal length and product number are engraved on the barrel of each lens for identification.

SPECIFICATIONS: LASER DIODE RECEIVER LENSES

Paraxial Focal Length: ±2% at 830nm
Design Wavelength: 830nm
Wavelength Range: 780 to 1550nm
Clear Aperture: ±0.5mm
Dimensions: ±0.25mm
Coating: Single layer MgF_2 for 830nm
Housing: Aluminum
Finish: Black anodized

Laser Diode Receiver Lenses

f_o (mm)	D (mm)	f_o/D	Field Of View (degrees)	Spot Size (microns)	Diameter ϕ (mm)	Length L (mm)	PRODUCT NUMBER
120.0	48.0	2.5	±1.0	30.0	63.5	27.3	06 GLR 001
211.0	48.0	4.4	±1.0	60.0	63.5	17.4	06 GLR 002

LASER DIODE PLANO-CONVEX GLASS SINGLETS

These diffraction limited singlets have been optimized for use with laser diodes for focusing and collection applications.

Made from a high-index glass, these plano-convex singlets provide better levels of correction than are normally available over this wavelength range with BK 7 singlets. Wavefront distortion is less than λ5 at 830nm.

High-efficiency antireflection coatings are available for three specific wavelength ranges as noted below. These coatings will reduce reflection losses to less than 0.25% per surface at their design wavelengths.

See page 90 for many more plano-convex glass singlets.

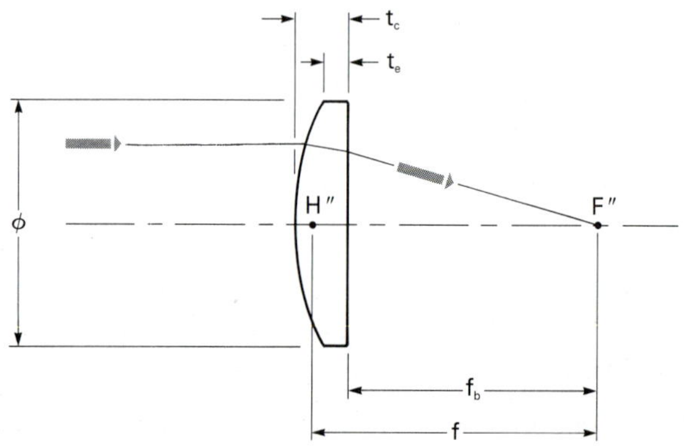

06 LXP

COATING OPTIONS

The Coating Suffix system should be used to request antireflection coatings for these lenses. If no Coating Suffix is appended to the Product Number, the lenses will be supplied uncoated.

Antireflection Coatings; Normal Incidence

Wavelength (nm)	Maximum Reflectance (%)	COATING SUFFIX
780-830	0.25	/076
1300	0.25	/168
1550	0.25	/169

SPECIFICATIONS: LASER DIODE PLANO-CONVEX GLASS SINGLETS

Paraxial Focal Length: ±2%
Design Wavelength: 830nm
Refractive Index: $n_{830}=1.76312$, $n_{1300}=1.74888$, $n_{1550}=1.74525$
Wavelength Range: 780 to 1550nm
Wavefront Distortion: $<\lambda/5$ at 830nm
Centration: 3 arc minutes
Diameter (ϕ): ±0.25mm
Center Thickness: (t_c): ±0.15mm
Materials: Schott SF 11, grade A
Cosmetic Surface Quality: Pitch-polished, 80-50 scratch and dig
Coating: Must be specified by suffix
Circumference is ground and outside edges have a slight protective bevel.

Laser Diode Plano-Convex Glass Singlets

for λ=830nm f	for λ=830nm f_b	for λ=830nm f/ϕ	ϕ	t_c	t_e	for λ=1300nm f	for λ=1300nm f_b	for λ=1550nm f	for λ=1550nm f_b	PRODUCT NUMBER
15.0	13.9	5.0	3.0	2.0	1.9	15.3	14.1	15.4	14.2	06 LXP 001
25.0	23.6	5.0	5.0	2.5	2.3	25.5	24.0	25.6	24.2	06 LXP 003
50.0	48.3	6.3	8.0	3.0	2.8	51.0	49.2	51.2	49.5	06 LXP 005
65.0	63.0	6.5	10.0	3.5	3.3	66.2	64.2	66.6	64.6	06 LXP 007
110.0	107.8	4.0	15.0	4.0	3.7	112.1	109.8	112.6	110.3	06 LXP 009

Supplied uncoated. For antireflection coated lenses, append coating suffix from the coating options chart.

LASER DIODE SILICON SINGLETS

These silicon singlet lenses have been designed specifically for use with laser diodes at 1300 and 1550nm. At these wavelengths, the high index of refraction of silicon allows excellent aberration correction. All these lenses are diffraction limited and exhibit less than λ/10 of wavefront distortion at 1550nm.

These singlet lenses have to be refocused when the wavelength domain is changed.

High-efficiency antireflection coatings are available for these lenses at two wavelengths (1300 and 1550nm). To request these coatings please append the coating suffix as specified below.

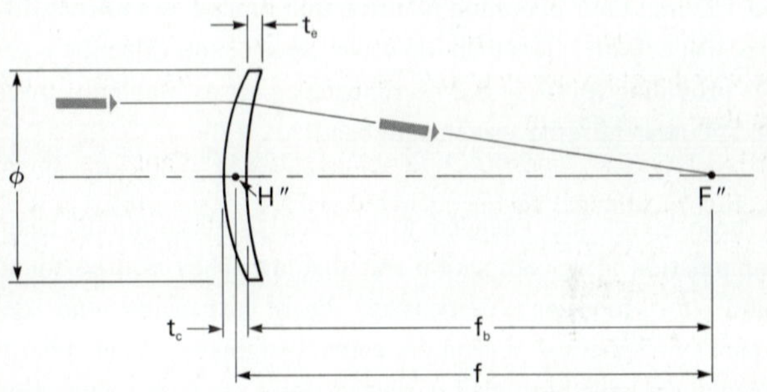

06 LSM

COATING OPTIONS

The Coating Suffix system should be used to request antireflection coatings for these lenses. If no Coating Suffix is appended to the Product Number the lenses will be supplied uncoated.

Antireflection coatings; Normal Incidence

Wavelength (nm)	Maximum Reflectance (%)	COATING SUFFIX
1300	0.25	/168
1550	0.25	/169

SPECIFICATIONS: LASER DIODE SILICON SINGLETS

Paraxial Focal Length: ±2%
Design Wavelength: 1550nm
Wavelength Range: 1300 to 1550nm
Wavefront Distortion: <λ/10 at 1550nm
Centration: 3 arc minutes
Diameter (ϕ): ±0.25mm
Center Thickness: (t_c): ±0.15mm
Materials: Optical Grade Silicon
Cosmetic Surface Quality: Pitch-polished, 80-50 scratch and dig
Coating: Must be specified by suffix
Circumference is ground and outside edges have a slight protective bevel.

Laser Diode Silicon Singlets

for λ= 1550nm							for λ= 1300nm		PRODUCT
f	f_b	f/ϕ	ϕ	t_c	t_e	B	f	f_b	NUMBER
15.0	13.5	3.0	5	2.3	2.2	13.4	14.8	13.3	06 LSM 001
25.0	23.3	3.1	8	2.6	2.5	23.1	24.7	22.8	06 LSM 003
35.0	33.1	3.5	10	2.8	2.7	32.8	34.6	32.7	06 LSM 005
60.0	57.6	4.0	15	3.4	3.2	57.3	59.3	56.9	06 LSM 007
90.0	87.2	4.5	20	4.0	3.7	86.8	89.0	86.2	06 LSM 009
120.0	116.4	4.8	25	5.0	4.7	116.0	118.7	115.1	06 LSM 011

Supplied uncoated. For antireflection coated lenses, append coating suffix from the coating options chart.

LASER DIODE GLASS DOUBLETS

These cemented glass doublets have been designed to perform well throughout the entire near-infrared spectrum. Between 780 and 1550nm their operation is diffraction limited; wavefront distortion at 830nm is less than $\lambda/5$ wave. Special antireflection coatings providing less than 0.25% reflectance are available for three specific near-infrared wavelength bands.

These lenses are intended to be used for laser diode beam manipulation after collimation and shaping. They will be found useful for focusing, defocusing, beam expansion and de-expansion. Spherical aberration, coma, astigmatism, and spherochromatism have been well corrected. Pure chromatic aberration (variation of focal length with wavelength) has been left uncorrected to provide greater freedoms in the other corrections. When used with monochromatic laser diodes this causes no aberration contribution, but will mean that systems have to be refocused for use with different wavelength laser diodes. The focal lengths over a range of 750 to 1550nm are shown graphically.

High-efficiency antireflection coatings are available for 780-830, 1300, and 1550nm which provide air-glass surface reflectivities of less than 0.25% per surface.

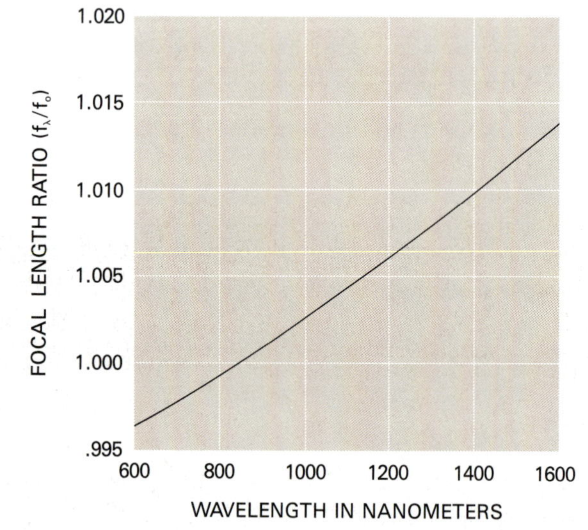

FOCAL LENGTH OF LASER DIODE GLASS DOUBLETS as a function of wavelength.

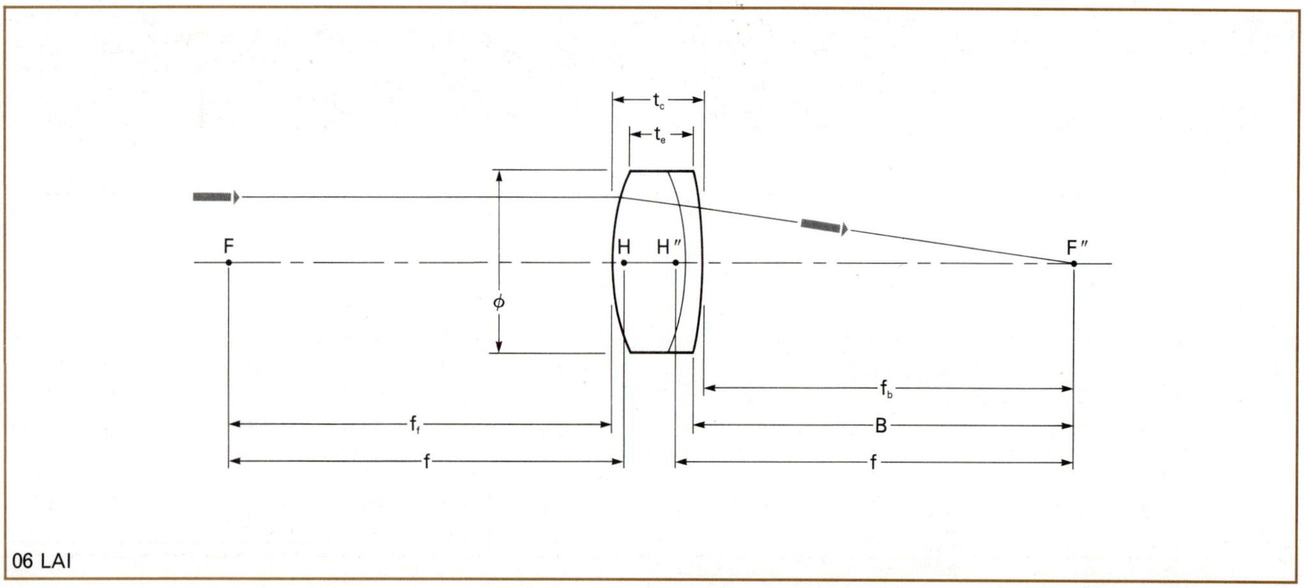

SPECIFICATIONS: LASER DIODE GLASS DOUBLETS

Paraxial Focal Length: ±2%
Design Wavelength: 830nm
Wavelength Range: 780 to 1550nm
Wavefront Distortion: $<\lambda/5$ at 830nm
Centration: 4 arc minutes
Diameter (ϕ): ±0.25mm
Center Thickness: (t_c): ±0.10mm
Materials: SK 11 and SF 5, grade A
Cement: Ultraviolet cured synthetic polyester
Cosmetic Surface Quality: Pitch-polished, 80-50 scratch and dig
Coating: Must be specified by suffix
Circumference is ground and outside edges have a slight protective bevel.

COATING OPTIONS

The Coating Suffix system should be used to request antireflection coatings for these lenses. If no Coating Suffix is appended to the Product Number, the lenses will be supplied uncoated.

Antireflection Coatings; Normal Incidence

Wavelength (nm)	Maximum Reflectance (%)	COATING SUFFIX
780-830	0.25	/076
1300	0.25	/168
1550	0.25	/169

We can offer assistance in incorporating stock lenses into your instrument to achieve a ready and cost-effective design.

Laser Diode Glass Doublets

for λ=830nm							for λ=1300nm		for λ=1550nm		PRODUCT
f	f_b	f/ϕ	ϕ	t_c	t_e	B	f	f_b	f	f_b	NUMBER
10.0	8.2	2.0	5.0	3.5	2.8	8	10.1	8.3	10.1	8.3	06 LAI 001
25.0	22.4	2.5	10.0	5.0	3.9	23	25.2	22.6	25.3	22.8	06 LAI 003
40.0	37.2	2.7	15.0	5.5	4.0	38	40.3	37.6	40.5	37.8	06 LAI 005
60.0	56.7	3.0	20.0	6.5	4.7	57	60.5	57.1	60.8	57.4	06 LAI 007
80.0	75.6	3.2	25.0	8.5	6.4	76	80.6	76.2	81.0	76.6	06 LAI 009
100.0	94.8	3.3	30.0	10.0	7.6	95	100.8	95.6	101.3	96.1	06 LAI 011
145.0	138.5	3.6	40.0	12.5	9.5	139	146.1	139.6	146.8	140.4	06 LAI 013
190.0	183.0	3.8	50.0	13.5	10.0	184	191.4	184.4	192.4	185.5	06 LAI 015

Supplied uncoated. For antireflection coated lenses, append coating suffix from the coating options chart.

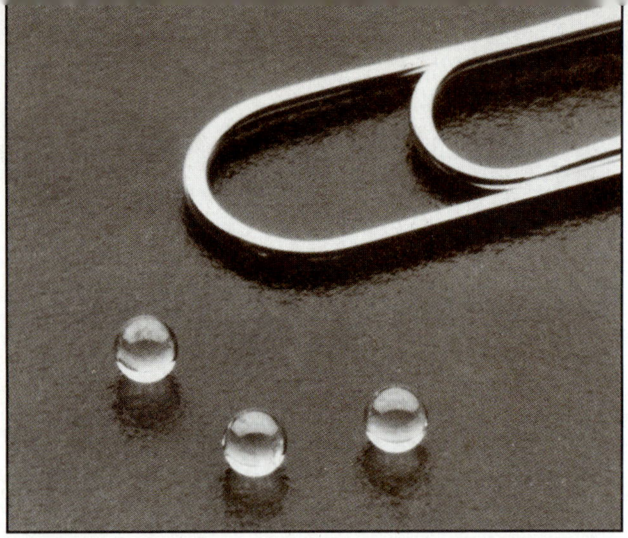

FIBER COUPLING SPHERES

Of the various techniques presently used to couple optical fibers to each other, (and to laser diodes and detectors) the ball lens method is one which allows the researcher to interact with the beam in the coupling process. Using this method, spheres are arranged such that the fiber end (or input/output device) is located at the focal point of the sphere. The output from the sphere is then a collimated beam. If two spheres are arranged in axial alignment with each other the beam will be transferred from one focal point to the other. This is exactly analagous to a pair of thin lenses relaying the object to the image space through a collimator/focusing lens combination. By enlarging the coupling beam, translational alignment sensitivity is reduced.

Spheres are much easier to manufacture than thin lenses in these small diameters. In the same way they are easy to align in experimental applications. Because of their simplicity, ball lens coupling systems tend to achieve better coupling efficiencies than other methods. Using communication fibers, coupling efficiencies up to 95% (0.2dB insertion loss) have been achieved.

These optical spheres are available in three diameters. The larger spheres are easier to handle and the diameter of the collimated output beam is greater to facilitate interaction with the beam and to minimize translational sensitivity. Conversely, larger spheres require substantially bigger packaging volumes.

Spherical aberration is the overriding defect in all coupling systems. Fiber diameters are so much smaller than the coupler's focal length that off-axis aberrations are absent. Evidence exists which indicates that a slight negative defocusing of the ball lens will substantially reduce the spherical aberration third-order contribution (see Lens Coupling in Fiber-Optics Devices: Efficiency Limits, A Nicia, *Applied Optics*, Vol. 20, No. 18, pp 3136-3145).

All of these spheres are coated on both hemispheres with a single layer MgF_2 antireflection coating. Because of the high index of these spheres, this results in a very effective coating with low reflectivity at the specified wavelengths. A narrow equatorial band is not coated due to fixturing but the optical axis may be located by a small black dot that is located on the optical axis. This dot should be used to locate the optical axis during mounting. Once aligned the black dot may be removed by applying acetone with a cotton swab.

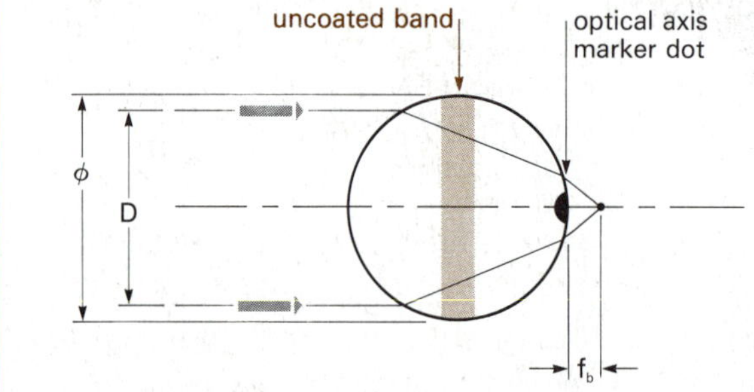

06 LMS

SPECIFICATIONS: FIBER COUPLING SPHERES

Wavelength Range: 600 to 1550nm

Refractive Index: $n_{633}=1.84491$, $n_{830}=1.83098$, $n_{1330}=1.81767$, $n_{1550}=1.81337$

Sphericity: $<\lambda/4$ at 632.8nm

Reflectance: <0.25% at specified wavelength except uncoated version

Diameter (ϕ): ±1 micron

Material: LaSF 9

Coating: Single layer MgF_2

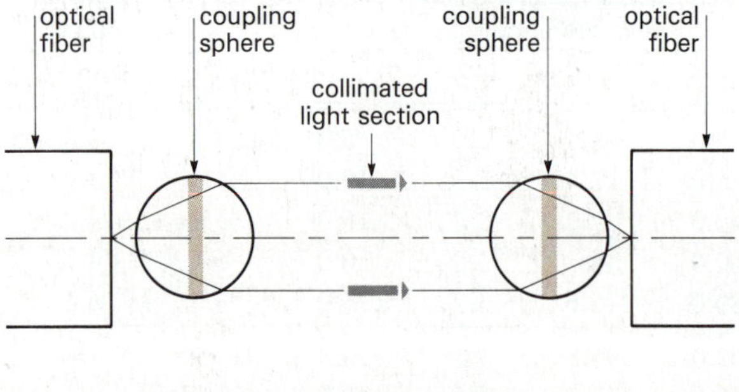

FIBER COUPLING USING SPHERES. Lateral positioning sensitivity is greatly reduced by this method.

Fiber Coupling Spheres; uncoated

ϕ (mm)	f_{830} (mm)	f_b (mm)	f_{830}/D	D (mm)	PRODUCT NUMBER
2.0	1.10	0.10	0.69	1.60	06 LMS 002
3.0	1.65	0.15	0.69	2.40	06 LMS 003
5.0	2.75	0.25	0.69	4.00	06 LMS 005

Fiber Coupling Spheres; antireflection coated for $\lambda = 633$nm

ϕ (mm)	f_{633} (mm)	f_b (mm)	f_{633}/D	D (mm)	PRODUCT NUMBER
2.0	1.09	0.09	0.68	1.60	06 LMS 102
3.0	1.64	0.14	0.68	2.40	06 LMS 103
5.0	2.73	0.23	0.68	4.00	06 LMS 105

Fiber Coupling Spheres; antireflection coated for $\lambda = 830$nm

ϕ (mm)	f_{830} (mm)	f_b (mm)	f_{830}/D	D (mm)	PRODUCT NUMBER
2.0	1.10	0.10	0.69	1.60	06 LMS 202
3.0	1.65	0.15	0.69	2.40	06 LMS 203
5.0	2.75	0.25	0.69	4.00	06 LMS 205

Fiber Coupling Spheres; antireflection coated for $\lambda = 1300$nm

ϕ (mm)	f_{1300} (mm)	f_b (mm)	f_{1300}/D	D (mm)	PRODUCT NUMBER
2.0	1.11	0.11	0.69	1.60	06 LMS 302
3.0	1.67	0.17	0.69	2.40	06 LMS 303
5.0	2.78	0.28	0.69	4.00	06 LMS 305

Fiber Coupling Spheres; antireflection coated for $\lambda = 1550$nm

ϕ (mm)	f_{1550} (mm)	f_b (mm)	f_{1550}/D	D (mm)	PRODUCT NUMBER
2.0	1.11	0.11	0.69	1.60	06 LMS 402
3.0	1.67	0.17	0.69	2.40	06 LMS 403
5.0	2.79	0.29	0.69	4.00	06 LMS 405

GRADIENT-INDEX LENSES

WHY GRADIENT INDEX?

In coventional lenses imaging is a result of discrete refractions occurring at the boundaries of homogeneous media of different refractive index. By using materials in which the refractive index varies in some controlled way it is possible to form images by continuous refraction.

Combining surface refraction with continuous refraction provides a number of advantages over conventional lens systems. The more notable of these advantages are:

- correction of aberrations without complex, multi-element systems or aspherics,
- simplification of the geometry of lenses,
- formation of real images at the lens surface.

Index variation may be axial, radial or a combination of the two. Gradient lens technology has, so far, been entirely governed by the supply of material. However, recent advances in ion diffusion techniques have enabled a variety of gradient index materials to be made available.

The gradient index lenses we offer are manufactured from SELFOC® which is a radial gradient index material. These are rod lenses combining refraction at the plane end surfaces with continuous refraction within the rod. They are extremely easy to use and are highly suitable for coupling the output of a laser diode into an optical fiber as well as numerous other applications.

IMAGING THEORY IN GRADIENT INDEX ROD LENSES

In SELFOC® material refractive index varies parabolically as a function of radius. The index variation may be expressed as:

$$n_r = n_{oo}\left(1 - \frac{A}{2} r^2\right)$$

where, n_r = refractive index at a distance r from the optical axis

n_{oo} = design index on the optical axis

A = a positive constant

r = radial distance from the optical axis.

As a result of this parabolic index variation a ray incident on the front surface follows a sinusoidal path along the rod lens. The period of this sinusoidal path is called the "pitch" of the lens and is an important parameter in gradient index imaging. It is given by:

$$P = 2\pi/\sqrt{A}$$

SELFOC® is a registered trademark of the Nippon Sheet Glass Company.

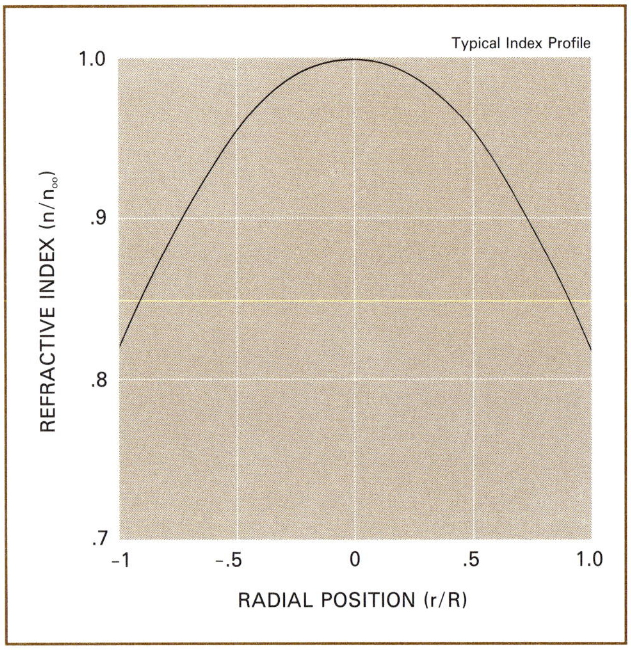

REFRACTIVE INDEX PROFILE IN SELFOC® gradient index material.

Another important parameter is the maximum acceptance angle,

$$\theta = \sin^{-1}(n_{oo} R \sqrt{A}) \text{ , where}$$

$$R = \text{radius of lens} = \phi/2$$

In order to determine the object and image position for gradient index rod lenses it is convenient to define the lens parameters in analogy with those of a conventional spherical surfaced, homogeneous lens. The accompanying diagram shows the input parameters and their relationships.

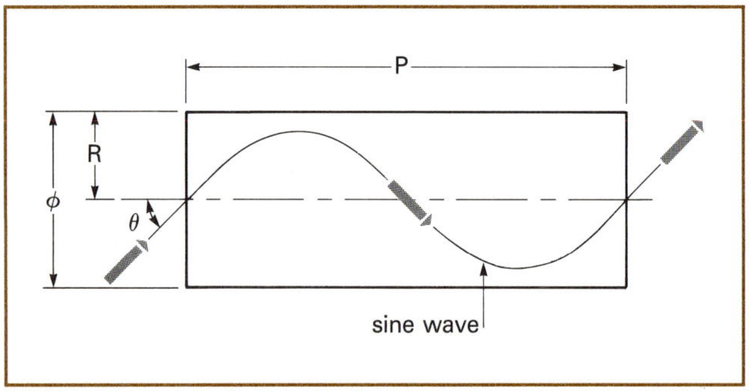

RAY PATH in gradient index rod lenses.

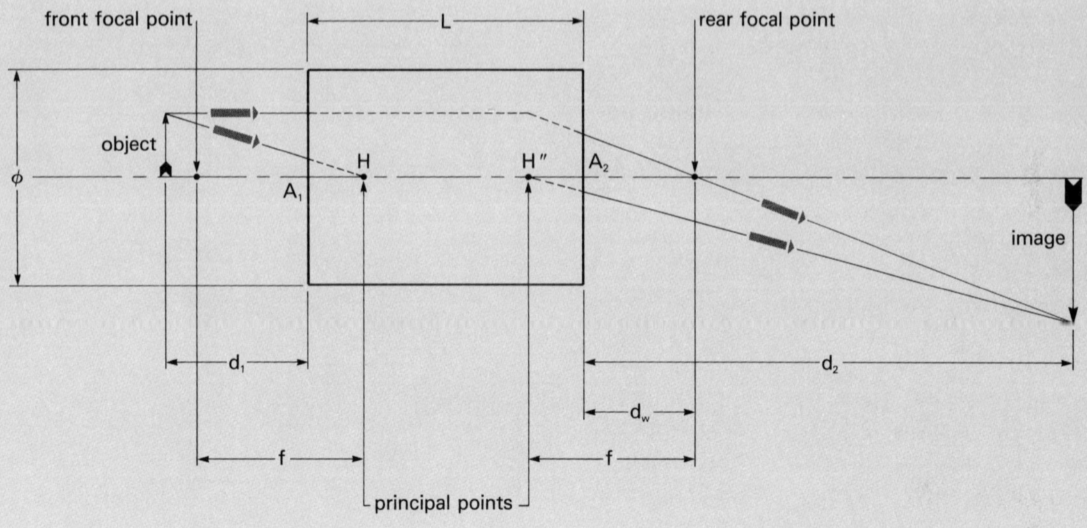

ϕ = Lens diameter

L = Lens length

n_o = Refractive index on axis

$\sqrt{A}$ = Quadratic constant (as previously defined)

$$\text{Pitch} = \frac{2\pi}{\sqrt{A}}$$

Normalized lens length = $L\sqrt{A}$

$$\text{Fractional pitch} = \frac{L\sqrt{A}}{2\pi}$$

$$f = \text{Effective focal length} = \frac{1}{n_o\sqrt{A}\,\sin(L\sqrt{A})}$$

d_w = Working distance (back focal length)

$$= \frac{1}{n_o\sqrt{A}}\cot(L\sqrt{A})$$

$$A_2H'' = \text{Principal plane position} = \frac{1}{n_o\sqrt{A}}\tan\left(\frac{L\sqrt{A}}{2}\right)$$

d_1 = Object to lens distance

d_2 = Image to lens distance

$$= \frac{1}{n_o\sqrt{A}} \cdot \frac{n_o\sqrt{A}\,d_1\cos(L\sqrt{A}) + \sin(L\sqrt{A})}{n_o\sqrt{A}\,d_1\sin(L\sqrt{A}) - \cos(L\sqrt{A})}$$

$$m = \text{Magnification} = \frac{-1}{n_o\sqrt{A}\,d_1 - \cot(L\sqrt{A})}$$

IMAGING PROPERTIES of a gradient index rod lens at finite conjugate ratio.

GRADIENT INDEX LENS TYPES

Knowing the pitch (P) of gradient index rod material, it is possible to achieve various imaging characteristics simply by varying the length of the lens. Imaging properties of common fractional pitch lens types are shown in the accompanying diagram.

A quarter pitch lens forms an inverted, real image of an object at infinity on the opposite end of the lens. A half pitch lens transfers an inverted real image from one end of the lens to the other. Uneven fractional pitch lenses (0.23P and 0.29P) are useful for coupling laser diode outputs either to a collimated beam or into an optical fiber. Their advantage lies in the fact that the image is displaced from the end of the lens and is therefore accessible. A 0.29P lens may be used over a range of magnifications (see graph).

Both the index on-axis (n_0) and the quadratic gradient constant ($\sqrt{A}$) are wavelength dependent. This results in the pitch, and consequently the length, varying with wavelength. For this reason we list lenses separately for each wavelength of interest. We can, of course, supply lenses for other wavelengths on special request.

COATINGS

Since these lenses are wavelength specific we have chosen to offer them with an antireflection coating. This reduces the surface reflection losses to less than 1.5% per surface.

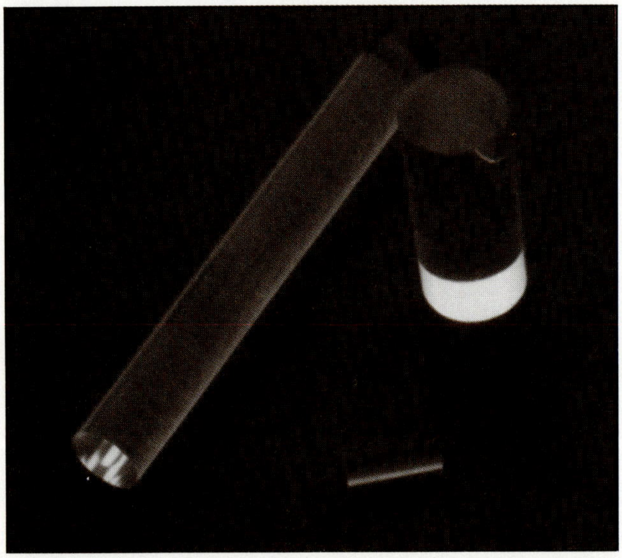

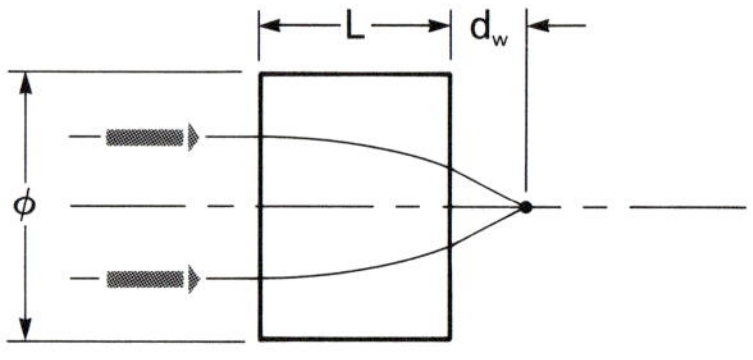

0.23P

06 LGT

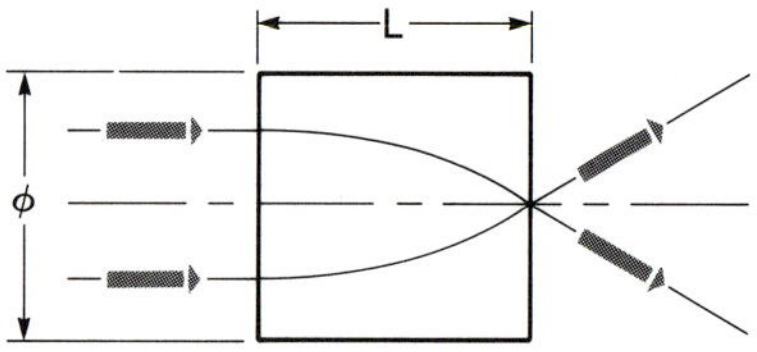

0.25P

06 LGS

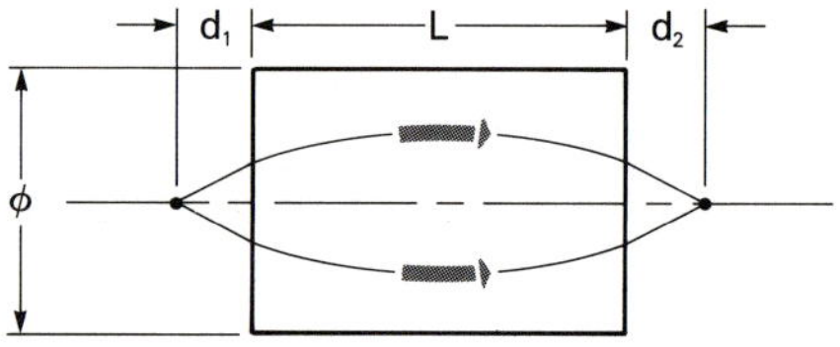

0.29P

06 LGE

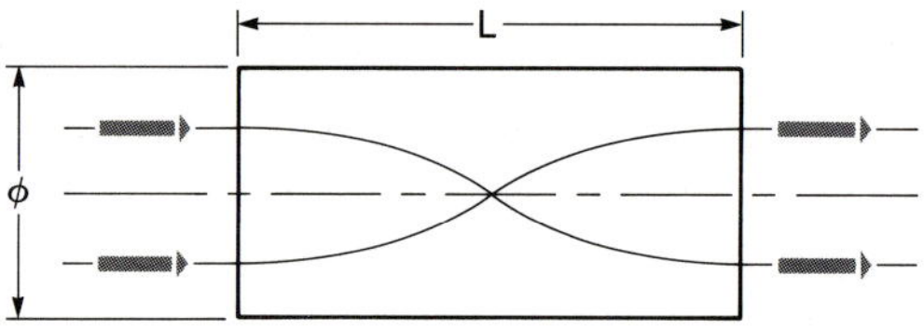

0.5P

06 LGD

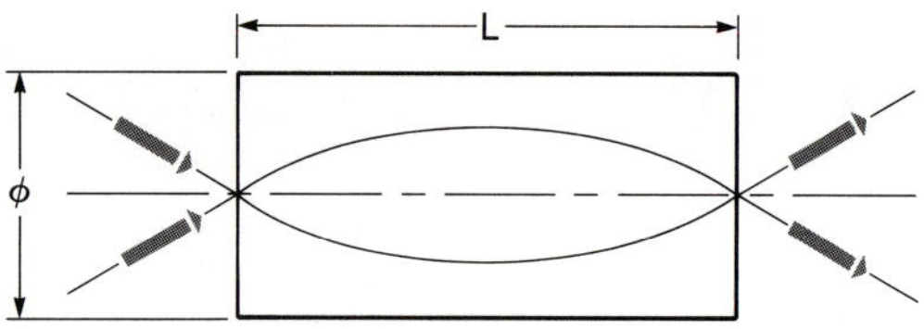

0.5P

06 LGD

SPECIFICATIONS: GRADIENT-INDEX ROD LENSES

Design Wavelength: 633, 830, 1300 or 1550nm
Gradient: ±1.5%
Working distance (d_w): for 0.23 pitch lenses
(at 633nm) = 0.15mm
(at 830nm) = 0.21mm
(at 1300nm) = 0.26mm
(at 1550nm) = 0.26mm
Working distance: (d_1, d_2): for 0.29 pitch lenses — see graph.
Length: ±3%
Diameter (ϕ): 1mm ±0.01mm
1.8 and 2mm ±0.02mm

Gradient index lenses for other wavelengths can be supplied to OEM's on special order.

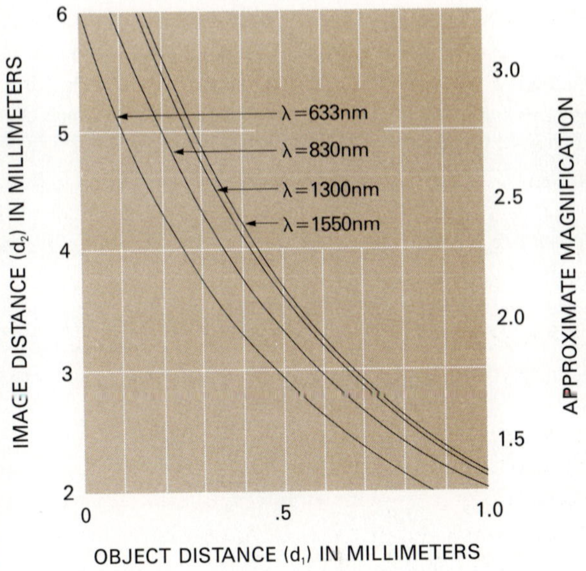

OBJECT AND IMAGE DISTANCE for 0.29 pitch lenses.

Gradient Index Rod Lenses; for λ=633nm

Fractional Pitch	N.A.	Diameter ϕ (mm)	Length L (mm)	Maximum Acceptance Angle, 2θ (deg)	Design Index On-axis n_{oo}	Quadratic Constant $\sqrt{A}$	PRODUCT NUMBER
0.23	0.46	1.8	4.3	55	1.608	0.339	06 LGT 114
0.25	0.37	1.0	3.2	43	1.564	0.499	06 LGS 111
	0.37	2.0	6.4	43	1.564	0.247	06 LGS 112
	0.46	1.0	2.6	55	1.608	0.608	06 LGS 113
	0.46	1.8	4.6	55	1.608	0.339	06 LGS 114
	0.46	2.0	5.2	55	1.608	0.304	06 LGS 115
	0.60	1.8	3.7	72	1.658	0.428	06 LGS 116
0.29	0.46	1.8	5.4	55	1.608	0.339	06 LGE 114
0.50	0.37	1.0	6.3	43	1.564	0.499	06 LGD 111
	0.37	2.0	12.7	43	1.564	0.247	06 LGD 112
	0.46	1.0	5.2	55	1.608	0.608	06 LGD 113
	0.46	1.8	9.3	55	1.608	0.339	06 LGD 114
	0.46	2.0	10.3	55	1.608	0.304	06 LGD 115
	0.60	1.8	7.3	72	1.658	0.428	06 LGD 116

Gradient Index Rod Lenses; for λ=830nm

Fractional Pitch	N.A.	Diameter ϕ (mm)	Length L (mm)	Maximum Acceptance Angle, 2θ (deg)	Design Index On-axis n_{oo}	Quadratic Constant $\sqrt{A}$	PRODUCT NUMBER
0.23	0.46	1.8	4.4	55	1.600	0.332	06 LGT 214
0.25	0.37	1.0	3.2	43	1.557	0.489	06 LGS 211
	0.37	2.0	6.5	43	1.557	0.243	06 LGS 212
	0.46	1.0	2.6	55	1.600	0.601	06 LGS 213
	0.46	1.8	4.7	55	1.600	0.332	06 LGS 214
	0.46	2.0	5.3	55	1.600	0.298	06 LGS 215
	0.60	1.8	3.7	72	1.646	0.423	06 LGS 216
0.29	0.46	1.8	5.5	55	1.600	0.332	06 LGE 214
0.50	0.37	1.0	6.4	43	1.557	0.489	06 LGD 211
	0.37	2.0	13.0	43	1.557	0.243	06 LGD 212
	0.46	1.0	5.2	55	1.600	0.601	06 LGD 213
	0.46	1.8	9.5	55	1.600	0.332	06 LGD 214
	0.46	2.0	10.5	55	1.600	0.298	06 LGD 215
	0.60	1.8	7.4	72	1.646	0.423	06 LGD 216

Gradient Index Rod Lenses; for λ=1300nm

Fractional Pitch	N.A.	Diameter ϕ (mm)	Length L (mm)	Maximum Acceptance Angle, 2θ (deg)	Design Index On-axis n_{oo}	Quadratic Constant $\sqrt{A}$	PRODUCT NUMBER
0.23	0.46	1.8	4.4	55	1.592	0.327	06 LGT 314
0.25	0.37	1.0	3.2	43	1.551	0.483	06 LGS 311
	0.37	2.0	6.6	43	1.551	0.283	06 LGS 312
	0.46	1.0	2.6	55	1.592	0.597	06 LGS 313
	0.46	1.8	4.8	55	1.592	0.327	06 LGS 314
	0.46	2.0	5.3	55	1.592	0.295	06 LGS 315
	0.60	1.8	3.7	72	1.636	0.418	06 LGS 316
0.29	0.46	1.8	5.6	55	1.592	0.327	06 LGE 314
0.50	0.37	1.0	6.5	43	1.551	0.483	06 LGD 311
	0.37	2.0	13.2	43	1.551	0.238	06 LGD 312
	0.46	1.0	5.3	55	1.592	0.597	06 LGD 313
	0.46	1.8	9.6	55	1.592	0.327	06 LGD 314
	0.46	2.0	10.6	55	1.592	0.295	06 LGD 315
	0.60	1.8	7.5	72	1.636	0.418	06 LGD 316

Gradient Index Rod Lenses; for $\lambda = 1550nm$

Fractional Pitch	N.A.	Diameter ϕ (mm)	Length L (mm)	Maximum Acceptance Angle, 2θ (deg)	Design Index On-axis n_{oo}	Quadratic Constant $\sqrt{A}$	PRODUCT NUMBER
0.23	0.46	1.8	4.4	55	1.590	0.326	06 LGT 414
0.25	0.37	1.0	3.3	43	1.550	0.481	06 LGS 411
	0.37	2.0	6.6	43	1.550	0.237	06 LGS 412
	0.46	1.0	2.6	55	1.592	0.596	06 LGS 413
	0.46	1.8	4.8	55	1.592	0.326	06 LGS 414
	0.46	2.0	5.3	55	1.592	0.294	06 LGS 415
	0.60	1.8	3.8	72	1.634	0.417	06 LGS 416
0.29	0.46	1.8	5.5	55	1.590	0.326	06 LGE 414
0.50	0.37	1.0	6.5	43	1.550	0.481	06 LGD 411
	0.37	2.0	13.3	43	1.550	0.237	06 LGD 412
	0.46	1.0	5.3	55	1.592	0.596	06 LGD 413
	0.46	1.8	9.6	55	1.592	0.326	06 LGD 414
	0.46	2.0	10.7	55	1.592	0.294	06 LGD 415
	0.60	1.8	7.5	72	1.634	0.417	06 LGD 416

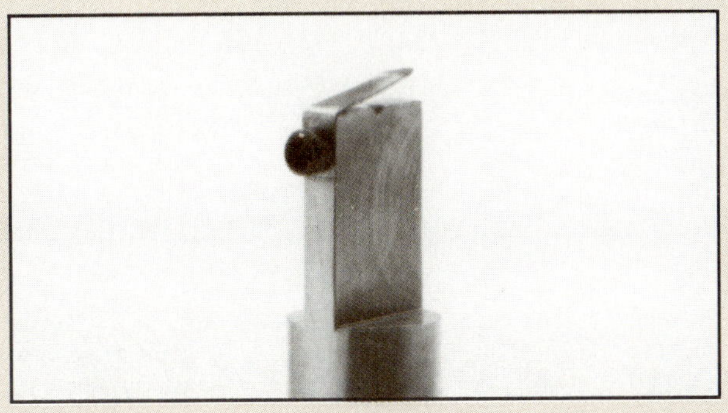

GRADIENT INDEX ROD LENS HOLDER

A simple holder for gradient index rod lenses is offered for your convenience together with many other accessories and positioning devices which you will find useful in laser diode/fiber optics systems. See page 471.

PRODUCT NUMBER
07 HGI 001

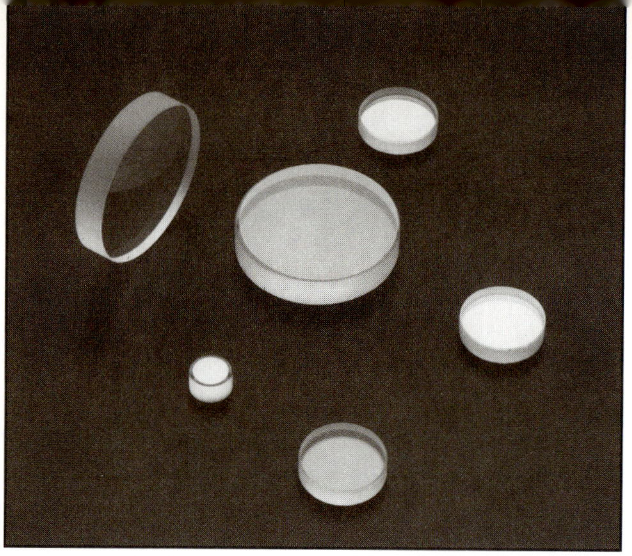

LASER DIODE MIRRORS

One of the most commonly used elements in an optical system is a mirror. Laser diode systems are no exception. The reflectance of protected aluminum coated mirrors starts to decline in the near-infrared making these less suitable for use with laser diodes. The best way to maximize the efficiency of laser diode reflectance is to use a dielectric maximum reflectance coating. These coatings are optimized for a specific wavelength and angle.

We offer laser diode MAX-R mirrors in two flatnesses, λ/6 and λ/30. These mirrors are available from stock for normal incidence or 45° incidence in three sizes.

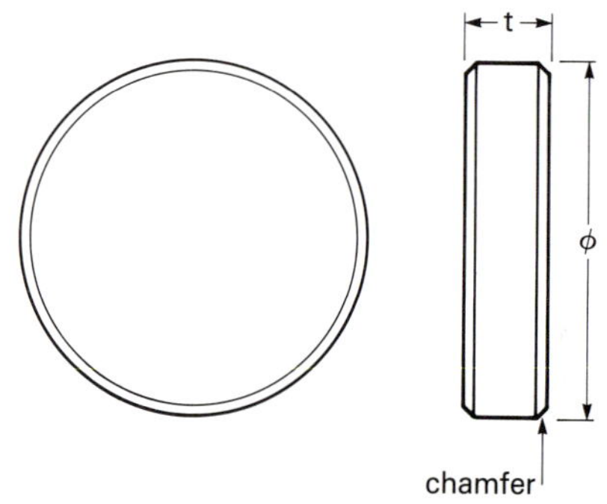

02 MFG / 02 MLQ

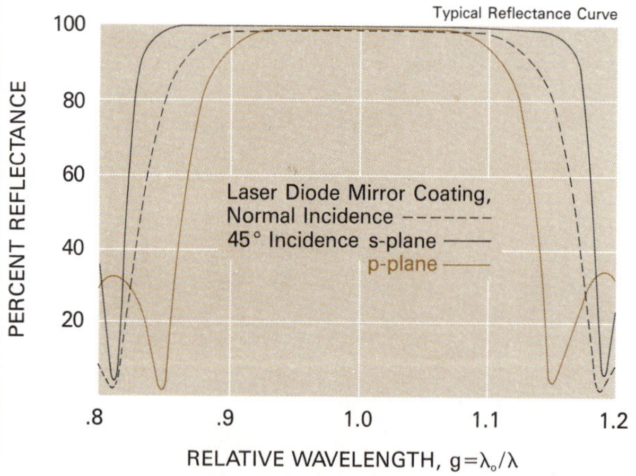

If you need mirror mounts, we have a broad selection. All of these have been designed to provide the stability and precision of adjustment you need. *See page 472 for further details.*

SPECIFICATIONS: LASER DIODE MIRRORS

Surface Flatness: at 830nm over central 90% of diameter
Material: Pyrex or synthetic fused silica
Dimensions: ±0.25mm
Cosmetic Surface Quality: 60-40 scratch and dig or better.
Complies With: MIL-C-675 and MIL-M-13508

Laser Diode Mirror Substrates

Flatness	Material	Diameter φ (mm)	Thickness t (mm)	PRODUCT NUMBER
λ/6	Pyrex	12.5	6	02 MFG 014
	Pyrex	25.0	6	02 MFG 015
	Pyrex	50.0	10	02 MFG 019
λ/30	Silica	12.5	6	02 MLQ 009
	Silica	25.0	6	02 MLQ 011
	Silica	50.0	10	02 MLQ 015

To order, select substrate product number above and append coating suffix from tables below.

Laser Diode Mirror Coatings; Normal Incidence

Wavelength (nm)	Minimum Reflectance, R_p (%) 0° incidence	0°±15° incidence	COATING SUFFIX
830	99.3	99.0	/237
1300	99.2	99.0	/245
1550	99.2	99.0	/247

Laser Diode Mirror Coatings; 45° Incidence

Wavelength (nm)	Minimum Reflectance, R_p (%) 45° incidence	45°±10° incidence	COATING SUFFIX
830	99.0	98.5	/287
1300	99.0	98.5	/295
1550	99.0	98.5	/297

LASER DIODE CUBE BEAMSPLITTERS

Cube beamsplitters are well suited for use with laser diodes. When used in a collimated output beam, they provide an ideal means of separating the beam into two components. Because of their symmetrical and robust design, cube beamsplitters are both thermally and mechanically stable. The lack of a second surface (as with plate type beamsplitters) eliminates ghosts.

If cube beamsplitters are used in convergent or divergent portions of an optical beam, they will contribute substantial amounts of unwanted aberration. These can be avoided or minimized by making sure these beamsplitters are used in collimated or nearly collimated laser diode beams. Conjugate distances which include cubes should be long. Alternatively, other elements of the system can be designed to compensate for any aberrations introduced by the cube in a convergent beam.

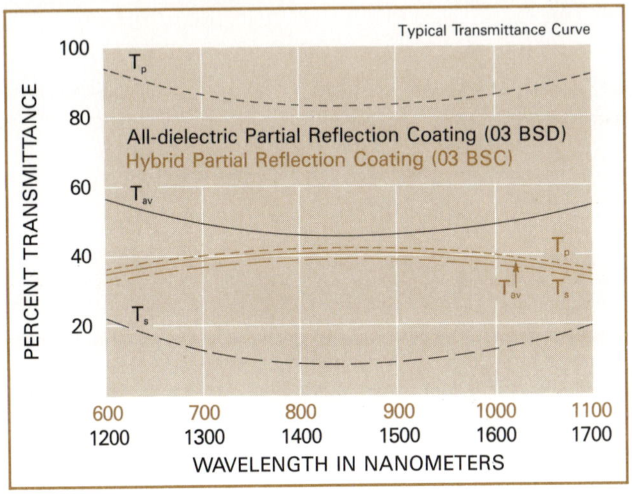

Cube beamsplitters are made from two identical right-angle prisms. The hypotenuse face on one of the prisms is coated with a vacuum-deposited beamsplitting film prior to being cemented to the other prism. (The particular prism half which is coated is marked with a small dot on one of the ground faces.) In the 780-830nm wavelength region these beamsplitters are coated with a metal-dielectric hybrid coating which will produce a 40% reflected (R) by 40% transmitted (T) beam. This coating is relatively polarization insensitive. Further into the infra-red the coating is all-dielectric with a nominal split ratio of 50/50 for unpolarized light. Because these latter coatings are all-dielectric, they are very sensitive to polarization of the input beam, so appropriate care should be taken in use.

SPECIFICATIONS: LASER DIODE CUBE BEAMSPLITTERS

780-830nm
- Reflective Coating: Hybrid, metal-dielectric hybrid
- Split Ratio (R/T): 40/40
- Split Ratio Tolerance: ±5%

1300-1550nm
- Reflective Coating: All-dielectric
- Split Ratio (R/T): 50/50
- Tolerance: ±5%

Dimensions: ±0.3mm
Angle Tolerance: 3 arc minutes
Material: BK 7 grade A fine annealed
Cosmetic Surface Quality: 80-50 scratch and dig
AR Coating: <0.25%R — all four faces
Unused faces are fine ground, and all edges are lightly beveled.

Laser Diode Cube Beamsplitters

A=B=C (mm)	PRODUCT NUMBER Wavelength Range (nm) 780-830	1300-1550
5.0	03 BSC 023	03 BSD 024
10.0	03 BSC 025	03 BSD 026
20.0	03 BSC 027	03 BSD 028

LASER DIODE POLARIZING BEAMSPLITTERS

Similar in design to the cube beamsplitters previously described, these polarizing beamsplitters separate the beam into two orthogonally polarized beams, separated from each other by 90° in angle. The beam which passes straight through the cube, emerges linearly polarized with a purity of at least 96%. The plane of the emerging electric vector is parallel (p-polarized) to the plane of incidence defined by the multilayer film. The beam which is reflected by the internal multilayer film emerges at right angles to the incident beam and it is polarized to a purity of at least 96% as well. In this case, the plane of the electric vector is perpendicular to the plane of incidence defined by the multilayer film (s-polarized). In both cases the residual transmitted beam is less than 2%.

An incident linearly polarized laser diode beam is similarly divided into two polarized beams. The ratio of output beam irradiances is dependent on the orientation of the electric vector of the initial beam. If the plane of polarization of the incident beam is rotated (by turning the laser diode or by use of a half wave plate), the output of the polarizing beamsplitter can be made continuously variable between 1:43 and 43:1.

Polarizing beamsplitters are available coated for four wavelengths: 780, 830, 1300, and 1550nm. Some other laser wavelengths are also available.

All four entrance and exit faces of the beamsplitter are coated with special high-efficiency antireflection coatings which reflect less than 0.25% per surface at their specified wavelengths.

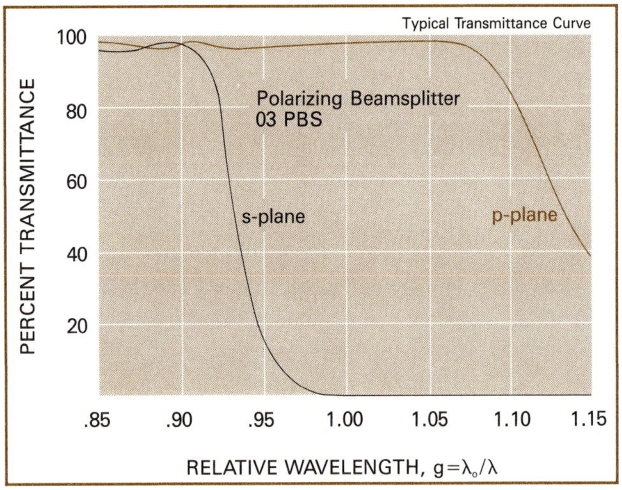

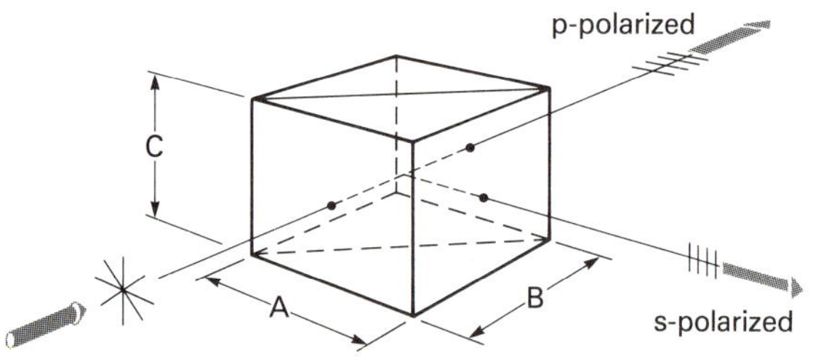

03 PBS

SPECIFICATIONS: LASER DIODE POLARIZING BEAMSPLITTERS

Principal Transmittance: >96%T for p-polarization; <2%T for s-polarization.
Principal Reflectance: >96%R for s-polarization; <2%R for p-polarization.
Beam Deviation: <3 arc minutes
Useful Angular Diameter: 10°
Entrance/Exit Surface Flatness: <λ/10 at 632.8nm
Wavefront Distortion: <λ/2 at 830nm
Dimensions: ±0.2mm
Material: BK 7 grade A fine annealed
Cosmetic Surface Quality: 40-20 scratch and dig
Coating: ±0.25%R — all four faces
Unused faces are fine ground, and all edges are lightly beveled.

Laser Diode Polarizing Beamsplitters

A=B=C (mm)	PRODUCT NUMBER Wavelength (nm) 780	830	1300	1550
3.2	03 PBS 051	03 PBS 061	03 PBS 071	03 PBS 081
5.0	03 PBS 052	03 PBS 062	03 PBS 072	03 PBS 082
10.0	03 PBS 053	03 PBS 063	03 PBS 073	03 PBS 083
20.0	03 PBS 055	03 PBS 065	03 PBS 075	03 PBS 085

CARBON DIOXIDE LASERS

In this section, we describe our completely new range of waveguide CO_2 lasers which have been developed by Melles Griot Laser Products in San Marcos, California.

Contents **Page Number**

SELF-CONTAINED CO_2 LASERS

These new carbon dioxide lasers are small, compact, lightweight, and completely self-contained. They require only 115 VAC/60 Hz or 230 VAC/50 Hz standard single-phase power input for operation. These portable, single-unit lasers will operate reliably in any orientation, in any normal laboratory environment without adjustment or realignment. The power output at 10.6 microns is dependable, stable and long-lived. Nominal output power is 8 watts. If, for any reason, these lasers should fail to have an output power of at least 5 watts within one year after the date of purchase, Melles Griot will refurbish the laser free of charge.

Three versions of this laser are currently available. The first has a randomly polarized output with a guaranteed minimum output power of 8 watts. The second has a linearly polarized output power of at least 7 watts (polarization is vertical to the base of the unit). The third unit, also with 7 watts polarized output, has a grating-controlled single-frequency output tuned by the factory. Beam diameter at the $1/e^2$ points for all models is 1.6mm; beam divergence is less than 8.4 milliradians. Plasma excitation is by direct current discharge. The internal operating temperature is stabilized to within a degree by a thermistor-controlled variable-speed fan. Operation is virtually silent. The solid-state electronics are completely contained in the black aluminum housing for portability. Overall length is only 40cm; weight is only 7.0 kilograms.

These lasers do not require an external gas supply or water supply. Both the laser and its power supply are contained in a single, very small, completely self-contained housing. No radio frequency discharge is emitted. No adjustments are required during startup or operation. No alignment is required to assure the user of continuous, stable output power.

These lasers are designed and manufactured by Melles Griot, Laser Products in San Marcos, California. They incorporate novel design features to achieve their reliable operation. Like the helium neon lasers which preceded them, they have been designed and built with the Original Equipment Manufacturer in mind. The OEM demands performance: long-life, reliable operation, stable output — his very reputation depends upon these features. Melles Griot's success in the helium neon laser business is the result of this dedication to quality. We believe these CO_2 lasers are deserving of the same reputation.

MEDICAL AND LABORATORY CO_2 LASER APPLICATIONS

Medical applications for carbon dioxide lasers are filled with promise and potential. Historically, CO_2 lasers have been difficult to use in a medical or surgical environment. The flowing gas systems required bulky, heavy, and expensive laser equipment;

counter-balanced and articulated-arm positining systems were required to guide and aim the laser beam; the clinical advantages gained by using 10.6 microns radiation were heavily outweighed by practical considerations.

Carbon dioxide surgical laser applications have included various orthopedic, arthroscopic, cardiovascular, gynecologic, otologic, otolaryngologic, and neurosurgical techniques. The CO_2 laser provides very precise control of tissue removal and brings to the surgeon a feel of command over the surgical procedure. Coronary artery clean-out techniques using carbon dioxide lasers have been explored with some success. Argon lasers (also used in a number of medical applications) have an output wavelength which is absorbed by the arterial walls surrounding the blockage in the artery; this absorption results in damage to the arterial wall. The output wavelength from carbon dioxide lasers is not absorbed by the surrounding tissue, and energy may be directly concentrated on the blockage itself.

With these new Melles Griot self-contained CO_2 lasers medical and surgical applications can be immediately made more practical. The small, portable size means that a complete laser system can be carried, easily, from room to room. Delivery of the energy beam to the specific clinical area of interest is convenient, inexpensive and reliable. The small, portable size reduces sterilization and shrouding problems considerably in an operating room environment. Gas sterilization methods may be used easily and conveniently.

Laser machining with carbon dioxide lasers is now recognized for performing a variety of useful and critical procedures. In microelectronic applications, ceramics can be scribed and drilled, resistors may be trimmed, and micro-solder joining may be performed. Paper, textiles, natural and synthetic fibers may be cut and drilled. Plastics, metals and glass may be micromachined with reliability and precision. The far-infrared wavelength is such that most materials totally absorb the CO_2 radiation resulting in an efficient and rapid energy transfer.

Materials which have been successfully machined by CO_2 lasers include: ABS, acrylic, Kapton, nylon, mylar, polyvinyl chloride, silicon, styrene, and vinyl. These materials respond to laser machining by melting or instantaneous vaporization. The output beam can be focused to direct the energy into highly localized areas. CO_2 lasers have been used to drill consistent diameter, burr-free holes from a few microns in diameter to many millimeters. Synthetic and natural organic materials used for wire insulation can be easily stripped from their metallic conductor without nicking or damaging the conductor itself. With higher power devices, even thin gauge metals may be cut, machined and/or drilled.

The grating controlled single-frequency CO_2 lasers will be found useful for communication, optical-radar, Doppler rangefinding and spectroscopic applications.

All these new Melles Griot CO_2 lasers are well suited for experimentation and product development. The small size, light weight, and self-contained features lend them to prototype testing and evaluation for applications which may ultimately require higher power, more complex production-level lasers.

OPERATION OF CARBON DIOXIDE LASERS

Carbon dioxide is a linear molecule. The three atoms of the molecule lie in a straight line with the carbon atom in the middle. Three modes of vibration can occur with a molecule of this geometry. In the first mode the central carbon atom remains motionless while the outer two oxygen atoms vibrate in an axial direction about their equilibrium positions. In the second vibrational mode all the atoms oscillate in the axial direction of the molecule — the outer two oxygen atoms moving in one direction while the central carbon atom moves in the opposite direction. The third vibrational mode allows the central carbon atom to move perpendicular to the molecular axis, while the outer oxygen atoms move in the opposite perpendicular direction. In each case the motion is a simple harmonic motion with all the atoms moving in phase with each other and oscillating about their own equilibrium positions.

Each vibrational mode has an energy level associated with it and, according to the rules of quantum mechanics, each energy level is quantized. When there are no vibrational motions, the molecule is said to be in the ground state. The excited states correspond to one or more quanta of vibrational energy and, since the carbon dioxide molecule can vibrate in more than one state at any one time, a number of vibrational quanta are possible in each mode.

While laser output at 10.6 microns is possible with pure CO_2, the power output is low. All CO_2 lasers today use a proprietary mixture of carbon dioxide, nitrogen, helium, and other gases. The highest energy level of carbon dioxide is only slightly above the first excited vibrational energy level of nitrogen, which can itself be excited by collisions with electrons having approximately two volts of energy in the electrical discharge. The nitrogen's energy is transfered to the carbon dioxide by collision and results in a population inversion between two primary states for the CO_2 molecule. Because the two transitions share a common upper state, the higher energy level (10.6 microns) tends to predominate over the other (9.6 microns) level. Therefore, CO_2 lasers tend to lase at 10.6 microns unless some other wavelength selecting mechanism, such as a grating, is employed. The addition of helium to the gas mixture tends to increase the laser's power output. The helium helps to deplete the lower population level collsions and tends to keep the gas mixture cool because of the high mobility of the helium atoms. Allowing the temperature of the cavity walls and the gas mixture to increase tends to decrease the laser's power output.

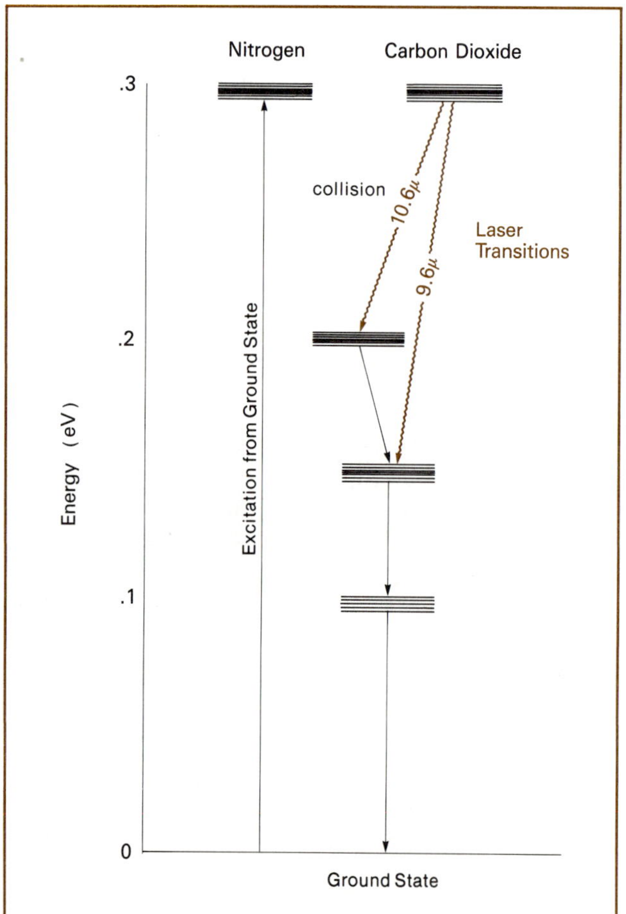

ENERGY LEVEL DIAGRAMS for nitrogen and carbon dioxide.

The CO_2 laser is notable for its high efficiency. Most other lasers have a low efficiency of conversion of electrical energy into optical energy. Efficiencies of a few tenths of a percent to a few percent are typical. CO_2 lasers, such as the new Melles Griot design, are capable of efficiencies of up to 20%. Part of this efficiency is the result of the waveguide design employed in the laser.

WAVEGUIDES FOR EFFICIENCY

Melles Griot CO_2 lasers utilize a small bore diameter which acts as a waveguide to contain the lasing volume and guide the laser radiation. The bore behaves much like an optical fiber to contain the plasma within the lasing volume. The light propagation in this hollow structure with a dielectric wall is well-controlled and has well-defined, low-loss modes of propagation. Thus, a power output of 8 watts and more is possible in a package 40cm long. Conventional, flowing gas, CO_2 lases confine the lasing medium within a fairly large diameter tube — typically 1 centimeter or larger in diameter. To cause the carbon dioxide contained within the bore to lase, the two mirrors on the cavity ends must be aligned to one another (and the large bore) very precisely. Waveguide lasers require less critical alignment because the cavity mirrors need to be aligned only to the local orientation of the waveguide itself. The end mirrors need only return the radiation into the waveguide to assure continued, stable lasing operation.

Several benefits accrue from this approach. The built-in stability of the waveguide design and the intimate association of the cavity mirrors assure both mechanical and thermal stability. System efficiency is greatly improved because the plasma is contained by the waveguide diffraction losses are minimized and high-gain devices with short overall length may be manufactured. Mode stability is greatly improved because the waveguide-style laser has a markedly lower Fresnel number than conventional laser systems. This results in better discrimination against transverse mode excitation and results in a stable EH_{11} output mode with less than 3 percent higher-order modes being present. This is essentially equivalent to a TEM_{00} Gaussian mode after the beam leaves the waveguide.

Early CO_2 lasers (and many higher powered units still today) utilized a flowing gas system. The gas flows through the laser and is exhausted. In a sealed-off laser the interaction between the electrical discharge and the gasses causes some decomposition and resulting harmful by-products. In a sealed system some method must be provided to remove these detrimental decomposed gasses. The Melles Griot CO_2 laser utilizes a proprietary "getter" within the system to trap the majority of these poisonous by-products and remove them from the operating gas supply. In time, however, the by-products will build up and lasing output power will decrease.

To overcome this problem Melles Griot has utilized a novel seal design which allows the CO_2 laser to be easily and quickly emptied, purged and re-filled with a fresh supply of proprietary gas mixture. Because of the simplicity of the design and the great portability of the unit, this re-filling operation is both inexpensive and rapid. Return of the fully refurbished and fully warranted unit is assured within two weeks of receipt by Melles Griot.

Beneath the plasma tube, there is a stainless steel gas reservoir. The reservoir acts as a ballasting supply of the gas mixture during operation thereby extending the useful lifetime of the laser. Furthermore, it provides a strong, stable and rigid internal support structure to the plasma tube itself. The integrity of this design aids the thermal and mechanical stability of the entire laser system.

NOVEL SEALS FOR ASSURED OPERATION

The Melles Griot CO_2 laser utilizes proprietory seals at both end-mirrors. These seals are flexible and yet non-permeable by the gas mixture. They form tight, reliable, long-lived seals at the cavity ends. Because of their flexible nature they will conform to the factory alignment of the end-mirrors and retain this position. Yet because the end-mirrors are not permanently epoxied or frit-sealed to the cavity, they may be easily removed, replaced and the system realigned for continued operation.

So confident of these seals is Melles Griot that these CO_2 lasers are guaranteed to have an output of at least 5 watts continuous power for 12 months after the date of purchase. There is no limit to operational hours during this time. The only exception to this operational warranty is obvious misuse, damage or breakage of the laser tube or housing. Should the output power of these lasers fall below this level within the time limit, Melles Griot will refurbish the laser for you free of charge.

Following the warranty period Melles Griot CO_2 lasers may be re-filled and refurbished quickly for a minimal charge. In either event please telephone your nearest Melles Griot office to obtain a Return Authorization Number prior to shipping the unit. This will help greatly in assuring you a rapid response.

DESIGN FEATURES FOR RELIABILITY

The capillary discharge bore in the Melles Griot CO_2 laser is made from high-purity ceramic. This material has a very low coefficient of thermal expansion. The capillary also acts as the waveguide to contain the lasing medium, thereby increasing stabilization and reducing overall sytem length. The capillary material has a high thermal conductivity and allows the laser to be air cooled. A combination of internal heat sink, cooling fins and a small, silent fan promptly stabilizes the operating temperature.

Warm-up to operating temperature is rapid. Fully stabilized operation is reached within 15 minutes. During the initial start-up time the output will oscillate through a number of power levels (as is typical for all CO_2 lasers) as the mechanical size of the cavity grows before reaching thermal equilibrium. Once operating temperature is reached, the variable-speed internally mounted fan will operate at whatever speed is required to keep the internally mounted thermistor, and hence the plasma tube, at a constant operating temperature.

Because of the various novel design features and the evenly distributed and symmetrical internal thermal mass, this cooling technique results in stable and trouble-free operation for extended or even continuous duty cycles.

Operational controls are simple and straightforward. The left-hand side panel contains a key-operated ON/OFF switch, and indicator light to show the laser is turned on and operational, and a current-limiting control which allows the user to decrease the output power from its nominal operating level. Reduction of the operating current will, in general, increase the operational life of the laser. At the front end there is a push-to-close shutter. When the shutter is open the control button acts as a visual indicator.

POLARIZATION

Both randomly polarized an linearly polarized versions of these CO_2 lasers are available. The randomly polarized version is preferable for maximum raw power output for applications such as cutting, soldering, and surgical applications. The linearly polarized model has its plane of polarization oriented vertically (perpendicular to the mounting plane of the laser). Output is through a ZnSe Brewster angle window. The window selects only one plane of polarization for amplification within the laser. The resulting output polarization ratio is 100 to 1, thereby effectively eliminating polarization noise within the system. The plane of polarization can be confirmed by observing the power level with an infrared power meter with the beam reflected (at Brewster's angle) from a dielectric surface such as a mirror or polished glass substrate.

The randomly polarized output beams contain both orthogonal output planes of polarization, but the ratio and phases of the two components vary rapidly with time. The output therefore appears to be non-polarized when integrated over a fairly short period of time. Even so, care should be exercised with these lasers in situations where polarization effects may occur further down the optical train.

POWER SUPPLY

The power supply for these lasers is completely built-in to the housing. The supplied three-wire, grounded power cord is detachable for convenience. Operation from either 115VAC or 230VAC is key-switch controlled on the side panel of the laser. Line frequencies of either 50Hz or 60Hz are acceptable. Power consumption for any of the three models is about 70 watts. The power supply itself is a true state-of-the-art design. The output is feedback regulated and frequency switching circuits provide a stable and reliable output. The power supply module is internally mounted on a separate printed circuit board adjacent to the plasma tube.

A side panel ON/OFF key switch controls the laser. A green indicator lamp on this same panel glows when the laser is on. The power supply (as well as the plasma tube) is covered by our 12-month "no-nonsense" warranty.

MOUNTING CONVENIENCE

These lasers have been designed to be convenient to use in the laboratory and workshop. The all-aluminum case is finished with a durable, baked black powder coating.

The front end of these lasers has a one-inch diameter 32 TPI threaded hole centered on the laser output beam. This may be used to mount the various far-infrared laser accessories which will become available from Melles Griot in the near future. Accessories manufactured by other companies may also be mounted into this standard interface. The front end has four 4-40 tapped holes in a square pattern for attaching special devices.

The center of the laser beam is 70mm above the mounting surface making these lasers easily interchangeable with earlier CO_2 laser designs from other manufacturers.

SAFETY DURING OPERATION

CO_2 laser radiation is potentially dangerous to people and animals. The output radiation is invisible. The intensity of either direct or scattered radiation is sufficient to burn skin and severely damage eyes. Wood, plastics, paper, particle board and similar materials can be ignited by the output beam. The United States Bureau of Radioligcal Health rates all these lasers as Class IV Laser Products under 21-CFR-1040. Please use these lasers with care and foresight to avoid injury to yourself or others. Melles Griot Applications Engineers are always available to assist or advise you on safety procedures to assure hazard free operation.

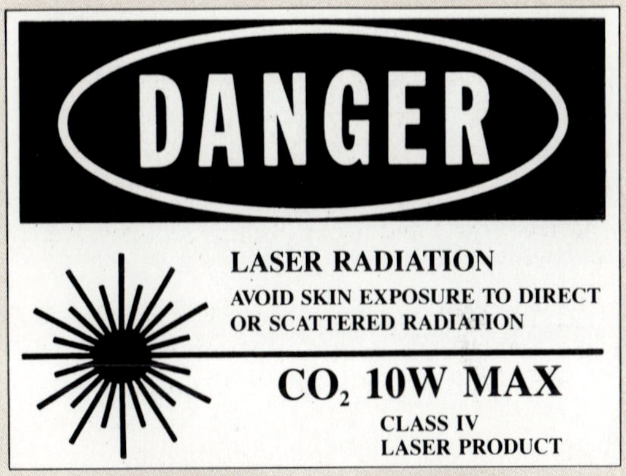

05 COR 080
05 COP 080
05 COP 081

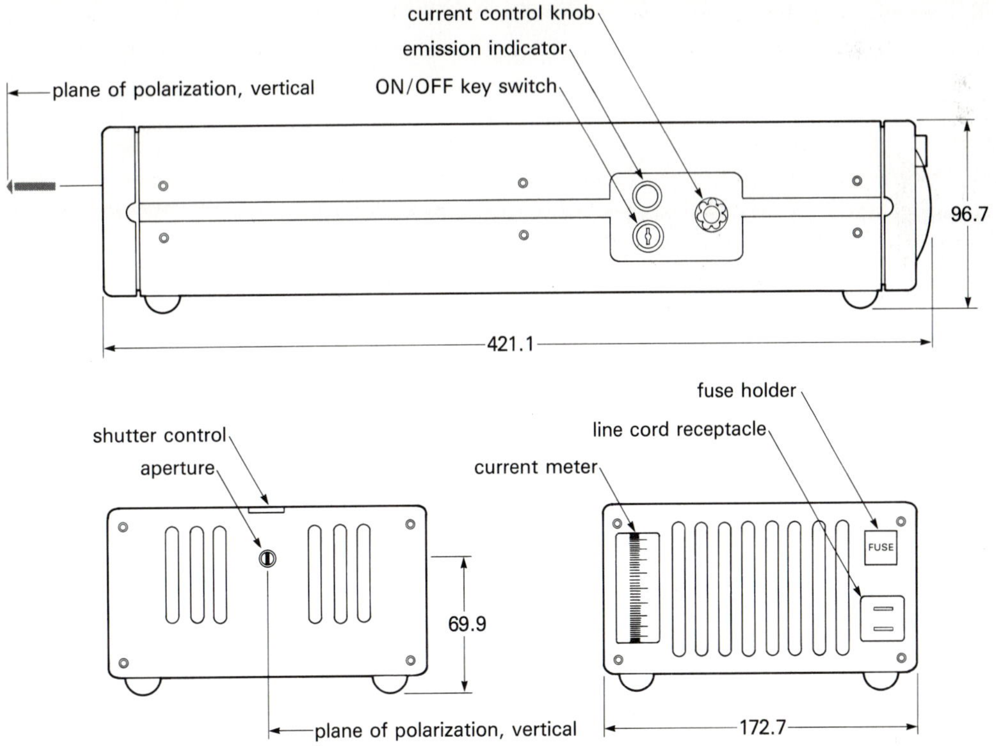

Carbon Dioxide Lasers

SPECIFICATIONS	PRODUCT NUMBERS		
	05 COR 080	05 COP 080	05 COP 081*
Wavelength of operation: (microns)	10.6	10.6	9.6 to 10.6*
Minimum Initial CW Power Output EH_{11}: (watts)	8	7	6 to 7
Operating Wavelength: (microns)	10.6	10.6	9.4 to 10.8
Higher Order Mode Population	>5%	>5%	>1%
Pressure Broadened Linewidth: (MHz)	500	500	100
Beam Diameter $1/e^2$: (mm)	1.6	1.6	1.6
Beam Divergence: (mrad, full)	8.4	8.4	8.4
Polarization:	Random	100:1	100:1
Warm-up Time: (min)	15	15	15
Amplitude Stability after Warm-up			
Short Term: (minutes)	±2%	±2%	±2%
Long Term: (hours)	±5%	±5%	±5%
Input Power Consumption at 110/220VAC, 50/60Hz: (watts)	70	70	70
Weight: (kg)	7.4	7.4	7.4

*When ordering this grating controlled, single frequency model, please specify the exact wavelength in the the range 9.6 to 10.6 microns.

Common To All Product Numbers

Static Alignment:	Beam centered within 0.20mm and 2.0mrad
Angular Drift:	<0.05 mrad after 15 minutes
Power Cord:	3-wire, grounded, 2.8 meter
Dimensions:	±0.5mm

Environmental Specifications	Operating	Non-Operating
Temperature:	-20°C to +50°C	-40°C to +80°C
Altitude:	0 to 3000 meters	0 to 5800 meters
Relative Humidity:	0 to 100%	0 to 100%
Shock:	12g for 11msec	12g for 11msec

OPTICAL TABLES AND RAILS

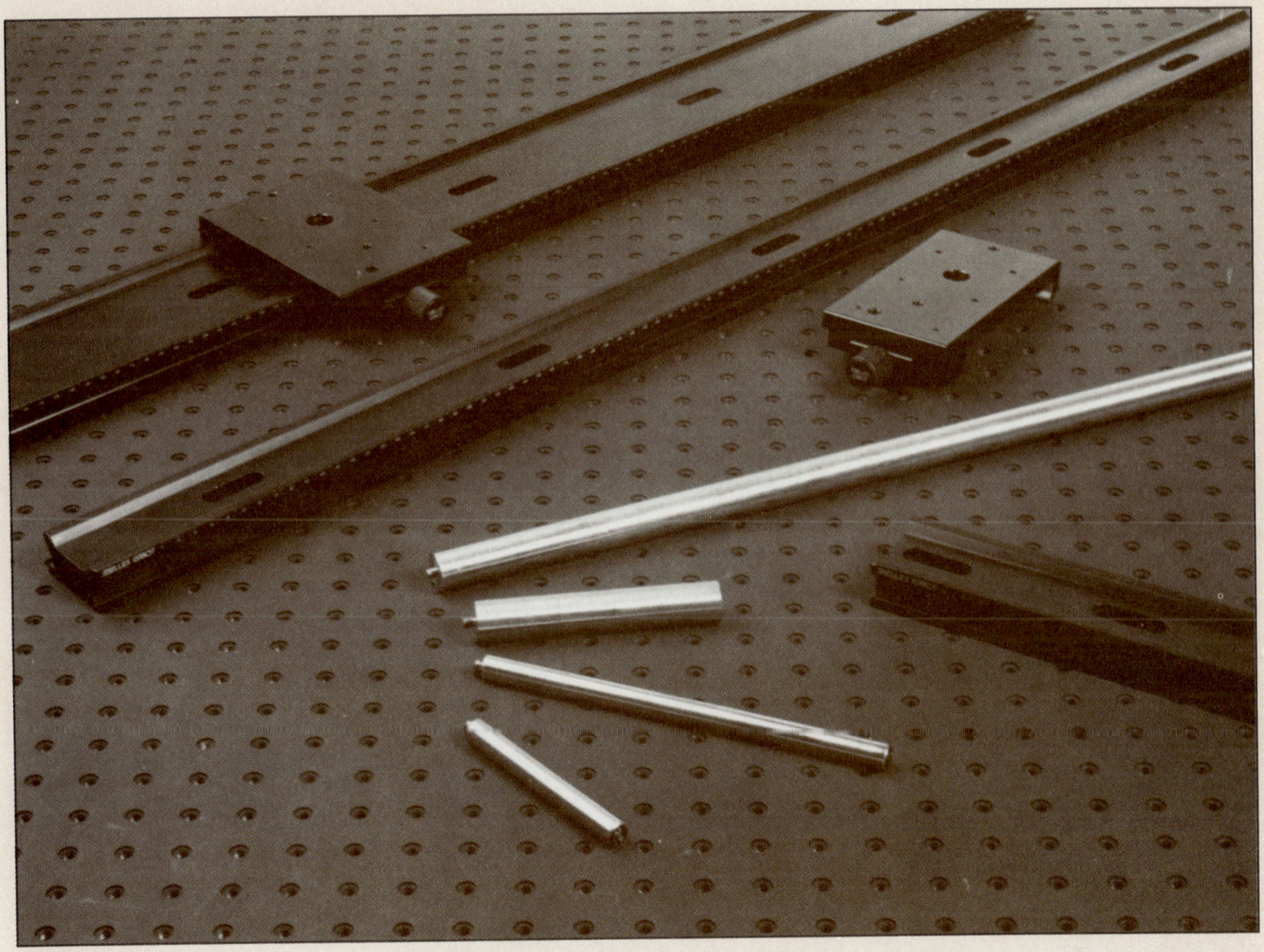

Melles Griot is proud to announce a completely new range of optical tables, rails and associated accessories. This product line represents a major advance in optical table technology. Our tables are demonstrably superior in stiffness and damping to those in common use today. Furthermore, Melles Griot is the only company which offers you a stock choice between the two most commonly used mounting hole patterns.

Contents **Page Number**

OPTICAL TABLES

Optical tables provide the rigid, stable base required for optical and laser system development. To be effective, an optical table must provide a firm, reliable, vibration-free environment isolated from external disturbances. When components are added or moved on the surface the table structure must be adequately rigid to resist bending or flexure; the structure must provide internal damping to dissipate vibrations generated on the table surface itself; isolation must be provided to separate the table from external vibrations or disturbances. Failure to achieve these requirements results in inaccuracies and misleading results during system development.

The range of optical tables described here is completely new. Their internal design is novel and is the result of considerable research by Melles Griot. Extensive industry experience with optical systems and laser systems has provided Melles Griot with the insight and know-how required to develop this truly superior series of tables. Not only are these tables more stable, more rigid and better isolated than any other units presently on the market, but they also incorporate conveniences and features which make them functionally superior and easier to use.

Furthermore, the design provides a strong and simple interface for joining tables together to suit specific applications and needs. After joining, the resultant structure is essentially as stable and rigid as it would have been had it been made as a single unit.

WHAT IS IMPORTANT IN AN OPTICAL TABLE?

In a word, the single most important attribute in an optical table is stability. Once positioned, it is critically important that an optical beam does not move during the remainder of the system development. Minor disturbances of an optical setup can have major detrimental effects on the overall operation of the prototype system. Focusing and alignment sensitivities of less than a micron are not unusual in many optical imaging systems. Interferometric systems depend on constant and unchanging positional stability for clear, readable, and unambiguous interferograms. When doing holography the relative positions of components must not change during an exposure or the encoded data in the hologram will be destroyed. High resolution spectrography demands constant positional stability for reliable results. Studies of high-speed phenomena require great positional stability to prevent losses of system

resolution. In most optical applications, a failure to assure a stable and reliable support to the entire system will result in a system unable to live up to its design goals.

Stability of an optical table involves a number of factors:

1. Vibrational resonances which amplify induced vibrations must be absent from the table design.
2. Broad-spectrum vibrational damping must be incorporated into the table design to rapidly dissipate any oscillations introduced by components mounted on the table surface or transmitted to the table from external acoustical sources.
3. The table must have great built-in static rigidity in order to resist surface flexure under varying loads and load positions.
4. Effective isolation must be provided from ground and building vibration disturbances.

The first three factors are controllable in the design of the table and its inherent structure; the fourth factor must be controlled by a separate table vibration isolation system. Using modern structural computer-aided design techniques, Melles Griot has created a table and isolation system which exceeds all previous design attempts; its stability performance is unsurpassed.

DESIGNING A BETTER OPTICAL TABLE

A variety of materials have been used in optical tables in the past. Cast iron, cast aluminum, concrete and granite have all been tried. The assumption in these designs was that a massive weight would provide large inertial resistance to external forces. Unfortunately all have basic and unavoidable flaws. Cast iron and cast aluminum have sharply apparent resonances in the audio range; when struck they "ring" like a bell. Further, their large masses create a long-term thermal stabilization problem resulting in constant shape changes during equalization. Concrete tables undergo very long-term internal chemical changes resulting in millimeter level motions for many decades. Granite (like concrete) absorbs water readily and warps unevenly, and sometimes permanently. These monolithic structures relied entirely on their internal composition for damping (which was frequently non-existent) and, of course, they were unwieldy because of their great weight. Honeycomb paper, aluminum and/or steel structures are all more practical and have gained wide acceptance during the past decade. Top and bottom skins of aluminum or steel are bonded (usually with epoxy) to the honeycomb under pressure. The result is a firm, rigid structure of relatively light weight. The weakness in all of these structures to date is their lack of edge constraint. The internal honeycomb structure provides great resistance to compression but far less bending resistance. In many honeycomb-technology applications this bending freedom is advantageous; airplane wings and skis make considerable use of this effect. Bendable optical tables are not, however, advantageous.

When composite honeycomb tables are assembled with epoxy under pressure, internal strains are built into the system. Because of the brittleness of this form of bonding and lack of resistance to tensional forces, separation of the skins from the honeycomb frequently occurs in localized areas after several months. Evidence of this problem is rarely obvious until large, unbalanced structures are mounted on the table surface. Furthermore, the relatively light weight of honeycomb structures denies the inertial advantages of massive structures. When conventional optical tables are manufactured, the tops are drilled and tapped after assembly. Both the drilling and tapping processes require lubrication to prolong tool life. Oil or water-emulsified waxes are used to provide this lubrication and, in the process, flow into the inside of the table. It is impossible to drain the residual material completely from the table core and capillary action will cause the oil to "seep" from the table-top throughout its life — thereby contaminating the optical set-up on the surface. Some manufacturers have attempted to "dry-cut" the top threads to avoid this problem. The resulting thread-form produced by quickly dulled tools is weak and spalled from cutter tear. This results in easily stripped threads and the permanent loss of a mounting hole. Inevitably, of course, these lost holes occur in the area of the table-top most used.

Clearly, what was needed was a better design.

Melles Griot optical tables are fabricated from multi-layer steel top and bottom skins with aluminum honeycomb core sections. High damping capacity visco-elastic compounds bond the components together and provide considerable and wide spread vibration absorption. The skin laminations provide enhanced rigidity without increased weight. System resonant frequencies are higher and the damping characteristics are improved. The composite lamination structure is permanently bonded into a rigid welded steel framework which completely encloses the table core. This framework provides immense structural integrity and continuity to the table edges and completely eliminates the bending function inherent in the honeycomb composite. This novel combination of a structural steel frame with laminated skins on a honeycomb composite results in a table design with greatly improved static and dynamic rigidity at a minimum weight.

The top and bottom skins are ferro-magnetic steel. The top skin is a two-layer sandwich bonded with a visco-elastic polymer. This bonding agent has considerable tensional strength and table skins do not separate from the honeycomb core even when subjected to high angular torque loads. Total top skin thickness is 8mm. This thick, laminated top skin is partly responsible for the table's extremely high damping factor which quickly attenuates oscillations introduced by components mounted on the table surface. Internal visco-elastic polymeric layers separate the cores into two or three further layers (depending on table's total thickness). These internal layers also contribute to the table's low-Q damping characteristics. The combination of thick skins and a multilayer core provides a more rigid structure capable of carrying very heavy loads without detrimental deflection. Using more conventional technology, competitive tables would need to be some 28% thicker to exhibit equivalent rigidity.

IMPROVED USER CONVENIENCE

A number of valuable user conveniences are designed into every Melles Griot optical table.

- Hardened steel inserts for long thread life.
- Matte black top finish for reduced glare.
- T-slots around the edge for convenience.
- Laser beam pass-through access hole at one end.
- Optional bottom mounted shelves for lasers and accessories.
- Both inch and metric versions available from stock.
- Special OEM configurations easily achieved.

Melles Griot optical tables are available with either 1/4-20 threaded holes on one-inch centers or with M6 threaded holes on 25mm centers. With either version hardened tool-steel inserts are used to provide the threads. The user gains a number of very real advantages with these threaded inserts. Neither cutting oil nor water-based wax lubricant is ever used on the table-top, so oil seepage is never a problem. Because the threaded portion of the insert is quite deep (12.5mm) and because the quality of the thread

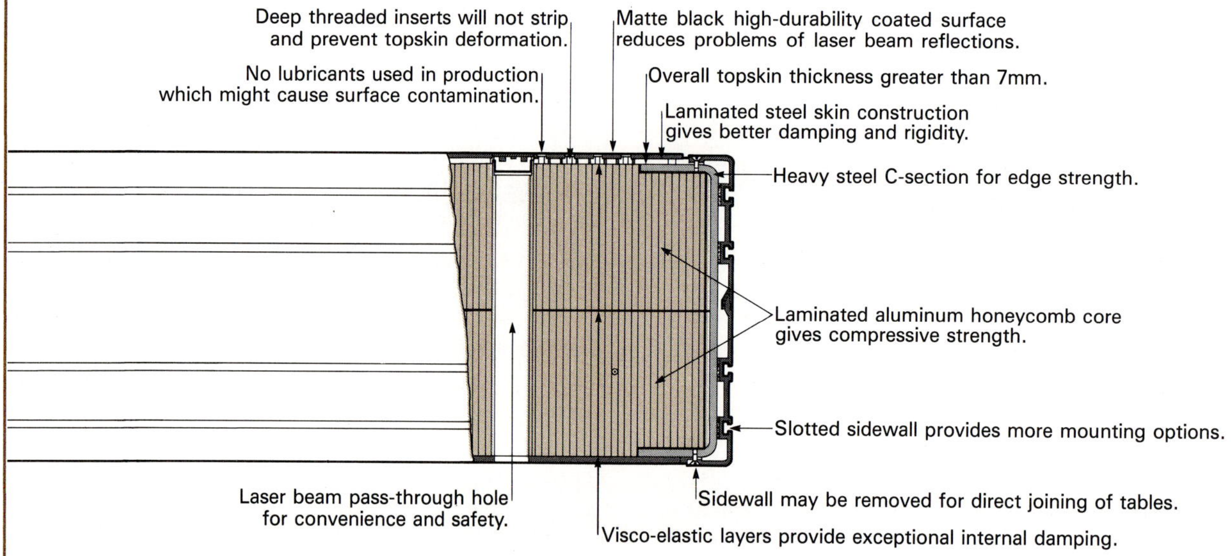

CROSS SECTION OF OPTICAL TABLE showing the many superior design features.

form is better than provided by a conventional tap, the insert resists both cross-threading and stripped threads better than a conventional table top. The base of the insert is quite broad (nearly 15mm) and spreads the pull-down load over nearly 250% greater surface area than a conventional tapped hole. This, combined with the markedly thicker top-skin prevents the all-too-common local "dimpling" of the top caused by overightening mounting hardware on a conventional optical table.

Table tops are available with a matte black epoxy finish. This black finish helps to reduce stray or scattered light reflected from the table surface, thereby improving both performance and safety. The black coating is extremely hard and resists scratching, abrasion and most laboratory liquids. The coating is thick enough that it has negligible effect on surface flatness and only a minor reduction of the magnetic holding power of the surface. Conventional textured stainless steel top skins are also available.

An extruded aluminum protective cover surrounds the welded steel frame of the table edge. It is attached to the table sides making contact through a polyurethane high-density foam which acts as a further damping compound to reduce table edge vibrations and higher order resonances. This cover is provided with a series of useful T-slots. Depending on table thickness, either 4 or 5 parallel T-slots are provided along both sides and both ends. Each slot will capture either 1/4-20 or M6 hexagonal bolt heads and prevent them from rotating while being tightened. A variety of useful accessories (see page 431) will mount to these T-slots. Alternately, they will be found convenient for mounting power outlets, switch boxes, cable guides, keepers for gas or liquid lines, and the myriad other devices normally used close to optical tables.

A laser access hole is provided, at 150mm from one end, in every Melles Griot optical table. The 32mm clear-diameter hole allows a laser beam to be passed through from the bottom of the optical table without having to resort to outboard mirror holders. Laser shelves (see page 422) may be conveniently mounted below the surface of the optical table to carry lasers, power supplies and other space-consuming equipment. When not in use, the laser pass-through hole may be covered with the threaded access cap provided, using the tool provided.

SPECIAL SHAPES AND SIZES

A large variety of standard shapes and sizes is offered to fill the majority of requirements. Special applications, however, frequently do arise and Melles Griot optical tables can be adapted easily to meet these situations. Special shapes and configurations may be most easily achieved by joining different sizes of table end-to-end, side-to-side, or side-to-end. Because of the novel welded steel framework incorporated into the basic table design, size and end joining techniques are both simplified and made more reliable than with more conventional optical tables. When the extruded aluminum sidewall is removed, the structural steel frame is exposed. By having this frame specially slotted, joining hardware may be easily attached which firmly and securely joins two or more table sections together. These adaptations can be inexpensively carried out by Melles Griot prior to shipping or, if your needs should require, we can supply you with adequate information to allow your own in-house shop to make the adaptations at a later date. Interconnected or modified vibration systems are also easily adapted to modified table systems.

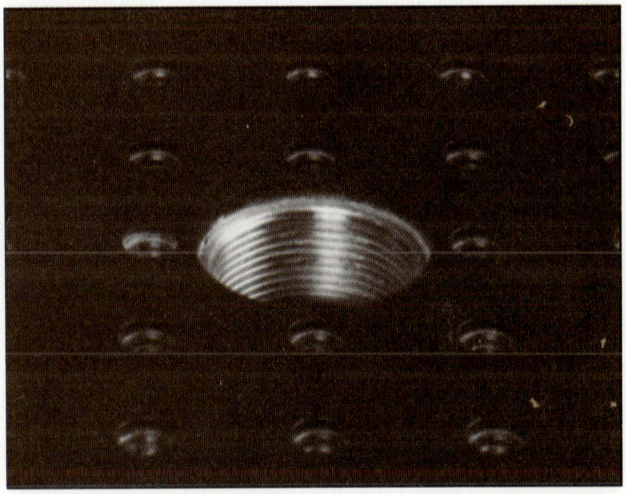

Tables in non-standard thicknesses from 50mm to 900mm are also available on special order.

TESTING PROVES PERFORMANCE

The value of the motions optical tables are designed to eliminate are so small and so transitory that it is difficult for a user to evaluate these products before, or even after, purchase. Furthermore the testing procedures used to describe the performance of the table system are always open to interpretation. One manufacturer will point to a series of attributes regarding his table design and use them to justify his claims of superiority; the next manufacturer will choose another set of parameters to refute those claims.

The problem is, of course, that there is not one set or group of specifications which will separately define the performance of an optical table. Nor is there one set of tests. (Although it might be noted here that there is one set of quite unscientific tests the customer may perform easily before buying an optical table. Sharply rap the top of the table with your knuckles or a nylon face hammer at several locations. Listen carefully. A truly "dead" sound is indicative of a well-damped system and shows that the manufacturer is at least trying to produce a satisfactory product. Melles Griot optical tables will surprise you by sounding extremely "dead.")

Very complex iterative design and testing techniques are required to develop an optical table. Finite element modeling techniques, as developed by the aerospace industry, are used to design a locally optimized structure. Prototypes are then built, tested and analyzed to determine whether that particular local optimum was in the appropriate direction and the results are fed back into the finite element model for further refinement. On-line, mainframe interfacing with the software systems used by NASA, Boeing, McDonnell-Douglas and other airframe designers provide Melles Griot engineers with true "state-of-the-art" design methodology. The prototyping process provides an opportunity to evaluate damping methods which do not lend themselves to accurate prediction by modeling.

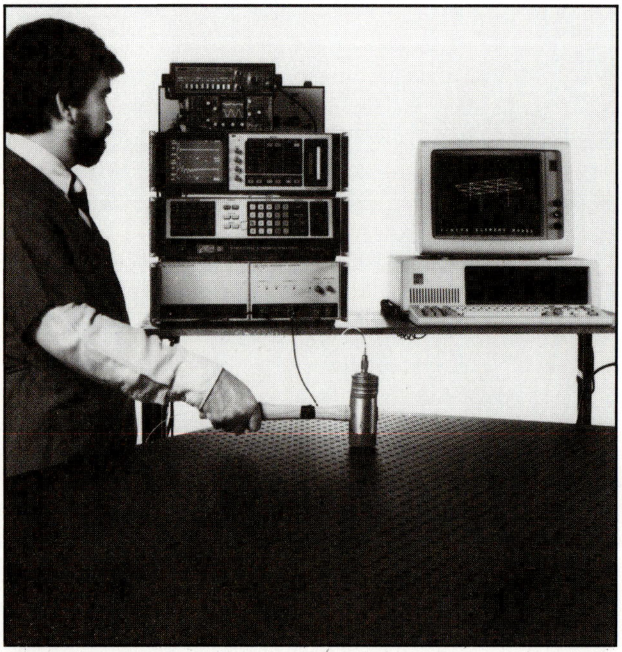

To illustrate the testing and analysis process that needs to be undertaken to evaluate the performance of various table designs, we compare data taken from actual tests of an alternate optical table and a Melles Griot optical table. The alternate table is lacking careful consideration of the static rigidity properties, in some areas.

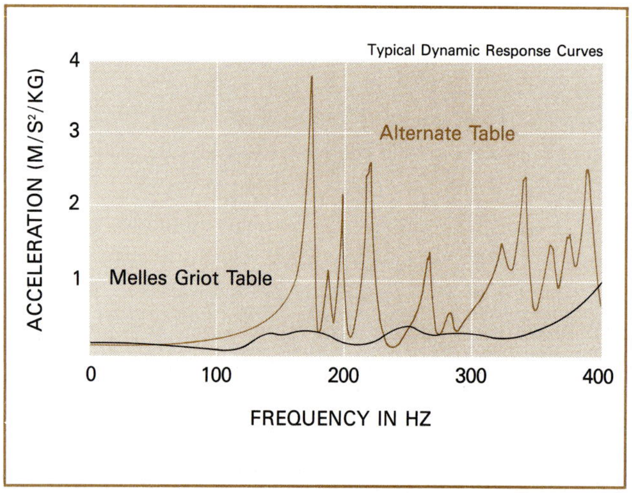

DYNAMIC CHARACTERISTICS of a Melles Griot table and an alternate table.

The accompanying figure illustrates the dynamic characteristics of a typical undamped optical table and the new generation Melles Griot optical table. The horizontal axis of the graph represents the frequency, in Hertz, of the vibration oscillations of the table. The vertical axis represents the acceleration, meters/sec/sec, of vibration due to a one kilogram force applied to one corner of the table. What is clear from the graph is that the alternate table is highly sensitive to an externally applied force when compared with the Melles Griot table.

To appreciate the information provided in the graph, let us assume that an object, such as a lens holder, weighing one kilogram was dropped from a height of 38mm onto the corner of the table. When the object hit the table a 141 kilogram force impulse of 2.5 milliseconds duration was imparted to the table. The table then responded to this force by vibrating at its natural resonant frequencies. We are not concerned with the displacement of the table isolator at this time. But we are concerned with the actual motion of the table at its flexural resonant frequencies. To determine the deflection we double integrate this data. The results are the associated compliance (flexibility) curves. We have truncated the com-

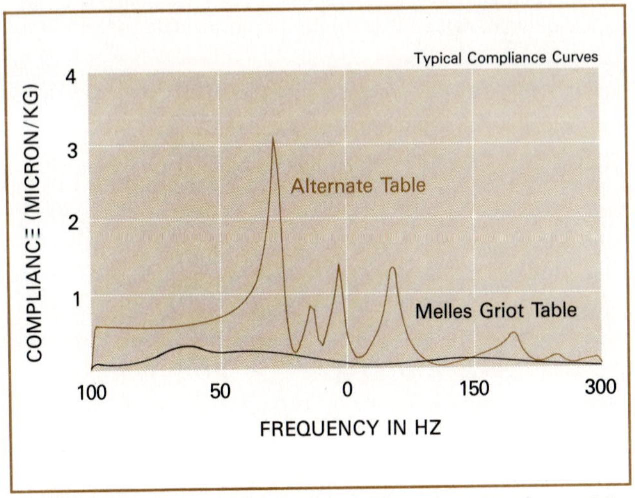

COMPLIANCE of a Melles Griot table and an alternate table.

pliance curves to lie within the range of 100Hz to 300Hz where the dominant flexural resonances for both tables occur. Above 300Hz both tables exhibit negligible compliance amplitude. Below 100Hz the table isolator system will eliminate most transient displacements in the present comparison.

The compliance functions shown in the figure illustrate the sensitivity of each table to forces applied at the corners of the table. Similar results would be demonstrated for disturbances introduced at other locations on the table. For example, 1.0 kilogram force at the corner of a typical undamped table would produce a deflection of 3.2 microns at the same corner at a frequency of 172Hz. The same magnitude force would produce only a 0.35 micron deflection of a Melles Griot table at its first resonance of frequency of 145Hz. These results, taken from actual data, illustrate the superior performance of a Melles Griot optical table.

The following impulse response curves compare the time deflection response between the same two tables when excited in this way. Both figures clearly illustrate the low-Q damping characteristics of the Melles Griot optical table when compared with the competitive unit. The undamped table vibration persists nearly ten times longer than the Melles Griot table. At 50 milliseconds after the impulse, the competitive table is vibrating at greater than 20 microns peak-to-peak deflection. The Melles Griot table vibration has decayed to less than 1 micron peak-to-peak deflection. The undamped table took nearly 400 milliseconds for the vibration to decay to an equivalent level.

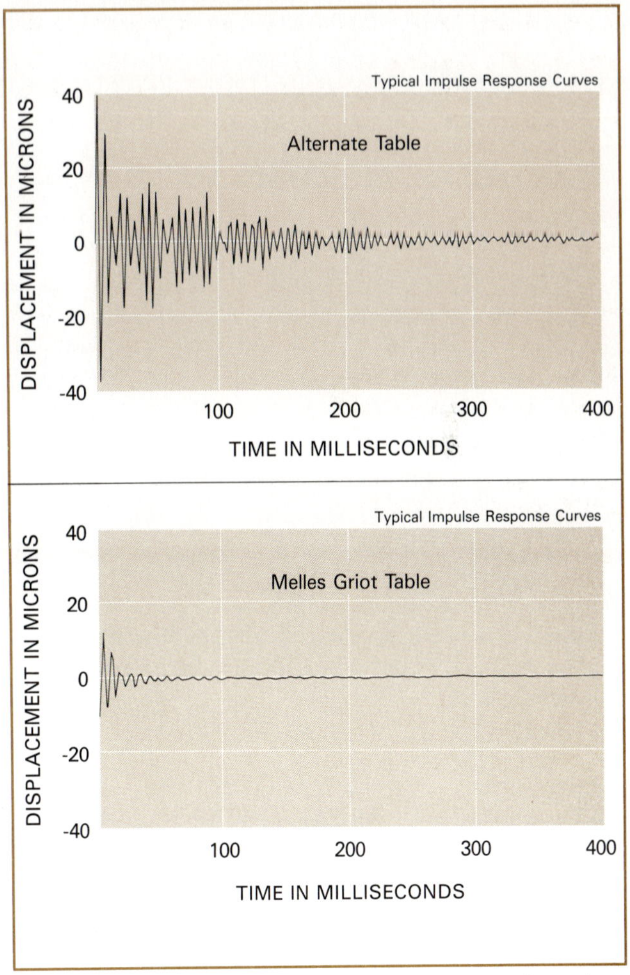

IMPULSE RESPONSE of a Melles Griot table and an alternate table.

The actual forces encountered by an optical table will, of course, vary greatly and will depend on the user's environment. However, with low damping characteristics and high structural compliance, a typical optical table may well be sensitive to acoustical noise and ground or building vibrations which are transmitted through the table's isolator mounting system. By having a compliance characteristic ten times smaller than other typical optical tables the Melles Griot table will, in general, exhibit an order of magnitude less vibration deflection. Furthermore, the duration of any transient noise or vibration condition would be ten times shorter for the Melles Griot table.

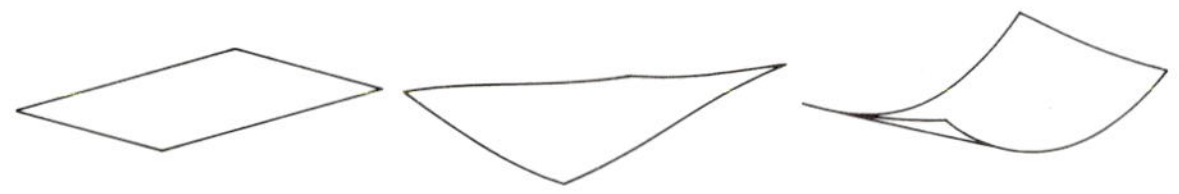

VIBRATIONAL MODES for an alternate table.

This superior performance is not accidental. The figure above illustrating the data-set geometry used, shows the structural resonance deflections of the competitive table. Included in the figure are two of the first five resonant vibration modes found in this table. The modes present illustrate classical structural resonance phenomena. The modes are well defined and are not significantly coupled. Because this table has relatively low damping, the resonant modes result in very sharply defined resonant peaks in the compliance curve.

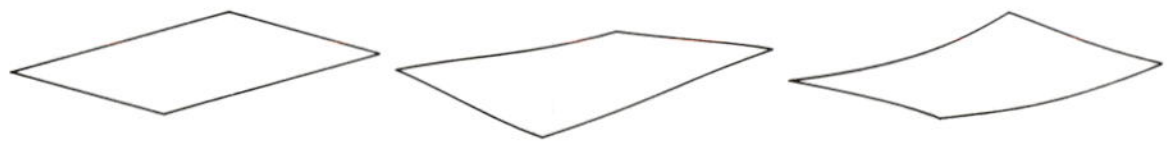

VIBRATIONAL MODES for a Melles Griot table.

The second figure illustrates the data-set geometry used to evaluate the structural resonance deflections of the Melles Griot optical table. Included are two of the first three modes found in the Melles Grtio table. In this case there is very strong coupling between the modes, and the classical resonant vibration modes are not present. The extremely high damping values achieved by the Melles Griot design are responsible for the low resonant peaks inthe compliance function.

It should be remembered that the curve and data shown here are not theoretical; they represent real data, taken in likely situations and represent, as best we understand, likely real-life performance.

ANALYSIS CREATES IMPROVEMENTS

By working with real performance data from a variety of existing and competitive optical tables, Melles Griot engineers were able to compare and optimize mathematical models of real structures. By creating adequate models of existing table designs, it was possible to modify the design within the "true-to-life" model by simple parameter modifications. This type of "parametric study" allowed rapid iteration based on error-free model simulations.

Mathematically modeled static and dynamic loads were applied to the model to determine the effect on changing honeycomb materials, density and thickness; top, bottom and visco-elastic skin thicknesses could be varied; side-beam cross sections and internal structures could be changed or added to; all other design features could be easily maximized. A finite element model was established which included all the critical table components and a final set of design parameters was generated.

The mode diagrams illustrated for twisting and bending were initially set by the model parameters. By manipulation, the first order twisting mode was forced to occur at a higher natural frequency. Fine adjustments then yielded a system that exhibited considerable coupling between the twisting mode and the various bending modes. The final table product, in manufacture, fitted the design parameters well; the high damping coefficient and the strong cross-coupling of the various modes resulted in the desired low amplification compliance function as shown. The final outcome is, we believe, the finest optical table available anywhere today.

We can supply non-standard sizes and configurations on special request.

SPECIFICATIONS: OPTICAL TABLES

Dimensions: ±5mm
Hole positions: ±0.25mm
Flatness; Black finish: ±0.20mm
Stainless Steel: ±0.125mm
Skin Material: Ferromagnetic steel
Top skin composite 8mm (0.310 inch) thick

Core: Multilayer aluminum honeycomb
0.04 g/cm³, 3.2 cm² cell area
Density of Table, 270mm tables: 15 g/cm² (32 lbs/ft²)
390mm tables: 17 g/cm² (37 lbs/ft²)
Mounting Holes; Inch version: 1/4-20 on 25.4mm (1 inch) centers
Metric version: M6 on 25mm centers
Finish, Skins: black epoxy coating or stainless steel
Sides: black epoxy coating

Optical Tables, Black Finish

Thickness (mm)	Width (m)	Width (ft)	Length (m)	Length (ft)	Approx. Weight (kg)	PRODUCT NUMBER Inch Version	PRODUCT NUMBER Metric Version
270	1.2	4	1.8	6	354	07 OTB 001	07 OTB 501
	1.2	4	2.4	8	473	07 OTB 002	07 OTB 502
	1.2	4	3.0	10	591	07 OTB 003	07 OTB 503
	1.2	4	3.7	12	709	07 OTB 004	07 OTB 504
	1.5	5	3.0	10	739	07 OTB 013	07 OTB 513
	1.5	5	3.7	12	886	07 OTB 014	07 OTB 514
390	1.2	4	2.4	8	535	07 OTB 022	07 OTB 522
	1.2	4	3.0	10	668	07 OTB 023	07 OTB 523
	1.2	4	3.7	12	802	07 OTB 024	07 OTB 524
	1.5	5	3.0	10	835	07 OTB 033	07 OTB 533
	1.5	5	3.7	12	1000	07 OTB 034	07 OTB 534

Order isolator separately

Optical Tables, Stainless Finish

Thickness (mm)	Width (m)	Width (ft)	Length (m)	Length (ft)	Approx. Weight (kg)	PRODUCT NUMBER Inch Version	PRODUCT NUMBER Metric Version
270	1.2	4	1.8	6	354	07 OTF 001	07 OTF 501
	1.2	4	2.4	8	473	07 OTF 002	07 OTF 502
	1.2	4	3.0	10	591	07 OTF 003	07 OTF 503
	1.2	4	3.7	12	709	07 OTF 004	07 OTF 504
	1.5	5	3.0	10	739	07 OTF 013	07 OTF 513
	1.5	5	3.7	12	886	07 OTF 014	07 OTF 514
390	1.2	4	2.4	8	535	07 OTF 022	07 OTF 522
	1.2	4	3.0	10	668	07 OTF 023	07 OTF 523
	1.2	4	3.7	12	802	07 OTF 024	07 OTF 524
	1.5	5	3.0	10	835	07 OTF 033	07 OTF 533
	1.5	5	3.7	12	1000	07 OTF 034	07 OTF 534

Order isolator separately

TABLE ISOLATORS AND STANDS

Almost as important as the optical table itself is the isolation system. For a table to perform correctly, it must be properly supported. A good isolator will support an optical table free from all ground disturbances. To do this it must act as a soft spring of very low natural frequency. Air springs have long been regarded as the most efficient form of isolator for optical tables. They have very soft characteristics without the need for very long range of travel.

Melles Griot isolators use a thin walled diaphragm to form an air chamber. Supported by the diaphragm is a piston which is connected to a load bearing plate. The air chamber is connected to a larger air reservoir by means of a small orifice. Any acceleration imparted at the base of the isolator by ground disturbance is transmitted to the air in the piston chamber. Compression of the piston chamber air causes air to flow through the damping orifice into the reservoir chamber. The energy lost by flow through the orifice causes the disturbing vibrations to dampen out quickly and maintain the static stability of the optical table. Sensors on the underside of the table increase or decrease the air supply to the reservoir to maintain a constant height setting of the table irrespective of load. These position control valves also act rapidly in order to stabilize the table quickly when a significant load change takes place during the set-up of an optical system.

Melles Griot isolators are mounted in black, square section, steel tubes. They are supplied as a set of four, complete with all necessary fittings and tools. Three legs are equipped with sensors and the fourth is slaved to one of the adjacent legs to provide an effective three point self-levelling system. To provide added stability, the isolators should be bolted together by means of tie-bars. These tie-bars are supplied in sets of four to suit individual table sizes and serve to position the isolators securely in the correct place before mounting the optical table. Another function of the tie-bars is to provide a safe and tidy location for the necessary interconnecting tubing.

These isolators operate from a compressed air supply or bottled gas. Very little air is used once the system has stabilized at its normal operating height.

Three different load capacities are offered. The light-duty isolator should only be used for low load tasks up to 500 Kg. The normal-duty isolator may be used in most situations and is recommended when the load conditions are uncertain. The heavy-duty isolator is only recommended for use with 390mm thick optical tables, due to their extra weight. This isolator is also shorter to accommodate for the difference in thickness of the tables. To determine which isolator is right for your application, first obtain

the weight of your optical table from the chart on page 411. Then estimate the additional weight of accessories and everything to be mounted on the table. Ideally, this load should equal one half of the isolator capacity, but anything between a quarter and full capacity is acceptable.

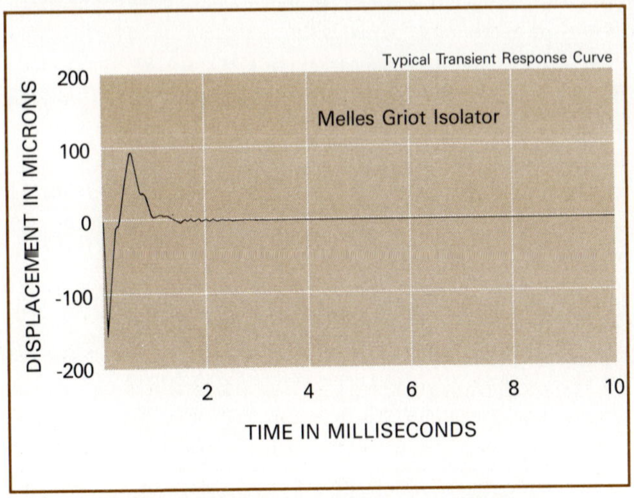

TRANSIENT RESPONSE of a Melles Griot isolator.

ANALYSIS OF ISOLATORS

The figure above presents the transient response of a Melles Griot optical table isolation system to a force applied to a corner. This illustrates the rapid attenuation of the rigid body motions of the table by the isolator damping mechanism and the fast acting self-levelling valves.

The figure in the next column represents the transmissibility curve for the Melles Griot optical table isolation system. This curve

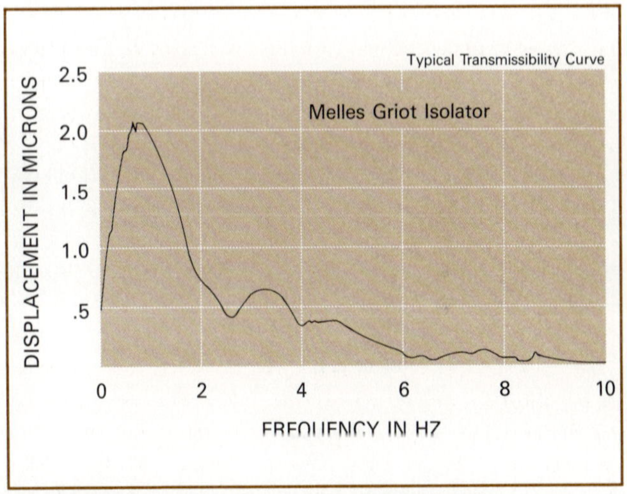

TRANSMISSIBILITY of a Melles Griot isolator.

demonstrates the high attenuation of floor disturbance frequencies. At the first resonance mode of the optical table, the transmissibility curve shows an attenuation of 0.001. This indicates that less than 0.1% of the floor vibration will feed into the optical table at its first resonant frequency. The very high damping of the Melles Griot optical table will quickly attenuate any vibration energy transmitted through to the table.

FIXED STANDS

In some situations, isolation is not necessary. In these cases a fixed frame stand may be used. Our fixed stands utilize the same tie-bar system as the isolators. However, they are simply made from the square section steel tubes without the isolation fittings. Instead, sturdy levelling bolts with tilt bearing plates are provided.

For special situations, combinations of six or more isolators may be used to increase the load capacity. Please discuss these situations with an Applications Engineer.

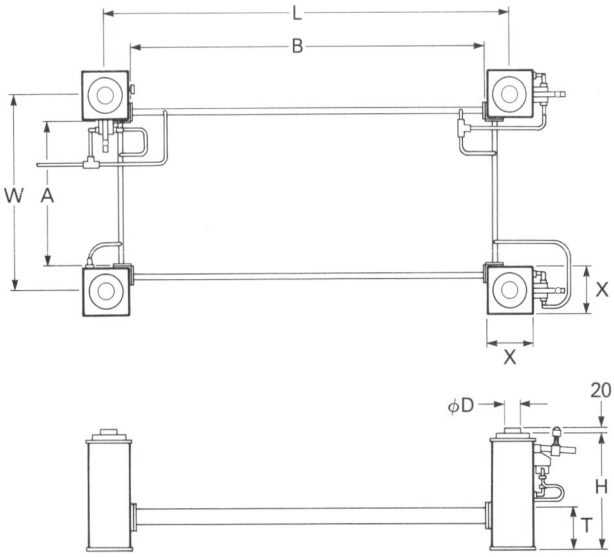

PRODUCT NUMBER	H	T	X	ϕD
07 OTI 002	653	270	160	76
with tie bars	A	B	W	L
07 OTJ 001	740	900	911	1071
07 OTJ 002	740	1140	911	1311

PRODUCT NUMBER	H	T	X	ϕD
07 OTI 003	653	270	160	119
with tie bars	A	B	W	L
07 OTJ 001	740	900	911	1071
07 OTJ 002	740	1140	911	1311
07 OTJ 003	740	1420	911	1591
07 OTJ 004	740	1810	911	1981

PRODUCT NUMBER	H	T	X	ϕD
07 OTI 004	523	140	222	155
with tie bars	A	B	W	L
07 OTJ 002	740	1140	973	1373
07 OTJ 003	740	1420	973	1653
07 OTJ 004	740	1810	973	2043

SPECIFICATIONS: TABLE ISOLATORS

Dimensions: $\pm$2mm

Air supply pressure: 2.5 to 5.0 atmospheres

Material: Thin walled rubber rolling diaphragm seal within steel containment vessels

Finish: Durable black paint

Table Stand

	Load Capacity (kg)	Height (mm)	PRODUCT NUMBER
Normal duty	2000	650	07 OTL 001

Please select tie bars below to suit table size.

Table Isolators; set of 4

	Load Capacity (kg)	Resonant Frequency (Hz)	Height (mm)	PRODUCT NUMBER
Light duty	500	1.2	650	07 OTI 002
Normal duty	2000	1.5	650	07 OTI 003
Heavy duty	4000	1.6	520	07 OTI 004

Please select tie bars from the list opposite to suit table size.

Tie Bars for Table Isolators and Stands; set of 4

For use with Table Length (m)	PRODUCT NUMBER
1.8	07 OTJ 001
2.4	07 OTJ 002
3.0	07 OTJ 003
3.7	07 OTJ 004

OPTICAL BREADBOARDS

In spite of being thinner and lighter than conventional optical tables, Melles Griot optical breadboards are exceptionally rigid, stiff and internally damped. They are ideally suited to the precision set-up of optical systems and sub-systems. Breadboards are extremely useful in building prototypes, setting up quality control systems, aligning and adjusting multi-component imaging systems and in interfacing optical and electronic systems. Many OEM's assemble, align and test sub-systems on a small breadboard and incorporate the entire structure into their final equipment. The large variety of available sizes means that virtually any prototype or experimental system can be accommodated comfortably and reliably.

Melles Griot optical breadboards owe their high standards of performance to a variety of unique design features which provide a superior level of performance when compared with previously available products. Further, the features of this new design provide far greater long-term serviceability than has been previously available.

The heavy-duty threaded inserts distribute pull-down loads over a large surface area of the table and also resist deformation of the top skin. The unique foam ring around the edges of the breadboard improves damping characteristics. The novel two-layer, sandwiched top skin provides great local rigidity with high resonant frequencies of low-Q value. This is a truly superior optical work surface. The edge attachment slots provide component mounting flexibility and a method of joining the breadboards together.

The stiffness of these optical breadboards is great enough that they may be used by themselves on a normal laboratory bench for many applications. They are designed, futhermore, for direct mounting on top of either 1/4-20, one inch spaced or M6, 25mm spaced optical tables which are in common use throughout the world. Breadboards may be loosely placed on optical table tops or firmly secured in place with four or five attaching screws, depending on the breadboard size. Alternatively, the breadboards may be attached by means of levelling devices or standoff posts which raise them 127, 254 or 508mm (5, 10 or 20 inches) above the optical table surface.

Breadboards are finished with an extremely durable and resistant flat black coating. This thin, epoxy-like surface reduces surface glare and reflection substantially without reducing the effectiveness of magnetic holders used on the breadboard top. For those who need an uncoated surface, the conventional, bright finished, ferromagnetic stainless steel tops are also available. These tops are also ground to provide superior flatness.

CONSTRUCTION

Melles Griot optical breadboards are constructed as a composite of top and bottom steel skins, aluminum honeycomb and polyurethane filled aluminum sides. The top skin is, in fact, a two-layer, ferromagnetic steel sandwich made up of an 11 gauge (3mm) skin and a heavy (5mm) back plate bonded together. This sandwich construction produces an extremely rigid top with very "dead" vibratory characteristics, the viscoelastic bond line providing internal damping of the surface. The aluminum honeycomb core is

also laminated in two sections with viscoelastic bonding.

Heavy-duty threaded inserts (1/4-20 or M6) are fitted into the back of the top skin. These hardened steel inserts provide firm, permanent, virtually unstripable threads for mounting components on the breadboard top. Because of the large diameter of the inserts (13mm) the pull-down load is distributed widely over the surface of the skin, thereby greatly reducing the possibility of local skin deformation. As no oil is used in tapping the top skin holes there is no problem of this affecting delicate optical components. The result is a far more dependable mounting surface than has been available previously. The clean construction methods make these breadboards ideal for use in a clean room environment.

All four breadboard edges have a 6.5mm (0.27 inch) slot around the outer edge. The slot will accommodate a 1/4-20 hex-head bolt and constrain the head from rotating. In this way, several bolts may be slipped into the slots and various components loosely placed over them and locked in position by tightening nuts over the bolts. Attachment devices for joining breadboards, side-to-side are also available.

Co-fabrication techniques allow the polyurethane foam filled rim to be assembled between the skins at the same time that the honeycomb is being bonded. This construction is stronger than an edge assembly which is post-fabricated. Further, the combination of aluminum extrusion edge pieces and foam filler provides a low-Q damping ring around the complete breadboard for increased internal stability.

STATIC AND DYNAMIC RIGIDITY

These Melles Griot optical breadboards have been designed to provide markedly improved static rigidity with a lower mechanical Q than any other optical breadboard currently available. This improved performance provides you, the user, with that extra margin of assurance that breadboard related problems will not degrade your work.

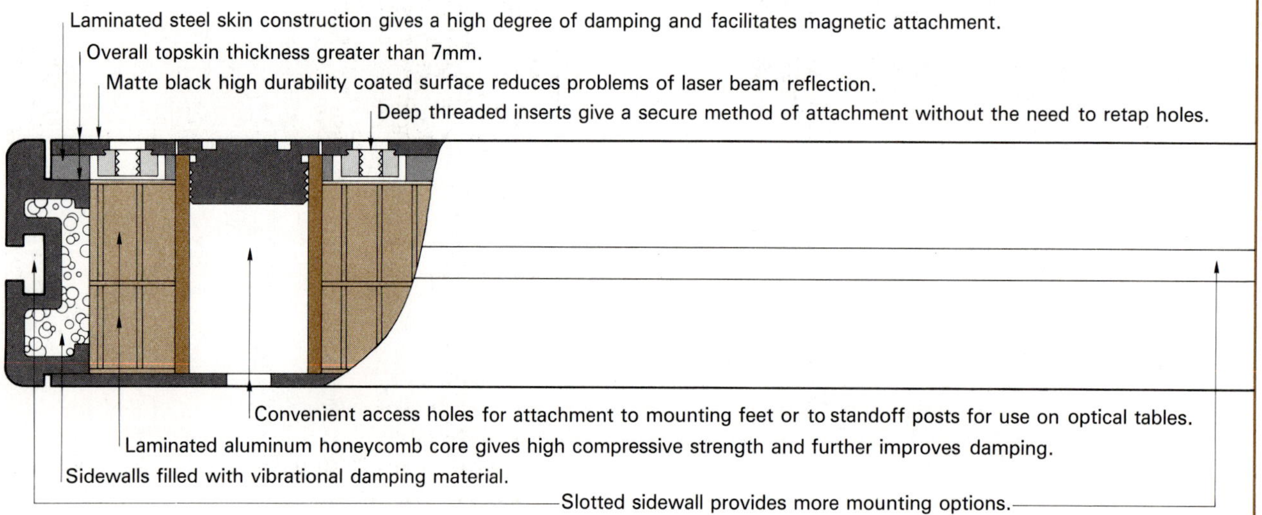

CROSSECTION OF OPTICAL BREADBOARD showing the many superior design features.

STATIC RIGIDITY

Melles Griot optical breadboards achieve their high static rigidity from a combination of design factors. The results are a more rigid surface than cast iron, granite or other honeycomb cores. Static rigidity (the ability to withstand bending or twisting when static loads are applied) is of particular importance to the optical researcher. During the course of an experiment or design the weight distribution on the top of a breadboard changes frequently as components, tools and optical hardware is moved, added or removed. Motion (bending) of the breadboard surface during such load change is, invariably, detrimental. Bending clearly varies not only with breadboard design but with thickness, size and induced load as well. A properly supported Melles Griot optical breadboard (07 OBB 002) will exhibit a maximum deflection rate of 3.5 x 10^{-4} arc seconds per kilogram (1.6 x 10^{-4} arc sec/lb) of applied load.

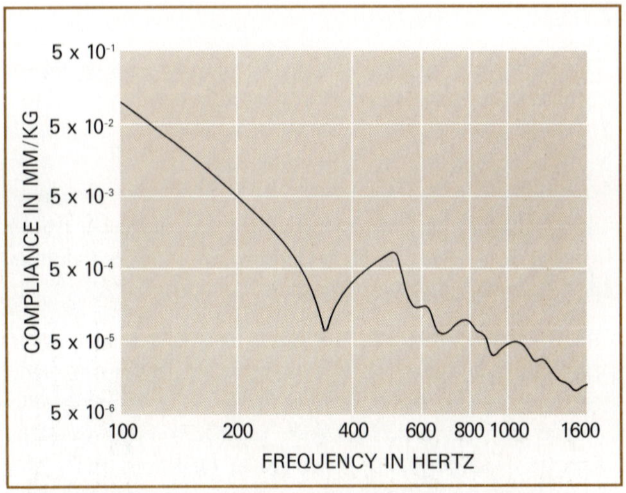

COMPLIANCE of a Melles Griot optical breadboard.

DYNAMIC RIGIDITY

Dynamic rigidity (the resistance to bending or twisting when time varying forces are applied) is also important in an optical breadbaord. High rigidity indicates a resistance to seismic, accoustic and local vibratory forces. Internal breadboard design considerations can improve the damping characteristics of honeycomb structures thereby improving their performance. Dynamic rigidity can be conveniently measured by analyzing the spectral time response of a breadboard to an impulse input force. The upper curve opposite indicates the compliance of a Melles Griot optical breadboard. The low peak values indicate the superiority of these breadboards. Peak values here are at least an order of magnitude lower than solid or non-damped optical tables and represent a marked improvement over other damped honeycomb breadboards.

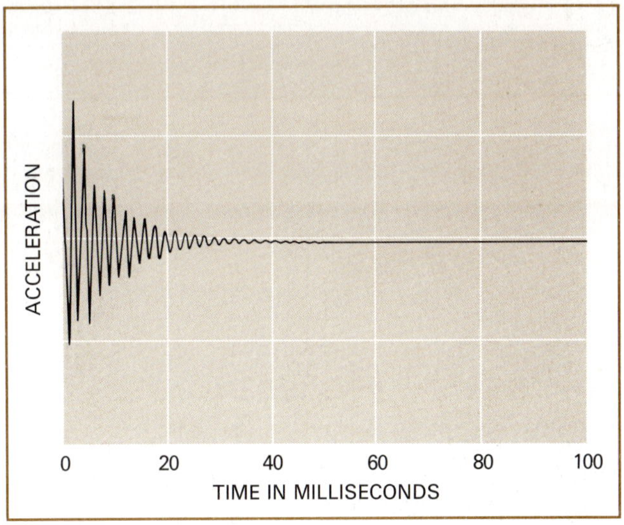

DAMPING of a Melles Griot optical breadboard.

DAMPING

The broad shouldered, non-peaky compliance curve indicates the superior damping characteristics of Melles Griot optical breadboards. The corner response (lower) curve for impulse corner input speaks for itself. The damping of a Melles Griot optical breadboard (07 OBB 007) is about 2.0%.

FLATNESS

Melles Griot black finish, optical breadboards are flat to within ±0.2mm which is ideal for most applications. For those who need even better flatness we recommend Melles Griot stainless steel, optical breadboards. These are carefully ground to ensure a flatness of ±0.125mm or better.

FINISH

The top skin of Melles Griot optical breadboards is finished with a thin flat black coating which is bonded into the metal surface. The result is the strongest, most abrasion-resistant coating available. This coating does not significantly affect the flatness nor the magnetic properties of the surface. Melles Griot breadboards are also available with conventional bright ferromagnetic stainless steel top skins. The rest of the breadboard is finished with dull black baked-on polyurethane paint.

Throughout the Melles Griot optical hardware product range, matte black finishes have been used predominantly. Our market research has shown that this type of finish is most desirable for use in laser systems experimentation in that it helps to reduce unwanted, and potentially troublesome reflections and scatter. Indeed, many of the problems involved in turning a prototype optical system into a production unit are caused by stray and scattered light.

MOUNTING AND LEVELLING

Melles Griot optical breadboards will mount on one, two and four inch spaced optical tables and 25, 50 and 100mm spaced optical tables. The optical table may be threaded with either 1/4-20 or M6. This is possible because the four corner attaching areas in Melles Griot optical breadboards have a slot to accommodate either format. Levelling devices and standoff posts (described later will also fit either inch or metric optical tables in this way. The standoff posts provide a great deal of flexibility and versatility in setting up prototypes on an otherwise crowded work surface.

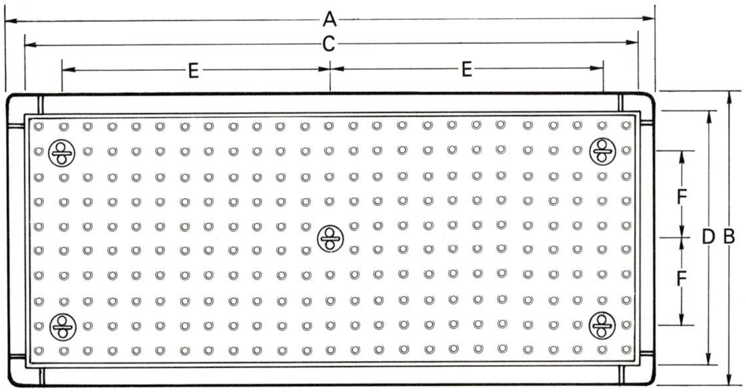

PRODUCT NUMBERS	A	B	C	D	E	F
*07 OBB 001 / 07 OBF 001	305	305	282	282	101.6	101.6
*07 OBB 002 / 07 OBF 002	610	305	587	282	254.0	101.6
07 OBB 003 / 07 OBF 003	914	305	892	282	406.4	101.6
*07 OBB 005 / 07 OBF 005	610	610	587	587	254.0	254.0
07 OBB 006 / 07 OBF 006	914	610	892	587	406.4	254.0
07 OBB 007 / 07 OBF 007	1219	610	1197	587	406.4	254.0
07 OBB 008 / 07 OBF 008	1524	610	1502	587	508.0	254.0
07 OBB 009 / 07 OBF 009	914	914	892	892	406.4	406.4
07 OBB 012 / 07 OBF 012	1219	914	1197	892	406.4	406.4
07 OBB 013 / 07 OBF 013	1524	914	1502	892	508.0	406.4
07 OBB 014 / 07 OBF 014	1219	1219	1197	1197	406.4	406.4
07 OBB 015 / 07 OBF 015	1524	1219	1502	1197	508.0	406.4

*Denotes breadboards with only 4 mounting holes. (Center hole omitted.)

PRODUCT NUMBERS	A	B	C	D	E	F
*07 OBB 501 / 07 OBF 501	305	305	282	282	100.0	100.0
*07 OBB 502 / 07 OBF 502	610	305	587	282	250.0	100.0
07 OBB 503 / 07 OBF 503	914	305	892	282	400.0	100.0
*07 OBB 505 / 07 OBF 505	610	610	587	587	250.0	250.0
07 OBB 506 / 07 OBF 506	914	610	892	587	400.0	250.0
07 OBB 507 / 07 OBF 507	1219	610	1197	587	400.0	250.0
07 OBB 508 / 07 OBF 508	1524	610	1502	587	500.0	250.0
07 OBB 509 / 07 OBF 509	914	914	892	892	400.0	400.0
07 OBB 512 / 07 OBF 512	1219	914	1197	892	400.0	400.0
07 OBB 513 / 07 OBF 513	1524	914	1502	892	500.0	400.0
07 OBB 514 / 07 OBF 514	1219	1219	1197	1197	400.0	400.0
07 OBB 515 / 07 OBF 515	1524	1219	1502	1197	500.0	400.0

*Denotes breadboards with only 4 mounting holes. (Center hole omitted.)

SPECIFICATIONS: OPTICAL BREADBOARDS

Dimensions: ±4mm

Flatness: Black finish ±0.20mm
Stainless steel ±0.125mm

Skin Material: Ferromagnetic steel (either black coated or stainless). Top skin composite 8mm (0.310 inch) thick.

Core: Double layer aluminum honeycomb $0.08 g/cm^3$, $1.61 cm^2$ cell area

Panel Thickness: 50.8mm (2 inches)

Weight: approx $7 g/cm^2$ ($15 lbs/ft^2$)

Mounting Holes: Inch version ¼-20 on 25.4mm (1 inch) centers
Metric version M6 on 25mm centers

Finish: Surface — black coating or stainless
Sides and underside — polyurethane paint

Damping: Composite top skin, core, and polyurethane foam outer ring.

Optical Breadboards, Black Finish

Width		Length		Weight	PRODUCT NUMBER	
(mm)	(ft)	(mm)	(ft)	(kg)	Inch Version	Metric Version
305	1	305	1	8	07 OBB 001	07 OBB 501
305	1	610	2	15	07 OBB 002	07 OBB 502
305	1	914	3	23	07 OBB 003	07 OBB 503
305	1	1219	4	31	07 OBB 004	07 OBB 504
610	2	610	2	31	07 OBB 005	07 OBB 505
610	2	914	3	46	07 OBB 006	07 OBB 506
610	2	1219	4	62	07 OBB 007	07 OBB 507
610	2	1524	5	77	07 OBB 008	07 OBB 508
914	3	914	3	70	07 OBB 009	07 OBB 509
914	3	1219	4	93	07 OBB 012	07 OBB 512
914	3	1524	5	116	07 OBB 013	07 OBB 513
1219	4	1219	4	124	07 OBB 014	07 OBB 514
1219	4	1524	5	155	07 OBB 015	07 OBB 515

Optical Breadboards, Stainless Finish

Width		Length		Weight	PRODUCT NUMBER	
(mm)	(ft)	(mm)	(ft)	(kg)	Inch Version	Metric Version
305	1	305	1	8	07 OBF 001	07 OBF 501
305	1	610	2	15	07 OBF 002	07 OBF 502
305	1	914	3	23	07 OBF 003	07 OBF 503
305	1	1219	4	31	07 OBF 004	07 OBF 504
610	2	610	2	31	07 OBF 005	07 OBF 505
610	2	914	3	46	07 OBF 006	07 OBF 506
610	2	1219	4	62	07 OBF 007	07 OBF 507
610	2	1524	5	77	07 OBF 008	07 OBF 508
914	3	914	3	70	07 OBF 009	07 OBF 509
914	3	1219	4	93	07 OBF 012	07 OBF 512
914	3	1524	5	116	07 OBF 013	07 OBF 513
1219	4	1219	4	124	07 OBF 014	07 OBF 514
1219	4	1524	5	155	07 OBF 015	07 OBF 515

RUBBER MOUNTING FEET

These grey rubber non-marking feet provide positional stability when optical breadboards are placed on other surfaces (such as an optical table). The feet raise the bottom of the breadboard 13.5mm from the mounting surface, thereby providing a convenient finger space for moving or lifting. Feet are provided with a 1/4-20 or M6 tapped hole for attachment directly to the bottom of Melles Griot optical breadboards. Finish is zinc plated steel. Furnished as a set of five, complete with all mounting hardware.

Rubber Mounting Feet

	PRODUCT NUMBER	
	Inch Version	Metric Version
Rubber Mounting Feet (set of five)	07 OBM 001	07 OBM 501

ACCESSORY KITS FOR OPTICAL BREADBOARDS

All required screws and tools are supplied with Melles Griot optical breadboards when originally shipped. For your convenience, we offer several kits of assorted fasteners and tools packaged in a closeable plastic box.

Accessory Kits

	PRODUCT NUMBER
Inch Version	07 AKT 001
Metric Version	07 AKT 501

LEVELLING DEVICES

Melles Griot levelling devices allow one to level or adjust the height of an optical breadboard. These units are easy to use and, with their metric lead threads of one millimeter per complete turn, allow one to set breadboards to specific heights.

The fully closed height of the leveler is 115mm. A vertical, upward adjustment of 25mm is provided. Adjustment is simply a matter of grasping the straight knurled outer shell and rotating. A yellow ring at the top of the leveler acts as a jam nut to lock the device in position.

The base of the levelling device is provided with holes so that it may be attached to the top of all threaded optical tables. The Melles Griot optical breadboard is placed on top (after removing the access covers in the breadboard with the tool provided) and attached to the levelling device with the included 1/4-20 or M6 Allen cap screws.

Levelling devices are supplied individually, complete with all necessary hardware. Levelling devices may be freely interchanged with the 127mm length standoff posts following.

SPECIFICATIONS: LEVELLING DEVICES

Nominal Height: 127mm
Adjustment Range: ±12mm
Lead: 1 thread/mm
Material: Steel; aluminum base
Finish: Flat black

Levelling Device

	PRODUCT NUMBER	
	Inch Version	Metric Version
Levelling Device, complete with hardware	07 OBL 001	07 OBL 501

Melles Griot offers both inch spaced (1/4-20 on one inch centers) and metric spaced (M6 on 25mm centers) optical hardware. Throughout this Optics Guide you will find both versions listed together. For compatibility be sure to keep all accessories within the same (inch or metric) system.

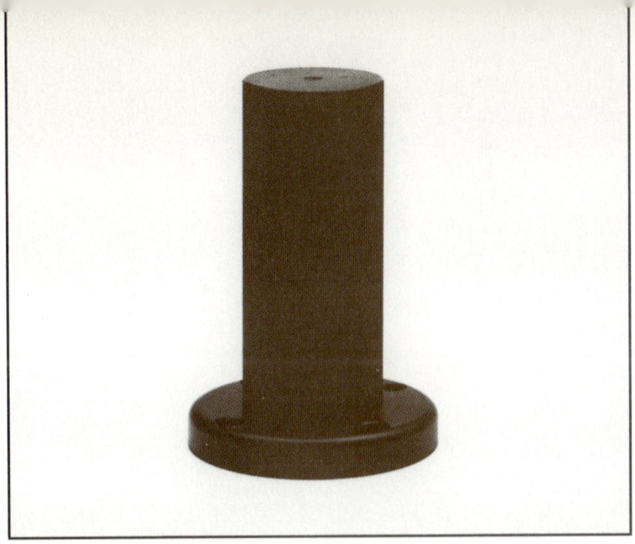

STANDOFF POSTS

Standoff posts for Melles Griot optical breadboards are available in 127, 254 and 508mm (5, 10 an 20 inch) lengths. They are made from black anodized aluminum extrusions and are filled with a damping material.

Standoff posts are tapped on their tops with 1/4-20 or M6 screw holes and are supplied complete with all hardware to attach to the four corner slots and the central mounting hole of Melles Griot optical breadboards. The bottoms of the standoff posts are flanged to provide a stable base. Holes in the flange allow the posts to be secured with cap screws to either inch-spaced or metric-spaced optical tables.

SPECIFICATIONS: STANDOFF POSTS

Material: Aluminum
Finish: Black anodized
Length: ±0.13mm

Standoff Posts

Length (mm)	PRODUCT NUMBER Inch Version	Metric Version
127.0	07 OBP 001	07 OBP 501
254.0	07 OBP 002	07 OBP 502
508.0	07 OBP 004	07 OBP 504

BREADBOARD SIDE JOINERS

These convenient devices allow one to firmly draw together two (or more) Melles Griot optical breadboards at their edges so that they are securely joined.

The side joiners consist of two flanged studs which slip into the slots along the sides of Melles Griot optical breadboards and two knurled rings. By rotating the knurled rings in opposite directions the breadboards are pulled together and locked into position.

Breadboard sides, it should be clearly noted, are not strong enough to freely support the breadboard by the side joiners alone. It is intended that the joiners be used as alignment and positioning devices and to keep the separation of one or more breadboards constant; they are not intended to support an optical breadboard in a cantilevered position.

Side joiners are made of flat black anodized aluminum. Four radial holes are drilled in the central knobs to accommodate a tightening bar. Supplied as a set of two, complete with tightening bars.

Breadboard Side Joiners

	PRODUCT NUMBER
Breadboard Side Joiners, set of two	07 OBJ 001

LASER AND ACCESSORY SHELVES

These shelves will be found useful in providing space for accessories or large lasers without occupying valuable table surface area. They may be mounted above the optical table or below it. In either case they are firmly attached to the table and if an isolator is used they are isolated from ground disturbances together with the table.

The shelves are similar in construction and appearance to Melles Griot optical breadboards. The main difference is that they do not have an array of mounting holes. The top and bottom skins are a single thickness of ferromagnetic steel and the core is of laminated cellulose honeycomb. The structure is somewhat lighter than a breadboard but retains many of the important breadboard features, including good structural damping.

These shelves may be mounted above an optical table by means of the standoff posts previously described (page 421). In this configuration they are ideal for supporting instruments. They may also be slung underneath a table by means of the shelf mounting hardware. In this alternate configuration the shelves are useful for supporting large lasers. The laser beam may be brought to the surface through the access hole in the optical table by means of a beamsteering device (see page 432). All Melles Griot optical tables have threaded inserts on the underside which are positioned to accept laser shelf hardware. However, if you are using an existing table of alternate manufacture it is a simple matter to drill and tap the appropriate holes.

It is important to specify an inch or metric version of these shelves in order that the mounting hardware will line up with the holes on the underside of your Melles Griot optical table.

SPECIFICATIONS: LASER AND ACCESSORY SHELVES

Skin Material: Ferromagnetic steel, black coated
Core: Double layer cellulose honeycomb
Panel Thickness: 50.8mm
Density: 6 g/cm^2

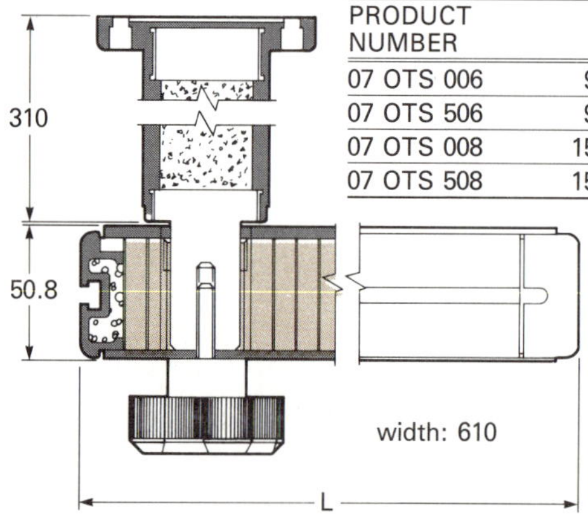

PRODUCT NUMBER	L
07 OTS 006	914
07 OTS 506	914
07 OTS 008	1524
07 OTS 508	1524

LASER SHELF MOUNTING HARDWARE

This hardware consists of four, 32cm long, internally damped suspension rods plus all screws and washers necessary to mount either a laser shelf or an optical breadboard of equivalent size beneath an optical table. The inch version is supplied with 1/4-20 screws and the metric version with M6 screws to match the appropriate tables.

After attaching the suspension rods to the underside of the table, the ends of these rods are fitted into the access holes on the shelf or breadboard. The shelf is then secured in place with the screws and washers provided.

Laser Shelf Mounting Hardware

PRODUCT NUMBER	
Inch Version	Metric Version
07 OTA 001	07 OTA 501

Laser and Accessory Shelves

Width		Length		Weight	PRODUCT NUMBER	
(mm)	(ft)	(mm)	(ft)	(kg)	Inch Version	Metric Version
610	2	914	3	36	07 OTS 006	07 OTS 506
610	2	1524	5	59	07 OTS 008	07 OTS 508

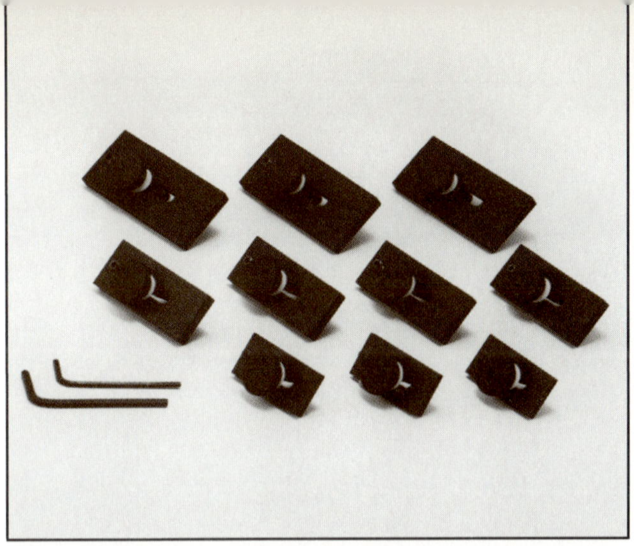

TABLE CLAMPS

Table clamps are extremely useful for clamping various components in position on top of an optical breadboard or table. The clamps are sold singly, but we recommend that you purchase them in economical packages of 10, 25 or 50 each.

Each clamp consists of a black anodized strip with an adjustable, one inch long, Allen set screw at one end for clamping height adjustment (0.0mm to 20.0mm) and a plastic knobbed 1/4-20 or M6 knurled tie-down thumb screw. Three different lengths are offered; 50, 72 and 90mm. The clamps have a slot for the tie-down screw to provide for adjustment of the clamp length. Economy packs contain a selection of the three sizes of the clamp plus Allen keys for height adjustment.

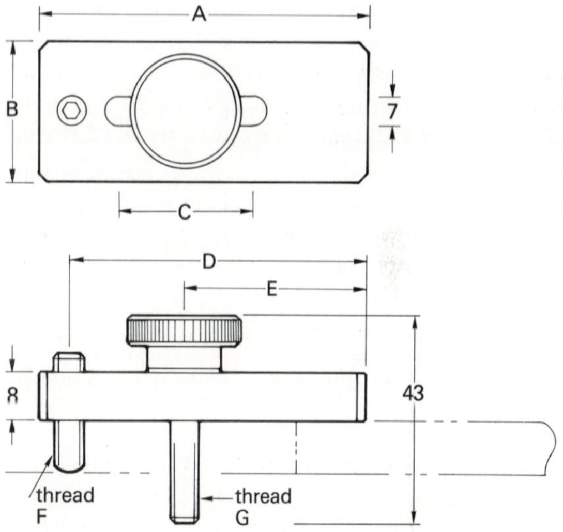

PRODUCT NUMBER	A	B	C	D	E	thread F	thread G
07 CTT 001	50	24	20	45	26	M6x20	1/4-20 UNC
07 CTT 002	72	30	30	65	40	M6x25	1/4-20 UNC
07 CTT 003	90	40	40	82	50	M8x25	1/4-20 UNC
07 CTT 501	50	24	20	45	26	M6x20	M6
07 CTT 502	72	30	30	65	40	M6x25	M6
07 CTT 503	90	40	40	82	50	M8x25	M6

Table Clamps, Individual

	Clamping Adjustment		PRODUCT NUMBER	
Length (mm)	Height (mm)	Length (mm)	Inch Version	Metric Version
50	0 to 15	0 to 22	07 CTT 001	07 CTT 501
72	0 to 20	0 to 51	07 CTT 002	07 CTT 502
90	0 to 20	0 to 66	07 CTT 003	07 CTT 503

Table Clamps, Packages

	PRODUCT NUMBER	
	Inch Version	Metric Version
Package of 10 Table Clamps (Contains three 50mm clamps, four 72mm clamps, three 90mm clamps and two Allen keys)	07 CTS 001	07 CTS 501
Package of 25 Table Clamps (Contains eight 50mm clamps, twelve 72mm clamps, five 90mm clamps and four Allen keys)	07 CTS 002	07 CTS 502
Package of 50 Table Clamps (Contains sixteen 50mm clamps, twenty four 72mm clamps, ten 90mm clamps and six Allen keys)	07 CTS 003	07 CTS 503

FASTENER KIT

All required fasteners and screws are supplied with Melles Griot optical hardware when originally shipped. These kits are designed specifically to replace lost screws and fasteners. They are packaged, complete with the specific tools, in a closeable plastic box.

PRODUCT NUMBER	
Inch Version	Metric Version
07 AKT 005	07 AKT 505

MAGNETIC BASES

Magnetic bases are extremely convenient for rapid set-ups on ferromagnetic optical breadboards and tables. The "on-off" feature of these bases allows them to be moved about on the table surface until properly located and then firmly locked in place. Melles Griot magnetic bases may be used upright or tipped onto one end for added versatility and convenience.

These bases are available in two sizes, standard and heavy-duty. The standard base produces a holding force of 75 kilograms; the heavy-duty base produces a 150 kilogram force. Both have an M16 tapped hole in the center of the top to accept all Melles Griot heavy-duty posts. Both also have four M4 tapped holes in the corners which will accept an appropriate top plate which is tapped with 1/4-20 or M6 holes on one inch or 25mm centers. These top plates will, in turn, accept all Melles Griot translation stages, rotational stages, tilt stages, mirror mounts and so forth.

Both bases are provided with a knob to control the on-off cycle. Both are painted flat black to reduce stray reflections and scatter.

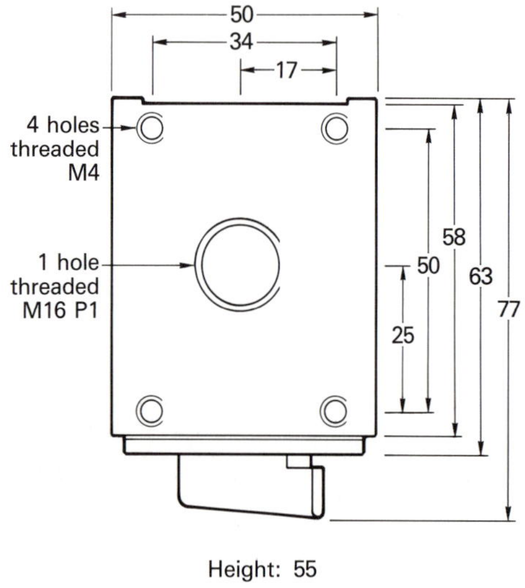

07 BMS 001

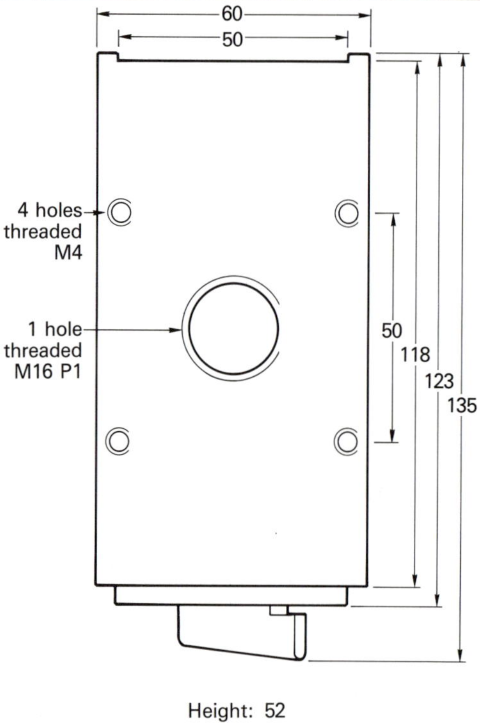

07 BMS 002

SPECIFICATIONS: MAGNETIC BASES

Material: Mild steel
Finish: Flat black paint
Holding Force: Standard-duty 75 kg
Heavy-duty 150 kg
Weight: Standard-duty 1.0 kg
Heavy-duty 2.4 kg

Magnetic Bases

	PRODUCT NUMBER
Standard-Duty	07 BMS 001
Heavy-Duty	07 BMS 002

TOP PLATES FOR MAGNETIC BASES

Aluminum top plates are available for both magnetic bases. The aluminum acts as an isolator to decouple the magnetic field from other components attached to the top plate. Both are drilled and tapped 1/4-20 on one inch centers or M6 on 25mm centers to accept a wide variety of accessories. Both top plates mount on their respective magnetic bases with four M4 countersunk Phillips screws. Top plates are supplied complete with mounting screws. Each top plate is black anodized.

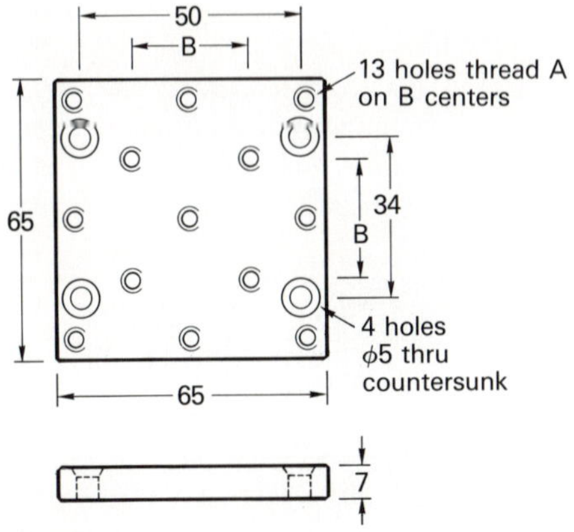

PRODUCT NUMBER	thread A	B
07 BMA 001	1/4-20 UNC	25.4
07 BMA 501	M6	25

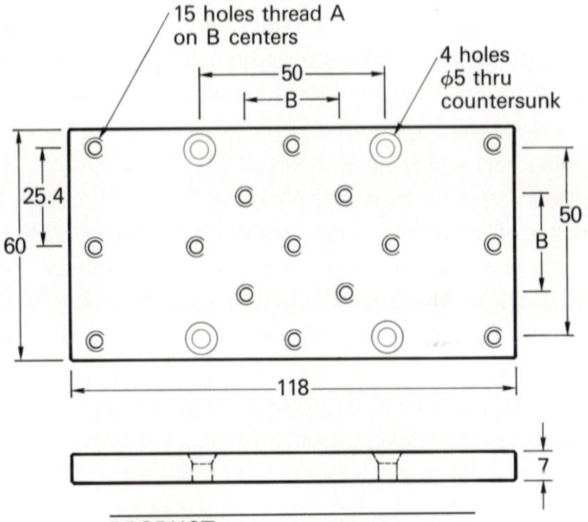

PRODUCT NUMBER	thread A	B
07 BMA 002	1/4-20 UNC	25.4
07 BMA 502	M6	25

Top Plates for Magnetic Bases

	PRODUCT NUMBER	
For Use With	Inch Version	Metric Version
07 BMS 001	07 BMA 001	07 BMA 501
07 BMS 002	07 BMA 002	07 BMA 502

ACCESSORY KITS FOR OPTICAL HARDWARE

All required screws and tools are supplied with Melles Griot optical hardware when shipped. For your convenience, we offer several kits of assorted fasteners and tools packaged in a closeable plastic box. See page 480 for a full list of contents.

	PRODUCT NUMBER
Inch Version	07 AKT 003
Metric Version	07 AKT 503

LOW PROFILE MAGNETIC BASES

These magnetic bases offer a broad, stable, low profile mounting base for a large variety of Melles Griot mechanical components. The permanent magnets provide a large holding force which may be continuously varied from off to its maximum value. Control of the holding force is provided by the conveniently located knurled ring which protrudes slightly from one side of the base.

Rotation of this ring positions a permanent magnet within the base, such that the magnetic field is contained within the table base system at all times. Whether on or off, these units may be used near photo-electric devices, which are sensitive to magnetic fields, without danger.

Low profile magnetic bases are provided in two sizes. The small base produces a maximum holding force of 10 kilograms when mounted on a Melles Griot optical table or breadboard; the large base produces a 45 kilogram force. The top mounting surface of both bases is non-magnetic. A central M16 tapped hole in the top of the large low profile magnetic bases will accept all Melles Griot heavy-duty posts. The top surfaces of both bases are also tapped with an array of 1/4-20 holes on one inch centers or M6 holes on 25mm centers; 8 tapped holes in the smaller base, 24 in the larger.

Both sizes of the low profile magnetic base are finished flat black to reduce stray reflections.

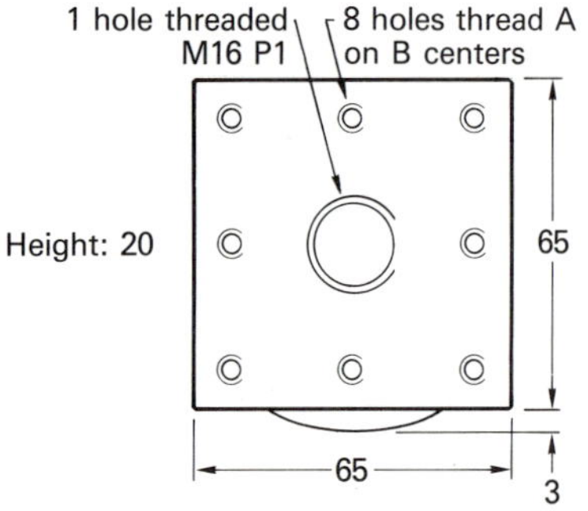

PRODUCT NUMBER	thread A	B
07 BMS 003	1/4-20 UNC	25.4
07 BMS 503	M6	25

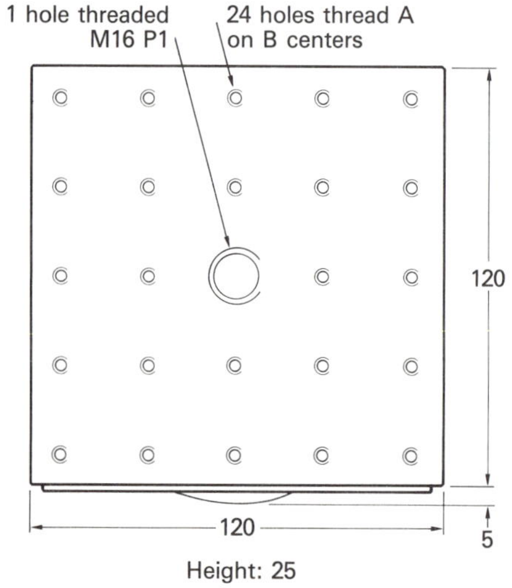

PRODUCT NUMBER	thread A	B
07 BMS 004	1/4-20 UNC	25.4
07 BMS 504	M6	25

SPECIFICATIONS: LOW PROFILE MAGNETIC BASES

Material: Steel
Finish: Flat black
Holding Force: Small base: 10 kg
Large base: 45 kg
Weight: Small base: 0.7 kg
Large base: 4.0 kg

Low Profile Magnetic Bases

	PRODUCT NUMBER	
	Inch Version	Metric Version
Small Base	07 BMS 003	07 BMS 503
Large Base	07 BMS 004	07 BMS 504

The translation stages on pages 490-493 may be attached directly to these magnetic bases.

KINEMATIC BASES

Frequently one wishes to remove a mounted optical component from a system and then return it to exactly the same place. These true kinematic mounts are constrained in five freedoms of motion; gravity (or the retaining clamp) provides the sixth restraint.

These bases consist of two plates with facility for 1/4-20 on one inch center, or M6 on 25mm center mounting of components on top. The bottom plate has clearance holes for mounting directly to an optical breadboard, table, bench carrier or magnetic base.

The underside of the top plate holds three precision ground steel spheres which fit into the bottom plate's kinematic locators. The top plate is threaded in the middle with an M16 hole which directly accepts all Melles Griot heavy-duty posts.

The top plate is also tapped with 1/4-20 holes on one inch centers or M6 holes on 25mm centers to carry all Melles Griot translation stages, rotational stages, tilt stages, and so forth.

Clamps are provided on the sides of the bottom plates to retain the top plate when mounted. The bottom plate and the top plate are black chrome plate ferromagnetic steel.

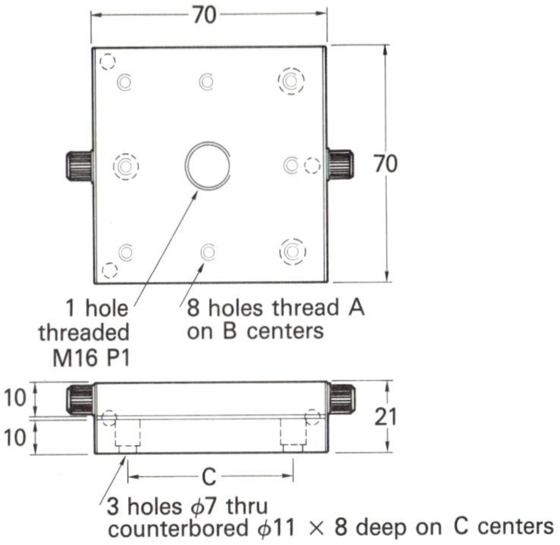

PRODUCT NUMBER	thread A	B	C
07 BKS 001	1/4-20 UNC	25.4	50.8
07 BKS 501	M6	25	50

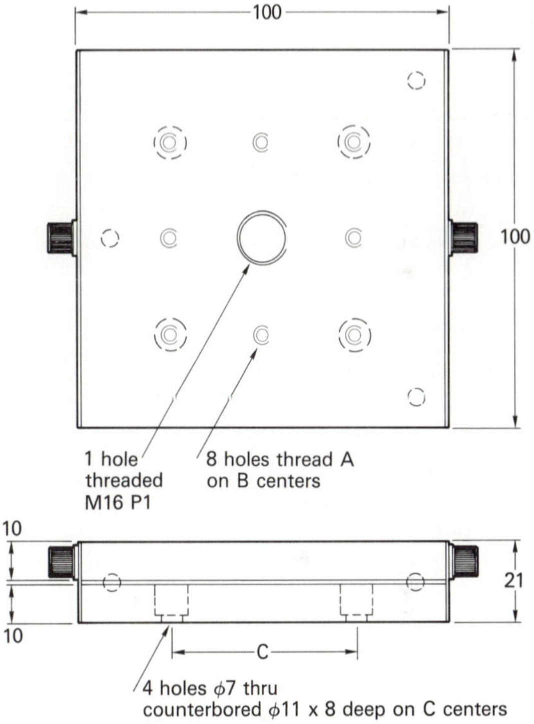

PRODUCT NUMBER	thread A	B	C
07 BKS 002	1/4-20 UNC	25.4	50.8
07 BKS 502	M6	25	50

SPECIFICATIONS: KINEMATIC BASES

Material: Ferromagnetic steel
Finish: Black chrome
Weight: Standard-duty 0.78 kg
Heavy-duty 1.50 kg

Kinematic Bases

	PRODUCT NUMBER	
	Inch Version	Metric Version
Standard-Duty	07 BKS 001	07 BKS 501
Heavy-Duty	07 BKS 002	07 BKS 502

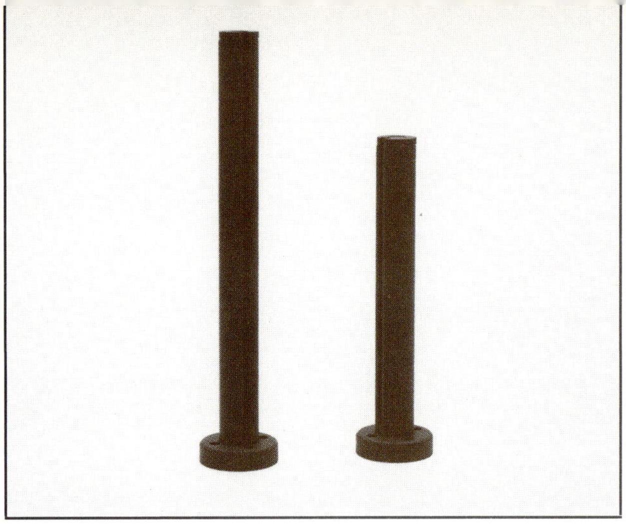

DAMPED MOUNTING POSTS

For stable mounting of optical components on an optical table it is necessary to use a broad-based, sturdy mounting post. Heavy metal rods, when used for this purpose, will have a tendency to "ring" at their resonant frequency. They may therefore set up or amplify disturbances in a way which can adversely affect the optical performance of the system.

These new, Melles Griot damped mounting posts (patent pending) overcome this problem through their novel, exceptionally highly damped construction. The posts consist of a 38mm diameter aluminum tube with a tee-slot section on one face. Inside the tube is suspended a large number of small masses within an elastometric damping compound. Any disturbance imparted to the post is damped out by relative motion of the masses to the outer tube. The effect of the masses is to work the damping medium and dissipate any induced vibration. This damping system is effective over a wide range of frequencies. The post is attached to a circular base having clearance holes to match both inch and metric mounting patterns.

Finite element modeling as verified by empirical measurement of dynamic response characteristics shows exceptionally high damping at the fundamental and higher order resonances. This results in instantaneous dissipation of any induced disturbance.

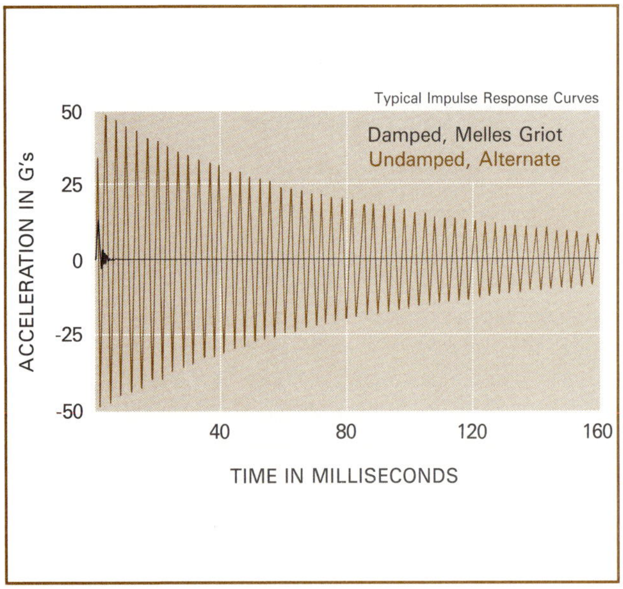

IMPULSE RESPONSE of a Melles Griot damped post compared with an undamped post.

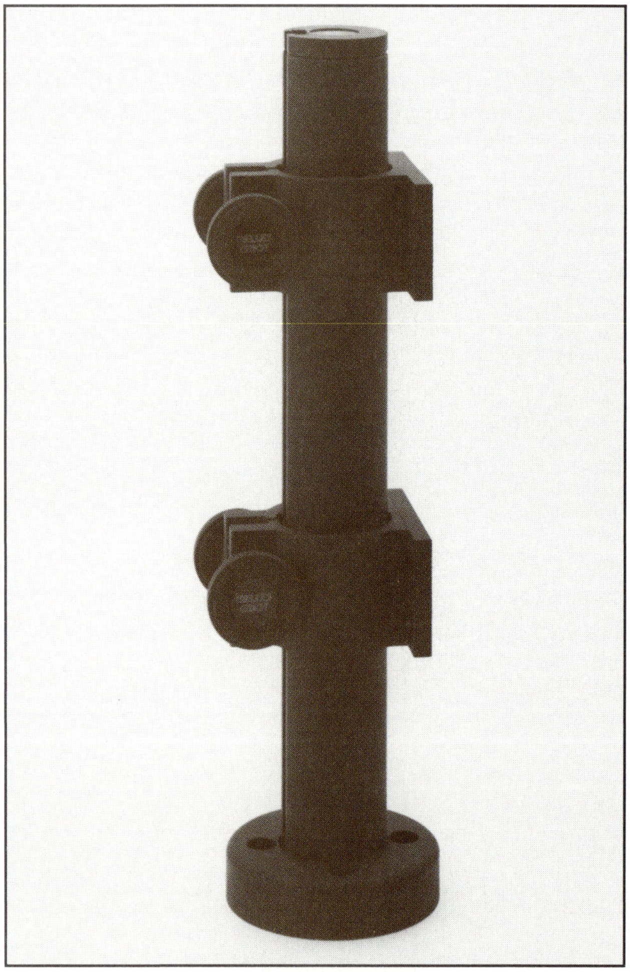

The impulse response graph compares a Melles Griot damped post with an undamped post, both in a cantilever mode, as normally used. Note that the Melles Griot post damps out all vibration within about 50msec. Damping of the first bending mode has been empirically determined to be as high as 20%.

Damped mounting posts are offered with or without a rack gear and in two heights, 300 and 400mm. A number of accessory mounts are also available. The rack gear enables pinion carriers to be used to facilitate vertical adjustment. However, since the rack is within the diameter of the post surface plain carriers may also be used and may be rotated without interference from the rack. Using a post without a rack, it is possible to mount accessories to the tee-slot by means of the tee-slot mounts listed later in this section (page 431).

Damped post carriers are of the split ring type for more secure clamping. They are offered with and without the pinion gear. They have a vertical mounting face with a pattern of 1/4-20 or M6 threaded holes for the attachment of other accessories. If it is desired to make use of a horizontal or angled mounting surface one of the Melles Griot angle brackets may be used to convert the vertical mounting face to the desired angle.

A locking collar is offered to facilitate height setting. Once a component height has been set the locking collar may be used to maintain this height even when the component clamp is released.

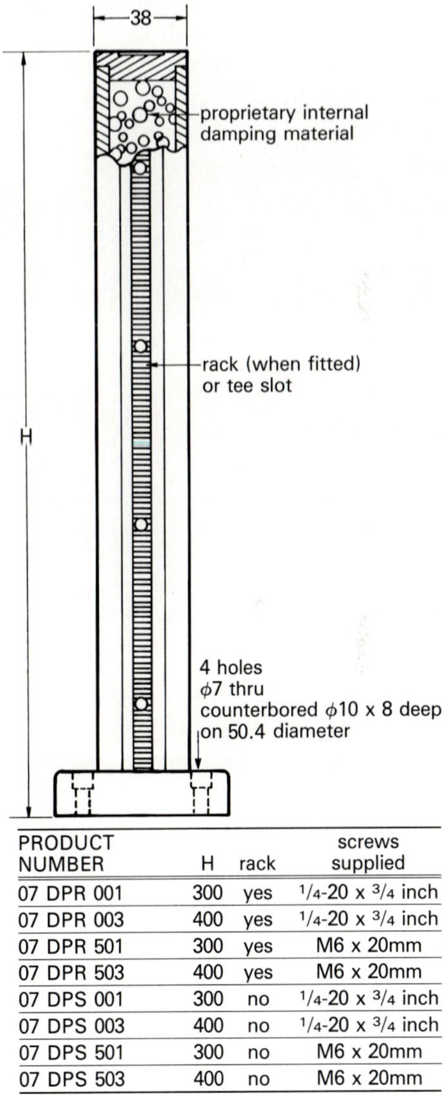

PRODUCT NUMBER	H	rack	screws supplied
07 DPR 001	300	yes	1/4-20 x 3/4 inch
07 DPR 003	400	yes	1/4-20 x 3/4 inch
07 DPR 501	300	yes	M6 x 20mm
07 DPR 503	400	yes	M6 x 20mm
07 DPS 001	300	no	1/4-20 x 3/4 inch
07 DPS 003	400	no	1/4-20 x 3/4 inch
07 DPS 501	300	no	M6 x 20mm
07 DPS 503	400	no	M6 x 20mm

SPECIFICATIONS: DAMPED MOUNTING POSTS

Dimensions: Height: ±0.2mm
Diameter: 38 ±0.2mm

Resonant Frequency, (cantilever mode) 300mm post: 200Hz
400mm post: 140Hz

Damping: Internally damped, 20%

Materials: Steel, Aluminum and proprietary damping materials

Finish: Matte Black Anodized

Damped Mounting Posts*

	Height (mm)	PRODUCT NUMBER Inch Version	Metric Version
with rack gear	300	07 DPR 001	07 DPR 501
	400	07 DPR 003	07 DPR 503
without rack gear	300	07 DPS 001	07 DPS 501
	400	07 DPS 003	07 DPS 503

*Patent Pending

We believe this damping system developed by Melles Griot engineers represents a major advance in composite damping technology. If you are interested in applying this system to other structures, please call us for further information.

CARRIERS FOR DAMPED POSTS

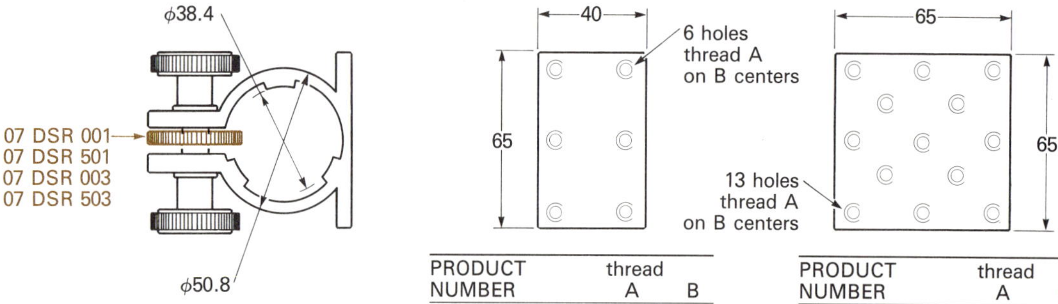

PRODUCT NUMBER	thread A	B
07 DSR 001	1/4-20	25.4
07 DSS 001	1/4-20	25.4
07 DSR 501	M6	25
07 DSS 501	M6	25

PRODUCT NUMBER	thread A	B
07 DSR 003	1/4-20	25.4
07 DSS 003	1/4-20	25.4
07 DSR 503	M6	25
07 DSS 503	M6	25

Mounting Post Carriers

	PRODUCT NUMBER	
	Inch Version	Metric Version
small carrier with pinion	07 DSR 001	07 DSR 501
small carrier without pinion	07 DSS 001	07 DSS 501
larege carrier with pinion	07 DSR 003	07 DSR 503
large carrier without pinion	07 DSS 003	07 DSS 503
locking collar	07 DPC 001	as Inch

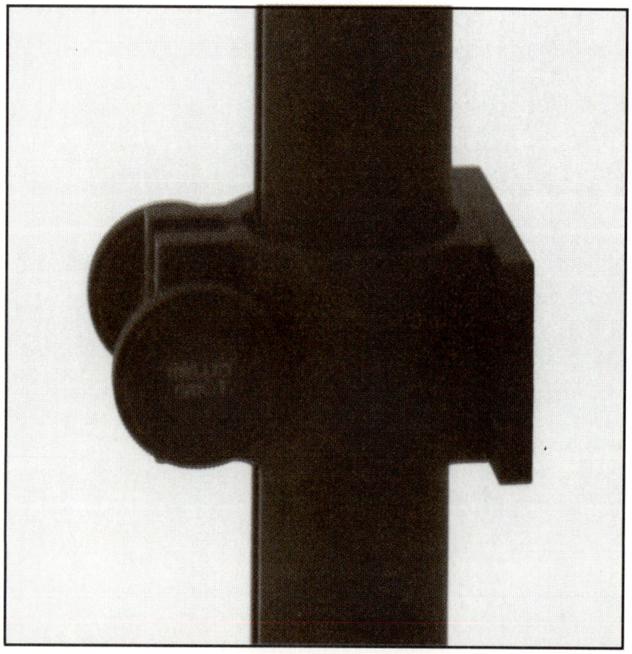

These damped posts may be adapted to carry various loads by using the angle brackets described on page 494.

TEE-SLOT MOUNTS

Melles Griot has been providing a tee-slot feature in several new products introduced over the past two years. We plan to expand on this concept as we believe it provides a simple and convenient means of attachment. These mounts increase the usefulness of tee-slots by providing a means for attachment of accessories.

They will be found particularly useful for attaching components to the damped mounting posts previously described but they may also be used for attachment to other products having tee-slots.

Two types of mount are offered. One is a 65mm square base plate with a threaded hole pattern and the other is a screw attachment for holding lens or mirror mounts. Both types of mount are attached to the tee-slot by inserting a round boss (allowing rotational freedom) inside the tee and then clamping it with a ring clamp. These devices are easy to attach and remove, yet they are secure when in place.

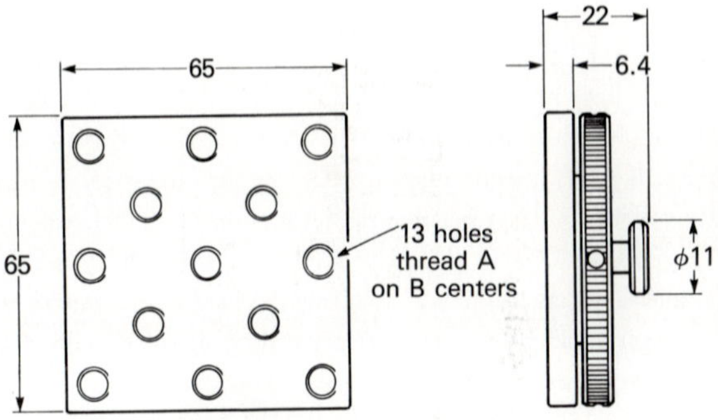

PRODUCT NUMBER	thread A	B
07 STC 001	1/4-20	25.4
07 STC 501	M6	25

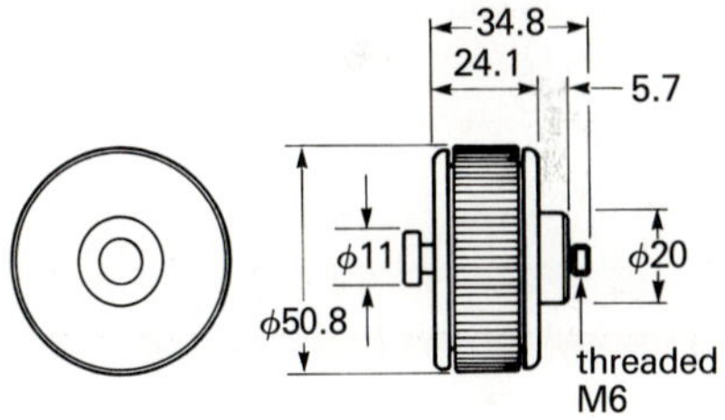

07 STC 003

Tee-slot Mounts

Type	PRODUCT NUMBER Inch Version	Metric Version
Square base	07 STC 001	07 STC 501
Component thread	07 STC 003	as Inch

Tee-slots will be found on all Melles Griot optical tables, optical breadboards, damped posts, and self-contained lasers. They will accept these tee-slot mounts or a 1/4-20 or M6 hex-head bolt.

BEAMSTEERING DEVICES

PRECISION BEAMSTEERING DEVICE

A commonly encountered requirement in laser systems is the steering of a laser beam from one part of an optical table to another. These precision beam-steering devices provide a universal means of beam adjustment. They are constructed from standard Melles Griot components including the damped mounting post, previously described and mirror mounts from the assortment on pages 474-475. In this way it is easy for you to make up alternate versions of the beamsteering device by selection of individual components. The beamsteering device we offer here as a package consists of the following components:

Damped Mounting Post with Rack, 07 DPR 001/07 DPR 501
Lower Mirror Arm Assembly (without pinion), 07 BSA 001/ 07 BSA 501
Upper Mirror Arm Assembly (with pinion), 07 BSA 003/ 07 BSA 503
Lower Mirror Mount, 07 BSM 001/07 BSM 501
Upper Mirror Mount, 07 MAS 001
Two 02 MFG 015 mirrors, mounted in mirror mounts.

You may make up your own version by selecting individual components. Any one of our post mounted mirror holders may be mounted by unscrewing the post and attaching the mirror mount directly to the arm. Our post mounted beamsplitter mount, 07 MSS 001, may be attached in the same manner.

The lower mirror mount, 07 BSM 001, has been devised so the center of the mirror may be placed within 33mm of the table top.

The damped posts 07 DPS 001 and 07 DPS 003 or metric equivalent without a rack, may be used as a beamsteering device by adding two 07 BSA 001s, the lower arms without pinions and your choice of mirror mounts and mirrors.

The mirror arm assemblies are comprised of a mounting block (07 BSB 001) bolted to one of the small carriers for the damped post. The mounting block can also be mounted directly to a table, breadboard or translation stage. It will fit either 1/4-20 on one-inch centers or M6 on 24mm centers hole patterns.

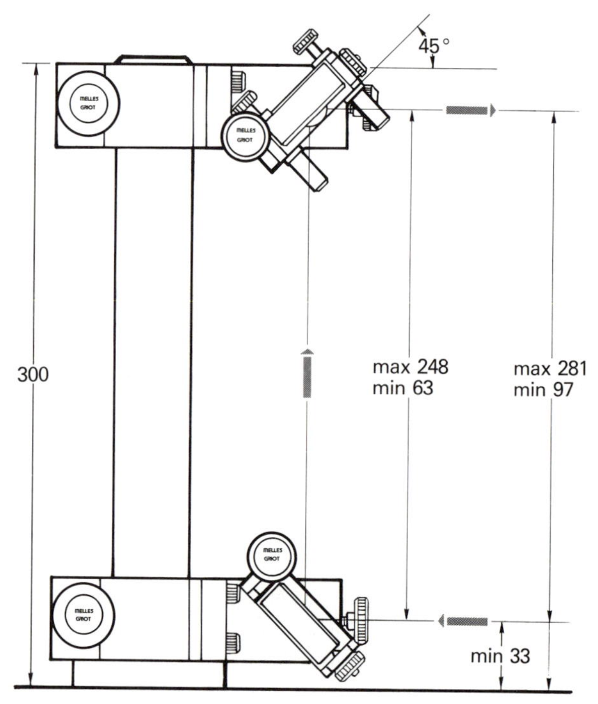

07 BSD

Precision Beamsteering Device

	PRODUCT NUMBER	
	Inch Version	Metric Version
Precision beamsteering device, complete	07 BSD 001	07 BSD 501
Lower arm, less mirror mount, no pinion	07 BSA 001	07 BSA 501
Upper arm, less mirror mount, with pinion	07 BSA 003	07 BSA 503
Lower mirror mount, less mirror	07 BSM 001	07 BMS 501
Upper mirror mount, less mirror	07 MAS 001	as Inch
Mounting block	07 BSB 001	07 BSB 501

Various combinations of beamsteering devices may be assembled from these components. A frequently used form of beamsteering device employs only a single mirror as in the photograph.

This device may also be mounted upside down under any Melles Griot optical table facilitate transfer of a laser beam through the access hole. Threaded inserts are provided in every table for this purpose.

SIMPLE BEAMSTEERING DEVICE

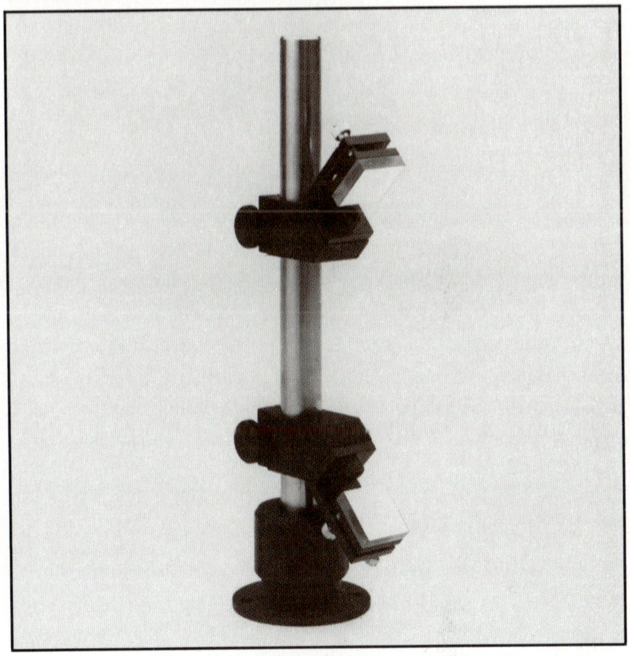

This simple beamsteering device may be used for positioning of laser beams when the ultimate precision and stability is not so important. It utilizes a standard 20mm diameter mounting post and two mounted mirrors which are attached to the post by means of thumb screws. For your convenience we offer a complete package of the following components:

Mounting Post 20mmϕ by 250mm long, 07 REL 003 or 07 RML 003

Mounted mirrors (qty. 2), 07 MMA 006

Pair of beamsteerer clamp brackets, 07 BSC 006.

To make up your own version select individual components.

Simple Beam-Steering Device

	PRODUCT NUMBER	
	Inch Version	Metric Version
Simple beamsteering device, complete	07 BSP 001	07 BSP 501
Pair of beamsteerer clamp brackets	07 BSC 006	as Inch

MINISYSTEM

The increasing use of fiber optics and laser diode sources has resulted in a demand for smaller and smaller experimental systems. In some situations even a small optical breadboard seems huge in comparison with the optical path under investigation. For those situations Melles Griot offers this "Minisystem."

"Minisystem" is a name we have given to a collection of mounts and component holders you will find spread throughout the remainder of this catalog. It is based on a small mounting plate and a miniature optical rail. It uses 8mm diameter post mounted component holders. Many of these, such as laser diode holders and gradient index rod lens holders, are specifically oriented toward fiber optic and laser diode applications. Melles Griot 25mm square translation slides form a part of the Minisystem. Together with the components listed here they make up the smallest commercially available optical support and positioning system.

MINI-MOUNTING PLATES

These Minisystem mounting plates are solid aluminum plates, 10mm thick, arranged with an array of 1/4-20 tapped holes on inch centers or M6 tapped holes on 25mm centers. They are black anodized to reduce surface reflections and glare. Each plate has four of its tapped holes replaced by counterbored clearance holes so that 1/4 inch or M6 screws can be used to attach the plate itself to some other device such as a breadboard or extender posts. Three sizes are offered to cover most applications.

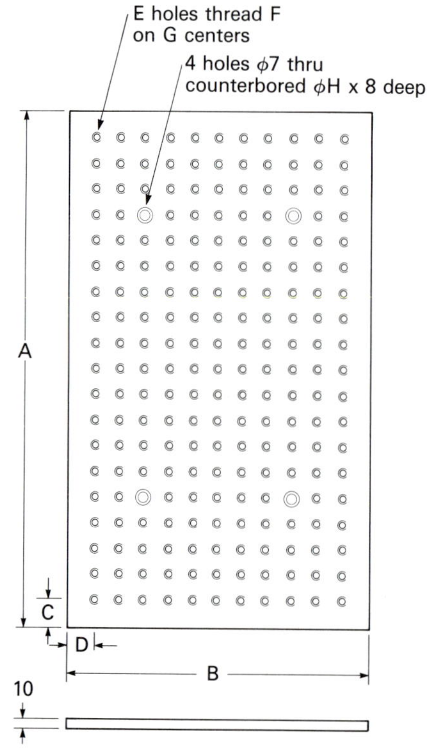

PRODUCT NUMBER	A	B	C	D	E	thread F	G	ϕH
*07 BLP 001	200	200	—	—	45	1/4-20	25.4	10
*07 BLP 002	300	300	—	—	77	1/4-20	25.4	10
07 BLP 003	500	300	21.4	23	205	1/4-20	25.4	11
*07 BLP 501	200	200	—	—	45	M6	25	11
*07 BLP 502	300	300	—	—	77	M6	25	11
07 BLP 503	500	300	25	25	205	M6	25	11

*See drawing on page 487 for specific hole patterns.

Mini-Mounting Plates

Length (mm)	Width (mm)	PRODUCT NUMBER Inch Version	Metric Version
200	200	07 BLP 001	07 BLP 501
300	300	07 BLP 002	07 BLP 502
500	300	07 BPM 003	07 BPM 503

MINI-RAILS

These Minisystem rails are simple dovetail section aluminum optical rails. They ar black anodized to reduce surface reflections and glare. These may be used in connection with an optical breadboard or mounting plate to provide a linear movement for optical components. The rails are milled with a number of counterbored slots so that they can be screwed down to either an inch or metric hole pattern at a wide range of angles. If it is desired to make up a three-dimensional structure, these rails may be mounted vertically or at an angle by attachment to a mini-rail mounting bracket.

Carriers are offered in two sizes for mounting posts, post holders or stages. Provided only 8mm post mounted accessories are used, a common center height of 75mm will be maintained.

Mini-rails

Length (mm)	Number of slots	PRODUCT NUMBER
50	1	07 ORM 001
100	2	07 ORM 003
250	3	07 ORM 005
500	5	07 ORM 007

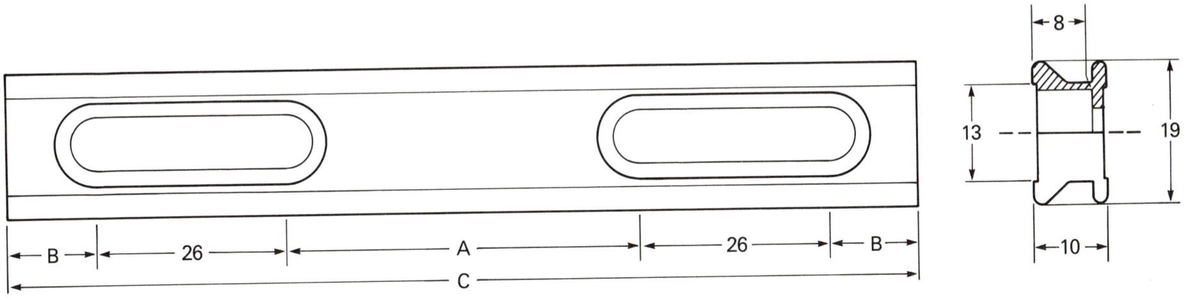

PRODUCT NUMBER	A	B	C	no. of slots
07 ORM 001	—	12	50	1
07 ORM 003	24	12	100	2
07 ORM 005	74	12	250	3
07 ORM 007	74	37	500	5

MINI-RAIL CARRIERS

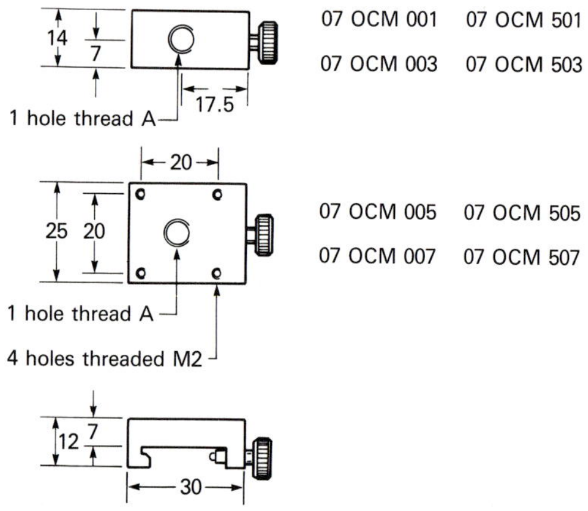

PRODUCT NUMBER	thread A
07 OCM 001	8-32 UNC
07 OCM 003	1/4-20 UNC
07 OCM 501	M4
07 OCM 503	M6
07 OCM 005	8-32 UNC
07 OCM 007	1/4-20 UNC
07 OCM 505	M4
07 OCM 507	M6

MINI-RAIL BRACKET

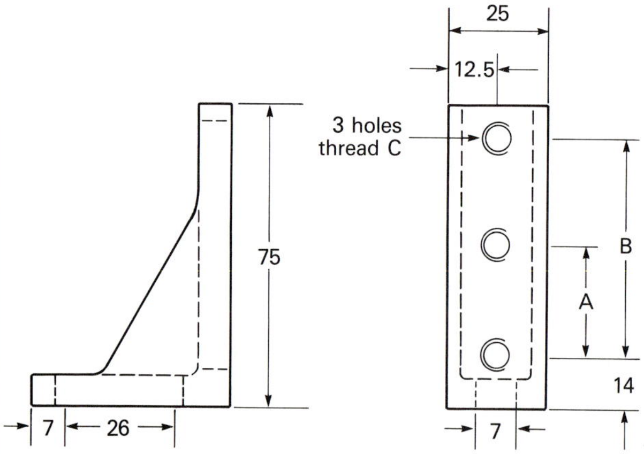

PRODUCT NUMBER	A	B	thread C
07 ORB 001	25.4	50.8	1/4-20 UNC
07 ORB 501	25.0	50.0	M6

Mini-rail Bracket

PRODUCT NUMBER	
Inch Version	Metric Version
07 ORB 001	07 ORB 501

Mini-rail Carriers

Width (mm)	Central Thread	PRODUCT NUMBER Inch Version	PRODUCT NUMBER Metric Version
14	8-32	07 OCM 001	—
14	M4	—	07 OCM 501
14	1/4-20	07 OCM 003	—
14	M6	—	07 OCM 503
25	8-32	07 OCM 005	—
25	M4	—	07 OCM 505
25	1/4-20	07 OCM 007	—
25	M6	—	07 OCM 507

See also the mini-translation stage on page 493 and the mini-mounted mirror on page 473.

OPTICAL RAILS

In the past many optical researchers made use of heavy and precise "optical benches." In today's optical systems, however, two- and three-dimensional arrays of components are common; hence the popularity of optical tables. Sometimes, it is desirable to align components in a way that will permit a variation of the spacing between them. For these situations an optical rail is the ideal solution.

Melles Griot optical rails are dovetail section aluminum extrusions, which are black anodized to reduce surface reflections. They may be mounted on an optical table or used in a free-standing mode. These rails are low in profile, enabling components to be mounted as close as possible to the table surface for maximum stability.

Counterbored slots, located in the center of the rails along their length, enable them to be bolted down to an optical table or other work surface. Since slots are used, and not simply clearance holes, the rails may be attached at a wide range of angles to the hole pattern and will also accommodate both inch and metric hole patterns. Cross sections of the rails are such that the outer edges fall in contact with the table surface and pressure exerted by screws on the center-line will clamp the rails securely in position. An alternate method of attachment is to use table clamps (listed on page 423) to clamp the outer ledge of the rails in place.

Melles Griot optical rails are offered in two widths, 50 and 100mm, with a variety of lengths. Rail width should be selected to suit the size of components to be mounted.

A selection of carriers is offered for each rail width. These carriers have a lip which engages with one side of the dovetail and a clamp on the other side. The clamp can be released to permit removal of the carrier from the rail or tightened to secure the carrier in position. All contacting surfaces are covered with a low friction tape for ease of translation. Carrier surfaces have a central M16 threaded hole for attachment of post holders (see pages 445-448), and an array of 1/4-20 or M6 threaded holes for attachment of positioning stages and other accessories. Short carriers should be used to mount components on the rails in close proximity to each other. Longer carriers offer greater stability and may be used with translation slides and other large objects.

A scale along the edge of the rail can be read against a fiducial mark on the carriers as an indication of their relative positions. This scale has millimeter divisions, enabling position to be estimated to an accuracy of 0.2mm.

SPECIFICATIONS: OPTICAL RAILS

Straightness: ±0.5mm/m

Scale Division: 1mm

Material: Aluminum, black anodized

Optical Rails

Width (mm)	Length (m)	Number of Slots	PRODUCT NUMBER
50	0.25	3	07 ORN 001
50	0.50	3	07 ORN 003
50	0.75	4	07 ORN 005
50	1.00	5	07 ORN 007
50	1.50	7	07 ORN 009
50	2.00	10	07 ORN 011
100	0.50	3	07 ORP 003
100	1.00	5	07 ORP 007
100	2.00	10	07 ORP 011
100	3.00	15	07 ORP 015

Optical Rail Carriers

For Rail Width (mm)	Carrier Length (mm)	PRODUCT NUMBER Inch Version	PRODUCT NUMBER Metric Version
50	25	07 OCN 001	07 OCN 501
50	65	07 OCN 003	07 OCN 503
100	40	07 OCP 001	07 OCP 501
100	65	07 OCP 003	07 OCP 503
100	120	07 OCP 005	07 OCP 505
100	160	07 OCP 007	07 OCP 507

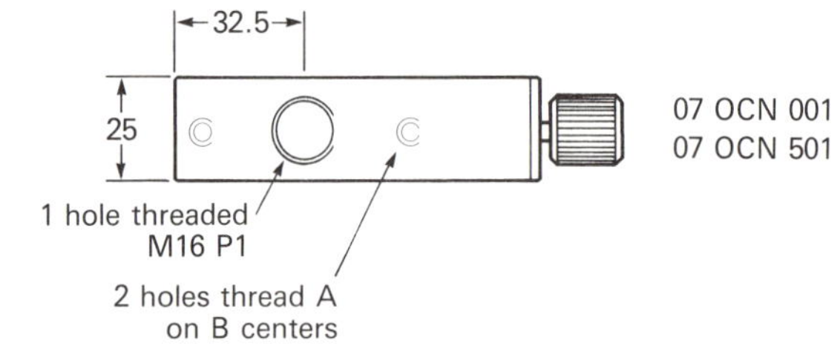

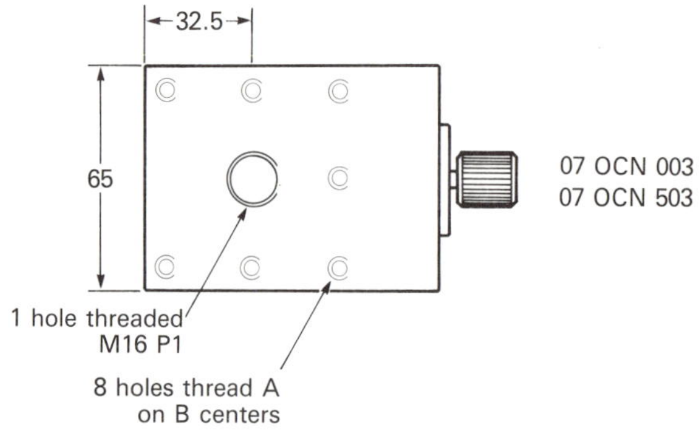

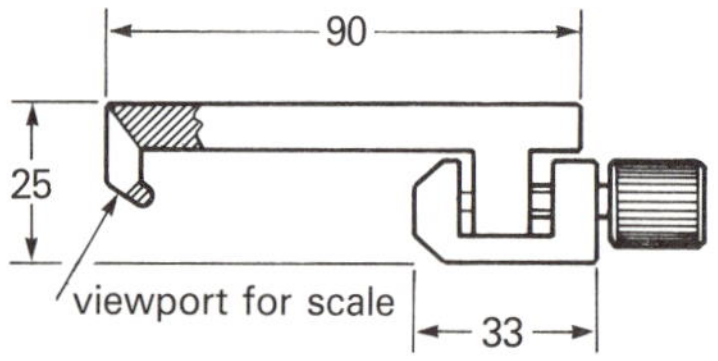

PRODUCT NUMBER	thread A	B
07 OCN 001	1/4-20 UNC	25.4
07 OCN 003	1/4-20 UNC	25.4
07 OCN 501	M6	25
07 OCN 503	M6	25

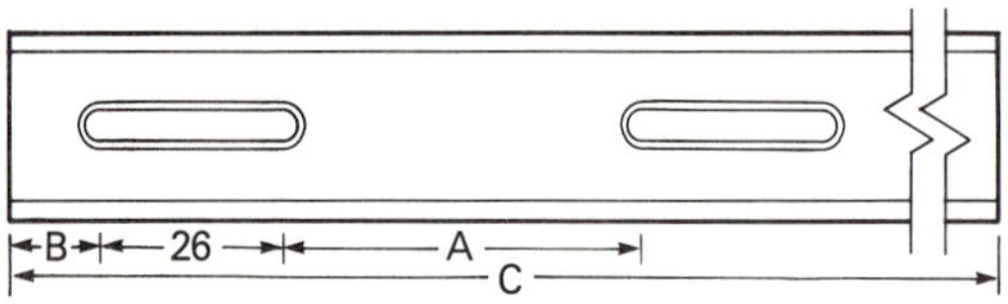

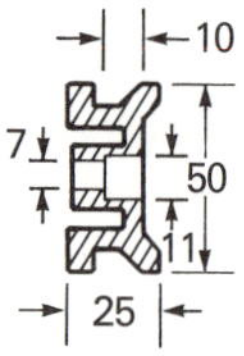

PRODUCT NUMBER	A	B	C	no. of slots
07 ORN 001	74	12	250	2
07 ORN 003	174	37	500	2
07 ORN 005	174	62	750	3
07 ORN 007	174	87	1000	4
07 ORN 009	174	37	1500	7
07 ORN 011	174	87	2000	9

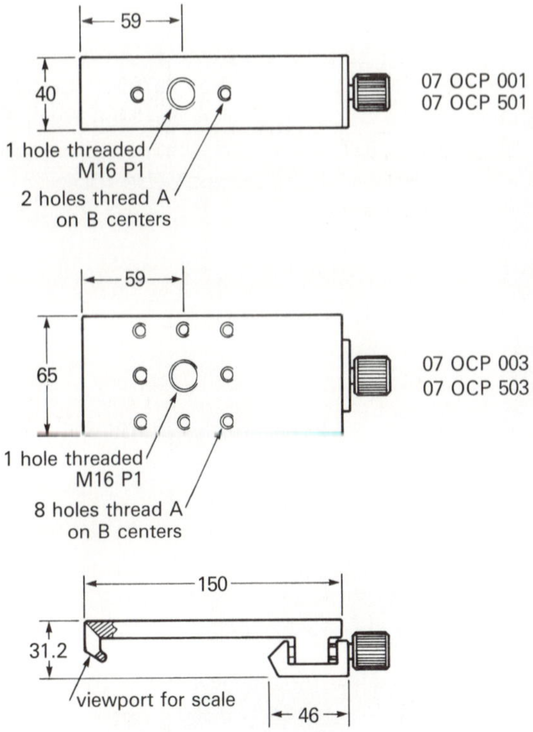

PRODUCT NUMBER	thread A	B
07 OCP 001	1/4-20 UNC	25.4
07 OCP 003	1/4-20 UNC	25.4
07 OCP 501	M6	25
07 OCP 503	M6	25

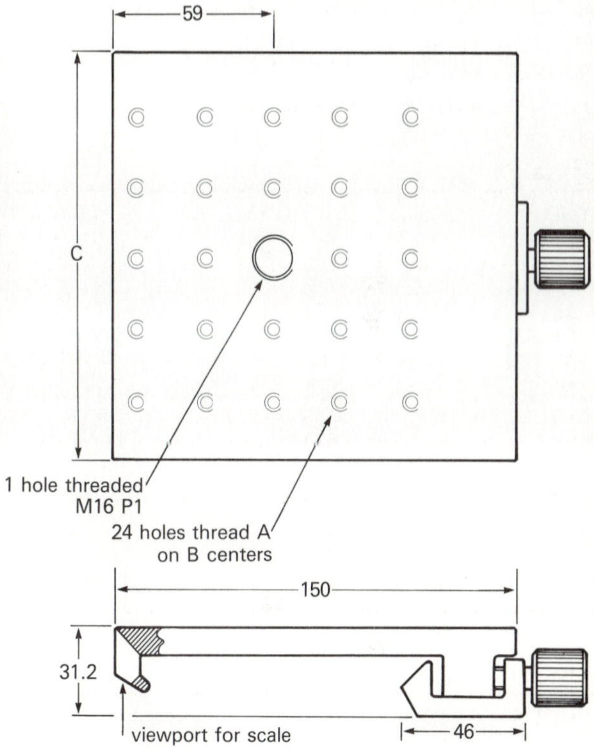

PRODUCT NUMBER	thread A	B	C
07 OCP 005	1/4-20 UNC	25.4	120
07 OCP 007	1/4-20 UNC	25.4	160
07 OCP 505	M6	25	120
07 OCP 507	M6	25	160

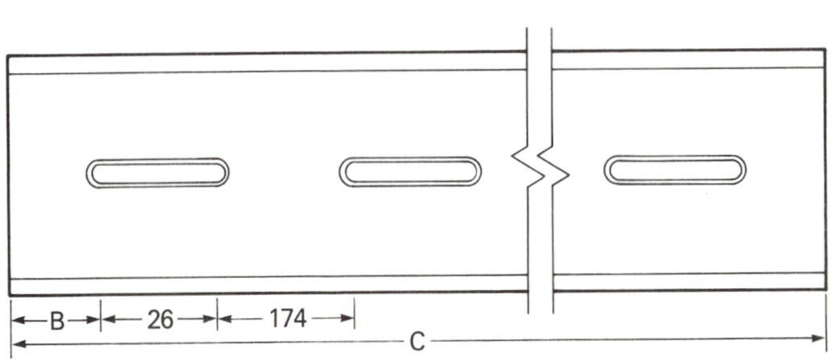

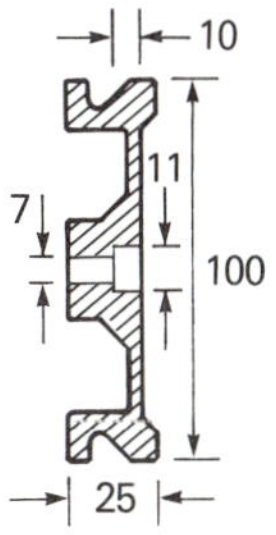

PRODUCT NUMBER	B	C	no. of slots
07 ORP 003	37	500	3
07 ORP 007	87	1000	5
07 ORP 011	87	2000	10
07 ORP 015	87	3000	15

OPTICAL TABLES

Melles Griot Composites in Tustin, California has been established with the sole purpose of providing better optical tables than have previously been available.

We make extensive use of technology developed by the aerospace industry to manufacture stable and highly damped composite tables.

From insertion of the threads, to bonding of the honeycomb core, to assembly of the table, to finishing of the top plates, we take great care to produce a better table.

Furthermore, we are able to respond quickly to your need for special sizes and configurations. So if you're equipping a new laboratory or refurbishing an existing one — call us first.

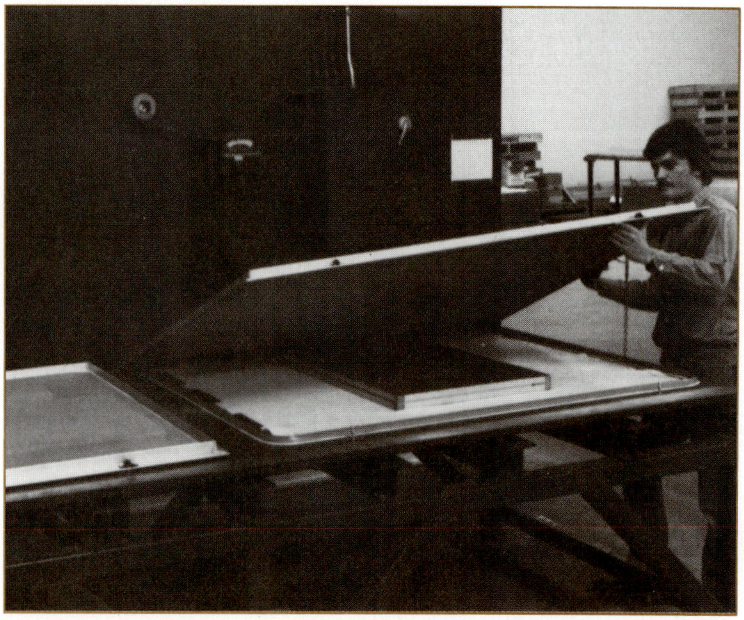

COMPONENT HOLDERS AND MOUNTS

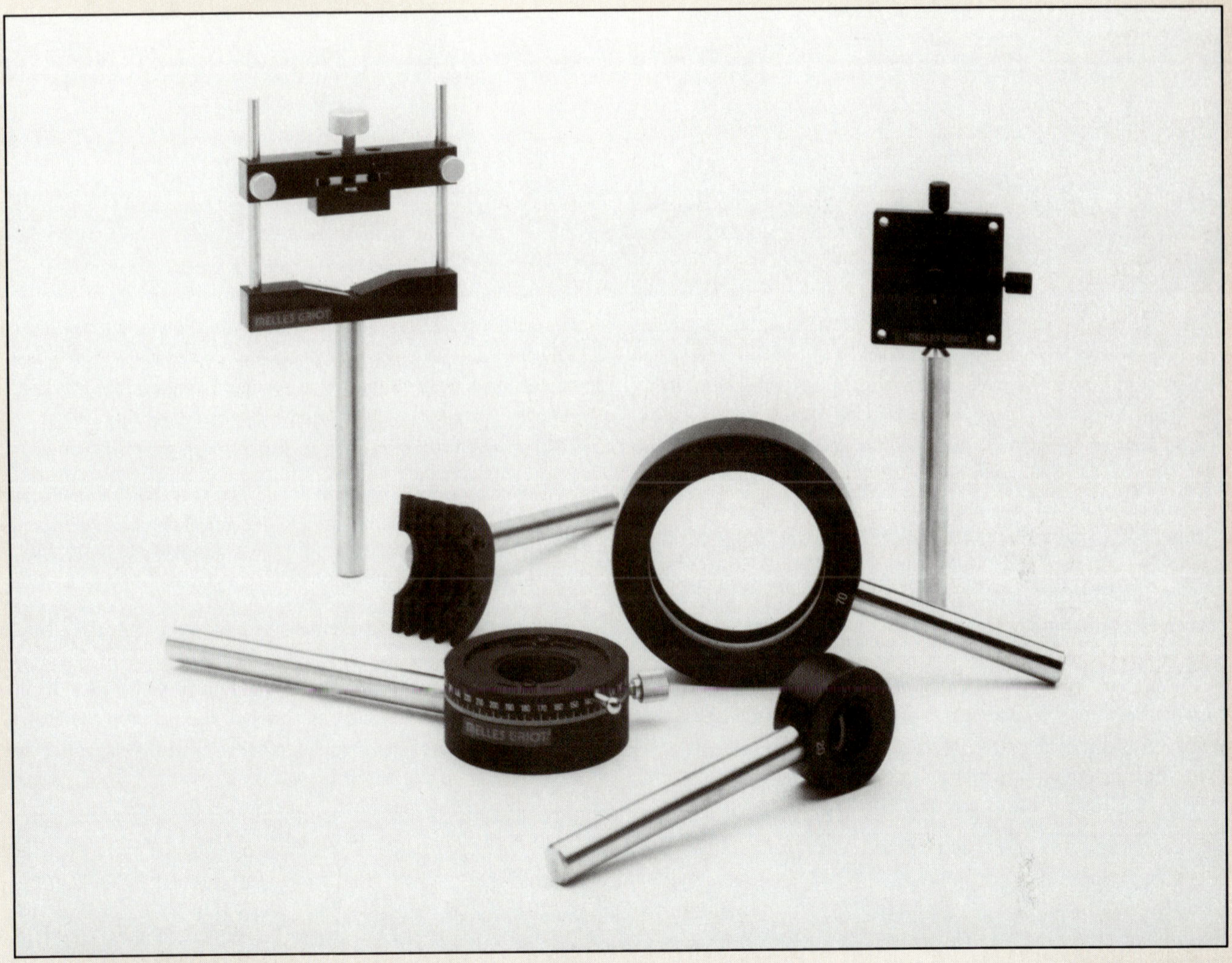

If you have ever searched around a laboratory to find something suitable to hold an optical component, you will appreciate the wide selection of holders we offer here. We have made every effort to supply holders for all the components we offer and for other commonly used devices.

Contents **Page Number**

POST-MOUNTING SYSTEMS

Post-mounting of optical components is rapid, convenient and cost effective. Components can be positioned, removed and accurately replaced. By placing post holders in-line on an optical breadboard, an optical axis can be quickly established and held. By combining post holders with translation and rotational stages, very precise motions and offsets may be achieved. By using post holders in conjunction with the carriers of an optical rail components may be moved in a straight line.

Because of this versatility a variety of post-mounted systems have evolved in the past. Many offer a bewildering array of special purpose mechanical components which have been designed to overcome shortcomings in the original system design. Others have sacrificed stability and reliability for versatility.

The Melles Griot post-mounting system has been designed primarily for the optical engineer, to provide him with the best possible mix of reliability, stability and versatility. Post lengths have been established on a component-by-component basis to assure that the optical center line will lie at a convenient height. Post systems are available in three diameters to allow a sensible match between the mechanical and optical components. Component holders have been designed and specified to match Melles Griot's extensive range of standard lenses, mirrors, prisms, filters and other components.

All items are carried in stock and are available for immediate delivery. This, too, is important to the optical engineer because, as we all know, when a piece of optical hardware is needed, it is needed immediately.

Further, most of these components are available in modified or simplified form for the OEM user. By making use of essentially identical mounting systems in manufacture as used during the initial design and prototyping, the original equipment manufacturer assures himself of dependable, reliable system performance as well as reduced engineering cost.

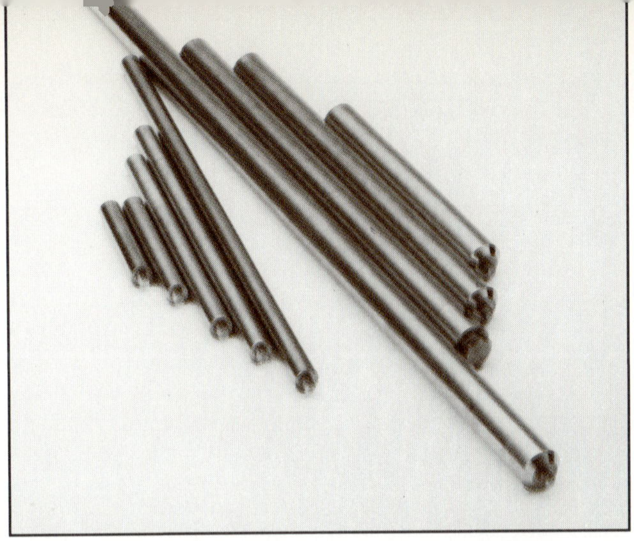

MOUNTING POSTS

Post-mounting of optical components on an optical table or rail is the most convenient and inexpensive method of setting up prototype optical systems. One has the freedom to establish quickly and accurately an optical axis at a given and convenient height above the reference surface and to align components to one another with ease and confidence.

Over the years a variety of post diameters have evolved from different manufacturers of this type of component, the most common sizes being 8.0mm, 12.0mm, 12.7mm, 13.7mm, 14.3mm and 20mm diameter.

Melles Griot post and post-mounted components are available in three standard sizes: 8mm, 12mm, and 20mm. Post holders and clamps are bored to accept these size posts smoothly and precisely. The result is a firm and controlled fit of all components with greatly improved accuracy during initial set-up. Melles Griot also offers a range of universal post holders which will acccommodate any post from 12.0mm to 14.5mm in diameter. These post holders have a machined groove down one side to provide two line contacts when a smaller post is clamped by the thumb screw.

Melles Griot 8mm posts are intended for use with Minisystem components in such applications as fiber optic or laser diode assemblies. Melles Griot 12mm posts are intended for use with lightweight components such as lenses, filters, and pinholes. The 20mm posts provide greater stability for mounting heavier components such as mirror mounts, translation stages and large optics.

Various lengths and terminations are available. The 1/4-20 or M6 threads will be found to be most convenient for mounting directly to an optical breadboard or table. The M16 threads provide for direct mounting of either the small or large posts to the complete range of Melles Griot magnetic bases, kinematic bases and translation and rotational stages. Posts with an M16 threaded end are provided with a hole (on the 20mm posts) or a pair of flats on the base flange (12mm posts) for tightening or loosening the posts during mounting. The 4mm diameter hole accepts a 3.5mm diameter breaker bar which is supplied in the 07 AKT 003 and 07 AKT 503 accessory kits for optical hardware (see page 480). The M6 threads are used when changing posts on all post-mounted accessories such as lens holders, mirror holders, polarizers and so forth.

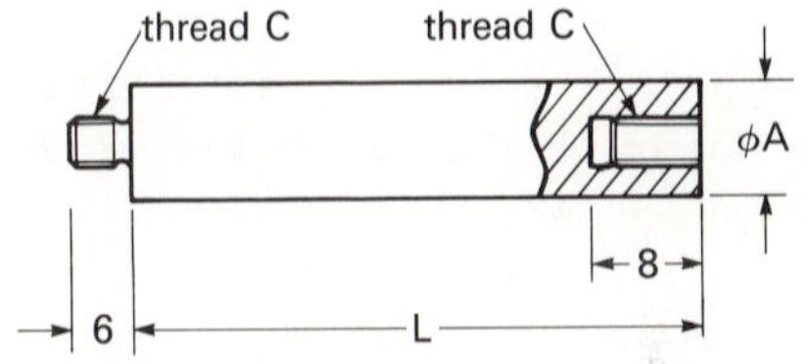

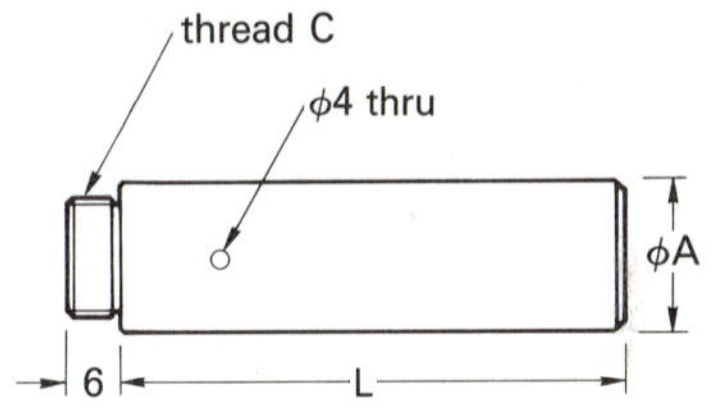

07 PML

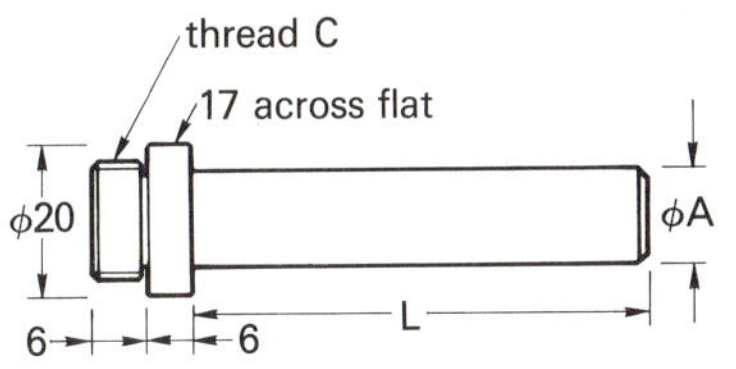

07 PMS

All Melles Griot mounting posts, except those within M16 termination, are supplied with matching male and female threads on either end. This enables the posts to be joined end to end for greater length.

SPECIFICATIONS: MOUNTING POSTS

Diameter (ϕ): + 0.00mm, -0.03mm
Length: ±0.02mm
Termination: 8-32, 1/4-20, M4, M6 or M16 male threaded stud
Material: Stainless steel
Finish: Centerless ground

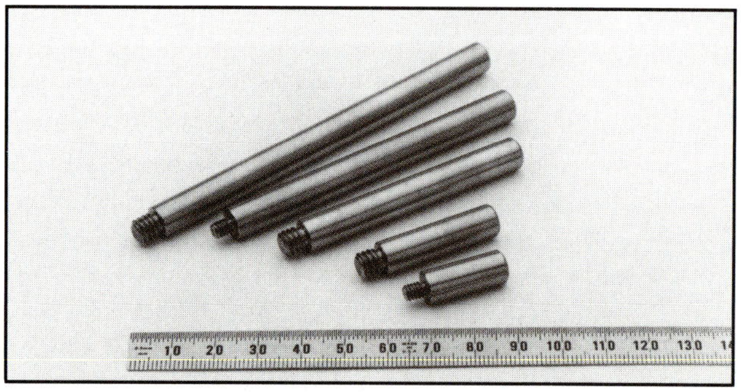

Mounting Posts

Diameter ϕA (mm)	Length L (mm)	PRODUCT NUMBER Thread, C 8-32	1/4-20	M4	M6	M16
8	20	07 REU 001	07 REW 001	07 RMU 001	07 RMW 001	—
8	30	07 REU 003	07 REW 003	07 RMU 003	07 RMW 003	—
8	40	07 REU 005	07 REW 005	07 RMU 005	07 RMW 005	—
8	60	07 REU 007	07 REW 007	07 RMU 007	07 RMW 007	—
8	80	07 REU 009	07 REW 009	07 RMU 009	07 RMW 009	—
8	100	07 REU 011	07 REW 011	07 RMU 011	07 RMW 011	—
12	40	—	07 RES 001	—	07 RMS 001	—
12	60	—	07 RES 002	—	07 RMS 002	—
12	80	—	07 RES 003	—	07 RMS 003	—
12	100	—	07 RES 004	—	07 RMS 004	—
12	150	—	07 RES 005	—	07 RMS 005	07 PMS 001
12	200	—	07 RES 006	—	07 RMS 006	—
12	250	—	07 RES 007	—	07 RMS 007	—
12	300	—	07 RES 008	—	07 RMS 008	07 PMS 002
12	500	—	07 RES 009	—	07 RMS 009	07 PMS 003
20	120	—	07 REL 001	—	07 RML 001	—
20	150	—	07 REL 002	—	07 RML 002	07 PML 001
20	250	—	07 REL 003	—	07 RML 003	—
20	300	—	—	—	—	07 PML 002
20	500	—	07 REL 004	—	07 RML 004	07 PML 003

To change the post length on any of the optics holders described on the following pages, select a mounting post with an M6 thread from the table above.

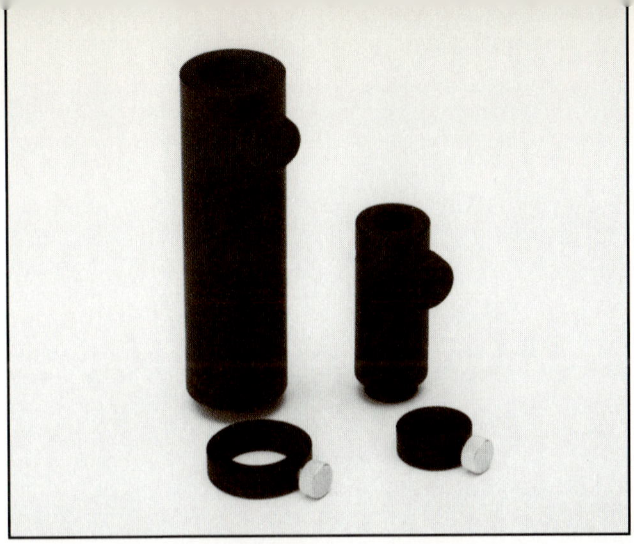

POST HOLDERS

Melles Griot post holders are machined from solid brass to provide stability, accuracy and reliability. Standard post holders have an 8mm, a 12mm or a 20mm bore to fit our complete range of post-mounted accessories precisely. A thumb screw locks the post firmly in position with only moderate pressure. Because of the exact fit, mounted accessories have a firm feel during adjustment with no tendency to rock side-to-side as is more commonly found with oversized holders.

For even greater positioning stability, collet-type clamping post holders are offered. These are available for both 12mm and 20mm posts.

For larger, non-standard mounting posts (from 12 to 14.5mm in diameter) a universal post holder is also available. This post holder has a cylindrical slot down one side to provide two-line contact, and improved stability, when the posts are clamped by the thumb screw.

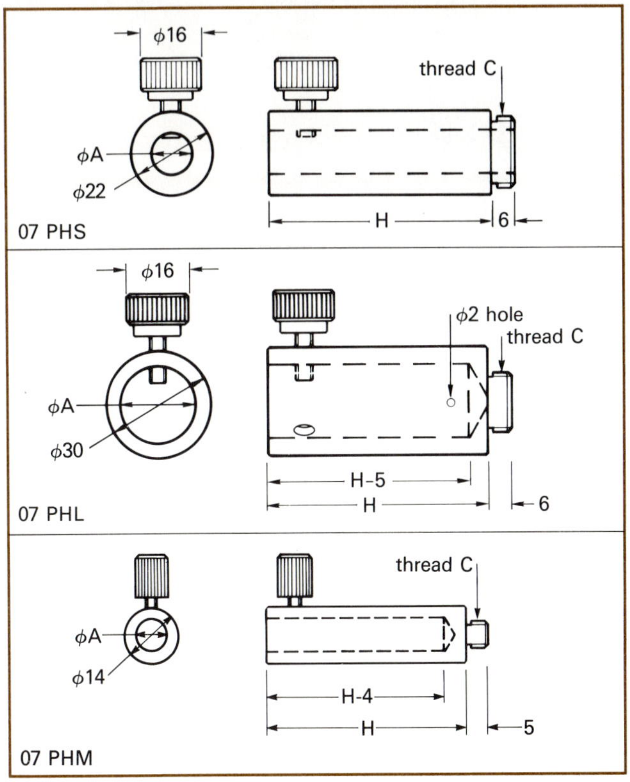

All standard and universal post holders terminate in a male thread on the bottom for mounting directly into our complete range of translation and rotation stages as well as the kinematic bases, magnetic bases and optical rail carriers.

For optical table mounting, accessory bases are available for the standard and universal post holders. The collet-type post holder is available with a similar integral mounting base.

All are chrome plated flat black to reduce possible stray reflections and scatter. All 8mm, 12mm and 20mm post holders are supplied with a clamping collar for attachment to post-mounted accessories to preserve the setting height when these items are removed and then replaced in the post holder.

SPECIFICATIONS: POST HOLDERS

Bore: +0.05mm, -0.00mm
Length: ±0.02mm
Termination: 1/4-20, M6 or M16 male threaded stud.
Material: Brass
Finish: Black chrome, bright chrome knobs

Standard Post Holders

Bore φA (mm)	Height H (mm)	Thread C	Load Capacity (kg)	PRODUCT NUMBER
8	20	1/4-20	10	07 PHM 001
8	30	1/4-20	10	07 PHM 003
8	50	1/4-20	10	07 PHM 005
8	80	1/4-20	10	07 PHM 007
8	20	M6	10	07 PHM 501
8	30	M6	10	07 PHM 503
8	50	M6	10	07 PHM 505
8	80	M6	10	07 PHM 507
12	30	M16	30	07 PHS 001
12	50	M16	30	07 PHS 003
12	80	M16	30	07 PHS 006
12	120	M16	30	07 PHS 008
20	30	M16	50	07 PHL 001
20	50	M16	50	07 PHL 003
20	80	M16	50	07 PHL 006
20	120	M16	50	07 PHL 008

UNIVERSAL POST HOLDERS

These post holders are intended for use with non-standard diameter mounting posts. They accept mounting posts from 12 to 14.5mm in diameter. These post holders have a cylindrical slot down one side of the central bore. When the thumb screw is tightened the post is forced into line contact along the two edges of the groove. In this way stable positioning of undersized mounting posts may be achieved. To avoid side-to-side motion during set-up it is prudent to locate the groove in the direction of the optical axis.

Universal post holders are made of brass and finished with black chrome. The bottom terminates in an M16 male thread for direct mounting on the complete range of Melles Griot translation and rotation stages.

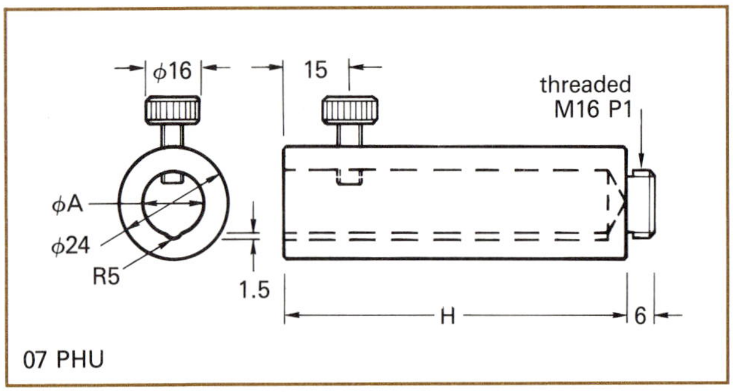

Universal Post Holders

Bore ϕA (mm)	Height H (mm)	Load Capacity (kg)	PRODUCT NUMBER
14.5	30	30	07 PHU 001
14.5	50	30	07 PHU 003
14.5	80	30	07 PHU 006
14.5	120	30	07 PHU 008

COLLAR FOR MOUNTING POSTS

An appropriate size collar is supplied with each 8mm, 12mm or 20mm Post Holder. These collars are, however, also available separately. Material is brass, finished with flat black chrome.

Collar For Mounting Posts

Bore (mm)	PRODUCT NUMBER
8	07 PHC 003
12	07 PHC 001
20	07 PHC 002

To hold mounting posts of other manufacturers, please use the universal post holders listed here

COLLET TYPE POST HOLDERS

Melles Griot collet type post holders firmly and securely grip 12 and 20mm diameter posts assuring precise location of optical components. Both bore sizes utilize a four jaw compression collet to grip the post. Collet pressure is controlled by the knurled top clamping ring.

Collet type post holders are available with either M16 male threaded studs on their bottom or with a flanged base which accepts either 1/4-20 or M6 screws. Holes are drilled on 50.4mm centers. This base suits both 1/4-inch holes on inch centers and M6 holes on 25mm centers. Both styles are available in 12mm and 20mm post sizes.

SPECIFICATIONS: COLLET TYPE POST HOLDERS

Dimensions (ϕ): $\pm$0.1mm
Bore: +0.05mm, -0.00mm
Material: Brass
Finish: Black Chrome

Collet Post Holders

Bore ϕB (mm)	Load Capacity (kg)	PRODUCT NUMBER Table Mounting	M16 Thread
12	50	07 PCS 001	07 PCS 002
20	75	07 PCL 001	07 PCL 002

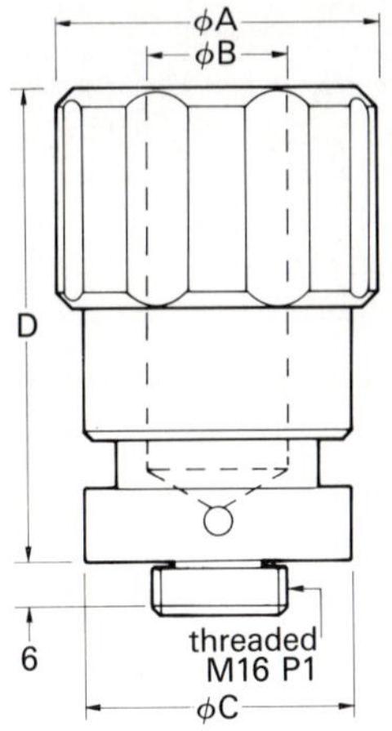

PRODUCT NUMBER	φA	φB	φC	D
07 PCL 002	45	20	38	65
07 PCS 002	36	12	28	60

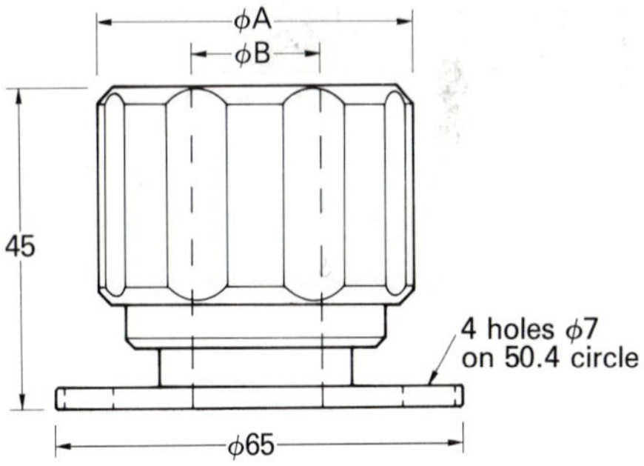

PRODUCT NUMBER	φA	φB
07 PCL 001	46	20
07 PCS 001	35	12

This type of post holder provides a far more secure and uniform clamp than the screw type post holders previously described.

ADJUSTABLE HEIGHT POST HOLDERS

These collet type post holders are provided with a precise and continuous height adjustment capability. A grip collar may be rotated to provide a vertical adjustment of 10mm to the post. Rotation of the collar acts on a broad but fine thread causing the inner collet to be raised or lowered.

The base of the post holders are fitted with an M16 male thread for attachment to any of the Melles Griot post holder bases or translation stages.

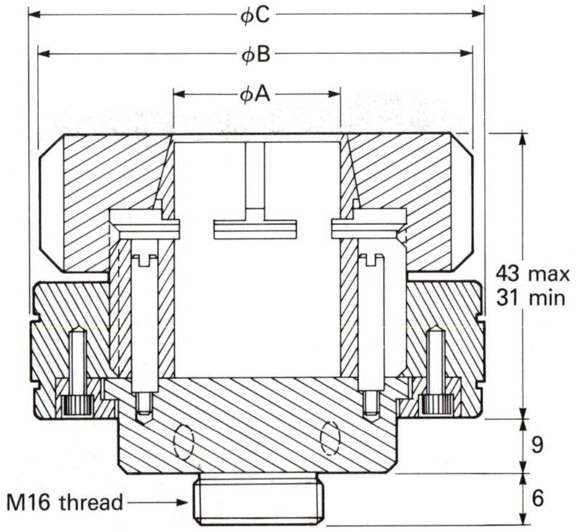

PRODUCT NUMBER	φA	φB	φC
07 PCA 001	12	50.8	54
07 PCA 003	20	50.8	54

Adjustable Height Post Holders

For Post Diameter (mm)	PRODUCT NUMBER
12	07 PCA 001
20	07 PCA 003

Supplied complete with 4mm tightening bar.

FASTENER KITS

All required fasteners and screws are supplied with Melles Griot optical hardware when originally shipped. This kit is designed specifically to replace lost screws and fasteners. It is packaged, complete with specific tools, in a closeable plastic box.

Fastener Kits

Contents	Quantity	PRODUCT NUMBER
Inch Version:		
1/4-20x3/8 inch Allen cap screws	50	
1/4-20x1/2 inch Allen cap screws	50	
1/4-20x3/4 inch Allen cap screws	50	
M6x12mm Allen cap screws	50	07 AKT 005
M6x20mm Allen cap screws	50	
1/4-20 Allen key	2	
M6 Allen key	2	
1/4-20x1/2 inch thumb screws	4	
M6x12mm thumb screws	4	
Metric Version:		
M6x10mm Allen cap screws	50	
M6x12mm Allen cap screws	50	
M6x20mm Allen cap screws	50	07 AKT 505
M6 Allen key	2	
M6x12mm thumb screws	4	

POST/POST HOLDER BASES

These bases facilitate the mounting of either posts or post holders on optical breadboards, tables or other surfaces. Each has a central threaded hole — which accepts the male stud of either a post holder or a post.

We offer three different types of mounting bases for posts and post holders. Each has its own special features and selection will depend on the restraints of your application.

SQUARE BASE

The square base is the most versatile in that it allows a great deal of flexibility in positioning of the post holder or accessory. By using the slots it is possible to position this base anywhere from one row of holes to the next in either direction.

RECTANGULAR BASE

This base is used when it is necessary to position components close together. It allows some lateral positioning but is also as small as possible.

CIRCULAR BASE

This base is used to position a holder above a mounting hole. It permits no positioning of the holder but it will accommodate both metric and inch hole spacing.

All bases are supplied complete with all necessary mounting hardware.

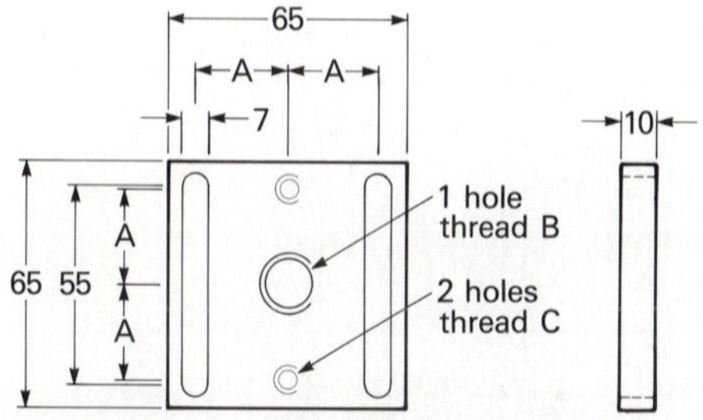

PRODUCT NUMBER	A	thread B	thread C
07 RPC 012	25.4	1/4-20	1/4-20
07 RPC 014	25.4	M16	1/4-20
07 RPC 512	25	M6	M6
07 RPC 514	25	M16	M6

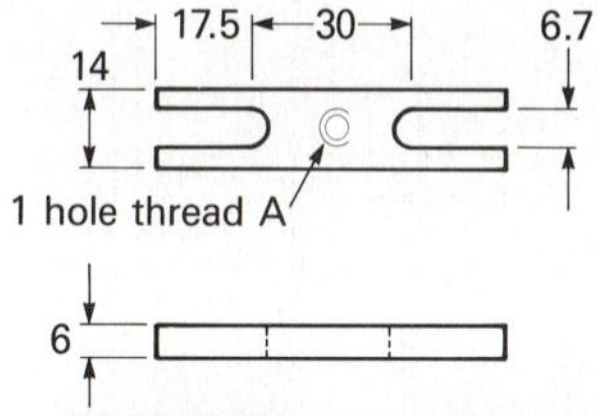

PRODUCT NUMBER	thread A
07 RPC 001	8-32 UNC
07 RPC 003	1/4-20 UNC
07 RPC 501	M4
07 RPC 503	M6

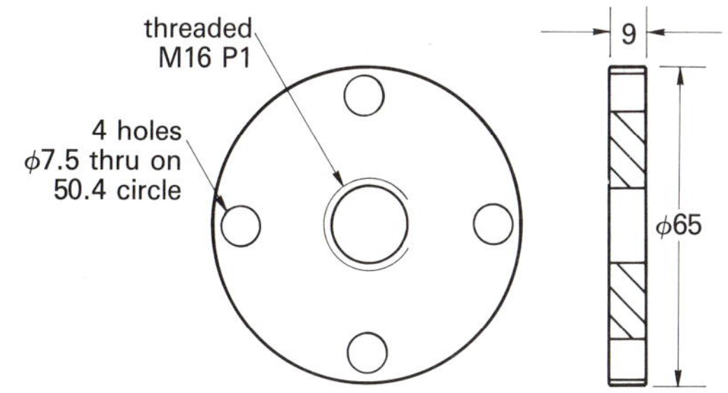

07 PHA 001

Post/Post Holder Bases

Shape	Central thread	PRODUCT NUMBERS Inch Version	Metric Version
Square	M16	07 RPC 014	07 RPC 514
	1/4-20	07 RPC 012	—
	M6	—	07 RPC 512
Rectangular	1/4-20	07 RPC 003	—
	M6	—	07 RPC 503
	8-32	07 RPC 001	—
	M4	—	07 RPC 501
Circular	M16	07 PHA 001	07 PHA 501

POST CLAMPS

Post clamps are used to join posts or post-mounted components together at either right angles or arbitrary angles.

Standard and heavy-duty clamps are available. The standard clamps utilize a knurled thumb screw to squeeze the post within the bore. The heavy-duty clamps have a split segment on one side which firmly grasps the post when the stainless steel screw is tightened.
on an optical breadboard or table. One post may be screwed directly and securely into the table top at an approximate location. Then by using two right angle clamps (or one adjustable clamp) post-mounted components or accessories may be positioned precisely on the required optical axis.

For increased stability, two posts may be mounted a small distance apart and a bridge made from two right angle clamps. A third clamp, mounted on the cross member, will carry the desired component while providing for both vertical and lateral motions.

All clamps are machined from solid brass and are flat black chrome plated, to reduce possible reflections and scatter. Moderate finger pressure on the clamp screw will hold posts firmly and securely.

SPECIFICATIONS: POST CLAMPS

Dimensions: (ϕ): ±0.1mm
Bore: +0.05mm; -0.00mm
Fixed Right Angle: ±5 arc minutes
Load Capacity, Standard-duty: 30 kg
Heavy-duty: 50 kg
Material: Brass; Stainless steel screw on heavy duty clamps, brass screw on light duty clamps
Finish: Body, flat black paint

Right Angle Post Clamps

Bore ϕA (mm)	Bore ϕB (mm)	PRODUCT NUMBER Standard-Duty	PRODUCT NUMBER Heavy-Duty
8	8	07 CFL 004	—
8	12	07 CFL 005	—
12	12	07 CFL 001	07 CFH 001
12	20	07 CFL 002	07 CFH 002
20	20	07 CFL 003	07 CFH 003

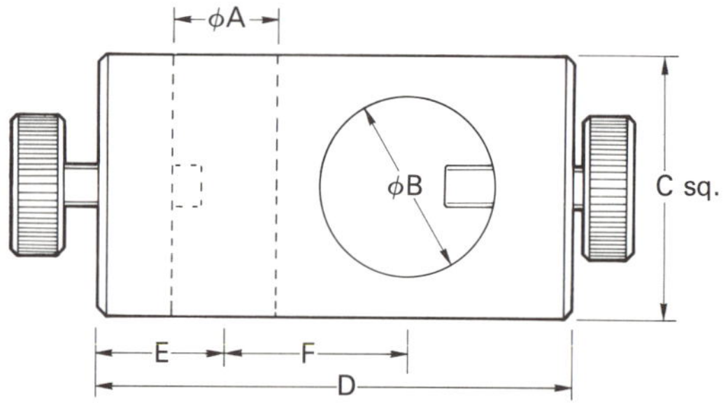

PRODUCT NUMBER	ϕA	ϕB	C	D	E	F
07 CFL 001	12	12	20	44	14	16
07 CFL 002	12	20	30	53	14	21
07 CFL 003	20	20	30	61	18	25
07 CFL 004	8	8	14	40	12	16
07 CFL 005	8	12	14	40	12	16

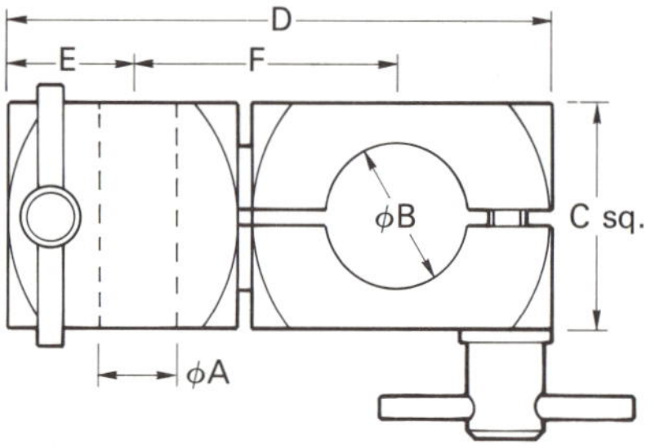

PRODUCT NUMBER	ϕA	ϕB	C	D	E	F
07 CFH 001	12	12	20	62	16	30
07 CFH 002	12	20	30	72	16	35
07 CFH 003	20	20	30	82	21	40

ADJUSTABLE POST CLAMPS

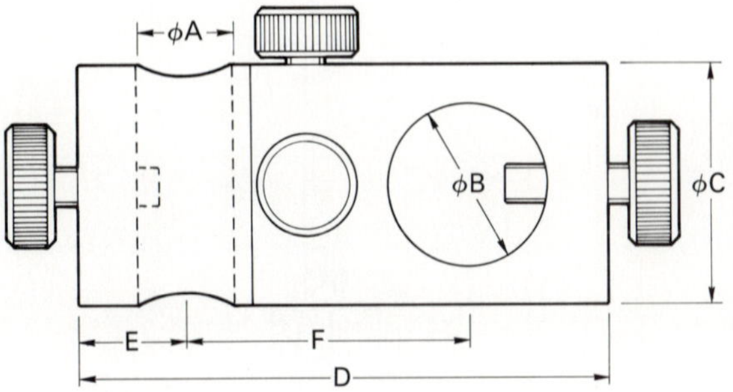

PRODUCT NUMBER	ϕA	ϕB	ϕC	D	E	F
07 CRL 001	12	12	22	00	14	32
07 CRL 002	12	20	30	68	14	36
07 CRL 003	20	20	30	76	18	40
07 CRL 004	8	8	15	50	14	25
07 CRL 005	8	12	22	60	14	32

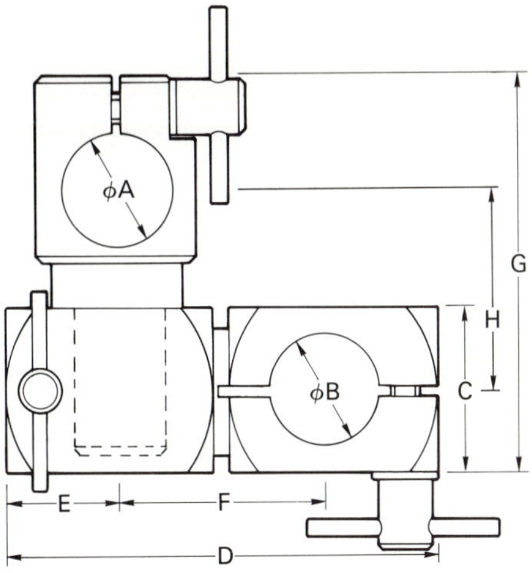

PRODUCT NUMBER	ϕA	ϕB	C	D	E	F	G	H
07 CRH 001	12	12	20	62	16	30	48	20
07 CRH 002	12	20	30	72	16	35	58	37
07 CRH 003	20	20	30	82	21	40	73	37

Adjustable Post Clamps

Bore ϕA (mm)	Bore ϕB (mm)	PRODUCT NUMBERS Standard-Duty	PRODUCT NUMBERS Heavy-Duty
8	8	07 CRL 004	—
8	12	07 CRL 005	—
12	12	07 CRL 001	07 CRH 001
12	20	07 CRL 002	07 CRH 002
20	20	07 CRL 003	07 CRH 003

CLAMP END

An 8mm, 12mm or 20mm bore post clamp with a short, same diameter stub which fits other post holders or clamps. Useful for converting right angle clamps to adjustable angle.

Clamp End

Bore (mm)	PRODUCT NUMBER
8	07 CFR 004
12	07 CFR 001
20	07 CFR 002

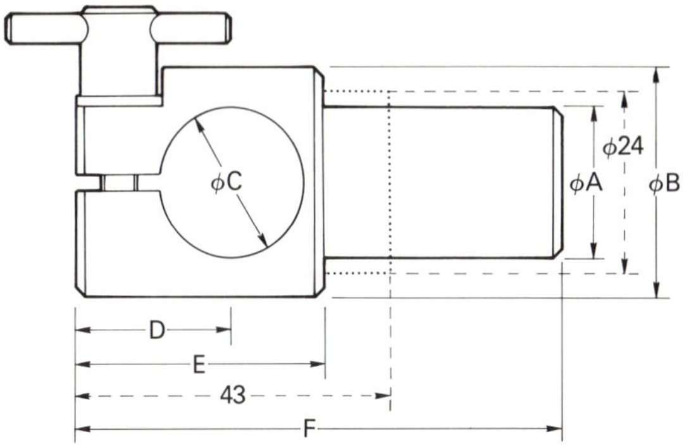

PRODUCT NUMBER	ϕA	ϕB	ϕC	D	E	F
07 CFR 001	12	20	12	18	27	48
07 CFR 002	20	30	20	21	34	66
07 CFR 004	8	20	8	18	27	36

Dotted dimension 07 CFR 002 only

LENS HOLDERS

Lens holders are all mounted on 12mm or 20mm diameter posts for use with Melles Griot simple concave and convex lenses from 10mm diameter to 153mm diameter. Adapters are available for lens diameters from 5.2mm to 20mm. These fit the 30mm lens holder.

Each lens holder is engraved with its clear aperture in millimeters (inches where even) so that lenses may be mounted rapidly and with assurance. The post lengths are adjusted for each lens holder so that the nominal optical axis is approximately constant at 150mm. This is of great convenience when setting up prototype systems as a common optical axis is readily settable for all components.

A nylon inner ring cushions the lens when the brass retainer ring is screwed in to mount the lens. A slot spanner, listed below, is required to tighten and remove the retainer ring.

SLOT SPANNER

An adjustable slot spanner for tightening and loosening the retainer ring used in Melles Griot lens holders, mirror holders and polarizer holders. Made of steel.

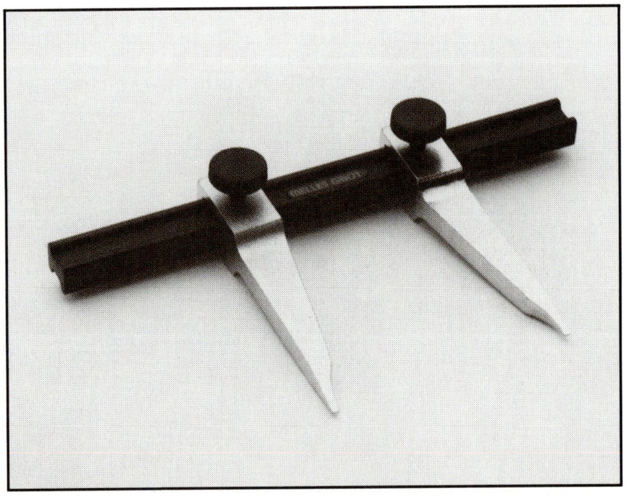

Slot Spanner

	PRODUCT NUMBER
	07 AKT 007

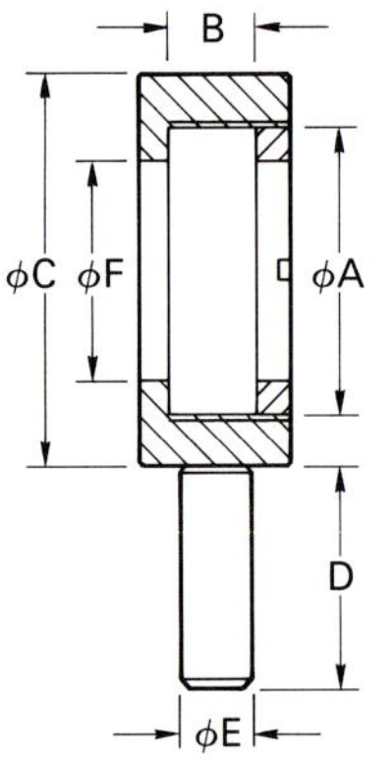

PRODUCT NUMBER	φA	B	φC	D	E	φF
07 LHF 001	10	6	30	100	12	7
07 LHF 002	13	6	30	100	12	10
07 LHF 003	15	6	30	100	12	12
07 LHF 004	18	6	36	100	12	15
07 LHF 005	20	6	40	100	12	17
07 LHF 006	23	6	40	100	12	19
07 LHF 007	25.4	6	45	100	12	21
07 LHF 008	27	7	50	100	12	23
07 LHF 009	30	7	50	100	12	26
07 LHF 011	32	10	50	100	12	28
07 LHF 012	35	10	50	100	12	31
07 LHF 013	38.1	10	60	80	12	34
07 LHF 014	40	10	60	80	12	36
07 LHF 015	42	12	65	80	12	37
07 LHF 016	45	12	65	80	12	40
07 LHF 017	47	12	65	80	12	42
07 LHF 018	50	12	70	80	12	45
07 LHF 019	50.8	12	80	80	12	46
07 LHF 022	52	15	75	80	12	47
07 LHF 023	58	15	75	80	12	53
07 LHF 024	60	15	80	80	12	55
07 LHF 025	63.5	15	85	80	12	56
07 LHF 026	66	15	85	80	12	59
07 LHF 027	70	15	90	80	12	63
07 LHF 028	76.2	15	100	80	12	70
07 LHF 029	80	15	100	60	20	73
07 LHF 033	82	15	100	60	20	75
07 LHF 034	90	15	108	60	20	83
07 LHF 035	95	15	114	60	20	88
07 LHF 036	100	20	120	60	20	93
07 LHF 037	101.6	20	120	60	20	94
07 LHF 038	105	20	125	60	20	97
07 LHF 039	120	20	140	40	20	112
07 LHF 044	130	20	150	40	20	122
07 LHF 045	140	20	160	40	20	132
07 LHF 046	145	20	170	40	20	137
07 LHF 047	152	22	175	40	20	145

Lens Holders

Lens Diameter (mm)	Max Lens Edge Thickness (mm)	Post Diameter (mm)	PRODUCT NUMBER
10	6	12	07 LHF 001
13	6	12	07 LHF 002
15	6	12	07 LHF 003
18	6	12	07 LHF 004
20	6	12	07 LHF 005
23	6	12	07 LHF 006
25.4	6	12	07 LHF 007
27	7	12	07 LHF 008
30	7	12	07 LHF 009
32	10	12	07 LHF 011
35	10	12	07 LHF 012
38.1	10	12	07 LHF 013
40	10	12	07 LHF 014
42	12	12	07 LHF 015
45	12	12	07 LHF 016
47	12	12	07 LHF 017
50	12	12	07 LHF 018
50.8	12	12	07 LHF 019
52	15	12	07 LHF 022
58	15	12	07 LHF 023
60	15	12	07 LHF 024
63.5	15	12	07 LHF 025
66	15	12	07 LHF 026
70	15	12	07 LHF 027
77	15	12	07 LHF 028
80	15	20	07 LHF 029
82	15	20	07 LHF 033
90	15	20	07 LHF 034
95	15	20	07 LHF 035
100	20	20	07 LHF 036
101.6	20	20	07 LHF 037
105	20	20	07 LHF 038
120	20	20	07 LHF 039
130	20	20	07 LHF 044
140	20	20	07 LHF 045
145	20	20	07 LHF 046
153	22	20	07 LHF 047

ADAPTERS FOR SMALL LENSES

Each adapter is made of aluminum, finished black anodized and is supplied complete with nylon cushion ring and threaded brass retaining ring similar to post mounted lens holders.

These adapters are most convenient for mounting small diameter stock Melles Griot lenses. The clear aperture is reduced by the retaining ring. (See table below.)

All adapters fit the 30mm Lens Holder (Product Number 07 LHF 009) by screwing them into the holder.

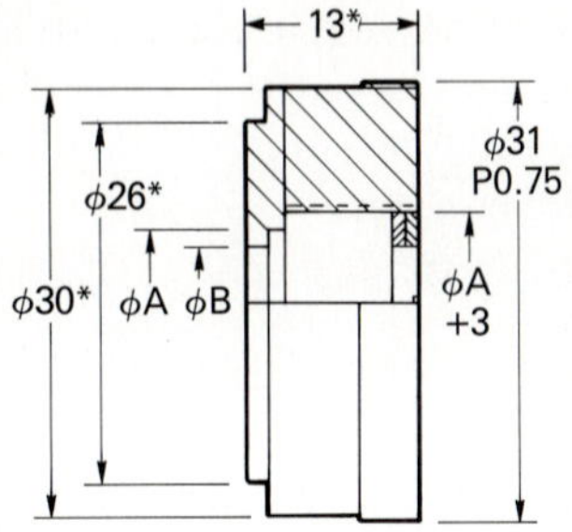

PRODUCT NUMBER	ϕA	ϕB
07 LAD 001	5.2	3.7
07 LAD 002	6.3	4.8
07 LAD 003	6.7	5.2
07 LAD 004	7.5	6.0
07 LAD 005	8.5	7.0
07 LAD 006	10.0	8.0
07 LAD 007	15.0	13.0
07 LAD 008	20.0	17.0
07 LAD 009	25.4	22.0

*subtract 1mm for 07 LAD 009

Lens Holder Adapters

Inside Diameter (mm)	PRODUCT NUMBER
5.2	07 LAD 001
6.3	07 LAD 002
6.7	07 LAD 003
7.5	07 LAD 004
8.5	07 LAD 005
10.0	07 LAD 006
15.0	07 LAD 007
20.0	07 LAD 008
25.4	07 LAD 009

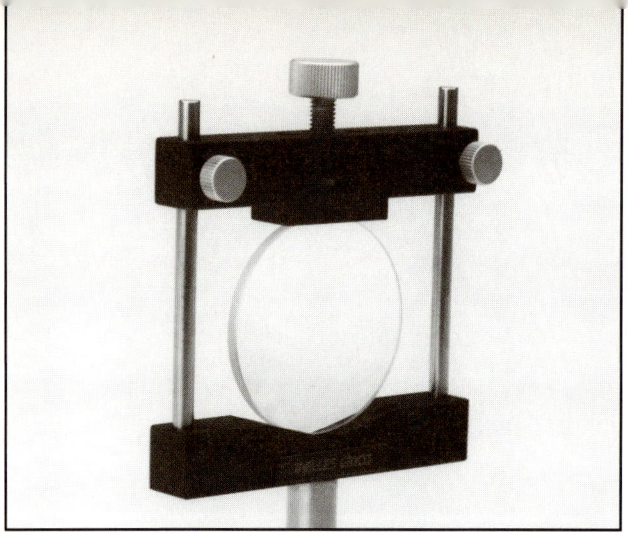

ADJUSTABLE LENS HOLDERS

These adjustable lens holders are designed to hold lenses from 6mm to 25.4mm, from 10mm to 60mm, from 20mm to 101.6mm, and from 60mm to 150mm in diameter. Lenses may be easily and quickly removed and replaced without moving the lens holder position.

Clamp screws on the two vertical side guides allow the cross bar to be lowered for approximately the correct position. The central thumb screw then may be adjusted to hold the lens firmly and softly in the correct location.

"V" grooves in both the top and the bottom center a simple lens both vertically and horizontally. When a barrel lens is mounted, the cross vee in the base and top clamp hold the lens firmly on the optical axis.

These adjustable lens holders are made of aluminum with stainless steel vertical pins and chrome-plated brass clamp screws. The mounting post is of stainless steel also. The body is black anodized to reduce scatter and stray reflections.

Post diameter is 8mm, 12mm or 20mm as appropriate to the size of lenses held.

Adjustable Lens Holders

Maximum Opening (mm)	Post Diameter (mm)	PRODUCT NUMBER
25.4	8	07 LHA 004
60.0	12	07 LHA 001
101.6	20	07 LHA 002
150.0	20	07 LHA 003

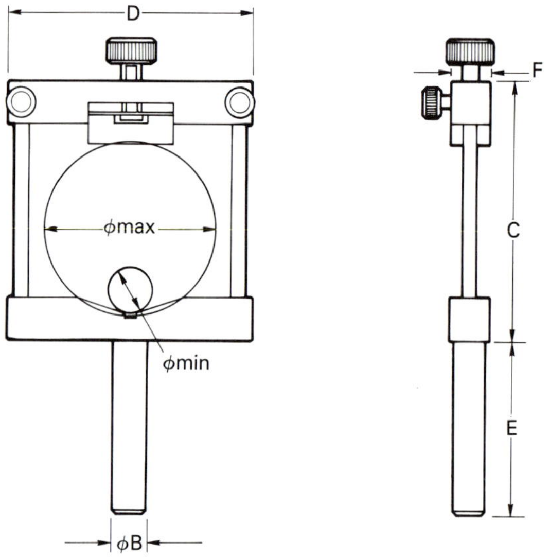

PRODUCT NUMBER	ϕmax	ϕmin	B	C	D	E	F
07 LHA 001	60	10	12	90	86	100	14
07 LHA 002	100	20	20	134	134	90	19
07 LHA 003	150	60	20	195	200	50	20
07 LHA 004	25	6	8	45	40	40	12

If you wish to change the length of any mounting post select the appropriate length post with an M6 thread from the mounting posts table on page 444. Using moderate torque these posts may be unscrewed and interchanged.

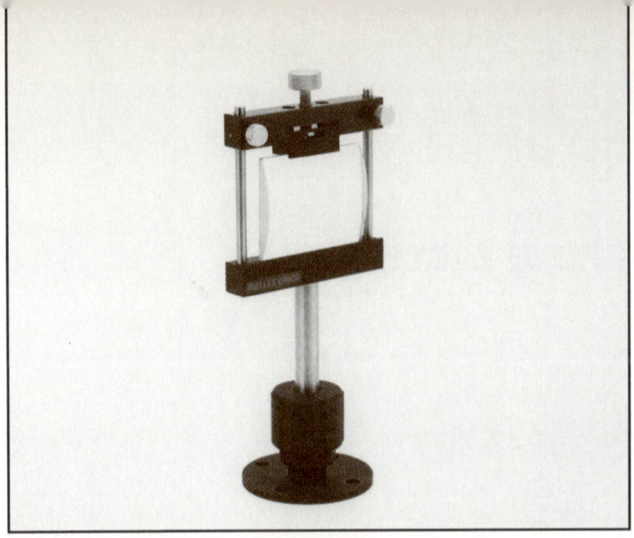

CYLINDRICAL LENS HOLDERS

These adjustable lens holders are intended to hold a wide variety of cylindrical (rectangular shape) lenses. Widths (y) from 10 to 60mm and lengths (x) from 10 to 130mm can be accommodated. They are suitable for both positive and negative focal length components. Positive elements will be held by a vee groove but negative elements will rest on the flat surface. In both cases a ledge stop serves to align and locate the lens. In addition to cylindrical lenses these holders are suitable for other rectangular shape components.

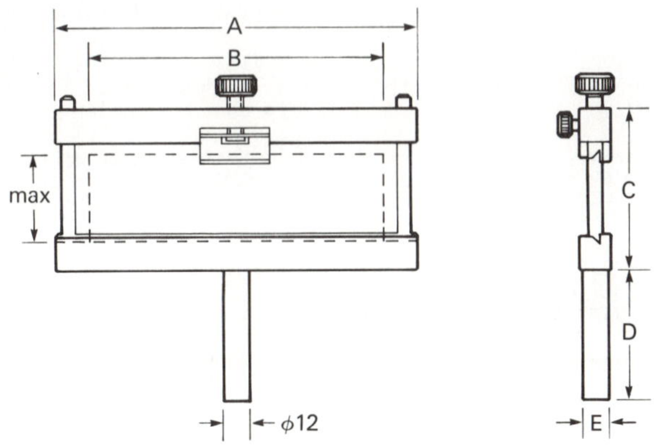

PRODUCT NUMBER	A	B	C	D	E	max
07 LHC 001	40	25	39	80	12	25
07 LHC 003	86	60	84	60	14	60
07 LHC 005	160	130	78	60	14	50

Cylindrical Lens Holders

Maximum Opening (mm)	Maximum Lens Length (mm)	Post Diameter (mm)	PRODUCT NUMBER
25	25	12	07 LHC 001
60	60	12	07 LHC 003
50	130	12	07 LHC 005

SELF CENTERING LENS HOLDERS

These lens holders will accept and center a wide selection of different size lenses. The mini-holder will accept lenses from 5 to 25mm diameter while the larger holder will accept lenses from 25 to 80mm diameter.

To operate simply rotate the knurled ring, insert the lens and gently release the ring. Operation is spring loaded. One ring rotates within another and three cam actuated arms are moved radially to hold the lens. A clamp screw permits the arms to be locked at any open or closed position.

These lens holders are made from brass with a stainless steel mounting post. The body is black chrome coated to reduce scatter and stray reflections.

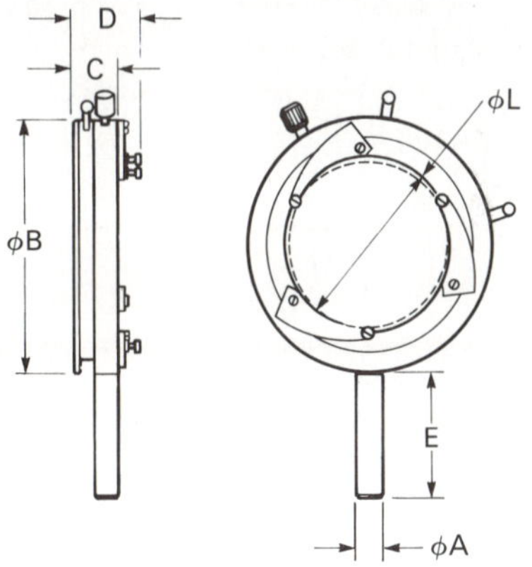

PRODUCT NUMBER	ϕA	ϕB	C	D	E	ϕL min	ϕL max
07 LHJ 001	8	48	17	23	40	5	25
07 LHJ 003	12	120	16	25	60	25	80

Self Centering Lens Holders

Maximum Opening (mm)	Minimum Opening (mm)	Post Diameter (mm)	PRODUCT NUMBER
25	5	8	07 LHJ 001
80	25	12	07 LHJ 003

PRISM TABLES

Three standard sized prism tables are available. They will accept right angle prisms, cube beamsplitters and similar components with overall heights of 26mm, 35mm, and 54mm.

Prisms are held in position on the table top by a cork-padded pressure thumb screw which is supported by two side posts and a cross member. The brass posts and cross member may be removed and relocated in the opposite 90° orientation to assure convenient location of the tilt and rotational control knobs. Maximum distance between these pins (and, therefore, maximum prism diagonal dimensions) are listed below.

When prisms less than the minimum height specification are to be mounted, spacers may be added above or below the prism to make up the difference.

θ_x and θ_y tilt of the table surface of $\pm 3°$ in each axis is provided by two spring-loaded thumb screws. Similarly, a rotation, θ_{29} about a vertical axis through the prism $\pm 3°$ is provided by a third thumb screw. These features provide a smooth and convenient beamsteering function through the prism.

The prism tables are provided with 1/4-20 or M6 clearance holes for mounting directly to optical breadboards or tables.

SPECIFICATIONS: PRISM TABLES

Tilt Travel: ±3° both axes
Rotational Travel: ±3°
Sensitivity: <5 arc seconds
Material: Aluminum body, brass knobs
Finish: Body, flat black anodized;
chrome plated controls

Prism Tables

Maximum Prism Height (mm)	Maximum Prism Diagonal (mm)	PRODUCT NUMBER Inch Version	Metric Version
26.0	36.0	07 TTA 001	07 TTA 501
35.0	54.0	07 TTA 002	07 TTA 502
54.0	76.0	07 TTA 003	07 TTA 503

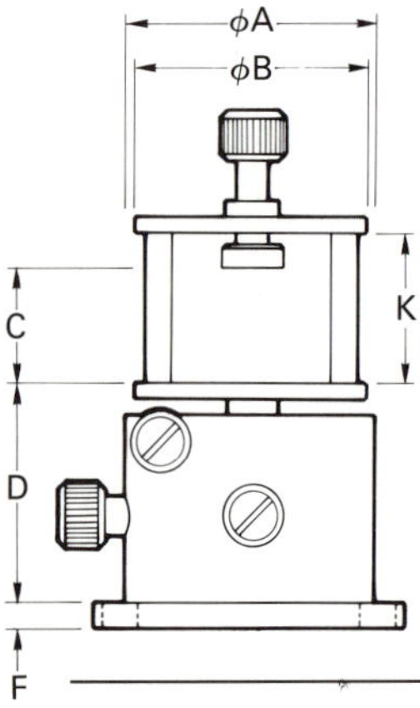

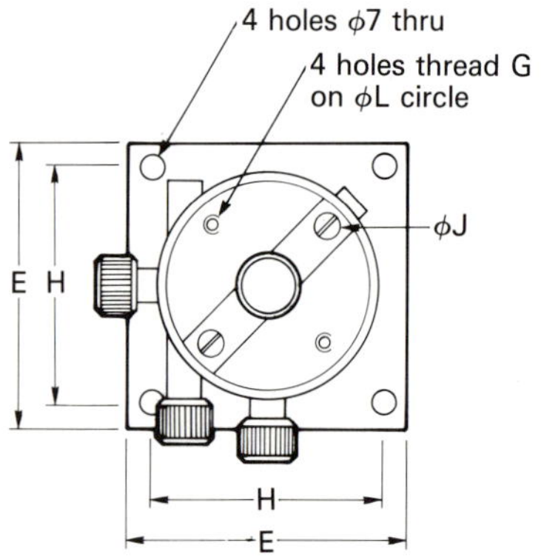

PRODUCT NUMBER	φA	φB	C	D	E	F	G	H	max cube size	φJ	K	φL
07 TTA 001	48	46	5-26	41	65	5	M2.6	50.8	25.4	4	32	36
07 TTA 002	68	68	13-35	60	90	5	M4	76.2	30	6	43	54
07 TTA 003	96	92	24-60	82	120	8	M4	101.6	50.8	10	63	84
07 TTA 501	48	46	5-26	41	65	5	M2.6	50.8	25.4	4	32	36
07 TTA 002	68	68	13-35	60	90	5	M4	75	30	6	43	54
07 TTA 503	96	92	24-60	82	120	8	M4	100	50.8	10	63	84

PRISM HOLDERS

These prism holders provide an easy means of mounting a variety of prisms or beamsplitters. The prism is placed on top of the platform and held in position by a spring retainer. Tilt adjustment about two axes is provided by means of the screws projecting from the underside of the platform.

Mini and standard sizes are offered to suit prisms and cubes up to 26mm or 51mm in height respectively.

The holders are made of aluminum with stainless steel posts. See also preceding prism tables if post mounting is not desired.

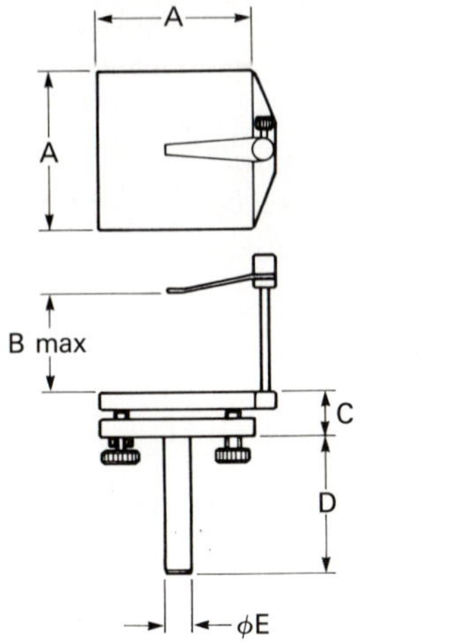

PRODUCT NUMBER	A	B max	C	D	ϕE
07 TTM 001	25	25	12	40	8
07 TTM 003	70	50.8	20	60	12

Prism Holders

Maximum Prism Height (mm)	Post Diameter (mm)	PRODUCT NUMBER
26	8	07 TTM 001
51	12	07 TTM 003

See page 223 for prisms and beamsplitters.

RETROREFLECTOR HOLDERS

These convenient holders simplify the mounting of retroreflectors (corner cubes) in an optical system. The retroreflector rests in a recess on one side of an aluminum disc. It is retained in position by means of three clamps which act on the flat faces of the cube. The disc is black anodized and engraved with the aperture of the retroreflector in millimeters. It is mounted on a stainless steel post.

All Melles Griot retroreflectors described on page 235 will fit these holders.

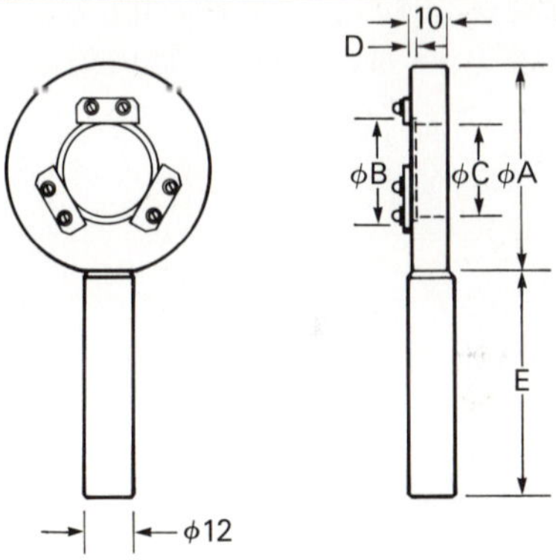

PRODUCT NUMBER	ϕA	ϕB	ϕC	D	E
07 HCC 001	40	15.2	12	1.2	100
07 HCC 003	50	25.2	22	1.4	100
07 HCC 005	70	38.2	34	1.8	80
07 HCC 007	80	50.2	45	2.2	80
07 HCC 008	100	70.2	64	3.4	60

Retroreflector Holders

Retroreflector Aperture (mm)	Post Diameter (mm)	PRODUCT NUMBER
15	12	07 HCC 001
25	12	07 HCC 003
38	12	07 HCC 005
50	12	07 HCC 007
70	12	07 HCC 008

FILTER HOLDERS

MULTIPLE FILTER HOLDERS

These post-mounted filter holders allow a number of round or square filters to be introduced and removed from an optical beam with ease and rapidity. Filter holders are available for three different widths of filter (7mm, 5mm and 3mm maximum) and for two different diameters or sides (25mm and 50mm).

The wide (7mm) filter holder will accommodate up to three individual filters at one time. It is specifically designed to accommodate Melles Griot interference filters and filter sets. These filters are all mounted in 25mm diameter by 6.9mm thick protective aluminum rings to enhance resistance to water vapor invasion. These bandpass filter sets, when used with the filter holders, allow rapid, precise change of the spectral region of interest.

The standard (5mm) filter holder will accommodate up to three filters 5mm thick plus one filter 3mm thick at one time. It is well suited for Melles Griot neutral density filter sets, color process filter set and various individual dichroic and colored glass filters.

All filter holders are made of flat black anodized aluminum and are mounted on stainless steel 12mm diameter posts.

Multiple Filter Holders

For use with filters		
Diameter or Side (mm)	Max. thickness (mm)	PRODUCT NUMBER
25	5	07 HFR 001
25	7	07 HFR 002
50	5	07 HFR 003
50	7	07 HFR 004

Melles Griot offers a large selection of colored glass, interference, dichroic, and neutral density filters. *See the section starting on page 243 for details.*

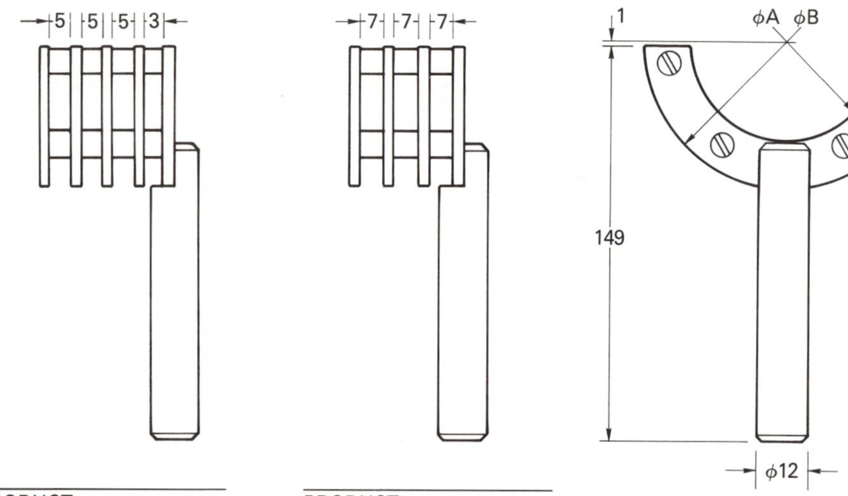

PRODUCT NUMBER	φA	φB
07 HFR 001	23	50
07 HFR 003	48	70

PRODUCT NUMBER	φA	φB
07 HFR 002	23	50
07 HFR 004	48	70

SINGLE FILTER HOLDERS

These spring-loaded filter holders are intended for the positioning of single filters, test targets or other flat components in an optical system. They will hold 25 or 50mm, round or square filters. Filters up to 6mm thickness may be accommodated.

To operate, simply insert the filter behind the flexible spring clips. It will be firmly and vertically held against the upright surface. If multiple reflections are a problem the filter can easily be rotated slightly off-axis using these holders.

These filter holders are made of aluminum with spring steel clips and stainless steel posts. They are black anodized to reduce scatter and surface reflections.

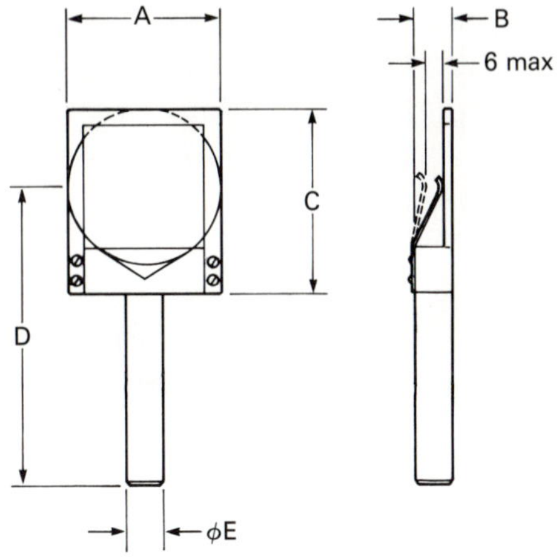

PRODUCT NUMBER	A	B	C	D	ϕE
07 HFU 001	28	8	—	60	8
07 HFU 002	25	8	28	120	12
07 HFU 003	50	12	60	120	12

Single Filter Holders

Filter Size (mm)	Max Filter Thickness (mm)	Post Diameter (mm)	PRODUCT NUMBER
25	6	8	07 HFU 001
25	6	12	07 HFU 002
50	6	12	07 HFU 003

CIRCULAR NEUTRAL-DENSITY WEDGE FILTER HOLDERS

Circular linear-wedge neutral-density filters are useful in many optical systems where light intensity needs to be continuously variable. These holders enable the user to mount Melles Griot wedge filters (see page 253), and to rotate it to vary the intensity throughput.

The holders are supplied in two parts: a boss with mounting post is clamped to the central hole in the filter and a knurled ring is fitted to the outer rim of the filter. Once mounted in this way, the short mounting post may be attached by means of a cross clamp to any vertical post such that a clear area of the disc is available to the optical beam. Clamps and vertical posts must be ordered additionally. (See pages 443-444 and 450-451.)

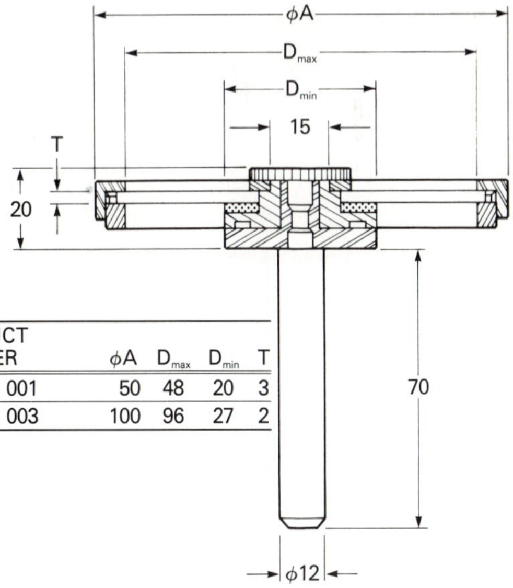

PRODUCT NUMBER	ϕA	D_{max}	D_{min}	T
07 HFD 001	50	48	20	3
07 HFD 003	100	96	27	2

Circular Neutral-Density Wedge Filter Holders

Filter Diameter (mm)	For Use With	PRODUCT NUMBER
50	03 FDC 001	07 HFD 001
100	03 FDC 003	07 HFD 003

POLARIZER HOLDERS

These post-mounted accessories are for holding and accurately rotating polarization components. They may be used with dichroic sheet polarizers, calcite prism polarizers, and quartz or mica retardation plates.

They feature a large, firm brass on stainless steel plain bearing surface to provide smooth, wobble-free rotation. The outer edge of the rotating portion has an engraved 360° scale marked every degree. With the thumb screw released, the holder portion may be rotated to any desired orientation and set. When the thumb screw is tightened, the holder is firmly locked in position. The rotational control is smooth enough that fractional degree angular positioning may be achieved.

Components are mounted in the polarizer holders with a threaded retainer ring.

As supplied the polarizer holders will accept the following sizes of polarizer directly: 20.0mm, 20.6mm, 30.00mm, 30.2mm, 50.8mm, 101.6mm and 103.0mm. All other sizes offered by Melles Griot, including the Glan-Thompson, Glan-Taylor and Wollaston prisms may be fitted by means of an adapter.

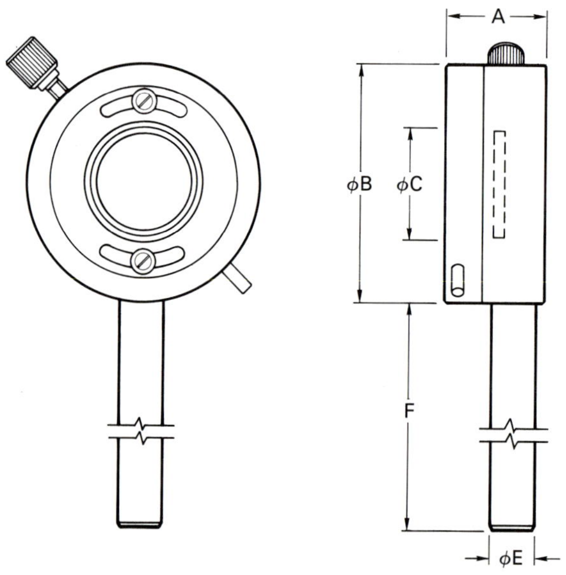

PRODUCT NUMBER	A	φB	φC	φE	F
07 HPR 001	26	64	30	12	120
07 HPR 003	20	50	20.6	8	25
07 HPR 005	26	84	50.8	12	60
07 HPR 007	30	148	103	20	60

Melles Griot offers a wide selection of polarization components including calcite birefringent polarizers, thin film polarizing beamsplitters, dichroic polarizers, near-infrared polarizers and a selection of retarders. *See the section starting on page 283 for full details of these products and a description of their function.*

Polarizer Holders

Polarizer Diameter (mm)	Post Diameter (mm)	PRODUCT NUMBER
20 or 20.6	8	07 HPR 003
30 or 30.2	12	07 HPR 001
50 or 50.8	12	07 HPR 005
100 or 103	20	07 HPR 007

POLARIZER HOLDER ADAPTERS

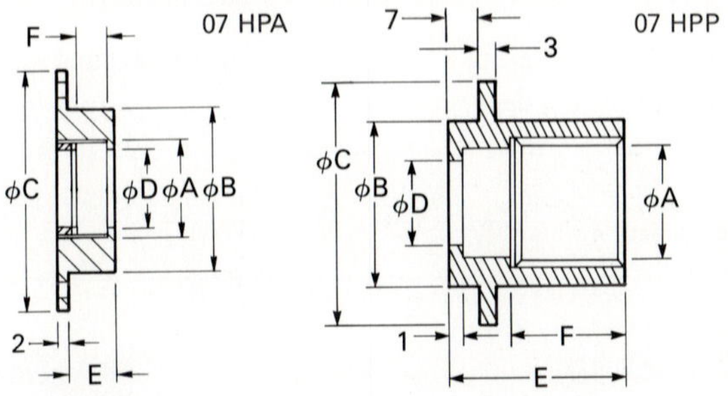

PRODUCT NUMBER	ϕA	ϕB	ϕC	ϕD	E	F
07 HPA 001	10	34	50	7	10	6.5
07 HPA 003	20.6	34	50	17.6	10	6.5
07 HPA 005	41.3	54	70	38	13	9
07 HPA 007	63.5	108	130	58	13	9
07 HPA 009	82.6	108	130	76	13	9
07 HPA 011	93	108	130	85	13	9
07 HPP 003	25	34	50	17.8	40	30
07 HPP 005	30	54	70	22	55	42
07 HPP 007	40	54	70	30	70	52

Polarizer Holder Adapters

Polarizer Diameter (mm)	fits Holder	PRODUCT NUMBER
for wave plates and dichroic filters		
10.0	07 HPR 001	07 HPA 001
20.0 or 20.6	07 HPR 001	07 HPA 003
40.0 or 41.3	07 HPR 005	07 HPA 005
63.5	07 HPR 007	07 HPA 007
82.6	07 HPR 007	07 HPA 009
93.0	07 HPR 007	07 HPA 011
for mounted calcite prism polarizers		
25.0	07 HPR 001	07 HPP 003
30.0	07 HPR 005	07 HPP 005
40.0	07 HPR 005	07 HPP 007

DIVIDED CIRCLE ROTATING MOUNT

This divided circle rotating mount is intended primarily for the support of Soleil-Babinet compensators. With its help, the orientation about the compensator optical axis can be accurately measured, or previously determined orientations accurately reproduced. The divided circle mount is made of non-magnetic metal, to minimize errors in measured orientations due to the Faraday effect.

Please note that the mounting post for this product is 13.7mm in diameter. It will fit the universal post holders described on page 446.

SPECIFICATIONS: DIVIDED CIRCLE ROTATING MOUNT

Diameter: 140mm
Calibration: ±90 degrees (from zero at top) in 1° steps
Vernier Precision: ±3 arc minutes
Post Diameter: 13.7mm

Divided Circle Rotating Mount

PRODUCT NUMBER
04 SBM 001

CYLINDRICAL LASER HOLDERS

We offer three different size fixed laser holders which are ideal for supporting helium neon laser heads. The separation of the supports matches the recommended mounting positions for Melles Griot HeNe lasers. The six nylon faced screws gently but firmly hold cylindrical laser tubes in such a way that the beam may be aligned precisely to the required axis.

The base of the laser holder is fixed permanently and firmly to a 20mm diameter mounting post. All laser holders are made of aluminum which is black anodized to reduce possible stray reflections. The 20mm diameter stainless steel posts fits all Melles Griot heavy-duty post holders and provides a rigid support.

All laser holders will accommodate lasers up to a maximum outside diameter of 55mm.

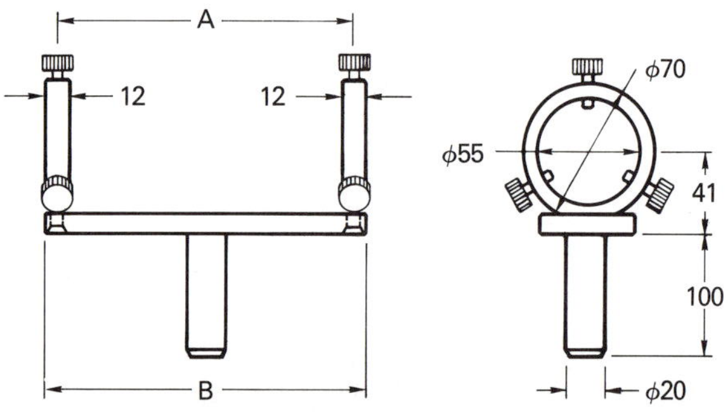

PRODUCT NUMBER	A	B
07 HLH 001	228	240
07 HLH 002	308.5	320.5
07 HLH 003	368.2	380.2

Cylindrical Laser Holders

For use with	Post Diameter (mm)	PRODUCT NUMBER
1 and 2mW HeNe lasers	20	07 HLH 001
4 and 5mW HeNe lasers	20	07 HLH 002
7mW HeNe lasers	20	07 HLH 003

LASER HOLDER WITH FINE ADJUSTMENT

This laser holder may be adjusted for horizontal and vertical tilt by adjusting the appropriate screws. Aiming accuracy to within a small fraction of an arc minute may be achieved in this way. In other respects it is identical to the laser holders previously described.

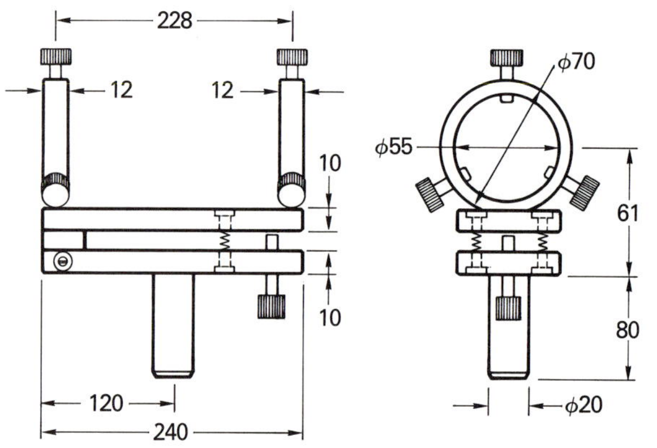

07 HLA 001

Cylindrical Laser Holder with Fine Adjustment

For use with	Post Diameter (mm)	PRODUCT NUMBER
All cylindrical HeNe lasers	20	07 HLA 001

Melles Griot helium neon lasers are now recognized as some of the most stable and reliable lasers available. Full specifications are listed on pages 356-359.

SELF-CONTAINED LASER HOLDER

This adjustable laser holder is intended for mounting all Melles Griot self-contained lasers. The laser is mounted by simply placing it on the horizontal platform. Two T-nuts are inserted into the underside T-slot on the laser and used to secure the laser to the platform.
alignment is facilitated by the adjusting screw on the underside of the holder.

This laser holder is made of aluminum with a stainless steel post. It is black anodized to reduce scatter and stray reflections.

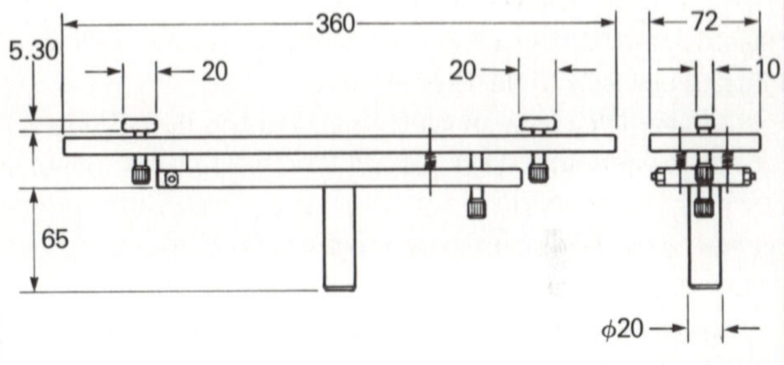

07 HLA 002

Self-Contained Laser Holder with Fine Adjustment

For use with	Post Diameter (mm)	PRODUCT NUMBER
All Melles Griot self-contained lasers	20	07 HLA 002

For details of the Melles Griot self-contained lasers see pages 345-355.

FIBER OPTIC HOLDERS

Two versions of this fiber optic holder are available: one has X-Y-Z positioning of the fiber only and the other has both X-Y-Z and two axes of tilt adjustment. They are both mounted on a 12mm diameter stainless steel post. Both provide firm fiber optic ferrule support and provide light spring-loaded fore and aft motion for proper positioning. Total axial motion is 2.75mm. The spring tension in the fiber's axial direction is approximately 0.5 kilogram.

X and Y positioning on the fiber, to maximize the beam coupling efficiently, is provided by two orthogonal, spring-loaded thumb screws. Total travel in each direction is 3.0mm.

In the tiltable version, angular orientation around both the X and Y axes is provided for beam coupling efficiency. Two spring-loaded, backlash-free thumb screws on the rear of the mount provide a total angular travel of ±5° in both axes. The thumb screws are kinematically positioned against the mounting plate to protect against mechanical over-constraint in these motions.

These holders have been designed to accept 6.35mm diameter fiber optic chucks primarily. Smaller sizes may be used, but locational firmness within the holder bore will be sacrificed as sizes become smaller than 3.5mm. The fiber ferrule is held in place by two metric Allen set screws, at right angles to each other, located at the rear of the holder's barrel.

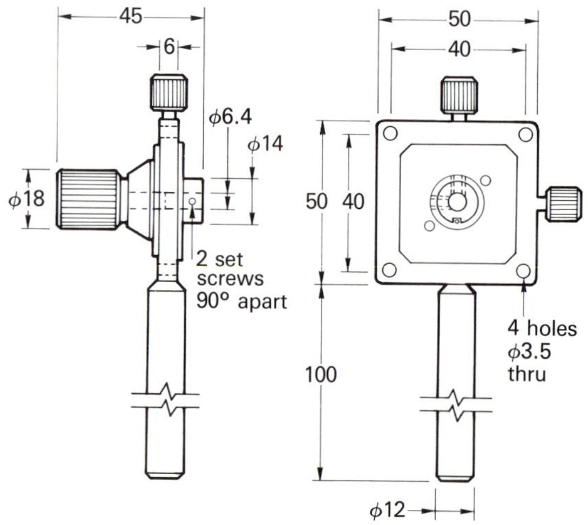

07 HFO 001

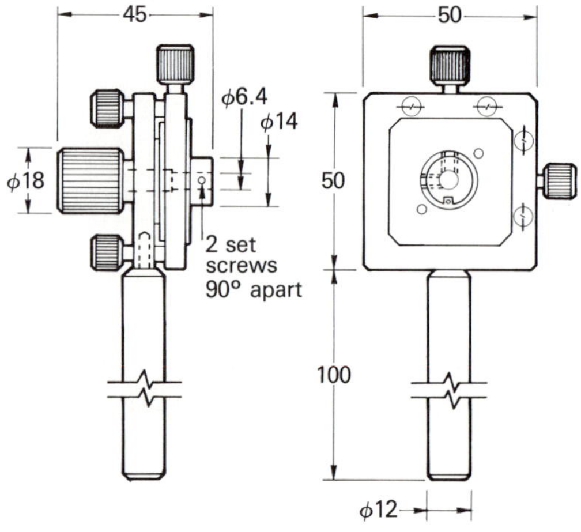

07 HFO 002

SPECIFICATIONS: FIBER OPTIC HOLDERS

X-Y Travel: 3.0mm by 3.0mm
θ_x–θ_y Tilt Travel: ±15°
Fiber Motion: 2.75mm
Bundle Diameter: 3.5mm to 6.35mm
Material: Aluminum
Finish: Black anodized

Fiber Optic Holders

Degrees of Freedom	Post Diameter (mm)	PRODUCT NUMBER
X-Y-Z	12	07 HFO 001
X-Y-X-θ_y–θ_z	12	07 HFO 002

FIBER OPTIC CHUCK

This fiber optic chuck is intended to hold most common types of optical fiber when these are to be mounted in a fiber optic holder. The outer diameter of the chuck is 6.35mm and will fit directly into all Melles Griot fiber optic holders. Internally, the chuck consists of two coaxial clamps.

The larger diameter clamp holds the outer sheath of the fiber and provides strain relief. This is very important when trying to mount optical fibers as the slightest movement of the heavy outer sheath with respect to the inner core can cause breakage. The strain relief collet accepts cable sheaths of 3.0mm diameter.

The smaller clamp is for holding the inner fiber. It operates on the outside of the cladding. This clamp consists of two split nylon ferrules separated by a spacer. These two ferrules can be closed by longitudinal pressure exerted by the screwed end cap. They lightly but firmly hold the fiber without scratching or causing other damage. A selection of ferrules are supplied with the chuck covering a range of cladding diameters from 50 to 300 microns.

In use it is important to clamp the outer sheath prior to the cladding in order to minimize the possibility of breaking the fiber. Once mounted in the chuck the fiber can be handled without risk and can easily be positioned in a fiber optic holder.

Fiber Optic Chuck

	PRODUCT NUMBER
for use with most common fibers	07 HFC 002

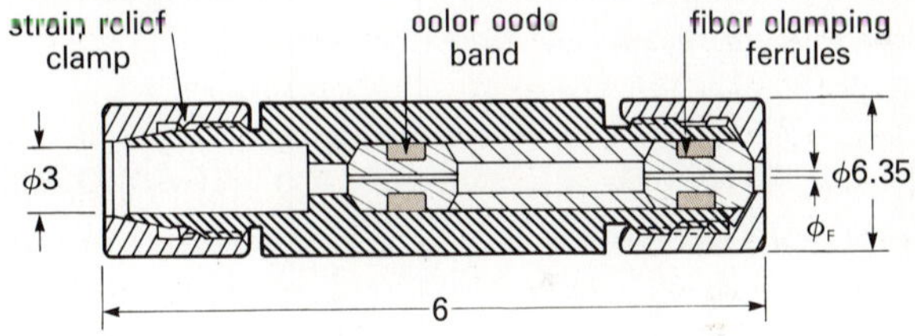

ferrule color code	ϕ_F (microns) min	max
black	50	100
yellow	100	250
red	250	400
blue	400	500

07 HFC

We offer a broad selection of optics for use in fiber optic systems. This selection includes coupling spheres, gradient index lenses and miniature beamsplitters. See pages 369-394 for details.

PRECISION X-Y POSITIONING ACCESSORY HOLDER

This holder is a precision, differential micrometer stage mounted on a stainless steel post. It may be used with accessory mounts to support microscope objectives, pinholes, fiber optics and laser diodes. Simple adapters may be made to mount a variety of other components also.

This holder will be found helpful when it is necessary to provide very fine x-y positioning together with a stable support. The differential micrometers have a range of 6mm and a sensitivity of 0.1 microns. They have a coarse control of 2 turns per millimeter which can be locked to provide a fine control of 0.4 microns per micrometer division.

The stage is mounted on a vertical aluminum plate having an 18mm aperture. Four M3 threaded holes are used to attach the accessory mounts. The holder is made of steel and is black chrome coated to reduce scatter and stray reflections.

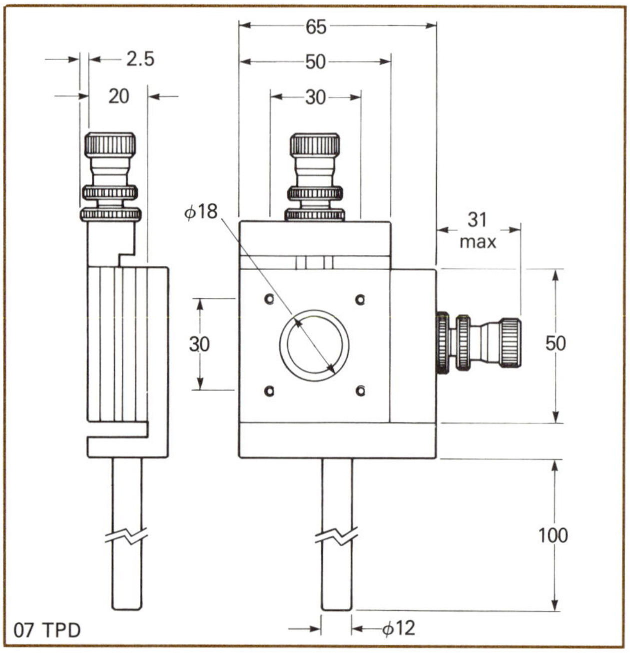

SPECIFICATIONS: PRECISION X-Y POSITIONING ACCESSORY HOLDER

Range; Coarse: ±3mm
fine: ±0.125mm
Sensitivity: 0.1 micron
Dimensions: ±0.1mm
Material: Steel
Finish: Black Chrome

Precision X-Y Positioning Accessory Holder

Post Diameter (mm)	PRODUCT NUMBER
12	07 TPD 003

ACCESSORY SUB-MOUNTS

These Accessory mounts are intended for use with the Precision X-Y Positioning Holder described above. They will hold a selection of commonly used components, as listed.

Accessory Mounts

for use with	PRODUCT NUMBER
Microscope Objective (see page 312)	07 HMO 003
Pinhole or Slit (see page 323)	07 HMP 001
Fiber Optic Chuck (see page 465)	07 HFC 001
Laser Diode	07 HLD 003

PINHOLE HOLDER

This holder with X-Y adjustment is convenient for the mounting and positioning of pinholes, slits and other apertures. A retaining ring secures the aperture in place in a 16mm diameter recess. A sub-plate on which the aperture is mounted can be moved by means of screws over a range of ±1.5mm.

This holder will accept directly any of the Melles Griot mounted pinholes or slits described on page 323.

The holder is made of aluminum and is black anodized to reduce scatter and stray reflections. It has a stainless steel post.

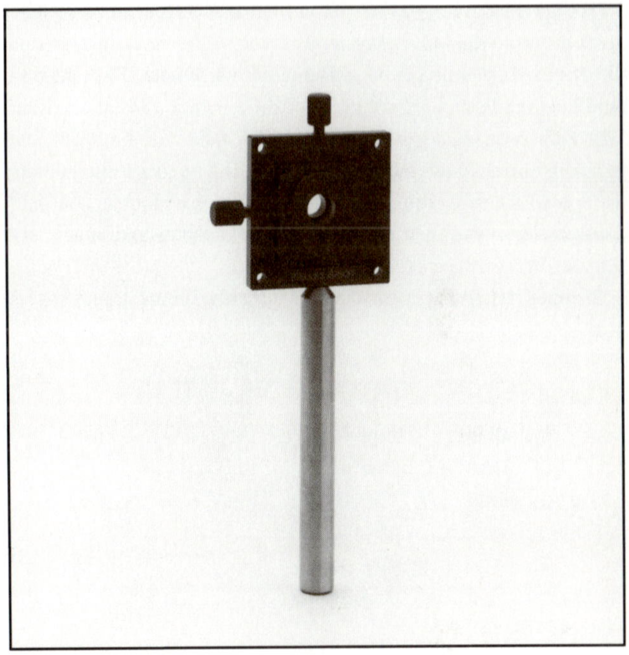

Pinhole Holder

Post Diameter (mm)	PRODUCT NUMBER
12	07 HPH 001

ACCESSORY KITS

For your convenience we offer the following kits of tools and hardware. See page 480 for a full list of contents.

	PRODUCT NUMBER
Inch Version	07 AKT 003
Metric Version	07 AKT 503

SPATIAL FILTER ASSEMBLY

This open laboratory-type spatial filter will appeal to those who need access to the laser beam as it traverses the filter. The spatial filter consists of a precision, differential micrometer X-Y stage to hold the pinhole and a focusing optic holder on a translation slide. You may select from the Melles Griot range of pinholes and focusing optics which are listed on the following page to complete the spatial filter. If you wish to use a microscope objective as a focusing lens, you may remove the inner adapter and attach any standard microscope objective directly.

The entire assembly is mounted on a sturdy 20mm stainless steel post.

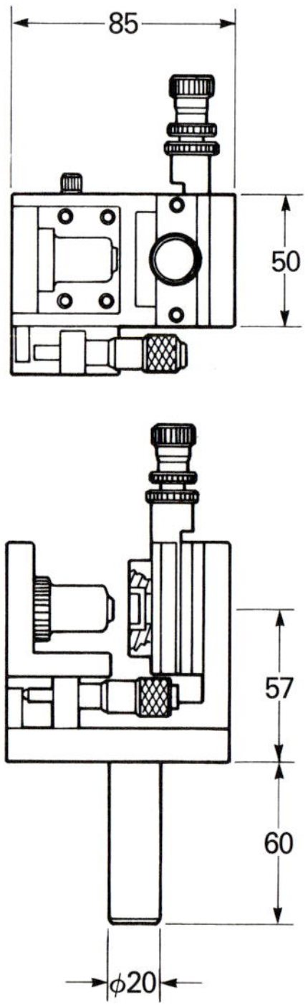

07 SFM 001

SPECIFICATIONS: SPATIAL FILTER ASSEMBLY

Range; x,y, Coarse: ±3mm
Fine: ±0.125mm
z: ±6mm
Sensitivity; x, y: 0.1 micron
z: 2 microns
Dimensions: ±0.1mm
Material: Steel
Finish: Black Chrome

Spatial Filter Assembly

Post Diameter (mm)	PRODUCT NUMBER
20	07 SFM 001

Please select pinholes and focusing optics from the next page.

For an enclosed spatial filter which fits the front mounting ring of our self-contained, laboratory, helium neon lasers *see page 362.*

FOCUSING OPTICS AND PINHOLES

FOCUSING OPTICS

These focusing optics are for use with the Melles Griot spatial filter assembly to provide a focused beam at the pinhole plane. Alternatively they may be used without the pinhole to create an unfiltered spherical wavefront of varying numerical aperture.

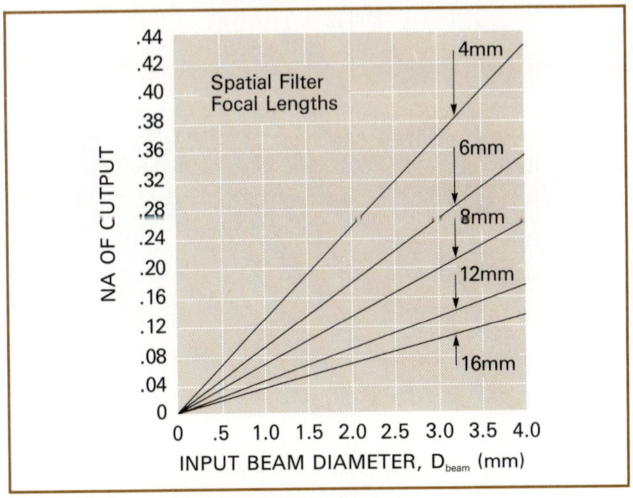

FOCUSING OPTICS selection graph.

Focusing optics are chosen by reference to the graph above and the $1/e^2$ focal spot radius (w_o) may be calculated from the following formula:

$w_o = \lambda f/\pi w$, where λ = wavelength
f = focal length of Focusing Optics
w = $1/e^2$ radius of input beam

Focusing Optics

Focal Length f_o (mm)	Maximum Aperture (mm)	Wavelength Range (nm)	PRODUCT NUMBER
4.0	1.6	450-750	09 LSL 001
6.0	2.4	450-750	09 LSL 003
8.0	3.2	450-750	09 LSL 005
12.0	4.8	450-750	09 LSL 007
16.0	5.0	450-750	09 LSL 009

PINHOLES

Thirteen different size pinholes are recommended for matching the aperture to the $1/e^2$ diameter of the focused laser beam. Each pinhole is mounted in an aluminum cell which fits the X-Y stage in the spatial filter assembly. The pinholes themselves are precisely laser-drilled in 0.013mm stainless steel (0.0025mm stainless foil in the small sizes). All pinholes are exceptionally round and free from edge defects.

Pinhole Selection Chart; for λ = 632.8nm

Focusing Optics	Pinhole Diameter (microns)		
	Input Beam Diameter		
f_o (mm)	0.6mm	0.8mm	1.0mm
4.0	10	8	6
6.0	15	12	10
8.0	20	15	12
12.0	30	25	20
16.0	40	30	25

Pinhole selection is directly related to the spot size. We recommended the use of a pinhole of diameter about 2 times the $1/e^2$ beam contour diameter at focus.

To select an appropriate pinhole for use with the most red helium neon lasers, refer to the chart above.

Mounted Pinholes

Diameter (microns)	PRODUCT NUMBER
2.0	04 PPM 001
3.0	04 PPM 027
4.0	04 PPM 029
5.0	04 PPM 003
6.0	04 PPM 033
8.0	04 PPM 035
10.0	04 PPM 005
12.0	04 PPM 037
15.0	04 PPM 007
20.0	04 PPM 009
25.0	04 PPM 011
30.0	04 PPM 039
40.0	04 PPM 045

ACCESSORY HOLDERS

MICROSCOPE OBJECTIVE HOLDER

Microscope objectives are frequently used in laser imaging systems because of their convenience as focusing lenses. This holder will accept any standard microscope objective having a standard RMS thread.

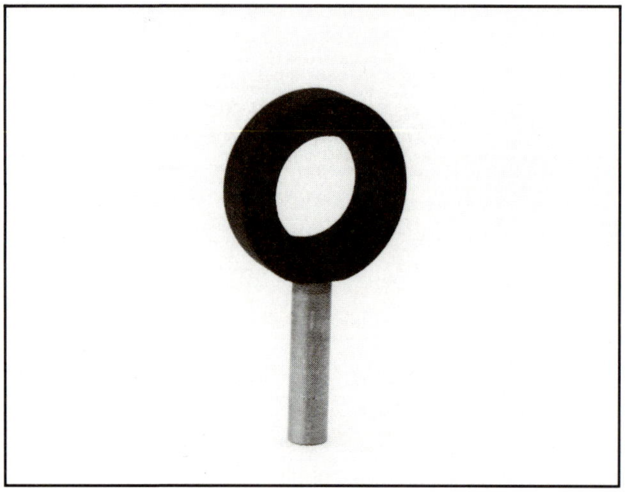

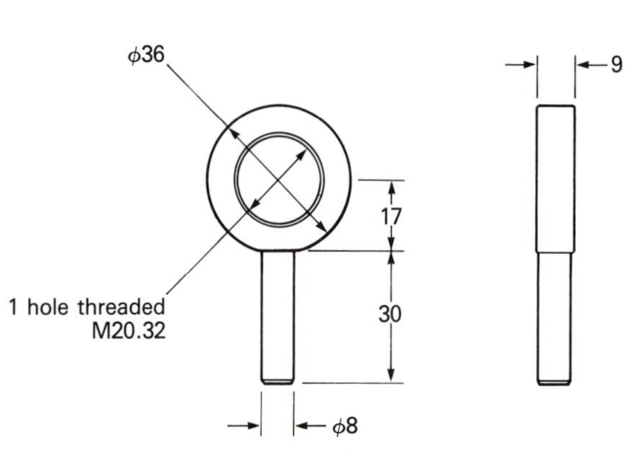

07 HMO 001

Microscope Objective Holder

Post Diameter (mm)	PRODUCT NUMBER
8	07 HMO 001

LASER ACCESSORY HOLDER

Melles Griot offers a selection of mounted laser accessories including a spatial filter, collimators and beam expanders (see pages 361-367). The holder listed here will accept any of these accessories if it is desired to mount them independently of a laser.

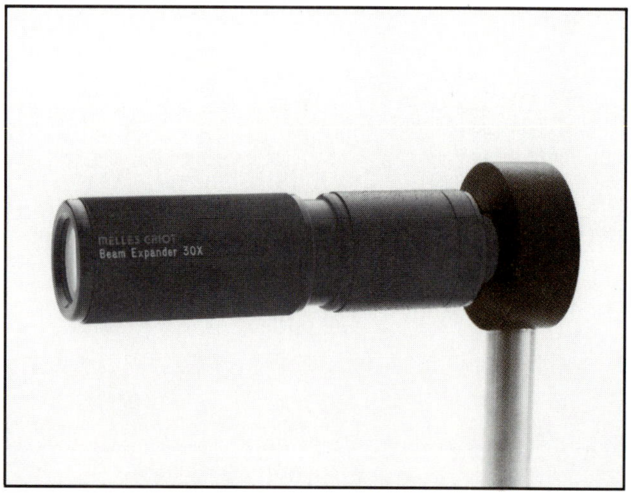

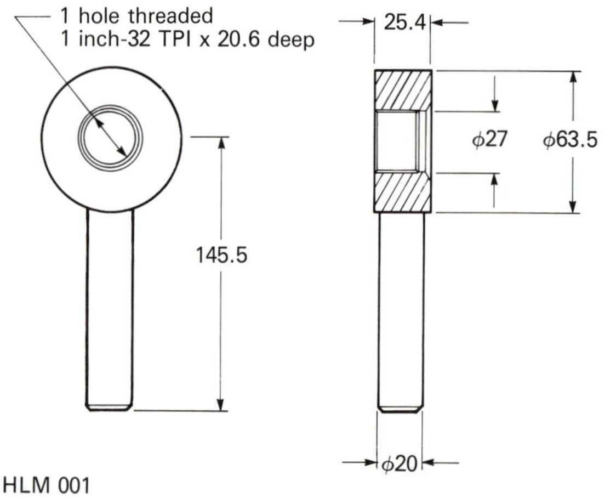

07 HLM 001

Laser Accessory Holder

Post Diameter (mm)	PRODUCT NUMBER
20	07 HLM 001

LASER DIODE HOLDER

This laser diode holder facilitates the mounting of at least two of the most common didoes in use today. It may easily be modified to suit many others. The holder has two screws to accommodate the rectancgular flange mounting (e.g. Hitachi HL 7801 E) and also a split ring for clamping the diameter of cylindrical flanges (e.g. Hitachi HL 7801 G).

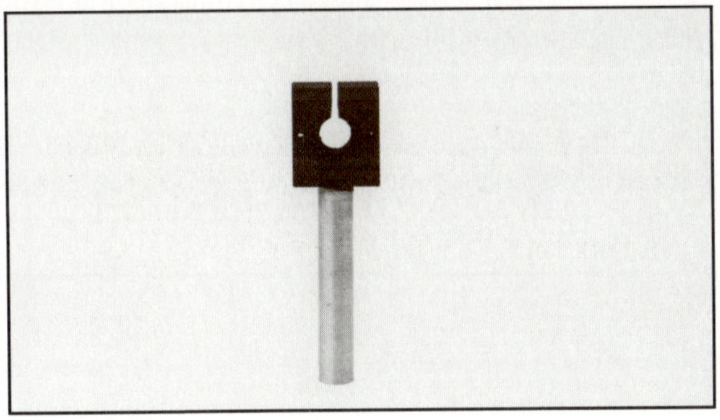

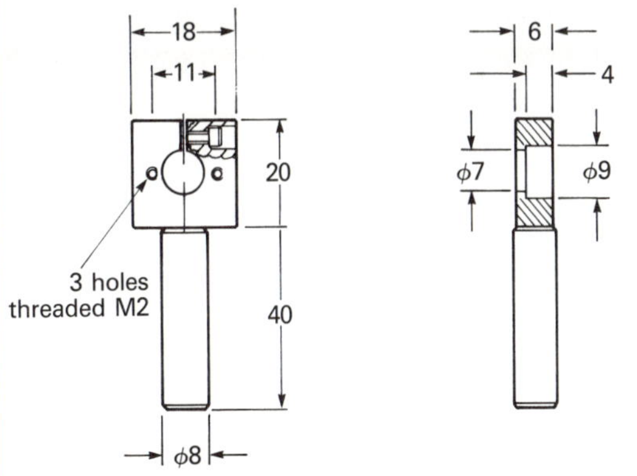

07 HLD 001

Laser Diode Holder

Post Diameter (mm)	PRODUCT NUMBER
8	07 HLD 001

GRADIENT INDEX ROD LENS HOLDER

Gradient index rod lenses (see page 386) are extremely useful for the coupling of light into or out of optical fibers or laser diodes. Their extremely small size is an advantage in these applications but it presents some problems in mounting. The lens holders described here provide a secure means of mounting these lenses, without themselves intruding into the optical path. They will be found extremely useful in experimental situations.

These lens holders consist of a mounting rod which has been reduced in width to 3mm at the top. A phosphor bronze spring is used to retain the gradient index lens in a groove across the top of the post. This will allow access to both ends of, even the smallest, gradient index lens supplied by Melles Griot.

Mini and standard sizes of gradient index rod lens holder are offered.

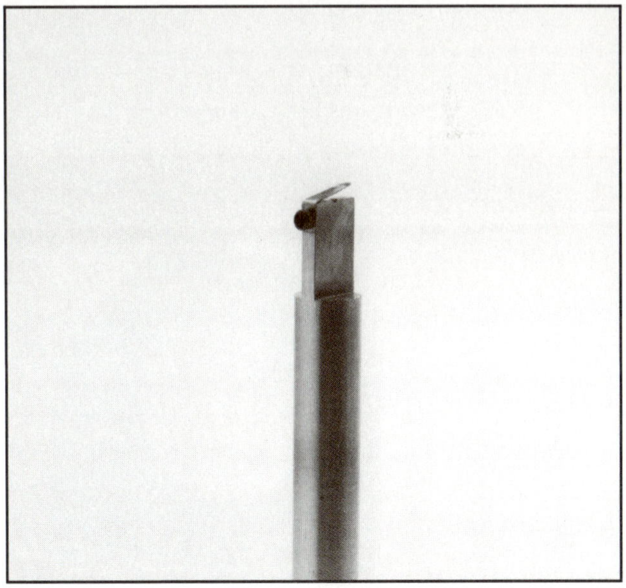

Gradient Index Rod Lens Holders

Post Diameter (mm)	PRODUCT NUMBER
8	07 HGI 001
12	07 HGI 003

MIRROR MOUNTS

Mounting a mirror is a more demanding requirement than one realizes at first thought. Because of the reflection, all angular tolerances immediately double at a mirror surface. So the first problem is simply being able to position a mirror in space so that its surface is normal to the optical axis.

If the mount distorts the surface, the effects are far more dramatic than in a transmissive element, and invariably the results are a reduction in system optical performance.

Even more distressing is rotating a mirror surface around an axis which is not coplanar with the surface of a flat mirror. Not only is the surface translated in space as well as being rotated, but in a convergent beam the conjugate distances are changed as well.

Melles Griot mirror mounts will firmly hold mirrors without distortion. The face of a flat mirror is coplanar with the rotational axes when properly assembled, and the rotational motions are smooth and precise.

A variety of different size and rotational resolution mirror mounts are available to fit a variety of needs. Ultra-precise units provide an angular adjustment sensitivity of 0.2 arc seconds. Large units are available for mirrors up to 254mm in diameter. All have been designed for high mechanical and thermal stability and will provide many years of trouble-free service.

Mirror mounts and adapters are available from stock to fit all Melles Griot standard mirror, window and beamsplitter sizes. Rapid delivery assures you of not being interrupted in the midst of a critical setup.

For the OEM, all mirror mounts and accessories are available in modified or simplified formats. Making use of pre-tested designs can save time and expense.

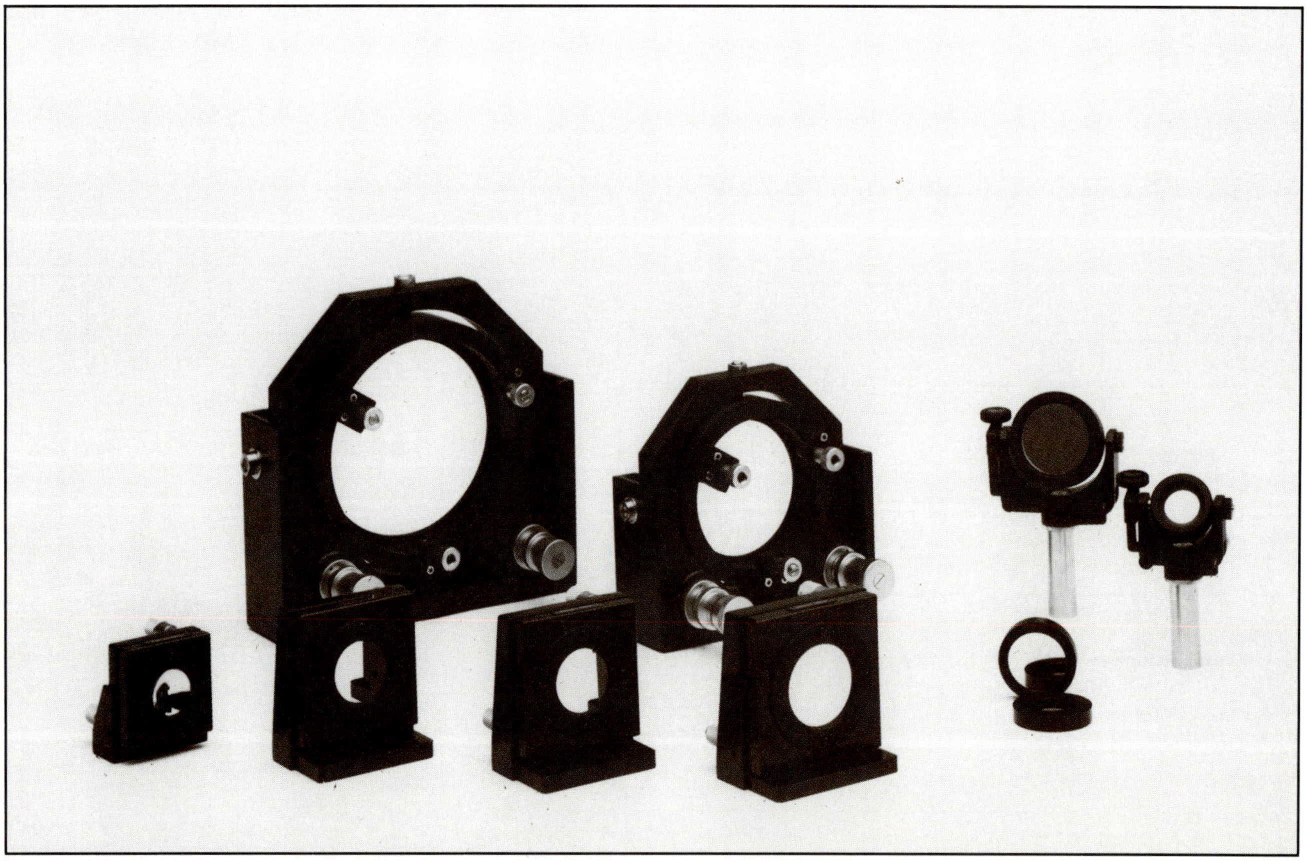

MINI-
MOUNTED MIRRORS

These mini-mounted mirrors are extremely useful, general purpose 25mm square flat mirrors, mounted on a smooth, precise angular motion stage. The glass mirror is flat to $\lambda/4$ overall and is aluminum coated with a silicon monoxide protective overcoat (/ 011).

The angular control is provided by two lapped, precision thumb screws. Total angular range is 6° in both horizontal and vertical rotational axes. Angular sensitivity is 2 arc minutes.

The base is offered with three alternate fixing means. They are a 1/4-20 female thread, an M6 female thread or a 6.5mm clearance slot to accept a 1/4-20 or M6 bolt. In each case the center of the fitting screw is directly under the center of the mirror surface permitting the mount to be rotated about the mirror surface. These mirror mounts may be attached to the end of a mounting post or may be screwed down to a table or other mounting surface.

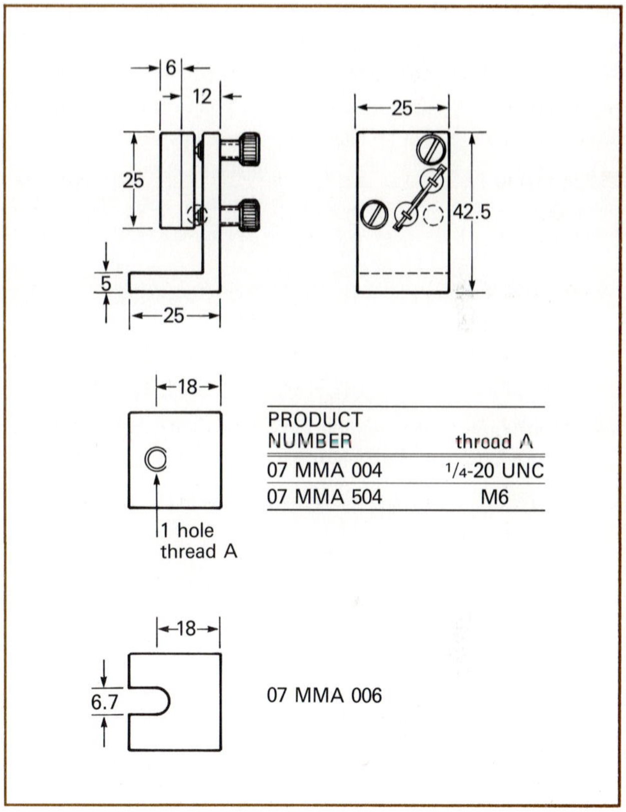

Mini Mounted Mirrors

Fixing Means	PRODUCT NUMBER
1/4-20 thread	07 MMA 004
M6 thread	07 MMA 504
slot	07 MMA 006

COMPONENTS FOR THE OEM

A number of Melles Griot mechanical components are well suited to OEM applications. The mini-mounted mirrors, described above, are a good example. These units can be supplied without the mirrors or with a variety of specially coated or different sized mirrors to fit your specific requirement.

The advantages to the OEM are many fold: design and reliability study costs are reduced, inventory costs are reduced and a pretested system is readily developed.

Please contact our Applications Engineers for further information.

POST-TYPE MIRROR MOUNTS

These 20mm post-type mirror mounts rotate flat mirrors about their surface, rather than some remote axis as is typical of other designs. This means that optical beams may be moved in an angular sense without changing the conjugate lengths of the reflected beam. This is of particular importance in set-ups involving imaging or relaying through a fold-mirror surface.

The mirror mounting plate is kinematically held against the frame with stainless steel balls. The spring-loaded attachment assures an absence of backlash and smooth, precise angular motion. Both the fine-pitch drive screws (0.5mm pitch) and the differential micrometers provide low stiction and push against the mount with hardened stainless steel balls.

These mirror holders are available in three sizes and a variety of adapters are offered to accommodate Melles Griot standard round mirror diameters (see page 187). Mirrors are held in place by a slotted retaining ring and a nylon cushion ring (between the mirror and retaining ring).

Coarse motion in both vertical and horizontal axes is continuous through 360°. After locking the coarse motion with the appropriate thumb screw, fine motion is provided by either precision lapped, spring-loaded lead screws or, for really precise control, by differential micrometers.

The total angular range for the lead-screw controls is 6°; for the differential micrometers it is 3°. Angular sensitivity is stated in the table below.

All three mirror holders will accept either 25, 50 or 100mm diameter mirrors or, respectively, 1, 2 or 4 inch diameter mirrors with equivalent accuracy.

These mirror mounts may each be used in the precision beam-steering device described on page 432.

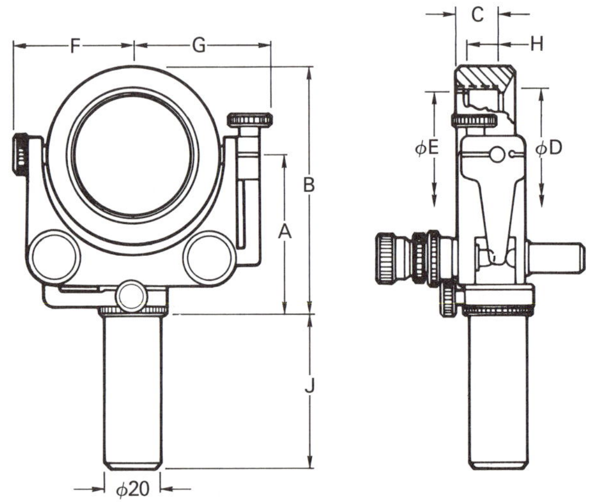

PRODUCT NUMBER	A	B	C	φD	φE	F	G	H	J
07 MAD 001	48	68	10	25.4	22.4	33	45	6	100
07 MAD 005	59	91	15	50.8	47.8	45	56	11	80
07 MAD 008	116	178	23	101.6	98.6	83	101	19	50

SPECIFICATIONS: POST-TYPE MIRROR MOUNTS

Dimensions: ±0.1mm

Angular Range; Lead screw: 6°

Differential micrometer: 3°

Material: Brass

Finish: flat black and satin chrome plate

Post-Type Mirror Mounts

Adjuster Type	Mirror Diameter (mm)	Maximum Mirror Thickness (mm)	Angular Sensitivity (arc sec)	Fine Range (arc min)	PRODUCT NUMBER
Screw	25 or 25.4	6.0	30.0	—	07 MAS 001
	50 or 50.8	11.0	22.0	—	07 MAS 005
	100 or 101.6	19.0	10.0	—	07 MAS 008
Differential Micrometer	25 or 25.4	6.0	0.6	24	07 MAD 001
	50 or 50.8	11.0	0.4	18	07 MAD 005
	100 or 101.6	19.0	0.2	8	07 MAD 008

BEAMSPLITTER MOUNTS

A post-type beamsplitter mount for precise angular control. These mounts are similar to our post-type mirror mounts, except that their profile has been thinned down to allow for a maximum diameter of both the reflected and transmitted beam at 45° angle of incidence.

Control is either by precision lead screws or by differential micrometers. The holder plate is kinematically held, spring-loaded and backlash-free. Only 50mm (or two inch) diameters are available, although mirror mount adapters may be used successfully to mount smaller diameters. Beamsplitters which fit this mount are described on page 236.

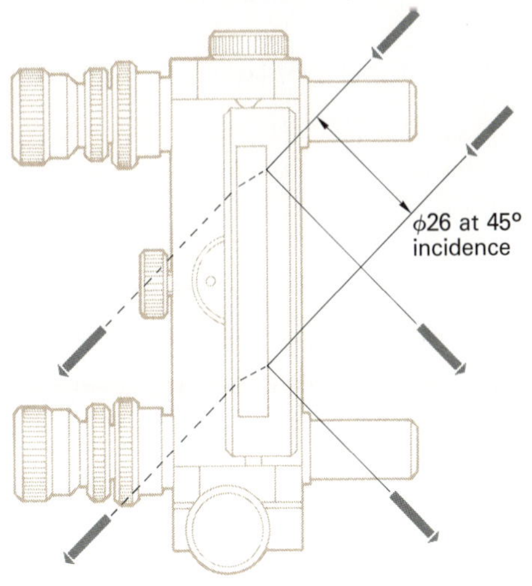

07 MSD 001

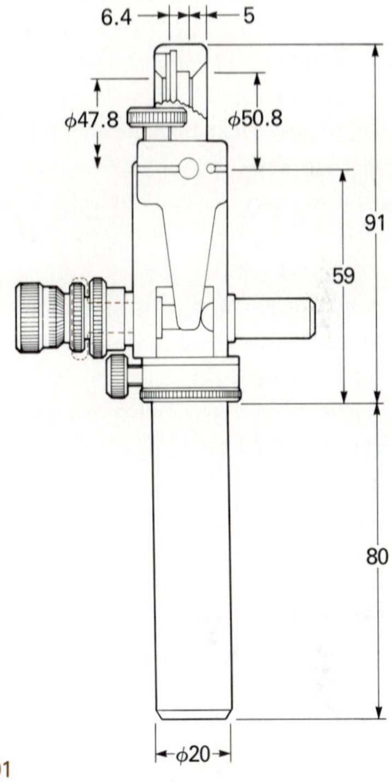

07 MSS 001
07 MSD 001

SPECIFICATIONS: BEAMSPLITTER MOUNTS

Angular Range; Screw controls: 6°
Differential micrometer controls: 3°

Post Length: 80mm

Material: Brass

Finish: Flat black and satin chrome

Beamsplitter Mounts

Adjuster Type	Beamsplitter Diameter (mm)	Maximum Beamsplitter Thickness (mm)	Angular Sensitivity (arc sec)	Fine Range (arc min)	PRODUCT NUMBER
Screw	50.8	6.5	22.0	—	07 MSS 001
Differential Micrometer	50.8	6.5	0.4	18	07 MSD 001

BRACKET-TYPE MIRROR MOUNTS

These mirror mounts are intended for mounting directly on an optical table or on a variety of Melles Griot translational and rotational stages. Coarse rotational motion is not incorporated into these mounts.

Mirror mounts are kinematically mounted on robust, reinforced aluminum cast angle brackets. The brackets are clearance drilled for 1/4-20 screws on one inch or M6 holes on 25mm centers for mounting.

Control is provided by either precision lapped stainless steel thumb screws (with a lead of 2 turns per mm) or by metric micrometer. Both press against the spring-loaded mirror holder plate with a hardened stainless steel ball. This provides a precise, backlash-free, control to the kinematically mounted mirror plate. The control screws are located conveniently for ease of operation.

Mirrors are held in the mount by slotted retaining rings and a nylon cushion to protect the mirror. The surface of the mounted mirror is coplanar with the axes of rotation such that no Z translation is imparted to the mirror when it is rotated. Adapters, which slide into the mount and held in place with the retainer ring, are available for other standard sized Melles Griot mirrors, and are listed on the following page.

Mirror mounts are made of solid brass and are finished flat black to reduce possible stray reflections.

SPECIFICATIONS: BRACKET-TYPE MIRROR MOUNTS

Angular Range; Lead screws: 6°
Micrometers: 6°
Material: Brass holder with cast aluminum frame
Finish: Body, flat black paint;
Controls, satin chrome plate

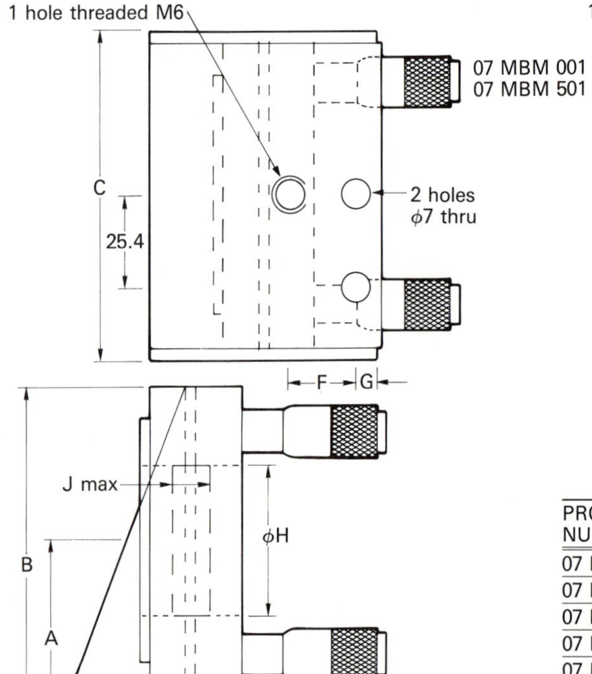

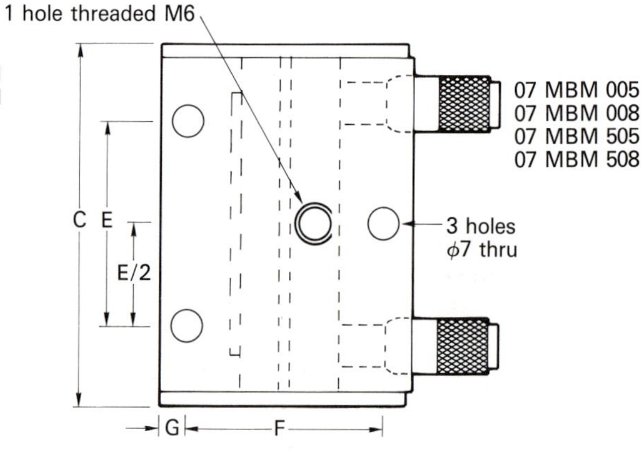

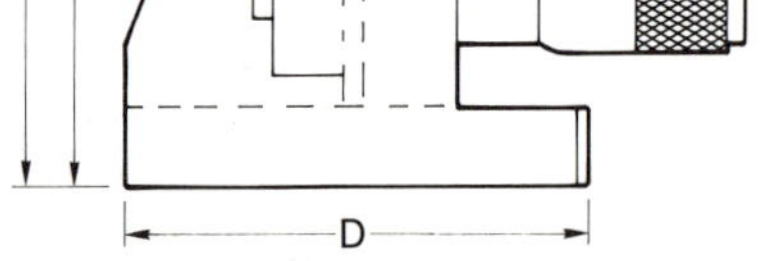

PRODUCT NUMBER	A	B	C	D	E	F	G	ϕH	J
07 MBM 001	40	70	70	44	–	12	10	25.4	10
07 MBM 005	55	100	100	64	50.8	50.8	7	50.8	15
07 MBM 008	80	150	152	64	101.6	50.8	7	101.6	20
07 MBM 501	40	70	70	44	–	12	10	25.4	10
07 MBM 505	55	100	100	64	50	50	7	50.8	15
07 MBM 508	80	150	152	64	100	50	7	101.6	20

Bracket-Type Mirror Mounts

Adjuster Type	Mirror Diameter (mm)	Maximum Mirror Thickness (mm)	Angular Sensitivity (arc sec)	PRODUCT NUMBER Inch Version	Metric Version
Screw	25 or 25.4	10	17	07 MBS 001	07 MBS 501
	50 or 50.8	15	10	07 MBS 005	07 MBS 505
	100 or 101.6	20	6	07 MBS 008	07 MBS 508
Micrometer	25 or 25.4	10	8	07 MBM 001	07 MBM 501
	50 or 50.8	15	5	07 MBM 005	07 MBM 505
	100 or 101.6	20	3	07 MBM 008	07 MBM 508

MIRROR MOUNT ADAPTERS

For mounting other circular mirrors directly into the mirror mounts. Adapters are provided complete with their own retaining ring and nylon cushion ring.

They are designed such that, when fully assembled, the mounted mirror surface is coplanar with the axes of rotation of the mount.

Mirror mount adapters are made of solid aluminum and are flat black anodized.

Mirror Mount Adapters

Mirror Diameter (mm)	Maximum Mirror Thickness (mm)	Outside Diameter (mm)	PRODUCT NUMBER
6.4	6.0	25.4	07 MMS 001
12.5	6.0	25.4	07 MMS 002
15.0	6.0	25.4	07 MMS 003
20.0	11.0	50.8	07 MMS 007
25.4	11.0	50.8	07 MMS 008
30.0	11.0	50.8	07 MMS 009
36.0	11.0	50.8	07 MMS 011
38.1	11.0	50.8	07 MMS 012
40.0	11.0	50.8	07 MMS 013
60.0	20.0	101.6	07 MMS 015
70.0	20.0	101.6	07 MMS 016
76.2	20.0	101.6	07 MMS 018
80.0	20.0	101.6	07 MMS 019

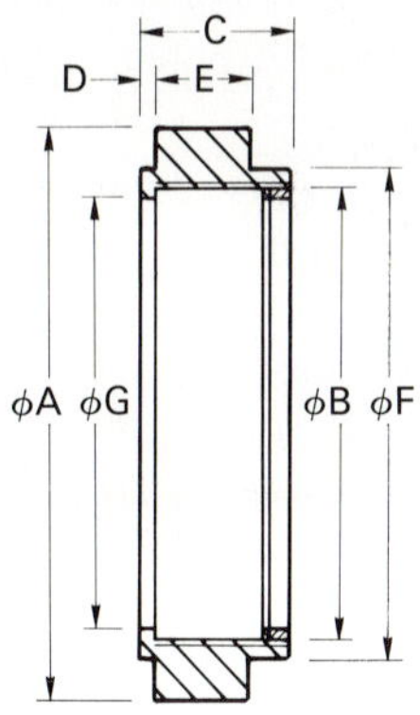

PRODUCT NUMBER	ϕA	ϕB	C	D	E	ϕF	ϕG
07 MMS 001	25.4	6.4	12	2	4	22	4.4
07 MMS 002	25.4	12.5	12	2	4	22	10.5
07 MMS 003	25.4	15	12	2	4	22	12
07 MMS 007	50.8	20	16	2	9	44	17
07 MMS 008	50.8	25.4	16	2	9	44	22.4
07 MMS 009	50.8	30	16	2	9	44	27
07 MMS 011	50.8	36	16	2	9	44	33
07 MMS 012	50.8	38.1	16	2	9	44	35
07 MMS 013	50.8	40	16	2	9	44	37
07 MMS 015	101.6	60	25	2	14	88	56
07 MMS 016	101.6	70	25	2	14	88	66
07 MMS 018	101.6	76.2	25	2	14	88	72.2
07 MMS 019	101.6	80	25	2	14	88	76

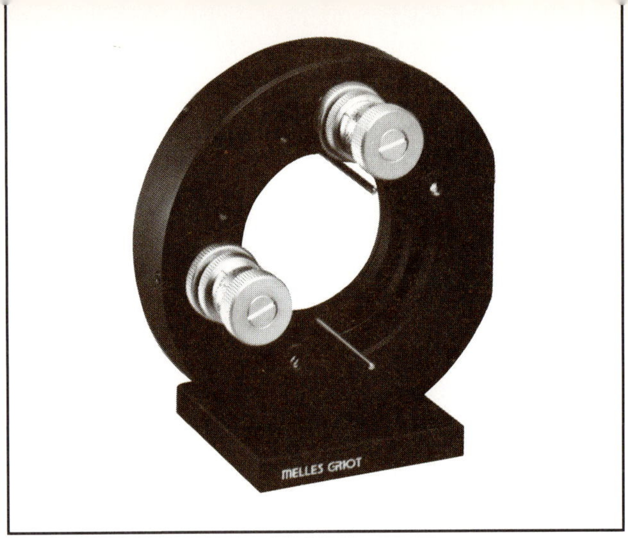

PRECISION MIRROR MOUNTS

A top-quality mirror mount for extremely precise angular beam control. These mirror mounts offer the most precise angular control available. Angular sensitivity is 0.5 arc seconds or better over a total angular range of 6°. This extremely stable holder is available for 50mm (and two inch) diameter mirrors only (see page 187). The differential micrometer has a total range of 8mm and a fine travel range of 200 microns. The fine travel has a motion of 0.4 microns per division.

The holder is made of solid brass for mechanical stability. The mirror plate is held firmly by springs against the kinematic interface with the base. The precision polished stainless steel balls of the kinematic support sit in precisely machined kinematic locations firmly and positively.

Motion is provided by two precision differential micrometers with polished steel spherical ends pressing against the motion plate. Motion is virtually stiction free and, of course, without backlash. Only light pressure is required to impart a smooth, assured, controlled motion.

The mirror is held in position, from the rear, by two spring-loaded thumb screws carried by a circular pressure plate. The open rear is essentially symmetrical to the front, thereby providing similar environments for thermal stability for the mirror.

This mirror mount has been especially designed for the most demanding and assured angular control of an optical beam as is possible. The reduced angular adjustment range presumes that it will be used in precision set-ups. The base is clearance drilled for four 1/4-20 screws on one inch (or M6 on 25mm) multiples for direct mounting to optical breadboards or tables and the complete range of Melles Griot translational, rotational and positional stages.

SPECIFICATION: PRECISION MIRROR MOUNTS

Angular Range: 6°
Angular Sensitivity: <0.5 arc seconds
Material: Solid steel with hardened stainless steel motion components
Finish: Flat black paint

Precision Mirror Mounts

Mirror Diameter (mm)	Mirror Thickness (mm)	Angular Sensitivity (arc sec)	PRODUCT NUMBER Inch Version	Metric Version
50.8	12.0	0.5	07 MCD 005	07 MCD 505

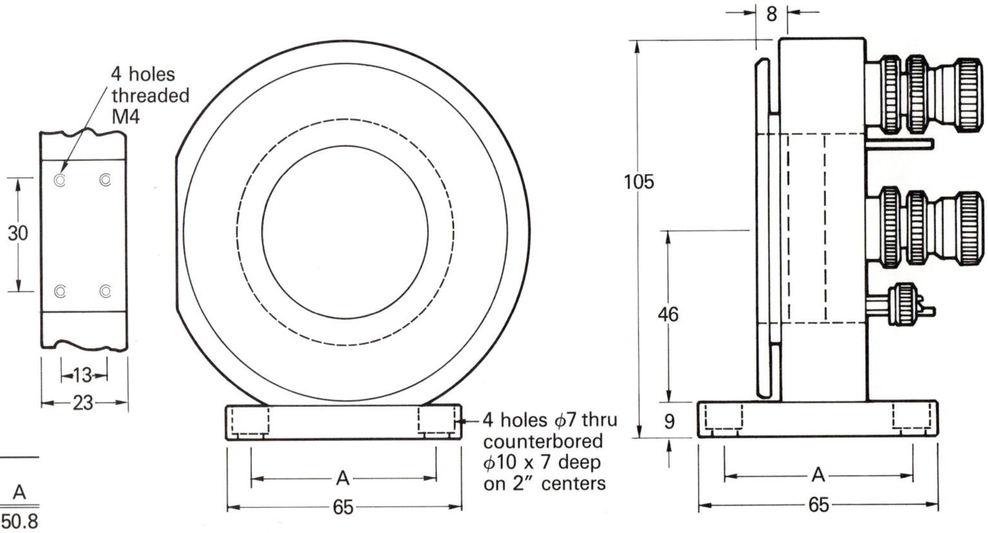

PRODUCT NUMBER	A
07 MCD 005	50.8
07 MCD 505	50

LARGE MIRROR MOUNTS

These large mirror mounts are extra robust to provide precision and stability to large optical mirrors. Motion to both axes is provided by differential micrometers for high resolution and precise motion. The differential micrometers provide this extremely precise angular control over a restricted operating range which is dependent on the mirror holder size.

Mirror mounts are kinematically held internally and are spring-loaded for backlash-free motion. The angle bracket-shaped bases are cast aluminum and are provided with a sub-plate that can either be clamped to a table or breadboard with the table clamps or screwed down, through clearance holes, with three 1/4-20 or M6 cap screws. The base hole pattern is available for one inch spaced or 25mm spaced multiples.

These mirror holders are extra sturdy for positive and reliable support and control of heavy mirrors. Both mechanical and thermal stability are excellent.

SPECIFICATIONS: LARGE MIRROR MOUNTS

Angular Range: 6°
Material: Brass Holders with cast aluminum brackets
Finish: Flat black paint

Large Mirror Mounts

Mirror Diameter (mm)	Maximum Mirror Thickness (mm)	Angular Sensitivity (arc sec)	PRODUCT NUMBER Inch Version	Metric Version
101.6	30	0.3	07 MDD 001	07 MDD 501
152.4	40	0.2	07 MDD 002	07 MDD 502
203.2	45	0.16	07 MDD 003	07 MDD 503
254.0	50	0.15	07 MDD 004	07 MDD 504

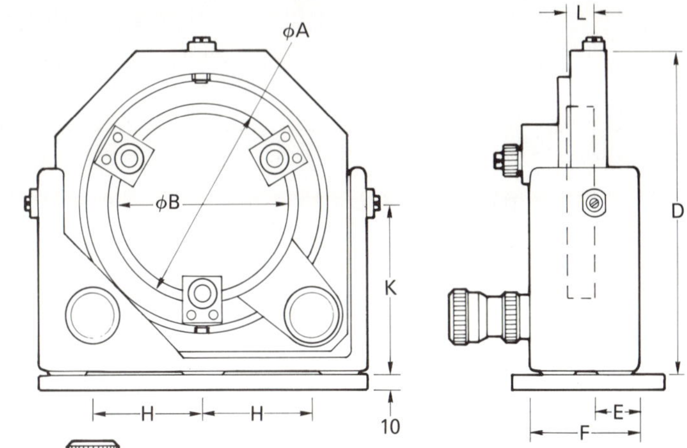

PRODUCT NUMBER	φA	φB
07 MDD 001	101.6	92
07 MDD 002	152.4	138
07 MDD 003	203.2	188
07 MDD 004	254	242
07 MDD 501	101.6	92
07 MDD 502	152.4	138
07 MDD 503	203.2	188
07 MDD 504	254	242

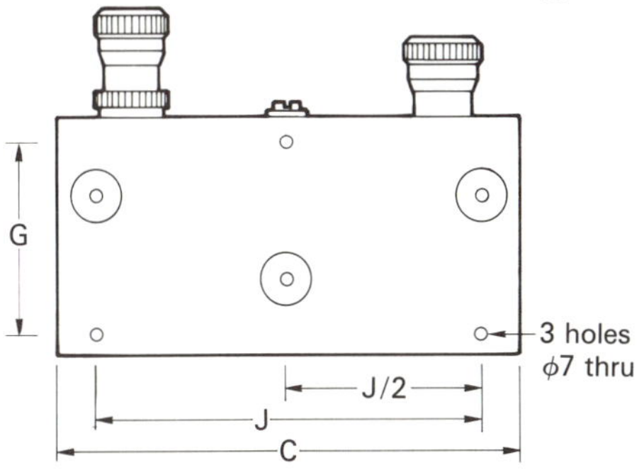

PRODUCT NUMBER	C	D	E	F	G	H	J	K	L
07 MDD 001	180	177	25	90	76.2	60	152.4	90	0-30
07 MDD 002	240	226	25	90	76.2	80	203.2	120	0-40
07 MDD 003	295	284	30	120	101.6	100	254	150	5-45
07 MDD 004	348	340	33	140	127	126	304.8	180	10-50
07 MDD 501	180	177	25	90	75	60	150	90	0-30
07 MDD 502	240	226	25	120	100	80	200	120	0-40
07 MDD 503	295	284	30	120	100	100	250	150	5-45
07 MDD 504	348	340	33	140	125	126	300	180	10-50

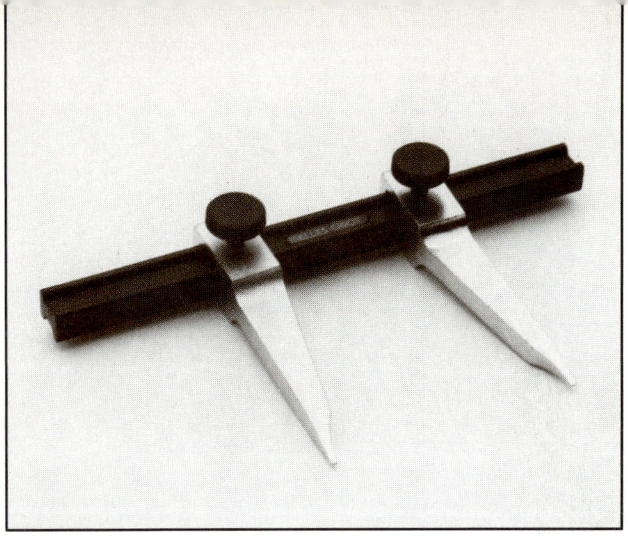

ACCESSORY KITS

FASTENER KITS

These kits are designed specifically to replace lost screws and fasteners. They are packaged, complete with specific tools, in a closeable plastic box.

Fastener Kits

Contents	Quantity	PRODUCT NUMBER
Inch Version:		
1/4-20x3/8 inch Allen screws	50	07 AKT 005
1/4-20x1/2 inch Allen screws	50	
1/4-20x3/4 inch Allen screws	50	
M6x12mm Allen screws	50	
M6x20mm Allen screws	50	
1/4-20 Allen key	2	
M6 Allen key	2	
1/4-20x1/2inch thumb screws	4	
M6x12mm thumb screws	4	
Metric Version:		
M6x10mm Allen screws	50	07 AKT 505
M6x12mm Allen screws	50	
M6x20mm Allen screws	50	
M6 Allen key	2	
M6x12mm thumb screws	4	

SLOT SPANNER

An adjustable slot spanner tool for tightening and loosening the retainer ring used in Melles Griot lens holders, mirror mounts and polarizer holders. Made of steel.

Slot Spanner

PRODUCT NUMBER
07 AKT 007

ACCESSORY KITS FOR OPTICAL HARDWARE

These kits contain a comprehensive set of tools and fasteners for use with all Melles Griot optical hardware packaged in a closeable plastic box.

Fastener Kits

Contents	Quantity	PRODUCT NUMBER
Inch Version:		
1/4-20x3/8 inch Allen screws	8	07 AKT 003
1/4-20x1/2 inch Allen screws	8	
1/4-20x3/4 inch Allen screws	8	
8-32x3/8 inch Allen screws	8	
M4x10mm Allen screws	8	
M6x12mm Allen screws	8	
M6x20mm Allen screws	8	
1/4-20x1/2 inch thumb screws	4	
M6x12mm thumb screws	4	
8-32 Allen key	1	
1/4-20 Allen key	1	
M4 Allen key	1	
M6 Allen key	1	
M8 Allen key	1	
12mm post wrench	1	
Tightening bars (4 sizes)	1 each	
Spanner wrench for Melles Griot lens holders, mirror mounts and polarizer holders	1	
Metric Version:		
M4x10mm Allen screws	8	07 AKT 503
M6x12mm Allen screws	8	
M6x20mm Allen screws	8	
M6x12mm thumb screws	4	
M4 Allen key	1	
M6 Allen key	1	
M8 Allen key	1	
12mm post wrench	1	
Tightening bars (4 sizes)	1 each	
Spanner wrench for Melles Griot lens holders, mirror mounts and polarizer holders	1	

POSITIONING DEVICES

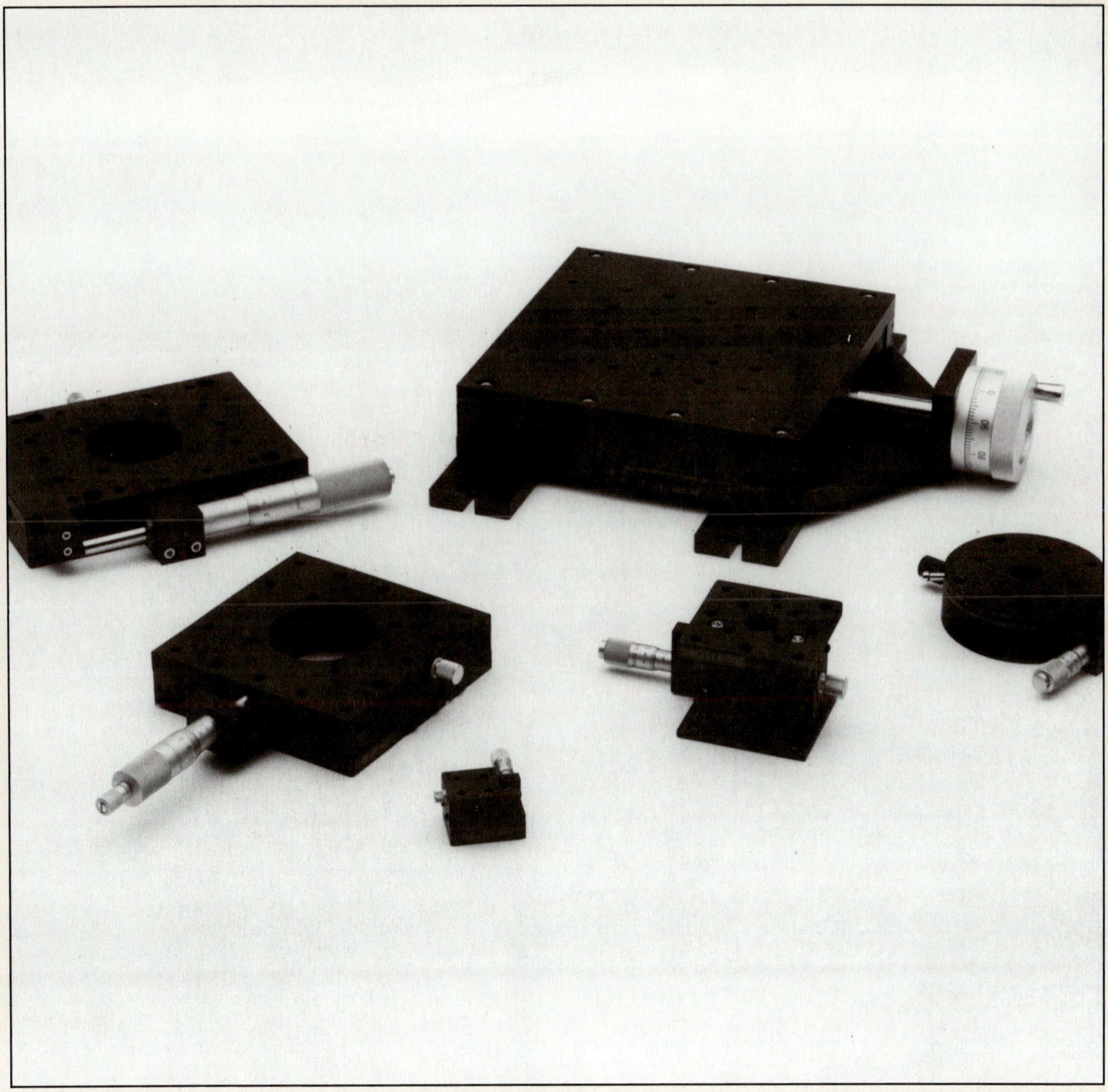

These positioning devices, used in combination, provide from one to six degrees of freedom for the placement of optical components.

Contents **Page Number**

POSITIONING DEVICES

The positioning of optical components can often be critical. As anyone who has attempted to align an optical system knows, lack of the appropriate mechanism can cause great inconvenience and delay. To facilitate your optical alignment tasks Melles Griot offers, from stock, a wide selection of devices, each having a specific function. Before looking at the individual product specification we suggest you consider your positioning problem in the light of the following device selection criteria.

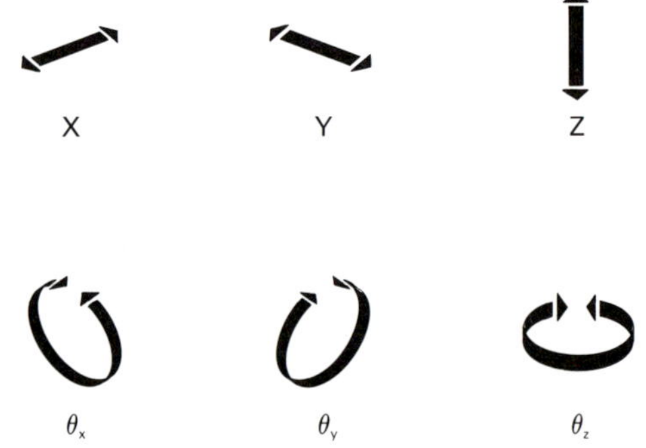

DEGREES OF FREEDOM

By using devices alone or in combination it is possible to achieve from one to six degrees of freedom. There are three translational degrees of freedom and three rotational degrees of freedom. All six, and the convention Melles Griot uses to refer to them, are illustrated in the accompanying diagram. Note that X is always in the direction of the optical axis, Y is horizontal/perpendicular to the axis and Z is vertical. θ_x, θ_y, and θ_z are rotations about the X, Y and Z axes, respectively. The devices listed in the following pages may be combined in many different ways to provide combinations of the six degrees of freedom.

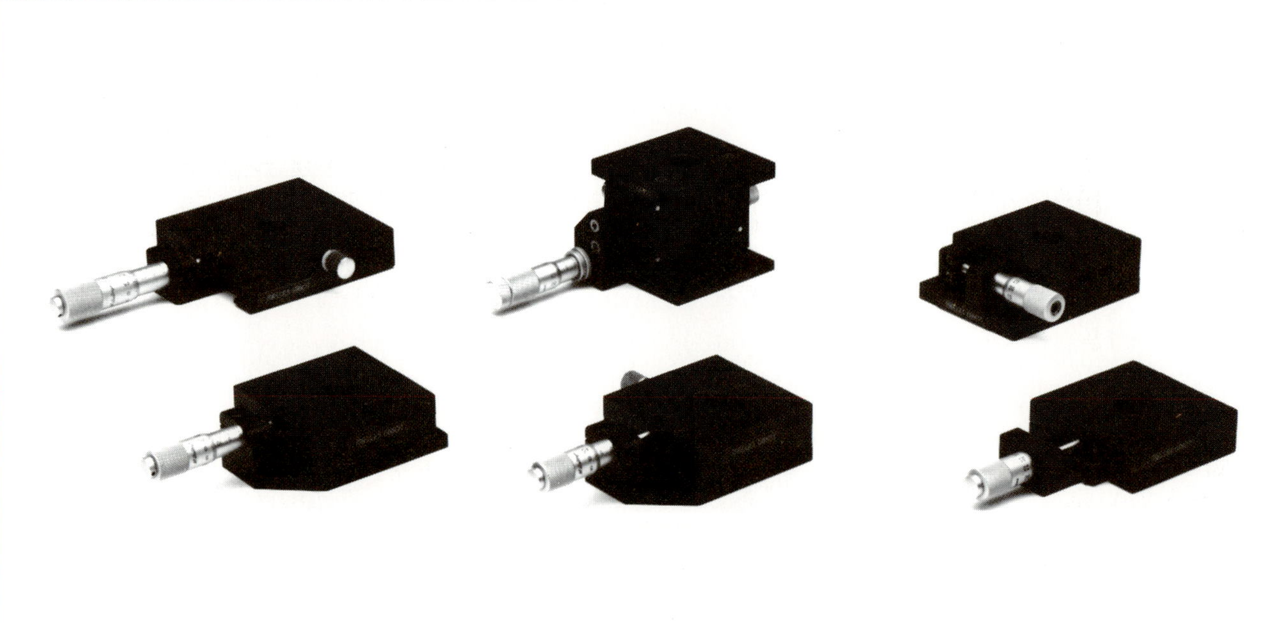

SIZE AND LOAD CAPACITY

The size of the mechanism is of importance. Often the smallest possible mechanism is the most desirable, but when heavy loads are to be positioned, larger and stronger devices are necessary. Melles Griot offers four basic sizes of positioning equipment relating to the following base sizes: 25mm, 65mm, 120mm, and 160mm. Load capacities range from one to 100 kilograms and are stated in the specifications.

MOUNTING PATTERN

Melles Griot is the *only* supplier who offers you a choice of two mounting patterns. We offer both the 1/4-20 thread on one inch (25.4mm) centers pattern, commonly used in the USA, and the M6 thread on 25mm centers pattern, commonly used in Europe and Japan. Most of our devices are available with either of these patterns. Provided you keep to the same pattern, the devices may be combined in many different ways.

RANGE OF TRAVEL

Sometimes you will need a long range of travel or continuous 360° rotatio. In other situations this is not necessary. A greater range usually requires a larger and more expensive device. We recommend that you consider the other advantages of shorter range devices whenever possible.

ACCURACY

The accuracy of an instrument is a measure of its deviation from the ideal. The accuracy of a positioning device is characterized by its linearity (straightness or circularity). All Melles Griot mechanisms are accurate. Even so, there are grades of accuracy depending on the type of bearing used. For example, we offer an inexpensive translation slide using a ball and rod bearing and a more accurate ground ball bearing slide.

PRECISION

The precision of an instrument or device is a measure of its ability to reproduce readings within a given accuracy. Precision of positioning is governed by the actuating mechanism. In most cases we use standard metric micrometers which may be read directly to 0.01mm and interpolated to 2 microns. When we state a sensitivity specification it relates to 1/5 micrometer division. It has been empirically determined that it is possible to reset a device to this degree of precision. When a very high degree of precision is needed, differential micrometer devices are recommended.

COMBINATIONS OF DEVICES

It is clear that the number of possible combinations of positioning devices is endless. We believe that it is better to explain the many options you have than try to put together standard combination packages which may, or may not, be applicable to your application. Therefore, we do not offer XY, XYZ, $YZ\theta_x$, or $\theta_y\theta_z$ packages, as others do. Putting together these packages is straight forward; all you do is select the individual component devices you need. All necessary fixing screws are supplied and assembly is easy. The accompanying photographs show some of the more commonly encountered combinations. These will serve to illustrate which components are needed to build a large variety of combinations. Detailed descriptions and specifications for these products will be found on the following pages. If you need further assistance in selection of correct devices please call one of our Applications Engineers who will be happy to assist you.

Many of the devices described here are suitable for incorporation in original equipment. If you have an OEM requirement, please call for a quotation.

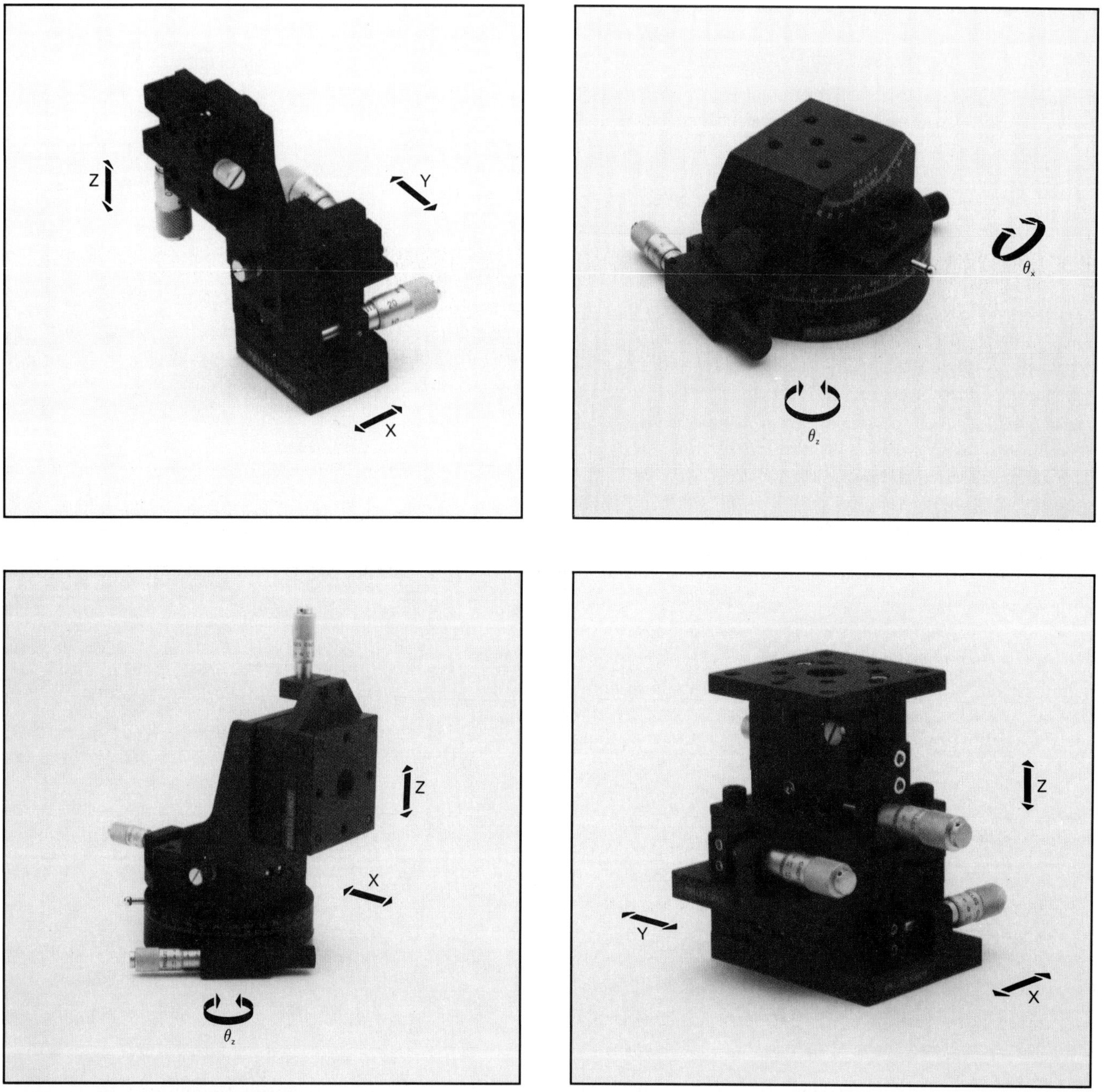

SOME COMMON COMBINATIONS OF MELLES GRIOT POSITIONERS; all six possible degrees of freedom can be achieved with the proper combination.

LAB JACKS

Melles Griot lab jacks are a secure and reliable means of accurately raising and lowering optical assemblies or sub-systems to the required axis. The classical dual pantograph design completely eliminates the stiction inherent in more complex, over-constrained designs. Non-parallelism between the top and bottom plates of all jacks is held to less than 50 microns throughout the complete range of travel.

A large knob is connected to a lead screw which draws the ends of the scissors together raising the jack. The metal-to-metal sleeve bearings in the four side arms on both sides provide great side-to-side and forward-to-rear rigidity while providing a smooth, controlled motion.

Lab jacks are available in three sizes: light-, standard-, and heavy-duty. All have 7mm clearance holes in a diamond pattern on their bases for directly mounting to an optical breadboard or table. The top mounting surface has an array of 1/4-20 or M6 threaded holes on one inch or 25mm centers for mounting a wide variety of translational and rotational stages as well as small post mounted components. The corners of both the top and bottom plates are tapped with four M6 holes to accept a platform extension.

Lab jacks may also be secured to optical tabletops with table clamps.

All three sizes of lab jacks are identical in design. The light-duty lab jack will carry a weight of 75 kilograms over an elevation range of 63mm to 85mm. The standard-duty lab jack will carry a weight of 125 kilograms over an elevation range of 82mm to 122mm. The heavy-duty lab jack will carry a weight of 210 kilograms over an elevation range of 100mm to 170mm above the table surface.

Oversize mounting plates are also available. These plates provide a greatly extended working surface on the top of the jacks.

All three jacks are made with steel top and bottom plates, for use with magnetic bases, and are finished flat black to reduce reflections and scattered light. The drive screw is stainless steel; linkages are of brass.

Throughout this Optics Guide we offer both inch and metric mounting hole patterns. Be sure to keep to a common system.

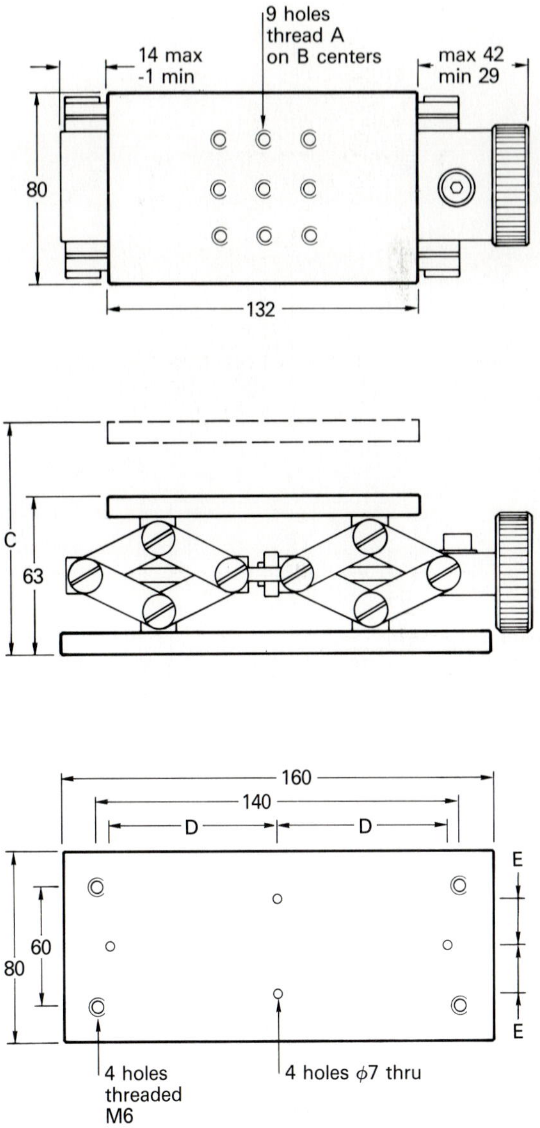

PRODUCT NUMBER	thread A	B	C	D	E
07 BLJ 001	1/4-20	25.4	87	63.5	25.4
07 BLJ 501	M6	25	85	62.5	25

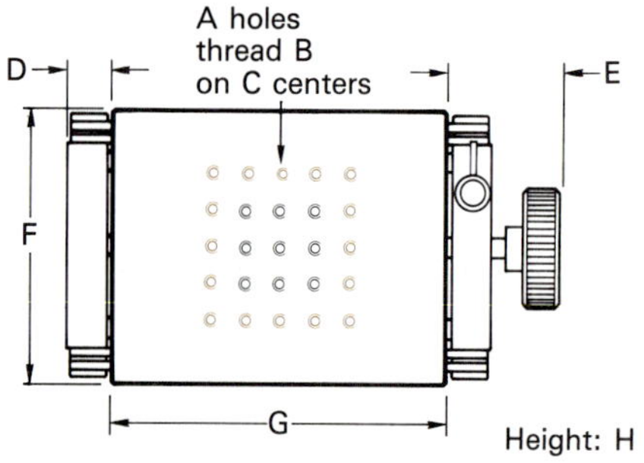

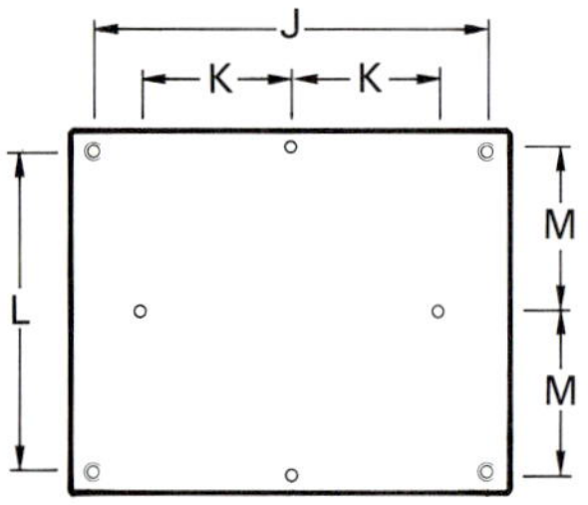

PRODUCT NUMBER	A	thread B	C	D min	D max	E min	E max	F	G
07 BLJ 002	9	1/4-20	25.4	-1	24	63	78	100	170
07 BLJ 003	25	1/4-20	25.4	11	48	87	124	170	210
07 BLJ 502	9	M6	25	-1	24	63	78	100	170
07 BLJ 503	25	M6	25	11	48	87	124	170	210

A Black hole pattern applies to 07 BLJ 002 and 07 BLJ 502 only.
Both black and brown hole patterns apply to 07 BLJ 003 and 07 BLJ 503.

PRODUCT NUMBER	H min	H max	J	K	L	M
07 BLJ 002	80	122	150	63.5	80	38.1
07 BLJ 003	100	170	190	76.2	150	76.2
07 BLJ 502	82	122	150	75	75	37.5
07 BLJ 503	100	170	190	75	150	75

SPECIFICATIONS: LAB JACKS

Load Capacity: Light-duty 75 kg
Standard-duty 125 kg
Heavy-duty 210 kg
Travel: Light-duty 63mm to 85mm
Standard-duty 80mm to 122mm
Heavy-duty 100mm to 170mm
Non-Parallelism Over Travel: $<50\mu m$
Weight: Light-duty 3.1 kg
Standard-duty 5.8 kg
Heavy-duty 10.0 kg
Material: See text
Finish: Flat black paint

Lab Jacks

	PRODUCT NUMBER Inch Version	PRODUCT NUMBER Metric Version
Light-Duty	07 BLJ 001	07 BLJ 501
Standard-duty	07 BLJ 002	07 BLJ 502
Heavy-Duty	07 BLJ 003	07 BLJ 503

PLATFORM EXTENSIONS

Platform extensions may be mounted below a lab jack to increase its starting height by a fixed amount.

Platform extensions are available for all sizes of lab jack. The platforms consist of a plate and three sets of four each 16mm diameter brass posts whose lengths are 20, 40, and 80mm. These have M6 threaded studs on one end which screw directly into the corners of the lab jacks and have M6 threaded holes at the other ends. The platform plate attaches to these posts with four M6 recessed Allen cap screws. The 11mm thick platform plate itself has a diamond pattern of clearance holes for direct attachment to an optical table or breadboard. By combining the extender posts it is possible to achieve extension heights of 31mm to 151mm in 20mm increments (28mm to 148mm in the case of the light-duty jacks).

Because scissors jacks are inherently most stable at their lowest extension, it is best to use the highest extension platform which is convenient. Extension platforms are supplied complete with mounting screws.

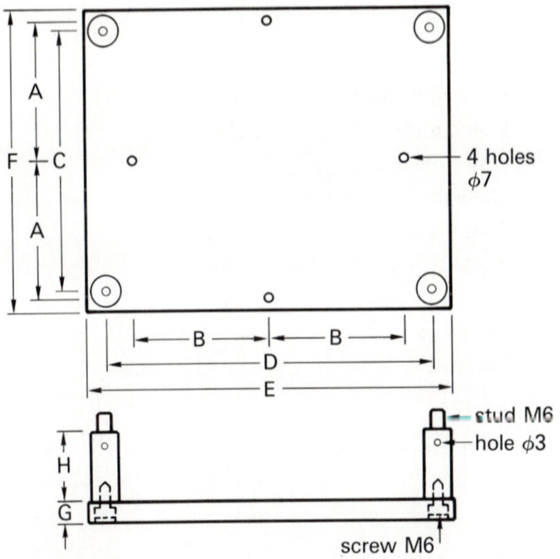

PRODUCT NUMBER	A	B	C	D	E	F	G
07 BLS 001	25.4	63.5	60	140	160	80	8
07 BLS 002	38.1	63.5	80	150	170	100	11
07 BLS 003	76.2	76.2	150	190	210	170	11
07 BLS 501	25	62.5	60	140	160	80	8
07 BLS 502	37.5	62.5	75	150	170	100	11
07 BLS 503	75	75	150	190	210	170	11

H — Height of spacers for all models are 20, 40 and 80mm. These may be combined for additional height.

Lab Jack Extension Platforms

for use with	PRODUCT NUMBER
Inch Version	
Light-Duty Jack (07 BLJ 001)	07 BLS 001
Standard-Duty Jack (07 BLJ 002)	07 BLS 002
Heavy-Duty Jack (07 BLJ 003)	07 BLS 003
Metric Version	
Light-Duty Jack (07 BLJ 501)	07 BLS 501
Standard-Duty Jack (07 BLJ 502)	07 BLS 502
Heavy-Duty Jack (07 BLJ 503)	07 BLS 503

MOUNTING PLATES

For mounting complete subassemblies or oversized loads on lab jacks or for general purpose mounting on an extended working surface. Each plate mounts directly on to the hole pattern on the top of the various jacks.

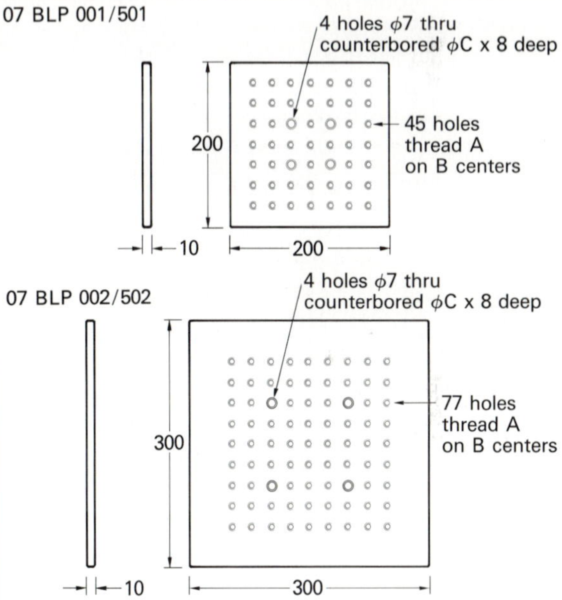

PRODUCT NUMBER	thread A	B	ϕC
07 BLP 001	1/4-20	25.4	10
07 BLP 002	1/4-20	25.4	10
07 BLP 501	M6	25	11
07 BLP 502	M6	25	11

Mounting Plates

Side (mm)	PRODUCT NUMBER Inch Version	Metric Version
200	07 BLP 001	07 BLP 501
300	07 BLP 002	07 BLP 502

GONIOMETERS

Melles Griot goniometers allow tilt motion around an axis either 30 or 50mm above the surface of the base. A vernier and scale on the base allow angles to be read directly to 1 degree, with a vernier reading of 1 arc minute.

The table is driven by a pinion and circular segment gear. The pinion may be controlled by knobs on either end for convenience. A thumb screw on the side will lock the tilt position.

The base is machined from solid brass and finished flat black. The dovetail circular-section bearings have an adjustable preload on one side to allow smooth, precise, controlled motion.

The goniometer base may be mounted directly on a breadboard or optical table top with 1/4-20 or M6 screw or clamped in position with table clamps (07 CTT 001). The tops of the goniometers are tapped with a pattern of holes for the mounting of additional accessories.

SPECIFICATIONS: GONIOMETERS

Vernier Scale: 1 arc minute
Capacity: 5 kg
Material: Brass
Finish: Black chrome

Goniometers

Angular Range (±deg)	Axis Height Above Base (mm)	PRODUCT NUMBER Inch Version	Metric Version
30	30	07 GOH 001	07 GOH 501
20	50	07 GOH 002	07 GOH 502

Tilt motion may also be achieved by using the mirror mounts described on pages 472-479.

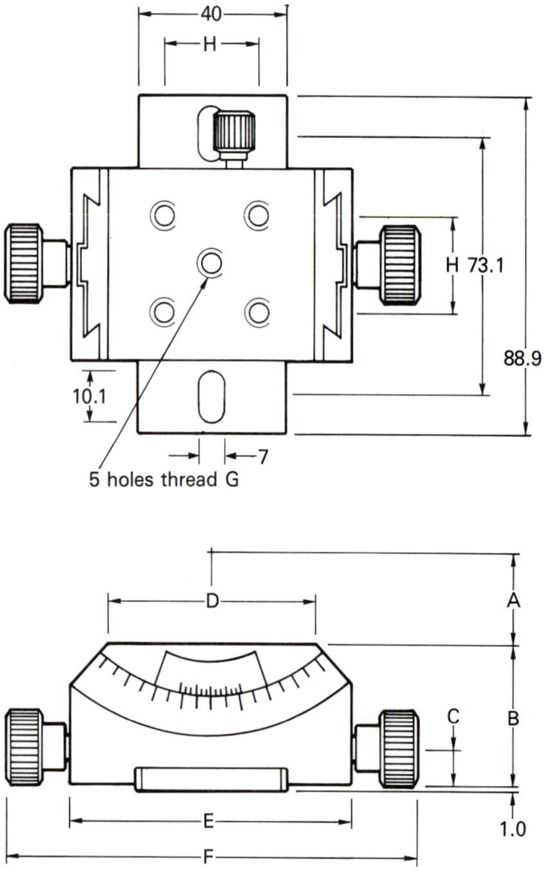

PRODUCT NUMBER	A	B	C	D	E	F	thread G	H
07 GOH 001	30	37	10	56	76	110	1/4-20	25
07 GOH 002	50	42	15	70	84	114	1/4-20	25
07 GOH 501	30	37	10	56	76	110	M6	25.4
07 GOH 502	50	42	15	70	84	114	M6	25.4

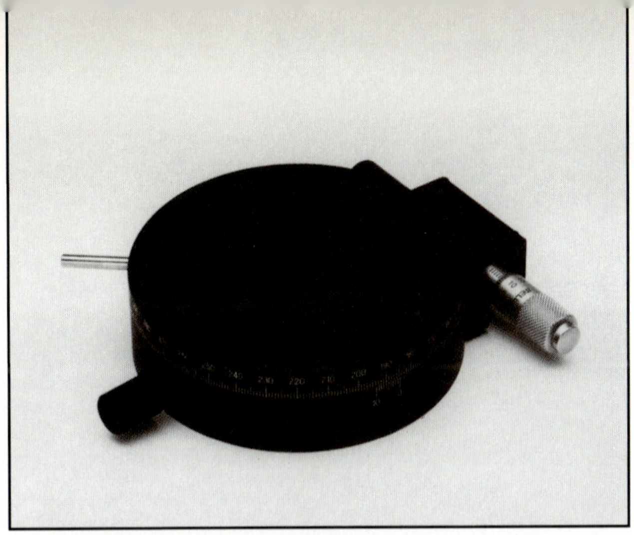

ROTATION STAGES

Melles Griot rotation stages provide a smooth and precise rotational control for optical components and subassemblies. The large, captive brass-on-steel sleeve bearing in the base provides a firm and reliable base.

Coarse motion of 360° rotation may be obtained by releasing the thumb locking screw on the base. When tightened the base is firmly locked to the micrometer driven anvil on the opposite side. Spring-loading against the micrometer hemispherical end eliminates backlash.

An angular scale, calibrated in degrees and readable to 6.0 arc minutes, surrounds the base. Total angular travel with micrometer control is ±5°; angular resolution is 40 arc seconds.

The bottoms of these rotation stages are clearance drilled for 1/4-20 screws on two inch centers or M6 screws on 50mm centers. These holes are accessible, in turn, by rotation of the stage top. The stage tops are tapped for 1/4-20 screws on one inch centers or M6 screws on 25mm centers and also have a central tapped M16 hole for mounting 20mm diameter Melles Griot posts directly or 12mm diameter post holders.

Rotation stages will mount on Melles Griot optical tables and breadboards as well as all of the larger Melles Griot translation stages. In a similar manner the tops of the rotation stages will accept a wide variety of translation stages, mirror mounts, lens holders and other accessories.

The bases are finished with a flat black chrome to reduce possible stray reflections and scatter.

SPECIFICATIONS: ROTATION STAGES

Angular Range: Coarse: 360°
Fine: ±5°
Fine Angular Resolution: 40 arc seconds
Load Capacity: 5 kg
Material: Brass and stainless steel
Finish: Flat black chrome

Rotation Stages

Stage Diameter (mm)	PRODUCT NUMBER Inch Version	Metric Version
84	07 TRT 001	07 TRT 501
98	07 TRT 002	07 TRT 502

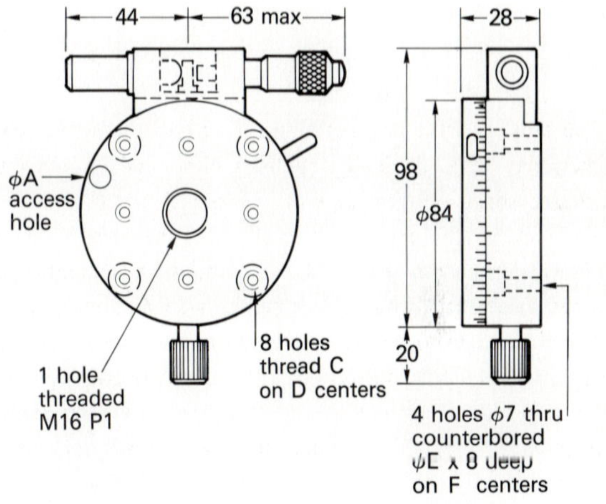

PRODUCT NUMBER	ϕA	thread C	D	ϕE	F
07 TRT 001	10	1/4-20	25.4	10	50.4
07 TRT 501	11	M6	25	11	50

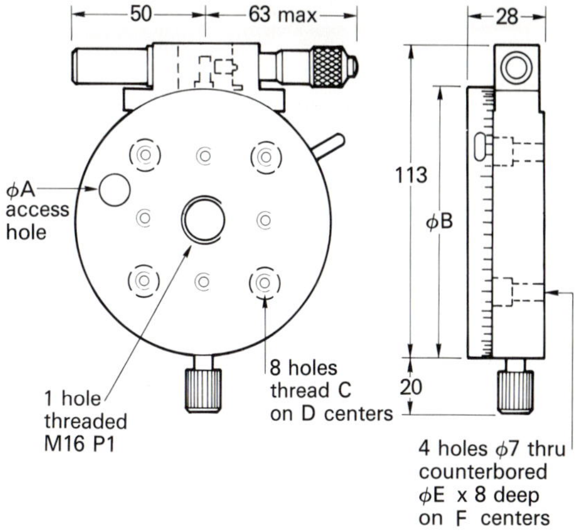

PRODUCT NUMBER	ϕA	ϕB	thread C	D	ϕE	F
07 TRT 002	10	100	1/4-20	25.4	10	50.8
07 TRT 502	11	98	M6	25	11	50

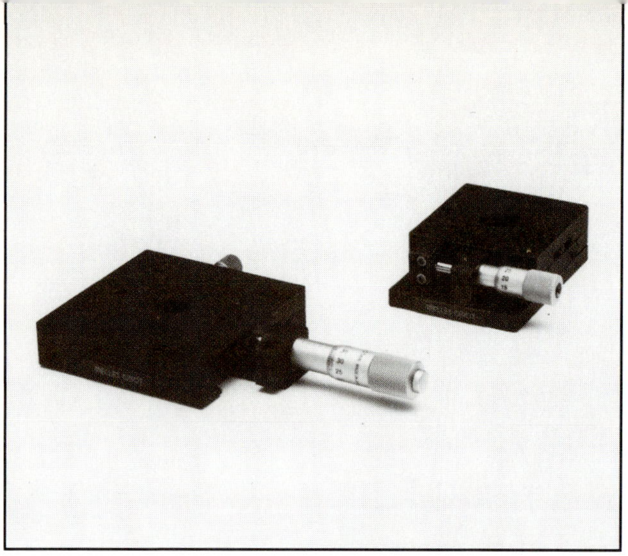

STEEL TRANSLATION STAGES

Melles Griot steel translation stages provide smooth, controlled, repeatable linear motion for optical components and subassemblies. Precision ground stainless steel balls ride between two pairs of precision ground "V" grooves and are slightly pre-loaded at assembly to assure freedom from side-to-side and up-to-down wobble. The motion stage is preloaded against the micrometer ball to provide a smooth motion without backlash.

These stages are all provided with metric micrometers. Positions may be read directly to 0.01mm and interpolated to 2 microns.

These translation stages are available in both side-mounted and center-mounted micrometer locations for convenience in set-up and operation. Both versions provide identical mechanical performance. These models are available in four different sizes with correspondingly different travels and load capacities.

All models are supplied with a slot-screw position locking device which, when tightened, will hold the stage firmly in position without affecting the location in any way. The locks are useful in assuring that positions are not accidentally changed after they have been set.

Stages may be stacked on top of one another to provide orthogonal (or parallel) travel. In the case of the mini-translation stage (25mm), the lower plate must be removed when stages are stacked. All stages will stack on top of all other stages of the same or greater size. Brackets (described on page 494) are available to allow X-Y-Z stacking configurations.

In a similar way, all stages may be mounted to all optical breadboards or other motion devices which have a 1/4-20 by one inch or M6 by 25mm tapped hole pattern. The tops of all translation stages are tapped with a 1/4-20 by one inch or M6 by 25mm hole pattern. Further, the top plates of the 65mm stages are centrally tapped with an M16 hole to mount the 20mm posts or post holders. The largest stages (120 and 160mm) have a 52mm diameter central hole for passing an optical beam. This hole is not obscured by the base plate when the stage is translated from end to end of its travel.

SPECIFICATIONS: STEEL TRANSLATION STAGES

Sensitivity: 2 microns
Material: Steel
Finish: Flat black chrome

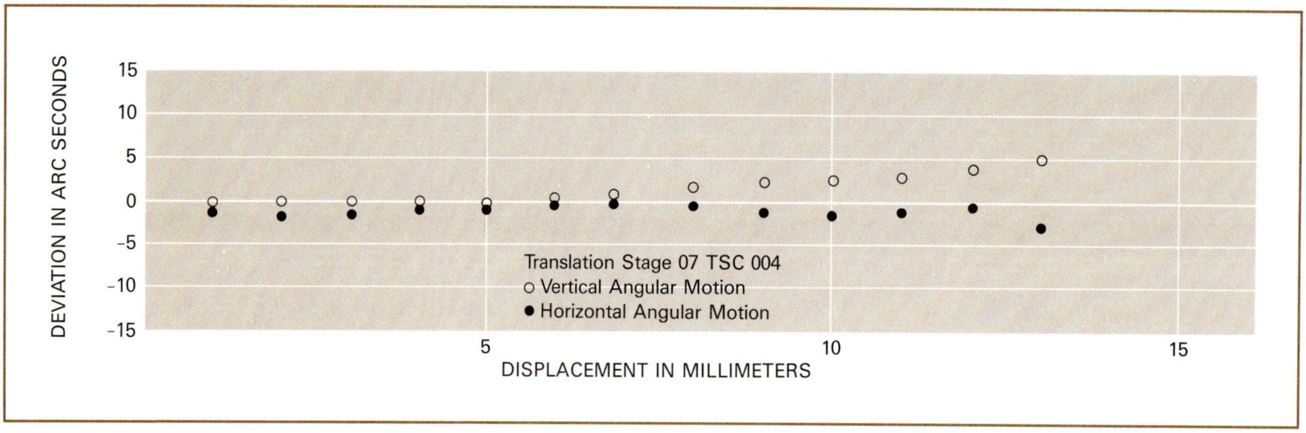

TRANSLATION SLIDE STRAIGHTNESS

Samples from every translation stage production batch are tested for straightness to assure continued high quality. Part of the test procedure calls for measuring the angular motion of the slide, in both horizontal and vertical planes, throughout the total travel. To do this, a mirror is mounted on the stage slide and a laser beam is eflected into a theodolite. The above graph is typical of the results. Within the 13mm travel total horizontal angular motion is about 3 arc seconds; vertical angular motion is about 5 arc seconds. These angular motions are equivalent to translational position errors of less than 0.2μm in both axes.

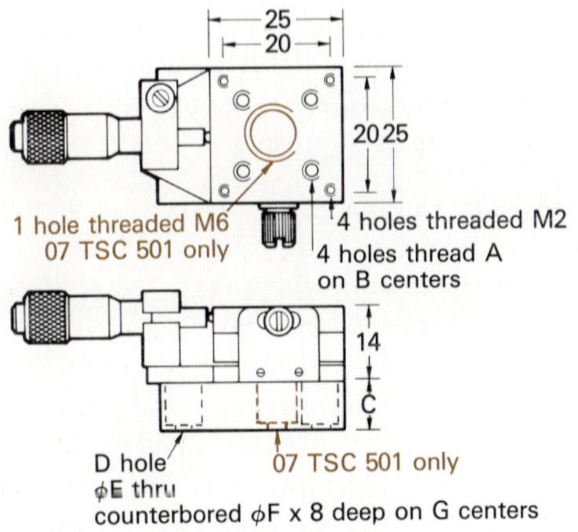

PRODUCT NUMBER	thread A	B	C	D	ϕE	ϕF	G
07 TSC 001	8-32	12.7	9	2	4.5	7	25.4
07 TSC 501	M4	12.5	10	1	7	11	–

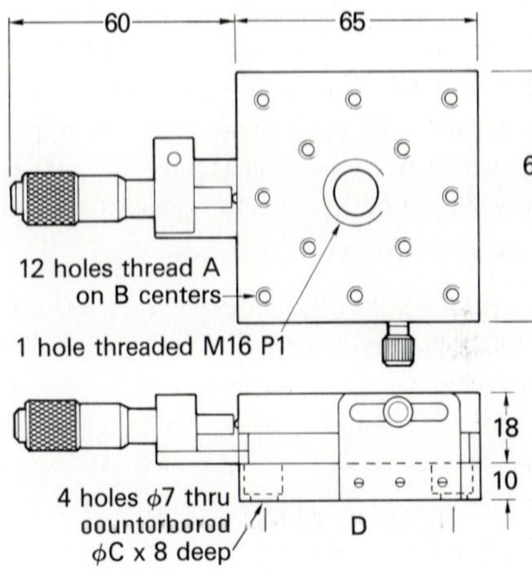

PRODUCT NUMBER	thread A	B	ϕC	D
07 TSC 004	1/4-20	25.4	10	50.8
07 TSC 504	M6	25	11	50

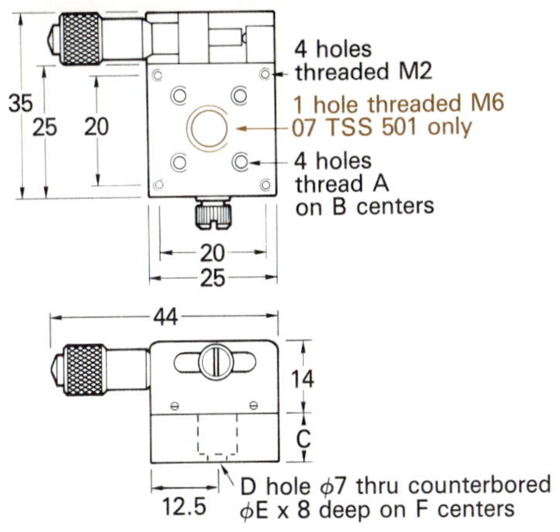

PRODUCT NUMBER	thread A	B	C	D	ϕE	F
07 TSS 001	8-32	12.7	9	2	7	25.4
07 TSS 501	M4	12.5	10	1	11	–

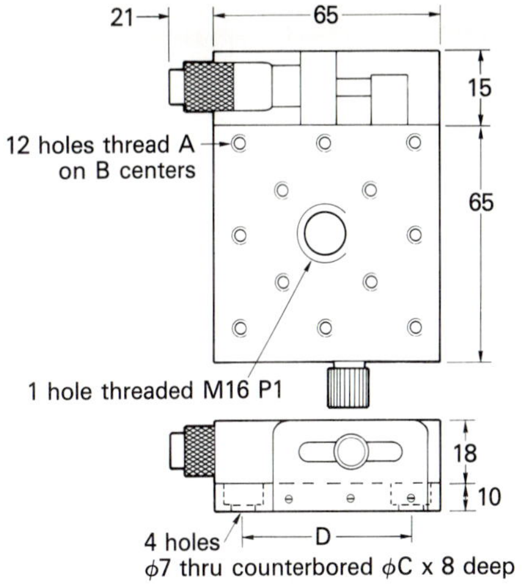

PRODUCT NUMBER	thread A	B	ϕC	D
07 TSS 004	1/4-20	25.4	10	50.8
07 TSS 504	M6	25	11	50

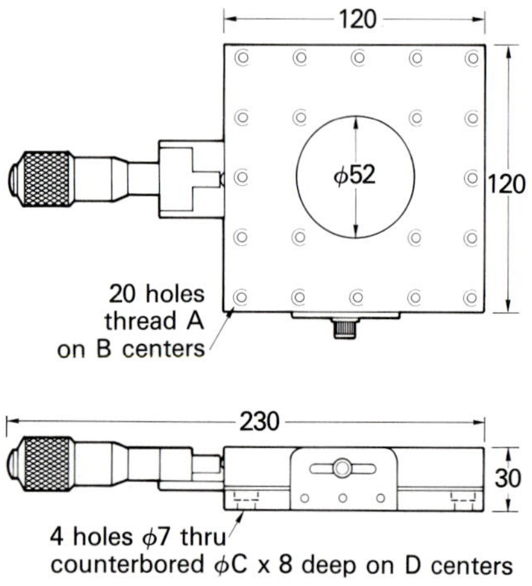

PRODUCT NUMBER	thread A	B	φC	D
07 TSC 007	1/4-20	25.4	10	101.6
07 TSC 507	M6	25	11	100

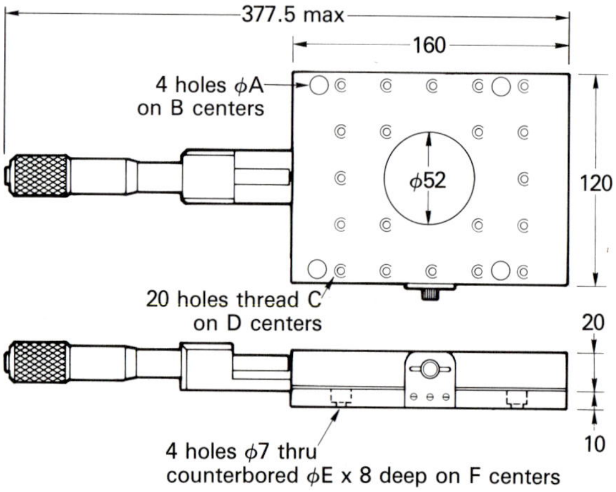

PRODUCT NUMBER	φA	B	thread C	D	φE	F
07 TSC 008	10	101.6	1/4-20	25.4	10	101.6
07 TSC 508	11	100	M6	25	11	100

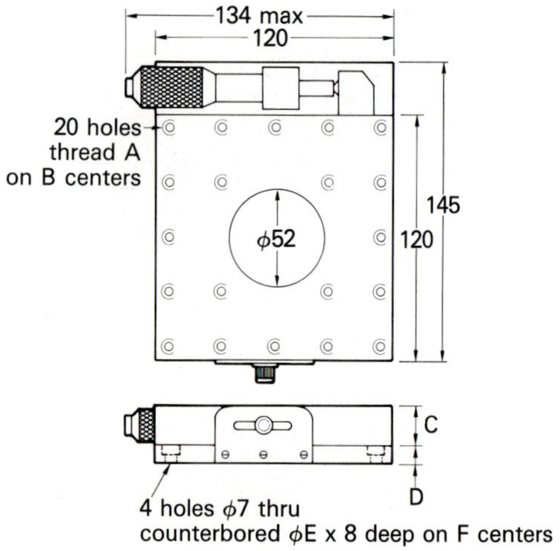

PRODUCT NUMBER	thread A	B	C	D	φE	F
07 TSS 007	1/4-20	25.4	19	8	10	101.6
07 TSS 507	M6	25	20	10	11	100

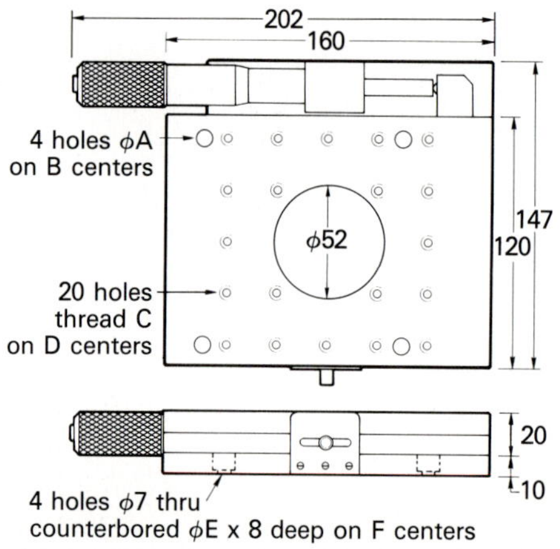

PRODUCT NUMBER	φA	B	thread C	D	φE	F
07 TSS 008	10	101.6	1/4-20	25.4	10	101.6
07 TSS 508	11	100	M6	25	11	100

Steel Translation Stages

Stage Length (mm)	Travel (mm)	Load Capacity (kg) (Normal)	Load Capacity (kg) (Maximum)	Straightness (μm)	PRODUCT NUMBER Inch Version Center Drive	Inch Version Side Drive	Metric Version Center Drive	Metric Version Side Drive
25	6	1.5	8	±1.0	07 TSC 001	07 TSS 001	07 TSC 501	07 TSS 501
65	13	6.0	30	±1.0	07 TSC 004	07 TSS 004	07 TSC 504	07 TSS 504
120	25	16.0	80	±1.5	07 TSC 007	07 TSS 007	07 TSC 507	07 TSS 507
160	50	20.0	100	±2.5	07 TSC 008	07 TSS 008	07 TSC 508	07 TSS 508

Z MOTION ADAPTERS

There are several ways to achieve Z-axis motion using these stages. A side drive stage mounted with its micrometer operating upwards is one of the best ways. However, if you wish to mount a center drive stage with its micrometer operating downwards, you will find that the load carrying capacity is drastically reduced. This is because gravity acts against the spring preload. To overcome this, the slide may be mounted the opposite way round with the slideway attached to the bracket. In this case an adapter plate is needed to mount further components on to the stage.

Z Motion Adapters

For Use With	PRODUCT NUMBER
Inch Version	
07 TSC 001	07 TSA 001
07 TSC 004	07 TSA 004
07 TSC 007	07 TSA 007
07 TSC 008	07 TSA 007
Metric version	
07 TSC 501	07 TSA 501
07 TSC 504	07 TSA 504
07 TSC 507	07 TSA 507
07 TSC 508	07 TSA 507

ANGLE BRACKETS

These cast aluminum brackets provide an angled attachment surface for connecting translation or rotation stages or other devices to each other or to the surface of an optical table or breadboard. They may also be used for attachment to the vertical mounting face of a damped post mounting plate. All surfaces are accurately machined to provide a smooth, strain-free surface, and the brackets are matte black anodized.

Brackets have either slots or clearance holes in their bases for mounting. Stages and other accessories can be attached directly to the perpendicular surface for Z-motion. The 65mm and 120mm brackets have a central aperture for passing an optical beam through the bracket. The 65mm bracket is available in two different heights.

A 65mm bracket is also offered with a circular slotted base enabling it to be used to swivel the mounting surface over an angular range of ±45 degrees.

SPECIFICATIONS: ANGLE BRACKETS

Angular Tolerance: < 5 arc minutes
Material: Cast aluminum
Finish: Matte black anodized

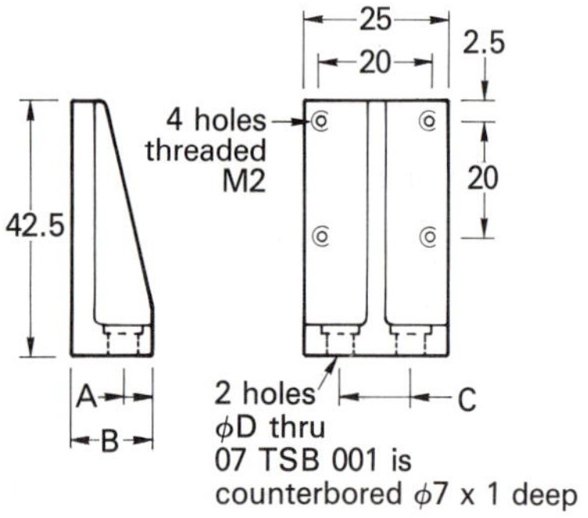

PRODUCT NUMBER	A	B	C	ϕD
07 TSB 001	5	14	12.7	4.5
07 TSB 501	2.5	12.5	20	2.5

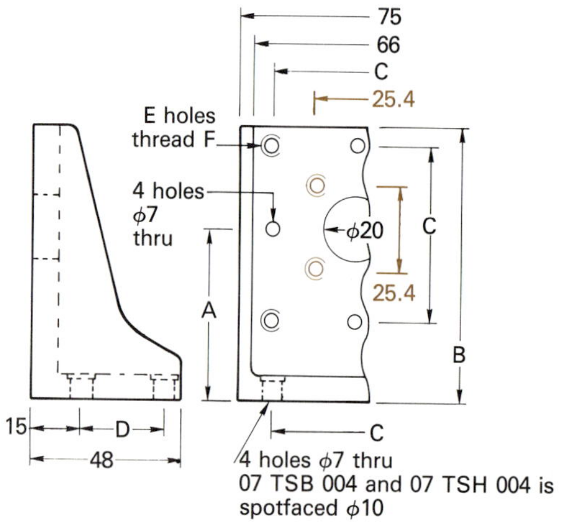

PRODUCT NUMBER	A	B	C	D	E	thread F
07 TSB 004	50	82.5	50.8	25.4	8	1/4-20
07 TSH 004	90	122.5	50.8	25.4	8	1/4-20
07 TSB 504	50	82.5	50	25	4	M6
07 TSH 504	90	122.5	50	25	4	M6

E: Black hole pattern applies to 07 TSB 504 and 07 TSH 504 only. Both black and brown hole patterns apply to 07 TSB 004 and 07 TSH 004.

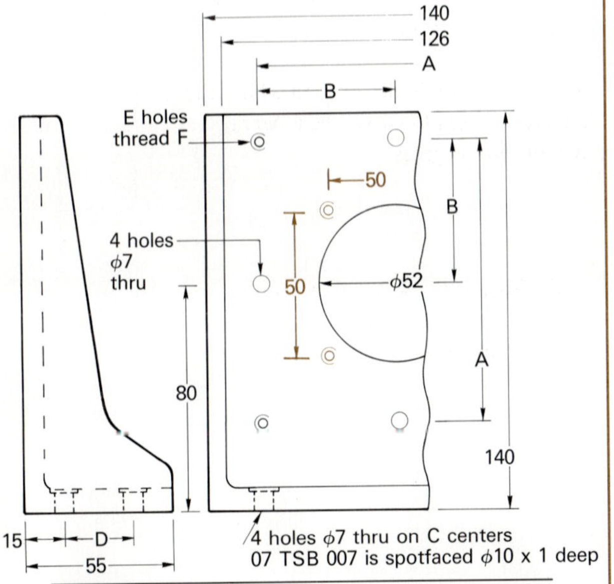

PRODUCT NUMBER	A	B	C	D	E	thread F
07 TSB 007	101.6	50.8	101.6	25.4	4	1/4-20
07 TSB 507	100	50	50	25	8	M6

E Black hole pattern applies to 07 TSB 007 only.
Both black and brown hole patterns apply to 07 TSB 507.

These brackets may be used as mounting platforms in conjunction with the damped posts on page 428.

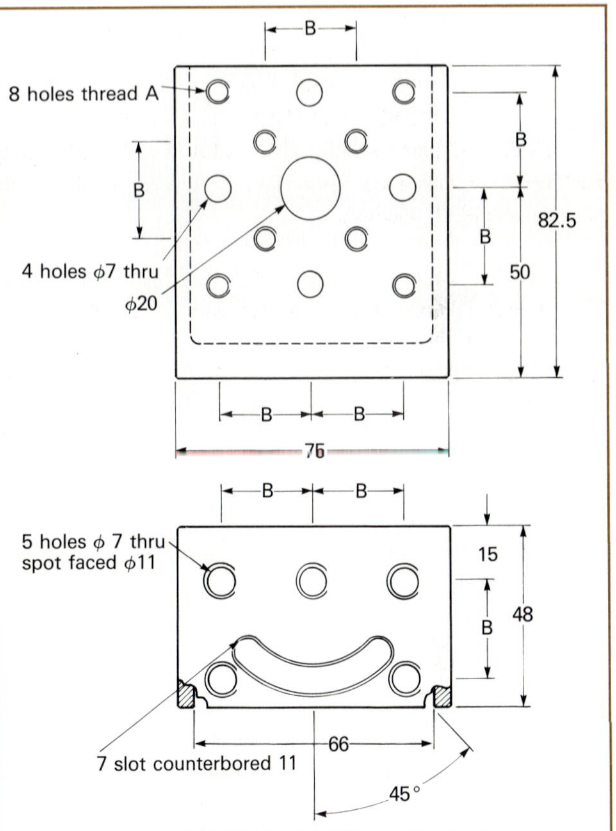

PRODUCT NUMBER	thread A	B
07 TSR 004	1/4-20 UNC	25.4
07 TSR 504	M6	25

Angle Brackets

Width (mm)	Center Height (mm)	Load Capacity (kg)	PRODUCT NUMBER Inch Version	Metric Version
25	30	45	07 TSB 001	07 TSB 501
65	50	130	07 TSB 004	07 TSB 504
*65	50	130	07 TSR 004*	07 TSR 504*
65	90	132	07 TSH 004	07 TSH 504
120	80	260	07 TSB 007	07 TSB 507

*The 07 TSR brackets have a swivel adjustment slot.

VERTICAL TRANSLATION STAGE

This vertical translation stage provide an alternate and simple method of precisely positioning components and holders in the vertical direction. The Melles Griot vertical translation stage is essentially a shortened version of the standard translation stage, mounted on its end and controlled by a micrometer actuated tangent arm.

Precision ground stainless steel balls ride between two precision ground "V" grooves to form the slide ways. The ways are preloaded to the stage by a light preloaded fit to prevent side play and end-to-end wobble.

The 10mm maximum vertical motion is imparted to the stage by a metric micrometer through a one-to-one right angle tangent arm. Angular displacement errors at the input end of the tangent arm are exactly cancelled by the congruent geometry at the output end of the arm. Two hardened steel hemispheres implanted in the ends of the arm ensure point contact.

Vertical preload (to prevent backlash) is provided in the downward direction by an internal tension spring. Vertical position may be locked with the knurled thumb screw.

Clearance holes in the base are provided for mounting to an optical table or breadboard or a variety of other translational and rotational stages. The top of the stage is provided with a central M16 tapped mounting hole for Melles Griot heavy duty posts and an array of 1/4-20 tapped holes on one inch centers for mounting and other mechanical components.

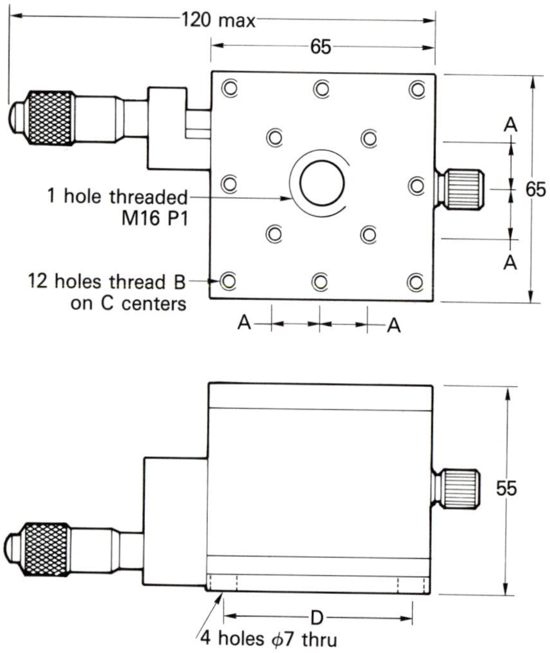

PRODUCT NUMBER	A	thread B	C	D
07 TSZ 004	12.7	1/4-20	25.4	50.8
07 TSZ 504	12.5	M6	25	50

FASTENER KIT

All required fasteners and screws are supplied with Melles Griot optical hardware when originally shipped. This kit is designed specifically to replace lost screws and fasteners. It is packaged, complete with the specific tools, in a closeable plastic box. For contents see page 499.

PRODUCT NUMBER	
Inch Version	Metric Version
07 AKT 005	07 AKT 505

SPECIFICATIONS: VERTICAL TRANSLATION STAGE

Straightness: ±2 microns over travel
Travel: 10mm
Load Capacity: 10 kg
Sensitivity: 2 microns
Material: Steel
Finish: Black chrome

Vertical translation stage

PRODUCT NUMBER	
Inch Version	Metric Version
07 TSZ 004	07 TSZ 504

SMALL ALUMINUM TRANSLATION STAGE

This inexpensive aluminum translation stage provides a useful and versatile method of moving optical components. The ways are made from two pairs of centerless-ground stainless steel rods, containing ball bearings, on either side of the slide. Bearings are pre-loaded.

Drive motion is imparted by a metric micrometer threaded through the frame and captured by the stage. The stage motion is protected from backlash by a spring which loads the stage against the micrometer.

It is supplied with a slot/screw position locking device which when tightened will hold the stage firmly without affecting the location in any way. The lock is useful in assuring that the positions are not changed after they have been set.

This aluminum stage is tapped so that it may be attached to any other stage in an X-Y, X-Z, Y-Z, or X-Y-Z configuration. The bottom of the stage is clearance drilled so that it may be mounted directly on optical tables and breadboards. The aluminum stage is matte black anodized. The micrometer is satin chrome plated.

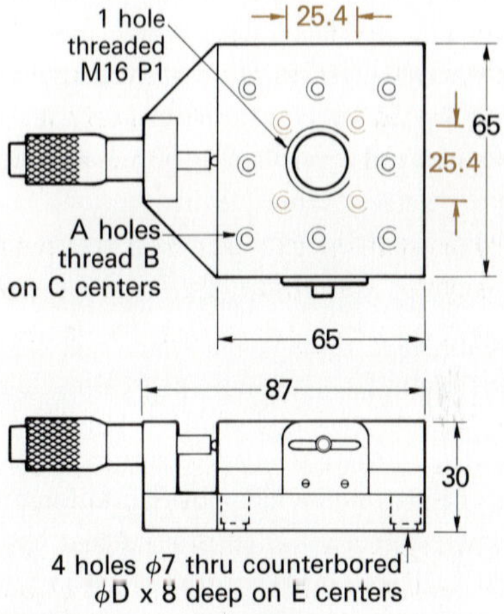

PRODUCT NUMBER	A	thread B	C	ϕD	E
07 TAC 004	12	1/4-20	25.4	10	50.8
07 TAC 504	8	M6	25	11	50

A: Black hole pattern applies to 07 TAC 504 only. Both black and brown hole patterns apply to 07 TAC 004.

OEM APPLICATIONS

The translation stages described here and on the next page and other components are well-suited to use by the OEM. Melles Griot carries all of these components in stock for rapid delivery, the designs are proven in response to real needs. Development and prototype costs can be largely avoided and you are assured of a high-level of quality control.

Modifications in controls (such as replacement of thumb screws with Allen set-screws) travel length, metric-to-inch conversions of drive controls, all can be accomplished quickly and economically.

For further information regarding OEM applications and modifications, please contact your Applications Engineer — who will be pleased to have an opportunity to work with you to meet your requirements.

SPECIFICATIONS: SMALL ALUMINUM TRANSLATION STAGE

Straightness: ±5 microns over travel
Travel: 13mm
Load Capacity: 20 kg
Sensitivity: 2 microns
Material: Aluminum body
Finish: Black anodized body; satin chrome micrometer

Small Aluminum Translation Stage

Body Length (mm)	PRODUCT NUMBER Inch Version	Metric Version
65	07 TAC 004	07 TAC 504

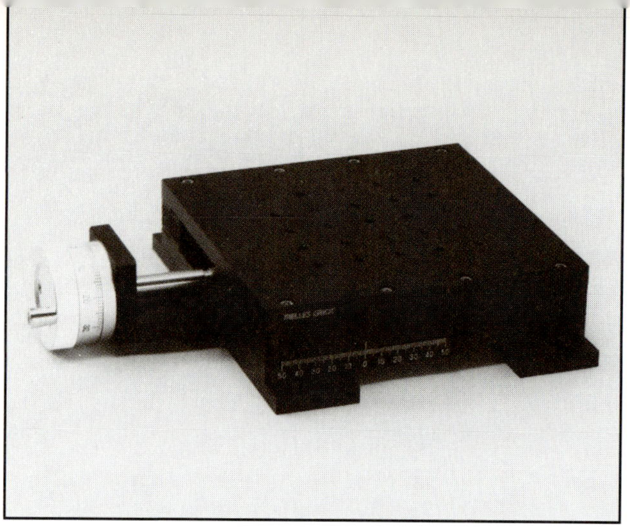

LARGE ALUMINUM TRANSLATION STAGE

Similar in construction to the small aluminum translational stage described on the previous page, these larger stages provide a useful, versatile, less-expensive means of moving optical components in linear directions. The ways consist of two pairs of centerless ground steel rods on either side of the slide. Ball-bearings are held between these rods under a pre-loading tension, set to eliminate side and end play.

Drive is provided by a fine pitch metric lead screw (2 threads per mm) which is threaded through the frame and captured by the stage. Backlash is eliminated by a pair of springs which load the stage against the lead screw ball-end.

Both models are provided with a position locking device which will hold the stage precisely in position after it has been set. A metric scale (readable to 1.0mm) is attached to the side of the stage to record position.

Sides and edges of the stages are drilled and tapped in such a way that they may be attached to each other in X-Y, X-Z, Y-Z or X-Y-Z configurations. The bottoms of both stages are clearance drilled to allow mounting to optical tables and breadboards.

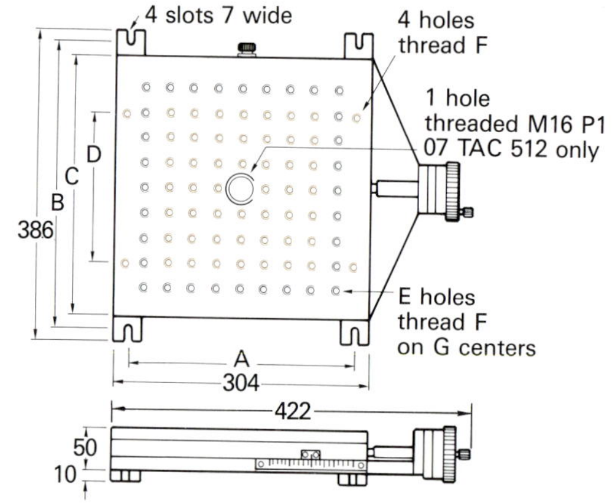

PRODUCT NUMBER	A	B	C	D	E	thread F	G
07 TAC 012	254	356	300	254	49	1/4-20	25.4
07 TAC 512	250	350	304	150	80	M6	25

E: Brown hole pattern applies to 07 TAC 012 only. Both black and brown hole patterns apply to 07 TAC 512.

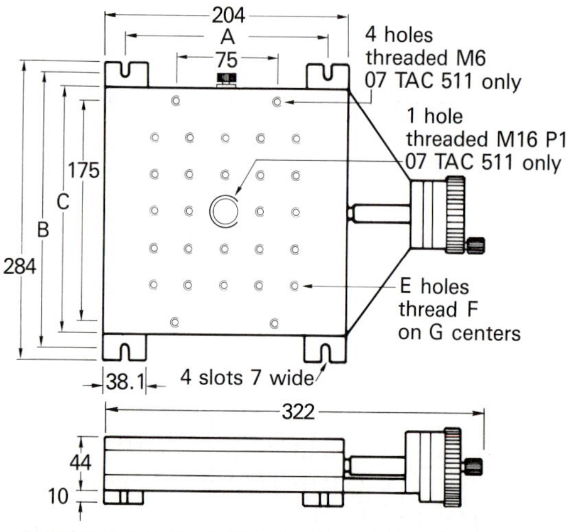

PRODUCT NUMBER	A	B	C	E	thread F	G
07 TAC 011	152.4	254	200	25	1/4-20	25.4
07 TAC 511	152	250	204	24	M6	25

SPECIFICATIONS: LARGE ALUMINUM TRANSLATION STAGES

Material: Aluminum body; Steel thumb screw
Finish: Black anodized body; Satin steel lead screw
Straightness: ±25 microns over travel
Sensitivity: 4 microns

Large Aluminum Translation Stages

Body Length (mm)	Travel (mm)	Load Capacity (kg)	PRODUCT NUMBER Inch Version	Metric Version
200	80	120	07 TAC 011	07 TAC 511
300	100	160	07 TAC 012	07 TAC 512

ACCESSORY KITS

ACCESSORY KIT FOR OPTICAL BREADBOARDS

This kit provides tools and fasteners for use with Melles Griot optical breadboards packaged in a closeable plastic box.

Accessory Kits for Optical Breadboards

Contents	Quantity	PRODUCT NUMBER
Inch Version:		
1/4-20x1/2 inch Allen screws	8	
1/4-20x3/4 inch Allen screws	8	
1/4-20x3/8 inch Allen screws	8	
1/4-20x1 inch hex head bolts	4	07 AKT 001
1/4-20x1/2 inch thumb screws	4	
nuts and washers (for side and track mounts)	4	
"T" handle Allen key (for table feet fasteners)	1	
Cover tool (for removing caps)	1	
Metric Version:		
M6x10mm Allen screws	10	
M6x20mm Allen screws	10	
M6x25mm hex head bolts	4	
M6x12mm thumb screws	4	07 AKT 501
nuts and washers (for side and track mounts)	4	
M6 tee nut	4	
"T" handle Allen key (for table fasteners)	1	
Cover tool (for removing caps)	1	

FASTENER KIT

This kit is designed specifically to replace lost screws and fasteners. It is packaged, complete with the specific tools, in a closeable plastic box.

Fastener Kit

Contents	Quantity	PRODUCT NUMBER
Inch Version:		
1/4-20x3/8 inch Allen cap screws	50	
1/4-20x1/2 inch Allen cap screws	50	
1/4-20x3/4 inch Allen cap screws	50	
M6x12mm Allen cap screws	50	07 AKT 005
M6x20mm Allen cap screws	50	
1/4-20 Allen key	2	
M6 Allen key	2	
1/4-20x1/2 inch thumb screws	4	
M6x12mm thumb screws	4	
Metric Version:		
M6x10mm Allen cap screws	50	
M6x12mm Allen cap screws	50	
M6x20mm Allen cap screws	50	07 AKT 505
M6 Allen key	2	
M6x12mm thumb screws	4	

SLOT SPANNER

An adjustable slot spanner tool for tightening and loosening the retainer ring used in Melles Griot lens holders, mirror holders and polarizer holders.

PRODUCT NUMBER
07 AKT 007

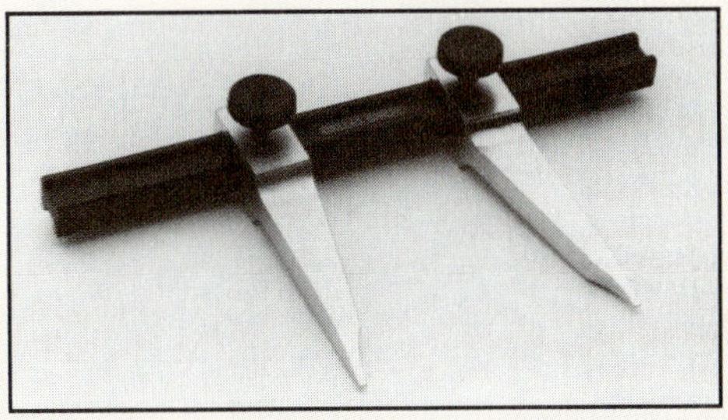

ACCESSORY KITS FOR OPTICAL HARDWARE

These kits contain a comprehensive set of tools and fasteners for use with all Melles Griot optical hardware packaged in a closeable plastic box.

Contents	Quantity	PRODUCT NUMBER
Metric Version:		
M4x10mm Allen screws	8	
M6x10mm Allen screws	8	
M6x12mm Allen screws	8	
M6x20mm Allen screws	8	
M6x12mm thumb screws	4	
Allen key (for M4 Allen cap screws	1	
Allen key (for M6 Allen cap screws	1	07 AKT 503
Allen key (for M8 Allen cap screws	1	
Wrench (for 12mm mounting posts)	1	
Tightening bars (3 sizes)	1	
Spanner wrench (for Melles Griot lens, mirror and polarizer holders)	1	

Accessory Kits for Optical Hardware

Contents	Quantity	PRODUCT NUMBER
Inch Version:		
1/4-20x1/2 inch Allen screws	8	
1/4-20x3/4 inch Allen screws	8	
1/4-20x3/8 inch Allen screws	8	
8-32x3/8 inch Allen screws	8	
M4x10mm Allen screws	8	
M6x12mm Allen screws	8	
M6x20mm Allen screws	8	
1/4-20x1/2 inch thumb screws	4	
M6x12mm thumb screws	4	
Allen key (for 8-32 Allen cap screws)	1	
Allen key (for 1/4-20 Allen cap screws)	1	07 AKT 003
Allen key (for M4 Allen cap screws)	1	
Allen key (for M6 Allen cap screws)	1	
Allen key (for M8 Allen cap screws)	1	
Wrench (for 12mm mounting posts)	1	
Tightening bars (3 sizes)	1	
Spanner wrench (for Melles Griot lens, mirror and polarizer holders)	1	

07 AKT 005

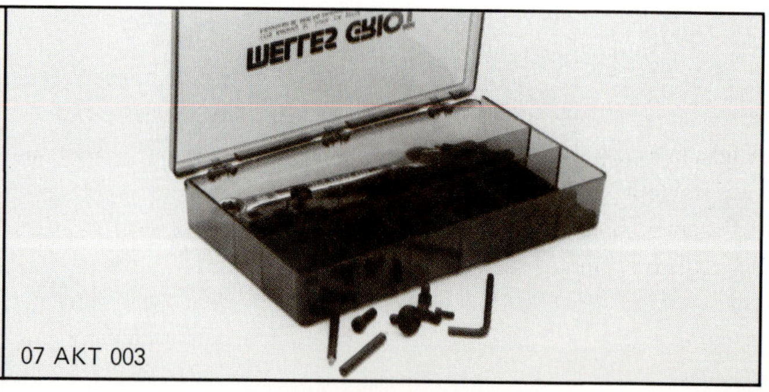
07 AKT 003

INDEXES

The two indexes which follow are designed to help you locate any of the products or technical sections contained within Optics Guide 3. The first index is a subject index. The subject index contains two types of entry; product entries, which are shown in black type, and technical entries, which are shown in brown type. The product entries list the pages where you will find that product's specifications and part number. The technical entries are generally discussions of product applications, sections on fundamental optics or relevant formulas. The product number index will help you find the pages on which a particular product is listed if the first five characters of the part number are known.

Contents **Page Number**

SUBJECT INDEX

PRODUCT NUMBER INDEX

NOTES

DO YOU KNOW SOMEONE WHO WANTS TO BORROW THIS OPTICS GUIDE?

If a friend or associate has frequent need to consult this useful reference book, let us supply another copy. We want everyone who uses lasers and optics to have benefit of the material published in Optics Guide 3.

ARE YOU ON OUR MAILING LIST?

If this Optics Guide was mailed directly to you, you are on our mailing list. If you obtained the catalog by another means, you are probably not on the list. We would like to keep you informed of our activities and new products and to supply you with occasional technical briefs. Be sure to sign up.

HAVE YOU MOVED OR CHANGED YOUR JOB?

If so, please let us know so we can update our records. We would like to keep in touch with you in your new position.

PLEASE COMPLETE AND RETURN THE CARD FOR YOUR AREA.

MELLES GRIOT

☐ Send Optics Guide 3 and price list.
☐ I have an Optics Guide — Please put me on mail list.
☐ Please note my new address listed below.

Name ______________________________
Position ______________________________
Company ______________________________
Address ______________________________
City ______________________________
State ________ Zip ________ Country ________
Phone ________ Ext. ________

Field of interest (check one)
CA ☐ Aerospace
CB ☐ Business Equipment
CC ☐ Optical Data Storage
CD ☐ Display Technology
CE ☐ Instrumentation
CF ☐ Laser-based Systems
CG ☐ Medical Instruments
CH ☐ Process Control
CI ☐ Semiconductors
CJ ☐ Optical Computing
CK ☐ Thermal Imaging
CL ☐ Optical Communications
CM ☐ Other ____________

Product Interest
EA ☐ Component Mounts & Holders
EB ☐ Laser Diode Optics
EC ☐ Filters
ED ☐ HeNe Lasers & Accessories
EE ☐ High Energy Optics
EF ☐ Mirrors, Prisms, Beamsplitters, Windows
EG ☐ Opto-Mechanical Accessories
EH ☐ Polarization Optics
EI ☐ Positioning Devices
EJ ☐ Single Element Lenses
EK ☐ Special Optical Components
EL ☐ Two & Three Element Lenses
EM ☐ Other ____________

Organizational Type (check one only)
AA ☐ Government/Military
AB ☐ Independent
AC ☐ Industrial
AD ☐ University/College

Job Function (check one)
BA ☐ Design/Development
BB ☐ Education
BC ☐ Field Application
BD ☐ Manufacturing
BE ☐ Marketing/Sales
BF ☐ Procurement
BG ☐ Quality Control
BH ☐ Research
BI ☐ Other ____________

This Card: For USA, Canada, Central and South America, the Far East, Australia.

Please complete and return one of these cards so we can be sure to keep you informed of new products and technical information.

For Europe, Africa, The Middle East and the Indian subcontinent: **This Card.**

MELLES GRIOT

☐ Send Optics Guide 3 and price list.
☐ I have an Optics Guide — Please put me on mail list.
☐ Please note my new address listed below.

Name ______________________________
Position ______________________________
Company ______________________________
Address ______________________________
City ______________________________
Country ______________________________
Phone ________ Ext. ________

Field of interest (check one)
☐ Aerospace
☐ Business Equipment
☐ Optical Data Storage
☐ Display Technology
☐ Instrumentation
☐ Laser-based Systems
☐ Medical Instruments
☐ Process Control
☐ Semiconductors
☐ Optical Computing
☐ Thermal Imaging
☐ Optical Communications
☐ Other ____________

Product Interest
☐ Lenses
☐ Mirrors/Prisms/Beamsplitters/Windows
☐ Filters
☐ Laser diode optics
☐ Infrared optics
☐ Polarizing optics
☐ Scanning lenses
☐ HeNe lasers & Accessories
☐ Optical benches/tables/breadboards
☐ Optical mounts and positioners
☐ Sapphire optics
☐ Microscope Accessories
☐ Low power sealed off CO_2 lasers
☐ 1.5 micron HeNe lasers
☐ Other ____________

Organizational Type (check one only)
AA ☐ Government/Military
AB ☐ Independent
AC ☐ Industrial
AD ☐ University/College

Job Function (check one)
BA ☐ Design/Development
BB ☐ Education
BC ☐ Field Application
BD ☐ Manufacturing
BE ☐ Marketing/Sales
BF ☐ Procurement
BG ☐ Quality Control
BH ☐ Research
BI ☐ Other ____________

Place
Stamp
Here

MELLES GRIOT

Optical Components

1770 Kettering St.

Irvine, California 92714-5670

Place
Stamp
Here

MELLES GRIOT

Edisonstraat 98

P.O. Box 272

6900 AG Zevenaar, Netherlands

DO YOU KNOW SOMEONE WHO WANTS TO BORROW THIS OPTICS GUIDE?

If a friend or associate has frequent need to consult this useful reference book, let us supply another copy. We want everyone who uses lasers and optics to have benefit of the material published in Optics Guide 3.

ARE YOU ON OUR MAILING LIST?

If this Optics Guide was mailed directly to you, you are on our mailing list. If you obtained the catalog by another means, you are probably not on the list. We would like to keep you informed of our activities and new products and to supply you with occasional technical briefs. Be sure to sign up.

HAVE YOU MOVED OR CHANGED YOUR JOB?

If so, please let us know so we can update our records. We would like to keep in touch with you in your new position.

PLEASE COMPLETE AND RETURN THE CARD FOR YOUR AREA.